NONLINEAR MECHANICS OF STRUCTURES

Nonlinear Mechanics of Structures

MICHAŁ KLEIBER AND CZESŁAW WOŹNIAK

Institute of Fundamental Technological Research,
Polish Academy of Sciences, Warsaw

KLUWER ACADEMIC PUBLISHERS
DORDRECHT/BOSTON/LONDON

PWN—POLISH SCIENTIFIC PUBLISHERS
WARSAW

Library of Congress Cataloging-in-Publication Data

Kleiber, Michał.
[Nieliniowa mechanika konstrukcji. English]
Nonlinear mechanics of structures/Michał Kleiber and Czesław Woźniak; [translated by the authors from the Polish edition].
p. cm. —
Translation of: Nieliniowa mechanika konstrukcji.
Includes bibliographical references.
ISBN-13: 978-94-010-6747-8
1. Mechanics Applied. 2. Nonlinear mechanics I. Woźniak, Czesław. II. Title.
TA350, K65313 1990
620. 1—dc20 90-4290

ISBN-13: 978-94-010-6747-8 e-ISBN-13: 978-94-009-0577-1
DOI: 10.1007/978-94-009-0577-1

Published by PWN—Polish Scientific Publishers,
Miodowa 10, 00-251 Warszawa, Poland
in co-edition with Kluwer Academic Publishers,
P.O. Box 17, 3300 AA Dordrecht, The Netherlands

Distributors:

for the United States and Canada: Kluwer Academic Publishers, 101 Philip Drive, Norwell, MA 02061, USA

for Albania, Bulgaria, Cuba, Czechoslovakia, Hungary, Korean People's Democratic Republic, People's Republic of Chile, Poland, Romania, the U.S.S.R., Vietnam and Yugoslavia: Ars Polona, Krakowskie Przedmiescie 7, 00-068 Warazawa, Poland

for all other countries: Kluwer Academic Publisher, P.O. Box 322, 3300 AH Dordrecht, The Netherlands

Contents

Introduction XI
Conventions and notations XIV

PART I: THEORETICAL FOUNDATIONS

1 General notions and principles
Purpose of the chapter 3
1.1 Introductory concepts 3
1.1.1 Space and time 3
1.1.2 Motion 3
1.1.3 Mass 5
1.1.4 Forces 6
1.1.5 Temperature, energy and entropy. 8
1.1.6 Heat supply 9
1.2 Fundamental principles 10
1.2.1 Balance of momentum 10
1.2.2 Balance of moment of momentum 10
1.2.3 Balance of energy 11
1.2.4 Dissipation principle 11
1.3 Basic fields 12
1.3.1 Displacements 12
1.3.2 Strains and strain rates 12
1.3.3 Stresses and stress rates 14
1.3.4 Heat fluxes 16
1.3.5 Matrix notation for the basic fields 16
1.4 Resulting relations. 19
1.4.1 Equation of motion 19
1.4.2 Energy equation 21
1.4.3 Dissipation inequality 22
1.4.4 Matrix form of the resulting equations 23
Summary 24
Problems 25

2 Materials, loadings and constraints

Purpose of the chapter 27
2.1 Materials 27
2.1.1 Real materials and ideal materials 27
2.1.2 Thermo-visco-elastic materials 28
2.1.3 Linear thermo-elastic materials 29
2.1.4 Thermo-elastic-plastic materials 30
2.1.5 Matrix form of constitutive relations 34
2.2 Loadings and heat supply 37
2.2.1 External agents 37
2.2.2 Body loadings and internal heat supply 38
2.2.3 Boundary loadings 38
2.2.4 Boundary heat supply. 39
2.3 Constraints 39
2.3.1 Boundary and internal constraints 39
2.3.2 Reactions 42
2.3.3 Bilateral constraints 44
2.3.4 Constraint functions 46
2.3.5 Constraints for stresses and heat fluxes 47
Summary 49
Problems 50

3 Formulation of engineering theories

Purpose of the chapter 52
3.1 General form of governing relations 52
3.1.1 Finite variational formulation 52
3.1.2 Incremental variational formulation 55
3.1.3 Variational equations for stresses and heat fluxes 58
3.1.4 Calculation of reactions 59
3.1.5 Matrix form of variational equations 60
3.2 Method of constraints: finite formulation 61
3.2.1 Foundations 61
3.2.2 Shell and plate theories 65
3.2.3 Rod and beam theories 71
3.2.4 Plane problems 74
3.3 Method of constraints: incremental formulation 77
3.3.1 General equations 77
3.3.2 Shells and rods 78
3.3.3 Plane problems 79

3.4 Method of constraints: discrete description of elements and structures 80
3.4.1 Discretization concept 80
3.4.2 Finite formulation 80
3.4.3 Incremental formulation 86
3.4.4 Shell and plate elements 87
3.4.5 Rod and beam elements 89
3.4.6 Special cases and matrix notation 91
3.4.7 Interaction between elements 96
3.5 Evaluation and correction of solutions 97
3.5.1 Residual reactions 97
3.5.2 Evaluation of solutions 99
3.5.3 Correction of solutions 101
3.5.4 On adaptive solution refinements 103
Summary 103
Problems 104

4 Extensions and specifications of the general theory

Purpose of the chapter 106
4.1 Domain discretization 107
4.2 Solving differential equations by weighted residual methods . . 109
4.3 Assembly of element matrices and vectors 113
4.4 Elastic pin-jointed element in space 118
4.5 Elastic frame element in space 123
4.6 Elastic lattice-type structures 131
4.6.1 Discrete models 131
4.6.2 Continuous models 141
4.7 Engineering shell theories 147
4.7.1 Foundations 147
4.7.2 General form of equations of motion 151
4.7.3 Special theories 158
4.7.4 Example: axisymmetric deformations of shells 162
4.8 Thin shell finite element for nonlinear axisymmetric analysis . . 164
4.9 Linear strain triangular element for nonlinear plane stress analysis 182
4.10 Constitutive matrices for thermo-elastic-plastic materials . . . 193
4.11 On inelastic analysis under non-proportionally varying loads . 197
4.11.1 Elastic-plastic structures subjected to mechanical loads . 197
4.11.2 Thermo-elastic-plastic structures subjected to mechanical and thermal loads 203

Summary 205
Problems 206

5 Numerical algorithms and software concepts

Purpose of the chapter 212
5.1 Introductory comments 212
5.2 Nonlinear quasi-statics 213
5.3 Integration of elastic-plastic constitutive law 225
5.4 Initial and linearized buckling 229
5.5 Nonlinear dynamics 232
5.6 Dynamic stability under non-periodic loads 237
5.7 Nonlinear heat transfer 242
5.8 Development of software 244
Summary 246
Problems 246

PART II: SELECTED APPLICATIONS

6 Trusses, frames, lattice-type shells

Purpose of the chapter 251
6.1 Trusses 251
6.1.1 Nonlinear effects in truss analysis — a model problem . . . 251
6.1.2 Accuracy of computations and further examples 261
6.1.3 Elastic-plastic truss under non-proportionally varying loads 272
6.2 Frames 277
6.2.1 Elastic-plastic constitutive matrices for beam elements . . . 277
6.2.2 Limit state conditions 283
6.2.3 Nonlinear analysis of curved beams 290
6.2.4 Examples of numerical frame analysis up to collapse 291
6.2.5 Elastic-plastic beam under non-proportionally varying loads . 298
6.2.6 Elastic buckling of plane grid 305
6.2.7 Approximate large displacement analysis of frames using buckling mode superposition 306
6.3 Lattice-type structures 312
6.3.1 Lattice shells — linear and "second-order" theories 312
6.3.2 Buckling of elastic grid 318
6.3.3 Perfectly-plastic lattice-type plates — discrete model 322
6.3.4 Limit load of polar grids — continuous model 339
6.3.5 Minimum weight design of ideally plastic rectangular dense grid 351
Summary 360
Problems 360

7 Thin plates loaded in-plane

Purpose of the chapter 363
7.1. Elastic-plastic bending of a cantilever beam 363
7.2 Limit analysis of perforated plates 369
7.3 Elastic-plastic analysis of a cantilever under non-proportionally varying loads 376
7.4 In-plane buckling of an elastic-plastic strip 379
7.5 Necking of an elastic-plastic strip 380
7.6 Extension of a thin rectangular plate with a central hole 383
Summary 384

8 Plate and shell problems .

Purpose of the chapter 385
8.1 Thin elastic-plastic axisymmetric shells 385
8.2 Elastic-viscoplastic analysis of axisymmetric shells 405
8.3 Buckling of elastic-plastic spatial plate assemblies 410
Summary 430

9 Heat conduction and thermal stress problems

Purpose of the chapter 431
9.1 Heat flow in flash and friction welding 431
9.2 Thermally induced stresses in elastic-plastic plate 440
9.3 Elastic-plastic cylindrical shell under thermo-mechanical load . 443
9.4 Elastic-plastic truss under nonproportionally varying temperature and mechanical load 445
Summary 448

Bibliography 449
Index 457

Introduction

The aim of this book is to provide a unified presentation of modern mechanics of structures in a form which is suitable for graduate students as well as for engineers and scientists working in the field of applied mechanics. Traditionally, students at technical universities have been taught subjects such as continuum mechanics, elasticity, plates and shells, frames or finite element techniques in an entirely separate manner. The authors' teaching experience clearly suggests that this situation frequently tends to create in students' minds an incomplete and inconsistent picture of the contemporary structural mechanics. Thus, it is very common that the fundamental laws of physics appear to students hardly related to simplified equations of different "technical" theories of structures, numerical solution techniques are studied independently of the essence of mechanical models they describe, and so on.

The book is intended to combine in a reasonably connected and unified manner all these problems starting with the very fundamental postulates of nonlinear continuum mechanics via different structural models of "engineering" accuracy to numerical solution methods which can effectively be used for solving boundary-value problems of technological importance.

The authors have tried to restrict the mathematical background required to that which is normally familiar to a mathematically minded engineering graduate.

Of course, no single book can fully embrace the broad field of nonlinear structural mechanics. Holding the book to reasonable length did not permit to include many interesting, and certainly important issues. In making our choice we have primarily aimed at logical presentation of the selected material and its anticipated usefulness in the long run. In particular, it is emphasized that the book is not intended to provide the reader with a great number of sophisticated nonlinear theories applicable to the refined analysis of specific structural configurations. Quite to the contrary, after giving a thorough continuum mechanics background we consistently develop merely relatively simple, or conventional, theories and apply them to solve numerous practical problems of engineering significance. It is our strong belief, however, that the general framework laid down in the present book will turn out to be a proper tool to develop more sophisticated theories in the near future.

A brief outline of the main portions of the book is as follows. The first part devoted to theoretical foundations of nonlinear mechanics of structures starts with the short presentation of the fundamental notions and laws of nonlinear continuum mechanics. We introduce the notation and get the reader acquainted with the description of motion of the material continuum, the states of strain and stress and finally with the conservation laws. The temperature effects are included into the formulation.

In Chapter 2 we outline the theory of constitutive equations and thoroughly discuss loadings and boundary constraints typical of structural configurations. Specific constitutive laws are given for materials known as linear elastic, visco-elastic and thermo-elastic-plastic.

Chapter 3 is of fundamental significance to the approach consequently followed in the book since it deals with the key concept of internal constraints imposed on the kinematics and/or state of stress of the continuum. By specifying the constraints we are able to systematically derive classical and non-classical equations governing different "engineering" theories which describe the nonlinear behaviour of beam-, plate- and shell-type configurations. The considerations of this chapter are based on a number of variational principles. Emphasis is placed on demonstrating the power of the principles in deriving the governing equations in a systematic way. Next, the method of constraints is used to introduce a discrete description of structural configurations. By this we mean setting up approximate models of structures described in terms of algebraic rather than differential equations. An interesting feature of the approach to discretization as presented in this book is that it is again based upon the internal constraint concept thus making the discussion fully compatible with that given previously in Sections 3.2, 3.3 of this chapter. Apart from some attractive error estimates offered by the way we look at these problems this portion of the book (Secs. 3.4, 3.5) essentially discusses the foundations and applications of the well-known and powerful finite element method in its displacement and mixed-type versions. Discretization with respect to space and time variables is discussed and final sets of nonlinear algebraic equations are derived.

Some useful extensions and specifications of the general theory proposed in Chapter 3 are dealt with in the next Chapter 4. We consider in it such problems as the domain discretization, assembly procedures to be used in summing up the element matrices and vectors, specific forms of typical element matrices as well as an explicit form of equations describing thermo-elastic-plastic materials. The theoretical background for the so-called shake-down analysis of structures is also presented to be used later while discussing specific boundary-value solutions to inelastic problems under variable loads.

The solution methods for the discretized problems are described in Chapter 5. Two basic sections in this chapter deal with numerical algorithms for static and dynamic nonlinear problems, respectively. The algorithms are based on repetitive solutions of sets of linear algebraic equations describing the instantaneous equilibrium of the discretized body and are additionally endowed with some Newton–Raphson iterative corrections to improve the accuracy of the results. Part I of the book is completed by a short discussion of software concepts which are typically employed while developing computer programs.

The contents of Part I of the book is made more specific in Part II, in which we consider different discretized models for beams, plates and shells. A number of effective finite elements is described in detail and test calculations are given to illustrate their performance in modelling different aspects of nonlinear structural behaviour. Trusses, frames, grids and shell-like structures made of regularly distributed beams are dealt with in Chapter 6. Chapter 7 is devoted to the analysis of thin plates loaded in-plane whereas Chapter 8 concentrates on shell-type problems. A short discussion of heat conduction and thermal stress problems given in Chapter 9 completes the book.

Problems for independent studies are provided at the end of most chapters so that students can check their understanding of the subjects. At places, problems serve also the purpose of extending the scope of the book into areas which could not have been covered for the sake of compactness of the presentation.

The course, as outlined here, is designed for a two-semester sequence at the graduate level. On the whole, it is believed that the treatment used in this book is in harmony with current trends toward a more fundamental approach in engineering education.

Conventions and notations

Throughout the book we shall use such mathematical objects as scalars, vectors, scalar and vector functions, vector and tensor components, coordinates of points, etc. We shall also perform certain operations on these objects such as scalar or vector products. We assume that the reader knows the elementary vector and tensor calculus both in the absolute and the component form. As far as possible we shall use the following conventions:

1. Sub- and superscripts i, j, k, l, m, n run over the sequence $1, 2, 3$ and are related to a cartesian coordinate system (so-called spatial coordinate system) in the physical space identified with the Euclidean 3-space $\boldsymbol{R}^3$ of spatial points. Sub- and superscripts $\alpha, \beta, \gamma, \delta$ run also over the sequence $1, 2, 3$ (unless a different range is clearly stated) and are related to a coordinate system introduced exclusively in a region Ω occupied by a body in a certain reference configuration (so-called material coordinate system). Ranges of any other indices appearing in the text will always be explained prior to their use.

2. Summation convention holds with respect to the indices $\alpha, \beta, \ldots$ and $i, j, \ldots$, provided they are repeated twice in a given expression and situated in it on the upper and lower level, respectively. Ricci symbol and Kronecker delta are denoted by ε^{ijk}, δ^{ij} respectively (position of i, j, k is immaterial).

3. Spatial points are denoted by $\mathbf{z}$, $\mathbf{z} \in \boldsymbol{R}^3$, and their (spatial) coordinates by z^k; points of Ω are denoted by $\mathbf{X}$, $\mathbf{X} \in \Omega$, and their (material) coordinates by X^α (instead of k or α we may be using any other pertinent Latin or Greek superscript, respectively). The symbols τ and t stand for time instants with t used to indicate a specific rather than an arbitrary time.

4. Scalars are represented mostly by small light-faced Greek italics: ϱ, α, π. Scalar-valued functions are denoted by $\varrho(\cdot)$, $\alpha(\cdot)$, $\pi(\cdot)$ etc.; domains of these functions are specified in the text.

5. Vectors and vector-valued functions are represented by small bold face Latin or Greek Roman letters; examples: $\mathbf{f}$, $\boldsymbol{\chi}$, $\hat{\mathbf{f}}(\cdot)$, $\hat{\boldsymbol{\chi}}(\cdot)$, respectively; their components are represented by small face Latin or Greek italics: f^k, χ^k, $\hat{f}^k(\cdot)$, $\hat{\chi}^A(\cdot)$.

6. Second order tensors and tensor-valued functions are represented by capital bold face Latin letters: $\mathbf{T}$, $\mathbf{S}$ and $\hat{\mathbf{T}}(\cdot)$, $\hat{\mathbf{S}}(\cdot)$; their components are denoted by T^{ij}, $S^{\alpha\beta}$ and $\hat{T}^{ij}(\cdot)$, $\hat{S}^{\alpha\beta}(\cdot)$, respectively.

7. Higher order tensors and tensor functions are represented by German capitals: $\mathfrak{C}$, $\mathfrak{H}$ and $\hat{\mathfrak{C}}(\cdot)$, $\hat{\mathfrak{H}}(\cdot)$, respectively; their components are indicated by light-faced Roman letters: C^{ijkl}, $\mathrm{H}_{\alpha\beta\gamma\delta}(\cdot)$ etc. Symbols $C[\cdot]$, $H[\cdot]$ stand for linear mappings of the form $\mathrm{C}^{ijkl}D_{kl}$, $\mathrm{H}_{\alpha\beta\gamma\delta}S^{\gamma\delta}$, respectively.

8. Comma denotes partial differentiation with respect to the spatial coordinates (example: $r_{,i} \equiv \frac{\partial r}{\partial x^i}$) or to the material coordinates (example: $T_{,\alpha} \equiv \frac{\partial T}{\partial X^\alpha}$). Dot denotes substantial differentiation with respect to the time coordinate (hence $\dot{g} = \frac{\partial g}{\partial \tau}$ provided $g = g(\mathbf{X}, \tau)$ and $\mathbf{X} = (X^\alpha)$ are the material coordinates). The gradient with respect to the material coordinates $\mathbf{X} = (X^\alpha)$ is denoted by ∇ (example: $\nabla\pi = (\pi_{,1}\pi_{,2}\pi_{,3})$, $\pi_{,\alpha} \equiv \frac{\partial \pi}{\partial X^\alpha}$, where $\pi = \pi(\mathbf{X}, \tau)$).

9. The component forms of products $\mathbf{a}\cdot\mathbf{b}$, $\mathbf{a}\times\mathbf{b}$, $\mathbf{Tn}$, $\mathbf{SE}$ are $a^k b_k$, $\varepsilon^{kl}{}_m a_l b^m$, $T^{kl}n_l$, $S^{\alpha\beta}E_{\beta\gamma}$, respectively.

10. Symbols $\operatorname{tr}\mathbf{A}$, $\det\mathbf{A}$ stand for trace and determinant of an arbitrary second order tensor $\mathbf{A}$. Symbols $\mathbf{A}^T$, $\mathbf{A}^{-1}$ stand for the transpose and the inverse of $\mathbf{A}$, respectively, $\mathbf{1}$ is a unit tensor.

11. Sets of points or vectors are represented by capital light face Greek letters (examples: Ω, Φ); sets of mappings are denoted by a light face script (examples: $\mathscr{P}$, $\mathscr{T}$).

PART I

THEORETICAL FOUNDATIONS

1

General notions and principles

Purpose of the chapter

The purpose of the chapter is to introduce and explain all general concepts and principles of thermo-mechanics which are common to all problems we are going to study throughout the book. We use the term "general" in order to underline the fact that these notions and rules are independent of such characteristic features of specific structural mechanics problems as the shape of the structure, the form of external agencies acting on it, materials the structure is made of as well as constraints imposed on possible deformations of the structure. The notions and principles developed in this chapter lead to a system of resulting relations which constitute the physical basis for all the problems of structural mechanics to be subsequently considered.

1.1 Introductory concepts

1.1.1 Space and time

We shall assume that a set of all events constitutes the Galilean space-time in which there is fixed once and for all an inertial coordinate system. Such a system assigns to every event a fourtouple (z_1, z_2, z_3, τ) of real numbers, $(z_1, z_2, z_3, \tau) \in \boldsymbol{R}^4$. Any triplet $\mathbf{z} = (z_1, z_2, z_3)$, $\mathbf{z} \in \boldsymbol{R}$, will be called a *spatial point*, reals z_i, $i = 1, 2, 3$, will be called *spatial coordinates* and real τ will be referred to as a *time coordinate* or a *time instant*. We shall always tacitly assume that a spatial distance $||\mathbf{z}-\bar{\mathbf{z}}|| \equiv \sqrt{(z_i-\bar{z}_i)(z_j-\bar{z}_j)\,\delta^{ij}}$ between any two spatial points $\mathbf{z}$, $\bar{\mathbf{z}}$ is determined in the fixed once for all unit length dimension $[L]$. Similarly, a time distance $|\tau-\bar{\tau}|$ between any two time instants $\tau, \bar{\tau}$ is assumed to be expressed in the fixed unit time dimension $[T]$. As a rule we shall use the International System of unit measures, setting $[L] = [\mathrm{m}]$, $[T] = [\mathrm{s}]$.

1.1.2 Motion

In order to describe a motion of a certain material (real) element of a solid or a structure under consideration, we introduce:

1. A region Ω in the Euclidean 3-space R^3 of spatial points, with a piecewise smooth boundary $\partial\Omega$. Region Ω is assumed to represent a certain reference configuration of a real piece of the structure considered. Points of Ω will be denoted by $\mathbf{X} \equiv (X^1, X^2, X^3)$, $\mathbf{X} \in \Omega$, and referred to as *material points* while real numbers X^α, $\alpha = 1, 2, 3$, will be called *material coordinates*. For the sake of simplicity set Ω will be referred to as the undeformed body[1].

2. A nonempty set $\mathscr{D}(\Omega)$ of invertible and continuously differentiable (as many times as required but possibly except at some surfaces, lines or points) mappings $\mathbf{z} = \boldsymbol{\varkappa}(\mathbf{X})$, $\mathbf{X} \in \Omega$, which will be called *deformations*, cf.

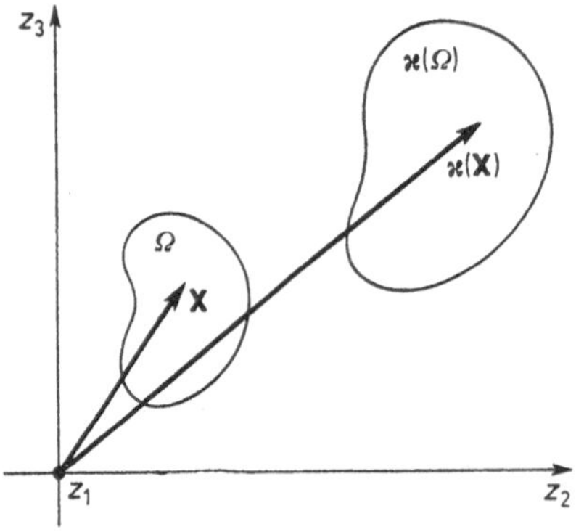

Fig. 1 Deformed and undeformed body

Fig. 1. Every $\boldsymbol{\varkappa}(\Omega)$ will be called a *deformed body* (with respect to Ω) provided $\boldsymbol{\varkappa}(\cdot)$ is not the identity mapping.

By a *motion of the body* in a time interval $[t_0, t_f]$ we shall mean any one-parameter family of deformations $\boldsymbol{\chi}(\cdot, \tau) \in \mathscr{D}(\Omega)$, $\tau \in [t_0, t_f]$ such that for every $\mathbf{X} \in \Omega$ the function $\mathbf{z} = \boldsymbol{\chi}(\mathbf{X}, \tau)$, $\tau \in [t_0, t_f]$, is continuous and has (except possibly at a finite number of time instants) continuous the first- and second-order time derivatives denoted by $\dot{\boldsymbol{\chi}}(\mathbf{X}, \tau)$ and $\ddot{\boldsymbol{\chi}}(\mathbf{X}, \tau)$, respectively (cf. Fig. 2, in which, for a fixed $\mathbf{X}$, the function $\mathbf{z} = \boldsymbol{\chi}(\mathbf{X}, \tau)$, $\tau \in [t_0, t_f]$, represents a smooth arc).

Every $\mathbf{z} = \boldsymbol{\chi}(\mathbf{X}, t)$ will be called a *position of the material point* $\mathbf{X}$, $\mathbf{X} \in \Omega$, at a time instant t, $t \in [t_0, t_f]$ in the motion $\boldsymbol{\chi}(\cdot, \tau) \in \mathscr{D}(\Omega)$, $\tau \in [t_0, t_f]$. Similarly, $\dot{\boldsymbol{\chi}}(\mathbf{X}, t)$ and $\ddot{\boldsymbol{\chi}}(\mathbf{X}, t)$ are said to be the *velocity* and the *acceleration* of $\mathbf{X}$ at a time instant t, respectively.

In continuum mechanics any invertible and continuously differentiable (except possibly at some surfaces, lines and points) mapping $\mathbf{z} = \boldsymbol{\varkappa}(\mathbf{X})$, $\mathbf{X} \in \Omega$,

[1] A term "body" is used here in order to underline the fact that the set Ω constitutes a mathematical description of a real solid which can be identified with a part of the material structure under consideration.

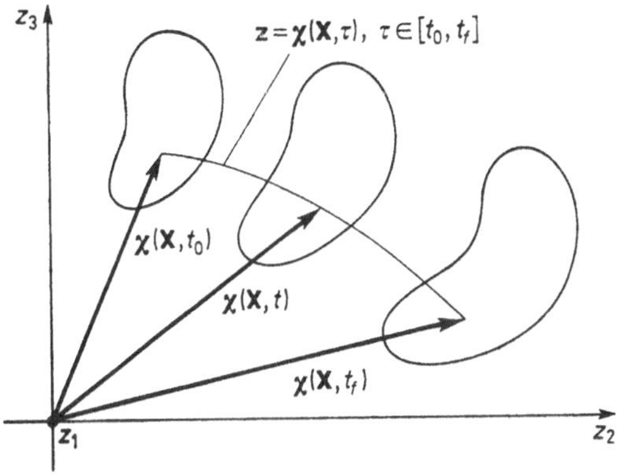

Fig. 2 Deformations of the body

can be taken as a deformation[1]. In structural mechanics problems, however, we confine ourselves to some special classes of deformations. For example, in the widely employed Kirchhoff plate theory we restrict our considerations to deformations in which material fibres normal to the midsurface of the plate in one configuration behave as rigid and remain normal to the midsurface in any other configuration. To quote another example: in the formulation of many beam theories we confine ourselves to deformations in which the beam cross-sections normal to the beam axis remain plane in every deformation. Further, in many special problems we assume that all the deformations are plane, i.e. for any fixed $\mathbf{X} \in \Omega$ and every pair (t_1, t_2), $t_1, t_2 \in [t_0, t_f]$, the relation $\chi_3(\mathbf{X}, t_1) = \chi_3(\mathbf{X}, t_2)$ holds. Generally speaking, to every class of problems in the field of structural mechanics corresponds as a rule a certain set $\mathscr{D}(\Omega)$ of deformations, which is assumed to be known a priori. In the sequel we are going to show that the choice of $\mathscr{D}(\Omega)$ constitutes a crucial point in the modelling of various engineering problems within mechanics and this choice is a typical feature of structural mechanics. Thus, from now on, we shall assume that the region Ω and the set $\mathscr{D}(\Omega)$ are known. The specific nature of both Ω and $\mathscr{D}(\Omega)$ depends on the character of the problem to be analysed and will be thoroughly discussed in the subsequent chapters of the book.

1.1.3 Mass

In order to describe a mass distribution in a certain element of a real solid or of a real structure we shall assume that there is known:

1. A region Ω in $\boldsymbol{R}^3$ (an undeformed body) introduced in Sec. 1.1.2,

[1] For particulars cf. [157]; the term "configuration" is there used instead of the term "deformation" introduced here.

2. A piecewise continuous positive real valued function $\varrho(\mathbf{X})$, $\mathbf{X} \in \Omega$, which will be called a *mass density* in the undeformed body Ω.

Let Π be an arbitrary subset of Ω which constitutes a region in a space R^3 with a piecewise smooth boundary. Every Π will be called a *part of a* (undeformed) *body*. The value of the integral

$$\mu(\Pi) = \int_{\Pi} \varrho(\mathbf{X}) \mathrm{d}V(\mathbf{X}), \qquad \mathrm{d}V(\mathbf{X}) \equiv \mathrm{d}X^1 \mathrm{d}X^2 \mathrm{d}X^3,$$

is assumed to represent the mass of the part Π of the body. The mass is expressed in the fixed once for all mass unit measure $[M]$ (as a rule, we shall assume $[M] = [\mathrm{kg}]$).

Now, let $\mathscr{D}(\Omega)$ be known and $\boldsymbol{\varkappa}(\mathbf{X})$, $\mathbf{X} \in \Omega$, be an arbitrary deformation, $\boldsymbol{\varkappa}(\cdot) \in \mathscr{D}(\Omega)$. Setting

$$\mathrm{d}v(\mathbf{z}) \equiv \mathrm{d}z_1 \mathrm{d}z_2 \mathrm{d}z_3, \qquad J(\mathbf{X}) \equiv \det \nabla \boldsymbol{\varkappa}(\mathbf{X}),$$

for $\mathbf{z} = \boldsymbol{\varkappa}(\mathbf{X})$ we have $\mathrm{d}v(\mathbf{z}) = J(\mathbf{X}) \mathrm{d}V(\mathbf{X})$. Denoting

$$\varrho_{\varkappa}(\mathbf{z}) \equiv \varrho(\mathbf{X}) J^{-1}(\mathbf{X}), \qquad \mathbf{X} = \boldsymbol{\varkappa}^{-1}(\mathbf{z}),$$

we obtain the formula

$$\mu(\Pi) = \int_{\boldsymbol{\varkappa}(\Pi)} \varrho_{\varkappa}(\mathbf{z}) \mathrm{d}v(\mathbf{z})$$

which holds for every part Π of the body and for every deformation $\boldsymbol{\varkappa}(\cdot) \in \mathscr{D}(\Omega)$. The function $\varrho_{\varkappa}(\mathbf{z})$, $\mathbf{z} \in \boldsymbol{\varkappa}(\Omega)$, is assumed to be the mass density in the deformed body[1].

In all problems of structural mechanics we are going to discuss, a mass density $\varrho(\mathbf{X})$, $\mathbf{X} \in \Omega$, in the undeformed body is assumed to be given. On the other hand, no mass density $\varrho_{\varkappa}(\mathbf{z})$, $\mathbf{z} \in \boldsymbol{\varkappa}(\Omega)$, in the deformed body is known, since deformations $\boldsymbol{\varkappa}(\cdot)$ of the body are not known a priori.

1.1.4 Forces

We shall assume that to every motion $\boldsymbol{\chi}(\cdot, \tau) \in \mathscr{D}(\Omega)$, $\tau \in [t_0, t_f]$, to every part Π of Ω and to every time instant $t \in [t_0, t_f]$ there is assigned a force resultant $\mathbf{f}_t(\Pi)$, $\mathbf{f}_t(\Pi) \in R^3$, acting at $\boldsymbol{\chi}(\Pi, t)$, given by

$$\mathbf{f}_t(\Pi) = \oint_{\partial \boldsymbol{\chi}(\Pi, t)} \mathbf{s}_{\partial \boldsymbol{\chi}(\Pi, t)}(\mathbf{z}, t) \mathrm{d}a(\mathbf{z}) + \int_{\boldsymbol{\chi}(\Pi, t)} \mathbf{b}_{\boldsymbol{\chi}}(\mathbf{z}, t) \mathrm{d}v(\mathbf{z}),$$

---

[1] The interpretation of $\varrho_{\varkappa}(\mathbf{z})$, from the physical point of view, as a mass density in the deformed body is equivalent to the principle of mass conservation.

where $da(\mathbf{z})$ is an element of the surface $\partial\boldsymbol{\chi}(\Pi, t)$. Here $\mathbf{s}_{\partial\boldsymbol{\chi}(\Pi,t)}(\mathbf{z}, t)$ are called *surface tractions* (acting on $\boldsymbol{\chi}(\Pi, t)$ across $\partial\boldsymbol{\chi}(\Pi, t)$) and $\mathbf{b}_{\boldsymbol{\chi}}(\mathbf{z}, t)$, $\mathbf{z} \in \boldsymbol{\chi}(\Omega, t)$, are called *body force densities*[1].

The aforementioned assumption restrict our considerations to two kinds of forces only, i.e. to forces due to the surface tractions $\mathbf{s}_{\partial\boldsymbol{\chi}(\Pi,t)}(\mathbf{z}, t)$, $\mathbf{z} \in \partial\boldsymbol{\chi}(\Pi, t)$, and to forces due to the body force densities $\mathbf{b}_{\boldsymbol{\chi}}(\mathbf{z}, t)$, $\mathbf{z} \in \boldsymbol{\chi}(\Omega, t)$. As a rule, in problems of structural mechanics both foregoing force densities are not known a priori. The surface tractions acting at an arbitrary part $\boldsymbol{\chi}(\Pi, t)$ of the deformed body (at a time instant t) are visualized in Fig. 3.

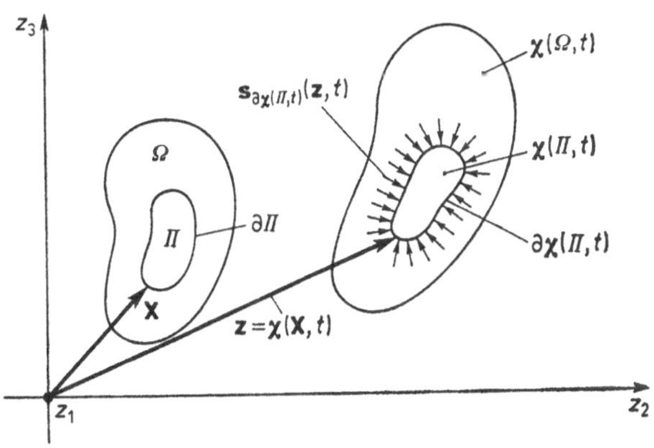

Fig. 3 Surface tractions

The vector $\mathbf{f}_t(\Pi)$ has components which will be denoted by $f_t(\Pi)^i$; setting $\|\mathbf{f}_t(\Pi)\| \equiv \sqrt{f_t(\Pi)^i f_t(\Pi)^j \delta_{ij}}$, we shall always assume that $\|\mathbf{f}_t(\Pi)\|$ is expressed in the force unit of measure $[F] = [MLT^{-2}]$.

Let $dA(\mathbf{X})$, $\mathbf{X} \in \partial\Pi$, be an element of the surface $\partial\Pi$ and $da(\mathbf{z})$ an element of $\partial\boldsymbol{\chi}(\Pi, t)$ at $\mathbf{z} = \boldsymbol{\chi}(\mathbf{X}, t)$. Define now the scalar densities $j(\mathbf{X}, t)$ setting $da(\mathbf{z}) = j(\mathbf{X}, t)dA(\mathbf{X})$ for $\mathbf{z} = \boldsymbol{\chi}(\mathbf{X}, t)$ and let $J(\mathbf{X}, t) \equiv \det \nabla\boldsymbol{\chi}(\mathbf{X}, t)$. Setting

$$\mathbf{s}_{\partial\Pi}(\mathbf{X}, t) \equiv \mathbf{s}_{\partial\boldsymbol{\chi}(\Pi,t)}(\boldsymbol{\chi}(\mathbf{X}, t), t)j(\mathbf{X}, t), \quad \mathbf{X} \in \partial\Pi,$$

$$\mathbf{b}(\mathbf{X}, t) \equiv \mathbf{b}_{\boldsymbol{\chi}}(\boldsymbol{\chi}(\mathbf{X}, t), t)J(\mathbf{X}, t), \quad \mathbf{X} \in \Omega,$$

for all values of arguments $\mathbf{X}$, t for which the right-hand sides (RHS) of the foregoing formulae are defined, we obtain

$$\mathbf{f}_t(\Pi) = \oint_{\partial\Pi} \mathbf{s}_{\partial\Pi}(\mathbf{X}, t)dA(\mathbf{X}) + \int_{\Pi} \mathbf{b}(\mathbf{X}, t)dV(\mathbf{X}). \tag{1.1}$$

[1] Instead of the body force densities $\mathbf{b}_{\boldsymbol{\chi}}(\mathbf{z}, t)$ in many textbooks body forces $\mathbf{b}(\mathbf{z}, t)$ are introduced, being interrelated with $\mathbf{b}_{\boldsymbol{\chi}}(\mathbf{z}, t)$ by means of the relation $\mathbf{b}_{\boldsymbol{\chi}}(\mathbf{z}, t) = \varrho_{\boldsymbol{\chi}}(\mathbf{z}, t)\mathbf{b}(\mathbf{z}, t)$, $\varrho_{\boldsymbol{\chi}}(\mathbf{z}, t) \equiv \varrho_{\boldsymbol{\chi}(\cdot,t)}(\mathbf{z})$. Functions $\mathbf{b}_{\boldsymbol{\chi}}(\cdot, t)$ may suffer discontinuities at some surfaces, lines or points in $\boldsymbol{\chi}(\Omega, t)$ while functions $\mathbf{s}_{\partial\boldsymbol{\chi}(\Pi,t)}(\cdot, t)$ may be not defined at some lines or points on $\partial\boldsymbol{\chi}(\Pi, t)$.

Here, $\mathbf{s}_{\partial\Pi}(\mathbf{X}, t)$ and $\mathbf{b}(\mathbf{X}, t)$ are the densities of surface tractions and body forces, respectively, related to the undeformed body Ω. We shall also define

$$\mathbf{s}(\mathbf{X}, t) \equiv \mathbf{s}_{\partial\Omega}(\mathbf{X}, t), \quad \mathbf{X} \in \partial\Omega,$$

and refer to it as the boundary tractions.

1.1.5 *Temperature, energy and entropy*

In order to describe a distribution of the absolute temperature in a material element of a real solid or structure we use again the notion of the undeformed body Ω, cf. Sec. 1.1.2, and introduce a set $\mathscr{T}(\Omega)$ of continuously differentiable (as many times as required but possibly except at some surfaces, lines or points) positive-valued functions $\theta(\mathbf{X})$, $\mathbf{X} \in \Omega$, which will be interpreted as the absolute temperature distributions. We tacitly assume that $\theta(\mathbf{X})$ is expressed in the unit of the measure of temperature $[K]$.

In all problems of structural mechanics in which the non-mechanical effects can not be neglected we shall assume that to every motion $\boldsymbol{\chi}(\cdot, \tau) \in \mathscr{D}(\Omega)$, $\tau \in [t_0, t_f]$, it is assigned an evolution of the temperature, given by a family of mappings $\theta(\cdot, \tau) \in \mathscr{T}(\Omega)$, $\tau \in [t_0, t_f]$, such that for every $\mathbf{X} \in \Omega$ function $\theta(\mathbf{X}, \tau) \in \boldsymbol{R}_+$, $\tau \in [t_0, t_f]$, is continuously differentiable except possibly at a finite number of time instants. We shall also define the relative temperature field $\vartheta_t(\mathbf{X})$, $\mathbf{X} \in \Omega$, setting $\vartheta_t(\mathbf{X}) \equiv \theta(\mathbf{X}, t) - \theta_0$ where θ_0 is the known absolute temperature. In the sequel the term "temperature" (without a specification) will always be understood as the relative temperature.

It has to be emphasized that the set $\mathscr{T}(\Omega)$ plays here a role analogous to that of a set $\mathscr{D}(\Omega)$ introduced earlier. It means that in order to simplify the mathematical description of structural mechanics problems we shall restrict ourselves to some special distributions of temperature only (for instance, to linear distributions, constant temperature fields, etc.), provided that such a restriction is motivated by a physical character of the problem.

We shall further assume in all non-purely mechanical problems of structural mechanics that in every motion $\boldsymbol{\chi}(\cdot, \tau) \in \mathscr{D}(\Omega)$, $\tau \in [t_0, t_f]$, and to every part Π of Ω as well as to every time instant $t \in [t_0, t_f]$, two scalars are assigned:

1. The total energy $\sigma_t(\Pi)$, $\sigma_t(\Pi) \in \boldsymbol{R}$, given by

$$\sigma_t(\Pi) = \int_{\Pi} [\tfrac{1}{2}\varrho(\mathbf{X})\dot{\boldsymbol{\chi}}(\mathbf{X}, t) \cdot \dot{\boldsymbol{\chi}}(\mathbf{X}, t) + \varrho(\mathbf{X})\,\varepsilon(\mathbf{X}, t)]\,\mathrm{d}V(\mathbf{X}) \tag{1.2}$$

where $\varepsilon(\mathbf{X}, t)$ is called the *internal energy*.

2. The entropy $\eta_t(\Pi)$, $\eta_t(\Pi) \in \boldsymbol{R}$, given by

$$\eta_t(\Pi) = \int_{\Pi} \varrho(\mathbf{X})\,\eta(\mathbf{X}, t)\,\mathrm{d}V(\mathbf{X}) \tag{1.3}$$

where $\eta(\mathbf{X}, t)$, $\mathbf{X} \in \Omega$, is called the *specific entropy*.

Values of $\sigma_t(\Pi)$ and $\eta_t(\Pi)$ are assumed to be expressed in the unit measures $[ML^2T^{-2}]$, $[MLT^{-1}K^{-1}]$, respectively. Functions $\varepsilon(\cdot, t)$, $\eta(\cdot, t)$ defined on Ω (except possibly at some surfaces, lines or points in Ω and at a finite number of time instants $t \in [t_0, t_f]$) are assumed to satisfy all regularity conditions which will be required in the sequel. It can easily be seen that a total energy $\sigma_t(\Pi)$ is a sum of the kinetic energy and of an energy which is due to the postulated internal energy $\varepsilon(\mathbf{X}, t)$, $\mathbf{X} \in \Omega$. As a rule, in problems of structural mechanics the functions $\varepsilon(\cdot, t)$ and $\eta(\cdot, t)$ are not known a priori, but in every special problem the set $\mathscr{T}(\Omega)$ is assumed to be given.

REMARK. Instead of the internal energy $\varepsilon(\mathbf{X}, t)$ we shall also use the function $\varphi(\mathbf{X}, \tau)$, $\mathbf{X} \in \Omega$, $\tau \in [t_0, t_f]$, defined by

$$\varphi(\mathbf{X}, t) \equiv \varepsilon(\mathbf{X}, t) - \eta(\mathbf{X}, t)\theta(\mathbf{X}, t)$$

which is said to be the *free energy*.

1.1.6 *Heat supply*

We shall assume that to every motion $\boldsymbol{\chi}(\cdot, \tau) \in \mathscr{D}(\Omega)$, $\tau \in [t_0, t_f]$, to every part Π of Ω and to every time instant $t \in [t_0, t_f]$, there is assigned a scalar $\delta_t(\Pi)$, $\delta_t(\Pi) \in \boldsymbol{R}$, which is called the *total heat supply to* $\boldsymbol{\chi}(\Pi, t)$, given by

$$\delta_t(\Pi) = \oint_{\partial\boldsymbol{\chi}(\Pi, t)} \pi_{\partial\boldsymbol{\chi}(\Pi, t)}(\mathbf{z}, t)\, da(\mathbf{z}) + \int_{\boldsymbol{\chi}(\Pi, t)} \alpha_{\boldsymbol{\chi}}(\mathbf{z}, t)\, dv(\mathbf{z}),$$

where $\pi_{\partial\boldsymbol{\chi}(\Pi, t)}(\mathbf{z}, t)$ is called the *surface heat supply* and $\alpha_{\boldsymbol{\chi}}(\mathbf{z}, t)$ is called the *heat absorption*. Every $\delta_t(\Pi)$ is assumed to be expressed in the unit of the measure of energy $[ML^2T^{-2}]$.

Now by setting, cf. Sec. 1.1.4,

$$\pi_{\partial\Pi}(\mathbf{X}, t) \equiv \pi_{\partial\boldsymbol{\chi}(\Pi, t)}(\boldsymbol{\chi}(\mathbf{X}, t), t) j(\mathbf{X}, t), \quad \mathbf{X} \in \partial\Pi,$$

$$\alpha(\mathbf{X}, t) \equiv \alpha_{\boldsymbol{\chi}}(\boldsymbol{\chi}(\mathbf{X}, t), t) J(\mathbf{X}, t), \quad \mathbf{X} \in \Omega,$$

for all values of the arguments $\mathbf{X}$, t for which the RHS of the foregoing formulae are defined, we obtain

$$\delta_t(\Pi) = \oint_{\partial\Pi} \pi_{\partial\Pi}(\mathbf{X}, t)\, dA(\mathbf{X}) + \int_{\Pi} \alpha(\mathbf{X}, t)\, dV(\mathbf{X}). \tag{1.4}$$

Here $\pi_{\partial\Pi}(\mathbf{X}, t)$ and $\alpha(\mathbf{X}, t)$ are the densities of the surface heat supply and of the internal heat supply, respectively, related to the undeformed body Ω. Setting

$$\pi(\mathbf{X}, t) \equiv \pi_{\partial\Omega}(\mathbf{X}, t), \quad \mathbf{X} \in \partial\Omega,$$

we shall refer to $\pi(\mathbf{X}, t)$ as the *boundary heat supply*.

The notion of the heat supply will be taken into account in all structural mechanics problems in which the thermal effects can not be neglected. As a rule, the functions $\pi_{\partial\Pi}(\cdot, t)$ and $\alpha(\cdot, t)$ are not known a priori. Moreover, the heat supply (1.4) implies a certain production of entropy (the rate of entropy) in every part Π of the body, given by

$$\oint_{\partial\Pi} \frac{\pi_{\partial\Pi}(\mathbf{X}, t)}{\theta(\mathbf{X}, t)}\, \mathrm{d}A(\mathbf{X}) + \int_{\Pi} \frac{\alpha(\mathbf{X}, t)}{\theta(\mathbf{X}, t)}\, \mathrm{d}V(\mathbf{X})$$

which is expressed in the unit of measure $[ML^2T^{-2}K^{-1}]$.

1.2 Fundamental principles

1.2.1 Balance of momentum

We postulate that in every motion $\boldsymbol{\chi}(\cdot, \tau) \in \mathscr{D}(\Omega)$, $\tau \in [t_0, t_f]$, for every part Π of a body and for every time instant t, $t \in [t_0, t_f]$, the following condition holds:

$$\frac{\mathrm{d}}{\mathrm{d}t} \int_{\Pi} \dot{\boldsymbol{\chi}}(\mathbf{X}, t)\varrho(\mathbf{X})\mathrm{d}V(\mathbf{X}) = \mathbf{f}_t(\Pi). \tag{1.5}$$

This postulate is called the *principle of balance of momentum* since the integral on the LHS of the foregoing equality represents the momentum of the part Π of the body at the time instant t.

1.2.2 Balance of moment of momentum

Let us define the vector moment $\mathbf{m}_t(\Pi, \mathbf{z}_0)$ (with respect to a spatial point $\mathbf{z}_0$) of all forces acting at a part Π of the body in its motion $\boldsymbol{\chi}(\cdot, \tau) \in \mathscr{D}(\Omega)$, $\tau \in [t_0, t_f]$, and at an arbitrary time instant t, $t \in [t_0, t_f]$:

$$\mathbf{m}_t(\Pi, \mathbf{z}_0) \equiv \oint_{\partial\Pi} [\boldsymbol{\chi}(\mathbf{X}, t) - \mathbf{z}_0] \times \mathbf{s}_{\partial\Pi}(\mathbf{X}, t)\mathrm{d}A(\mathbf{X}) + \int_{\Pi} [\boldsymbol{\chi}(\mathbf{X}, t) - \mathbf{z}_0] \times \mathbf{b}(\mathbf{X}, t)\mathrm{d}V(\mathbf{X}).$$

We postulate that in every motion $\boldsymbol{\chi}(\cdot, \tau) \in \mathscr{D}(\Omega)$, $\tau \in [t_0, t_f]$, the condition

$$\frac{\mathrm{d}}{\mathrm{d}t} \int_{\Pi} [\boldsymbol{\chi}(\mathbf{X}, t) - \mathbf{z}_0] \times \dot{\boldsymbol{\chi}}(\mathbf{X}, t)\varrho(\mathbf{X})\mathrm{d}V(\mathbf{X}) = \mathbf{m}_t(\Pi, \mathbf{z}_0) \tag{1.6}$$

holds for every part Π of the body and for every time instant t, $t \in [t_0, t_f]$. This postulate is known as the principle of balance of moment of momentum since the integral over Π on the LHS of the condition (1.6) represents the pertinent moment of momentum.

1.2.3 Balance of energy

Let us define the rate of work $\omega_t(\Pi)$ of forces acting at the part Π of the body in its motion $\boldsymbol{\chi}(\cdot, \tau) \in \mathscr{D}(\Omega)$, $\tau \in [t_0, t_f]$, and at an arbitrary time instant t, $t \in [t_0, t_f]$:

$$\omega_t(\Pi) \equiv \oint_{\partial\Pi} \mathbf{s}_{\partial\Pi}(\mathbf{X}, t) \cdot \dot{\boldsymbol{\chi}}(\mathbf{X}, t) \mathrm{d}A(\mathbf{X}) + \int_{\Pi} \mathbf{b}(\mathbf{X}, t) \cdot \dot{\boldsymbol{\chi}}(\mathbf{X}, t) \mathrm{d}V(\mathbf{X}).$$

We postulate that in every motion $\boldsymbol{\chi}(\cdot, \tau) \in \mathscr{D}(\Omega)$, $\tau \in [t_0, t_f]$, the condition

$$\frac{\mathrm{d}}{\mathrm{d}t} \int_{\Pi} [\tfrac{1}{2}\varrho(\mathbf{X}) \dot{\boldsymbol{\chi}}(\mathbf{X}, t) \cdot \dot{\boldsymbol{\chi}}(\mathbf{X}, t) + \varrho(\mathbf{X}) \varepsilon(\mathbf{X}, t)] \mathrm{d}V(\mathbf{X}) = \omega_t(\Pi) + \delta_t(\Pi) \quad (1.7)$$

holds for every part Π of the body and for every time instant t, $t \in [t_0, t_f]$.

The integral over Π on the LHS of the foregoing condition is referred to as the total energy, cf. Sec. 1.1.5, and hence the postulated condition is called the *principle of balance of energy*. This principle is also known as the *first law of thermodynamics*, provided only mechanical and thermal effects are taken into account.

1.2.4 Dissipation principle

Let us define the *production of entropy* $\gamma_t(\Pi)$ in the part Π of the body and at the time instant t, by setting

$$\gamma_t(\Pi) \equiv \frac{\mathrm{d}}{\mathrm{d}t} \eta_t(\Pi) - \left[\oint_{\partial\Pi} \frac{\pi_{\partial\Pi}(\mathbf{X}, t)}{\theta(\mathbf{X}, t)} \mathrm{d}A(\mathbf{X}) + \int_{\Pi} \frac{\alpha(\mathbf{X}, t)}{\theta(\mathbf{X}, t)} \mathrm{d}V(\mathbf{X}) \right].$$

The sum in the square brackets represents the production of entropy due to the heat supply to the part Π of the body, cf. Sec. 1.1.6. Hence, $\gamma_t(\Omega)$ is the total production of entropy at the time instant t.

We postulate that in every motion $\boldsymbol{\chi}(\cdot, \tau) \in \mathscr{D}(\Omega)$, $\tau \in [t_0, t_f]$, and for every evolution of temperature $\theta(\cdot, \tau) \in \mathscr{T}(\Omega)$, $\tau \in [t_0, t_f]$, the condition

$$\gamma_t(\Pi) \geqslant 0 \quad (1.8)$$

holds for every part Π of the body and for every time instant t, $t \in [t_0, t_f]$.

The postulate stated above is called the *dissipation principle* or the *Clausius–Duhem inequality*. This principle corresponds to the known second law of thermodynamics and describes the irreversibility of processes encountered in nature. It has to be emphasized that the dissipation principle has not the form of a balance principle like those formulated in Secs. 1.2.1–1.2.4. Hence also the role which the dissipation principle plays within mechanics is different from that of the balance principles.

1.3 Basic fields

1.3.1 Displacements

Let $\boldsymbol{\chi}(\cdot, \tau) \in \mathscr{D}(\Omega)$, $\tau \in [t_0, t_f]$, be a certain motion and t be a fixed time instant, $t \in [t_0, t_f]$. Deformations $\boldsymbol{\chi}(\cdot, t_0)$ and $\boldsymbol{\chi}(\cdot, t)$ will be referred to as the initial and actual deformations, respectively. By the *displacement field* we shall mean a function $\mathbf{u}_t \colon \Omega \to \boldsymbol{R}^3$ defined by

$$\mathbf{u}_t(\mathbf{X}) \equiv \boldsymbol{\chi}(\mathbf{X}, t) - \boldsymbol{\chi}(\mathbf{X}, t_0) \tag{1.9}$$

for every $\mathbf{X} \in \Omega$. Any $\mathbf{u}_t(\mathbf{X})$, $\mathbf{u}_t(\mathbf{X}) \in \boldsymbol{R}^3$, will be called the *displacement vector of the material point* $\mathbf{X}$ *at the time instant* t. The example of such vector is

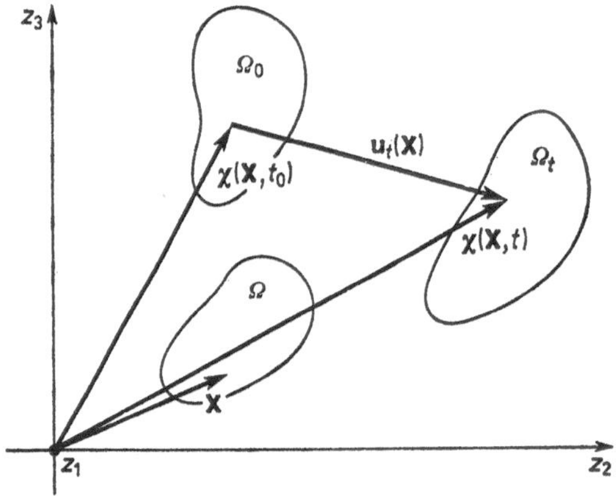

Fig. 4 Displacement vector

visualized in Fig. 4, where we have also used the notation $\Omega_0 \equiv \boldsymbol{\chi}(\Omega, t_0)$, $\Omega_t \equiv \boldsymbol{\chi}(\Omega, t)$.

Let $\mathbf{X}, \mathbf{X} \in \Omega$, be an arbitrary but fixed material point. The first and second (time) derivatives of functions $[t_0, t_f] \ni \tau \to \mathbf{u}_\tau(\mathbf{X}) \in \boldsymbol{R}^3$ at the time instant $\tau = t$ represent the velocity $\dot{\mathbf{u}}_t(\mathbf{X})$ and the acceleration $\ddot{\mathbf{u}}_t(\mathbf{X})$ respectively, of the material point $\mathbf{X}$ at the time instant t. Let us observe that $\dot{\mathbf{u}}_t(\mathbf{X}) = \dot{\boldsymbol{\chi}}(\mathbf{X}, t)$, $\ddot{\mathbf{u}}_t(\mathbf{X}) = \ddot{\boldsymbol{\chi}}(\mathbf{X}, t)$ for every $\mathbf{X}$ and $t \in (t_0, t_f)$.

1.3.2 Strains and strain rates

Let Γ be an arbitrary smooth arc in Ω. The material points of Γ will be denoted by $\mathbf{Z}$. Define $\Gamma_0 = \boldsymbol{\chi}(\Gamma, t_0)$, $\Gamma_t \equiv \boldsymbol{\chi}(\Gamma, t)$, where $\boldsymbol{\chi}(\cdot, \tau) \in \mathscr{D}(\Omega)$,

$\tau \in [t_0, t_f]$, is a certain motion and t, $t \in [t_0, t_f]$, is an arbitrary but fixed time instant. Let us also assume that both Γ_0, Γ_t are smooth arcs in Ω_0, Ω_t, respectively, cf. Fig. 5. Since $\mathbf{z} = \boldsymbol{\chi}(\mathbf{Z}, \tau)$, $\mathbf{Z} \in \Gamma$, $\tau \in [t_0, t_f]$, and $\mathrm{d}\mathbf{z} = \nabla\boldsymbol{\chi}(\mathbf{Z}, \tau)\mathrm{d}\mathbf{Z}$, then the length elements $\mathrm{d}s_0$, $\mathrm{d}s_t$ of Γ_0, Γ_t, respectively, are given by the simple formulae

$$\mathrm{d}s_0^2 = \mathrm{d}\mathbf{z}_0 \cdot \mathrm{d}\mathbf{z}_0, \qquad \mathrm{d}\mathbf{z}_0 = \nabla\boldsymbol{\chi}(\mathbf{Z}, t_0)\mathrm{d}\mathbf{Z},$$
$$\mathrm{d}s_t^2 = \mathrm{d}\mathbf{z}_t \cdot \mathrm{d}\mathbf{z}_t, \qquad \mathrm{d}\mathbf{z}_t = \nabla\boldsymbol{\chi}(\mathbf{Z}, t)\mathrm{d}\mathbf{Z}.$$

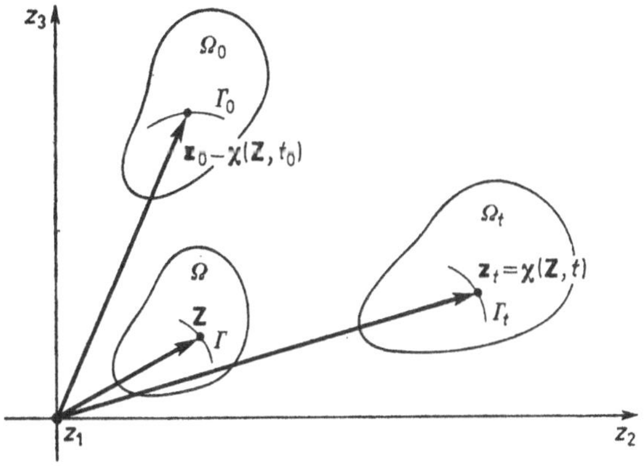

Fig. 5 Deformation of the line element

By denoting

$$\mathbf{C}(\mathbf{X}, \tau) \equiv \nabla\boldsymbol{\chi}^T(\mathbf{X}, \tau)\nabla\boldsymbol{\chi}(\mathbf{X}, \tau), \qquad \tau \in [t_0, t_f], \qquad \mathbf{X} \in \Omega,$$
$$\mathbf{G}(\mathbf{X}) \equiv \mathbf{C}(\mathbf{X}, t_0),$$

we obtain immediately

$$\mathrm{d}s_0^2 = \mathrm{d}\mathbf{Z} \cdot [\mathbf{G}(\mathbf{Z})\mathrm{d}\mathbf{Z}],$$
$$\mathrm{d}s_t^2 = \mathrm{d}\mathbf{Z} \cdot [\mathbf{C}(\mathbf{Z}, t)\mathrm{d}\mathbf{Z}].$$

Setting now

$$\mathbf{E}_t(\mathbf{X}) \equiv \tfrac{1}{2}[\mathbf{C}(\mathbf{X}, t) - \mathbf{G}(\mathbf{X})],$$

we arrive at the final relationship

$$\mathrm{d}s_t^2 - \mathrm{d}s_0^2 = 2\mathrm{d}\mathbf{Z} \cdot [\mathbf{E}_t(\mathbf{Z})\mathrm{d}\mathbf{Z}].$$

Hence we see that the tensor field $\mathbf{E}_t(\mathbf{X})$, $\mathbf{X} \in \Omega$, can be used as a strain measure if we pass from the initial deformation $\boldsymbol{\chi}(\cdot, t_0)$ to the actual deformation $\boldsymbol{\chi}(\cdot, t)$. Every $\mathbf{E}_t(\mathbf{X})$ is called the *Green–Saint Venant strain tensor*. By virtue of $\boldsymbol{\chi}(\mathbf{X}, t) = \boldsymbol{\chi}(\mathbf{X}, t_0) + \mathbf{u}_t(\mathbf{X})$, we arrive at the following strain-displacement relation

$$\mathbf{E}_t(\mathbf{X}) = \tfrac{1}{2}[\nabla\boldsymbol{\chi}^T(\mathbf{X}, t_0)\nabla\mathbf{u}_t(\mathbf{X}) + \nabla\mathbf{u}_t^T(\mathbf{X})\nabla\boldsymbol{\chi}(\mathbf{X}, t_0) + \nabla\mathbf{u}_t^T(\mathbf{X})\nabla\mathbf{u}_t(\mathbf{X})]. \quad (1.10)$$

Now define $\hat{\mathbf{C}}(\mathbf{X}, t) \equiv \mathbf{G}^{-1}(\mathbf{X})\mathbf{C}(\mathbf{X}, t)$ and take into account the secular equation

$$\det(\hat{\mathbf{C}} - \lambda\mathbf{1}) = 0, \quad \hat{\mathbf{C}} \equiv \hat{\mathbf{C}}(\mathbf{X}, t),$$

where $\mathbf{1}$ stands for the unit matrix. The principal values $\lambda_1, \lambda_2, \lambda_3$ of the symmetric matrix $\hat{\mathbf{C}} = \hat{\mathbf{C}}(\mathbf{X}, t)$ are called the *principal strains*; they are the roots of the aforementioned equation which can also be written in the form

$$-\lambda^3 + I_1\lambda^2 - I_2\lambda + I_3 = 0$$

where

$$I_1 \equiv \operatorname{tr}\hat{\mathbf{C}}, \quad I_2 \equiv \operatorname{tr}(\hat{\mathbf{C}}^{-1})I_3, \quad I_3 \equiv \det\hat{\mathbf{C}},$$

are called the *principal strain invariants.*

By the *strain rate tensor* we shall mean the value of the time derivative of functions $[t_0, t_f] \ni \tau \to \mathbf{E}_\tau(\mathbf{X}) \in R^{3\times3}$ at the time instant $\tau = t$ ($\mathbf{X}$ is here an arbitrary but fixed). The strain rate tensor for the material point $\mathbf{X}$ and at the time instant t will be denoted by $\dot{\mathbf{E}}_t(\mathbf{X})$. Hence we shall obtain the following relation

$$\dot{\mathbf{E}}_t(\mathbf{X}) = \tfrac{1}{2}[\nabla\boldsymbol{\chi}^T(\mathbf{X}, t)\nabla\dot{\mathbf{u}}_t(\mathbf{X}) + \nabla\dot{\mathbf{u}}_t^T(\mathbf{X})\nabla\boldsymbol{\chi}(\mathbf{X}, t)]. \tag{1.11}$$

The strain rate tensor defined by eq. (1.11) can also be called the *material strain rate tensor.* Setting

$$\mathbf{A}(\mathbf{X}, t) \equiv [\nabla\boldsymbol{\chi}(\mathbf{X}, t)]^{-1}$$

and

$$\mathbf{D}(\mathbf{X}, t) \equiv \mathbf{A}^T(\mathbf{X}, t)\dot{\mathbf{E}}_t(\mathbf{X})\mathbf{A}(\mathbf{X}, t),$$

we shall refer $\mathbf{D}(\mathbf{X}, t)$ to as the *spatial strain rate tensor.*

The fields of displacements $\mathbf{u}_t(\cdot)$, velocities $\dot{\mathbf{u}}_t(\cdot)$, accelerations $\ddot{\mathbf{u}}_t(\cdot)$, strain tensors $\mathbf{E}_t(\cdot)$ and strain rate tensors $\dot{\mathbf{E}}_t(\cdot)$ (or $\mathbf{D}(\cdot, t)$), defined on Ω, constitute the *basic kinematic fields* in solid and structural mechanics. It has to be emphasized that the kinematic fields are defined for a given motion of the body and for an arbitrary time instant.

1.3.3 Stresses and stress rates

Let $\boldsymbol{\chi}(\cdot, \tau) \in \mathscr{D}(\Omega)$, $\tau \in [t_0, t_f]$, be an arbitrary motion and $\Pi, \Pi \subset \Omega$, an arbitrary part of Ω such that $\bar{\Pi} \subset \Omega$. By $\mathbf{n}(\mathbf{z})$, $\mathbf{n}(\mathbf{z}) \in R^3$, $\mathbf{z} \in \partial\boldsymbol{\chi}(\Pi, t)$, we define the unit outer normal to the surface $\partial\boldsymbol{\chi}(\Pi, t)$ at the point $\mathbf{z}$ provided that such a normal is uniquely defined. The following lemma constitutes the basis for all considerations in which surface tractions are involved, cf. Sec. 1.1.4:

FUNDAMENTAL LEMMA. *In every motion* $\boldsymbol{\chi}(\cdot, \tau) \in \mathscr{D}(\Omega)$, $\tau \in [t_0, t_f]$, *and for every time instant* t, $t \in [t_0, t_f]$, *there exists a second order tensor field* $\mathbf{T}(\cdot, t)$ *defined on* Ω_t *(except possibly at some surfaces, lines or points) such that the relation*

$$\mathbf{s}_{\partial\boldsymbol{\chi}(\Pi,t)}(\mathbf{z}, t) = \mathbf{T}(\mathbf{z}, t)\mathbf{n}(\mathbf{z}),$$

holds for every part Π *of* Ω *and every* $\mathbf{z} \in \partial\boldsymbol{\chi}(\Pi, t)$ *for which* $\mathbf{T}(\mathbf{z}, t)$ *and* $\mathbf{n}(\mathbf{z})$ *are uniquely defined*[1].

Tensor $\mathbf{T}(\mathbf{z}, t)$, where $\mathbf{z} \in \Omega_t$, is called the *Cauchy stress tensor* or, briefly, the *stress tensor*. We assume that the stress tensor field $\mathbf{T}(\cdot, t)$ is continuous in Ω (except possibly at some surfaces, lines or points), is defined on $\bar{\Omega}_t$[2] and some additional smoothness conditions may be imposed upon the stress field.

The eigenvalues of the linear mapping given by the matrix $\mathbf{T}(\mathbf{z}, t)$ are called the *principal stresses* and the corresponding eigenvectors are called the *principal directions of stress*. The quantity $\sigma \equiv \frac{1}{3}\operatorname{tr}\mathbf{T}$ is called the *hydrostatic pressure* (for the sake of compactness we neglect here the dependence of $\mathbf{T}$ on its arguments). The symmetric tensor defined by $\mathbf{T}_D \equiv \mathbf{T} - \sigma\mathbf{1}$, $\mathbf{1}$ being the unit tensor, is called the *stress deviator*. It is seen that $\operatorname{tr}\mathbf{T}_D = 0$. Hence the stress deviator has only two independent invariants $I_D \equiv \operatorname{tr}(\mathbf{T}_D\mathbf{T}_D)$ and $II_D \equiv \operatorname{tr}(\mathbf{T}_D\mathbf{T}_D\mathbf{T}_D)$. For a more detailed discussion of the state of stress the reader is referred to any textbook on continuum mechanics.

Define now $J(\mathbf{X}, t) \equiv \det\nabla\boldsymbol{\chi}(\mathbf{X}, t)$, $\mathbf{X} \in \Omega$, $t \in [t_0, t_f]$. Moreover, let $\mathbf{n}(\mathbf{X})$, $\mathbf{X} \in \partial\Pi$, stands for the unit outer normal to $\partial\Pi$ at $\mathbf{X}$. Bearing in mind the definition $\mathbf{A}(\mathbf{X}, t) \equiv [\nabla\boldsymbol{\chi}(\mathbf{X}, t)]^{-1}$, we introduce

$$\mathbf{S}_t(\mathbf{X}) \equiv J(\mathbf{X}, t)\mathbf{A}(\mathbf{X}, t)\mathbf{T}(\boldsymbol{\chi}(\mathbf{X}, t), t)\mathbf{A}^T(\mathbf{X}, t), \quad \mathbf{X} \in \Omega, \ t \in [t_0, t_f].$$

Then it can be shown that the densities of the surface tractions $\mathbf{s}_{\partial\Pi}(\mathbf{X}, t)$, $\mathbf{X} \in \partial\Pi$, cf. Sec. 1.1.4, related to the undeformed body Ω, are given by

$$\mathbf{s}_{\partial\Pi}(\mathbf{X}, t) = \nabla\boldsymbol{\chi}(\mathbf{X}, t)\mathbf{S}_t(\mathbf{X})\mathbf{n}(\mathbf{X}), \quad \mathbf{X} \in \partial\Pi, t \in [t_0, t_f]. \tag{1.12}$$

It follows from the above relation that the tensor $\mathbf{S}_t(\mathbf{X})$ plays a role analogous to that of the stress tensor $\mathbf{T}(\mathbf{z}, t)$, $\mathbf{z} \in \Omega_t$. Tensor $\mathbf{S}_t(\mathbf{X})$ is called the *second Piola–Kirchhoff stress tensor*. It has to be emphasized that the tensor field $\mathbf{S}_t(\cdot)$ is defined on Ω, i.e. it is related to an undeformed body Ω.

Let $\mathbf{X}$ be an arbitrary material point, $\mathbf{X} \in \Omega$. By the *material stress rate tensor at* $\mathbf{X}$ *and at the time instant* t, we shall mean the value of the time derivative

---

[1] For the proof of the lemma the reader is referred to [158].

[2] It means that $\mathbf{T}(\mathbf{z}, t)\mathbf{n}(\mathbf{z})$ are defined also on $\partial\Omega_t$ except possibly at a finite number of lines or points, $\mathbf{n}(\mathbf{z})$ being the unit normal to $\partial\Omega_t$.

$\dot{\mathbf{S}}_t(\mathbf{X})$ of a function $[t_0, t_f] \ni \tau \to \mathbf{S}_\tau(\mathbf{X}) \in R^{3\times 3}$ at $\tau = t$ (provided it exists). The spatial stress rate will be defined by

$$\mathring{\mathbf{T}}(\mathbf{X}, t) \equiv \nabla\boldsymbol{\chi}(\mathbf{X}, t)\dot{\mathbf{S}}_t(\mathbf{X})\nabla\boldsymbol{\chi}^T(\mathbf{X}, t)J^{-1}(\mathbf{X}, t).$$

The second Piola–Kirchhoff stress tensor fields $\mathbf{S}_t(\cdot)$ and the stress rate tensor fields $\dot{\mathbf{S}}_t(\cdot)$ defined on Ω are the basic dynamic fields in problems which are the subject of the present book. The final remark given in Sec. 1.3.2 is valid also for dynamic fields.

1.3.4 Heat fluxes

Using the notation introduced at the beginning of Sec. 1.3.3 the following lemma is the basis for all the problems in which a surface heat supply is involved:

FUNDAMENTAL LEMMA. *In every motion $\boldsymbol{\chi}(\cdot, \tau) \in \mathscr{D}(\Omega)$, $\tau \in [t_0, t_f]$, and for every fixed time instant t, $t \in [t_0, t_f]$, there exists a vector field $\mathbf{h}(\cdot, t)$ defined on Ω_t (except possibly at some surfaces, lines or points) such that the relation*

$$\pi_{\partial\boldsymbol{\chi}(\Pi, t)}(\mathbf{z}, t) = \mathbf{h}(\mathbf{z}, t) \cdot \mathbf{n}(\mathbf{z})$$

holds for every part Π of Ω and every $\mathbf{z} \in \partial\boldsymbol{\chi}(\Pi, t)$ for which $\mathbf{h}(\mathbf{z}, t)$ and $\mathbf{n}(\mathbf{z})$ are defined [1].

The vector $\mathbf{h}(\mathbf{z}, t)$, where $\mathbf{z} \in \Omega_t$, is called the *heat flux vector.* We assume that the vector field $\mathbf{h}(\cdot, t)$ satisfies regularity conditions which are analogous to those imposed on the stress field $\mathbf{T}(\cdot, t)$. Setting

$$\mathbf{h}_t(\mathbf{X}) \equiv J(\mathbf{X}, t)\mathbf{A}(\mathbf{X}, t)\mathbf{h}(\boldsymbol{\chi}(\mathbf{X}, t), t), \qquad \mathbf{X} \in \Omega, t \in [t_0, t_f],$$

it can be shown that (cf. Sec. 1.1.6 for the definition of $\pi_{\partial\Pi}(\mathbf{X}, t)$)

$$\pi_{\partial\Pi}(\mathbf{X}, t) = \mathbf{h}_t(\mathbf{X}) \cdot \mathbf{n}(\mathbf{X}). \tag{1.13}$$

The vector $\mathbf{h}_t(\mathbf{X})$ will be called the *material heat flux.* The vector fields $\mathbf{h}_t(\cdot)$, defined on Ω, together with the temperature fields $\vartheta_t(\cdot) \equiv \theta(\cdot, t) - \theta_0$ (note that θ_0 is a certain reference temperature), are the basic thermal fields in all the non-mechanical problems to be discussed throughout the book.

1.3.5 Matrix notation for the basic fields

The theoretical foundations of thermo-mechanics, which are introduced and explained throughout the first part of the book, are formulated in terms of the absolute notation of the tensor calculus. Such a notation constitutes a proper and convenient mathematical tool for the general continuum theory.

[1] The proof of the lemma can be found in [158].

However, in all applications of the general theory to problems of structural mechanics we shall use another notation known as the matrix formalism. Such a notation enables us to express the governing relations of structural mechanics in a form most suitable for numerical calculations. This is why we shall be switching from tensor to matrix notation whenever a specific structural mechanics problem is encountered. The way in which we transform the tensor equations into their equivalent matrix forms is outlined below.

Let t be an arbitrary but fixed time instant, $t \in [t_0, t_f]$; for the sake of simplicity the dependence of all functions on the time variable t will not be explicitly indicated in the matrix notation. Hence an arbitrary displacement vector defined by eq. (1.9) is denoted by a column matrix (i.e. a vector)

$$\mathbf{u} \equiv \begin{Bmatrix} u_1(\mathbf{X}) \\ u_2(\mathbf{X}) \\ u_3(\mathbf{X}) \end{Bmatrix}.$$

Similarly, by $\dot{\mathbf{u}}$ and $\ddot{\mathbf{u}}$ we denote column matrices representing the velocity vector $\dot{\mathbf{u}}_t(\mathbf{X})$ and the acceleration vector $\ddot{\mathbf{u}}_t(\mathbf{X})$, respectively. On passing from the displacement fields to the strain measures we represent an arbitrary Green–Saint Venant strain tensor in the following matrix form

$$\boldsymbol{\epsilon} \equiv \{E_{11}(\mathbf{X}) \;\; E_{22}(\mathbf{X}) \;\; E_{33}(\mathbf{X}) \;\; \sqrt{2}E_{12}(\mathbf{X}) \;\; \sqrt{2}E_{23}(\mathbf{X}) \;\; \sqrt{2}E_{31}(\mathbf{X})\}.$$

Similarly, by $\dot{\boldsymbol{\epsilon}}$ we denote a vector representing the strain rate tensor

$$\dot{\boldsymbol{\epsilon}} \equiv \{\dot{E}_{11}(\mathbf{X}) \;\; \dot{E}_{22}(\mathbf{X}) \;\; \dot{E}_{33}(\mathbf{X}) \;\; \sqrt{2}\dot{E}_{12}(\mathbf{X}) \;\; \sqrt{2}\dot{E}_{23}(\mathbf{X}) \;\; \sqrt{2}\dot{E}_{31}(\mathbf{X})\}.$$

It has to be emphasized that in the aforementioned definitions we have tacitly taken into account the fact that $\mathbf{E}(\mathbf{X})$ and $\dot{\mathbf{E}}(\mathbf{X})$ are symmetric second order tensors and hence they can be represented by merely six independent components. This is why the vectors $\boldsymbol{\epsilon}$ and $\dot{\boldsymbol{\epsilon}}$ consist of six components only.

Let $\nabla\mathbf{u}$ be a 3×3 matrix with components $u_{i,\beta}(\mathbf{X})$. The interrelation between $\boldsymbol{\epsilon}$ and $\nabla\mathbf{u}$ follows directly from eq. (1.10) and is given as

$$E_{\alpha\beta}(\mathbf{X}) = \tfrac{1}{2}[\chi^k{}_{,\alpha}(\mathbf{X}, t_0)u_{k,\beta}(\mathbf{X}) + \chi^k{}_{,\beta}(\mathbf{X}, t_0)u_{k,\alpha}(\mathbf{X}) + u^k{}_{,\alpha}(\mathbf{X})u_{k,\beta}(\mathbf{X})].$$

Similarly, let $\nabla\dot{\mathbf{u}}$ be a 3×3 matrix with components $\dot{u}_{i,\beta}(\mathbf{X})$. Then the interrelation between $\dot{\boldsymbol{\epsilon}}$ and $\nabla\dot{\mathbf{u}}$ is determined by eq. (1.11) as

$$\dot{E}_{\alpha\beta}(\mathbf{X}) = \tfrac{1}{2}[\chi^k{}_{,\alpha}(\mathbf{X})\dot{u}_{k,\beta}(\mathbf{X}) + \chi^k{}_{,\beta}(\mathbf{X})\dot{u}_{k,\alpha}(\mathbf{X})],$$

where we have by means of eq. (1.9)

$$\chi^k{}_{,\alpha}(\mathbf{X}) = \chi^k{}_{,\alpha}(\mathbf{X}, t_0) + u^k{}_{,\alpha}(\mathbf{X})$$

keeping in mind that all the basic kinematic fields are dependent on the time variable t, $t \in [t_0, t_f]$. It can be easily seen that the interrelation between $\dot{\boldsymbol{\epsilon}}$

and $\nabla\dot{\mathbf{u}}$ is linear (but dependent on the displacement gradient matrix $\nabla\mathbf{u}$) while the interrelation between $\boldsymbol{\epsilon}$ and $\nabla\mathbf{u}$ is nonlinear.

On passing to the matrix notation for stresses and stress rates we take into account the fact (to be proved in Sec. 1.4.1) that the second Piola–Kirchhoff stress tensor $\mathbf{S}(\mathbf{X})$ as well as the corresponding stress rate tensor $\dot{\mathbf{S}}(\mathbf{X})$ are symmetric second order tensors and hence they are determined by six independent components only. Thus, in the matrix notation they are represented by means of the column matrices

$$\boldsymbol{\sigma} \equiv \left\{S^{11}(\mathbf{X}) \;\; S^{22}(\mathbf{X}) \;\; S^{33}(\mathbf{X}) \;\; \sqrt{2}S^{12}(\mathbf{X}) \;\; \sqrt{2}S^{23}(\mathbf{X}) \;\; \sqrt{2}S^{31}(\mathbf{X})\right\},$$
$$\dot{\boldsymbol{\sigma}} \equiv \left\{\dot{S}^{11}(\mathbf{X}) \;\; \dot{S}^{22}(\mathbf{X}) \;\; \dot{S}^{33}(\mathbf{X}) \;\; \sqrt{2}\dot{S}^{12}(\mathbf{X}) \;\; \sqrt{2}\dot{S}^{23}(\mathbf{X}) \;\; \sqrt{2}\dot{S}^{31}(\mathbf{X})\right\}.$$

Let $\mathbf{s}_{\partial\Pi}$ be the surface traction vector defined by (cf. Sec. 1.1.4)

$$\mathbf{s}_{\partial\Pi} \equiv \{s^1_{\partial\Pi}(\mathbf{X}) \;\; s^2_{\partial\Pi}(\mathbf{X}) \;\; s^3_{\partial\Pi}(\mathbf{X})\}.$$

The interrelation between $\boldsymbol{\sigma}$ and $\mathbf{s}_{\partial\Pi}$ is given by eq. (1.12):

$$s^k_{\partial\Pi}(\mathbf{X}) = \chi^k{}_{,\alpha}(\mathbf{X})S^{\alpha\beta}(\mathbf{X})n_\beta(\mathbf{X}), \qquad \mathbf{X} \in \partial\Pi,$$

where $\chi^k{}_{,\alpha}(\mathbf{X}) = \chi^k{}_{,\alpha}(\mathbf{X}, t_0)+u^k{}_{,\alpha}(\mathbf{X})$ in accordance with eq. (1.9).

Finally, the material heat flux is denoted by the column matrix

$$\mathbf{h} \equiv \begin{Bmatrix} h^1(\mathbf{X}) \\ h^2(\mathbf{X}) \\ h^3(\mathbf{X}) \end{Bmatrix}$$

which is interrelated with the surface heat supply $\pi_{\partial\Pi} = \pi_{\partial\Pi}(\mathbf{X})$ by means of eq. (1.13)

$$\pi_{\partial\Pi} = h^\alpha(\mathbf{X})n_\alpha(\mathbf{X}), \qquad \mathbf{X} \in \partial\Pi.$$

We conclude that in problems of structural mechanics the values of the basic fields for any $\mathbf{X} \in \Omega$ and for an arbitrary but fixed time instant t, $t \in [t_0, t_f]$, can be represented by:

1. The displacement vector $\mathbf{u}$ together with the velocity and acceleration vectors $\dot{\mathbf{u}}, \ddot{\mathbf{u}}$, respectively, and the 3×3 gradient matrices $\nabla\mathbf{u}$, $\nabla\dot{\mathbf{u}}$,
2. The strain vector $\boldsymbol{\epsilon}$ and the strain rate vector $\dot{\boldsymbol{\epsilon}}$,
3. The (second Piola–Kirchhoff) stress vector $\boldsymbol{\sigma}$ and the stress rate vector $\dot{\boldsymbol{\sigma}}$,
4. The material heat flux vector $\mathbf{h}$,
5. The values of scalar fields representing temperature $\vartheta = \vartheta_t(\mathbf{X})$, internal energy $\varepsilon = \varepsilon(\mathbf{X}, t)$, specific entropy $\eta = \eta(\mathbf{X}, t)$ and free energy $\varphi = \varphi(\mathbf{X}, t)$, cf. Sec. 1.1.5.

1.4 Resulting relations

1.4.1 Equation of motion

Taking into account the results of Sec. 1.1.4 by virtue of $\ddot{\mathbf{u}}_t(\mathbf{X}) = \ddot{\boldsymbol{\chi}}(\mathbf{X}, t)$, let us rewrite the condition representing the balance of momentum in the form

$$\int_{\Pi} \ddot{\mathbf{u}}_t(\mathbf{X})\varrho(\mathbf{X})\,\mathrm{d}V(\mathbf{X}) = \oint_{\partial\Pi} \nabla\boldsymbol{\chi}(\mathbf{X}, t)\mathbf{S}_t(\mathbf{X})\mathbf{n}(\mathbf{X})\,\mathrm{d}A(\mathbf{X}) + \int_{\Pi} \mathbf{b}(\mathbf{X}, t)\,\mathrm{d}V(\mathbf{X}).$$

Assume now that the function $\nabla\boldsymbol{\chi}(\cdot, t)\mathbf{S}_t(\cdot)$ is continuous together with its first derivative in $\bar{\Omega}$. Then for $\bar{\Pi} \subset \Omega$ the condition

$$\oint_{\partial\Pi} \nabla\boldsymbol{\chi}(\mathbf{X}, t)\mathbf{S}_t(\mathbf{X})\mathbf{n}(\mathbf{X})\,\mathrm{d}A(\mathbf{X}) = \int_{\Pi} \mathrm{Div}[\nabla\boldsymbol{\chi}(\mathbf{X}, t)\mathbf{S}_t(\mathbf{X})]\,\mathrm{d}V(\mathbf{X})$$

can easily be proved by means of the known divergence theorem[1].

Let us further assume that $\mathbf{b}(\cdot, t) - \ddot{\mathbf{u}}_t(\cdot)\varrho(\cdot)$ is continuous in Ω. Substituting $\Pi = B(\mathbf{X}_0, r)$ where $B(\mathbf{X}_0, r)$ is a ball with the center $\mathbf{X}_0$ and the radius r, dividing the expressions on both sides of the equation

$$\int_{B(\mathbf{X}_0, r)} \{\mathrm{Div}[\nabla\boldsymbol{\chi}(\mathbf{X}, t)\mathbf{S}_t(\mathbf{X})] + \mathbf{b}(\mathbf{X}, t) - \varrho(\mathbf{X})\ddot{\mathbf{u}}_t(\mathbf{X})\}\,\mathrm{d}V(\mathbf{X}) = 0$$

by the volume $\mathrm{vol}\, B(\mathbf{X}_0, r)$ of $B(\mathbf{X}_0, r)$ and passing to the limit as $r \to 0$, by means of the continuity of the integrand and after replacing $\mathbf{X}_0$ by $\mathbf{X}$ we arrive at the equation

$$\mathrm{Div}[\nabla\boldsymbol{\chi}(\mathbf{X}, t)\mathbf{S}_t(\mathbf{X})] + \mathbf{b}(\mathbf{X}, t) = \varrho(\mathbf{X})\ddot{\mathbf{u}}_t(\mathbf{X})$$

which holds in Ω. The equation obtained is known as the *Cauchy law of motion.*

Let now Π be a layer as indicated in Fig. 6. Applying the balance of momentum to a region Π, bearing in mind that $\mathbf{s}(\mathbf{X}, t) = \mathbf{s}_{\partial\Pi}(\mathbf{X}, t)$, cf. Sec. 1.1.4, passing to the limit as $h \to 0$ and $l \to 0$ and employing the assumption that in $\bar{\Pi}$ all the functions considered are continuous, we arrive at the condition

$$\mathbf{s}(\mathbf{X}, t) = \nabla\boldsymbol{\chi}(\mathbf{X}, t)\mathbf{S}_t(\mathbf{X})\mathbf{n}(\mathbf{X}),$$

which has to hold on $\partial\Omega$ (except at points and lines where the unit outer normal $\mathbf{n}(\mathbf{X})$ is not uniquely defined, cf. points A, B, C in Fig. 6 at which $\mathbf{S}_t(\mathbf{X})$ suffers discontinuities).

Let $\mathbf{v}(\mathbf{X}) \in R^3$, $\mathbf{X} \in \bar{\Omega}$, be an arbitrary vector function which is continuous in $\bar{\Omega}$ together with its first derivatives $\nabla\mathbf{v}(\mathbf{X})$. In other words, let $\mathbf{v}(\cdot) \in C_1^3(\bar{\Omega})$.

---

[1] In the sequel we shall assume that $\partial\Pi$ and $\partial\Omega$ are the Lipschitz boundaries, for which the divergence theorem holds.

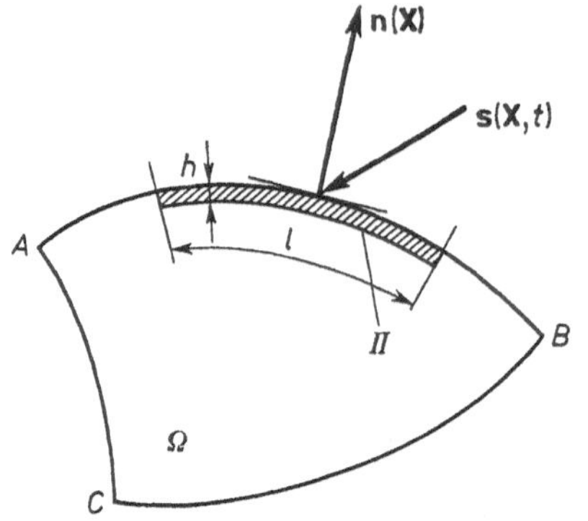

Fig. 6 Material layer of the body

Then

$$\int_{\Omega} \{\mathrm{Div}[\nabla\boldsymbol{\chi}(\mathbf{X},t)\mathbf{S}_t(\mathbf{X})] + \mathbf{b}(\mathbf{X},t) - \varrho(\mathbf{X})\ddot{\mathbf{u}}_t(\mathbf{X})\} \cdot \mathbf{v}(\mathbf{X})\,dV(\mathbf{X}) +$$
$$+ \oint_{\partial\Omega} [\mathbf{s}(\mathbf{X},t) - \nabla\boldsymbol{\chi}(\mathbf{X},t)\mathbf{S}_t(\mathbf{X})\mathbf{n}(\mathbf{X})] \cdot \mathbf{v}(\mathbf{X})\,dA(\mathbf{X}) = 0.$$

Using the divergence theorem we arrive finally at the following condition

$$\int_{\Omega} \mathrm{tr}\{\mathbf{S}_t(\mathbf{X})[\nabla\boldsymbol{\chi}^T(\mathbf{X},t)\nabla\mathbf{v}(\mathbf{X})]\}\,dV(\mathbf{X})$$
$$= \oint_{\partial\Omega} \mathbf{s}(\mathbf{X},t) \cdot \mathbf{v}(\mathbf{X})\,dA(\mathbf{X}) + \int_{\Omega} [\mathbf{b}(\mathbf{X},t) - \varrho(\mathbf{X})\ddot{\mathbf{u}}_t(\mathbf{X})] \cdot \mathbf{v}(\mathbf{X})\,dV(\mathbf{X}) \tag{1.14}$$

which has to hold for every $\mathbf{v}(\cdot) \in C_1^3(\bar{\Omega})$ and where $\boldsymbol{\chi}(\mathbf{X},t) = \boldsymbol{\chi}(\mathbf{X},t_0) + \mathbf{u}_t(\mathbf{X})$. In the following the relation (1.14) will be referred to as the (integral or weak) equation of motion. Functions $\mathbf{v}(\cdot)$ for which eq. (1.14) has to hold are called the *test functions.*

Let us now take into account the principle of balance of moment of momentum. By means of $\mathbf{s}_{\partial\Pi}(\mathbf{X},t) = \nabla\boldsymbol{\chi}(\mathbf{X},t)\mathbf{S}_t(\mathbf{X})\mathbf{n}(\mathbf{X})$ and noting that $\dot{\boldsymbol{\chi}}(\mathbf{X},t) \times \dot{\boldsymbol{\chi}}(\mathbf{X},t) = \mathbf{0}$, we obtain

$$\int_{\Pi} [\boldsymbol{\chi}(\mathbf{X},t) - \mathbf{z}_0] \times \ddot{\boldsymbol{\chi}}(\mathbf{X},t)\varrho(\mathbf{X})\,dV(\mathbf{X})$$
$$= \oint_{\partial\Pi} [\boldsymbol{\chi}(\mathbf{X},t) - \mathbf{z}_0] \times [\nabla\boldsymbol{\chi}(\mathbf{X},t)\mathbf{S}_t(\mathbf{X})\mathbf{n}(\mathbf{X})]\,dA(\mathbf{X}) +$$
$$+ \int_{\Pi} [\boldsymbol{\chi}(\mathbf{X},t) - \mathbf{z}_0] \times \mathbf{b}(\mathbf{X},t)\,dV(\mathbf{X}).$$

Applying the divergence theorem to the surface integral, taking into account the Cauchy law of motion and using the procedure outlined above we arrive at the condition[1]

$$\operatorname{ant}[\nabla\boldsymbol{\chi}(\mathbf{X}, t)\mathbf{S}_t(\mathbf{X})\nabla\boldsymbol{\chi}^T(\mathbf{X}, t)] = 0.$$

It follows that the second Piola–Kirchhoff stress tensor is symmetric, that is

$$\mathbf{S}_t(\mathbf{X}) = \mathbf{S}_t^T(\mathbf{X}) \tag{1.15}$$

for every $\mathbf{X} \in \Omega$ for which $\mathbf{S}_t(\mathbf{X})$ exists. Thus we also conclude that the Cauchy stress tensor $\mathbf{T}(\mathbf{X}, t)$ is symmetric, cf. Sec. 1.3.3.

1.4.2 Energy equation

Introducing the notions of the stress tensor and the heat flux vector, cf. Sec. 1.3.4, into the principle of balance of energy postulated in Sec. 1.2.3, we obtain

$$\int_{\Pi} [\varrho(\mathbf{X})\ddot{\boldsymbol{\chi}}(\mathbf{X}, t)\cdot\dot{\boldsymbol{\chi}}(\mathbf{X}, t)+\varrho(\mathbf{X})\dot{\varepsilon}(\mathbf{X}, t)]\,\mathrm{d}V(\mathbf{X})$$
$$= \oint_{\partial\Pi} [\dot{\boldsymbol{\chi}}(\mathbf{X}, t)\cdot\nabla\boldsymbol{\chi}(\mathbf{X}, t)\mathbf{S}_t(\mathbf{X})\mathbf{n}(\mathbf{X})+\mathbf{h}_t(\mathbf{X})\cdot\mathbf{n}(\mathbf{X})]\,\mathrm{d}A(\mathbf{X})+$$
$$+\int_{\Pi} [\dot{\boldsymbol{\chi}}(\mathbf{X}, t)\cdot\mathbf{b}(\mathbf{X}, t)+\alpha(\mathbf{X}, t)]\,\mathrm{d}V(\mathbf{X}) = 0.$$

Using the divergence theorem to the surface integral (under the assumption that $\dot{\boldsymbol{\chi}}(\mathbf{X}, t)\cdot\nabla\boldsymbol{\chi}(\mathbf{X}, t)\mathbf{S}_t(\mathbf{X})$ and $\mathbf{h}_t(\mathbf{X})$, $\mathbf{X} \in \bar{\Omega}$, are continuous with their first derivatives in $\bar{\Omega}$), recalling the notion of the strain rate tensor $\dot{\mathbf{E}}_t(\mathbf{X})$ and taking into account the Cauchy law of motion we arrive at the condition

$$\int_{\Pi} \{\operatorname{Div}\mathbf{h}_t(\mathbf{X})+\operatorname{tr}[\mathbf{S}_t(\mathbf{X})\dot{\mathbf{E}}_t(\mathbf{X})]+\alpha(\mathbf{X}, t)-\varrho(\mathbf{X})\dot{\varepsilon}(\mathbf{X}, t)\}\,\mathrm{d}V(\mathbf{X}) = 0$$

which has to hold for every part Π of the body. By means of the approach analogous to that used in Sec. 1.3.3 we obtain in Ω the condition which is called the *local form of the energy balance* as

$$\operatorname{Div}\mathbf{h}_t(\mathbf{X})+\operatorname{tr}[\mathbf{S}_t(\mathbf{X})\dot{\mathbf{E}}_t(\mathbf{X})]+\alpha(\mathbf{X}, t) = \varrho(\mathbf{X})\dot{\varepsilon}(\mathbf{X}, t).$$

Similarly as in Sec. 1.3.3 we obtain on $\partial\Omega$ (except at points and lines where the unit normal $\mathbf{n}(\mathbf{X})$ and $\mathbf{h}_t(\mathbf{X})$ are not uniquely defined):

$$\pi(\mathbf{X}, t) = \mathbf{h}_t(\mathbf{X})\cdot\mathbf{n}(\mathbf{X}).$$

[1] The symbol ant $\mathbf{M}$, where $\mathbf{M}$ is an arbitrary square matrix, stands for the antisymmetric part of $\mathbf{M}$, i.e. $\operatorname{ant}\mathbf{M} \equiv \frac{1}{2}(\mathbf{M}-\mathbf{M}^T)$, where as usual $\mathbf{M}^T$ is the transpose of $\mathbf{M}$.

Let $\zeta(\mathbf{X}) \in \boldsymbol{R}$, $\mathbf{X} \in \bar{\Omega}$, be an arbitrary scalar function which is continuous in $\bar{\Omega}$ with its first derivatives $\nabla\zeta(\mathbf{X})$, i.e. let $\zeta(\cdot) \in C_1(\bar{\Omega})$. Then

$$\int_{\Omega} \{\mathrm{Div}\,\mathbf{h}_t(\mathbf{X}) + \mathrm{tr}[\mathbf{S}_t(\mathbf{X})\dot{\mathbf{E}}_t(\mathbf{X})] + \alpha(\mathbf{X}, t) - \varrho(\mathbf{X})\dot{\varepsilon}(\mathbf{X}, t)\}\zeta(\mathbf{X})\mathrm{d}V(\mathbf{X}) +$$

$$+ \oint_{\partial\Omega} [\pi(\mathbf{X}, t) - \mathbf{h}_t(\mathbf{X}) \cdot \mathbf{n}(\mathbf{X})]\zeta(\mathbf{X})\mathrm{d}A(\mathbf{X}) = 0.$$

Using the divergence theorem we finally obtain the condition

$$\int_{\Omega} \mathbf{h}_t(\mathbf{X}) \cdot \nabla\zeta(\mathbf{X})\mathrm{d}V(\mathbf{X})$$

$$= \oint_{\partial\Omega} \pi(\mathbf{X}, t)\zeta(\mathbf{X})\mathrm{d}A(\mathbf{X}) + \int_{\Omega} \alpha(\mathbf{X}, t)\zeta(\mathbf{X})\mathrm{d}V(\mathbf{X}) +$$

$$+ \int_{\Omega} \{\mathrm{tr}[\mathbf{S}_t(\mathbf{X})\dot{\mathbf{E}}_t(\mathbf{X})] - \varrho(\mathbf{X})\,\dot{\varepsilon}(\mathbf{X}, t)\}\zeta(\mathbf{X})\mathrm{d}V(\mathbf{X}) \tag{1.16}$$

which has to hold for every $\zeta(\cdot) \in C_1(\bar{\Omega})$. The relation obtained will be called the (integral or weak) *energy equation*.

1.4.3 Dissipation inequality

The dissipation principle $\gamma_t(\Pi) \geqslant 0$ postulated in Sec. 1.2.4 can also be written by using the relation $\pi_{\partial\Pi}(\mathbf{X}, t) = \mathbf{h}_t(\mathbf{X}) \cdot \mathbf{n}(\mathbf{X})$ in the form of a condition

$$\int_{\Pi} \varrho(\mathbf{X})\dot{\eta}(\mathbf{X}, t)\mathrm{d}V(\mathbf{X}) - \oint_{\partial\Pi} \frac{\mathbf{h}_t(\mathbf{X}) \cdot \mathbf{n}(\mathbf{X})}{\theta(\mathbf{X}, t)}\mathrm{d}A(\mathbf{X}) + \int_{\Pi} \frac{\alpha(\mathbf{X}, t)}{\theta(\mathbf{X}, t)}\mathrm{d}V(\mathbf{X}) \geqslant 0$$

which has to hold for every part Π of the body. Applying the divergence theorem to the surface integral (under the assumption that the function $\mathbf{h}_t(\mathbf{X})/\theta(\mathbf{X}, t)$, $\mathbf{X} \in \bar{\Omega}$, is continuous in $\bar{\Omega}$ together with its first order derivatives) and using the procedure analogous to that presented in Sec. 1.4.1 and 1.4.2 we arrive at the inequality

$$\varrho(\mathbf{X})\dot{\eta}(\mathbf{X}, t)\theta(\mathbf{X}, t) - \mathrm{Div}\,\mathbf{h}_t(\mathbf{X}) + \frac{\mathbf{h}_t(\mathbf{X}) \cdot \nabla\theta(\mathbf{X}, t)}{\theta(\mathbf{X}, t)} - \alpha(\mathbf{X}, t) \geqslant 0$$

which has to hold in Ω provided the functions $\alpha(\cdot, t)$, $\theta(\cdot, t)$, $\varrho(\cdot)$ and $\dot{\eta}(\cdot, t)$ are continuous in Ω. This condition is known as the *local form of the dissipation inequality*. Taking into account that

$$\mathrm{Div}\,\mathbf{h}_t(\mathbf{X}) + \alpha(\mathbf{X}, t) = \varrho(\mathbf{X})\,\dot{\varepsilon}(\mathbf{X}, t) - \mathrm{tr}[\mathbf{S}_t(\mathbf{X})\dot{\mathbf{E}}_t(\mathbf{X})],$$

which was derived in Sec. 1.4.2, we arrive at the alternative local form of the dissipation inequality

$$\varrho(\mathbf{X})[\dot{\eta}(\mathbf{X}, t)\theta(\mathbf{X}, t)-\dot{\varepsilon}(\mathbf{X}, t)]+\mathrm{tr}[\mathbf{S}_t(\mathbf{X})\dot{\mathbf{E}}_t(\mathbf{X})]+\frac{\mathbf{h}_t(\mathbf{X})\cdot\nabla\theta(\mathbf{X}, t)}{\theta(\mathbf{X}, t)} \geqslant 0.$$

Introducing now the free energy $\varphi(\mathbf{X}, t)$, cf. Sec. 1.1.5, and using the relationships

$$\dot{\varphi}(\mathbf{X}, t) = \dot{\varepsilon}(\mathbf{X}, t)-\dot{\eta}(\mathbf{X}, t)\theta(\mathbf{X}, t)-\eta(\mathbf{X}, t)\dot{\theta}(\mathbf{X}, t),$$

$$\dot{\theta}(\mathbf{X}, t) = \dot{\vartheta}_t(\mathbf{X}), \quad \nabla\theta(\mathbf{X}, t) = \nabla\vartheta_t(\mathbf{X}),$$

we finally obtain

$$-\varrho(\mathbf{X})[\dot{\varphi}(\mathbf{X}, t)+\eta(\mathbf{X}, t)\dot{\vartheta}_t(\mathbf{X})]+\mathrm{tr}[\mathbf{S}_t(\mathbf{X})\dot{\mathbf{E}}_t(\mathbf{X})]+\frac{\mathbf{h}_t(\mathbf{X})\cdot\nabla\vartheta_t(\mathbf{X})}{\theta(\mathbf{X}, t)} \geqslant 0 \tag{1.17}$$

where $\theta(\mathbf{X}, t) = \theta_0+\vartheta_t(\mathbf{X})$. The meaning of the dissipation inequality (1.17) will be explained in Sec. 2.1.2.

1.4.4 *Matrix form of the resulting equations*

Bearing in mind the matrix notation introduced in Sec. 1.3.5 we shall now rewrite the integral equation of motion (1.14) and the integral energy equation (1.16) using this notation. To this aim let us introduce a column vector

$$\mathbf{v} \equiv \begin{Bmatrix} v_1(\mathbf{X}) \\ v_2(\mathbf{X}) \\ v_3(\mathbf{X}) \end{Bmatrix}$$

defined for every $\mathbf{X} \in \bar{\Omega}$, such that $\mathbf{v}(\cdot)$ is an arbitrary test function, cf. Sec. 1.4.1. Define next

$$\mathbf{L_F v} \equiv \{(L_F v)_{11} \;\; (L_F v)_{22} \;\; (L_F v)_{33} \;\; 2(L_F v)_{12} \;\; 2(L_F v)_{23} \;\; 2(L_F v)_{31}\},$$

setting $\mathbf{F} \equiv \nabla\boldsymbol{\chi}(\mathbf{X}, t)$ and

$$(L_F v)_{\alpha\beta} \equiv \tfrac{1}{2}[F^k{}_{,\alpha}(\mathbf{X})v_{k,\beta}(\mathbf{X})+F^k{}_{,\beta}(\mathbf{X})v_{k,\alpha}(\mathbf{X})],$$

where

$$F^k{}_{,\alpha} \equiv \chi^k{}_{,\alpha}(\mathbf{X}) = \chi^k{}_{,\alpha}(\mathbf{X}, t_0)+u^k{}_{,\alpha}(\mathbf{X}),$$

and where the simplified notation of Sec. 1.3.5 has been employed. Mind that every $\mathbf{L_F v}$ is an element of the six dimensional vector space $\boldsymbol{R}^6$. Let $\mathbf{a}, \mathbf{b} \in \boldsymbol{R}^6$ and define $\mathbf{a} \circ \mathbf{b} = \sum a_k b_k$ where the summation index "k" runs over the sequence $1, \dots, 6$. Then it can be easily verified that

$$\mathrm{tr}\{\mathbf{S}_t(\mathbf{X})[\nabla\boldsymbol{\chi}^T(\mathbf{X}, t)\nabla\mathbf{v}(\mathbf{X})]\} = \boldsymbol{\sigma} \circ \mathbf{L_F v}$$

with the meaning of the symbol $\boldsymbol{\sigma}$ explained in Sec. 1.3.5. Replacing for simplicity $\mathrm{d}V(\mathbf{X})$, $\mathrm{d}A(\mathbf{X})$ by $\mathrm{d}V$, $\mathrm{d}A$, respectively, we rewrite the integral equation of motion (1.14) in the form

$$\int_{\Omega} \boldsymbol{\sigma} \circ \mathbf{L}_{\mathbf{F}} \mathbf{v} \mathrm{d}V = \oint_{\partial\Omega} \mathbf{s} \cdot \mathbf{v} \mathrm{d}A + \int_{\Omega} (\mathbf{b} - \varrho \ddot{\mathbf{u}}) \cdot \mathbf{v} \mathrm{d}V \tag{1.18}$$

which is assumed to hold for every test function $\mathbf{v}(\cdot)$.

Proceeding along the similar lines the integral energy equation (1.16) is rewritten as

$$\int_{\Omega} \mathbf{h} \cdot \nabla \zeta \mathrm{d}V = \oint_{\partial\Omega} \pi \zeta \mathrm{d}A + \int_{\Omega} \alpha \zeta \mathrm{d}V + \int_{\Omega} (\boldsymbol{\sigma} \circ \dot{\boldsymbol{\epsilon}} - \varrho \dot{\varepsilon}) \zeta \mathrm{d}V \tag{1.19}$$

which has to hold for every test function $\zeta(\cdot)$.

The conditions given in eqs. (1.18), (1.19) are called the *matrix forms of the integral equation of motion* and *energy equation*, respectively. It should be emphasized that $\mathbf{L}_{\mathbf{F}}$ in $\mathbf{L}_{\mathbf{F}} \mathbf{v}$ represents the linear first order differential operator which depends on the displacement gradient $\nabla \mathbf{u}(\mathbf{X}, t)$. Let us also observe that using the operator $\mathbf{L}_{\mathbf{F}}$ and employing eq. (1.11) we obtain that $\dot{\boldsymbol{\epsilon}} = \mathbf{L}_{\mathbf{F}} \dot{\mathbf{u}}$.

The derivation of the matrix form for the governing relations of nonlinear thermo-mechanics constitutes the first step towards obtaining the system of relations which is convenient in studying specific boundary-value problems of structural mechanics from the viewpoint of computer-oriented solution methods.

Summary

The general notions, fundamental principles and fundamental lemmas of mechanics lead to the relationships which constitute the physical foundations of the theories and methods to be studied in the remainder of the book. The resulting relations are:

1. Equation of motion (1.14) which upon substituting the formula $\boldsymbol{\chi}(\mathbf{X}, t) = \boldsymbol{\chi}(\mathbf{X}, t_0) + \mathbf{u}_t(\mathbf{X})$ interrelates the displacement $\mathbf{u}_t(\cdot)$, the acceleration $\ddot{\mathbf{u}}_t(\cdot)$, the second Piola–Kirchhoff stress $\mathbf{S}_t(\cdot)$, the body force $\mathbf{b}(\cdot, t)$ (all these fields are defined on Ω) and the boundary tractions $\mathbf{s}(\cdot, t)$ (which is the field defined on $\partial\Omega$).
2. Energy equation (1.16) which interrelates the heat flux $\mathbf{h}_t(\cdot)$, the heat absorption $\alpha(\cdot, t)$, the rate of internal energy $\dot{\varepsilon}(\cdot, t)$, the second Piola–Kirchhoff stress $\mathbf{S}_t(\cdot)$, the strain rate $\dot{\mathbf{E}}_t(\cdot)$ (all these fields are defined on Ω) and the boundary heat flux $\pi(\cdot, t)$ (which is the field defined on $\partial\Omega$).

3. Dissipation inequality (1.17) which apart from the objects $\mathbf{S}_t(\mathbf{X})$, $\dot{\mathbf{E}}_t(\mathbf{X})$, $\mathbf{h}_t(\mathbf{X})$, $\dot{\varepsilon}(\mathbf{X}, t)$, mentioned above involves also, upon substituting the formula $\theta(\mathbf{X}, t) = \theta_0 + \vartheta_t(\mathbf{X})$, the temperature $\vartheta_t(\mathbf{X})$ its gradient $\nabla\vartheta_t(\mathbf{X})$ and the rate of entropy $\dot{\eta}(\mathbf{X}, t)$.
4. Rate compatibility relation (1.18) which by using the formula $\boldsymbol{\chi}(\mathbf{X}, t) = \boldsymbol{\chi}(\mathbf{X}, t_0) + \mathbf{u}_t(\mathbf{X})$ interrelates strain rate field $\dot{\mathbf{E}}_t(\cdot)$ with the displacement $\mathbf{u}_t(\cdot)$ and the velocity $\dot{\mathbf{u}}_t(\cdot)$.

All the aforementioned fields used in any specific problem to be discussed are considered at an arbitrary but fixed time instant t, $t \in [t_0, t_f]$. As a rule, these fields and their increments (which will be introduced in Sec. 3.1) are not known a priori in structural mechanics problems and therefore have to be determined in the course of the solution procedure. On the other hand, we may usually assume that the domain of definition Ω (the undeformed body) of these fields is known from the beginning, and that there is known a mass density function $\varrho(\mathbf{X})$, $\mathbf{X} \in \Omega$, of the undeformed body as well as the reference temperature θ_0. Moreover, in every problem of structural mechanics we shall assume that the following objects are known a priori:

1. the set $\mathscr{D}(\Omega)$ of all admissible deformations $\boldsymbol{\varkappa}(\cdot)$: $\Omega \to \boldsymbol{R}^3$,
2. the set $\mathscr{T}(\Omega)$ of all admissible temperature fields $\theta(\cdot)$: $\Omega \to \boldsymbol{R}_+$.

Thus we shall deal with the additional restrictions imposed on deformation and temperature fields given by $\boldsymbol{\chi}(\cdot, t) \in \mathscr{D}(\Omega)$, $\theta(\cdot, t) \in \mathscr{T}(\Omega)$. These restrictions are postulated a priori for every specific problem of structural mechanics in order to simplify the mathematical description of the problem and hence to make it possible for us to obtain the solution to it, cf. Secs. 1.1.2 and 1.1.5.

Problems

1.1 By taking into account the results of Sec. 1.3.2 find the interrelations between the strain invariants I_1, I_2, I_3 and the components of the Green–Saint Venant strain tensor $\mathbf{E}$.
1.2 The interrelation between the densities of surface tractions $\mathbf{s}_{\partial\Pi}(\mathbf{X}, t)$ and $\mathbf{s}_{\partial\chi(\Pi,t)}(\mathbf{z}, t)$ referred to undeformed and deformed body, respectively, cf. Sec. 1.1.4, depends on the deformation $\mathbf{z} = \boldsymbol{\chi}(\mathbf{X}, t)$, $\mathbf{X} \in \Omega$. Treating the mapping $\mathbf{z} = \boldsymbol{\chi}(\mathbf{X}, t)$, $\mathbf{X} \in \Omega$, as the transformation of coordinates (with the fixed t) find the exact form of this dependence.
1.3 Under the additional assumption that the surface tractions $\mathbf{s}_{\partial\Pi}(\mathbf{X}, t)$, $\mathbf{X} \in \partial\Pi$, satisfy the conditions$_{\partial\Pi}(\mathbf{X}, t) = \mathbf{s}(\mathbf{X}, t, \mathbf{n}(\mathbf{X}))$, where $\mathbf{n}(\mathbf{X})$ is the unit outer normal to $\partial\Pi$ at $\mathbf{X}$, prove the fundamental lemma of Sec. 1.3.3. Hint: apply the principle of balance of momentum (1.5) to a region Π which is a tetrahedron $OA_1A_2A_3$ with three edges OA_1, OA_2, OA_3 parallel to the coordinate axes and assume $\mathbf{n}(\mathbf{X})$ as the outer normal to $A_1A_2A_3$; pass to the limit with the volume of $OA_1A_2A_3$, $\mathrm{vol}(OA_1A_2A_3) \to 0$, holding $\mathbf{n}(\mathbf{X})$ constant.
1.4 Consider the principle of balance of moment of momentum and perform all the calculations leading to eq. (1.15).

1.5 Starting with eq. (1.14) derive the equations

$$\mathrm{Div}[\nabla\boldsymbol{\chi}(\mathbf{X}, t)\mathbf{S}_t(\mathbf{X})]+\mathbf{b}(\mathbf{X}, t) = \varrho(\mathbf{X})\ddot{\mathbf{u}}_t(\mathbf{X}), \qquad \mathbf{X}\in\Omega,$$

$$\mathbf{s}(\mathbf{X}, t) = \nabla\boldsymbol{\chi}(\mathbf{X}, t)\mathbf{S}_t(\mathbf{X})\mathbf{n}(\mathbf{X}), \qquad \mathbf{X}\in\partial\Omega.$$

Hint: apply the known du Bois–Raymond lemma (the fundamental lemma of variational calculus) to eqs. (1.14).

1.6 Consider the equations mentioned in Problem 1.4 and obtain the principle of balance of momentum (1.5).

1.7 Starting with the energy equation (1.16) derive the principle of balance of energy (1.7).

1.8 Starting with eqs. (1.15) and the equations of Problem 1.4 derive the principle of balance of moment of momentum.

1.9 Consider the strain-displacement relation (1.10) for the case in which $\mathbf{z} = \boldsymbol{\chi}(\mathbf{X}, t)$, $\mathbf{X}\in\Omega$, is the identity mapping for the fixed t.

1.10 Perform the exact calculations leading from the dissipation inequality (1.8) to its local form (1.17).

1.11 Let Γ be a smooth oriented surface in Ω across which the second Piola–Kirchhoff stress tensor may suffer a jump $[\mathbf{S}_t(\mathbf{X})]$, $\mathbf{X}\in\Gamma$. Starting from eq. (1.5) prove that $[\mathbf{S}_t(\mathbf{X})]\mathbf{n}(\mathbf{X}) = 0$ has to hold on Γ where $\mathbf{n}(\mathbf{X})$, $\mathbf{X}\in\Gamma$, is the unit outer normal to Γ.

Hint: use an approach similar to that employed in Sec. 1.4.1 leading to $\mathbf{s}(\mathbf{X}, t) = \nabla\boldsymbol{\chi}(\mathbf{X}, t)\mathbf{S}_t(\mathbf{X})\mathbf{n}(\mathbf{X})$, $\mathbf{X}\in\partial\Omega$.

2

Materials, loadings and constraints

Purpose of the chapter

By taking into account merely the resulting relations formulated in Sec. 1.4 we are not able to develop a well determined analytical description of problems typical of structural and solid mechanics. To this aim we have to know also what materials is the structure made of and how it is loaded and supported. Moreover, the mathematical models of structural components as well as those of loadings and supports have to be simple enough in order to facilitate the analysis and effective derivation of solutions. In this chapter we shall show that this can be accomplished by considering certain a priori restrictions imposed on deformation and temperature distributions. Such restrictions will be called *constraints.*

The aim of the chapter is to propose a mathematical description of physical concepts such as materials, loadings, supports and constraints, which are crucial for every structural mechanics problem.

2.1 Materials

2.1.1 Real materials and ideal materials

Attempts to describe in the precise language of mathematics various real materials constitute a very important branch of continuum mechanics known as the theory of constitutive modelling. The characteristic feature of every constitutive relation is that it is determined independently for every $\mathbf{X} \in \Omega$ (except possibly at some surfaces across which material properties suffer discontinuity). In thermomechanics constitutive relations can be interpreted as the equations involving the following histories: $\tau \to \mathbf{S}_\tau(\mathbf{X})$, $\tau \to \mathbf{E}_\tau(\mathbf{X})$, $\tau \to \vartheta_t(\mathbf{X})$, $\tau \to \nabla\vartheta_t(\mathbf{X})$, $\tau \to \mathbf{h}_t(\mathbf{X})$, $\tau \to \eta(\mathbf{X}, \tau)$ and $\tau \to \varphi(\mathbf{X}, \tau)$ which are defined for $\tau \leqslant t$ where t is an arbitrary but fixed instant of time ($\mathbf{X}$ is considered fixed as well). The functional form of the constitutive relation does not depend on t and has to ensure the fulfilment of the dissipation inequality (1.17).

Every real material is represented within the framework of mechanics by a certain constitutive relation and this relation is identified with the so-called ideal material, i.e. with a mathematical description of the real material considered. In the sequel the term "material" will always be used in the sense of the "ideal material". It should be emphasized that one real material can be represented by means of various ideal materials, and the different idealized mathematical representations may turn out useful in different problems of mechanics.

Throughout the book we shall restrict ourselves to constitutive relations which are local[1] in time in the sense that they do not involve the whole histories but only the values $\mathbf{S}_t(\mathbf{X})$, $\dot{\mathbf{S}}_t(\mathbf{X})$ (or $\mathbf{T}(\mathbf{X},t)$, $\dot{\mathbf{T}}(\mathbf{X},t)$, respectively), $\mathbf{E}_t(\mathbf{X})$, $\dot{\mathbf{E}}_t(\mathbf{X})$, $\vartheta_t(\mathbf{X})$, $\dot{\vartheta}_t(\mathbf{X})$, $\nabla\vartheta_t(\mathbf{X})$, $\mathbf{h}_t(\mathbf{X})$, $\varphi(\mathbf{X},t)$, $\dot{\varphi}(\mathbf{X},t)$, $\eta(\mathbf{X},t)$, $\dot{\eta}(\mathbf{X},t)$. In particular, in the present section we shall confine our considerations to the three important classes of materials, namely to thermo-visco-elastic materials, linear thermo-elastic materials and thermo-elastic-plastic materials.

For a more detailed discussion of the theory of constitutive modelling the reader is referred to [47], [53], [157], for instance.

2.1.2 *Thermo-visco-elastic materials*

The constitutive relations defining thermo-visco-elastic materials are assumed in the form

$$\begin{aligned}
\mathbf{S} &= \hat{\mathbf{S}}(\mathbf{X};\mathbf{E},\vartheta)+\mathfrak{H}(\mathbf{X};\mathbf{E},\vartheta)[\dot{\mathbf{E}}],\\
\mathbf{h} &= \mathbf{K}(\mathbf{X};\mathbf{E},\vartheta)\nabla\vartheta,\\
\varphi &= \hat{\varphi}(\mathbf{X};\mathbf{E},\vartheta),\\
\eta &= \hat{\eta}(\mathbf{X};\mathbf{E},\vartheta),
\end{aligned}$$

where the point $\mathbf{X}\in\Omega$ is arbitrary but fixed and $\hat{\mathbf{S}}(\mathbf{X};\,\cdot)$, $\mathfrak{H}(\mathbf{X};\,\cdot)$, $\mathbf{K}(\mathbf{X};\,\cdot)$, $\hat{\varphi}(\mathbf{X};\,\cdot)$, $\hat{\eta}(\mathbf{X};\,\cdot)$ are assumed to be the known functions satisfying any required regularity conditions. The aforementioned form of the constitutive relation is obviously based on some physical premises. Roughly speaking, it constitutes an acceptable mathematical idealization of a large class of real structural materials. Substituting the RHS of these equations into the dissipation inequality (1.17) we obtain

$$\begin{aligned}
\operatorname{tr}\left\{\left[\hat{\mathbf{S}}(\mathbf{X};\mathbf{E},\vartheta)-\varrho(\mathbf{X})\frac{\partial\hat{\varphi}(\mathbf{X};\mathbf{E},\vartheta)}{\partial\mathbf{E}}\right]\dot{\mathbf{E}}\right\}-\left[\hat{\eta}(\mathbf{X};\mathbf{E},\vartheta)+\frac{\partial\hat{\varphi}(\mathbf{X};\mathbf{E},\vartheta)}{\partial\vartheta}\right]\dot{\vartheta}+\\
+\operatorname{tr}\{\mathfrak{H}(\mathbf{X};\mathbf{E},\vartheta)[\dot{\mathbf{E}}]\dot{\mathbf{E}}\}+\mathbf{K}(\mathbf{X};\mathbf{E},\vartheta)\nabla\vartheta/\theta\geqslant 0,\qquad \theta=\theta_0+\vartheta.
\end{aligned}$$

[1] The constitutive relations which are local in time can still involve certain functions defined on $[t_0, t_f]$ (and their first order derivatives), which are called the *internal state variables.* As examples of internal state variables may serve plastic strains, strain hardening parameters, etc., cf. Sec. 2.1.4.

We conclude that the dissipation inequality will be satisfied identically if

$$\hat{\mathbf{S}}(\mathbf{X}; \mathbf{E}, \vartheta) = \varrho(\mathbf{X}) \frac{\partial \hat{\varphi}(\mathbf{X}; \mathbf{E}, \vartheta)}{\partial \mathbf{E}},$$

$$\hat{\eta}(\mathbf{X}; \mathbf{E}, \vartheta) = -\frac{\partial \hat{\varphi}(\mathbf{X}; \mathbf{E}, \vartheta)}{\partial \vartheta},$$

$$\mathbf{K}(\mathbf{X}; \mathbf{E}, \vartheta)\nabla\vartheta \geqslant 0 \text{ for every } \nabla\vartheta,$$

and if $\mathfrak{H}(\mathbf{X}; \mathbf{E}, \vartheta)$ is a positive definite mapping, i.e. if

$$\operatorname{tr}\{\mathbf{M}\mathfrak{H}(\mathbf{X}; \mathbf{E}, \vartheta)[\mathbf{M}]\} \geqslant 0$$

holds for every 3×3 matrix $\mathbf{M}$. Thus the bodies made of thermo-visco-elastic materials considered are assumed to be defined by the equations

$$\begin{aligned} \mathbf{S}_t(\mathbf{X}) &= \varrho(\mathbf{X}) \frac{\partial \hat{\varphi}(\mathbf{X}; \mathbf{E}_t(\mathbf{X}), \vartheta_t(\mathbf{X}))}{\partial \mathbf{E}_t(\mathbf{X})} + \mathfrak{H}[\mathbf{X}; \mathbf{E}_t(\mathbf{X}), \vartheta_t(\mathbf{X})][\dot{\mathbf{E}}_t(\mathbf{X})], \\ \mathbf{h}_t(\mathbf{X}) &= \mathbf{K}(\mathbf{X}; \mathbf{E}_t(\mathbf{X}), \vartheta_t(\mathbf{X}))\nabla\vartheta_t(\mathbf{X}), \\ \varphi(\mathbf{X}, t) &= \hat{\varphi}(\mathbf{X}; \mathbf{E}_t(\mathbf{X}), \vartheta_t(\mathbf{X})), \end{aligned} \tag{2.1}$$

which holds for $\mathbf{X} \in \Omega$ (except possibly at some surfaces, lines or points) and where $\hat{\varphi}(\mathbf{X};\ \cdot)$, $\mathfrak{H}(\mathbf{X};\ \cdot)$, $\mathbf{K}(\mathbf{X};\ \cdot)$ are assumed to be the known functions.

2.1.3 Linear thermo-elastic materials

Consider now the special physical situation in which the principal values of the (symmetric) strain tensor $\mathbf{E}(\mathbf{X}, t)$ are sufficiently small with respect to the principal values of the tensor $\mathbf{G}(\mathbf{X})$ and the relative temperatures $\vartheta_t(\mathbf{X}) = \theta(\mathbf{X}, t) - \theta_0$ are sufficiently small with respect to the reference temperature θ_0. For the sake of simplicity let us also consider the case in which the effects of viscosity can be neglected, which corresponds to setting $\mathfrak{H}(\mathbf{X};\ \cdot) \equiv 0$. Expanding the functions $\mathbf{K}(\mathbf{X};\ \cdot)$, $\hat{\varphi}(\mathbf{X};\ \cdot)$ into power series near $\mathbf{E} = 0$, $\vartheta = 0$ and neglecting in eq. (2.1) the nonlinear terms (under the condition that $\mathbf{K}(\mathbf{X}; \mathbf{E}, \vartheta)$ and $\hat{\varphi}(\mathbf{X}; \mathbf{E}, \vartheta)$ are analytical at $\mathbf{E} = \mathbf{0}$, $\vartheta = 0$) we arrive at

$$\begin{aligned} \mathbf{S}_t(\mathbf{X}) &= \mathfrak{C}(\mathbf{X})[\mathbf{E}_t(\mathbf{X})] + \mathbf{B}(\mathbf{X})\vartheta_t(\mathbf{X}), \\ \mathbf{h}_t(\mathbf{X}) &= \mathbf{K}_0(\mathbf{X})\nabla\vartheta_t(\mathbf{X}), \\ \eta(\mathbf{X}, t) &= -\beta(\mathbf{X})\vartheta_t(\mathbf{X}) - \operatorname{tr}[\mathbf{B}(\mathbf{X})\mathbf{E}_t(\mathbf{X})]/\varrho(\mathbf{X}), \\ \varphi(\mathbf{X}, t) &= \frac{1}{2\varrho(\mathbf{X})} \operatorname{tr}\{\mathbf{E}_t(\mathbf{X})\mathfrak{C}(\mathbf{X})[\mathbf{E}_t(\mathbf{X})]\} + \\ &\quad + \frac{1}{\varrho(\mathbf{X})} \operatorname{tr}[\mathbf{B}(\mathbf{X})\mathbf{E}_t(\mathbf{X})]\vartheta_t(\mathbf{X}) + \tfrac{1}{2}\beta(\mathbf{X})[\vartheta_t(\mathbf{X})]^2 \end{aligned} \tag{2.2}$$

where $\mathfrak{C}(\mathbf{X})$ is a fourth order tensor called the tensor of elastic modulae, $\mathbf{B}(\mathbf{X})$ is a second order symmetric tensor, $\beta(\mathbf{X})$ is a scalar and $\mathbf{K}_0(\mathbf{X})$

$\equiv \mathbf{K}(\mathbf{X}; 0, 0)$ is a second order symmetric tensor called the thermal conductivity tensor. From the symmetry of $\mathbf{E}_t(\mathbf{X})$ and from eq. $(2.2)_1$ we conclude that the components $C^{\alpha\beta\gamma\delta}(\mathbf{X})$ of $\mathfrak{C}(\mathbf{X})$ can be assumed in the form satisfying the symmetry conditions: $C^{\alpha\beta\gamma\delta}(\mathbf{X}) = C^{\beta\alpha\gamma\delta}(\mathbf{X}) = C^{\beta\alpha\delta\gamma}(\mathbf{X})$ and $C^{\alpha\beta\gamma\delta}(\mathbf{X}) = C^{\gamma\delta\alpha\beta}(\mathbf{X})$; hence, there are at most 21 independent components of $\mathfrak{C}(\mathbf{X})$. Eqs. (2.2) are the constitutive relations of the linear thermo-elastic materials. We also assume that the free energy $\varphi(\mathbf{X}, t)$ is always non-negative; thus, the RHS of eq. $(2.2)_4$ represents a positive-definite quadratic form. Moreover, if $\mathfrak{C}(\mathbf{X})$, $\mathbf{B}(\mathbf{X})$ and $\mathbf{K}_0(\mathbf{X})$ are given by

$$\begin{aligned}
C^{\alpha\beta\gamma\delta}(\mathbf{X}) &= \lambda(\mathbf{X})G^{\alpha\beta}(\mathbf{X})G^{\gamma\delta}(\mathbf{X})+\mu(\mathbf{X})[G^{\alpha\gamma}(\mathbf{X})G^{\beta\delta}(\mathbf{X})+G^{\alpha\delta}(\mathbf{X})G^{\beta\gamma}(\mathbf{X})],\\
B^{\alpha\beta}(\mathbf{X}) &= -[3\lambda(\mathbf{X})+2\mu(\mathbf{X})]\alpha(\mathbf{X})G^{\alpha\beta}(\mathbf{X}),\\
K_0^{\alpha\beta}(\mathbf{X}) &= \varkappa(\mathbf{X})G^{\alpha\beta}(\mathbf{X}),
\end{aligned}$$

where $G^{\alpha\beta}(\mathbf{X})$ are elements of the inverse to the matrix $\mathbf{G}(\mathbf{X}) = \mathbf{C}(\mathbf{X}, t_0)$, cf. Sec. 1.3.2, $\lambda(\mathbf{X})$, $\mu(\mathbf{X})$ are called the Lamé modulae and $\alpha(\mathbf{X})$, $\varkappa(\mathbf{X})$ are said to be the linear thermal expansion and conductivity coefficient, then the linear thermo-elastic material is said to be *isotropic*. For particulars the reader is referred to any textbook on the theory of linear thermoelasticity.

2.1.4 *Thermo-elastic-plastic materials*

This section is devoted to the discussion of the theory of thermo-elastic-plastic materials. The symbols $\mathbf{S}$ and $\mathbf{T}$ stand, as before, for the 3×3 symmetric matrices representing the second Piola–Kirchhoff and the Cauchy stress tensors, respectively. By $\mathbf{q} \equiv (q_1, \ldots, q_r)$ we define the r-tuple of real numbers called the *internal state variables* the meaning of which will be explained below. In order to define thermo-elastic-plastic materials we assume that for points $\mathbf{X}$ of Ω (except possibly at some surfaces in Ω) the following mathematical objects are known (the argument $\mathbf{X}$ is arbitrary but fixed):

1. A real valued differentiable function $f(\mathbf{X}; \mathbf{T}, \vartheta, \mathbf{q})$ defined for every $\mathbf{T}$, ϑ and $\mathbf{q}$. The functions $f(\mathbf{X}; \cdot, \vartheta, \mathbf{q})$ are called the *plastic potentials* while the equations $f(\mathbf{X}; \mathbf{T}, \vartheta, \mathbf{q}) = 0$ are called the *yield conditions*. By virtue of the stress symmetry property $\mathbf{T} = \mathbf{T}^T$ the yield condition for every fixed $\mathbf{X}$, ϑ and $\mathbf{q}$ may be given an interpretation of the so-called loading surfaces in the 6th dimensional stress space of vectors $(T^{11}, T^{22}, T^{33}, \sqrt{2}T^{12}, \sqrt{2}T^{23}, \sqrt{2}T^{31})$. Moreover, the region in this space bounded by any loading surface is assumed to be convex and to comprise the origin, i.e. the point representing zero stress, cf. Fig. 7 in which $\mathbf{T}$ is treated as the vector with components $(T^{11}, T^{22}, T^{33}, \sqrt{2}T^{12}, \sqrt{2}T^{23}, \sqrt{2}T^{31})$.

2. An invertible linear mapping $\mathfrak{C}(\mathbf{X})[\cdot]$ and a symmetric tensor $\mathbf{B}(\mathbf{X})$ which assign to every spatial strain rate tensor $\mathbf{D}$ and every temperature

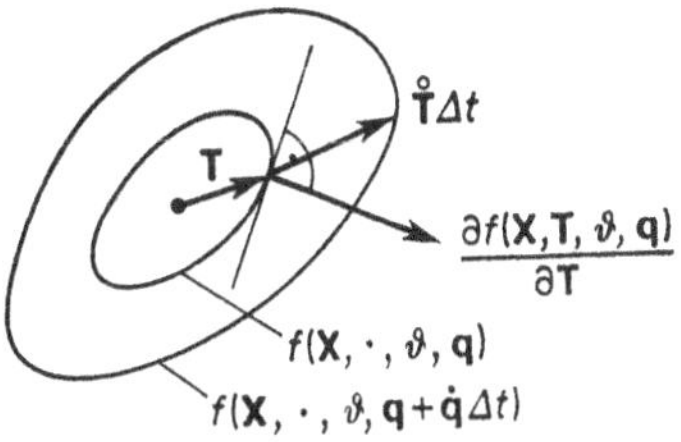

Fig. 7 Loading surfaces

rate $\dot{\vartheta}$ the value of the spatial stress rate $\mathring{\mathbf{T}}$ as $\mathring{\mathbf{T}} = \mathfrak{C}(\mathbf{X})[\mathbf{D}]+\mathbf{B}(\mathbf{X})\dot{\vartheta}$ (cf. Secs. 1.3.2 and 1.3.3 for the definitions of $\mathbf{D}$ and $\mathring{\mathbf{T}}$). We also assume that the formula $\mathbf{h}_t = \mathbf{K}(\mathbf{X};\mathbf{E},\vartheta)\nabla\vartheta$ holds true. Here $\mathbf{K}(\mathbf{X};\ \cdot)$, $\mathfrak{C}(\mathbf{X})$, $\mathbf{B}(\mathbf{X})$ describe the thermoelastic properties of the material, cf. Sec. 2.1.3.

3. A sequence of r symmetric continuous functions $\mathbf{A}_\varrho(\mathbf{X};\mathbf{T},\vartheta,\mathbf{q})$, $\varrho = 1,\dots,r$, defined for every $\mathbf{T}$, ϑ and every $\mathbf{q} \equiv (q_1,\dots,q_r)$ such that $\mathbf{A}_\varrho(\mathbf{X};\mathbf{T},\vartheta,\mathbf{q})$ are symmetric 3×3 matrices; these functions are called the *internal state parameters* and used to describe the so-called hardening properties of the material.

We shall now introduce the following fundamental postulates which define the class of elastic-plastic materials:

1. The total strain rate $\mathbf{D}$ is the sum $\mathbf{D} = \mathbf{D}_e+\mathbf{D}_p$, where $\mathbf{D}_e$ and $\mathbf{D}_p$ are called the elastic and plastic parts of the total strain rate $\mathbf{D}$.

2. The plastic part $\mathbf{D}_p$ of $\mathbf{D}$ is given by

$$\mathbf{D}_p = \delta\lambda\frac{\partial f(\mathbf{X};\mathbf{T},\vartheta,\mathbf{q})}{\partial\mathbf{T}};$$

$$f' \equiv \operatorname{tr}\left(\frac{\partial f}{\partial\mathbf{T}}\mathring{\mathbf{T}}\right)+\frac{\partial f}{\partial\vartheta}\dot{\vartheta},$$

$$f \equiv f(\mathbf{X};\mathbf{T},\vartheta,\mathbf{q}),$$

$$\delta \equiv \begin{cases} 0 & \text{if} \quad f<0 \quad \text{or} \quad \text{if } f=0 \text{ and } f' \leqslant 0,\\ 1 & \text{if} \quad f=0 \quad \text{and} \quad f'>0,\end{cases}$$

where λ is an arbitrary non-negative constant. The above condition is called the *associated flow rule.*

3. The stress rate $\mathring{\mathbf{T}}$ is interrelated with the elastic part $\mathbf{D}_e$ of the total strain rate $\mathbf{D}$ and the temperature rate $\dot{\vartheta}$ by means of the equation $\mathring{\mathbf{T}} = \mathfrak{C}(\mathbf{X})[\mathbf{D}_e]+\mathbf{B}(\mathbf{X})\dot{\vartheta}$.

4. The rates $\dot{\mathbf{q}} = (\dot{q}_1,\dots,\dot{q}_r)$ of the internal state variables $\mathbf{q} = (q_1,\dots$ $\dots,q_r)$ are interrelated with the plastic part $\mathbf{D}_p$ of the total strain rate by means of the equations

$$\dot{q}_\varrho = \operatorname{tr}[\mathbf{A}_\varrho(\mathbf{X};\mathbf{T},\vartheta,\mathbf{q})\mathbf{D}_p], \qquad \varrho = 1,\dots,r.$$

If q_ϱ, $\varrho = 1, \ldots, 6$, are interpreted as plastic strains then $r \geqslant 6$ and the above equations are identities for $\varrho = 1, \ldots, 6$ since $\mathbf{D}_p$ represents the rate of the plastic strain.

Let $\mathring{\mathbf{T}}\Delta t$ represents a certain small stress increment. Taking into account the aforementioned assumptions we see that the plastic part $\mathbf{D}_p$ of the strain rate $\mathbf{D}$ is not equal to zero only if

$$f(\mathbf{X}; \mathbf{T}+\mathring{\mathbf{T}}\Delta t, \vartheta+\dot{\vartheta}\Delta t, \mathbf{q}+\dot{\mathbf{q}}\Delta t) = 0$$

provided $\delta = 1$. Hence, in the limit as $\Delta t \to 0$ we obtain the condition (here and below summation over $\varrho = 1, \ldots, r$ is assumed to hold)

$$\delta\left[\operatorname{tr}\left(\frac{\partial f}{\partial \mathbf{T}}\mathring{\mathbf{T}}\right)+\frac{\partial f}{\partial \vartheta}\dot{\vartheta}\right]+\frac{\partial f}{\partial q_\varrho}\dot{q}_\varrho = 0$$

which has to be satisfied in all situations considered (for $\delta = 0$ we have $\dot{q}_\varrho = 0$!).

Substituting

$$\mathring{\mathbf{T}} = \mathfrak{C}(\mathbf{X})[\mathbf{D}_e]+\mathbf{B}(\mathbf{X})\dot{\vartheta} = \mathfrak{C}(\mathbf{X})[\mathbf{D}]-\delta\lambda\mathfrak{C}(\mathbf{X})\left[\frac{\partial f(\mathbf{X}; \mathbf{T}, \vartheta, \mathbf{q})}{\partial \mathbf{T}}\right]+\mathbf{B}(\mathbf{X})\dot{\vartheta},$$

$$\dot{q}_\varrho = \operatorname{tr}[\mathbf{A}_\varrho(\mathbf{X}; \mathbf{T}, \vartheta, \mathbf{q})\mathbf{D}_p] = \delta\lambda \operatorname{tr}\left[\mathbf{A}_\varrho(\mathbf{X}; \mathbf{T}, \vartheta, \mathbf{q})\frac{\partial f(\mathbf{X}; \mathbf{T}, \vartheta, \mathbf{q})}{\partial \mathbf{T}}\right]$$

into the aforementioned condition we can calculate $\delta\lambda$. Denoting

$$\alpha \equiv \operatorname{tr}\left\{\frac{\partial f(\mathbf{X}; \mathbf{T}, \vartheta, \mathbf{q})}{\partial \mathbf{T}}\mathfrak{C}(\mathbf{X})\left[\frac{\partial f(\mathbf{X}; \mathbf{T}, \vartheta, \mathbf{q})}{\partial \mathbf{T}}\right]\right\},$$

$$\beta \equiv \frac{\partial f}{\partial q_\varrho}\operatorname{tr}\left[\mathbf{A}_\varrho(\mathbf{X}; \mathbf{T}, \vartheta, \mathbf{q})\frac{\partial f(\mathbf{X}; \mathbf{T}, \vartheta, \mathbf{q})}{\partial \mathbf{T}}\right]$$

we arrive at

$$\delta\lambda = \frac{\delta}{\alpha-\beta}\left\{\operatorname{tr}\left[\frac{\partial f(\mathbf{X}; \mathbf{T}, \vartheta, \mathbf{q})}{\partial \mathbf{T}}\mathfrak{C}(\mathbf{X})[\mathbf{D}]\right]+\frac{\partial f(\mathbf{X}; \mathbf{T}, \vartheta, \mathbf{q})}{\partial \vartheta}\dot{\vartheta}\right\}.$$

At the same time

$$\mathring{\mathbf{T}} = \mathfrak{C}(\mathbf{X})[\mathbf{D}_e] = \mathfrak{C}(\mathbf{X})[\mathbf{D}]-\delta\lambda\mathfrak{C}(\mathbf{X})\left[\frac{\partial f(\mathbf{X}; \mathbf{T}, \vartheta, \mathbf{q})}{\partial \mathbf{T}}\right]+\mathbf{B}(\mathbf{X})\dot{\vartheta},$$

$$\dot{q}_\varrho = \delta\lambda \operatorname{tr}\left[\mathbf{A}_\varrho(\mathbf{X}; \mathbf{T}, \vartheta, \mathbf{q})\frac{\partial f(\mathbf{X}; \mathbf{T}, \vartheta, \mathbf{q})}{\partial \mathbf{T}}\right],$$

so that finally

$$\overset{\circ}{\mathbf{T}} = \mathfrak{C}(\mathbf{X})[\mathbf{D}] - \frac{\delta}{\alpha-\beta}\, \mathfrak{C}(\mathbf{X})\left[\frac{\partial f(\mathbf{X};\mathbf{T},\vartheta,\mathbf{q})}{\partial \mathbf{T}}\right] \times$$
$$\times \left\{\mathrm{tr}\left[\frac{\partial f(\mathbf{X};\mathbf{T},\vartheta,\mathbf{q})}{\partial \mathbf{T}}\, \mathfrak{C}(\mathbf{X})[\mathbf{D}]\right] + \frac{\partial f(\mathbf{X};\mathbf{T},\vartheta,\mathbf{q})}{\partial \vartheta}\,\dot{\vartheta}\right\} + \mathbf{B}(\mathbf{X})\dot{\vartheta}, \qquad (2.3)$$
$$\dot{q}_\varrho = \frac{\delta}{\alpha-\beta}\,\mathrm{tr}\left[\mathbf{A}_\varrho(\mathbf{X};\mathbf{T},\vartheta,\mathbf{q})\, \frac{\partial f(\mathbf{X};\mathbf{T},\vartheta,\mathbf{q})}{\partial \mathbf{T}}\right].$$

Eqs. (2.3), together with the equations for α, β and δ given above, form the constitutive relations which define a class of thermo-elastic-plastic materials. The RHS of eq. $(2.3)_1$ determines linear mapping $\tilde{\mathfrak{C}}(\mathbf{X};\mathbf{T},\vartheta,\mathbf{q},\delta)[\cdot]$ and a tensor $\tilde{\mathbf{B}}(\mathbf{X};\mathbf{T},\vartheta,\mathbf{q},\delta)$ defined by

$$\tilde{\mathfrak{C}}(\mathbf{X};\mathbf{T},\vartheta,\mathbf{q},\delta)[\mathbf{D}]$$
$$\equiv \mathfrak{C}(\mathbf{X})[\mathbf{D}] - \frac{\delta}{\alpha-\beta}\, \mathfrak{C}(\mathbf{X})\left[\frac{\partial f(\mathbf{X};\mathbf{T},\vartheta,\mathbf{q})}{\partial \mathbf{T}}\right]\mathrm{tr}\left\{\frac{\partial f(\mathbf{X};\mathbf{T},\vartheta,\mathbf{q})}{\partial \mathbf{T}}\, \mathfrak{C}(\mathbf{X})[\mathbf{D}]\right\},$$
$$\tilde{\mathbf{B}}(\mathbf{X};\mathbf{T},\vartheta,\mathbf{q},\delta) \equiv \mathbf{B}(\mathbf{X}) - \frac{\delta}{\alpha-\beta}\, \mathfrak{C}(\mathbf{X})\left[\frac{\partial f(\mathbf{X};\mathbf{T},\vartheta,\mathbf{q})}{\partial \mathbf{T}}\right] \frac{\partial f(\mathbf{X};\mathbf{T},\vartheta,\mathbf{q})}{\partial \vartheta}$$

for every symmetric 3×3 matrix $\mathbf{D}$. Thus

$$\overset{\circ}{\mathbf{T}} = \tilde{\mathfrak{C}}(\mathbf{X};\mathbf{T},\vartheta,\mathbf{q},\delta)[\mathbf{D}] + \tilde{\mathbf{B}}(\mathbf{X};\mathbf{T},\vartheta,\mathbf{q},\delta)\dot{\vartheta}$$

is an equivalent form of eq. $(2.3)_1$; the components $\tilde{C}^{ijkl}$ of $\tilde{\mathfrak{C}}$ and $\tilde{B}^{ij}$ of $\tilde{\mathbf{B}}$ are given by

$$\tilde{C}^{ijkl}(\mathbf{X};\mathbf{T},\vartheta,\mathbf{q},\delta)$$
$$= C^{ijkl}(\mathbf{X}) - \frac{\delta}{\alpha-\beta}\, C^{ijmn}(\mathbf{X})\, \frac{\partial f(\mathbf{X};\mathbf{T},\vartheta,\mathbf{q})}{\partial T^{mn}}\, \frac{\partial f(\mathbf{X};\mathbf{T},\vartheta,\mathbf{q})}{\partial T^{pr}}\, C^{prkl}(\mathbf{X}),$$
$$\tilde{B}^{ij}(\mathbf{X};\mathbf{T},\vartheta,\mathbf{q},\delta) = B^{ij}(\mathbf{X}) - \frac{\delta}{\alpha-\beta}\, C^{ijmn}(\mathbf{X})\, \frac{\partial f(\mathbf{X};\mathbf{T},\vartheta,\mathbf{q})}{\partial T^{mn}}\, \frac{\partial f(\mathbf{X};\mathbf{T},\vartheta,\mathbf{q})}{\partial \vartheta}.$$

Now taking into account the interrelations between spatial and material stress and strain rates (cf. Secs. 1.3.2 and 1.3.3) and defining a linear mapping $\tilde{\mathfrak{D}}(\mathbf{X};\mathbf{S},\vartheta,\mathbf{q},\delta)[\cdot]$ and a tensor $\tilde{\mathbf{G}}(\mathbf{X};\mathbf{S},\vartheta,\mathbf{q},\delta)$ such that

$$\tilde{D}^{\alpha\beta\gamma\delta}(\mathbf{X};\mathbf{S},\vartheta,\mathbf{q},\delta)$$
$$\equiv J(\mathbf{X},t)A^{\alpha}{}_{i}(\mathbf{X},t)A^{\beta}{}_{j}(\mathbf{X},t)\tilde{C}^{ijkl}(\mathbf{X};\mathbf{T},\mathbf{q},\delta)A^{\gamma}{}_{k}(\mathbf{X},t)A^{\delta}{}_{l}(\mathbf{X},t),$$
$$\tilde{G}^{\alpha\beta}(\mathbf{X};\mathbf{S},\vartheta,\mathbf{q},\delta) \equiv J(\mathbf{X},t)A^{\alpha}{}_{i}(\mathbf{X},t)A^{\beta}{}_{j}(\mathbf{X},t)\tilde{B}^{ij}(\mathbf{X};\mathbf{T},\vartheta,\mathbf{q},\delta)$$

for

$$\mathbf{T} = J^{-1}(\mathbf{X}, t)\nabla\boldsymbol{\chi}^T(\mathbf{X}, t)\mathbf{S}\nabla\boldsymbol{\chi}(\mathbf{X}, t),$$

we obtain

$$\dot{\mathbf{S}} = \tilde{\mathfrak{D}}(\mathbf{X}; \mathbf{S}, \vartheta, \mathbf{q}, \delta)[\dot{\mathbf{E}}] + \tilde{\mathbf{G}}(\mathbf{X}; \mathbf{S}, \vartheta, \mathbf{q}, \delta)\dot{\vartheta},$$

where $\tilde{\mathfrak{D}}(\cdot)$, $\tilde{\mathbf{G}}(\cdot)$ depend also on the material gradient $\nabla\boldsymbol{\chi}(\mathbf{X}, t)$. If $\boldsymbol{\chi}(\cdot, t)$ for a fixed time instant t is an identity mapping then $\tilde{\mathfrak{D}}(\cdot) = \tilde{\mathfrak{C}}(\cdot)$, $\tilde{\mathbf{G}}(\cdot) = \tilde{\mathbf{B}}(\cdot)$ and we can replace $\mathbf{T}$ in the constitutive equations (2.3) by $\mathbf{S}$. The requirements imposed on the relations obtained by the dissipation inequality are rather complicated and are not discussed here.

If $f(\mathbf{X};\ \cdot)$ are independent of temperature and internal variables and the functions $\mathbf{A}_\varrho(\mathbf{X},\ \cdot)$ drop out from considerations then we deal with the so-called elastic-ideally plastic materials. Eqs. (2.3) hold then under an additional assumption that $\alpha_q \equiv 0$, $f(\mathbf{X}; \mathbf{T}, \vartheta, \mathbf{q}) = f(\mathbf{X}, \mathbf{T})$ and

$$\delta = \begin{cases} 0 & \text{if } f(\mathbf{X};\mathbf{T}) < 0 \text{ or if } f(\mathbf{X};\mathbf{T}) = 0 \text{ and } \operatorname{tr}\left[\dfrac{\partial f(\mathbf{X};\mathbf{T})}{\partial \mathbf{T}}\overset{\circ}{\mathbf{T}}\right] < 0, \\ 1 & \text{if } f(\mathbf{X};\mathbf{T}) = 0 \quad \text{and} \quad \operatorname{tr}\left[\dfrac{\partial f(\mathbf{X};\mathbf{T})}{\partial \mathbf{T}}\overset{\circ}{\mathbf{T}}\right] = 0. \end{cases}$$

Hence, the mappings $\tilde{\mathfrak{C}}[\cdot]$ and $\tilde{\mathfrak{D}}[\cdot]$ are independent of $\mathbf{q}$.

In most cases we assume that $f(\mathbf{X}; \mathbf{T}, \vartheta, \mathbf{q})$ and $\mathbf{A}_\varrho(\mathbf{X}; \mathbf{T}, \vartheta, \mathbf{q})$ depend on $\mathbf{T}$ as a function of invariants I_D, II_D of the stress deviator $\mathbf{T}_D = \mathbf{T} - \sigma\mathbf{1}$, cf. Sec. 1.3.3. We also assume that $\mathbf{C}(\mathbf{X})$ coincides with the tensor of elastic modulae for an isotropic linear material, i.e.

$$C^{ijkl} = \lambda\delta^{ij}\delta^{kl} + \mu(\delta^{ik}\delta^{jl} + \delta^{il}\delta^{jk})$$

where λ, μ are the Lamé modulae, cf. Sec. 2.1.3. Some specific forms of the functions $f(\mathbf{X}; I_D, II_D, \vartheta, \mathbf{q})$ and $\mathbf{A}_\varrho(\mathbf{X}; I_D, II_D, \vartheta, \mathbf{q})$ will be detailed later in the book.

If $\delta \equiv 0$, i.e., if there are no plastic deformations, then we arrive at the constitutive relation $\dot{\mathbf{S}} = \mathfrak{D}(\mathbf{X}; \mathbf{S})[\dot{\mathbf{E}}]$. Under the assumption that $\mathfrak{D}(\mathbf{X}; \mathbf{S})[\cdot]$ is invertible we also obtain $\dot{\mathbf{E}} = \mathfrak{H}(\mathbf{X}; \mathbf{S})[\dot{\mathbf{S}}]$.

Such materials are referred to as the rate-type materials.

2.1.5 *Matrix form of constitutive relations*

Throughout this section let τ be an arbitrary but fixed time instant, $\tau \in [t_0, t_f]$, whereas $\boldsymbol{\chi}(\cdot, \tau)$ be the identity mapping, i.e. $\mathbf{X} = \boldsymbol{\chi}(\mathbf{X}, \tau)$ for every $\mathbf{X} \in \Omega$ and for the fixed τ. Thus, the region Ω coincides with the region

$\chi(\Omega, \tau)$ and $\nabla\chi(\mathbf{X}, \tau)$ is the 3×3 unit matrix. In Summary of Chapter 1 we have assumed that Ω is known a priori and that the deformations $\chi(\cdot, t) \in \mathscr{D}(\Omega)$ for $t \in [t_0, t_f]$ are unknown; hence it follows that we deal now with a certain special situation in which the deformation at the time instant $t = \tau$ has been previously determined. It follows from eq. (1.11) that

$$\dot{\mathbf{E}}_\tau(\mathbf{X}) = \tfrac{1}{2}[\nabla\dot{\mathbf{u}}_\tau(\mathbf{X}) + \nabla\dot{\mathbf{u}}_\tau^T(\mathbf{X})]$$

and the definition of $\mathbf{D}(\mathbf{X}, t)$ in Sec. 1.3.2 leads to the conclusion that $\dot{\mathbf{E}}_\tau(\mathbf{X}) = \mathbf{D}(\mathbf{X}, \tau)$, i.e. that material and spatial strain rate tensors coincide for $t = \tau$. Similarly, the formulae given in Sec. 1.3.3 imply that $\mathbf{S}_\tau(\mathbf{X}) = \mathbf{T}(\mathbf{X}, \tau)$ and $\mathring{\mathbf{T}}(\mathbf{X}, \tau) = \dot{\mathbf{S}}_\tau(\mathbf{X})$.

Moreover, let us also assume that the strains are small for every $t \in [t_0, t_f]$; it means that every matrix $\mathbf{C}(\mathbf{X}, t)$ can be approximated by $\mathbf{C}(\mathbf{X}, t_0) = \mathbf{G}(\mathbf{X})$ (cf. Sec. 1.3.2 for the definitions of $\mathbf{C}(\mathbf{X}, t)$ and $\mathbf{G}(\mathbf{X})$).

We start with the constitutive relations for linear thermo-elastic materials which were introduced in Sec. 2.1.3. We confine ourselves to the case of isotropic materials using the equations discussed at the end of that section in which $G^{\alpha\beta}$ is approximated by $\delta^{\alpha\beta}$. Introducing the 6×6 matrix

$$\mathbf{C} \equiv \begin{bmatrix} \lambda+2\mu & \lambda & \lambda & & & \\ \lambda & \lambda+2\mu & \lambda & & 0 & \\ \lambda & \lambda & \lambda+2\mu & & & \\ & & & 2\mu & 0 & 0 \\ & 0 & & 0 & 2\mu & 0 \\ & & & 0 & 0 & 2\mu \end{bmatrix}$$

and the column vector

$$\mathbf{b} \equiv \{(3\lambda+2\mu)\alpha \quad (3\lambda+2\mu)\alpha \quad (3\lambda+2\mu)\alpha \quad 0 \quad 0 \quad 0\}$$

and accounting for notation introduced in Sec. 1.3.5, we obtain the following form of the constitutive relations for the materials under considerations

$$\boldsymbol{\sigma} = \mathbf{C}\boldsymbol{\epsilon} - \mathbf{b}\vartheta,$$
$$\mathbf{h} = \varkappa\nabla\vartheta,$$

where $\mathbf{C} = \mathbf{C}(\mathbf{X}), \ldots, \varkappa = \varkappa(\mathbf{X})$. The matrix $\mathbf{C}$ is said to be the constitutive matrix of linear elasticity. Since $\mathbf{C}$ is nonsingular, it has the inverse $\mathbf{C}^{-1}$. By virtue of

$$\mathbf{b} = \mathbf{C}\mathbf{C}^{-1}\mathbf{b}$$

employing additionally the notation $\mathbf{a} \equiv \mathbf{C}^{-1}\mathbf{b}$ we arrive at the following alternative form of the constitutive relations for the linear thermo-elastic materials

$$\boldsymbol{\sigma} = \mathbf{C}(\boldsymbol{\epsilon}-\boldsymbol{\epsilon}^{\vartheta}), \quad \boldsymbol{\epsilon}^{\vartheta} = \mathbf{a}\vartheta, \tag{2.4}$$

$$\mathbf{h} = \varkappa\nabla\vartheta. \tag{2.5}$$

In many specific problems the linear thermal expansion coefficient $\alpha(\mathbf{X})$, the thermal conductivity coefficient $\varkappa(\mathbf{X})$ as well as the elastic modulae $\lambda(\mathbf{X})$, $\mu(\mathbf{X})$ depend on the temperature ϑ. Consequently, the values of $\mathbf{a}$, $\varkappa$ and $\mathbf{C}$ depend then not only on $\mathbf{X}$ but also on $\vartheta = \vartheta(\mathbf{X}, t)$. This is the case of what are called the thermo-linear elastic materials. From a formal point of view the linear thermo-elastic materials can be treated as a special case of the thermo-linear elastic materials.

We shall now pass to the case of thermo-elastic-plastic materials, the general theory of which was outlined in Sec. 2.1.4. Under the requirements formulated at the beginning of the present section (all the relations hold for the fixed time instant $t = \tau$ for which $\boldsymbol{\chi}(\cdot, \tau)$ is the identity mapping) and taking into account the notation introduced in Sec. 1.3.5, we shall now repeat and specify the assumptions listed in Sec. 2.1.4.:

1. The total strain rate $\dot{\boldsymbol{\epsilon}}$ is the sum $\dot{\boldsymbol{\epsilon}} = \dot{\boldsymbol{\epsilon}}^{(p)}+\dot{\boldsymbol{\epsilon}}^{(e)}$ of the elastic $\dot{\boldsymbol{\epsilon}}^{(e)}$ and plastic $\dot{\boldsymbol{\epsilon}}^{(p)}$ strain rates.

2. The plastic part $\dot{\boldsymbol{\epsilon}}^{(p)}$ is given by the plastic flow rule

$$\dot{\boldsymbol{\epsilon}}^{(p)} = \delta\lambda\left(\frac{\partial f}{\partial \boldsymbol{\sigma}}\right)^{T},$$

where $f \equiv f(\boldsymbol{\sigma}, \vartheta, \mathbf{q})$ is assumed to be the known scalar function.

For the present purpose it is enough to specify the vector of internal state parameters $\mathbf{q}$ as the one which has only one component (a scalar). We also postulate the plastic potentials $f(\cdot, \vartheta, \mathbf{q})$ in the form which corresponds to the yield condition given by

$$f(\boldsymbol{\sigma}, \vartheta, \mathbf{q}) = \sqrt{\tfrac{3}{2}(\boldsymbol{\sigma}^{D})^{T}\boldsymbol{\sigma}^{D}}-\hat{\sigma}(\vartheta, \mathbf{q}) = 0$$

where $\boldsymbol{\sigma}^{D}$ is the deviatoric stress (cf. Sec. 1.3.3) reading in matrix notation as

$$\boldsymbol{\sigma}^{D} \equiv \{\sigma_{11}-\sigma \quad \sigma_{22}-\sigma \quad \sigma_{33}-\sigma \quad \sqrt{2}\sigma_{12} \quad \sqrt{2}\sigma_{23} \quad \sqrt{2}\sigma_{31}\},$$

$$\sigma \equiv \tfrac{1}{3}(\sigma_{11}+\sigma_{22}+\sigma_{33}),$$

The above form of the condition $f(\boldsymbol{\sigma}, \vartheta, \mathbf{q}) = 0$ is known as the *Huber–Mises yield condition* while the function $\hat{\boldsymbol{\sigma}}(\cdot)$ is called the *tensile yield stress*. The plastic flow rule takes the form

$$\dot{\boldsymbol{\epsilon}}^{(p)} = \delta\lambda\left(\frac{\partial f}{\partial \boldsymbol{\sigma}}\right)^{T} = \delta\lambda\frac{3}{2}\frac{\boldsymbol{\sigma}^{D}}{\boldsymbol{\sigma}}.$$

3. The stress rate $\dot{\boldsymbol{\sigma}}$ is interrelated with the elastic part $\dot{\boldsymbol{\epsilon}}^{(e)}$ of the total strain rate $\dot{\boldsymbol{\epsilon}}$ and with a temperature rate $\dot{\vartheta}$ by means of the relation implied by the constitutive relation of the linear thermo-elastic material:

$$\dot{\boldsymbol{\sigma}} = \mathbf{C}(\dot{\boldsymbol{\epsilon}}-\dot{\boldsymbol{\epsilon}}^{(\vartheta)})+\dot{\mathbf{C}}(\boldsymbol{\epsilon}^{(e)}-\boldsymbol{\epsilon}^{(\vartheta)}).$$

Setting $\dot{\boldsymbol{\epsilon}}^{(e)} = \dot{\boldsymbol{\epsilon}}-\dot{\boldsymbol{\epsilon}}^{(p)}$ and denoting $'\boldsymbol{\sigma} \equiv \dot{\mathbf{C}}(\boldsymbol{\epsilon}^{(e)}-\boldsymbol{\epsilon}^{(\vartheta)})$, we obtain the formula

$$\dot{\boldsymbol{\sigma}} = \mathbf{C}(\dot{\boldsymbol{\epsilon}}-\dot{\boldsymbol{\epsilon}}^{(p)}-\dot{\boldsymbol{\epsilon}}^{\vartheta})+{}'\boldsymbol{\sigma}.$$

Here, $'\boldsymbol{\sigma}$ is the contribution of the stress rate due to the change in the temperature dependent elastic modulae. We also assume that $\boldsymbol{\epsilon}^{(e)} = \boldsymbol{\epsilon}-\boldsymbol{\epsilon}^{(p)}$ where $\boldsymbol{\epsilon}^{(p)}$ is the vector of plastic strains. It must be emphasized, however, that the decomposition $\boldsymbol{\epsilon} = \boldsymbol{\epsilon}^{(e)}+\boldsymbol{\epsilon}^{(p)}$ holds for the small strain case only.

The further discussion of thermo-elastic-plastic materials will be given later in the book. We shall also show the way the theory can be implemented in the computer and discuss some numerical solutions to specific boundary-value problems.

2.2 Loadings and heat supply

2.2.1 *External agents*

In every problem of structural mechanics we deal with certain interactions between the piece of the structure considered and its exterior. The mathematical description (or modelling) of such interactions is given in the subsequent Sections 2.2.2–2.3.1. The description is based on the assumption that the interactions are represented by loadings and heat sources as well as by what is called the boundary constraints. The loadings yield at every time instant t the force resultant $\mathbf{f}_t$ given by

$$\mathbf{f}_t = \int_{\Omega} \mathbf{b}_t(\mathbf{X})\,\mathrm{d}V(\mathbf{X})+ \oint_{\partial\Omega} \mathbf{s}_t(\mathbf{X})\,\mathrm{d}A(\mathbf{X}).$$

Similarly, the heat sources yield at every t the total heat supply δ_t given by

$$\delta_t = \int_{\Omega} \alpha_t(\mathbf{X})\,\mathrm{d}V(\mathbf{X})+ \oint_{\partial\Omega} \pi_t(\mathbf{X})\,\mathrm{d}A(\mathbf{X}).$$

Functions $\mathbf{b}_t(\cdot)$ and $\mathbf{s}_t(\cdot)$ defined for every $t \in [t_0, t_f]$ will be called the *body* and *boundary loadings*, respectively. Similarly, functions $\alpha_t(\cdot)$ and $\pi_t(\cdot)$ are said to be the *body* and *boundary heat sources*, respectively. All integrands in the above formulae will be referred to as the external agents acting upon the body.

Now, the basic assumption in the mathematical modelling of interactions between the body and its exterior states that all the exterior agents are either known a priori or are uniquely determined by the motion of the body and by the evolution of the temperature. The pertinent mappings will be referred to as the defining equations of the external agents. Throughout this section we shall confine ourselves to a certain class of loadings and heat sources mostly encountered in structural mechanics problems. However, it has to be stressed that the external agents considered do not represent all possible interactions between the body and its exterior. Therefore, we shall also introduce (in Sec. 2.3.1) certain restrictions imposed on motion and temperature distributions of the boundary of the body. These restrictions will be termed the boundary constraints of the kinematic- and thermal-type, respectively. It is assumed that the external agents and the boundary constraints are sufficient, at every time instant, to determine the structure interactions with the outside world.

2.2.2 Body loadings and internal heat supply

We shall restrict ourselves to the following defining equations for the body loadings:

$$\mathbf{b}_t(\mathbf{X}) = \hat{\mathbf{b}}(\mathbf{X}; t, \boldsymbol{\chi}(\mathbf{X}, t), \nabla\boldsymbol{\chi}(\mathbf{X}, t)), \quad \mathbf{X} \in \Omega, \tag{2.6}$$

where $\hat{\mathbf{b}}(\cdot)$ is for every specific situation a known function. The RHS of eq. (2.6) usually represents the force of gravity which is given by

$$\hat{\mathbf{b}} = \varrho(\mathbf{X})[\det(\nabla\boldsymbol{\chi}(\mathbf{X}, t))]^{-1}\mathbf{e},$$

where $\mathbf{e}$ is the unit vector determining the (constant) direction of the gravitational field. If the weight of the body can be neglected in the problem under cosideration (i.e. when the body loadings are sufficiently small in comparison to boundary loadings) then we can assume that $\hat{\mathbf{b}} = \mathbf{0}$.

Similarly, we shall confine ourselves to the defining equation for the internal heat supply in the form

$$\alpha_t(\mathbf{X}) = \hat{\alpha}(\mathbf{X}; t, \theta(\mathbf{X}, t)), \quad \mathbf{X} \in \Omega, \tag{2.7}$$

where $\hat{\alpha}(\cdot)$ is a known function for every specific problem considered. Functions $\hat{\mathbf{b}}(\cdot)$, $\hat{\alpha}(\cdot)$ are assumed to satisfy the regularity conditions required in the given problem.

2.2.3 Boundary loadings

In structural mechanics we deal as a rule with situations in which the boundary of the body (or a part of this boundary) is loaded. The defining equations for the boundary loadings is assumed in the form

$$\mathbf{s}_t(\mathbf{X}) = \tilde{\mathbf{s}}(\mathbf{X}, t) + \hat{\mathbf{s}}(\mathbf{X}, t, \boldsymbol{\chi}(\mathbf{X}, t), \nabla\boldsymbol{\chi}(\mathbf{X}, t)) + \\ + \mathbf{G}(\mathbf{X}, t, \boldsymbol{\chi}(\mathbf{X}, t), \nabla\boldsymbol{\chi}(\mathbf{X}, t))\dot{\mathbf{u}}_t(\mathbf{X}), \quad \mathbf{X} \in \partial\Omega, \tag{2.8}$$

where $\tilde{\mathbf{s}}(\cdot)$, $\hat{\mathbf{s}}(\cdot)$ are known vector functions and $\mathbf{G}(\cdot)$ is a known tensor function, all of which have to satisfy the regularity conditions required. Eq. (2.8) is assumed to be determined for every $\mathbf{X} \in \partial\Omega$ except possibly at a finite number of lines and points, arguments $\boldsymbol{\chi}(\mathbf{X}, t)$, $\nabla\boldsymbol{\chi}(\mathbf{X}, t)$ are the boundary values of the deformation $\boldsymbol{\chi}(\cdot, t)$ and its gradient $\nabla\boldsymbol{\chi}(\cdot, t)$, respectively. The first term on the RHS of eq. (2.8) represents what are called the dead loadings whereas the last term represents the so-called damping forces which depend on the velocity $\dot{\mathbf{u}}_t(\mathbf{X})$. If the part Δ_t of the boundary is not loaded at the time instant t then all the functions of the RHS of eq. (2.8) attain the zero value for every $\mathbf{X} \in \Delta_t$. From the general form of defining equation (2.8) we can easily pass to various specific boundary loadings needed in the analysis considered.

2.2.4 *Boundary heat supply*

The defining equations for the boundary heat supply is assumed in the form

$$\pi_t(\mathbf{X}) = \tilde{\pi}(\mathbf{X}, t) + \hat{\pi}(\mathbf{X}, t, \boldsymbol{\chi}(\mathbf{X}, t), \theta(\mathbf{X}, t)), \quad \mathbf{X} \in \partial\Omega \tag{2.9}$$

where $\tilde{\pi}(\cdot)$, $\hat{\pi}(\cdot)$ are known scalar functions satisfying the regularity conditions required. Eq. (2.8) is assumed to be determined for every $\mathbf{X} \in \Omega$ except possibly at a finite number of lines or points. Arguments $\boldsymbol{\chi}(\mathbf{X}, t)$, $\theta(\mathbf{X}, t)$ should be interpreted as the boundary values of the deformation and temperature fields $\boldsymbol{\chi}(\cdot, t)$ and $\theta(\cdot, t)$. It can happen that on a certain part Ξ_t of $\partial\Omega$ there is no boundary heat supply; the RHS of eq. (2.9) are then zero for every t.

2.3 Constraints

2.3.1 *Boundary and internal constraints*

In structural mechanics problems we always deal with bodies which are supported on certain parts of their boundaries. Such supports imply the necessity to impose certain restrictions on the values of deformations at the boundary $\partial\Omega$ of Ω. These restrictions will be referred to as the kinematic boundary constraints. The boundary value of deformation $\boldsymbol{\chi}(\cdot, \tau)$ on $\partial\Omega$, where $\boldsymbol{\chi}(\cdot, \tau) \in \mathscr{D}(\Omega)$, will be denoted by $\boldsymbol{\chi}_{\partial\Omega}(\cdot, \tau)$; every $\boldsymbol{\chi}_{\partial\Omega}(\cdot, \tau)$ is a mapping defined on $\partial\Omega$ with values in R^3. Thus the kinematic boundary constraints can be written in the following general form

$$\boldsymbol{\chi}_{\partial\Omega}(\cdot, \tau) \in \mathscr{D}_\tau(\partial\Omega),$$

where every $\mathscr{D}_\tau(\partial\Omega)$, $\tau \in [t_0, t_f]$, is the known set of all boundary values of the deformations satisfying the restrictions required by the presence of supports. For example, if a part Δ of $\partial\Omega$ at every time instant $\tau \in [t_0, t_f]$ coincides with a moving surface $\mathbf{z} = \boldsymbol{\xi}(\mathbf{X}, \tau)$, $\mathbf{X} \in \Omega$, then

$$\mathscr{D}_\tau(\partial\Omega) := \{\chi_{\partial\Omega}(\cdot, \tau) | \ \chi_{\partial\Omega}(\mathbf{X}, \tau) = \boldsymbol{\xi}(\mathbf{X}, \tau), \mathbf{X} \in \Delta\}.$$

Independently of the kinematic boundary constraints we introduced in Sec. 1.1.2 conditions of the form

$$\chi(\cdot, \tau) \in \mathscr{D}(\Omega), \qquad \tau \in [t_0, t_f],$$

the meaning of which was there explained. These conditions will be referred to as the kinematic internal constraints. We have stated before that the existence of internal constraints can be treated as a characteristic feature of structural mechanics problems for which the set $\mathscr{D}(\Omega)$ plays an essential role in formulating various engineering theories.

Similarly, in problems involving heat conduction we often deal with the so-called thermal boundary constraints which are restrictions imposed on boundary values of temperature fields $\theta(\cdot, \tau)$, $\tau \in [t_0, t_f]$, where $\theta(\cdot, \tau) \in \mathscr{T}(\Omega)$. The boundary values of $\theta(\cdot, \tau)$ will be denoted by $\theta_{\partial\Omega}(\cdot, \tau)$; every $\theta_{\partial\Omega}(\cdot, \tau)$ is a real valued function defined on $\partial\Omega$ (except possibly at some lines or points). Thus the thermal boundary constraints can be assumed in the form

$$\theta_{\partial\Omega}(\cdot, \tau) \in \mathscr{T}_\tau(\partial\Omega), \qquad \tau \in [t_0, t_f],$$

where every $\mathscr{T}_\tau(\partial\Omega)$ is the known set of all boundary values of the temperature fields which are admissible in the problem considered. For example, if at every time instant $\tau \in [t_0, t_f]$ the temperature at the part Δ of the boundary is known and equal $\bar{\theta}(\mathbf{X}, \tau)$, $\mathbf{X} \in \Delta$, then

$$\mathscr{T}_\tau(\partial\Omega) = \{\theta_{\partial\Omega}(\cdot, \tau) | \ \theta_{\partial\Omega}(\mathbf{X}, \tau) = \bar{\theta}(\mathbf{X}, \tau), \ \mathbf{X} \in \Delta\}.$$

Apart from the thermal boundary constraints in structural mechanics we can also deal with the known restrictions imposed on the temperature fields inside the body, defined by the relation

$$\theta(\cdot, \tau) \in \mathscr{T}(\Omega), \qquad \tau \in [t_0, t_f],$$

which were introduced in Sec. 1.1.5 and will be termed *thermal internal constraints.*

It has to be emphasized that both kinematic and thermal boundary constraints are due to the interactions between the material element and its exterior. However, these constraints are introduced only after the internal constraints have been determined. Thus, by setting

$$\mathscr{D}_\tau(\Omega) = \{\chi(\cdot, \tau) \in \mathscr{D}(\Omega) | \ \chi_{\partial\Omega}(\cdot, \tau) \in \mathscr{D}_\tau(\partial\Omega)\},$$

$$\mathscr{T}_\tau(\Omega) = \{\theta(\cdot, \tau) \in \mathscr{T}(\Omega) | \ \theta_{\partial\Omega}(\cdot, \tau) \in \mathscr{T}_\tau(\partial\Omega)\},$$

for every $\tau \in [t_0, t_f]$ we define the total constraints by means of the conditions

$$\begin{aligned} &\chi(\cdot, \tau) \in \mathscr{D}_\tau(\Omega), \\ &\theta(\cdot, \tau) \in \mathscr{T}_\tau(\Omega), \qquad \tau \in [t_0, t_f], \end{aligned} \tag{2.10}$$

where the sets $\mathscr{D}_\tau(\Omega)$, $\mathscr{T}_\tau(\Omega)$ are assumed to be known. We have here tacitly assumed that every deformation $\chi(\cdot, \tau) \in \mathscr{D}(\Omega)$ and every temperature field $\theta(\cdot, \tau) \in \mathscr{T}(\Omega)$ has a well defined trace $\chi_{\partial\Omega}(\cdot, \tau)$ and $\theta_{\partial\Omega}(\cdot, \tau)$, respectively, on the boundary $\partial\Omega$ of Ω. In the sequel in order to simplify the terminology we shall identify the internal, boundary and total kinematic constraints with the sets $\mathscr{D}(\Omega)$, $\mathscr{D}_\tau(\partial\Omega)$ and $\mathscr{D}_\tau(\Omega)$ respectively. Similarly, we shall identify the internal, boundary and total thermal constraints with the sets $\mathscr{T}(\Omega)$, $\mathscr{T}_\tau(\partial\Omega)$ and $\mathscr{T}_\tau(\Omega)$, respectively.

Now let t be an arbitrary but fixed time instant $t \in (t_0, t_f)$ and $\varkappa(\cdot)$, $\varkappa(\cdot) \in \mathscr{D}_\tau(\Omega)$ be an arbitrary fixed deformation. Moreover, let $\chi(\cdot, \tau) \in \mathscr{D}_\tau(\Omega)$, $\tau \in [t_0, t_f]$, represent an arbitrary motion such that $\chi(\cdot, t) = \varkappa(\cdot)$. Then $\dot{\chi}(\cdot, t)$ will be called the *velocity field* admissible by the constraints $\mathscr{D}_\tau(\Omega)$. The set of all velocity fields $\mathbf{v}(\cdot)$ admissible by (kinematic) constraints $\mathscr{D}_t(\Omega)$ at $\varkappa(\cdot)$ will be denoted by the symbol $\mathscr{V}_t(\varkappa(\cdot))$.

Let $\Theta(\cdot)$, $\Theta(\cdot) \in \mathscr{T}_t(\Omega)$ be an arbitrary but fixed temperature field and $\theta(\cdot, \tau) \in \mathscr{T}_\tau(\Omega)$, $\tau \in [t_0, t_f]$, represent an evolution of the temperature field during an arbitrary motion. Let us also assume that $\theta(\cdot, t) = \Theta(\cdot)$ and that the temperature rate field $\dot{\theta}(\cdot, t)$ is well defined. Then $\dot{\theta}(\cdot, t)$ will be called the temperature rate field admissible by (thermal) constraints $\mathscr{T}_\tau(\Omega)$ at the temperature field $\Theta(\cdot)$. The set of all temperature rate fields admissible by $\mathscr{T}_\tau(\Omega)$ at $\Theta(\cdot)$ will be denoted by $\mathscr{W}_t(\Theta(\cdot))$.

Observe that referring to $\mathscr{D}_\tau(\Omega)$, $\mathscr{T}_\tau(\Omega)$ as the total constraints we have assumed that formulae (2.10) represent all a priori restrictions imposed on motions $\chi(\cdot, \tau)$, $\tau \in [t_0, t_f]$, and on evolutions of temperature fields $\theta(\cdot, \tau)$, $\tau \in [t_0, t_f]$. In other words, there are no additional restrictions imposed on, say, the rates $\dot{\chi}(\cdot, \tau)$, $\dot{\theta}(\cdot, \tau)$. Thus the families of sets $\mathscr{V}_\tau(\varkappa(\cdot))$, $\varkappa(\cdot) \in \mathscr{D}_\tau(\Omega)$ and $\mathscr{W}_\tau(\Theta(\cdot))$, $\Theta(\cdot) \in \mathscr{T}_\tau(\Omega)$, define all the possible velocities and temperature rates in all admissible deformations $\varkappa(\cdot) \in \mathscr{D}_\tau(\Omega)$ and for all admissible temperature fields $\Theta(\cdot) \in \mathscr{T}_\tau(\Omega)$, respectively. We shall assume that every $\mathscr{V}_\tau(\varkappa(\cdot))$, $\varkappa(\cdot) \in \mathscr{D}_\tau(\Omega)$, is a convex set, i.e. if $\mathbf{v}_1(\cdot) \in \mathscr{V}_\tau(\varkappa(\cdot))$ and $\mathbf{v}_2(\cdot) \in \mathscr{V}_\tau(\varkappa(\cdot))$ then also $\mathbf{v}(\cdot) \equiv \mu_1 \mathbf{v}_1(\cdot) + \mu_2 \mathbf{v}_2(\cdot) \in \mathscr{V}_\tau(\varkappa(\cdot))$ for every $\mu_1, \mu_2 \in (0, 1)$ such that $\mu_1 + \mu_2 = 1$. Similarly, we shall also assume that every $\mathscr{W}_\tau(\Theta(\cdot))$, $\Theta(\cdot) \in \mathscr{T}_\tau(\Omega)$, is a convex set as well, which means that from $w_1(\cdot) \in \mathscr{W}_\tau(\Theta(\cdot))$ and $w_2(\cdot) \in \mathscr{W}_\tau(\Theta(\cdot))$ follows that $w(\cdot) \equiv \mu_1 w_1(\cdot) + \mu_2 w_2(\cdot) \in \mathscr{W}_\tau(\Theta(\cdot))$ for every $\mu_1, \mu_2 \in (0, 1)$ such that $\mu_1 + \mu_2 = 1$. These conditions are fulfilled in most problems of structural mechanics. More general situations can be taken into account as well but will not be considered here for the sake of book compactness; for a more detailed discussion of problems with constraints the reader is referred to [169], [170].

So far, the concept of constraints was analysed from the purely kinematic point of view. Considering the general formulations of problems within the framework of solid mechanics we shall assume that the constraints are maintained by certain supplementary forces called reactions of the constraints. Below we shall show how to describe mathematically the concept of reactions to both the kinematic and thermal constraints.

2.3.2 Reactions

Taking into account the concepts of forces and heat supply, cf. eqs. (1.1), (1.4), as well as the concept of external agents introduced in Sec. 2.2.1, we introduce now the following definitions

$$
\begin{aligned}
&\bar{\mathbf{b}}_t(\mathbf{X}) \equiv \mathbf{b}(\mathbf{X}, t)-\mathbf{b}_t(\mathbf{X}), \\
&\bar{\alpha}_t(\mathbf{X}) \equiv \alpha(\mathbf{X}, t)-\alpha_t(\mathbf{X}), \quad \mathbf{X} \in \Omega, \\
&\bar{\mathbf{s}}_t(\mathbf{X}) \equiv \mathbf{s}(\mathbf{X}, t)-\mathbf{s}_t(\mathbf{X}), \\
&\bar{\pi}_t(\mathbf{X}) \equiv \pi(\mathbf{X}, t)-\pi_t(\mathbf{X}), \quad \mathbf{X} \in \partial\Omega,
\end{aligned}
\tag{2.11}
$$

provided that the RHS of the above equalities are well defined in the given problem. The vector density fields $\bar{\mathbf{b}}_t(\cdot)$, $\bar{\mathbf{s}}_t(\cdot)$ will be interpreted as reactions due to the presence of the kinematic constraints $\mathscr{D}_t(\Omega)$ whereas the scalar density fields $\bar{\alpha}_t(\cdot)$, $\bar{\pi}_t(\cdot)$ will be interpreted as reactions due to the presence of the thermal constraints $\mathscr{T}_t(\Omega)$. Obviously, the reactions depend on the form of constraints. In order to interrelate the reactions $\bar{\mathbf{b}}_t(\cdot)$, $\bar{\mathbf{s}}_t(\cdot)$ and the total kinematic constraints $\mathscr{D}_t(\Omega)$ we shall follow the general line of approach leading to the concept of ideal constraints. Firstly, we shall introduce the important concept of what we call the virtual displacements. To this aim for every $\boldsymbol{\varkappa}(\cdot) \in \mathscr{D}_t(\Omega)$ and every $\mathbf{v}(\cdot) \in \mathscr{V}_t(\boldsymbol{\varkappa}(\cdot))$ we define a set $\mathscr{U}_t(\boldsymbol{\varkappa}(\cdot), \mathbf{v}(\cdot))$ of functions $\mathbf{u}(\cdot) \in C_1^3(\bar{\Omega})$ such that $\mathbf{u}(\cdot) = \lambda[\bar{\mathbf{v}}(\cdot)-\mathbf{v}(\cdot)]$ for some $\bar{\mathbf{v}}(\cdot) \in \mathscr{V}_t(\boldsymbol{\varkappa}(\cdot))$ and for some $\lambda \geqslant 0$:

$$
\mathscr{U}_t(\boldsymbol{\varkappa}(\cdot), \mathbf{v}(\cdot)) = \left\{\mathbf{u}(\cdot) \in C_1^3(\bar{\Omega}) \,\middle|\, \mathbf{u}(\cdot) = \lambda[\bar{\mathbf{v}}(\cdot)-\mathbf{v}(\cdot)], \bar{\mathbf{v}}(\cdot) \in \mathscr{V}_t(\boldsymbol{\varkappa}(\cdot)), \lambda \geqslant 0\right\}.
$$

The functions $\mathbf{u}(\cdot) \in \mathscr{U}_t(\boldsymbol{\varkappa}(\cdot), \mathbf{v}(\cdot))$ will be called the *virtual displacements* for $\boldsymbol{\varkappa}(\cdot) \in \mathscr{D}_t(\Omega)$ and for $\mathbf{v}(\cdot) \in \mathscr{V}_t(\boldsymbol{\varkappa}(\cdot))$. Let us observe that if $\mathbf{u}(\cdot) \in \mathscr{U}_t(\boldsymbol{\varkappa}(\cdot), \mathbf{v}(\cdot))$ then also $\alpha\mathbf{u}(\cdot) \in \mathscr{U}_t(\boldsymbol{\varkappa}(\cdot), \mathbf{v}(\cdot))$ for every non-negative number α, $\alpha \geqslant 0$.

Now, let us assume that the rate of work of all the reaction forces on every virtual displacement is non-negative. Thus

$$\oint_{\partial\Omega} \bar{\mathbf{s}}_t(\mathbf{X}) \cdot \mathbf{u}(\mathbf{X}) \mathrm{d}A(\mathbf{X}) + \int_{\Omega} \bar{\mathbf{b}}_t(\mathbf{X}) \cdot \mathbf{u}(\mathbf{X}) \mathrm{d}V(\mathbf{X}) \geqslant 0 \tag{2.12}$$

is assumed to hold for every $\mathbf{u}(\cdot) \in \mathscr{U}_t(\boldsymbol{\chi}(\cdot, t), \dot{\boldsymbol{\chi}}(\cdot, t))$.

The reactions $(2.11)_{1,3}$ to kinematic constraints for which inequality (2.12) is satisfied are called *ideal*. It can be easily shown that the aforementioned assumption is equivalent to the statement that at every time instant t and every deformation $\boldsymbol{\chi}(\cdot, t)$ the rate of work (cf. Sec. 1.2.3) of all the reaction forces acting in the body attains its minimum on the set $\mathscr{V}_t(\boldsymbol{\chi}(\cdot, t))$, where $\boldsymbol{\chi}(\cdot, t) \in \mathscr{D}_t(\Omega)$ (i.e. on the set of all velocity fields admissible by the constraints). Thus we see that the concepts of virtual displacements and ideal constraints have a simple physical interpretation in terms of the minimum principle.

Similarly, to interrelate the thermal reactions $\bar{\alpha}_t(\cdot)$, $\bar{\pi}_t(\cdot)$ and the total thermal constraints $\mathscr{T}_t(\Omega)$ we shall introduce what are called the virtual temperature increments. For every $\Theta(\cdot) \in \mathscr{T}_t(\Omega)$ and every $w(\cdot) \in \mathscr{W}_t(\Theta(\cdot))$ we define a set $\mathscr{Z}_t(\Theta(\cdot), w(\cdot))$ of functions $\zeta(\cdot) \in C_1(\bar{\Omega})$ such that $\zeta(\cdot) = \lambda[\bar{w}(\cdot) - w(\cdot)]$ for some $\bar{w}(\cdot) \in \mathscr{W}_t(\Theta(\cdot))$ and for some non-negative number λ, $\lambda \geqslant 0$. It can be seen that if $\zeta(\cdot) \in \mathscr{Z}_t(\Theta(\cdot), w(\cdot))$ then also $\alpha\zeta(\cdot) \in \mathscr{Z}_t(\Theta(\cdot), w(\cdot))$ for every α, $\alpha \geqslant 0$. The functions $\zeta(\cdot) \in \mathscr{Z}_t(\Theta(\cdot), w(\cdot))$ are called the *virtual temperature increments*[1]. If the inequality

$$\oint_{\partial\Omega} \bar{\pi}_t(\mathbf{X}) \zeta(\mathbf{X}) \mathrm{d}A(\mathbf{X}) + \int_{\Omega} \bar{\alpha}_t(\mathbf{X}) \zeta(\mathbf{X}) \mathrm{d}V(\mathbf{X}) \geqslant 0 \tag{2.13}$$

is assumed to hold for every $\zeta(\cdot) \in \mathscr{Z}_t(\theta(\cdot, t), \dot{\theta}(\cdot, t))$, $t \in [t_0, t_f]$, then reactions $(2.11)_{2,4}$ will be termed ideal[2].

In the sequel we shall assume that the reactions to both kinematic and thermal constraints are ideal. In order to simplify the terminology we shall also say that the constraints under consideration are ideal; for by the constraints (both kinematic and thermal) we mean not only families of sets $\mathscr{D}_\tau(\Omega)$, $\mathscr{T}_\tau(\Omega)$,

[1] The definitions of virtual displacements and virtual relative temperatures are valid only if the sets $\mathscr{V}_t(\boldsymbol{\varkappa}(\cdot))$, $\mathscr{W}_t(\Theta(\cdot))$ are convex, which was assumed at the end of Sec. 2.3.1. All the readers familiar with the basic concepts of functional analysis will observe that the sets $\mathscr{U}_t(\boldsymbol{\varkappa}(\cdot), \mathbf{v}(\cdot))$ and $\mathscr{Z}_t(\Theta(\cdot), w(\cdot))$ have to be non-empty closed convex cones in the pertinent linear topological spaces of sufficiently regular functions defined on Ω (having well defined traces on $\partial\Omega$) with values in $\boldsymbol{R}^3$ and $\boldsymbol{R}$, respectively.

[2] Under the condition that $|\vartheta_t(\mathbf{X})|/\theta_0 \ll 1$ assumption (2.13) states that the rate of the total production of entropy due to the thermal constraints and treated as a functional defined on $\mathscr{W}_t(\theta(\cdot, t))$ attains on $\mathscr{W}_t(\theta(\cdot, t))$ its minimum (cf. Sec. 1.2.4 for the notion of the total production of entropy $\gamma_t(\Omega)$).

$\tau \in [t_0, t_f]$, but also conditions (2.12), (2.13). For the detailed discussion of the problem of constraints in continuum and structural mechanics the reader is referred to [174]–[178]. The approach to constraints exploited in Secs. 2.3.1 and 2.3.2 will be taken as the basis for some specific considerations in the next section of the book.

2.3.3 *Bilateral constraints*

It follows from formulae (2.12), (2.13) that problems with the ideal constraints can be governed not only by equations but also by certain integral inequalities. However, for many structural mechanics problems it is sufficient to consider some rather simple forms of ideal constraints only, examples of which will be given below.

Let us assume that for every $\boldsymbol{\varkappa}(\cdot) \in \mathcal{D}_\tau(\Omega)$, $\Theta(\cdot) \in \mathcal{T}_\tau(\Omega)$, $\tau \in [t_0, t_f]$, sets $\mathcal{U}_\tau(\boldsymbol{\varkappa}(\cdot), \mathbf{v}(\cdot))$ and $\mathcal{Z}_\tau(\Theta(\cdot), w(\cdot))$ are independent of $\mathbf{v}(\cdot) \in \mathcal{V}_\tau(\boldsymbol{\varkappa}(\cdot))$ and $w(\cdot) \in \mathcal{W}_\tau(\Theta(\cdot))$, respectively. It means that

$$\mathcal{U}_\tau(\boldsymbol{\varkappa}(\cdot)) = \mathcal{U}_\tau(\boldsymbol{\varkappa}(\cdot), \mathbf{v}(\cdot)) \quad \text{for every } \mathbf{v}(\cdot) \in \mathcal{V}_\tau(\boldsymbol{\varkappa}(\cdot)),$$
$$\mathcal{Z}_\tau(\Theta(\cdot)) = \mathcal{Z}_\tau(\Theta(\cdot), w(\cdot)) \text{ for every } w(\cdot) \in \mathcal{W}_\tau(\Theta(\cdot)),$$

where $\mathcal{U}_\tau(\boldsymbol{\varkappa}(\cdot))$ and $\mathcal{Z}_\tau(\Theta(\cdot))$ are referred to as the sets of virtual displacements and virtual temperatures, respectively.

If the above conditions hold then, remembering that all sets $\mathcal{V}_\tau(\boldsymbol{\varkappa}(\cdot))$, $\mathcal{W}_\tau(\Theta(\cdot))$ are convex, cf. Sec. 2.3.1, the constraints under consideration will be termed bilateral.

Taking into account the definitions of sets $\mathcal{U}_\tau(\boldsymbol{\varkappa}(\cdot), \mathbf{v}(\cdot))$ and $\mathcal{Z}_\tau(\Theta(\cdot), w(\cdot))$ which were given in Sec. 2.3.2, we obtain for the bilateral constraints

$$\mathcal{U}_t(\boldsymbol{\varkappa}(\cdot)) = \{\mathbf{u}(\cdot) \in C_1^3(\bar{\Omega}) | \ \mathbf{u}(\cdot) = \lambda[\mathbf{v}_1(\cdot) - \mathbf{v}_2(\cdot)];$$
$$\mathbf{v}_1(\cdot), \mathbf{v}_2(\cdot) \in \mathcal{V}_t(\boldsymbol{\varkappa}(\cdot)), \lambda \geqslant 0\},$$
$$\mathcal{Z}_t(\Theta(\cdot)) = \{\zeta(\cdot) \in C_1(\bar{\Omega}) | \ \zeta(\cdot) = \lambda[w_1(\cdot) - w_2(\cdot)];$$
$$w_1(\cdot), w_2(\cdot) \in \mathcal{W}_t(\Theta(\cdot)), \lambda \geqslant 0\},$$

for every $\boldsymbol{\varkappa}(\cdot) \in \mathcal{D}_t(\Omega)$, $\Theta(\cdot) \in \mathcal{T}_t(\Omega)$ and $t \in [t_0, t_f]$. Let us observe that if condition (2.11) holds for some $\mathbf{u}(\cdot) \in \mathcal{U}_t(\boldsymbol{\varkappa}(\cdot))$ then it also holds for $\mathbf{u}_1(\cdot) = -\mathbf{u}(\cdot)$. Thus, it follows that condition (2.11) is satisfied by every $\mathbf{u}(\cdot) \in \mathcal{U}_t(\boldsymbol{\varkappa}(\cdot))$ if the equality

$$\oint_{\partial\Omega} \bar{\mathbf{s}}_t(\mathbf{X}) \cdot \mathbf{u}(\mathbf{X}) \mathrm{d}A(\mathbf{X}) + \int_{\Omega} \bar{\mathbf{b}}_t(\mathbf{X}) \cdot \mathbf{u}(\mathbf{X}) \mathrm{d}V(\mathbf{X}) = 0 \tag{2.14}$$

holds for every $\mathbf{u}(\cdot) \in \mathcal{U}_t(\boldsymbol{\varkappa}(\cdot))$. Similarly, if condition (2.13) holds for some $\zeta(\cdot) \in \mathcal{Z}_t(\Theta(\cdot))$ then it also holds for $\zeta_1(\cdot) \equiv -\zeta(\cdot)$. In this way the bilateral constraints condition (2.13) implies that the equality

$$\oint_{\partial\Omega} \bar{\pi}_t(\mathbf{X})\zeta(\mathbf{X})\,\mathrm{d}A(\mathbf{X}) + \int_{\Omega} \bar{\alpha}_t(\mathbf{X})\zeta(\mathbf{X})\,\mathrm{d}V(\mathbf{X}) = 0 \tag{2.15}$$

holds for every $\zeta(\cdot) \in \mathscr{Z}_t(\Theta(\cdot))$.

REMARK. It can be shown that if $\mathbf{u}_1(\cdot) \in \mathscr{U}_t(\boldsymbol{\varkappa}(\cdot))$ and $\mathbf{u}_2(\cdot) \in \mathscr{U}_t(\boldsymbol{\varkappa}(\cdot))$ then also $\alpha_1\mathbf{u}_1(\cdot) + \alpha_2\mathbf{u}_2(\cdot) \in \mathscr{U}_t(\boldsymbol{\varkappa}(\cdot))$ for any two real numbers α_1, α_2. Similarly, if $\zeta_1(\cdot) \in \mathscr{Z}_t(\Theta(\cdot))$ and $\zeta_2(\cdot) \in \mathscr{Z}_t(\Theta(\cdot))$ then also $\alpha_1\zeta_1(\cdot) + \alpha_2\zeta_2(\cdot) \in \mathscr{Z}_t(\Theta(\cdot))$ for any two reals α_1, α_2. These properties of sets $\mathscr{U}_t(\boldsymbol{\varkappa}(\cdot))$, $\mathscr{Z}_t(\Theta(\cdot))$ are due to the fact that $\mathscr{V}_t(\boldsymbol{\varkappa}(\cdot))$ and $\mathscr{Z}_t(\Theta(\cdot))$ are assumed to be convex sets, cf. the remark at the end of Sec. 2.3.1[1].

The constraints which are bilateral and ideal are called the *ideal bilateral constraints*. For such constraints the reactions have to satisfy conditions (2.14), (2.15). In the sequel we shall confine ourselves to problems with the bilateral ideal constraints. For the sake of simplicity we shall also use the notation

$$\mathscr{V}_t \equiv \mathscr{V}_t(\boldsymbol{\chi}(\cdot, t)), \quad \mathscr{U}_t \equiv \mathscr{U}_t(\boldsymbol{\chi}(\cdot, t)),$$
$$\mathscr{W}_t \equiv \mathscr{W}_t(\theta(\cdot, t)), \quad \mathscr{Z}_t \equiv \mathscr{Z}_t(\theta(\cdot, t)),$$

for every $\boldsymbol{\chi}(\cdot, t) \in \mathscr{D}_t(\Omega)$, $\theta(\cdot, t) \in \mathscr{T}_t(\Omega)$, $t \in [t_0, t_f]$.

In order to illustrate condition (2.14) let us assume that, cf. Sec. 2.3.1,

$$\mathscr{D}_\tau(\Omega) = \{\boldsymbol{\chi}(\cdot, \tau) \in \mathscr{D}(\Omega) \mid \boldsymbol{\chi}_{\partial\Omega}(\mathbf{X}, \tau) = \boldsymbol{\xi}(\mathbf{X}, \tau), \mathbf{X} \in \Delta\}, \qquad \tau \in [t_0, t_f],$$

where Δ is a part of $\partial\Omega$ and $\boldsymbol{\xi}(\mathbf{X}, \tau)$, $\mathbf{X} \in \Delta$, is the known function representing configurations of this part of the boundary. To further simplify the presentation of the example let us also assume that there are no internal constraints, i.e. the set $\mathscr{D}(\Omega)$ comprises all the possible deformations. In such a case we can take $\mathbf{v}(\mathbf{X})$, $\mathbf{X} \in \Omega$, as an arbitrary velocity field satisfying conditions $\mathbf{v}(\mathbf{X}) = \dot{\boldsymbol{\xi}}(\mathbf{X}, \tau)$ for $\mathbf{X} \in \Delta$. Hence, for every $t \in [t_0, t_f]$ we get $\mathscr{U}_t = \{\mathbf{u}(\cdot) \in C_1^3(\bar{\Omega}) \mid \mathbf{u}(\mathbf{X}) = \mathbf{0}$ for every $\mathbf{X} \in \Delta\}$ and formula (2.14) leads to the condition

$$\oint_{\partial\Omega\setminus\Delta} \bar{\mathbf{s}}_t(\mathbf{X}) \cdot \mathbf{u}(\mathbf{X})\,\mathrm{d}A(\mathbf{X}) + \int_{\Omega} \bar{\mathbf{b}}_t(\mathbf{X}) \cdot \mathbf{u}(\mathbf{X})\,\mathrm{d}V(\mathbf{X}) = 0$$

which has to hold for every $\mathbf{u}(\cdot) \in \mathscr{U}_t$. Using next the du Bois–Raymond lemma (the fundamental lemma of variational calculus) we obtain

$$\bar{\mathbf{s}}_t(\mathbf{X}) = \mathbf{0}, \quad \mathbf{X} \in \partial\Omega\setminus\bar{\Delta},$$
$$\bar{\mathbf{b}}_t(\mathbf{X}) = \mathbf{0}, \quad \mathbf{X} \in \Omega,$$

[1] To be more precise, the sets $\mathscr{U}_t(\boldsymbol{\varkappa}(\cdot))$, $\mathscr{Z}_t(\Theta(\cdot))$ constitute the closed linear subspaces in pertinent linear topological spaces of (sufficiently regular) functions defined on Ω (having well defined traces on $\partial\Omega$), with values in $\boldsymbol{R}^3$ and $\boldsymbol{R}$, respectively.

and $\bar{\mathbf{s}}_t(\mathbf{X}) \in R^3$ for $\mathbf{X} \in \Delta$. Thus we have arrived at the obvious conclusion that the reactions in the problem considered can be different from zero on the part Δ of the boundary $\partial\Omega$ only.

2.3.4 Constraint functions

We shall now be more specific by considering such ideal bilateral constraints in which $\boldsymbol{\chi}(\cdot, t) \in \mathscr{D}_t(\Omega)$ if and only if

$$\alpha_p(\mathbf{X}, \nabla\boldsymbol{\chi}(\mathbf{X}, t)) = 0, \quad \mathbf{X} \in \Omega, \quad p = 1, \ldots, P,$$
$$\beta_r(\mathbf{X}, t, \boldsymbol{\chi}_{\partial\Omega}(\mathbf{X}, t)) = 0, \quad \mathbf{X} \in \partial\Omega, \quad r = 1, \ldots, R,$$

where $\alpha_p(\cdot)$, $\beta_r(\cdot)$ are known differentiable real valued functions. It follows that $\mathbf{v}(\cdot) \in \mathscr{V}_t \equiv \mathscr{V}_t(\boldsymbol{\chi}(\cdot, t))$ if and only if

$$\operatorname{tr}\left[\frac{\partial\alpha_p}{\partial\nabla\boldsymbol{\chi}(\mathbf{X}, t)} \nabla\mathbf{v}(\mathbf{X})\right] = 0, \quad \mathbf{X} \in \Omega, \quad p = 1, \ldots, P,$$

$$\frac{\partial\beta_r}{\partial t} + \frac{\partial\beta_r}{\partial\boldsymbol{\chi}(\mathbf{X}, t)} \cdot \mathbf{v}(\mathbf{X}) = 0, \quad \mathbf{X} \in \partial\Omega, \quad r = 1, \ldots, R,$$

and the sets $\mathscr{U}_t \equiv \mathscr{U}_t(\boldsymbol{\chi}(\cdot, t))$ comprise all functions $\mathbf{u}(\cdot) \in C_1^3(\bar{\Omega})$ such that

$$\operatorname{tr}\left[\frac{\partial\alpha_p}{\partial\nabla\boldsymbol{\chi}(\mathbf{X}, t)} \nabla\mathbf{u}(\mathbf{X})\right] = 0, \quad \mathbf{X} \in \Omega, \quad p = 1, \ldots, P,$$

$$\frac{\partial\beta_r}{\partial\boldsymbol{\chi}(\mathbf{X}, t)} \cdot \mathbf{u}(\mathbf{X}) = 0, \quad \mathbf{X} \in \partial\Omega, \quad r = 1, \ldots, R.$$

The indices p and r here and below run over the sequences $1, \ldots, P$ and $1, \ldots, R$ respectively, and the summation convention with respect to these indices is assumed to apply. Let $\lambda^p(\cdot, t)$, $p = 1, \ldots, P$, and $\mu^r(\cdot, t)$, $r = 1, \ldots, R$, be arbitrary (sufficiently regular) real valued functions defined (almost everywhere) on Ω and $\partial\Omega$, respectively. The conditions

$$\lambda^p(\mathbf{X}, t)\operatorname{tr}\left[\frac{\partial\alpha_p}{\partial\nabla\boldsymbol{\chi}(\mathbf{X}, t)} \nabla\mathbf{u}(\mathbf{X})\right] = 0, \quad \mathbf{X} \in \Omega,$$

$$\mu^r(\mathbf{X}, t)\frac{\partial\beta_r}{\partial\boldsymbol{\chi}(\mathbf{X}, t)} \cdot \mathbf{u}(\mathbf{X}) = 0, \quad \mathbf{X} \in \partial\Omega,$$

have to hold for every $\mathbf{u}(\cdot) \in \mathscr{U}_t \equiv \mathscr{U}_t(\boldsymbol{\chi}(\cdot, t))$. Integrating these conditions over Ω and $\partial\Omega$, respectively, and using the divergence theorem we arrive at the conclusion stating that the condition

$$\oint_{\partial\Omega}\left[\mu^r(\mathbf{X},t)\frac{\partial\beta_r}{\partial\boldsymbol{\chi}(\mathbf{X},t)}-\lambda^p(\mathbf{X},t)\frac{\partial\alpha_p}{\partial\nabla\boldsymbol{\chi}(\mathbf{X},t)}\mathbf{n}(\mathbf{X})\right]\cdot\mathbf{u}(\mathbf{X})\mathrm{d}A(\mathbf{X})+$$

$$+\int_{\Omega}\mathrm{Div}\left[\lambda^p(\mathbf{X},t)\frac{\partial\alpha_p}{\partial\nabla\boldsymbol{\chi}(\mathbf{X},t)}\right]\cdot\mathbf{u}(\mathbf{X})\mathrm{d}V(\mathbf{X})=0,$$

where $\mathbf{n}(\mathbf{X})$ stands for the unit outer normal to $\partial\Omega$ at $\mathbf{X}$, has to hold for every $\mathbf{u}(\cdot)\in\mathscr{U}_t$. Taking into account condition (2.13) it may be seen that the reactions $\bar{\mathbf{s}}_t(\mathbf{X})$, $\bar{\mathbf{b}}_t(\mathbf{X})$ can be assumed in the form

$$\bar{\mathbf{s}}_t(\mathbf{X})=\mu^r(\mathbf{X},t)\frac{\partial\beta_r}{\partial\boldsymbol{\chi}(\mathbf{X},t)}-\lambda^p(\mathbf{X},t)\frac{\partial\alpha_p}{\partial\nabla\boldsymbol{\chi}(\mathbf{X},t)}\mathbf{n}(\mathbf{X}),\qquad \mathbf{X}\in\partial\Omega,$$

$$\bar{\mathbf{b}}_t(\mathbf{X})=\mathrm{Div}\left[\lambda^p(\mathbf{X},t)\frac{\partial\alpha_p}{\partial\nabla\boldsymbol{\chi}(\mathbf{X},t)}\right],\qquad \mathbf{X}\in\Omega,$$

where $\mu^r(\cdot,t)$, $r=1,\ldots,R$, and $\lambda^p(\cdot,t)$, $p=1,\ldots,P$, are arbitrary (sufficiently regular) real valued functions defined on $\partial\Omega$ and Ω, respectively. These functions are termed the *constraint functions* and represent $R+P$ additional unknowns. Their existence is due to $R+P$ additional equations $\beta^r(\cdot)=0$, $\lambda^p(\cdot)=0$. The approach indicated above can be applied to any ideal constraints provided that they are determined by equations similar to those stated at the beginning of this subsection.

2.3.5 *Constraints for stresses and heat fluxes*

In dealing with problems of structural mechanics we may need to introduce a priori certain hypotheses concerning the stress and/or heat flux distributions. From the viewpoint of mechanics, such hypotheses may be seen as constraints for stresses and/or heat fluxes. To have a closer look at such situations let us consider in a given problem only such stress fields which satisfy the condition

$$\mathbf{S}_\tau(\cdot)\in\mathscr{S}(\Omega),\qquad \tau\in[t_0,t_f],$$

where $\mathscr{S}(\Omega)$ is the known set of all admissible stress fields. Similarly, let us assume that heat flux fields $\mathbf{h}_\tau(\cdot)$, $\tau\in[t_0,t_f]$, are restricted by the condition

$$\mathbf{h}_\tau(\cdot)\in\mathscr{H}(\Omega),\qquad \tau\in[t_0,t_f],$$

with $\mathscr{H}(\Omega)$ as the known set of all admissible heat flux fields in the given problem. We shall also introduce, for every $\mathbf{S}(\cdot)\in\mathscr{S}(\Omega)$ and every $\mathbf{h}(\cdot)\in\mathscr{H}(\Omega)$, sets $\mathscr{S}^0(\mathbf{S}(\cdot))$, $\mathscr{H}^0(\mathbf{h}(\cdot))$ of all stress rates fields and heat flux rate fields which are continuous and admissible by constraints for the fixed $\mathbf{S}(\cdot)$ and $\mathbf{h}(\cdot)$, respectively. The sets $\mathscr{S}^0(\mathbf{S}(\cdot))$, $\mathscr{H}^0(\mathbf{h}(\cdot))$ have the meaning analogous to that of the sets $\mathscr{V}_t(\boldsymbol{\varkappa}(\cdot))$, $\mathscr{W}_t(\Theta(\cdot))$ considered in Sec. 2.3.1.

Following the general line of approach exploited in Sec. 2.3.2, we can define the sets of virtual stress increments and virtual heat flux increments by using the similar arguments to those used in defining the virtual displacements. We shall assume that the constraints to be considered are bilateral in the sense explained before. At the same time we shall now consider merely special cases of constraints, in which the sets of virtual stress increments coincide with the pertinent sets $\mathscr{S}^0(\mathbf{S}(\cdot))$, $\mathbf{S}(\cdot) \in \mathscr{S}(\Omega)$, of all stress rate fields. Similarly, we shall confine ourselves to the situation in which the sets of virtual heat flux increments coincide with the pertinent sets $\mathscr{H}^0(\mathbf{h}(\cdot))$, $\mathbf{h}(\cdot) \in \mathscr{H}(\Omega)$, of all heat flux rate fields. It can be shown that such situations are typical of constraints which are time independent. Note also our implicit assumption that both the set of all admissible stress fields $\mathscr{S}(\Omega)$ and the set of all admissible heat flux fields $\mathscr{H}(\Omega)$ are constant in time.

In order to define what will be called the ideal reactions to constraints for stresses and heat fluxes we introduce the following decompositions of strain rates $\dot{\mathbf{E}}_t(\mathbf{X})$ and temperature gradients $\nabla\vartheta_t(\mathbf{X})$:

$$
\begin{aligned}
\dot{\mathbf{E}}_t(\mathbf{X}) &= \mathbf{L}_t(\mathbf{X})+\bar{\mathbf{L}}_t(\mathbf{X}),\\
\nabla\vartheta_t(\mathbf{X}) &= \mathbf{g}_t(\mathbf{X})+\bar{\mathbf{g}}_t(\mathbf{X}),
\end{aligned}
\tag{2.16}
$$

where $\mathbf{L}_t(\mathbf{X})$, $\mathbf{g}_t(\mathbf{X})$ are said to be the constitutive part of $\dot{\mathbf{E}}_t(\mathbf{X})$ and $\nabla\vartheta_t(\mathbf{X})$, respectively, and $\bar{\mathbf{L}}_t(\mathbf{X})$, $\bar{\mathbf{g}}_t(\mathbf{X})$ are called the reactions to the constraints. It means that in the presence of constraints for stresses and heat fluxes only the constitutive parts $\mathbf{L}_t(\mathbf{X})$ and $\mathbf{g}_t(\mathbf{X})$ are to be taken into account in postulating the constitutive relationships (instead of the total strain rates $\dot{\mathbf{E}}_t(\mathbf{X})$ and temperature gradients $\nabla\vartheta_t(\mathbf{X})$). At the same time we assume that the conditions

$$
\begin{aligned}
&\int_\Omega \operatorname{tr}[\bar{\mathbf{L}}_t(\mathbf{X})\mathbf{T}(\mathbf{X})]\,\mathrm{d}V(\mathbf{X}) = 0,\\
&\int_\Omega \bar{\mathbf{g}}_t(\mathbf{X})\cdot\mathbf{j}(\mathbf{X})\,\mathrm{d}V(\mathbf{X}) = 0,
\end{aligned}
\tag{2.17}
$$

hold for every $\mathbf{T}(\cdot) \in \mathscr{S}^0(\mathbf{S}_t(\cdot))$, $\mathbf{j}(\cdot) \in \mathscr{H}^0(\mathbf{h}_t(\cdot))$.

Eqs. (2.16), (2.17) together with the restrictions $\mathbf{S}_t(\cdot) \in \mathscr{S}(\Omega)$, $\mathbf{h}_t(\cdot) \in \mathscr{H}(\Omega)$ determine the ideal constraints for stresses and heat fluxes. It has to be emphasized that in the presence of the ideal constraints for stresses and heat fluxes the appropriate constitutive relations have to be postulated in the form which involves the fields $\mathbf{L}_t(\cdot)$, $\mathbf{g}_t(\cdot)$ instead of $\dot{\mathbf{E}}_t(\cdot)$, $\nabla\vartheta_t(\cdot)$ respectively.

As an illustrative example let us consider the uniaxial state of stress in a rate-type material, cf. Sec. 2.1.4. Let all stress components different from S^{11} be equal zero. Then from conditions (2.17) we obtain (via the du Bois–

Raymond lemma) that $\bar{L}_{11}(\mathbf{X}, t) = 0$ and that all other components of $\bar{\mathbf{L}}_t(\mathbf{X})$ can assume arbitrary values. Thus, the constitutive relation yields[1]

$$\begin{aligned} &\dot{E}_{11}(\mathbf{X}, t) = H_{1111}(\mathbf{X}; S^{11}(\mathbf{X}, t))\dot{S}^{11}(\mathbf{X}, t), \\ &\dot{E}_{\alpha\beta}(\mathbf{X}, t) \in \boldsymbol{R}^{(3\times 3)} \quad \text{for } (\alpha, \beta) \neq (1, 1), \\ &S^{\alpha\beta}(\mathbf{X}, t) = 0 \quad \text{for } (\alpha, \beta) \neq (1, 1). \end{aligned}$$

This result means that the strain rates $\dot{E}_{\alpha\beta}(\mathbf{X}, t)$ for $(\alpha, \beta) \neq (1, 1)$ remain undetermined by the constitutive equation.

It has to be emphasized that in the recent literature on continuum and structural mechanics the concept of constraints is mainly understood in the sense of the kinematic constraints. Here we have also introduced the thermal constraints as well as the constraints for stresses and heat fluxes using the formal procedure similar to that used for the kinematic constraints. Such a generalization is motivated by the fact that the thermal constraints and the constraints for stresses and/or heat fluxes can also be encountered in many specific problems of structural mechanics.

Summary

In Chapter 1 the physical foundations of continuum mechanics have been laid out. In Chapter 2 we have introduced all the additional informations about structural elements necessary to develop the complete mathematical models of structural mechanics problems. This has come down to specifying materials the structure is made of, interactions between the structural element and its exterior, and restrictions (or constraints) imposed on deformations and temperature fields.

In Section 2.1 we have introduced three classes of materials: thermo-visco-elastic materials, linear thermo-elastic materials and elastic-plastic materials. In Section 2.2 we have discussed the concept of external agents acting on the body; the resulting formulae are given by eqs. (2.6)–(2.9). Finally, in Section 2.3 the concept of constraints has been exposed followed by the discussion of the so-called ideal reactions to constraints. From now on we shall mainly deal with the bilateral constraints by concentrating upon the conditions (2.14), (2.15) which determine reactions to constraints (2.10). The concept of constraints involves two elements. Firstly, we use it to describe support conditions for the structural element considered which imposes certain restrictions on the motion of its boundary (the boundary constraints).

[1] The components of $\dot{\mathbf{E}}_t(\mathbf{X})$, $\mathbf{S}_t(\mathbf{X})$ and $\dot{\mathbf{S}}_t(\mathbf{X})$ are denoted by $\dot{E}_{\alpha\beta}(\mathbf{X}, t)$, $S^{\alpha\beta}(\mathbf{X}, t)$ and $\dot{S}^{\alpha\beta}(\mathbf{X}, t)$, respectively. Moreover, $\boldsymbol{R}^{(3\times 3)}$ is the set of all real 3×3 symmetric matrices.

Secondly, we introduce certain, postulated a priori, restrictions imposed on possible deformations of the body in order to simplify the mathematical model of the problem (the internal constraints). The latter type of constraints is typical of structural mechanics. It is the purpose of the next chapter to show how the internal constraints make it possible for us to formulate different engineering theories of mechanics such as shell, plate or beam theories.

The fundamental, or resulting, relations obtained in Chapter 1 are valid in all problems of structural mechanics. On the other hand, the choice of particular relations (and of their specific forms) from those developed in the present chapter depends on the character of every problem to be dealt with. In other words, different physical situations are described by different constitutive relations of the type (2.1), (2.4), by different equations for external agents, eqs. (2.6)–(2.9) and by different boundary and internal constraints, eqs. (2.10), (2.14) and (2.15).

Problems

2.1 Consider the special case of eqs. (2.1) in the form

$$\mathbf{S}_t(\mathbf{X}) = \varrho(\mathbf{X}) \frac{\partial \hat{\varphi}(\mathbf{X}; \mathbf{E}(\mathbf{X}, t))}{\partial \mathbf{E}(\mathbf{X}, t)} .$$

The ideal material defined by this equation is called the *hyperelastic material*. If $\hat{\varphi}(\mathbf{X}; \mathbf{E}(\mathbf{X}, t)) = \tilde{\varphi}(\mathbf{X}; I_1(\mathbf{X}, t), I_2(\mathbf{X}, t), I_3(\mathbf{X}, t))$ where $I_i(\mathbf{X}, t)$ are the strain invariants and $\tilde{\varphi}(\mathbf{X}; \cdot)$ is a known function, then the hyperelastic material is said to be *isotropic*. Derive the general and linearized forms of constitutive relation for such a hyperelastic isotropic material. Hint: consider Problem 1.1 and eqs. (2.2).

2.2 Let eqs. (2.1) hold for a certain motion $\chi(\cdot, \tau) \in \mathscr{D}(\Omega)$, $\tau \in [t_0, t_f]$. Prove that these equations do not change their form under superimposing on the motion an arbitrary rigid motion.

2.3 Consider the definition of the rate-type material (cf. end of Sec. 2.1.4) and specify the conditions under which it can also describe a certain hyperelastic material.

2.4 Apply the results of Problem 2.3 to eqs. (2.3) for the hyperelastic nonlinear and linear isotropic materials.

2.5 Under the assumption that the body loadings $\mathbf{b}_t(\mathbf{X})$, $\mathbf{X} \in \Omega$ depend on the force of gravity only, prove that $\hat{\mathbf{b}}(\mathbf{X}, \cdot)$ is a function of the Green–Saint Venant stress tensor $\mathbf{E}_t(\mathbf{X})$.

2.6 Consider eqs. (2.10), (2.11) and (2.14). Find the reaction forces for the case of a rigid body.

2.7 Assume that the internal constraints are given by the incompressibility condition

$$\det \mathbf{C}(\mathbf{X}, t) = \det \mathbf{G}(\mathbf{X}),$$

$$\mathbf{C}(\mathbf{X}, t) \equiv \nabla\chi^T(\mathbf{X}, t)\nabla\chi(\mathbf{X}, t), \qquad \mathbf{G}(\mathbf{X}) \equiv \nabla\chi^T(\mathbf{X}, t_0)\nabla\chi(\mathbf{X}, t_0).$$

Find the reaction forces.

2.8 Let the internal constraints be given by

$$\alpha^p(\mathbf{X}; t, \nabla\chi(\mathbf{X}, t)) = 0, \qquad \mathbf{X} \in \Omega, \qquad p = 1, \ldots, P.$$

Prove that the resultant of all the reaction forces is equal to zero.

2.9 Let the internal constraints be given by

$$\alpha^p(\mathbf{X}; t, \nabla\chi^T(\mathbf{X}, t)\nabla\chi(\mathbf{X}, t)) = 0, \qquad \mathbf{X} \in \Omega, \qquad p = 1, \ldots, P.$$

Prove that the resultant moment af all the reaction forces is equal to zero.

2.10 Let the internal and boundary constraints given in Problem 2.9 do not explicitly depend on the time coordinate. Prove that the total rate of work of all the reaction forces is equal to zero.

2.11 Consider the internal constraints for stresses. Obtain the form of the constitutive relations for linear elastic materials in the case of plane stress. Compare the results obtained with those of the classical elasticity theory. Discuss the similarities and differences.

2.12 Prove that if we deal exclusively with boundary constraints then $\overline{\mathbf{b}}(\mathbf{X}, t) = 0$ holds for every $\mathbf{X} \in \Omega$ and every $t \in [t_0, t_f]$ (reactions to boundary constraints occur in the form of the boundary tractions only).

2.13 Prove that reactions to internal constraints are represented by the body forces $\overline{\mathbf{b}}(\mathbf{X}, t)$, $\mathbf{X} \in \Omega$, as well as by the surface tractions $\overline{\mathbf{s}}(\mathbf{X}, t)$, $\mathbf{X} \in \partial\Omega$.

3

Formulation of engineering theories

Purpose of the chapter

The purpose of the present chapter is twofold. First, by referring to the results of Chapters 1 and 2 we shall outline a general method of formulating governing equations of structural mechanics. By using the concept of internal constraints we shall then develop the so-called method of constraints which will make it possible to derive equations of various engineering theories. In this way we shall specifically show how to consistently derive the governing relations of various shell and rod theories. The results of this chapter will complete the theoretical foundations of structural mechanics and will enable us to pass over to more specific topics and problems.

3.1 General form of governing relations

3.1.1 Finite variational formulation

The idea of variational formulation of the governing relations of structural mechanics is essentially based on eliminating from our considerations the reactions to constraints. To this aim we combine eqs. (1.14) and (2.14) taking into account that[1]

$$\mathbf{b}(\mathbf{X}, t) = \mathbf{b}_t(\mathbf{X}) + \bar{\mathbf{b}}_t(\mathbf{X}), \qquad \mathbf{X} \in \Omega,$$
$$\mathbf{s}(\mathbf{X}, t) = \mathbf{s}_t(\mathbf{X}) + \bar{\mathbf{s}}_t(\mathbf{X}), \qquad \mathbf{X} \in \partial\Omega,$$

by virtue of definitions (2.10). Hence, we obtain the condition

$$\int_{\Omega} \operatorname{tr}[\mathbf{S}_t(\mathbf{X}) \nabla \boldsymbol{\chi}^T(\mathbf{X}, t) \nabla \mathbf{v}(\mathbf{X})] \, \mathrm{d}V(\mathbf{X})$$
$$= \oint_{\partial\Omega} \mathbf{s}_t(\mathbf{X}) \cdot \mathbf{v}(\mathbf{X}) \, \mathrm{d}A(\mathbf{X}) + \int_{\Omega} [\mathbf{b}_t(\mathbf{X}) - \varrho(\mathbf{X}) \ddot{\mathbf{u}}_t(\mathbf{X})] \cdot \mathbf{v}(\mathbf{X}) \, \mathrm{d}V(\mathbf{X}) \qquad (3.1)$$

---

[1] The decompositions may not be defined on some surfaces, lines or points in Ω and on some lines or points on $\partial\Omega$, cf. Sec. 2.3.2.

which has to hold for every $\mathbf{v}(\cdot) \in \mathscr{U}_t(\boldsymbol{\chi}(\cdot, t))$, $\boldsymbol{\chi}(\cdot, t) \in \mathscr{D}_t(\Omega)$. It has been here assumed that the kinematic constraints are bilateral. Assume further that the given element can be treated a thermo-visco-elastic. By means of eqs. (2.1) we have

$$\mathbf{S}_t(\mathbf{X}) = \hat{\mathbf{S}}(\mathbf{X}; \mathbf{E}_t(\mathbf{X}), \vartheta_t(\mathbf{X})) + \mathfrak{H}(\mathbf{X}; \mathbf{E}_t(\mathbf{X}), \vartheta_t(\mathbf{X}))[\dot{\mathbf{E}}_t(\mathbf{X})] \tag{3.2}$$

with

$$\hat{\mathbf{S}}(\mathbf{X}; \mathbf{E}_t(\mathbf{X}), \vartheta_t(\mathbf{X})) = \varrho(\mathbf{X}) \frac{\partial \hat{\varphi}(\mathbf{X}; \mathbf{E}_t(\mathbf{X}), \vartheta_t(\mathbf{X}))}{\partial \mathbf{E}_t(\mathbf{X})},$$

whereas by means of eqs. (1.10), (1.11) we get

$$\begin{aligned} &\mathbf{E}_t(\mathbf{X}) = \tfrac{1}{2}[\nabla\boldsymbol{\chi}^T(\mathbf{X}, t)\nabla\boldsymbol{\chi}(\mathbf{X}, t) - \mathbf{G}(\mathbf{X}, t)], \\ &\dot{\mathbf{E}}_t(\mathbf{X}) = \tfrac{1}{2}[\nabla\boldsymbol{\chi}^T(\mathbf{X}, t)\nabla\dot{\boldsymbol{\chi}}(\mathbf{X}, t) + \nabla\dot{\boldsymbol{\chi}}^T(\mathbf{X}, t)\nabla\boldsymbol{\chi}(\mathbf{X}, t)], \\ &\boldsymbol{\chi}(\mathbf{X}, t) = \boldsymbol{\chi}(\mathbf{X}, t_0) + \mathbf{u}_t(\mathbf{X}), \quad \mathbf{G}(\mathbf{X}) = \nabla\boldsymbol{\chi}^T(\mathbf{X}, t_0)\nabla\boldsymbol{\chi}(\mathbf{X}, t_0). \end{aligned} \tag{3.3}$$

At the same time we restrict the class of loadings acting on the structure to those which can be described by eqs. (2.6), (2.8):

$$\begin{aligned} &\mathbf{b}_t(\mathbf{X}) = \hat{\mathbf{b}}(\mathbf{X}, \boldsymbol{\chi}(\mathbf{X}, t), \nabla\boldsymbol{\chi}(\mathbf{X}, t)), \\ &\mathbf{s}_t(\mathbf{X}) = \tilde{\mathbf{s}}(\mathbf{X}, t) + \hat{\mathbf{s}}(\mathbf{X}, t, \boldsymbol{\chi}(\mathbf{X}, t), \nabla\boldsymbol{\chi}(\mathbf{X}, t)) + \mathbf{G}(\mathbf{X}, t, \boldsymbol{\chi}(\mathbf{X}, t))\dot{\boldsymbol{\chi}}(\mathbf{X}, t). \end{aligned} \tag{3.4}$$

Substituting now the RHS of eqs. (3.2), (3.4) into condition (3.1) and remembering eqs. (3.3) we obtain the first variational governing relation for the class of problem under consideration. This variational relation has to hold for every $\mathbf{v}(\cdot) \in \mathscr{U}_t(\boldsymbol{\chi}(\cdot, t))$ and for an arbitrary time instant t in the interval $[t_0, t_f]$. If the temperature field $\theta(\cdot, t) = \theta_0 + \vartheta_t(\cdot)$ is known a priori (for instance, if $\theta(\cdot, \tau) = \theta_0$ for every $\tau \in [t_0, t_f]$ where θ_0 is the known constant reference temperature, cf. Sec. 1.1.5), then the governing relation becomes a single variational equation for the motion $\boldsymbol{\chi}(\cdot, \tau) \in \mathscr{D}_\tau(\Omega)$, $\tau \in [t_0, t_f]$, given in terms of the family $\mathbf{u}_\tau(\cdot) = \boldsymbol{\chi}(\cdot, \tau) - \boldsymbol{\chi}(\cdot, t_0)$ of the displacement fields. It has to be remembered that the sets $\mathscr{D}_\tau(\Omega)$, $\mathscr{U}_\tau(\boldsymbol{\chi}(\cdot, \tau))$ as well as the functions on the RHS of eqs. (3.2), (3.4) have to be known in every problem considered.

Assume now that the temperature fields $\theta(\cdot, \tau)$ for $\tau \in [t_0, t_f]$ are not known in advance. Then, in order to determine the temperature we have to formulate the second governing relation. To this aim we eliminate the thermal reactions from eqs. (1.16), (2.14) using the equalities[1]

$$\begin{aligned} &\alpha(\mathbf{X}, t) = \alpha_t(\mathbf{X}) + \bar{\alpha}_t(\mathbf{X}), \quad \mathbf{X} \in \Omega, \\ &\pi(\mathbf{X}, t) = \pi_t(\mathbf{X}) + \bar{\pi}_t(\mathbf{X}), \quad \mathbf{X} \in \partial\Omega, \end{aligned}$$

[1] Cf. the previous footnote.

which are implied by the definitions (2.11). Hence we arrive at the condition

$$\int_{\Omega} \mathbf{h}_t(\mathbf{X})\cdot\nabla\zeta(\mathbf{X})\,dV(\mathbf{X}) = \oint_{\partial\Omega} \pi_t(\mathbf{X})\zeta(\mathbf{X})\,dA(\mathbf{X}) + \int_{\Omega} \alpha_t(\mathbf{X})\zeta(\mathbf{X})\,dV(\mathbf{X}) +$$
$$+\int_{\Omega} \{\mathrm{tr}[\mathbf{S}_t(\mathbf{X})\dot{\mathbf{E}}_t(\mathbf{X})] - \varrho(\mathbf{X})\dot{\varepsilon}(\mathbf{X}, t)\}\zeta(\mathbf{X})\,dV(\mathbf{X}) \qquad (3.5)$$

which has to hold for every $\zeta(\cdot)\in\mathscr{Z}_t(\theta(\cdot, t))$, $\theta(\cdot, t)\in\mathscr{T}_t(\Omega)$.

From eqs. (2.1) and from the relation, cf. Sec. 1.1.5,

$$\varepsilon(\mathbf{X}, t) = \varphi(\mathbf{X}, t) + \eta(\mathbf{X}, t)\theta(\mathbf{X}, t),$$

we obtain the following thermal constitutive equations

$$\begin{aligned} \mathbf{h}_t(\mathbf{X}) &= \mathbf{K}(\mathbf{X}; \mathbf{E}_t(\mathbf{X}), \theta(\mathbf{X}, t))\nabla\theta(\mathbf{X}, t), \\ \varepsilon(\mathbf{X}, t) &= \hat{\varepsilon}(\mathbf{X}; \mathbf{E}_t(\mathbf{X}), \theta(\mathbf{X}, t)), \end{aligned} \qquad (3.6)$$

in which it has been defined

$$\hat{\varepsilon}(\mathbf{X}; \mathbf{E}, \theta) \equiv \hat{\varphi}(\mathbf{X}; \mathbf{E}, \theta) - \frac{\partial\hat{\varphi}(\mathbf{X}; \mathbf{E}, \theta)}{\partial\theta}\,\theta.$$

Let us also rewrite eqs. (2.7), (2.9) in the form

$$\begin{aligned} \alpha_t(\mathbf{X}) &= \hat{\alpha}(\mathbf{X}; t, \theta(\mathbf{X}, t)), \\ \pi_t(\mathbf{X}) &= \tilde{\pi}(\mathbf{X}, t) + \hat{\pi}(\mathbf{X}, t, \boldsymbol{\chi}(\mathbf{X}, t), \theta(\mathbf{X}, t)). \end{aligned} \qquad (3.7)$$

Eqs. (3.7) determine the internal and boundary heat supply.

Substituting next the RHS of eqs. (3.6), (3.7) into condition (3.5) we arrive at the second variational governing relation for the class of problems under consideration. The relation has to hold for every $\zeta(\cdot)\in\mathscr{Z}_t(\theta(\cdot, t))$ and for $\theta(\cdot, t)\in\mathscr{T}_t(\Omega)$.

If the motion of the body is known a priori (for instance, if we deal with a rigid body being at rest in the time interval $[t_0, t_f]$), then the equation obtained is a single variational equation for the evolution of temperature $\theta(\cdot, \tau)\in\mathscr{T}_\tau(\Omega)$, $\tau\in[t_0, t_f]$. The sets $\mathscr{T}_t(\Omega)$, $\mathscr{Z}_t(\theta(\cdot, t))$ as well as the functions on the RHS of eqs. (3.6), (3.7) have to be known in every given problem.

We conclude by saying that for thermo-visco-elastic materials and bilateral (ideal) constraints the governing relations are given by the two following variational equations:

1. Eq. (3.1) in which $\mathbf{S}_t(\mathbf{X})$ and $\mathbf{b}_t(\mathbf{X})$, $\mathbf{s}_t(\mathbf{X})$ are given by eqs. (3.2) and (3.4), respectively.
2. Eq. (3.5) in which $\mathbf{h}_t(\mathbf{X})$, $\varepsilon(\mathbf{X}, t)$ and $\mathbf{S}_t(\mathbf{X})$ are given by eqs. (3.6) and (3.2), respectively.

Taking into account eqs. (3.3) and bearing in mind that

$$\chi(\cdot, t) = \chi(\cdot, t_0)+\mathbf{u}_t(\cdot) \in \mathscr{D}_t(\Omega),$$
$$\theta(\cdot, t) = \theta_0+\vartheta_t(\cdot) \in \mathscr{T}_t(\Omega), \qquad t \in [t_0, t_f],$$

it can easily be seen that we have arrived at the system of two coupled variational equations for the unknown displacement fields $\mathbf{u}_t(\cdot)$ and temperature fields $\vartheta_t(\cdot)$, $t \in [t_0, t_f]$.

In the whole analysis it was assumed that the values of $\chi(\mathbf{X}, t_0)$, $\dot{\chi}(\mathbf{X}, t_0)$ and $\theta(\mathbf{X}, t_0)$ are known, i.e. that there are known the appropriate initial conditions for the unknown fields.

Both the variational equations derived above represent what is called the general finite variational formulation of the governing relations of mechanics. It applies to problems in which we deal with thermo-visco-elastic materials but the transition to the case of linear thermo-elastic materials is straightforward.

Here and in the sequel we confine ourselves to the bilateral, ideal constraints so that $\mathscr{U}_t = \mathscr{U}_t(\chi(\cdot, t))$, $\mathscr{Z}_t = \mathscr{Z}_t(\theta(\cdot, t))$. More general cases, governed by inequalities (2.12) (which hold for $\mathbf{u}(\cdot) \in \mathscr{U}_t(\chi(\cdot, t), \dot{\chi}(\cdot, t))$ and by inequalities (2.13) (which hold for $\zeta(\cdot) \in \mathscr{Z}_t(\theta(\cdot, t), \dot{\theta}(\cdot, t))$, can also be considered within the framework of the approach just presented.

3.1.2 *Incremental variational formulation*

Let $\psi_\tau(\cdot)$, $\tau \in [t_0, t_f]$, be a one-parameter family of (sufficiently regular) scalar, vector or tensor fields defined in Ω or on $\partial\Omega$. Let further the functions $\tau \to \psi_\tau(\mathbf{X})$ of an argument $\tau \in [t_0, t_f]$ be continuous except possibly at a finite number of time instants in $[t_0, t_f]$[1]. Moreover, let t be an arbitrary but fixed time instant from the interval $[t_0, t_f]$ and Δt be a known positive time increment (as a rule, sufficiently small with respect to $t_f - t_0$). We shall consistently use the following notation

$$\Delta\psi(\mathbf{X}) \equiv \psi_{t+\Delta t}(\mathbf{X})-\psi_t(\mathbf{X}),$$

keeping in mind that $\Delta\psi(\mathbf{X})$ is an increment of $\psi_t(\mathbf{X})$ at the time instant t. It can readily be seen that

$$\Delta[\psi(\mathbf{X})\varphi(\mathbf{X})] = \psi_t(\mathbf{X})\nabla\varphi(\mathbf{X})+\varphi_t(\mathbf{X})\nabla\psi(\mathbf{X})+\underline{\Delta\psi(\mathbf{X})\Delta\varphi(\mathbf{X})}.$$

For sufficiently small increments $\Delta\psi(\mathbf{X})$, $\Delta\varphi(\mathbf{X})$ the underlined term can be neglected; if such a situation takes place the increments $\Delta\varphi(\mathbf{X})$, $\Delta\psi(\mathbf{X})$ will be called *linear*.

---

[1] The functions $\tau \to \psi_\tau(\mathbf{X})$ are defined either for every $\mathbf{X} \in \Omega$ (except possibly on some surfaces, lines or points in Ω), or for every $\mathbf{X} \in \partial\Omega$ (except possibly on some lines or points on $\partial\Omega$).

Using the operator Δ we shall obtain from eq. (3.1) the alternative form of the variational principle. This new form will be valid under the additional condition that $\boldsymbol{\chi}(\cdot, t)$ is the identity mapping for the fixed time instant $t, t \in [t_0, t_f]$ (but it is not identity for $\tau = t+\Delta t$!). We have then $\boldsymbol{\chi}(\mathbf{X}, t) = \mathbf{X}$ and the following condition follows

$$\int_{\Omega} \operatorname{tr}\{[\Delta\mathbf{S}(\mathbf{X})+\mathbf{S}_t(\mathbf{X})\nabla\Delta\mathbf{u}^T(\mathbf{X})+\Delta\mathbf{S}(\mathbf{X})\nabla\mathbf{u}^T(\mathbf{X})]\nabla\mathbf{v}(\mathbf{X})\}dV(\mathbf{X})$$

$$= \oint_{\partial\Omega} \Delta\mathbf{s}(\mathbf{X})\cdot\mathbf{v}(\mathbf{X})dA(\mathbf{X})+\int_{\Omega}[\Delta\mathbf{b}(\mathbf{X})-\varrho(\mathbf{X})\Delta\ddot{\mathbf{u}}(\mathbf{X})]\cdot\mathbf{v}(\mathbf{X})dV(\mathbf{X}). \tag{3.8}$$

Eq. (3.8) is assumed to hold for every $\mathbf{v}(\cdot)\in\mathscr{U}_t(\boldsymbol{\chi}(\cdot, t))$. It has to be emphasized that now $\Omega = \boldsymbol{\chi}(\Omega, t)$, $\boldsymbol{\chi}(\mathbf{X}, t) = \mathbf{X}$, $\mathbf{X}\in\Omega$.

Let us assume that for the given class of problems the material can be thought of as elastic-plastic. Let us also accept the approximate assumption that $\dot{\mathbf{S}} \simeq \Delta\mathbf{S}/\Delta t$, $\dot{\mathbf{E}} \simeq \Delta\mathbf{E}/\Delta t$, $\vartheta \simeq \Delta\theta/\Delta t$ which makes it possible to replace the constitutive relation $\dot{\mathbf{S}} = \tilde{\mathfrak{D}}\dot{\mathbf{E}}+\tilde{\mathbf{G}}\dot{\vartheta}$ by the following one

$$\Delta\mathbf{S}(\mathbf{X}) = \tilde{\mathfrak{D}}(\mathbf{X};\mathbf{S}_t(\mathbf{X}),\vartheta_t(\mathbf{X}),\mathbf{q}(\mathbf{X},t),\delta(\mathbf{X},t))[\Delta\mathbf{E}(\mathbf{X})]+$$
$$+\tilde{\mathbf{G}}(\mathbf{X};\mathbf{S}_t(\mathbf{X}),\vartheta_t(\mathbf{X}),\mathbf{q}(\mathbf{X},t),\delta(\mathbf{X},t))\Delta\theta(\mathbf{X}), \tag{3.9}$$

where $\tilde{\mathfrak{D}}(\cdot)$, $\tilde{\mathbf{G}}(\cdot)$ are defined by the same formula as in Sec. 2.1.4. We draw the reader's attention to the fact that "incremental" constitutive equation (3.9) is in general only an approximation to the old "rate" constitutive equation. For the elastic material we obtain from eq. (2.1) that

$$\Delta\mathbf{S}(\mathbf{X}) = \mathfrak{G}(\mathbf{X};\mathbf{E}_t(\mathbf{X}))[\Delta\mathbf{E}_t(\mathbf{X})],$$
$$\mathfrak{G}(\mathbf{X};\mathbf{E}_t(\mathbf{X})) = \varrho(\mathbf{X})\frac{\partial^2\hat{\varphi}(\mathbf{X};\mathbf{E}_t(\mathbf{X}))}{\partial\mathbf{E}_t(\mathbf{X})\,\partial\mathbf{E}_t(\mathbf{X})}. \tag{3.10}$$

By taking into account the definition of Δ and using eq. $(3.3)_1$ we arrive at the formula

$$\Delta\mathbf{E}(\mathbf{X}) = \tfrac{1}{2}[\Delta\mathbf{u}(\mathbf{X})+\Delta\mathbf{u}^T(\mathbf{X})+\Delta\mathbf{u}^T(\mathbf{X})\Delta\mathbf{u}(\mathbf{X})]. \tag{3.11}$$

It has to be remembered that eq. (3.11) as well as constitutive relation (3.9) hold under the condition that $\boldsymbol{\chi}(\cdot, t)$ is the identity mapping; such a situation is shown in Fig. 8. It has also to be recalled that t is an arbitrary but fixed time instant taken from the interval $[t_0, t_f)$ and that the incremental description of the problem is restricted to the (sufficiently small) time interval $[t, t+\Delta t]$ such that $[t, t+\Delta t] \subset [t_0, t_f]$.

The incremental description just developed bears in the recent computational mechanics literature the name of the Updated Lagrangean Description of the nonlinear structural response.

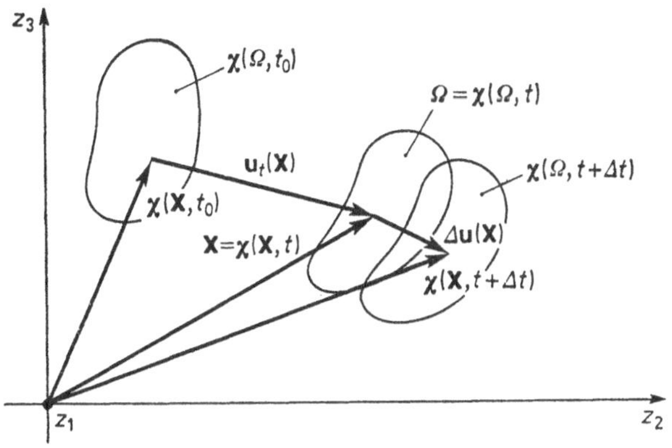

Fig. 8 Increments of displacements

Consider now eqs. (3.4). For an arbitrary function $\varphi(\mathbf{X}, t, \psi(\mathbf{X}, t))$ and fixed t, $\psi(\mathbf{X}, t)$, define

$$\Delta\varphi(\mathbf{X}, \Delta\psi(\mathbf{X})) \equiv \varphi(\mathbf{X}, t+\Delta t, \psi(\mathbf{X}, t)+\Delta\psi(\mathbf{X}, t)) - \varphi(\mathbf{X}, t, \psi(\mathbf{X}, t)).$$

Eqs. (3.4) then yield

$$\begin{aligned} \Delta\mathbf{b}(\mathbf{X}) &= \Delta\hat{\mathbf{b}}(\mathbf{X}, \Delta\mathbf{u}(\mathbf{X}), \nabla\Delta\mathbf{u}(\mathbf{X})), \\ \Delta\mathbf{s}(\mathbf{X}) &= \Delta\tilde{\mathbf{s}}(\mathbf{X}) + \Delta\hat{\mathbf{s}}(\mathbf{X}, \Delta\mathbf{u}(\mathbf{X}), \nabla\Delta\mathbf{u}(\mathbf{X})) + \\ &\quad + \frac{\Delta\mathbf{G}(\mathbf{X}, \Delta\mathbf{u}(\mathbf{X}))}{\Delta t}\Delta\mathbf{u}(\mathbf{X}) + \mathbf{G}(\mathbf{X}, t, 0)\Delta\dot{\mathbf{u}}(\mathbf{X}) + \Delta\mathbf{G}(\mathbf{X}, \Delta\mathbf{u}(\mathbf{X}))\Delta\dot{\mathbf{u}}(\mathbf{X}) \end{aligned} \tag{3.12}$$

under the additional assumption that $\dot{\mathbf{u}}(\mathbf{X})$ can be approximated by $\Delta\mathbf{u}(\mathbf{X})/\Delta t$, remembering that $\Delta\mathbf{u}(\mathbf{X}) = \Delta\boldsymbol{\chi}(\mathbf{X})$, $\boldsymbol{\chi}(\mathbf{X}, t) = \mathbf{X}$. Substituting the RHS of eqs. (3.12), (3.9) (or the RHS of eqs. (3.10) if we deal with the elastic or a rate-type material) into condition (3.8) and using eq. (3.11) we arrive at the condition which interrelates the increments $\Delta\mathbf{u}(\mathbf{X})$, $\Delta\dot{\mathbf{u}}(\mathbf{X})$ and $\Delta\ddot{\mathbf{u}}(\mathbf{X})$, provided the stress field $\mathbf{S}_t(\mathbf{X})$, $\mathbf{X} \in \Omega$, is known. The aforementioned condition represents what is called the general incremental variational formulation of the governing relations of structural mechanics. It concerns problems in which we deal either with the elastic-plastic or with the rate-type materials (in particular with the elastic or linear elastic materials). In the case of elasto-plasticity we have to account for the formulae of Sec. 2.1.3 in which the linear mappings $\tilde{\mathfrak{D}}(\mathbf{X}; \mathbf{S}, \vartheta, \mathbf{q}, \delta)$, $\tilde{\mathfrak{G}}(\mathbf{X}; \mathbf{S}, \vartheta, \mathbf{q}, \delta)$ are involved. The increments $\Delta\mathbf{u}(\mathbf{X})$, $\Delta\dot{\mathbf{u}}(\mathbf{X})$, $\mathbf{X} \in \Omega$, have to satisfy the conditions

$$\begin{aligned} &\boldsymbol{\chi}(\cdot, t) + \Delta\mathbf{u}(\cdot) \in \mathscr{D}_{t+\Delta t}(\Omega), \\ &\dot{\mathbf{u}}(\cdot, t) + \Delta\dot{\mathbf{u}}(\cdot) \in \mathscr{V}_{t+\Delta t}(\boldsymbol{\chi}(\cdot, t) + \Delta\mathbf{u}(\cdot)), \end{aligned}$$

whereas the sets on the RHS of these conditions are assumed to be known, cf. Sec. 2.3.1. For the sake of simplicity we have confined ourselves to the isothermal problems. However, using the same lines of reasoning the incremental form of the variational equation for the temperature can be developed.

3.1.3 *Variational equations for stresses and heat fluxes*

Let us now confine the considerations to problems in which the materials of the structural element can be treated as the rate-type (or elastic) ones. Eq. (2.16) yields then the condition

$$\int_{\Omega} \operatorname{tr}\{[\dot{\mathbf{E}}_t(\mathbf{X}) - \mathfrak{H}(\mathbf{X}; \mathbf{S}_t(\mathbf{X}))[\dot{\mathbf{S}}_t(\mathbf{X})]]\mathbf{T}(\mathbf{X})\}\, dV(\mathbf{X}) = 0 \tag{3.13}$$

which has to hold for every $\mathbf{T}(\cdot) \in \mathscr{S}^0$. Here, $\mathscr{S}^0$ is the known set the significance of which was delineated in Sec. 2.3.5[1]. In thc finite variational formulation condition (3.13), after substituting, cf. eq. (3.3)

$$\dot{\mathbf{E}}_t(\mathbf{X}) = \tfrac{1}{2}[\nabla\boldsymbol{\chi}^T(\mathbf{X}, t)\nabla\dot{\mathbf{u}}_t(\mathbf{X}) + \nabla\dot{\mathbf{u}}_t^T(\mathbf{X})\nabla\boldsymbol{\chi}(\mathbf{X}, t)],$$

has to be considered instead of the rate-type constitutive relation $\dot{\mathbf{E}}_t(\mathbf{X}) = \mathfrak{H}(\mathbf{X}; \mathbf{S}_t(\mathbf{X}))\,[\dot{\mathbf{S}}_t(\mathbf{X})]$. Thus we deal now with the system of two variational equations for the two unknown fields $\mathbf{u}_t(\cdot)$, $\mathbf{S}_t(\cdot)$ defined in Ω, such that conditions

$$\boldsymbol{\chi}(\cdot\,, t_0) + \mathbf{u}_t(\cdot) \in \mathscr{D}_t(\Omega),$$
$$\mathbf{S}_t(\cdot) \in \mathscr{S}(\Omega),$$

have to hold for every $t \in [t_0, t_f]$.

In the incremental variational formulation condition (3.13) is replaced by the relation

$$\int_{\Omega} \operatorname{tr}\{(\Delta\mathbf{E}(\mathbf{X}) - \mathfrak{H}(\mathbf{X}; \mathbf{S}_t(\mathbf{X}))[\Delta\mathbf{S}(\mathbf{X})])\mathbf{T}(\mathbf{X})\}\, dV(\mathbf{X}) = 0 \tag{3.14}$$

which has to hold for every $\mathbf{T}(\cdot) \in \mathscr{S}^0$ where $\mathbf{S}_t(\cdot) \in \mathscr{S}(\Omega)$ and $\Delta\mathbf{E}(\mathbf{X})$ is given by eq. (3.11). The aforementioned condition together with condition (3.8) constitute the system of two variational equations which interrelate the increments $\Delta\mathbf{u}(\mathbf{X})$, $\Delta\dot{\mathbf{u}}(\mathbf{X})$, $\Delta\ddot{\mathbf{u}}(\mathbf{X})$, $\Delta\mathbf{S}(\mathbf{X})$. At the same time we have

$$\boldsymbol{\chi}(\cdot\,, t) + \Delta\mathbf{u}(\cdot) \in \mathscr{D}_{t+\Delta t}(\Omega),$$
$$\dot{\mathbf{u}}_t(\cdot) + \Delta\dot{\mathbf{u}}(\cdot) \in \mathscr{V}_{t+\Delta t}(\boldsymbol{\chi}(\cdot\,, t+\Delta t)),$$
$$\mathbf{S}_t(\cdot) + \Delta\mathbf{S}(\cdot) \in \mathscr{S}(\Omega),$$

where the sets on the RHS of the foregoing relations are known, cf. Sec. 2.3.1.

---

[1] $\mathscr{S}^0$ has to be the linear subspace in the space of all continuous in $\bar{\Omega}$ stress rate fields, cf. Sec. 2.3.5.

The approach presented here can also be employed for constraints imposed on heat fluxes. Taking into account the results of Sec. 2.3.5 and assuming that every matrix $\mathbf{K}(\mathbf{X}; \mathbf{E}_t(\mathbf{X}), \vartheta_t(\mathbf{X}))$ has the inverse $\mathbf{J}(\mathbf{X}; \mathbf{E}_t(\mathbf{X}), \vartheta_t(\mathbf{X}))$ we obtain from eq. $(2.17)_2$ the condition

$$\int_{\Omega} [\nabla\vartheta_t(\mathbf{X}) - \mathbf{J}(\mathbf{X}; \mathbf{E}_t(\mathbf{X}), \vartheta_t(\mathbf{X}))\mathbf{h}_t(\mathbf{X})] \cdot \mathbf{j}(\mathbf{X}) \mathrm{d}V(\mathbf{X}) = 0$$

which has to hold for every $\mathbf{j}(\cdot) \in \mathscr{H}^0$. This condition should be used instead of eqs. (2.1).

3.1.4 Calculation of reactions

In the finite variational formulation of the problem the basic unknowns are: The displacement field $\mathbf{u}_t(\cdot)$ and the field of temperature $\vartheta_t(\cdot)$, $t \in [t_0, t_f]$. Let us assume for the time being that the problem under consideration has been solved and $\mathbf{u}_t(\cdot)$, $\vartheta_t(\cdot)$ are therefore known for every $t \in [t_0, t_f]$. From eqs. (3.3) we can then calculate $\mathbf{E}_t(\cdot)$, $\dot{\mathbf{E}}_t(\cdot)$ and, by eqs. (2.1), obtain $\mathbf{S}_t(\cdot)$, $\mathbf{h}_t(\cdot)$, $\eta(\cdot, t)$, $\varphi(\cdot, t)$. Similarly, from eqs. (3.4), (3.7) and (3.6) we derive the values of external agents $\mathbf{b}_t(\cdot)$, $\mathbf{s}_t(\cdot)$, $\alpha_t(\cdot)$, $\pi_t(\cdot)$ and of the internal energy $\varepsilon(\cdot, t)$, $t \in [t_0, t_f]$.

In order to compute the reactions to constraints we shall take into account the local forms of the equation of motion and of the energy equation as well as the pertinent conditions at the boundary, cf. Secs. 1.4.1, 1.4.2. By using also the definitions of the reactions (2.11) we obtain

$$\begin{aligned} \bar{\mathbf{b}}_t(\mathbf{X}) &= \varrho(\mathbf{X})\ddot{\mathbf{u}}_t(\mathbf{X}) - \mathbf{b}_t(\mathbf{X}) - \mathrm{Div}[(\nabla\boldsymbol{\chi}(\mathbf{X}, t_0) + \nabla\mathbf{u}_t(\mathbf{X}))\mathbf{S}_t(\mathbf{X})], \\ \bar{\alpha}_t(\mathbf{X}) &= \varrho(\mathbf{X})\dot{\varepsilon}(\mathbf{X}, t) - \alpha_t(\mathbf{X}) - \mathrm{Div}\,\mathbf{h}_t(\mathbf{X}) - \mathrm{tr}[\mathbf{S}_t(\mathbf{X})\dot{\mathbf{E}}_t(\mathbf{X})] \end{aligned} \tag{3.15}$$

in Ω and

$$\begin{aligned} \bar{\mathbf{s}}_t(\mathbf{X}) &= \mathbf{s}_t(\mathbf{X}) - [\nabla\boldsymbol{\chi}(\mathbf{X}, t_0) + \nabla\mathbf{u}_t(\mathbf{X})]\mathbf{S}_t(\mathbf{X})\mathbf{n}(\mathbf{X}), \\ \bar{\pi}_t(\mathbf{X}) &= \pi_t(\mathbf{X}) - \mathbf{h}_t(\mathbf{X}) \cdot \mathbf{n}(\mathbf{X}) \end{aligned} \tag{3.16}$$

on $\partial\Omega$. Moreover, if the constraints for stresses and heat fluxes are also accounted for, then, cf. Sec. 1.3.4,

$$\begin{aligned} \bar{\mathbf{L}}_t(\mathbf{X}) &= \dot{\mathbf{E}}_t(\mathbf{X}) - \mathbf{L}(\mathbf{X}, t), \\ \bar{\mathbf{g}}_t(\mathbf{X}) &= \nabla\vartheta_t(\mathbf{X}) - \mathbf{g}(\mathbf{X}, t) \end{aligned}$$

in Ω. The above equations determine, for every $t \in [t_0, t_f]$, the complete system of the ideal reactions to constraints for the class of problems which can be stated and solved by means of the finite variational formulation of Sec. 3.1.1.

To simplify the presentation, in the case of the incremental variational formulation we have neglected all thermal effects. Let us assume for the moment that after having solved the incremental problem (i.e. after having passed from the time instant t to the "next" time instant $t+\Delta t$) we have calculated $\Delta\mathbf{u}(\cdot)$, $\Delta\dot{\mathbf{u}}(\cdot)$, $\Delta\ddot{\mathbf{u}}(\cdot)$ and $\Delta\mathbf{S}(\cdot)$, employing on the way eqs. (3.11), (3.9). Eqs. (3.12) enable us then to compute $\Delta\mathbf{b}(\cdot)$, $\Delta\mathbf{s}(\cdot)$. Using now eqs. $(3.15)_1$ and $(3.16)_1$ we arrive at

$$\begin{aligned}\Delta\bar{\mathbf{b}}(\mathbf{X}) &= \varrho(\mathbf{X})\Delta\ddot{\mathbf{u}}(\mathbf{X})-\Delta\mathbf{b}(\mathbf{X})-\mathrm{Div}\{[\nabla\boldsymbol{\chi}(\mathbf{X},t_0)+\nabla\mathbf{u}(\mathbf{X})]\Delta\mathbf{S}(\mathbf{X})+\\ &\qquad +\nabla\Delta\mathbf{u}(\mathbf{X})\Delta\mathbf{S}(\mathbf{X})\}, \qquad (3.17)\\ \Delta\bar{\mathbf{s}}(\mathbf{X}) &= \Delta\mathbf{s}(\mathbf{X})-\{[\nabla\boldsymbol{\chi}(\mathbf{X},t_0)+\mathbf{u}_t(\mathbf{X})+\Delta\mathbf{u}(\mathbf{X})]\Delta\mathbf{S}(\mathbf{X})+\Delta\mathbf{u}(\mathbf{X})\Delta\mathbf{S}(\mathbf{X})\}\mathbf{n}(\mathbf{X})\end{aligned}$$

in Ω and on $\partial\Omega$, respectively. If the constraints for stresses have also been used then, assuming $\Delta\mathbf{E}(\mathbf{X}) \simeq \dot{\mathbf{E}}_t(\mathbf{X})\Delta t$ and $\Delta\mathbf{S}(\mathbf{X}) \simeq \dot{\mathbf{S}}_t(\mathbf{X})\Delta t$, we obtain

$$\Delta\mathbf{L}(\mathbf{X}) = \Delta\mathbf{E}(\mathbf{X})-\mathfrak{H}(\mathbf{X};\mathbf{S}_t(\mathbf{X}))[\Delta\mathbf{S}(\mathbf{X})].$$

The formulae given here determine the increments of the reactions produced during the deformation process extending from the time to the time.

3.1.5 Matrix form of variational equations

Taking into account the notation introduced in Secs. 1.3.5 and 1.4.4 we shall now rewrite the variational equations (3.1) and (3.5) in the matrix form. Eq. (3.1) can be written down as

$$\int_\Omega \boldsymbol{\sigma}\circ\mathbf{L}_\mathbf{F}\mathbf{v}\,dV = \oint_{\partial\Omega}\mathbf{s}\cdot\mathbf{v}\,dA+\int_\Omega(\mathbf{b}-\varrho\ddot{\mathbf{u}})\cdot\mathbf{v}\,dV,$$

where $\mathbf{F} = \nabla\boldsymbol{\chi}(\mathbf{X},t)$; it has to hold for every $\mathbf{v}\equiv\mathbf{v}(\cdot)\in\mathscr{U}_t(\boldsymbol{\chi}(\cdot,t))$, where $\boldsymbol{\chi}(\cdot,t)\in\mathscr{D}_t(\Omega)$. Similarly, eq. (3.5) is equivalent to the equation

$$\int_\Omega \mathbf{h}\cdot\nabla\zeta\,dV = \oint_{\partial\Omega}\pi\zeta\,dA+\int_\Omega\alpha\zeta\,dV+\int_\Omega(\boldsymbol{\sigma}\circ\dot{\boldsymbol{\epsilon}}-\varrho\dot{\varepsilon})\zeta\,dV$$

which is assumed to hold for every $\zeta\equiv\zeta(\cdot)\in\mathscr{Z}_t(\theta(\cdot,t))$, where $\theta(\cdot,t)\in\mathscr{T}_t(\Omega)$.

Passing to the incremental formulation we see that

$$\Delta(\mathbf{L}_\mathbf{F}\mathbf{v}) = \mathbf{L}_{\Delta\mathbf{F}}\mathbf{v},$$

where $\Delta\mathbf{F} = \Delta[\nabla\boldsymbol{\chi}(\mathbf{X},t)] = \nabla[\Delta\boldsymbol{\chi}(\mathbf{X},t)]$. Thus we obtain the condition

$$\int_\Omega(\Delta\boldsymbol{\sigma}\circ\mathbf{L}_\mathbf{F}\mathbf{v}+\boldsymbol{\sigma}\circ\mathbf{L}_{\Delta\mathbf{F}}\mathbf{v}+\Delta\boldsymbol{\sigma}\circ\mathbf{L}_{\Delta\mathbf{F}}\mathbf{v})\,dV = \oint_{\partial\Omega}\Delta\mathbf{s}\cdot\mathbf{v}\,dA+\int_\Omega(\Delta\mathbf{b}-\varrho\Delta\ddot{\mathbf{u}})\cdot\mathbf{v}\,dV,$$

which has to be satisfied for every $\mathbf{v}\equiv\mathbf{v}(\cdot)\in\mathscr{U}_t(\boldsymbol{\chi}(\cdot,t))$, $\boldsymbol{\chi}(\cdot,t)\in\mathscr{D}_t(\Omega)$ Similarly we get

$$\begin{aligned}\int_\Omega\Delta\mathbf{h}\cdot\nabla\zeta\,dV &= \oint_{\partial\Omega}\Delta\pi\zeta\,dA+\int_\Omega\Delta\alpha\zeta\,dV+\\ &\quad+\int_\Omega(\boldsymbol{\sigma}\circ\Delta\dot{\boldsymbol{\epsilon}}+\Delta\boldsymbol{\sigma}\circ\dot{\boldsymbol{\epsilon}}+\Delta\boldsymbol{\sigma}\circ\Delta\dot{\boldsymbol{\epsilon}}-\varrho\Delta\dot{\varepsilon})\zeta\,dV\end{aligned}$$

for every $\zeta \equiv \zeta(\cdot) \in \mathscr{Z}_t(\theta(\cdot, t))$, $\theta(\cdot, t) \in \mathscr{T}_t(\Omega)$. Taking into account that $\dot{\boldsymbol{\epsilon}} = \mathbf{L_F}\dot{\mathbf{u}}$, we have here

$$\Delta\dot{\boldsymbol{\epsilon}} = \mathbf{L_F}\Delta\dot{\mathbf{u}} + \mathbf{L}_{\Delta\mathbf{F}}\dot{\mathbf{u}} + \mathbf{L}_{\Delta\mathbf{F}}\Delta\dot{\mathbf{u}}.$$

All the aforementioned conditions have to hold for every $t \in [t_0, t_f]$. If for some $t = \tau$ the mapping $\boldsymbol{\chi}(\cdot, t)$ is the identity mapping then $\mathbf{F} = \nabla\boldsymbol{\chi}(\mathbf{X}, t)$ is the unit matrix **1**. Setting $\mathbf{Lv} \equiv \mathbf{L_F v}$ for $\mathbf{F} = \mathbf{1}$, we shall obtain from the definition of the linear operator $\mathbf{L_F}$ given in Sec. 1.4.4 that

$$(\mathbf{Lv})_{\alpha\beta} = \tfrac{1}{2}[\delta^k_\alpha v_{k,\beta}(\mathbf{X}) + \delta^k_\beta v_{k,\alpha}(\mathbf{X})].$$

At the same time the increment $\Delta\boldsymbol{\chi}(\mathbf{X}, \tau)$ in $\mathbf{L}_{\Delta\mathbf{F}}$, where $\Delta\mathbf{F} = \nabla[\Delta\boldsymbol{\chi}(\mathbf{X}, \tau)]$ approximates the increment of the displacement vector.

The above variational equations have to be considered jointly with the constitutive relations (cf. Sec. 2.1.5) and with the defining equations of the external agencies (cf. Sec. 2.2).

3.2 Method of constraints: finite formulation

3.2.1 Foundations

We shall now introduce the special class of internal and boundary constraints which will lead from the general form of governing relations discussed in Sec. 3.1 to various engineering theories of structural mechanics. To this aim we shall introduce the following mathematical objects:

1. Set Ω^* which is a certain surface or curve in Ω and a differentiable projection $(\cdot)^*$: $\Omega \ni \mathbf{X} \to (\mathbf{X})^* \in \Omega^*$ of Ω onto Ω^*. An example of such

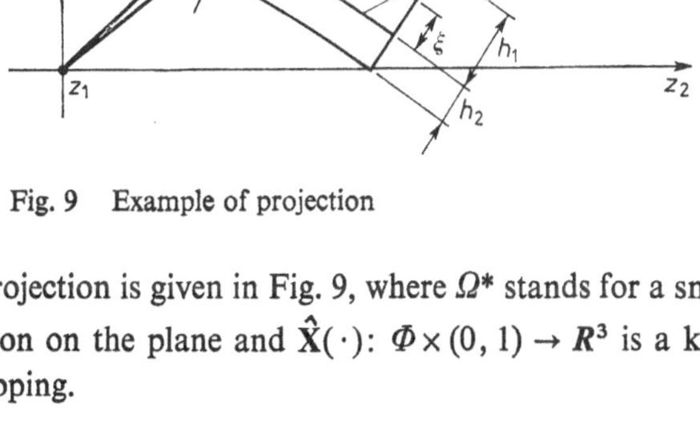

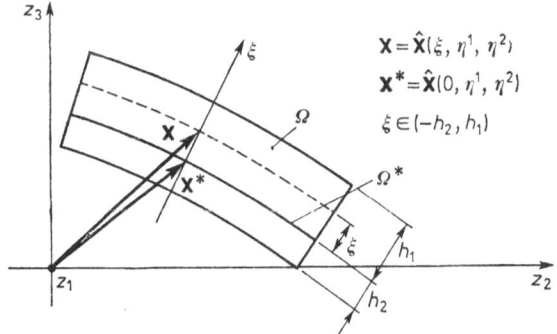

Fig. 9 Example of projection

a projection is given in Fig. 9, where Ω^* stands for a smooth arc, Φ is a regular region on the plane and $\hat{\mathbf{X}}(\cdot)$: $\Phi \times (0, 1) \to \boldsymbol{R}^3$ is a known invertible smooth mapping.

2. Family $\mathscr{D}_\tau^*(\Omega^*)$, $\tau \in [t_0, t_f]$, of sufficiently regular mappings $\boldsymbol{\varkappa}^*(\cdot)$: $\Omega^* \to \boldsymbol{R}^m$.

3. Family $\mathbf{F}(\mathbf{X})$, $\mathbf{X} \in \Omega$, of $3 \times m$ matrices that define linear mappings $\mathfrak{F}(\mathbf{X})[\cdot]$: $\boldsymbol{R}^m \to \boldsymbol{R}^3$ and a function $\boldsymbol{\chi}_0(\cdot, \tau)$: $\Omega \times (t_0, t_f) \to \boldsymbol{R}^3$. We assume that the functions $\mathbf{F}(\cdot)$ and $\boldsymbol{\chi}_0(\cdot, \tau)$ satisfy the regularity conditions required and specified below.

For the sake of compactness we shall write $\mathbf{X}^* \equiv (\mathbf{X})^*$ and use the notation $\mathfrak{F}(\mathbf{X})[\mathbf{q}] \equiv \mathbf{F}^a(\mathbf{X})q_a$, $\mathbf{q} = (q_1, \ldots, q_m) \in \boldsymbol{R}^m$. Index "$a$" runs over $1, \ldots$ $\ldots, m$; summation convention holds.

Assume that $\boldsymbol{\chi}(\cdot, \tau) \in \mathscr{D}_\tau(\Omega)$, $\tau \in [t_0, t_f]$, represents a motion of the body if and only if

$$\begin{aligned} \boldsymbol{\chi}(\mathbf{X}, \tau) &= \mathfrak{F}(\mathbf{X})[\boldsymbol{\chi}^*(\mathbf{X}^*, \tau)] + \boldsymbol{\chi}_0(\mathbf{X}, \tau) \\ &= \mathbf{F}^a(\mathbf{X})\boldsymbol{\chi}_a^*(\mathbf{X}^*, \tau) + \boldsymbol{\chi}_0(\mathbf{X}, \tau), \end{aligned} \tag{3.18}$$

for some

$$\boldsymbol{\chi}^*(\cdot, \tau) \in \mathscr{D}_\tau^*(\Omega^*), \qquad \tau \in [t_0, t_f].$$

Let functions $[t_0, t_f] \ni \tau \to \boldsymbol{\chi}^*(\mathbf{X}^*, \tau) \in \boldsymbol{R}^m$ for every $\mathbf{X}^* \in \Omega^*$ be continuous and have continuous first and second order time derivatives in $[t_0, t_f]$ except possibly at a finite number of time instants. The functions $\boldsymbol{\chi}^*(\cdot, \tau)$, $\dot{\boldsymbol{\chi}}^*(\cdot, \tau)$, $\ddot{\boldsymbol{\chi}}^*(\cdot, \tau)$ will be referred to as the generalized deformations, generalized velocities and generalized accelerations, respectively. Moreover, let $\mathscr{V}_t^* = \mathscr{V}_t^*(\boldsymbol{\chi}^*(\cdot, t))$ stands for a set of all possible generalized velocities at the time instant t and for a generalized deformation $\boldsymbol{\chi}^*(\cdot, t)$, $\boldsymbol{\chi}^*(\cdot, t) \in \mathscr{D}_t^*(\Omega)$. We shall assume that every set $\mathscr{V}_t^*$ is convex. We shall also define sets

$$\mathscr{U}_t^* := \{\mathbf{u}^*(\cdot) \in C_1^m(\bar{\Omega}^*) |\; \mathbf{u}^*(\cdot) = \lambda[\mathbf{v}_1^*(\cdot) - \mathbf{v}_2^*(\cdot)], \mathbf{v}_1^*, \mathbf{v}_2^* \in \mathscr{V}_t^*, \lambda \geqslant 0\}.$$

It can easily be seen that the line of approach proposed here leads to the specification of certain bilateral constraints and bears an analogy to that employed in Secs. 2.3.2 and 2.3.3. The elements of $\mathscr{U}_t^*$ will be called the generalized virtual displacements. Since $\mathscr{V}_t^* = \mathscr{V}_t^*(\boldsymbol{\chi}^*(\cdot, t))$ then also $\mathscr{U}_t^* = \mathscr{U}_t^*(\boldsymbol{\chi}^*(\cdot, t))$.

Let $X^{*K} = \pi^K(\mathbf{X})$, where $K = 1, 2$ or $K = 1$ and $\mathbf{X} \in \Omega$ be a coordinate form of a differentiable projection $\mathbf{X}^* = (\mathbf{X})^*$ of Ω onto Ω^*. Then, for an arbitrary differentiable real valued function $\varphi(\cdot)$ defined on Ω^*, $\varphi(\mathbf{X}^*)$, $\mathbf{X}^* \in \Omega^*$, we define $\varphi_{,K}(\mathbf{X}^*) \equiv \partial\varphi(\mathbf{X}^*)/\partial X^{*K}$ and

$$\nabla\varphi(\mathbf{X}^*) = \varphi_{,K}(\mathbf{X}^*)\nabla\pi^K(\mathbf{X}), \qquad \mathbf{X}^* = (\mathbf{X})^*,$$

for every $\mathbf{X} \in \Omega$ (summation convention holds). The variational equation (3.1) has now to hold for every differentiable function $\mathbf{v}(\cdot)$ such that

$$\mathbf{v}(\mathbf{X}) = \mathfrak{F}(\mathbf{X})[\mathbf{v}^*(\mathbf{X}^*)] = \mathbf{F}^a(\mathbf{X})v_a^*(\mathbf{X}^*),$$

$$\mathbf{v}^*(\cdot) \equiv (v_1^*(\cdot), \ldots, v_m^*(\cdot)) \in \mathscr{U}_t^*.$$

Taking into account that

$$\nabla[\mathbf{F}^a(\mathbf{X})v_a^*(\mathbf{X}^*)] = \nabla\mathbf{F}^a(\mathbf{X})v_a^*(\mathbf{X}^*)+\mathbf{F}^a(\mathbf{X})\nabla\pi^K(\mathbf{X})v_{a,K}^*(\mathbf{X}^*),$$

and setting

$$\mathbf{F}^{aK}(\mathbf{X}) \equiv \mathbf{F}^a(\mathbf{X})\otimes\nabla\pi^K(\mathbf{X})$$

where every $\mathbf{F}^{aK}(\mathbf{X})$, $a = 1, \ldots, m$, $\mathbf{X}\in\Omega$, is the known 3×3 matrix, we arrive finally at the condition

$$\begin{aligned}\int_\Omega \{&\mathrm{tr}[\mathbf{S}_t(\mathbf{X})\nabla\boldsymbol{\chi}^T(\mathbf{X}, t)\nabla\mathbf{F}^a(\mathbf{X})]v_a^*(\mathbf{X}^*)+\\ &+\mathrm{tr}[\mathbf{S}_t(\mathbf{X})\nabla\boldsymbol{\chi}^T(\mathbf{X}, t)\mathbf{F}^{aK}(\mathbf{X})]v_{a,K}^*(\mathbf{X}^*)\}\mathrm{d}V(\mathbf{X})\\ = &\oint_{\partial\Omega}\mathbf{s}_t(\mathbf{X})\cdot\mathbf{F}^a(\mathbf{X})v_a^*(\mathbf{X}^*)\mathrm{d}A(\mathbf{X})+\\ &+\int_\Omega[\mathbf{b}_t(\mathbf{X})-\varrho(\mathbf{X})\ddot{\mathbf{u}}_t(\mathbf{X})]\cdot\mathbf{F}^a(\mathbf{X})v_a^*(\mathbf{X}^*)\mathrm{d}V(\mathbf{X})\end{aligned}\tag{3.19}$$

which has to be satisfied by every $\mathbf{v}^*(\cdot)\in\mathscr{U}_t^*$, $t\in[t_0, t_f]$. Here

$$\boldsymbol{\chi}^*(\cdot, t)\in\mathscr{D}_t^*(\Omega^*)$$

and

$$\begin{aligned}\nabla\boldsymbol{\chi}(\mathbf{X}, t) &= \nabla[\mathbf{F}^a(\mathbf{X})\chi_a^*(\mathbf{X}^*, t)+\boldsymbol{\chi}_0(\mathbf{X}, t)]\\ &= \nabla\mathbf{F}^a(\mathbf{X})\chi_a^*(\mathbf{X}^*, t)+\mathbf{F}^{aK}(\mathbf{X})\chi_{a,K}^*(\mathbf{X}^*, t)+\nabla\boldsymbol{\chi}_0(\mathbf{X}, t),\end{aligned}$$

$$\ddot{\mathbf{u}}_t(\mathbf{X}) = \mathbf{F}^a(\mathbf{X})\ddot{\chi}_a^*(\mathbf{X}^*, t)+\ddot{\boldsymbol{\chi}}_0(\mathbf{X}, t).$$

Eqs. (3.19) represent the variational equations which are implied by the bilateral, ideal kinematic constraints of the form (3.18).

Using the procedure similar to that employed above, we are able to account for the constraints imposed on temperature. To this aim we assume that $\theta(\cdot, \tau) = \theta_0+\vartheta_\tau(\cdot)$ belongs to $\mathscr{T}_\tau(\Omega)$, $\tau\in[t_0, t_f]$, if and only if[1]

$$\theta(\mathbf{X}, t) = \mathbf{f}(\mathbf{X})[\theta^*(\mathbf{X}^*, t)]+\theta_0 = f^A(\mathbf{X})\theta_A^*(\mathbf{X}^*, t)+\theta_0, \quad \mathbf{X}\in\Omega, \tag{3.20}$$

for some $\theta^*(\cdot, t)\in\mathscr{T}_t^*(\Omega^*)$, where $\mathscr{T}_t^*(\Omega^*)$ are the known sets. Here $f^A(\mathbf{X})$, $\mathbf{X}\in\Omega$, are the known scalar differentiable functions and $\theta_A^*(\cdot, \tau)$ are unknown functions defined on Ω^* (for every $\tau\in[t_0, t_f]$). Every $\theta^*(\cdot, \tau)$ will be called the generalized temperature. We shall also use the notation

$$\mathbf{f}^{AK}(\mathbf{X}) \equiv f^A(\mathbf{X})\nabla\pi^K(\mathbf{X}), \quad K = 1, 2 \quad \text{or} \quad K = 1,$$

---

[1] Here $\mathscr{T}_t^*(\Omega^*)$, $t\in[t_0, t_f]$, are families of differentiable mappings $v^*(\cdot): \Omega^*\to R^n$, $\mathbf{f}(\mathbf{X})[\cdot]$: $R^n\to R$ is a linear mapping for every $\mathbf{X}\in\Omega$ and every $\mathbf{f}(\mathbf{X})$ is a vector with n components $f^A(\mathbf{X})$; hence $\mathbf{f}(\mathbf{X})[\mathbf{p}]\equiv f^A(\mathbf{X})p_A$ holds for every $\mathbf{p}\equiv(p_1, \ldots, p_n)\in R^n$. Index "$A$" runs over $1, \ldots, n$; summation convention holds. Moreover, the vector function $\mathbf{f}(\cdot)$ is assumed to be differentiable in Ω as many times as required.

where every $\mathbf{f}^{AK}(\mathbf{X})$, $A = 1, \ldots, m$, $\mathbf{X} \in \Omega$, is the known vector. Let for every $\mathbf{X}^* \in \Omega^*$ functions $[t_0, t_f] \ni \tau \to \theta^*(X^*, \tau)$ be continuous and have continuous first order time derivatives except possibly at a finite number of time instants. Let $\mathscr{W}_t^* \equiv \mathscr{W}_t^*(\theta^*(\cdot, t))$ be a set of all time derivatives $\dot{\theta}^*(\cdot, t) \equiv (\dot{\theta}_1^*(\cdot, t), \ldots$ $\ldots, \dot{\theta}_n^*(\cdot, t))$ which are admissible by the functions $[t_0, t_f] \ni \tau \to \theta^*(\mathbf{X}^*, \tau)$ at $\theta^*(\cdot, t)$. We shall assume that every $\mathscr{W}_t^*$ is a convex set. Setting then

$$\mathscr{Z}_t^* = \{\boldsymbol{\zeta}^*(\cdot) \in C_1^n(\overline{\Omega^*}) | \ \boldsymbol{\zeta}^*(\cdot) = \lambda[\mathbf{w}_1^*(\cdot) - \mathbf{w}_2^*(\cdot)], \quad \mathbf{w}_1^*(\cdot), \mathbf{w}_2^*(\cdot) \in \mathscr{W}_t^*, \lambda \geqslant 0\}$$

we obtain from eq. (3.5)

$$\begin{aligned}
&\int_\Omega [\mathbf{h}_t(\mathbf{X}) \cdot \nabla f^A(\mathbf{X}) \zeta_A^*(\mathbf{X}^*) + \mathbf{h}_t(\mathbf{X}) \cdot \mathbf{f}^{AK}(\mathbf{X}) \zeta_{A,K}^*(\mathbf{X}^*)] \mathrm{d}V(\mathbf{X}) \\
&= \oint_{\partial\Omega} \pi_t(\mathbf{X}) f^A(\mathbf{X}) \zeta_A^*(\mathbf{X}^*) \mathrm{d}A(\mathbf{X}) + \int_\Omega \alpha_t(\mathbf{X}) f^A(\mathbf{X}) \zeta_A^*(\mathbf{X}^*) \mathrm{d}V(\mathbf{X}) + \\
&\quad + \int_\Omega \{\mathrm{tr}[\mathbf{S}_t(\mathbf{X}) \dot{\mathbf{E}}_t(\mathbf{X})] - \varrho(\mathbf{X}) \dot{\varepsilon}(\mathbf{X}, t)\} f^A(\mathbf{X}) \zeta_A^*(\mathbf{X}^*) \mathrm{d}V(\mathbf{X}). \qquad (3.21)
\end{aligned}$$

Eq. (3.21) represents a variational equation which is assumed to hold for every $\boldsymbol{\zeta}^*(\cdot) = (\zeta_1^*(\cdot), \ldots, \zeta_n^*(\cdot))$ and for every time instant t. By virtue of $\mathscr{W}_t^* = \mathscr{W}_t^*(\theta(\cdot, t))$ we also have $\mathscr{Z}_t^* = \mathscr{Z}_t^*(\theta(\cdot, t))$. Taking into account definition (3.3) of the strain rate as well as the notation $\mathbf{F}^{aK}(\mathbf{X}) \equiv \mathbf{F}^a(\mathbf{X}) \otimes \nabla \pi^K(\mathbf{X})$, $X^{*K} = \pi^K(\mathbf{X})$, introduced earlier in this section, we obtain

$$\dot{\mathbf{E}}_t(\mathbf{X}) = \mathrm{sym}\{[\mathbf{F}^{aK}(\mathbf{X}) \chi_{a,K}^*(\mathbf{X}^*, t) + \nabla \mathbf{F}^a(\mathbf{X}) \boldsymbol{\chi}_a^*(\mathbf{X}^*, t)]^T \times \times [\mathbf{F}^{bL}(\mathbf{X}) \dot{\chi}_{b,L}^*(\mathbf{X}^*, t) + \nabla \mathbf{F}^b(\mathbf{X}) \dot{\boldsymbol{\chi}}_b^*(\mathbf{X}^*, t)]\},$$

where sym $\mathbf{M}$ stands for a symmetric part of an arbitrary square matrix $\mathbf{M}$, i.e. $\mathrm{sym}\,\mathbf{M} \equiv \frac{1}{2}(\mathbf{M} + \mathbf{M}^T)$ and the indices b, L have the same meaning as the indices a, K respectively.

We conclude that the constraints introduced above lead from problems for the functions $\boldsymbol{\chi}(\cdot, \tau)$, $\theta(\cdot, \tau)$ defined in Ω (for every $\tau \in [t_0, t_f]$) to problems for the functions $\boldsymbol{\chi}^*(\cdot, \tau)$, $\theta^*(\cdot, \tau)$ defined in Ω^*. The procedure presented above leading to the aforementioned result is said to be the *method of constraints*. It is a generalization of the known method of internal constraints, [160], and was developed in [174]–[178].

We have assumed so far in this section that the constraints are bilateral. For non-bilateral constraints we have to consider the sets $\mathscr{U}_t^*(\boldsymbol{\chi}^*(\cdot, t), \dot{\boldsymbol{\chi}}^*(\cdot, t))$ and $\mathscr{Z}_t^*(\theta^*(\cdot, t), \dot{\theta}^*(\cdot, t))$ instead of the sets $\mathscr{U}_t^*(\boldsymbol{\chi}^*(\cdot, t))$, $\mathscr{Z}_t^*(\theta^*(\cdot, t))$. The LHS of formulae (3.19) and (3.21) will then attain non-negative values only, cf. Sec. 2.3.1, for every $\mathbf{v}^*(\cdot) \in \mathscr{U}_t^*(\boldsymbol{\chi}^*(\cdot, t), \dot{\boldsymbol{\chi}}(\cdot, t))$ and $\boldsymbol{\zeta}^*(\cdot) \in \mathscr{Z}_t^*(\theta^*(\cdot, t), \dot{\theta}^*(\cdot, t))$, respectively.

The general approach given here will be illustrated in the remaining part of the chapter by a series of examples leading to various theories of engineering importance.

3.2.2 *Shell and plate theories*

Assume now that Ω^* is a known region on the plane and that $\mathbf{X}^* = (X^1, X^2)$. Let also $X^3 \in (-\delta, \delta)$ where δ is a known positive number. An example of the region Ω^* for which the above conditions are satisfied is shown in Fig. 10. We see that the projection $(\cdot)^*\colon \Omega \ni \mathbf{X} \to (\mathbf{X})^* \in \Omega^*$

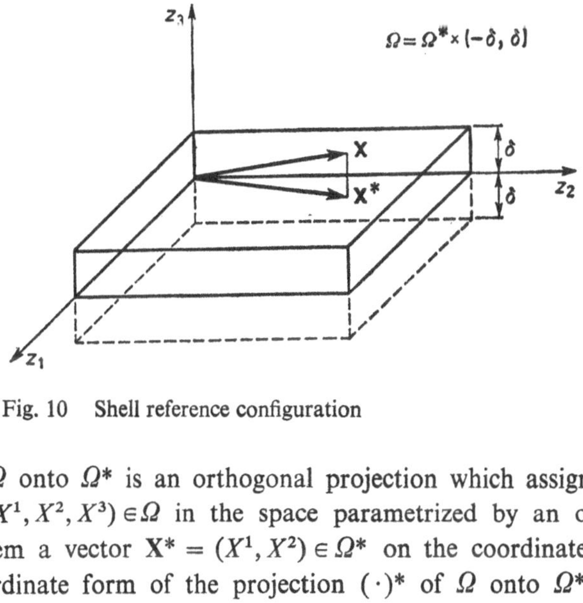

Fig. 10 Shell reference configuration

of Ω onto Ω^* is an orthogonal projection which assigns to every vector $\mathbf{X} = (X^1, X^2, X^3) \in \Omega$ in the space parametrized by an orthogonal Cartesian system a vector $\mathbf{X}^* = (X^1, X^2) \in \Omega^*$ on the coordinate plane $X^3 = 0$. The coordinate form of the projection $(\cdot)^*$ of Ω onto Ω^* is given by

$$X^K \equiv X^{*K} = \pi^K(\mathbf{X}) = \delta^K_\alpha X^\alpha, \qquad K = 1, 2, \qquad \alpha = 1, 2, 3$$

and hence $\nabla\pi^K(\mathbf{X}) = (\delta^K_1, \delta^K_2, 0)$. It follows that 3×3 matrices

$$\mathbf{F}^{aK}(\mathbf{X}) = \mathbf{F}^a(\mathbf{X}) \otimes \nabla\pi^K(\mathbf{X})$$

have components $F^{aK}_{K\alpha} = F^a_K \delta^K_\alpha$. Define $\eta \equiv X^3$, $\eta \in [-\delta, \delta]$, and assume $\mathbf{F}^a_{,K}(\mathbf{X}) = \mathbf{0}$. Since $\mathbf{X} = (\mathbf{X}^*, \eta)$, then

$$\boldsymbol{\chi}(\mathbf{X}, t) \equiv \boldsymbol{\chi}(\mathbf{X}^*, \eta, t) = \mathbf{F}^a(\eta)\boldsymbol{\chi}^*_a(\mathbf{X}^*, t) + \boldsymbol{\chi}_0(\mathbf{X}^*, \eta, t),$$

and variational equation (3.19) can be written in the form[1]

[1] In the sequal index "K" runs over 1,2; summation convention holds. By means of $\mathbf{F}^a_{,K}(\mathbf{X}) = \mathbf{0}$ we shall write $\mathbf{F}^a(\eta)$ and $\mathbf{F}^{aK}(\eta)$ instead of $\mathbf{F}^a(\mathbf{X})$ and $\mathbf{F}^{aK}(\mathbf{X})$, respectively. Moreover, by $dA(\mathbf{X}^*)$ we denote a line segment of $\partial\Omega^*$ at a point $\mathbf{X}^*$ and define $dV(\mathbf{X}^*) \equiv dX^1 dX^2$.

$$\int_{\Omega^*}\int_{-\delta}^{\delta} \{\mathrm{tr}[\mathbf{S}_t(\mathbf{X}^*, \eta)\nabla\boldsymbol{\chi}^T(\mathbf{X}^*, \eta, t)\nabla\mathbf{F}^a(\eta)]v_a^*(\mathbf{X}^*) +$$

$$+\mathrm{tr}[\mathbf{S}_t(\mathbf{X}^*, \eta)\nabla\boldsymbol{\chi}^T(\mathbf{X}^*, \eta, t)\mathbf{F}^{aK}(\eta)]v_{a,K}^*(\mathbf{X}^*)\}\mathrm{d}V(\mathbf{X}^*)\mathrm{d}\eta$$

$$= \int_{\Omega^*} \mathbf{s}_t(\mathbf{X}^*, \delta)\cdot\mathbf{F}^a(\delta)v_a^*(\mathbf{X}^*)\mathrm{d}V(\mathbf{X}^*) +$$

$$+\int_{\Omega^*} \mathbf{s}_t(\mathbf{X}^*, -\delta)\cdot\mathbf{F}^a(-\delta)v_a^*(\mathbf{X}^*)\mathrm{d}V(\mathbf{X}^*) +$$

$$+\oint_{\partial\Omega^*}\int_{-\delta}^{\delta} \mathbf{s}_t(\mathbf{X}^*, \eta)\cdot\mathbf{F}^a(\eta)v_a^*(\mathbf{X}^*)\mathrm{d}\eta\,\mathrm{d}A(\mathbf{X}^*) +$$

$$+\int_{\Omega^*}\int_{-\delta}^{\delta} [\mathbf{b}_t(\mathbf{X}^*, \eta) - \varrho(\mathbf{X}^*, \eta)\ddot{\mathbf{u}}_t(\mathbf{X}^*, \eta)]\cdot\mathbf{F}^a(\eta)v_a^*(\mathbf{X}^*)\mathrm{d}\eta\,\mathrm{d}V(\mathbf{X}^*)$$

where

$$\boldsymbol{\chi}(\mathbf{X}^*, \eta, t) = \mathbf{F}^a(\eta)\boldsymbol{\chi}_a^*(\mathbf{X}^*, t) + \boldsymbol{\chi}_0(\mathbf{X}^*, \eta, t),$$

$$\ddot{\mathbf{u}}_t(\mathbf{X}^*, \eta) = \ddot{\boldsymbol{\chi}}(\mathbf{X}^*, \eta, t) = \mathbf{F}^a(\eta)\ddot{\boldsymbol{\chi}}_a^*(\mathbf{X}^*, t) + \ddot{\boldsymbol{\chi}}_0(\mathbf{X}^*, \eta, t),$$

$$\nabla\boldsymbol{\chi}(\mathbf{X}^*, \eta, t) = \nabla\mathbf{F}^a(\eta)\boldsymbol{\chi}_a^*(\mathbf{X}^*, t) + \mathbf{F}^{aK}(\eta)\boldsymbol{\chi}_{a,K}^*(\mathbf{X}^*, t) + \nabla\boldsymbol{\chi}_0(\mathbf{X}^*, \eta, t).$$

Define now

$$H^{aK}(\mathbf{X}^*, t) \equiv \int_{-\delta}^{\delta} \mathrm{tr}[\mathbf{S}_t(\mathbf{X}^*, \eta)\nabla\boldsymbol{\chi}^T(\mathbf{X}^*, \eta, t)\mathbf{F}^{aK}(\eta)]\mathrm{d}\eta,$$

$$h^a(\mathbf{X}^*, t) \equiv -\int_{-\delta}^{\delta} \mathrm{tr}[\mathbf{S}_t(\mathbf{X}^*, \eta)\nabla\boldsymbol{\chi}^T(\mathbf{X}^*, \eta, t)\nabla\mathbf{F}^a(\eta)]\mathrm{d}\eta, \tag{3.22}$$

$$f^a(\mathbf{X}^*, t) \equiv \int_{-\delta}^{\delta} \mathbf{b}_t(\mathbf{X}^*, \eta)\cdot\mathbf{F}^a(\eta)\mathrm{d}\eta +$$

$$+\mathbf{s}_t(\mathbf{X}^*, \delta)\cdot\mathbf{F}^a(\delta) + \mathbf{s}_t(\mathbf{X}^*, -\delta)\cdot\mathbf{F}^a(-\delta),$$

$$M^{ab}(\mathbf{X}^*) \equiv \int_{-\delta}^{\delta} \varrho(\mathbf{X}^*, \eta)\mathbf{F}^a(\eta)\cdot\mathbf{F}^b(\eta)\mathrm{d}\eta,$$

for $\mathbf{X}^* \in \Omega^*$, and

$$s^a(\mathbf{X}^*, t) \equiv \int_{-\delta}^{\delta} \mathbf{s}_t(\mathbf{X}^*, \eta)\cdot\mathbf{F}^a(\eta)\mathrm{d}\eta \tag{3.23}$$

for $\mathbf{X}^* \in \partial\Omega^*$. Then variational equation (3.19) may be written as

$$\int_{\Omega^*} H^{aK}(\mathbf{X}^*, t) v^*_{a,K}(\mathbf{X}^*) \mathrm{d}V(\mathbf{X}^*)$$

$$= \oint_{\partial\Omega^*} s^a(\mathbf{X}^*, t) v^*_a(\mathbf{X}^*) \mathrm{d}A(\mathbf{X}^*) +$$

$$+ \int_{\Omega^*} [f^a(\mathbf{X}^*, t) + h^a(\mathbf{X}^*, t) - M^{ab}(\mathbf{X}^*) \ddot{\chi}^*_b(\mathbf{X}^*, t)] v^*_a(\mathbf{X}^*) \mathrm{d}V(\mathbf{X}^*) \quad (3.24)$$

which has to hold for every $\mathbf{v}^*(\cdot) \equiv (v^*_1(\cdot), \ldots, v^*_m(\cdot)) \in \mathscr{U}^*_t$.

Let us now consider constraints for temperature (3.20) setting

$$\theta(\mathbf{X}, t) = \theta(\mathbf{X}^*, \eta, t) = f^A(\eta)\theta^*_A(\mathbf{X}^*, t) + \theta_0, \qquad \mathbf{X}^* \in \Omega^*,$$

bearing in mind that $f^{A3}(\mathbf{X}) = 0$ and defining $f^{AK}(\eta)$ as a value of $f^{Ak}(\mathbf{X}^*, \eta)$ for $k = K$. Variational equation (3.21) takes the form

$$\int_{\Omega^*} \int_{-\delta}^{\delta} [\mathbf{h}_t(\mathbf{X}^*, \eta) \cdot \nabla f^A(\eta) \zeta^*_A(\mathbf{X}^*) + \mathbf{h}_t(\mathbf{X}^*, \eta) \cdot f^{AK}(\eta) \zeta^*_{A,K}(\mathbf{X}^*)] \mathrm{d}\eta \, \mathrm{d}V(\mathbf{X}^*)$$

$$= \int_{\Omega^*} [\pi_t(\mathbf{X}^*, \delta) f^A(\delta) + \pi_t(\mathbf{X}^*, -\delta) f^A(-\delta)] \zeta^*_A(\mathbf{X}^*) \mathrm{d}V(\mathbf{X}^*) +$$

$$+ \int_{\Omega^*} \int_{-\delta}^{\delta} \alpha_t(\mathbf{X}^*, \eta) f^A(\eta) \zeta^*_A(\mathbf{X}^*) \mathrm{d}\eta \, \mathrm{d}V(\mathbf{X}^*) +$$

$$+ \oint_{\partial\Omega^*} \int_{-\delta}^{\delta} \pi_t(\mathbf{X}^*, \eta) f^A(\eta) \zeta^*_A(\mathbf{X}^*) \mathrm{d}\eta \, \mathrm{d}A(\mathbf{X}^*) +$$

$$+ \int_{\Omega^*} \int_{-\delta}^{\delta} \{\mathrm{tr}[\mathbf{S}_t(\mathbf{X}^*, \eta) \dot{\mathbf{E}}_t(\mathbf{X}^*, \eta)] -$$

$$- \varrho(\mathbf{X}^*, \eta) \dot{\varepsilon}(\mathbf{X}^*, \eta, t)\} f^A(\eta) \zeta^*_A(\mathbf{X}^*) \mathrm{d}\eta \, \mathrm{d}V(\mathbf{X}^*),$$

where now, cf. Sec. 3.2.1,

$$\dot{\mathbf{E}}_t(\mathbf{X}^*, \eta) = \mathrm{sym}\{[\mathbf{F}^{aK}(\eta) \dot{\chi}^*_{a,K}(\mathbf{X}^*, t) +$$
$$+ \nabla \mathbf{F}^a(\eta) \chi^*_a(\mathbf{X}^*, t)]^T [\mathbf{F}^{bL}(\eta) \dot{\chi}^*_{b,L}(\mathbf{X}^*, t) + \nabla \mathbf{F}^b(\eta) \dot{\chi}^*_b(\mathbf{X}^*, t)]\};$$

the indices K, L and a, b run over 1, 2 and $1, \ldots, m$, respectively. Define further

$$G^{AK}(\mathbf{X}^*, t) \equiv \int_{-\delta}^{\delta} \mathbf{h}_t(\mathbf{X}^*, \eta) \cdot f^{AK}(\eta) \mathrm{d}\eta,$$

$$g^A(\mathbf{X}^*, t) \equiv - \int_{-\delta}^{\delta} \mathbf{h}_t(\mathbf{X}^*, \eta) \cdot \nabla f^A(\eta) \mathrm{d}\eta,$$

$$e^A(\mathbf{X}^*, t) \equiv \int_{-\delta}^{\delta} \alpha_t(\mathbf{X}^*, \eta) f^A(\eta) d\eta + \pi_t(\mathbf{X}^*, \delta) f^A(\delta) + \pi_t(\mathbf{X}^*, -\delta) f^A(-\delta),$$

$$\varepsilon^A(\mathbf{X}^*, t) \equiv \int_{-\delta}^{\delta} \varrho(\mathbf{X}^*, \eta)\, \varepsilon(\mathbf{X}^*, \eta, t) f^A(\eta) d\eta, \tag{3.25}$$

$$L^A(\mathbf{X}^*, t) \equiv \int_{-\delta}^{\delta} \mathrm{tr}[\mathbf{S}_t(\mathbf{X}^*, \eta) \dot{\mathbf{E}}_t(\mathbf{X}^*, \eta)] f^A(\eta) d\eta,$$

for $\mathbf{X}^* \in \Omega^*$ and

$$\pi^A(\mathbf{X}^*, t) \equiv \int_{-\delta}^{\delta} \pi_t(\mathbf{X}^*, \eta) f^A(\eta) d\eta \tag{3.26}$$

for $\mathbf{X}^* \in \partial\Omega^*$. Variational equation (3.21) takes then the final form

$$\int_{\Omega^*} G^{AK}(\mathbf{X}^*, t) \zeta^*_{A,K}(\mathbf{X}^*) dV(\mathbf{X}^*)$$
$$= \oint_{\partial\Omega^*} \pi^A(\mathbf{X}^*, t) \zeta^*_A(\mathbf{X}^*) dA(\mathbf{X}^*) +$$
$$+ \int_{\Omega^*} [e^A(\mathbf{X}^*, t) + g^A(\mathbf{X}^*, t) + L^A(\mathbf{X}^*, t) - \dot{\varepsilon}^A(\mathbf{X}^*, t)] \zeta^*_A(\mathbf{X}^*) dV(\mathbf{X}^*), \tag{3.27}$$

which has to hold for every $\boldsymbol{\zeta}^*(\cdot) \equiv (\zeta^*_1(\cdot), \ldots, \zeta^*_n(\cdot)) \in \mathscr{Z}^*_t$.

Eqs. (3.24) and (3.27) represent the variational equations of motion and heat conduction of what are commonly called the *theories of shells*. If the region Ω shown in Fig. 10 coincides with the region occupied by the structural element at a certain time instant, say $\tau = t_0$, then the theories of shells are transformed into the theories of plates. The terms "shell theories" and "plate theories" are used here to indicate that they can potentially be employed in many special engineering problems as the mathematical models of shell and plate structural elements.

Let us now confine ourselves to the case in which the material of the shell or plate can be treated as the thermo-visco-elastic. Substituting then the RHS of eqs. (3.2), (3.4) into the RHS of definitions (3.22), (3.23) and observing that

$$\begin{aligned} \boldsymbol{\chi}(\mathbf{X}^*, \eta, t) &= \mathbf{F}^a(\eta) \boldsymbol{\chi}^*_a(\mathbf{X}^*, t) + \boldsymbol{\chi}_0(\mathbf{X}^*, \eta, t), \\ \theta(\mathbf{X}^*, \eta, t) &= f^A(\eta) \theta^*_A(\mathbf{X}^*, t) + \theta_0, \end{aligned} \tag{3.28}$$

we shall arrive at the following equations

$$H^{aK}(\mathbf{X}^*, t) = \hat{H}^{aK}(\mathbf{X}^*, \nabla\boldsymbol{\chi}^*(\mathbf{X}^*, t), \boldsymbol{\chi}^*(\mathbf{X}^*, t), \nabla\dot{\boldsymbol{\chi}}^*(\mathbf{X}^*, t), \dot{\boldsymbol{\chi}}^*(\mathbf{X}^*, t), \theta^*(\mathbf{X}^*, t)),$$

$$
\begin{aligned}
h^a(\mathbf{X}^*, t) &= \hat{h}^a(\mathbf{X}^*, \nabla\boldsymbol{\chi}^*(\mathbf{X}^*, t), \boldsymbol{\chi}^*(\mathbf{X}^*, t), \nabla\dot{\boldsymbol{\chi}}(\mathbf{X}^*, t),\\
&\qquad \dot{\boldsymbol{\chi}}^*(\mathbf{X}^*, t), \theta^*(\mathbf{X}^*, t)),\\
f^a(\mathbf{X}^*, t) &= \hat{f}^a(\mathbf{X}^*, t, \nabla\boldsymbol{\chi}^*(\mathbf{X}^*, t), \dot{\boldsymbol{\chi}}^*(\mathbf{X}^*, t)), \quad \mathbf{X}^* \in \Omega^*, \qquad (3.29)\\
s^a(\mathbf{X}^*, t) &= \hat{s}^a(\mathbf{X}^*, t, \nabla\boldsymbol{\chi}^*(\mathbf{X}^*, t), \boldsymbol{\chi}^*(\mathbf{X}^*, t), \dot{\boldsymbol{\chi}}^*(\mathbf{X}^*, t)), \quad \mathbf{X}^* \in \partial\Omega^*.
\end{aligned}
$$

Analogously, substituting the RHS of eqs. (3.6), (3.7) into the RHS of definitions (3.25) and (3.26) we obtain

$$
\begin{aligned}
G^{AK}(\mathbf{X}^*, t) &= \hat{G}^{AK}(\mathbf{X}^*, \nabla\theta^*(\mathbf{X}^*, t), \theta^*(\mathbf{X}^*, t), \nabla\boldsymbol{\chi}^*(\mathbf{X}^*, t), \boldsymbol{\chi}^*(\mathbf{X}^*, t)),\\
g^A(\mathbf{X}^*, t) &= \hat{g}^A(\mathbf{X}^*, \nabla\theta^*(\mathbf{X}^*, t), \theta^*(\mathbf{X}^*, t), \nabla\boldsymbol{\chi}^*(\mathbf{X}^*, t), \boldsymbol{\chi}^*(\mathbf{X}^*, t)),\\
e^A(\mathbf{X}^*, t) &= \hat{e}^A(\mathbf{X}^*, t, \theta^*(\mathbf{X}^*, t), \boldsymbol{\chi}^*(\mathbf{X}^*, t)), \qquad (3.30)\\
\varepsilon^A(\mathbf{X}^*, t) &= \hat{\varepsilon}^A(\mathbf{X}^*, \theta^*(\mathbf{X}^*, t), \boldsymbol{\chi}^*(\mathbf{X}^*, t), \nabla\boldsymbol{\chi}^*(\mathbf{X}^*, t)),\\
L^A(\mathbf{X}^*, t) &= \hat{L}^A(\mathbf{X}^*, \nabla\boldsymbol{\chi}^*(\mathbf{X}^*, t), \boldsymbol{\chi}^*(\mathbf{X}^*, t), \nabla\dot{\boldsymbol{\chi}}^*(\mathbf{X}^*, t),\\
&\qquad \dot{\boldsymbol{\chi}}^*(\mathbf{X}^*, t), \theta^*(\mathbf{X}^*, t)); \quad \mathbf{X}^* \in \Omega^*,\\
\pi^A(\mathbf{X}^*, t) &= \hat{\pi}^A(\mathbf{X}^*, t, \boldsymbol{\chi}^*(\mathbf{X}^*, t), \theta^*(\mathbf{X}^*, t)), \quad \mathbf{X}^* \in \partial\Omega^*.
\end{aligned}
$$

It has to be stressed that in every problem of thermo-visco-elasticity with known material characteristics the given external agents and, for postulated a priori constraints (3.28), the RHS of eqs. (3.29), (3.30) can easily be calculated. Thus, eqs. (3.29), (3.30) represent in every special problem the constitutive relations of shell and plate theories.

Summing up, by the governing equations of shell or plate theories defined by various constraints (3.28) we shall understand both the variational equations (3.24), (3.27) and the constitutive relations (3.29), (3.30). The basic unknowns are:

$$
\begin{aligned}
\boldsymbol{\chi}^*(\mathbf{X}^*, \tau) &\equiv (\chi_1^*(\mathbf{X}^*, \tau), \ldots, \chi_m^*(\mathbf{X}^*, \tau)),\\
\theta^*(\mathbf{X}^*, \tau) &\equiv (\theta_1^*(\mathbf{X}^*, \tau), \ldots, \theta_n^*(\mathbf{X}^*, \tau)), \quad \mathbf{X}^* \in \Omega^*, \tau \in [t_0, t_f].
\end{aligned}
$$

Having solved the pertinent initial-boundary value problem of a shell theory we can calculate displacements $\mathbf{u}_t(\mathbf{X}) = \boldsymbol{\chi}(\mathbf{X}, t) - \boldsymbol{\chi}(\mathbf{X}, t_0)$ and temperatures $\theta(\mathbf{X}, t)$ by means of eqs. (3.28) and then compute all the other unknowns using formulae given in Secs. 3.3.1 and 3.1.4.

The approach to shell and plate theories based on the method of constraints and outlined in this section constitutes merely a starting point to formulate various special theories and problems. In most engineering applications of this approach we usually assume that eqs. (3.28) have the form

$$
\begin{aligned}
\boldsymbol{\chi}(\mathbf{X}^*, \eta, t) &= \mathbf{p}(\mathbf{X}^*, t) + \eta\mathbf{d}(\mathbf{X}^*, t),\\
\theta(\mathbf{X}^*, \eta, t) &= \psi(\mathbf{X}^*, t) + \eta\xi(\mathbf{X}^*, t),
\end{aligned} \qquad (3.31)
$$

where $\mathbf{p}(\cdot, t)$, $\mathbf{d}(\cdot, t)$ are unknown vector fields and $\psi(\cdot, t)$, $\xi(\cdot, t)$ are unknown scalar fields. We have assumed here that $\boldsymbol{\chi}_0(\mathbf{X}^*, \eta) \equiv 0$, $\theta_0 \equiv 0$, $m = 6$, $n = 2$ and

$$\mathbf{p}(\mathbf{X}^*, t) = \left(\chi_1^*(\mathbf{X}^*, t), \chi_2^*(\mathbf{X}^*, t), \chi_3^*(\mathbf{X}^*, t)\right),$$

$$\mathbf{d}(\mathbf{X}^*, t) = \left(\chi_4^*(\mathbf{X}^*, t), \chi_5^*(\mathbf{X}^*, t), \chi_6^*(\mathbf{X}^*, t)\right),$$

$$\psi(\mathbf{X}^*, t) = \theta_1^*(\mathbf{X}^*, t), \quad \xi(\mathbf{X}^*, t) = \theta_2^*(\mathbf{X}^*, t).$$

Surface $\mathbf{z} = \mathbf{p}(\mathbf{X}^*, \tau)$, $\mathbf{X}^* \in \Omega^*$, is called the *midsurface* of the shell at the time instant τ. Moreover, if the conditions $\boldsymbol{\chi}^*(\cdot, \tau) \in \mathscr{D}_\tau^*(\Omega^*)$ represent only the boundary constraints, i.e. if they do not impose any additional interrelations between the functions $\mathbf{p}(\cdot, \tau)$, $\mathbf{d}(\cdot, \tau)$ for every $\tau \in [t_0, t_f]$, then the pertinent shell theories are referred to as the six-parameter shell theories. If the condition $\boldsymbol{\chi}^*(\cdot, \tau) \in \mathscr{D}_\tau^*(\Omega^*)$ imposes on eq. (3.31) additional constraints of the form

$$\mathbf{d}(\mathbf{X}^*, \tau) \cdot \mathbf{d}(\mathbf{X}^*, \tau) = 1, \quad \mathbf{d}(\mathbf{X}^*, \tau) \cdot \mathbf{p}_{,K}(\mathbf{X}^*, \tau) = 0, \tag{3.32}$$

then we usual deal with the so-called Love–Kirchhoff shell theory. In this case every $\mathbf{d}(\mathbf{X}^*, \tau)$ is the unit vector normal to the midsurface $\mathbf{z} = \mathbf{p}(\mathbf{X}^*, \tau)$, $\mathbf{X}^* \in \Omega^*$, of the shell at the point with parameters $\mathbf{X}^* = (X^1, X^2)$ and at the time instant τ.

In the six-parameter shell theories the condition $\mathbf{v}^*(\cdot) = (v_1^*(\cdot), \ldots$ $\ldots, v_6^*(\cdot)) \in \mathscr{U}_\tau^*$ does not impose any restrictions on the values of $v_a^*(\cdot)$ inside of the region Ω^*. On the other hand, such restrictions are imposed in Love–Kirchhoff shell theories. Setting $\mathbf{v}^*(\cdot) \equiv (v_1(\cdot), v_2(\cdot), v_3(\cdot), w_1(\cdot), w_2(\cdot), w_3(\cdot))$ and taking into account the aforementioned internal constraints imposed on $\mathbf{p}(\cdot, \tau)$, $\mathbf{d}(\cdot, \tau)$ we obtain that $\mathbf{v}^*(\cdot) \in \mathscr{U}_\tau^*$ if

$$\mathbf{w}(\mathbf{X}^*) \cdot \mathbf{d}(\mathbf{X}^*, \tau) = 0,$$

$$\mathbf{w}(\mathbf{X}^*) \cdot \mathbf{p}_{,K}(\mathbf{X}^*, \tau) + \mathbf{d}(\mathbf{X}^*, \tau) \cdot \mathbf{v}_{,K}(\mathbf{X}^*) = 0. \tag{3.33}$$

This means that in Love–Kirchhoff shell theories in which the internal constraints are given by eqs. (3.31), (3.32) the variational equation of motion (3.24) is assumed to hold for $v_a^*(\mathbf{X}^*)$, $a = 1, \ldots, 6$, satisfying the conditions (3.33). At the same time the functions $\mathbf{p}(\cdot, \tau)$, $\mathbf{d}(\cdot, \tau)$ have to satisfy the boundary restrictions. For example, if the following conditions are prescribed on the boundary $\partial\Omega^*$

$$\gamma_\pi(\mathbf{X}^*, t, \mathbf{p}(\mathbf{X}^*, t), \mathbf{d}(\mathbf{X}^*, t)) = 0, \quad \mathbf{X}^* \in \partial\Omega^*, \quad \pi = 1, \ldots, P,$$

where $\gamma_\pi(\cdot)$ are known differentiable functions, then the functions $\mathbf{v}^*(\cdot)$ $\equiv (v_1(\cdot), v_2(\cdot), v_3(\cdot), w_1(\cdot), w_2(\cdot), w_3(\cdot))$ in the variational equation of motion (3.24) have also to satisfy the condition

$$\frac{\partial \gamma_\pi}{\partial \mathbf{p}(\mathbf{X}^*, \tau)} \cdot \mathbf{v}(\mathbf{X}^*) + \frac{\partial \gamma_\pi}{\partial \mathbf{d}(\mathbf{X}^*, \tau)} \cdot \mathbf{w}(\mathbf{X}^*) = 0, \qquad \pi = 1, \ldots, P,$$

on $\partial\Omega^*$. Remember that on a certain part of $\partial\Omega^*$ the functions $\gamma_\pi(\cdot)$ can attain the values identically equal to zero. For instance, if the conditions $\mathbf{p}(\mathbf{X}^*, \tau) - \mathbf{p}_0(\mathbf{X}^*, \tau) = \mathbf{0}$ hold for every $\mathbf{X}^* \in \Gamma$, where Γ is the known part of $\partial\Omega^*$ and $\mathbf{p}_0(\mathbf{X}^*, \tau)$ are known, then $\mathbf{v}(\mathbf{X}^*) = \mathbf{0}$ on Γ.

The engineering shell and plate theories are often based not only on the kinematic constraints (and on the constraints for the temperature fields), but also on certain constraints for stresses and heat fluxes. The form of these constraints depends on the character of the specific problem under consideration. In some problems we can postulate the plane state of stress in the shell setting $S_t^{33}(\mathbf{X}) = 0$, $S_t^{31}(\mathbf{X}) = 0$, $S_t^{32}(\mathbf{X}) = 0$ for every $\mathbf{X} \in \Omega$ and every $t \in [t_0, t_f]$. In other problems we can assume that $S_t^{33}(\mathbf{X}) = 0$ holds for every $\mathbf{X} \in \Omega$, $t \in [t_0, t_f]$ and that $S_t^{31}(\mathbf{X})$, $S_t^{32}(\mathbf{X})$ are uniquely determined by $\nabla\boldsymbol{\chi}^*(\mathbf{X}^*, t)$, being equal to zero for $X^3 = \pm\delta$. The general approach given in Sec. 2.3.5 can also be employed in all the cases in which constraints for stresses and heat fluxes are involved.

3.2.3 Rod and beam theories

Let Ω^* be the known segment $(0, l)$ of the z_1-coordinate axis, $\Omega^* = (0, l)$, and $X^* \equiv X^1 \in [0, l]$. Let also Φ be the known region on the coordinate plane $z_1 = 0$ such that $0 \in \Phi$ and assume that $\Omega = (0, l) \times \Phi$. Then the projection $(\cdot)^*$: $\Omega \ni \mathbf{X} \to (\mathbf{X}^*) \in \Omega^*$ of Ω onto $\Omega^* = (0, l)$ is an orthogonal projection which assigns to every vector $\mathbf{X} \equiv (X^1, X^2, X^3) \in \Omega$ a number $X^1 \in \Omega^*$

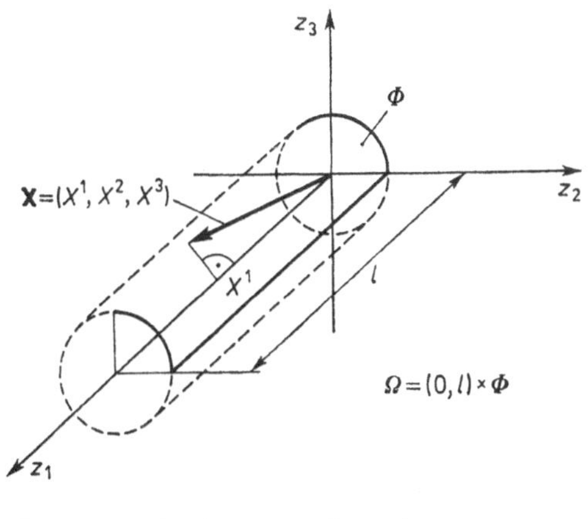

Fig. 11 Rod reference configuration

(the first coordinate of the vector **X**). Such a situation is illustrated in Fig. 11. The coordinate form of the projection $(\cdot)^*$ of Ω onto $\Omega^* = (0, l)$ is given by

$$X^* = X^1 = \pi(X^1, X^2, X^3) = \delta^1_\alpha X^\alpha.$$

Hence $\nabla\pi(\mathbf{X}) = (1, 0, 0)$ for every $\mathbf{X} \in \Omega$. For the sake of simplicity we shall use the notation: $\xi \equiv X^1$, $\boldsymbol{\eta} \equiv (\eta^1, \eta^2) \equiv (X^1, X^2)$. Then $\mathbf{X} = (\xi, \boldsymbol{\eta})$, $X^* = \xi$, and eqs. (3.18), (3.20) under the assumption that $\boldsymbol{\chi}_0(\mathbf{X}) \equiv \mathbf{0}$, $\theta_0(\mathbf{X}) \equiv 0$, yield[1]

$$\begin{aligned} \boldsymbol{\chi}(\mathbf{X}, \tau) &= \boldsymbol{\chi}(\xi, \boldsymbol{\eta}, \tau) = \mathbf{F}^a(\boldsymbol{\eta})\chi^*_a(\xi, \tau), \quad \boldsymbol{\chi}^*(\cdot, \tau) \in \mathscr{D}^*_t([0, l]), \\ \theta(\mathbf{X}, \tau) &= \theta(\xi, \boldsymbol{\eta}, \tau) = f^A(\boldsymbol{\eta})\theta^*_A(\xi, \tau), \quad \theta^*(\cdot, \tau) \in \mathscr{T}^*_t([0, l]). \end{aligned} \tag{3.34}$$

The formal procedure proceeds now in the same way as before in Sec. 3.2.2. Denoting[2]

$$\begin{aligned} H^a(\xi, t) &\equiv \int_\Phi \mathrm{tr}[\mathbf{S}_t(\xi, \boldsymbol{\eta})\nabla\boldsymbol{\chi}^T(\xi, \boldsymbol{\eta}, t)\mathbf{F}^a{}_{,1}(\boldsymbol{\eta})]\mathrm{d}\eta^1\mathrm{d}\eta^2, \\ h^a(\xi, t) &\equiv -\int_\Phi \mathrm{tr}[\mathbf{S}_t(\xi, \boldsymbol{\eta})\nabla\boldsymbol{\chi}^T(\xi, \boldsymbol{\eta}, t)\nabla\mathbf{F}^a(\boldsymbol{\eta})]\mathrm{d}\eta^1\mathrm{d}\eta^2, \\ f^a(\xi, t) &\equiv \int_\Phi \mathbf{b}_t(\xi, \boldsymbol{\eta}) \cdot \mathbf{F}^a(\boldsymbol{\eta})\mathrm{d}\eta^1\mathrm{d}\eta^2 + \oint_{\partial\Phi} \mathbf{s}_t(\xi, \boldsymbol{\eta}) \cdot \mathbf{F}^a(\eta)\mathrm{d}A(\boldsymbol{\eta}), \\ M^{ab}(\xi) &\equiv \int_\Phi \varrho(\xi, \boldsymbol{\eta})\mathbf{F}^a(\boldsymbol{\eta}) \cdot \mathbf{F}^b(\boldsymbol{\eta})\mathrm{d}\eta^1\mathrm{d}\eta^2 \end{aligned} \tag{3.35}$$

for $\xi \in (0, l)$ and

$$\mathbf{s}^a(t; 0) \equiv \int_\Phi \mathbf{s}_t(0, \boldsymbol{\eta}) \cdot \mathbf{F}^a(\boldsymbol{\eta})\mathrm{d}\eta^1\mathrm{d}\eta^2,$$

$$\mathbf{s}^a(t; l) \equiv \int_\Phi \mathbf{s}_t(l, \boldsymbol{\eta}) \cdot \mathbf{F}^a(\boldsymbol{\eta})\mathrm{d}\eta^1\mathrm{d}\eta^2,$$

we obtain from variational equation (3.19) the condition

$$\int_0^l H^a(\xi, t)v^*_{a,1}(\xi)\mathrm{d}\xi = \mathbf{s}^a(t; 0)v^*_a(0) + \mathbf{s}^a(t; l)v^*_a(l) + \\ + \int_0^l [f^a(\xi, t) + h^a(\xi, t) - M^{ab}(\xi)\ddot{\chi}^*_b(\xi, t)]v^*_a(\xi)\mathrm{d}\xi, \tag{3.36}$$

which has to hold for every $\mathbf{v}^*(\cdot) \equiv (v^*_1(\cdot), \ldots, v^*_m(\cdot)) \in \mathscr{U}^*_t$.

---

[1] We assume that $\mathbf{F}^a(\cdot)$, $f^A(\cdot)$ do not depend on X^1.

[2] The symbol $\mathrm{d}A(\boldsymbol{\eta})$ stands for a line element of $\partial\Phi$ at $\boldsymbol{\eta} \in \partial\Phi$. We also define $\varphi_{,1}(\xi, \boldsymbol{\eta}, t) \equiv \partial\varphi/\partial\xi$ for an arbitrary differentiable function $\varphi(\cdot)$.

In analogy, denoting

$$G^A(\xi, t) \equiv \int_{\Phi} \mathbf{h}_t(\xi, \boldsymbol{\eta}) \cdot f^{A1}(\boldsymbol{\eta}) \mathrm{d}\eta^1 \mathrm{d}\eta^2,$$

$$g^A(\xi, t) \equiv -\int_{\Phi} \mathbf{h}_t(\xi, \boldsymbol{\eta}) \cdot \nabla f^A(\boldsymbol{\eta}) \mathrm{d}\eta^1 \mathrm{d}\eta^2,$$

$$e^A(\xi, t) \equiv \int_{\Phi} \alpha_t(\xi, \boldsymbol{\eta}) f^A(\boldsymbol{\eta}) \mathrm{d}\eta^1 \mathrm{d}\eta^2 + \oint_{\partial\Phi} \pi_t(\xi, \boldsymbol{\eta}) f^A(\boldsymbol{\eta}) \mathrm{d}A(\boldsymbol{\eta}), \tag{3.37}$$

$$\varepsilon^A(\xi, t) \equiv \int_{\Phi} \varrho(\xi, \boldsymbol{\eta})\, \varepsilon(\xi, \boldsymbol{\eta}, t) f^A(\boldsymbol{\eta}) \mathrm{d}\eta^1 \mathrm{d}\eta^2,$$

$$L^A(\xi, t) \equiv \int_{\Phi} \mathrm{tr}[\mathbf{S}_t(\xi, \boldsymbol{\eta}) \dot{\mathbf{E}}_t(\xi, \boldsymbol{\eta})] f^A(\boldsymbol{\eta}) \mathrm{d}\eta^1 \mathrm{d}\eta^2,$$

for $\xi \in (0, l)$ and

$$\pi^A(t; 0) \equiv \int_{\Phi} \pi_t(0, \boldsymbol{\eta}) f^A(\boldsymbol{\eta}) \mathrm{d}\eta^1 \mathrm{d}\eta^2,$$

$$\pi^A(t; l) \equiv \int_{\Phi} \pi_t(l, \boldsymbol{\eta}) f^A(\boldsymbol{\eta}) \mathrm{d}\eta^1 \mathrm{d}\eta^2,$$

eq. (3.21) yields the condition

$$\int_0^l G^A(\xi, t) \zeta^*_{A.1}(\xi) \mathrm{d}\xi = \pi^A(t; 0) \zeta^*_A(0) + \pi^A(t; l) \zeta^*_A(l) +$$

$$+ \int_0^l [e^A(\xi, t) + g^A(\xi, t) + L^A(\xi, t) - \dot{\varepsilon}^A(\xi, t)] \zeta^*_A(\xi) \mathrm{d}\xi \tag{3.38}$$

which has to be satisfied by every $\boldsymbol{\zeta}^*(\cdot) \equiv (\zeta^*_1(\cdot), \ldots, \zeta^*_n(\cdot)) \in \mathscr{Z}^*_t$.

Eqs. (3.36) and (3.38) will be referred to as the *variational equations of motion* and the *variational heat conduction equation*, respectively, for the rod theories. If the region Ω coincides with the region $\boldsymbol{\chi}(\Omega, t_0)$, i.e. if $\boldsymbol{\chi}(\cdot, t_0)$ is the identity mapping, then the rod theory will be termed the beam theory. The constitutive relations will be obtained by an approach similar to that given in Sec. 3.2.2. In this way we arrive at

$$\begin{aligned}
H^a(\xi, t) &= \hat{H}^a(\xi, \nabla\boldsymbol{\chi}^*(\xi, t), \boldsymbol{\chi}^*(\xi, t), \nabla\dot{\boldsymbol{\chi}}^*(\xi, t), \dot{\boldsymbol{\chi}}^*(\xi, t), \theta^*(\xi, t)),\\
h^a(\xi, t) &= \hat{h}^a(\xi, \nabla\boldsymbol{\chi}^*(\xi, t), \boldsymbol{\chi}^*(\xi, t), \nabla\dot{\boldsymbol{\chi}}^*(\xi, t), \dot{\boldsymbol{\chi}}^*(\xi, t), \theta^*(\xi, t)),\\
f^a(\xi, t) &= \hat{f}^a(\xi, t, \nabla\boldsymbol{\chi}^*(\xi, t), \boldsymbol{\chi}^*(\xi, t), \dot{\boldsymbol{\chi}}^*(\xi, t)),\\
s^a(t; 0) &= \hat{s}^a(0; t, \nabla\boldsymbol{\chi}^*(0, t), \boldsymbol{\chi}^*(0, t), \dot{\boldsymbol{\chi}}^*(0, t)),\\
s^a(t; l) &= \hat{s}^a(l; t, \nabla\boldsymbol{\chi}^*(l, t), \boldsymbol{\chi}^*(l, t), \dot{\boldsymbol{\chi}}^*(l, t)),
\end{aligned} \tag{3.39}$$

and

$$
\begin{aligned}
G^A(\xi,t) &= \hat{G}^A(\xi, \nabla\theta^*(\xi,t), \theta^*(\xi,t), \nabla\boldsymbol{\chi}^*(\xi,t), \boldsymbol{\chi}^*(\xi,t)),\\
g^A(\xi,t) &= \hat{g}^A(\xi, \nabla\theta^*(\xi,t), \theta^*(\xi,t), \nabla\boldsymbol{\chi}^*(\xi,t), \boldsymbol{\chi}^*(\xi,t)),\\
e^A(\xi,t) &= \hat{e}^A(\xi, t, \theta^*(\xi,t), \boldsymbol{\chi}^*(\xi,t)),\\
\varepsilon^A(\xi,t) &= \hat{\varepsilon}^A(\xi, \theta^*(\xi,t), \nabla\boldsymbol{\chi}^*(\xi,t), \boldsymbol{\chi}^*(\xi,t)),\\
L^A(\xi,t) &= \hat{L}^A(\xi, \nabla\boldsymbol{\chi}^*(\xi,t), \boldsymbol{\chi}^*(\xi,t), \nabla\dot{\boldsymbol{\chi}}^*(\xi,t), \theta^*(\xi,t));\\
&\qquad\qquad\qquad\qquad \xi \in (0,l),\\
\pi^A(t;0) &= \hat{\pi}^A(0; t, \boldsymbol{\chi}^*(0,t), \theta^*(0,t)),\\
\pi^A(t;l) &= \hat{\pi}^A(l; t, \boldsymbol{\chi}^*(l,t), \theta^*(l,t)).
\end{aligned}
\tag{3.40}
$$

In every given special problem the RHS of eqs. (3.39) and (3.40) are known, cf. Sec. 3.2.2. Eqs. (3.39), (3.40) are the constitutive relations for thermo-visco-elastic materials and together with the conditions determined by eqs. (3.36), (3.38) constitute the governing relations of problems described within the framework of rod and beam theories.

The general discussion of the governing relations obtained is similar to that presented at the end of Sec. 3.2.2 and will be omitted here. It should be noted, however, that in the engineering theories of rods and beams we often postulate, apart from the constraints (3.35) introduced here, certain constraints for stresses by assuming that the stress components normal to the material planes which in Ω are parallel to the z_1-axis, are equal to zero.

The rod and beam theories are usually formulated under additional assumptions concerning stresses and heat fluxes. In some special cases we can introduce constraints for stresses setting $S_t^{\alpha\beta}(\mathbf{X}) = 0$ for every (α, β) different from $(1, 1)$ and for every $\mathbf{X} \in \Omega$, $t \in [t_0, t_f]$. Other specifications of the stress constraints can also be employed; the form of postulated constraints will depend on the character of the problem at hand. In the face of constraints for stresses and/or heat fluxes we should proceed as indicated in Sec. 2.3.5.

3.2.4 Plane problems

Let Ω^* be the known region on the coordinate plane $z_3 = 0$ and $\mathbf{X}^* = (X^1, X^2)$. Let also $X^3 \in (-\delta, \delta)$, where δ is a certain positive number and a projection $(\cdot)^*$ of Ω onto Ω^* be an orthogonal projection which assigns to every vector $\mathbf{X} = (X^1, X^2, X^3) \in \Omega$ its projection $\mathbf{X}^* = (X^1, X^2) \in \Omega^*$. Thus we start here from the situation similar to that described at the begining of Sec. 3.2.2 and shown in Fig. 10.

Let $\mathbf{e}_3$ be the vector axis in the Cartesian (spatial) coordinate system $Oz_1z_2z_3$. We shall now specify the constraints (3.18) by setting $m = 2$, $\mathbf{F}^1$

$= (1, 0, 0)$, $\mathbf{F}^2 = (0, 1, 0)$, $\boldsymbol{\chi}_0(\mathbf{X}) = X^3\mathbf{e}_3$. Thus

$$\boldsymbol{\chi}(\mathbf{X}, \tau) = \mathbf{F}^K\chi_K^*(\mathbf{X}^*, \tau) + X^3\mathbf{e}_3, \qquad K = 1, 2,$$

for some

$$\boldsymbol{\chi}^*(\cdot, \tau) = \big(\chi_1(\cdot, \tau), \chi_2(\cdot, \tau)\big) \in \mathscr{D}_\tau^*(\Omega^*), \qquad \tau \in [t_0, t_f].$$

Hence $\boldsymbol{\chi}(\mathbf{X}, t) = \boldsymbol{\chi}^*(\mathbf{X}^*, \tau) + X^3\mathbf{e}_3$, cf. Fig. 12.

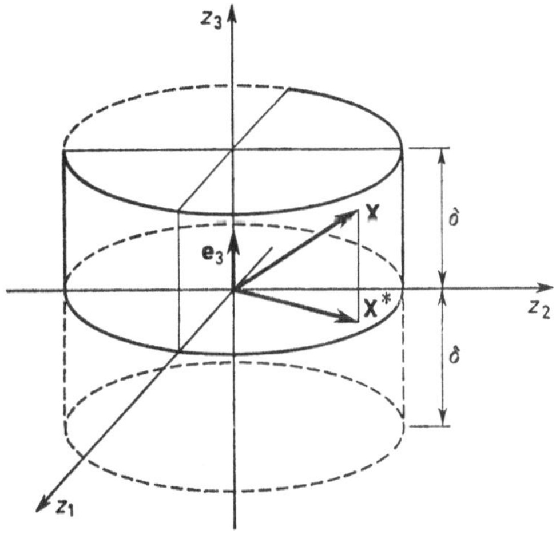

Fig. 12 Reference configuration for plane problem

Similarly, we shall specify the constraints (3.20) by setting $n = 1, f^1 = 1$, $\theta_0(\mathbf{X}) = 0$. Thus

$$\theta(\mathbf{X}, \tau) = \theta^*(\mathbf{X}^*, \tau)$$

for some $\theta^*(\cdot, \tau) \in \mathscr{T}_\tau^*(\Omega^*)$, $\tau \in [t_0, t_f]$.

The aforementioned constraints constitute the basis for the so-called plane problems of the deformation and temperature distributions. Such problems occur in practice when:

1. The coordinate plane $z_3 = 0$ can be treated as the plane of symmetry (both geometrical and physical) of the problem.

2. All material properties, external agents, boundary constraints (for $\mathbf{X}^* \in \partial\Omega^*$) and mass density can be treated as not dependent on the material coordinate X^3.

Taking into account the aforementioned constraints and conditions and using the approach described in Sec. 3.2.1 we obtain a special case of variational equations (3.19) and (3.21). Every $\mathbf{F}^{KL}(\mathbf{X}) = \mathbf{F}^K \otimes \nabla\pi^L(\mathbf{X})$, where $\pi^L(\mathbf{X}) = X^L$, $L = 1, 2$ (cf. Sec. 3.2.3) is now a 3×3 matrix with the components $\delta^{kK}\delta^{lL}$ (k, l run over $1, 2, 3$). Similarly $f^{1k}(\mathbf{X}) = f^1\nabla\pi^k(\mathbf{X}) = (\delta_1^k, \delta_2^k, 0)$.

For an arbitrary $\mathbf{f} = (f^1, f^2, f^3)$ and for an arbitrary 3×3 matrix $\mathbf{S}$ with the components $S^{\alpha\beta}$ (α, β run over the sequence $1, 2, 3$) we shall define

$$\mathbf{f}^* \equiv (f^1, f^2), \quad \mathbf{S}^* \equiv \begin{bmatrix} S^{11} & S^{12} \\ S^{21} & S^{22} \end{bmatrix}.$$

Under the assumptions introduced above we obtain from eq. (3.19) the following condition

$$\int_{\Omega^*} \operatorname{tr}[\mathbf{S}_t^*(\mathbf{X}^*)\nabla\boldsymbol{\chi}^{*T}(\mathbf{X}^*, t)\nabla\mathbf{v}^*(\mathbf{X}^*)]\,\mathrm{d}V(\mathbf{X}^*)$$

$$= \oint_{\partial\Omega^*} \mathbf{s}_t^*(\mathbf{X}^*)\cdot\mathbf{v}^*(\mathbf{X}^*)\,\mathrm{d}A(\mathbf{X}^*) +$$

$$+ \int_{\Omega^*} [\mathbf{b}_t^*(\mathbf{X}^*) - \varrho(\mathbf{X}^*)\ddot{\mathbf{u}}_t^*(\mathbf{X}^*)]\cdot\mathbf{v}^*(\mathbf{X}^*)\,\mathrm{d}V(\mathbf{X}^*),$$

which has to hold for every $\mathbf{v}^*(\cdot) = (v_1^*(\cdot), v_2^*(\cdot)) \in \mathscr{U}_t^*$ and where $\mathrm{d}V(\mathbf{X}^*) \equiv \mathrm{d}X^1\mathrm{d}X^2$ while $\mathrm{d}A(\mathbf{X}^*)$ is an element of $\partial\Omega^*$.

Similarly, eq. (3.21) implies the condition

$$\int_{\Omega^*} \mathbf{h}_t^*(\mathbf{X}^*)\cdot\nabla\boldsymbol{\zeta}^*(\mathbf{X}^*)\,\mathrm{d}V(\mathbf{X}^*)$$

$$= \oint_{\partial\Omega^*} \pi_t(\mathbf{X}^*)\boldsymbol{\zeta}^*(\mathbf{X}^*)\,\mathrm{d}A(\mathbf{X}^*) + \int_{\Omega^*} \alpha_t(\mathbf{X}^*)\boldsymbol{\zeta}^*(\mathbf{X}^*)\,\mathrm{d}V(\mathbf{X}^*) +$$

$$+ \int_{\Omega^*} \{\operatorname{tr}[\mathbf{S}_t^*(\mathbf{X}^*)\dot{\mathbf{E}}_t^*(\mathbf{X}^*)] - \varrho(\mathbf{X}^*)\dot{\varepsilon}(\mathbf{X}^*, t)\}\boldsymbol{\zeta}^*(\mathbf{X}^*)\,\mathrm{d}V(\mathbf{X}^*)$$

to hold for every $\boldsymbol{\zeta}^*(\cdot) \in \mathscr{Z}_t^*$, where the functions $\boldsymbol{\zeta}^*(\cdot)$, $\theta^*(\cdot, t)$ are defined on Ω^* and

$$\dot{\mathbf{E}}_t^*(\mathbf{X}^*) = \tfrac{1}{2}[\nabla\boldsymbol{\chi}^{*T}(\mathbf{X}^*, t)\nabla\dot{\mathbf{u}}_t^*(\mathbf{X}^*) + \nabla\dot{\mathbf{u}}^{*T}(\mathbf{X}^*)\nabla\boldsymbol{\chi}^*(\mathbf{X}^*, t)],$$

cf. eq. (3.3). It can be easily observed that the variational equations obtained have the form analogous to the form of eqs. (3.1) and (3.5), respectively. It means that all the results of Sec. 3.1.1 can also be applied to plane problems provided Ω is a certain region of the plane (hence $\mathbf{X} = (X^1, X^2)$), and all the vectors and tensors encountered in the pertinent formulae are represented by pairs or by 2×2 matrices of real numbers, respectively. Therefore, there is no need to separately develop an independent theory for plane problems.

If a length parameter δ in Fig. 12 is small then we often deal with the plane problem of a thin plate subjected to the action of external agents situated paralelly to the plane $z_3 = 0$. The stress components $S_t^{33}(\mathbf{X})$, $S_t^{31}(\mathbf{X})$, $S_t^{32}(\mathbf{X})$ and the heat flux component $h_t^3(\mathbf{X})$ are in such a case small as compared with the principal values of the matrix $\mathbf{S}_t(\mathbf{X})$ and the absolute value of the vector

$\mathbf{h}_t(\mathbf{X})$, respectively, provided that the boundaries $X^3 = \pm\delta$ of the plate are free of forces and heat fluxes. Mathematical modelling of such problems can be based on the a priori assumption that the problem is plane and that we also deal with constraints for stresses $S_t^{33}(\mathbf{X}) = 0$, $S_t^{31}(\mathbf{X}) = 0$, $S_t^{32}(\mathbf{X}) = 0$ and heat fluxes $h_t^3(\mathbf{X}) = 0$. In this case we can interpret all the non-vanishing stress and heat flux components as their values averaged through the thickness 2δ of the plate. Moreover, taking into account conditions (3.15) and (3.16) we immediately obtain that the reactions $L_t^{33}(\mathbf{X})$, $L_t^{31}(\mathbf{X})$, $L_t^{32}(\mathbf{X})$ and $l_t^3(\mathbf{X})$ are arbitrary real numbers and all other reactions (for stress and heat flux constraints) are equal to zero. The non-zero reactions can be computed from the constitutive relations, cf. Sec. 2.3.4 only after the solution to the problem has been obtained, cf. Sec. 3.1.4.

3.3 Method of constraints: incremental formulation

3.3.1 General equations

The basic notions of the incremental variational formulation were introduced in Sec. 3.1.2. Setting in eq. (3.19)

$$\mathbf{T}_t(\mathbf{X}) \equiv \mathbf{S}_t(\mathbf{X})\nabla\boldsymbol{\chi}^T(\mathbf{X}, t),$$

we obtain

$$\begin{aligned}\int_\Omega \{\mathrm{tr}[\Delta\mathbf{T}(\mathbf{X})\nabla\mathbf{F}^a(\mathbf{X})]v_a^*(\mathbf{X}^*) + \mathrm{tr}[\Delta\mathbf{T}(\mathbf{X})\mathbf{F}^{ak}(\mathbf{X})]v_{a,k}^*(\mathbf{X}^*)\}\mathrm{d}V(\mathbf{X}) \\ = \oint_{\partial\Omega} \Delta\mathbf{s}(\mathbf{X})\cdot\mathbf{F}^a(\mathbf{X})v_a^*(\mathbf{X}^*)\mathrm{d}A(\mathbf{X}) + \\ + \int_\Omega [\mathbf{b}(\mathbf{X}) - \varrho(\mathbf{X})\Delta\ddot{\mathbf{u}}(\mathbf{X})]\cdot\mathbf{F}^a(\mathbf{X})v_a^*(\mathbf{X}^*)\mathrm{d}V(\mathbf{X})\end{aligned} \tag{3.41}$$

for every $\mathbf{v}^*(\cdot) \in \mathscr{U}_t^* = \mathscr{U}_t^*(\boldsymbol{\chi}(\cdot, t))$, where $\boldsymbol{\chi}^*(\cdot, t) + \Delta\boldsymbol{\chi}^*(\cdot) \in \mathscr{D}_t^*(\Omega^*)$. Similarly, starting from eq. (3.21) we arrive at the condition

$$\begin{aligned}\int_\Omega [\Delta\mathbf{h}(\mathbf{X})\cdot\nabla f^A(\mathbf{X})\zeta_A^*(\mathbf{X}^*) + \Delta\mathbf{h}(\mathbf{X})\cdot f^{AK}(\mathbf{X})\zeta_{A,K}^*(\mathbf{X}^*)]\mathrm{d}V(\mathbf{X}) \\ = \oint_{\partial\Omega} \Delta\pi(\mathbf{X})f^A(\mathbf{X})\zeta_A^*(\mathbf{X}^*)\mathrm{d}A(\mathbf{X}) + \int_\Omega \Delta\alpha(\mathbf{X})f^A(\mathbf{X})\zeta_A^*(\mathbf{X}^*)\mathrm{d}V(\mathbf{X}) + \\ + \int_\Omega \{\mathrm{tr}[\Delta\mathbf{S}(\mathbf{X})\dot{\mathbf{E}}_t(\mathbf{X}) + \mathbf{S}_t(\mathbf{X})\Delta\dot{\mathbf{E}}(\mathbf{X}) + \Delta\mathbf{S}(\mathbf{X})\Delta\dot{\mathbf{E}}(\mathbf{X})] - \\ - \varrho(\mathbf{X})\Delta\dot{\varepsilon}(\mathbf{X})\}f^A(\mathbf{X})\zeta_A^*(\mathbf{X}^*)\mathrm{d}V(\mathbf{X}),\end{aligned} \tag{3.42}$$

which has to hold for every $\boldsymbol{\zeta}^*(\cdot) \in \mathscr{Z}_t^* = \mathscr{Z}_t^*(\theta(\cdot, t))$, where $\theta^*(\cdot, t) + \Delta\theta^*(\cdot, t) \in \mathscr{T}_t(\Omega^*)$. Eqs. (3.41), (3.42) represent the incremental formulation of variational equations of thermomechanics. From eqs. (2.6)–(2.9) we obtain, cf. also eqs. (3.12)

$$
\begin{aligned}
&\Delta\mathbf{b}(\mathbf{X}) = \Delta\hat{\mathbf{b}}(\mathbf{X}, \Delta\mathbf{u}(\mathbf{X}), \nabla\Delta\mathbf{u}(\mathbf{X})),\\
&\Delta\mathbf{s}(\mathbf{X}) = \Delta\tilde{\mathbf{s}}(\mathbf{X})+\Delta\hat{\mathbf{s}}(\mathbf{X}, \Delta\mathbf{u}(\mathbf{X}), \nabla\Delta\mathbf{u}(\mathbf{X}))+\frac{\Delta\mathbf{G}(\mathbf{X}, \Delta\mathbf{u}(\mathbf{X}))}{\Delta t}\Delta\mathbf{u}(\mathbf{X})+\\
&\qquad +\mathbf{G}(\mathbf{X}, t)\Delta\dot{\mathbf{u}}(\mathbf{X})+\Delta\mathbf{G}(\mathbf{X}, \Delta\mathbf{u}(\mathbf{X}))\Delta\dot{\mathbf{u}}(\mathbf{X}), \qquad (3.43)\\
&\Delta\alpha(\mathbf{X}) = \Delta\hat{\alpha}(\mathbf{X}, \Delta\theta(\mathbf{X})),\\
&\Delta\pi(\mathbf{X}) = \Delta\tilde{\pi}(\mathbf{X})+\Delta\hat{\pi}(\mathbf{X}, \Delta\mathbf{u}(\mathbf{X}), \Delta\theta(\mathbf{X})).
\end{aligned}
$$

Similarly, eqs. (2.1) yield

$$
\begin{aligned}
&\Delta\mathbf{S}(\mathbf{X}) = \Delta\hat{\mathbf{S}}(\mathbf{X}, \Delta\mathbf{E}(\mathbf{X}), \Delta\theta(\mathbf{X}))+\Delta\mathbf{H}(\mathbf{X}, \Delta\mathbf{E}(\mathbf{X}), \Delta\theta(\mathbf{X}))\frac{\Delta\mathbf{E}(\mathbf{X})}{\Delta t}+\\
&\qquad +\mathbf{H}(\mathbf{X}, 0, \theta(\mathbf{X}, t))\Delta\dot{\mathbf{E}}(\mathbf{X})+\Delta\mathbf{H}(\mathbf{X}, \Delta\mathbf{E}(\mathbf{X}), \Delta\theta(\mathbf{X}))\Delta\dot{\mathbf{E}}(\mathbf{X}), \qquad (3.44)\\
&\Delta\mathbf{h}(\mathbf{X}) = \Delta\mathbf{K}(\mathbf{X}, \Delta\mathbf{E}(\mathbf{X}), \Delta\theta(\mathbf{X}))\nabla\theta(\mathbf{X})+\mathbf{K}(\mathbf{X}, 0, \theta(\mathbf{X}, t))\Delta\nabla\theta(\mathbf{X})+\\
&\qquad +\Delta\mathbf{K}(\mathbf{X}, \Delta\mathbf{E}(\mathbf{X}), \Delta\theta(\mathbf{X}))\Delta\nabla\theta(\mathbf{X}),\\
&\Delta\varphi(\mathbf{X}) = \Delta\hat{\varepsilon}(\mathbf{X}, \Delta\mathbf{E}(\mathbf{X}), \Delta\theta(\mathbf{X})),
\end{aligned}
$$

and $\Delta\mathrm{E}(\mathbf{X})$ is given by eq. (3.11). The above equations are valid under the assumption that $\boldsymbol{\chi}(\cdot, t)$ is the identity mapping, cf. Fig. 8 in Sec. 3.1.2.

Substituting the RHS of eqs. (3.43), (3.44) into eqs. (3.41), (3.42) and taking into account the incremental form of the constraints (3.18), (3.20) given by

$$
\Delta\mathbf{u}(\mathbf{X}) = \mathbf{F}^a(\mathbf{X})\Delta\chi^*_a(\mathbf{X}^*), \qquad \Delta\theta(\mathbf{X}) = f^A(\mathbf{X})\Delta\theta^*_A(\mathbf{X}^*), \qquad \mathbf{X}^* \in \Omega^*,
$$

we arrive at the general of the incremental formulation for problems in structural mechanics of thermo-visco-elastic materials. For elastic-plastic materials we have to consider instead of eqs. (3.44) equations (3.9).

3.3.2 Shells and rods

The governing equations of thermomechanics of shells and plates are given by the two variational equations (3.24), (3.27) and by eqs. (3.29), (3.30). The incremental form of the aforementioned variational equations is given by

$$
\begin{aligned}
\int_{\Omega^*} \Delta H^{aK}(\mathbf{X}^*)v^*_{a,K}(\mathbf{X}^*)\,\mathrm{d}V(\mathbf{X}^*) &= \oint_{\partial\Omega^*} \Delta s^a(\mathbf{X}^*)v^*_a(\mathbf{X}^*)\,\mathrm{d}A(\mathbf{X}^*)+\\
&+\int_{\Omega^*} [\Delta f^a(\mathbf{X}^*)+\Delta h^a(\mathbf{X}^*)-M^{ab}(\mathbf{X}^*)\Delta\ddot{\chi}^*_b(\mathbf{X}^*)]v^*_a(\mathbf{X}^*)\,\mathrm{d}V(\mathbf{X}^*), \qquad (3.45)
\end{aligned}
$$

$$\int_{\Omega^*} \Delta G^{AK}(\mathbf{X}^*)\zeta^*_{A,K}(\mathbf{X}^*)\mathrm{d}V(\mathbf{X}^*) = \oint_{\partial\Omega^*} \Delta\pi^A(\mathbf{X}^*)\zeta^*_A(\mathbf{X}^*)\mathrm{d}A(\mathbf{X}^*)+$$

$$+\int_{\Omega^*} [\Delta e^A(\mathbf{X}^*)+\Delta g^A(\mathbf{X}^*)+\Delta L^A(\mathbf{X}^*)-\Delta\dot{\varepsilon}^A(\mathbf{X}^*)]\zeta^*_A(\mathbf{X}^*)\mathrm{d}V(\mathbf{X}^*),$$

and has to hold for every $\mathbf{v}^*(\cdot) = (v^*_1(\cdot), \dots, v^*_m(\cdot)) \in \mathscr{U}^*_t$ and $\boldsymbol{\zeta}^*(\cdot) = (\zeta^*_1(\cdot), \dots, \zeta^*_n(\cdot)) \in \mathscr{Z}^*_t$. The increments in eqs. (3.45) have to be expressed in terms of the increments $\Delta\mathbf{u}(\mathbf{X}^*)$, $\Delta\theta(\mathbf{X}^*)$ by means of eqs. (3.29), (3.30).

Similarly, the variational equations of thermomechanics of rods and beams are represented by eqs. (3.36), (3.38). The incremental form of these equations is given by

$$\int_0^l \Delta H^a(\xi)v^*_{a,1}(\xi)\mathrm{d}\xi = \Delta s^a(0)v^*_a(0)+\Delta s^a(l)v^*_a(l)+$$

$$+\int_0^l [\Delta f^a(\xi)+\Delta h^a(\xi)-M^{ab}(\xi)\Delta\ddot{\chi}^*_b(\xi)]v^*_a(\xi)\mathrm{d}\xi,$$

$$\int_0^l \Delta G^A(\xi)\zeta^*_{A,1}(\xi)\mathrm{d}\xi = \Delta\pi^A(0)\zeta^*_A(0)+\Delta\pi^A(l)\zeta^*_A(l)+ \tag{3.46}$$

$$+\int_0^l [\Delta e^A(\xi)+\Delta g^A(\xi)+\Delta L^A(\xi)-\Delta\dot{\varepsilon}^A(\xi)]\zeta^*_A(\xi)\mathrm{d}\xi$$

and has to be satisfied by every $\mathbf{v}^*(\cdot) = (v^*_1(\cdot), \dots, v^*_m(\cdot)) \in \mathscr{U}^*_t$ and every $\boldsymbol{\zeta}^*(\cdot) = (\zeta^*_1(\cdot), \dots, \zeta^*_n(\cdot)) \in \mathscr{Z}^*_t$. The increments in eqs. (3.46) have to be calculated from eqs. (3.39), (3.40).

3.3.3 Plane problems

The constraints and the additional assumptions leading to plane problems have been discussed in Sec. 3.2.4. It has been shown that from a purely formal point of view the governing equations for plane problems coincide with the governing relations of the finite variational formulation given in Sec. 3.1.1. Thus, it follows that from the formal point of view the incremental variational formulation of plane problems is also identical with the general form of the variational formulation discussed in Sec. 3.1.2. It must be remembered that both the finite and incremental forms of plane problems are physically valid only if the conditions of Sec. 3.2.4 hold and the pertinent constraints for deformation and temperature distributions have been postulated.

The incremental formulation of plane problems can be applied to the problem of thin plates subjected to external agents acting parallely to the plane

$z_3 = 0$. To describe such problems we have introduced in Sec. 3.2.4 the additional constraints for certain stress and heat flux components. In this case we deal with increments of the reactions not only for constraints imposed on deformation and temperature distributions but also for constraints imposed on stresses and heat fluxes.

3.4 Method of constraints: discrete description of elements and structures

3.4.1 Discretization concept

The concept of discretization is quite a natural one to every structural engineer since it is closely related to building up the structure by fitting together a number of separately treated structural elements. The main purpose of this section is the translation of this fundamental idea into the precise language of solid and structural mechanics. We shall start the considerations with the discrete description of an arbitrary single structural element. To this aim we shall use again the method of constraints. On this basis we shall show how to construct a discretized model of the whole structure by dividing it into special structural elements joined together by means of certain nodal points. This approach will bring us later to the so-called finite element formulation of structural mechanics problems.

3.4.2 Finite formulation

In order to construct a discretized model of the structural mechanics problem we have first to divide the region Ω into a finite number of disjoined subregions called finite elements. In this section we shall confine ourselves to one, arbitrary but fixed, solid finite element Ω_e. Using the approach introduced at the beginning of Sec. 1.1.2, we shall assume that the region Ω_e represents a reference configuration of a certain part of the structure; this part will be subsequently referred to as the structural element. It can be seen that the variational form of the governing relations (3.1) can also be applied to the part Ω_e provided $\mathbf{s}_t(\mathbf{X})$, $\mathbf{X} \in \partial\Omega_e$, are interpreted as the boundary tractions acting on the structural element considered and $\boldsymbol{\chi}(\cdot, \tau) \in \mathscr{D}(\Omega_e)$, $\tau \in [t_0, t_f]$, stands for a motion of this element. Hence

$$\int_{\Omega_e} \mathrm{tr}[\mathbf{S}_t(\mathbf{X})\nabla\boldsymbol{\chi}^T(\mathbf{X}, t)\nabla\mathbf{u}(\mathbf{X})]dV(\mathbf{X})$$

$$= \oint_{\partial\Omega_e} \mathbf{s}_t(\mathbf{X}) \cdot \mathbf{u}(\mathbf{X})dA(\mathbf{X}) + \int_{\Omega_e} [\mathbf{b}_t(\mathbf{X}) - \varrho(\mathbf{X})\ddot{\mathbf{u}}_t(\mathbf{X})] \cdot \mathbf{u}(\mathbf{X})dV(\mathbf{X}) \quad (3.47)$$

has to hold for every $\mathbf{u}(\cdot) \in \mathscr{U}_t^e$, where now

$$\mathscr{U}_t^e := \{\mathbf{u}(\cdot) \in C_1^3(\bar{\Omega}_e) \mid \mathbf{u}(\cdot) = \lambda[\mathbf{v}_1(\cdot) - \mathbf{v}_2(\cdot)], \mathbf{v}_1(\cdot), \mathbf{v}_2(\cdot) \in \mathscr{V}_t^e, \lambda \geqslant 0\}$$

and where $\mathscr{V}_t^e = \mathscr{V}_t^e(\boldsymbol{\chi}(\cdot, t))$ stands for a set of all velocity fields admissible by constraints. Simultaneously, eqs. (3.2), (3.4) are assumed to hold in the unchanged form for $\mathbf{X} \in \Omega_e$. Observe that the constraints for the deformation are imposed exclusively upon the selected structural element, i.e., they have the form $\boldsymbol{\chi}(\cdot, \tau) \in \mathscr{D}_\tau(\Omega_e)$, $\tau \in [t_0, t_f]$, while $\mathscr{U}_t^e = \mathscr{U}_t^e(\boldsymbol{\chi}(\cdot, t))$.

Assume now that $\boldsymbol{\chi}(\cdot, \tau) \in \mathscr{D}(\Omega_e)$ if $\boldsymbol{\chi}(\cdot, \tau)$ is of the form

$$\boldsymbol{\chi}(\mathbf{X}, \tau) = \mathbf{F}^a(\mathbf{X}) q_a(\tau) + \boldsymbol{\chi}_0(\mathbf{X}), \quad \mathbf{X} \in \bar{\Omega}_e, \tau \in [t_0, t_f], \tag{3.48}$$

where $\mathbf{F}^a(\mathbf{X})$, $a = 1, \ldots, m$, $\boldsymbol{\chi}_0(\mathbf{X})$, $\mathbf{X} \in \bar{\Omega}_e$, are known vectors and $\mathbf{q}(\tau) \equiv (q_1(\tau), \ldots, q_m(\tau)) \in \boldsymbol{R}^m$, $\tau \in [t_0, t_f]$, is an arbitrary vector function which is assumed to be continuous and to have the continuous first and second order derivatives. Then, $\mathbf{v}(\cdot) \in \mathscr{V}_t^e$ if and only if

$$\mathbf{v}(\mathbf{X}) = \mathbf{F}^a(\mathbf{X}) v_a, \quad \mathbf{X} \in \bar{\Omega}_e,$$

for every $\mathbf{v} = (v_1, \ldots, v_m) \in \boldsymbol{R}^m$. It follows that $\mathbf{u}(\cdot) \in \mathscr{U}_t^e$ if and only if

$$\mathbf{u}(\mathbf{X}) = \mathbf{F}^a(\mathbf{X}) u_a, \quad \mathbf{X} \in \bar{\Omega}^e,$$

for an arbitrary $\mathbf{u} \equiv (u_1, \ldots, u_m) \in \boldsymbol{R}^m$. Substituting the RHS of eqs. (3.48) into eq. (3.47) we obtain the condition

$$\int_{\Omega_e} \operatorname{tr}[\mathbf{S}_t(\mathbf{X}) \nabla \boldsymbol{\chi}^T(\mathbf{X}, t) \nabla \mathbf{F}^a(\mathbf{X})] \, dV(\mathbf{X}) u_a$$

$$= \oint_{\partial \Omega_e} \mathbf{s}_t(\mathbf{X}) \cdot \mathbf{F}^a(\mathbf{X}) \, dA(\mathbf{X}) u_a + \int_{\Omega_e} [\mathbf{b}_t(\mathbf{X}) - \varrho(\mathbf{X}) \ddot{\mathbf{u}}_t(\mathbf{X})] \cdot \mathbf{F}^a(\mathbf{X}) \, dV(\mathbf{X}) u_a$$

which has to be satisfied for every $\mathbf{u} = (u_1, \ldots, u_n) \in \boldsymbol{R}^n$. Hence, we conclude that

$$\int_{\Omega_e} \operatorname{tr}[\mathbf{S}_t(\mathbf{X}) \nabla \boldsymbol{\chi}^T(\mathbf{X}, t) \nabla \mathbf{F}^a(\mathbf{X})] \, dV(\mathbf{X})$$

$$= \oint_{\partial \Omega_e} \mathbf{s}_t(\mathbf{X}) \cdot \mathbf{F}^a(\mathbf{X}) \, dA(\mathbf{X}) + \int_{\Omega_e} [\mathbf{b}_t(\mathbf{X}) - \varrho(\mathbf{X}) \ddot{\mathbf{u}}_t(\mathbf{X})] \cdot \mathbf{F}^a(\mathbf{X}) \, dV(\mathbf{X}) \tag{3.49}$$

hold for $a = 1, \ldots, m$; it has to be remembered that by virtue of eq. (3.48) we have

$$\nabla \boldsymbol{\chi}(\mathbf{X}, t) = \nabla \mathbf{F}^a(\mathbf{X}) q_a(t),$$

$$\ddot{\mathbf{u}}_t(\mathbf{X}) = \ddot{\boldsymbol{\chi}}(\mathbf{X}, t) = \mathbf{F}^a(\mathbf{X}) \ddot{q}_a(t).$$

Note that now the sets $\mathscr{V}_t^e$ and $\mathscr{U}_t^e$ are independent of $\boldsymbol{\chi}(\cdot, t)$ and t.

A similar line of approach may be employed for the analysis of the heat conduction problem. To this aim we shall take into account eq. (3.5) and assume that

$$\int_{\Omega_e} \mathbf{h}_t(\mathbf{X}) \cdot \nabla \boldsymbol{\zeta}(\mathbf{X}) \, \mathrm{d}V(\mathbf{X})$$

$$= \oint_{\partial \Omega_e} \pi_t(\mathbf{X}) \boldsymbol{\zeta}(\mathbf{X}) \, \mathrm{d}A(\mathbf{X}) + \int_{\Omega_e} \alpha_t(\mathbf{X}) \boldsymbol{\zeta}(\mathbf{X}) \, \mathrm{d}V(\mathbf{X}) +$$

$$+ \int_{\Omega_e} \{\operatorname{tr}[\mathbf{S}_t(\mathbf{X}) \dot{\mathbf{E}}_t(\mathbf{X})] - \varrho(\mathbf{X}) \dot{\varepsilon}(\mathbf{X}, t)\} \boldsymbol{\zeta}(\mathbf{X}) \, \mathrm{d}V(\mathbf{X}), \quad (3.50)$$

holds for every $\boldsymbol{\zeta}(\cdot) \in \mathscr{Z}_t^e$, where

$$\mathscr{Z}_t^e := \{\boldsymbol{\zeta}(\cdot) \in C_1(\bar{\Omega}_e) | \ \boldsymbol{\zeta}(\cdot) = \lambda[w_1(\cdot) - w_2(\cdot)]; \ w_1(\cdot), w_2(\cdot) \in \mathscr{W}_t^e, \ \lambda \geqslant 0\},$$

and where $\mathscr{W}_t^e = \mathscr{W}_t^e(\theta(\cdot, t))$ is a set of all the temperature rate fields in the structural element which are admissible by the constraints $\theta(\cdot, \tau) \in \mathscr{T}_\tau(\bar{\Omega}_e)$, $\tau \in [t_0, t_f]$. Hence also $\mathscr{Z}_t^e = \mathscr{Z}_t^e(\theta(\cdot, t))$ for $\theta(\cdot, t) \in \mathscr{T}_t(\bar{\Omega}_e)$. It must be stressed that $\pi_t(\mathbf{X})$, $\mathbf{X} \in \partial \Omega_e$, is now a heat supply to the structural element across its boundary. Eqs. (3.6), (3.7) are simultaneously assumed to hold for every $\mathbf{X} \in \partial \Omega_e$.

Assume now that $\theta(\cdot, t) \in \mathscr{T}_t(\bar{\Omega}_e)$ if $\theta(\cdot, t)$ is of the form

$$\theta(\mathbf{X}, t) = f^A(\mathbf{X}) p_A(t) + \theta_0, \quad \mathbf{X} \in \bar{\Omega}_e, \ t \in [t_0, t_f], \quad (3.51)$$

where $f^A(\mathbf{X})$, $A = 1, \ldots, n$, θ_0 are known scalars and $\mathbf{p}(t) = (p_1(t), \ldots, p_n(t)) \in \boldsymbol{R}^n$ is an arbitrary vector function having continuous first derivatives. Then $w(\cdot) \in \mathscr{W}_t^e$ if and only if

$$\mathbf{w}(\mathbf{X}) = f^A(\mathbf{X}) w_A, \quad \mathbf{X} \in \bar{\Omega}_e,$$

for every $\mathbf{w} \equiv (w_1, \ldots, w_n) \in \boldsymbol{R}^n$ and hence $\boldsymbol{\zeta}(\cdot) \in \mathscr{Z}_t^e$ if and only if

$$\boldsymbol{\zeta}(\mathbf{X}) = f^A(\mathbf{X}) \zeta_A, \quad \mathbf{X} \in \bar{\Omega}_e. \quad (3.52)$$

Note that now the sets $\mathscr{W}_t^e$ and $\mathscr{Z}_t^e$ are independent of $\theta(\cdot, t)$ and t. Substituting the RHS of eqs. (3.52) into eq. (3.50) we obtain

$$\int_{\Omega_e} \mathbf{h}_t(\mathbf{X}) \cdot \nabla f^A(\mathbf{X}) \, \mathrm{d}V(\mathbf{X}) \zeta_A$$

$$= \Big[\oint_{\partial \Omega_e} \pi_t(\mathbf{X}) f^A(\mathbf{X}) \, \mathrm{d}A(\mathbf{X}) + \int_{\Omega_e} \alpha_t(\mathbf{X}) f^A(\mathbf{X}) \, \mathrm{d}V(\mathbf{X}) +$$

$$+ \int_{\Omega_e} \{\operatorname{tr}[\mathbf{S}_t(\mathbf{X}) \dot{\mathbf{E}}_t(\mathbf{X})] - \varrho(\mathbf{X}) \dot{\varepsilon}(\mathbf{X}, t)\} f^A(\mathbf{X}) \, \mathrm{d}V(\mathbf{X}) \Big] \zeta_A$$

for every $\boldsymbol{\zeta} \equiv (\zeta_1, \ldots, \zeta_n) \in R^n$. Hence the conditions

$$\int_{\Omega_e} \mathbf{h}_t(\mathbf{X}) \cdot \nabla f^A(\mathbf{X}) \mathrm{d}V(\mathbf{X})$$
$$= \oint_{\partial\Omega_e} \pi_t(\mathbf{X}) f^A(\mathbf{X}) \mathrm{d}A(\mathbf{X}) + \int_{\Omega_e} \alpha_t(\mathbf{X}) f^A(\mathbf{X}) \mathrm{d}V(\mathbf{X}) +$$
$$+ \int_{\Omega_e} \{\mathrm{tr}[\mathbf{S}_t(\mathbf{X})\dot{\mathbf{E}}_t(\mathbf{X})] - \varrho(\mathbf{X})\dot{\varepsilon}(\mathbf{X}, t)\} f^A(\mathbf{X}) \mathrm{d}V(\mathbf{X}) \qquad (3.53)$$

hold for $A = 1, \ldots, n$.

Eqs. (3.49), (3.53) represent what is called the *discrete description of a structural element* since the basic unknowns are the functions $\mathbf{q}(\tau) \equiv (q_1(\tau), \ldots, q_m(\tau))$, $\mathbf{p}(\tau) = (p_1(\tau), \ldots, p_n(\tau))$, $\tau \in [t_0, t_1]$, which are independent of the material coordinates $\mathbf{X} = (X^\alpha)$. In order to present eqs. (3.49) in a more convenient form we shall define the following quantities

$$h^a(t) \equiv -\int_{\Omega_e} \mathrm{tr}[\mathbf{S}_t(\mathbf{X})\nabla\boldsymbol{\chi}^T(\mathbf{X}, t)\nabla\mathbf{F}^a(\mathbf{X})] \mathrm{d}V(\mathbf{X}),$$
$$f^a(t) \equiv \oint_{\partial\Omega_e} \mathbf{s}_t(\mathbf{X}) \cdot \mathbf{F}^a(\mathbf{X}) \mathrm{d}A(\mathbf{X}) + \int_{\Omega_e} \mathbf{b}_t(\mathbf{X}) \cdot \mathbf{F}^a(\mathbf{X}) \mathrm{d}V(\mathbf{X}), \qquad (3.54)$$
$$M^{ab} \equiv \int_{\Omega_e} \varrho(\mathbf{X}) \mathbf{F}^a(\mathbf{X}) \cdot \mathbf{F}^b(\mathbf{X}) \mathrm{d}V(\mathbf{X}).$$

By virtue of the relation

$$\ddot{\mathbf{u}}_t(\mathbf{X}) = \ddot{\boldsymbol{\chi}}(\mathbf{X}, t) = \mathbf{F}^b(\mathbf{X})\ddot{q}_b(t),$$

we then obtain

$$h^a(t) + f^a(t) = M^{ab}\ddot{q}_b(t). \qquad (3.55)$$

Quite similarly, by setting

$$g^A(t) \equiv -\int_{\Omega_e} \mathbf{h}_t(\mathbf{X}) \cdot \nabla f^A(\mathbf{X}) \mathrm{d}V(\mathbf{X}),$$
$$e^A(t) \equiv \oint_{\partial\Omega_e} \pi_t(\mathbf{X}) f^A(\mathbf{X}) \mathrm{d}A(\mathbf{X}) + \int_{\Omega_e} \alpha_t(\mathbf{X}) f^A(\mathbf{X}) \mathrm{d}V(\mathbf{X}),$$
$$\varepsilon^A(t) \equiv \int_{\Omega_e} \varrho(\mathbf{X})\, \varepsilon(\mathbf{X}, t) f^A(\mathbf{X}) \mathrm{d}V(\mathbf{X}), \qquad (3.56)$$
$$L^{Aab}(t) \equiv -\int_{\Omega_e} \mathrm{tr}[\mathbf{S}_t(\mathbf{X})\nabla\mathbf{F}^a(\mathbf{X})^T\nabla\mathbf{F}^b(\mathbf{X})] f^A(\mathbf{X}) \mathrm{d}V(\mathbf{X}),$$

and remembering that

$$\dot{\mathbf{E}}_t(\mathbf{X}) = \tfrac{1}{2}[\nabla\boldsymbol{\chi}^T(\mathbf{X}, t)\nabla\dot{\boldsymbol{\chi}}(\mathbf{X}, t) + \nabla\dot{\boldsymbol{\chi}}^T(\mathbf{X}, t)\nabla\boldsymbol{\chi}(\mathbf{X}, t)]$$
$$= \nabla\mathbf{F}^a(\mathbf{X})^T\nabla\mathbf{F}^b(\mathbf{X}) q_a(t)\dot{q}_b(t),$$

we arrive at

$$g^A(t)+e^A(t)+L^{Aab}(t)q_a(t)\dot{q}_b(t) = \dot{\varepsilon}^A(t), \qquad A = 1, \ldots, n. \tag{3.57}$$

Eqs. (3.55), (3.57) represent the discrete description of motion and heat conduction, respectively, for a single structural element. Assuming that the material of the element is thermo-visco-elastic, substituting the RHS of eqs. (3.2) and (3.4) into eqs. (3.54), remembering that

$$\begin{aligned}
\mathbf{u}_t(\mathbf{X}) &= \boldsymbol{\chi}(\mathbf{X}, t)-\boldsymbol{\chi}(\mathbf{X}, t_0) = \mathbf{F}^a(\mathbf{X})q_a(t)-\boldsymbol{\chi}(\mathbf{X}, t_0),\\
\mathbf{E}_t(\mathbf{X}) &= \tfrac{1}{2}[\nabla\boldsymbol{\chi}^T(\mathbf{X}, t)\nabla\boldsymbol{\chi}(\mathbf{X}, t)-\nabla\boldsymbol{\chi}^T(\mathbf{X}, t_0)\nabla\boldsymbol{\chi}(\mathbf{X}, t_0)]\\
&= \tfrac{1}{2}[\nabla\mathbf{F}^a(\mathbf{X})^T\nabla\mathbf{F}^b(\mathbf{X})q_a(t)q_b(t)-\nabla\boldsymbol{\chi}^T(\mathbf{X}, t_0)\nabla\boldsymbol{\chi}(\mathbf{X}, t_0)],\\
\theta(\mathbf{X}, t) &= \theta_0+\vartheta_t(\mathbf{X}) = \theta_0+f^A(\mathbf{X})p_A(t),
\end{aligned} \tag{3.58}$$

and introducing the useful notation[1]

$$\hat{h}^a(\mathbf{q}(t), \mathbf{p}(t)) \equiv -\int_{\Omega_e} \operatorname{tr}[\hat{\mathbf{S}}_t(\mathbf{X})\nabla\boldsymbol{\chi}^T(\mathbf{X}, t)\nabla\mathbf{F}^a(\mathbf{X})]\,\mathrm{d}V(\mathbf{X}),$$

$$H^{ab}(\mathbf{q}(t), \mathbf{p}(t))$$

$$\equiv \int_{\Omega_e} [\operatorname{tr}\mathbf{H}(\mathbf{X}, \mathbf{E}_t(\mathbf{X}), \vartheta_t(\mathbf{X}))\nabla\mathbf{F}^b(\mathbf{X})^T\nabla\mathbf{F}^c(\mathbf{X})\nabla\boldsymbol{\chi}^T(\mathbf{X}, t)\nabla\mathbf{F}^a(\mathbf{X})]\,\mathrm{d}V(\mathbf{X})q_c(t),$$

$$\tilde{f}^a(t) \equiv \oint_{\partial\Omega_e} \tilde{\mathbf{s}}(\mathbf{X}, t)\cdot\mathbf{F}^a(\mathbf{X})\,\mathrm{d}A(\mathbf{X}),$$

$$\begin{aligned}
\hat{f}^a(t, \mathbf{q}(t)) \equiv &\oint_{\partial\Omega_e} \hat{\mathbf{s}}(\mathbf{X}, t, \boldsymbol{\chi}(\mathbf{X}, t), \nabla\boldsymbol{\chi}(\mathbf{X}, t))\cdot\mathbf{F}^a(\mathbf{X})\,\mathrm{d}A(\mathbf{X})+\\
&+\int_{\Omega_e} \hat{\mathbf{b}}(\mathbf{X}, \boldsymbol{\chi}(\mathbf{X}, t), \nabla\boldsymbol{\chi}(\mathbf{X}, t))\cdot\mathbf{F}^a(\mathbf{X})\,\mathrm{d}V(\mathbf{X}),
\end{aligned}$$

$$G^{ab}(t, \mathbf{q}(t)) \equiv \oint_{\partial\Omega_e} [G(\mathbf{X}, t, \boldsymbol{\chi}(\mathbf{X}, t))\mathbf{F}^b(\mathbf{X})]\cdot\mathbf{F}^a(\mathbf{X})\,\mathrm{d}A(\mathbf{X}),$$

we arrive at

$$\begin{aligned}
h^a(t) &= \hat{h}^a(\mathbf{q}(t), \mathbf{p}(t))+H^{ab}(\mathbf{q}(t), \mathbf{p}(t))\dot{q}_b(t),\\
f^a(t) &= \tilde{f}^a(t)+\hat{f}^a(t, \mathbf{q}(t))+G^{ab}(t, \mathbf{q}(t))\dot{q}_b(t).
\end{aligned} \tag{3.59}$$

Similarly, by substituting the RHS of eqs. (3.6), (3.7), (3.2) into the definitions (3.56), accounting for eqs. (3.58) and denoting (cf. the previous footnote)

$$\hat{g}^A(t, \mathbf{q}(t), \mathbf{p}(t)) \equiv -\int_{\Omega_e} [\mathbf{K}(\mathbf{X}, \mathbf{E}_t(\mathbf{X}), \vartheta_t(\mathbf{X}))\nabla\vartheta_t(\mathbf{X})]\cdot\nabla f^A(\mathbf{X})\,\mathrm{d}V(\mathbf{X}),$$

---

[1] It is here assumed that $\nabla\boldsymbol{\chi}(\mathbf{X}, t)$, $\mathbf{E}_t(\mathbf{X})$, $\vartheta_t(\mathbf{X})$ are all expressed in terms of $\mathbf{q}(t)$, $\mathbf{p}(t)$.

$$\tilde{e}^A(t) \equiv \oint_{\partial\Omega_e} \tilde{\pi}(\mathbf{X}, t) f^A(\mathbf{X}) \mathrm{d}A(\mathbf{X}),$$

$$\hat{e}^A(t, \mathbf{q}(t), \mathbf{p}(t)) \equiv \oint_{\partial\Omega_e} \hat{\pi}(\mathbf{X}, t, \mathbf{u}_t(\mathbf{X}), \vartheta_t(\mathbf{X})) f^A(\mathbf{X}) \mathrm{d}A(\mathbf{X}) + $$

$$+ \int_{\Omega_e} \hat{\alpha}(\mathbf{X}, t, \mathbf{u}_t(\mathbf{X}), \vartheta_t(\mathbf{X})) f^A(\mathbf{X}) \mathrm{d}V(\mathbf{X}),$$

$$\hat{\varepsilon}^A(\mathbf{q}(t), \mathbf{p}(t)) \equiv \int_{\Omega_e} \varrho(\mathbf{X}) \hat{\varepsilon}(\mathbf{X}; \mathbf{E}_t(\mathbf{X}), \vartheta_t(\mathbf{X})) f^A(\mathbf{X}) \mathrm{d}V(\mathbf{X}),$$

$$\hat{L}^{Aab}(\mathbf{q}(t), \dot{\mathbf{q}}(t), \mathbf{p}(t)) \equiv -\int_{\Omega_e} \mathrm{tr}\{\hat{\mathbf{S}}_t(\mathbf{X}; \mathbf{E}_t(\mathbf{X}), \vartheta_t(\mathbf{X})) +$$

$$+ \mathbf{H}(\mathbf{X}; \mathbf{E}_t(\mathbf{X}), \vartheta_t(\mathbf{X}))[\dot{\mathbf{E}}_t(\mathbf{X})] \nabla \mathbf{F}^a(\mathbf{X})^T \nabla \mathbf{F}^b(\mathbf{X})\} f^A(\mathbf{X}) \mathrm{d}V(\mathbf{X}),$$

we obtain

$$\begin{aligned} \varepsilon^A(t) &= \hat{\varepsilon}^A(\mathbf{q}(t), \mathbf{p}(t)), \\ L^{Aab}(t) &= \hat{L}^{Aab}(\mathbf{q}(t), \mathbf{p}(t)), \\ g^A(t) &= \hat{g}^A(t, \mathbf{q}(t), \mathbf{p}(t)), \\ e^A(t) &= \tilde{e}^A(t) + \hat{e}^A(t, \mathbf{q}(t), \mathbf{p}(t)). \end{aligned} \tag{3.60}$$

Eqs. (3.59), (3.60) represent the discrete description of constitutive relations in a structural element. Similarly, eqs. (3.59), (3.60) constitute the discrete description of the external agents (of both the mechanical and thermal type) acting on the structural element considered.

We conclude by pointing out again that the discrete description of a structural element is governed by eqs. (3.10), (3.12) and eqs. (3.13), (3.14). The specific form of governing equations depends on the choice of constraints, i.e., on the choice of vector functions $\mathbf{F}^a(\cdot)$, $a = 1, \dots, m$, and real-valued functions $f^A(\cdot)$, $A = 1, \dots, n$, defined on Ω_e. The basic unknowns are real valued functions $q_a(\cdot)$, $a = 1, \dots, m$ and $p_A(\cdot)$, $A = 1, \dots, n$, defined on the time interval $[t_0, t_f]$. It can be seen that the governing equations lead to the system of ordinary differential equations for $q_a(\cdot)$, $p_A(\cdot)$. If $q_a(\cdot)$, $p_A(\cdot)$ are known, then we can calculate $\mathbf{u}_t(\cdot)$, $\vartheta_t(\cdot)$ from eqs. (3.13) and on this basis all the remaining unknowns including reactions to constraints, cf. Secs. 3.1.1 and 3.1.4.

For the class of problems in which all the thermal effects can be neglected (in isothermal situations, for instance) the following discrete description of a visco-elastic structural element is obtained from eqs. (3.55), (3.59):

$$\begin{aligned} h^a(t)+f^a(t) &= M^{ab}\ddot{q}_b(t), \\ h^a(t) &= \hat{h}^a(\mathbf{q}(t))+H^{ab}(\mathbf{q}(t))\dot{q}_b(t), \\ f^a(t) &= \tilde{f}^a(t)+\hat{f}^a(t,\mathbf{q}(t))+G^{ab}(t,\mathbf{q}(t))\dot{q}_b(t). \end{aligned} \tag{3.61}$$

Similarly, if a structural element may be treated as rigid in a heat conduction problem, then

$$\begin{aligned} g^A(t)+e^A(t) &= \dot{\varepsilon}^A(t), \\ g^A(t) &= \hat{g}^A(t,\mathbf{p}(t)), \\ e^A(t) &= \tilde{e}^A(t)+\hat{e}^A(t,\mathbf{p}(t)), \\ \varepsilon^A(t) &= \hat{\varepsilon}^A(\mathbf{p}(t)). \end{aligned} \tag{3.62}$$

Eqs. (3.62) form the basis of the discrete heat conduction theory.

On the other hand, if eqs. (3.62) are valid while the functions $\hat{h}^a(\cdot)$, $H^{ab}(\cdot)$ in eqs. (3.61) depend also on $\mathbf{p}(t)$:

$$h^a(t) = \hat{h}^a(\mathbf{q}(t),\mathbf{p}(t))+H^{ab}(\mathbf{q}(t),\mathbf{p}(t))\dot{q}_b(t),$$

then we arrive at the discrete description of thermal stresses in the structural element.

We note at the end of this section that the method of constraints has been used here merely as a formal, purely mathematical tool in the discrete modelling of thermo-visco-elastic structural elements. The question concerning the choice of constraints (i.e., the functions $\mathbf{F}^a(\cdot)$, $f^A(\cdot)$) in a specific situation will be answered later in this chapter.

3.4.3 Incremental formulation

Using the incremental form of eqs. (3.55), (3.59) we are able to obtain the relations

$$\begin{aligned} \Delta h^a+\Delta f^a &= M^{ab}\Delta\ddot{q}_b, \\ \Delta h^a &= \Delta\hat{h}^a(\Delta\mathbf{q},\Delta\mathbf{p})+\delta H^{ab}(\Delta\mathbf{q},\Delta\mathbf{p})\Delta q_b+H^{ab}(\Delta\mathbf{q},\Delta\mathbf{p})\Delta\dot{q}_b, \\ \Delta f^a &= \Delta\tilde{f}^a+\Delta\hat{f}^a(\Delta\mathbf{q})+\delta G^{ab}(\Delta\mathbf{q})\Delta q_b+G^{ab}(\Delta\mathbf{q})\Delta\dot{q}_b, \end{aligned} \tag{3.63}$$

where, cf. Sec. 3.1.2

$$\begin{aligned} \delta H^{ab}(\Delta\mathbf{q},\Delta\mathbf{p}) &\equiv [H^{ab}(\Delta\mathbf{q},\Delta\mathbf{p})-H^{ab}(0,0)]\Delta t, \\ \delta G^{ab}(\Delta\mathbf{q}) &\equiv [G^{ab}(\Delta\mathbf{q})-G^{ab}(0)]\Delta t. \end{aligned}$$

Thermo-visco-elastic material properties have been tacitly assumed in the above equations. For elastic-plastic materials we consider instead of eqs. (3.63) the relation

$$\Delta h^a = D^{ab}\Delta q_b$$

where

$$D^{ab} \equiv -\int_{\Omega_e} \mathrm{tr}[\Delta \mathbf{S}_t(\mathbf{X})\nabla \mathbf{F}^a(\mathbf{X})]\mathbf{F}^b \,\mathrm{d}V(\mathbf{X})$$

with $\Delta \mathbf{S}(\mathbf{X})$ given by eqs. (3.9). Similarly, starting from eqs. (3.57), (3.60) we arrive at

$$\begin{aligned}
&\Delta g^A + \Delta e^A + \Delta[\hat{L}^{Aab}(\mathbf{q},\mathbf{p})q_a\dot{q}_b] = \Delta\dot{\varepsilon}^A,\\
&\Delta g^A = \Delta\hat{g}^A(\Delta\mathbf{q},\Delta\mathbf{p}),\\
&\Delta e^A = \Delta\hat{e}^A(\Delta\mathbf{q},\Delta\mathbf{p}),\\
&\Delta\varepsilon^A = \Delta\hat{\varepsilon}^A(\Delta\mathbf{q},\Delta\mathbf{p}).
\end{aligned} \tag{3.64}$$

Eqs. (3.63), (3.64) represent the incremental form of the discrete description of thermo-mechanics of a structural element.

3.4.4 Shell and plate elements

The approach taken at the beginning of Sec. 3.4.1 was based on dividing the region Ω (in the Euclidean 3-space $\boldsymbol{R}^3$) into a finite number of disjoined subregions called the finite elements and then on restricting the considerations to a single finite element Ω_e. Every finite element represents a certain part of the structure and has been referred to as the structural element. An analogous approach can be employed for structures described in terms of a certain plate or shell theory, as it was done in Sec. 3.2.2. By dividing the plane region Ω^* into a finite number of disjoined subregions we then obtain the system of finite elements representing pertinent shell or plate elements. Let Ω_e^* be an arbitrary but fixed plane finite element. Eq. (3.24) specified to this element Ω_e^* of Ω^* leads to the condition

$$\begin{aligned}
&\int_{\Omega_e^*} H^{aK}(\mathbf{X}^*, t)v^*_{a,K}(\mathbf{X}^*)\mathrm{d}V(\mathbf{X}^*)\\
&\quad = \oint_{\partial\Omega_e^*} s^a(\mathbf{X}^*, t)v_a^*(\mathbf{X}^*)\mathrm{d}A(\mathbf{X}^*) +\\
&\quad + \int_{\Omega_e^*} [f^a(\mathbf{X}^*, t) + h^a(\mathbf{X}^*, t) - M^{ab}(\mathbf{X}^*)\ddot{\chi}_b(\mathbf{X}^*, t)]v_a^*(\mathbf{X}^*)\mathrm{d}V(\mathbf{X}^*)
\end{aligned} \tag{3.65}$$

which has to hold for every $\mathbf{v}^*(\cdot) \equiv (v_1^*(\cdot), \ldots, v_m^*(\cdot)) \in \mathscr{U}_t^{*e}$, where the set

$\mathscr{U}_t^{*e}$ of virtual displacements is defined in analogy to the set $\mathscr{U}_t^*$ in Sec. 3.2.1, but the domain of all functions is restricted to $\bar{\Omega}_e^*$. Similarly, eq. (3.27) now yields

$$\int_{\Omega_{e\tau}^*} G^{AK}(\mathbf{X}^*, t)\zeta_{A,K}^*(\mathbf{X}^*)\mathrm{d}V(\mathbf{X}^*)$$
$$= \oint_{\partial\Omega_e^*} \pi^A(\mathbf{X}^*, t)\zeta_A^*(\mathbf{X}^*)\mathrm{d}A(\mathbf{X}^*) +$$
$$+ \int_{\Omega_e^*} [e^A(\mathbf{X}^*, t) + g^A(\mathbf{X}^*, t) + L^A(\mathbf{X}^*, t) - \dot{\varepsilon}^A(\mathbf{X}^*, t)]\zeta_A^*(\mathbf{X}^*)\mathrm{d}V(\mathbf{X}^*) \tag{3.66}$$

and has to hold for every $\boldsymbol{\zeta}^*(\cdot) \equiv (\zeta_1^*(\cdot), \ldots, \zeta_n^*(\cdot)) \in \mathscr{Z}_t^{*e}$, where $\mathscr{Z}_t^{*e}$ is the set of all virtual temperature increments restricted to $\bar{\Omega}_e^*$. Both aforementioned variational equations are considered together with the definitions (3.22), (3.23), (3.25) and (3.26). The functions $s^a(\mathbf{X}^*, t)$, $\pi^A(\mathbf{X}^*, t)$, $\mathbf{X}^* \in \partial\Omega_e^*$, represent now the external agents acting on the shell or plate element across its boundary which is determined by the boundary $\partial\Omega_e^*$ of the (plane) finite element. It should be remembered that the constraints

$$\boldsymbol{\chi}^*(\cdot, \tau) \in \mathscr{D}_\tau^*(\Omega_e^*), \quad \theta^*(\cdot, \tau) \in \mathscr{T}_\tau(\Omega_e^*), \quad \tau \in [t_0, t_f],$$

are imposed now exclusively on the shell or plate element.

In order to obtain the discrete description of such an element we have to specify the sets $\mathscr{D}_\tau^*(\Omega_e^*)$ and $\mathscr{T}_\tau^*(\Omega_e^*)$ by following the lines indicated in Sec. 4.1. To this aim we assume that $\boldsymbol{\chi}^*(\cdot, \tau) \in \mathscr{D}_\tau^*(\Omega_e^*)$ if

$$\boldsymbol{\chi}^*(\mathbf{X}^*, \tau) = \mathbf{F}^{\bar{a}}(\mathbf{X}^*)q_{\bar{a}}(\tau) + \boldsymbol{\chi}_0^*(\mathbf{X}^*), \quad \mathbf{X}^* \in \bar{\Omega}_e^*, \ \tau \in [t_0, t_f], \tag{3.67}$$

where $\mathbf{F}^{\bar{a}}(\mathbf{X}^*)$, $\bar{a} = \bar{1}, \ldots, \bar{m}$, $\boldsymbol{\chi}_0^*(\mathbf{X}^*)$ are known and $q_{\bar{a}}(\tau)$, $\tau \in [t_0, t_f]$, are arbitrary (sufficiently regular) unknown functions. Similarly, we assume that $\theta^*(\cdot, \tau) \in \mathscr{T}_\tau(\Omega_e^*)$ if and only if

$$\theta^*(\mathbf{X}^*, \tau) = f^{\bar{A}}(\mathbf{X}^*)p_{\bar{A}}(\tau) + \theta_0^*, \quad \mathbf{X}^* \in \bar{\Omega}_e^*, \ \tau \in [t_0, t_f], \tag{3.68}$$

where $f^{\bar{A}}(\mathbf{X}^*)$, $\bar{A} = \bar{1}, \ldots, \bar{n}$, θ_0^* are known and $p_{\bar{A}}(\tau)$, $\tau \in [t_0, t_f]$, are unknown functions. Bearing in mind the procedure employed in Sec. 3.4.2 which led from eqs. (3.47), (3.50) to eqs. (3.49), (3.53), respectively, we shall get the discrete description of the shell or plate element. We shall see that this description can be expressed by a system of $\bar{m}+\bar{n}$ equations for unknown functions $q_{\bar{a}}(\cdot)$, $\bar{a} = \bar{1}, \ldots, \bar{m}$, $p_{\bar{A}}(\cdot)$, $\bar{A} = \bar{1}, \ldots, \bar{n}$, defined on the time interval $[t_0, t_f]$. To this aim let us observe that the constraints postulated above written now as

$$\chi_a^*(\mathbf{X}^*, \tau) = F_a^{\bar{a}}(\mathbf{X}^*)q_{\bar{a}}(\tau) + \chi_{0a}^*(\mathbf{X}^*), \quad a = 1, \ldots, m,$$
$$\theta_A^*(\mathbf{X}^*, \tau) = f_A^{\bar{A}}(\mathbf{X}^*)p_{\bar{A}}(\tau) + \theta_{0A}^*, \quad A = 1, \ldots, n,$$

imply that

$$\mathbf{v}^*(\cdot) \equiv (v_1^*(\cdot), \ldots, v_m^*(\cdot)) \in \mathscr{U}_t^{*e}, \qquad \boldsymbol{\zeta}^*(\cdot) \equiv (\zeta_1^*(\cdot), \ldots, \zeta_n^*(\cdot)) \in \mathscr{Z}_t^{*e}$$

if and only if

$$v_a^*(\mathbf{X}^*) = F_a^{\bar{a}}(\mathbf{X}^*) v_{\bar{a}}, \qquad \zeta_A^*(\mathbf{X}^*) = f_A^{\bar{A}}(\mathbf{X}^*) \zeta_{\bar{A}},$$

for every $\bar{\mathbf{v}} \equiv (v_{\bar{1}}, \ldots, v_{\bar{m}}) \in R^{\bar{m}}$, $\bar{\boldsymbol{\zeta}} \equiv (\zeta_{\bar{1}}, \ldots, \zeta_{\bar{n}}) \in R^{\bar{n}}$. The sets $\mathscr{U}_t^{*e}$ and $\mathscr{Z}_t^{*e}$ are independent of $\boldsymbol{\chi}^*(\cdot, t)$ and $\theta^*(\cdot, t)$, respectively, as well as of the variable X^3. Eqs. (3.65), (3.66) lead to

$$\int_{\Omega_e^*} H^{aK}(\mathbf{X}^*, t) F_{a,K}^{\bar{a}}(\mathbf{X}^*) \mathrm{d}V(\mathbf{X}^*) = \oint_{\partial\Omega_e^*} s^a(\mathbf{X}^*, t) F_a^{\bar{a}}(\mathbf{X}^*) \mathrm{d}A(\mathbf{X}^*) +$$

$$+ \int_{\Omega_e^*} [f^a(\mathbf{X}^*, t) + h^a(\mathbf{X}^*, t) - M^{ab}(\mathbf{X}^*) \ddot{\chi}_b^*(\mathbf{X}^*, t)] F_a^{\bar{a}}(\mathbf{X}^*) \mathrm{d}V(\mathbf{X}^*),$$

$$\bar{a} = \bar{1}, \ldots, \bar{m}, \tag{3.69}$$

$$\int_{\Omega_e^*} G^{AK}(\mathbf{X}^*, t) f_{A,K}^{\bar{A}}(\mathbf{X}^*) \mathrm{d}V(\mathbf{X}^*) = \oint_{\partial\Omega_e^*} \pi^A(\mathbf{X}^*, t) f_A^{\bar{A}}(\mathbf{X}^*) \mathrm{d}A(\mathbf{X}^*) +$$

$$+ \int_{\Omega_e^*} [e^A(\mathbf{X}^*, t) + g^A(\mathbf{X}^*, t) + L^A(\mathbf{X}^*, t) - \dot{\varepsilon}^A(\mathbf{X}^*, t)] f_A^{\bar{A}}(\mathbf{X}^*) \mathrm{d}V(\mathbf{X}^*),$$

$$\bar{A} = \bar{1}, \ldots, \bar{n}.$$

Substituting into eqs. (3.69) the RHS of the definitions (3.22), (3.23), (3.25), (3.26) and taking into account constraints (3.65), (3.66) we can easily see that all integrands in eqs. (3.65) are the known functions of the argument $\mathbf{X}^* \equiv (X^1, X^2)$ and hence all integrals over Ω_e^* and $\partial\Omega_e^*$ in eqs. (3.69) can be computed. In this way we get from eqs. (3.69) the system of equations for $\bar{m} + \bar{n}$ unknown functions $q_{\bar{a}}(\tau)$, $\bar{a} = \bar{1}, \ldots, \bar{m}$, $p_{\bar{A}}(\tau)$, $\bar{A} = \bar{1}, \ldots, \bar{n}$, $\tau \in [t_0, t_f]$. This system represents the discrete description of a single shell or plate element.

3.4.5 Rod and beam elements

Assume that a part of the structure considered represents a rod or a beam element, the reference configuration of which can be seen in Fig. 11. For the sake of simplicity we confine ourselves to the situation in which all the thermal effects are negligible. The governing equations can be assumed in the form of eqs. (3.36), and (3.38) (with the argument $\theta^*(\xi, t)$ dropping out from eq. (3.38)) with $\boldsymbol{\chi}^*(\cdot, \tau) \equiv (\chi_1^*(\cdot, \tau), \ldots, \chi_m^*(\cdot, \tau)) \in \mathscr{D}_\tau^*([0, l])$. In order to obtain the discrete description of this problem we shall apply the same procedure as above assuming that

$$\boldsymbol{\chi}^*(\cdot, \tau) \equiv (\chi_1^*(\cdot, \tau), \ldots, \chi_m^*(\cdot, \tau)) \in \mathscr{D}_\tau^*([0, 1])$$

holds if

$$\chi^*(\xi, \tau) = \bar{\mathbf{F}}^1(\xi)\mathbf{q}_1(\tau) + \bar{\mathbf{F}}^2(\xi)\mathbf{q}_2(\tau) + \chi_0^*(\xi, \tau), \quad \xi \in [0, l], \tag{3.70}$$

where $\mathbf{q}_1(\tau)$, $\mathbf{q}_2(\tau)$ are arbitrary p-dimensional vectors (p-tuples of real numbers), $\bar{\mathbf{F}}^1(\xi)$, $\bar{\mathbf{F}}^2(\xi)$ are known $m \times p$ matrices with components $\bar{F}_a^{1k}(\xi)$, $\bar{F}_a^{2k}(\xi)$ respectively, and $\chi_0^*(\xi, \tau)$ is the known function. The vector functions $\mathbf{q}_1(\cdot)$, $\mathbf{q}_2(\cdot)$ defined on $[t_0, t_f]$ and continuous together with their first and second derivatives are now the basic unknowns. The values $\mathbf{q}_1(\tau)$ and $\mathbf{q}_2(\tau)$ will be interpreted as certain generalized displacements of the boundary cross sections of the rod, i.e., the cross sections determined by $\xi = 0$ and $\xi = l$, respectively, cf. Fig. 11. Using the procedure employed in Sec. 3.4.1 we shall derive the system of equations for $\mathbf{q}_1(\cdot)$, $\mathbf{q}_2(\cdot)$. To this aim we have to substitute into eqs. (3.36), (3.38) the RHS of eqs. (3.70) and to assume that eq. (3.36) has to hold for every $\mathbf{v}^*(\xi) = \bar{\mathbf{F}}^1(\xi)\mathbf{u}_1 + \bar{\mathbf{F}}^2(\xi)\mathbf{u}_2$, where $\mathbf{u}_1$, $\mathbf{u}_2$ are arbitrary p-tuples of real numbers[1]. From eq. (3.36) there follows

$$\int_0^l H^a(\xi, t)\bar{F}^K_{a,1}(\xi)d\xi = s^a(t, 0)\bar{F}^K_a(0) + s^a(t, l)\bar{F}^K_a(l) + $$
$$+ \int_0^l \{f^a(\xi, t) + h^a(\xi, t) - M^{ab}(\xi)[\bar{F}_b^1(\xi)\ddot{q}_1(\tau) + $$
$$+ \bar{F}_b^2(\xi)\ddot{q}_2(\tau)]\}\bar{F}_a^K(\xi)d\xi, \tag{3.71}$$

where $K = 1, 2$. Eqs. (3.39) combined with eq. (3.70) yields

$$\begin{aligned}
H^a(\xi, t) &= \bar{H}^a(\xi, \mathbf{q}(t), \dot{\mathbf{q}}(t)),\\
h^a(\xi, t) &= \bar{h}^a(\xi, \mathbf{q}(t), \dot{\mathbf{q}}(t)),\\
f^a(\xi, t) &= \bar{f}^a(\xi, t, \mathbf{q}(t), \dot{\mathbf{q}}(t)),\\
s^a(t; 0) &= \bar{s}^a(0; t, \mathbf{q}(t), \dot{\mathbf{q}}(t)),\\
s^a(t; l) &= \bar{s}^a(l; t, \mathbf{q}(t), \dot{\mathbf{q}}(t)),
\end{aligned} \tag{3.72}$$

where $\mathbf{q}(t) \equiv (\mathbf{q}_1(t), \mathbf{q}_2(t))$ and all the functions on the RHS of eqs. (3.72) are known.

We conclude that by substituting the RHS of eqs. (3.72) into eq. (3.71) we can compute all the integrals and obtain the system of ordinary differential equations of the second order for the unknowns $\mathbf{q}_1(\tau)$, $\mathbf{q}_2(\tau)$, $\tau \in [t_0, t_f]$.

[1] Recalling the results of Sec. 3.2.3 we have to assume that $\mathbf{v}^*(\cdot) \in \mathscr{U}_t^*$ if and only if $\mathbf{v}^*(\cdot) = \bar{\mathbf{F}}^1(\cdot)\mathbf{u}_1 + \bar{\mathbf{F}}^2(\cdot)\mathbf{u}_2$ for every $\mathbf{u}_1, \mathbf{u}_2 \in \mathbf{R}^k$. The sets $\mathscr{U}_t^*$ are independent of $\chi^*(\cdot, t)$ and t, i.e., all such sets coincide.

Thus eqs. (3.71), (3.72) together with formula (3.70) represent the discrete description of a rod or beam element, separated from the rest of the structure by the two end cross sections. In the elementary beam theory based on the Bernoulli hypothesis of the plane beam cross sections we can assume that $\mathbf{q}_K(\tau) = (u_K^1(\tau), u_K^2(\tau), u_K^3(\tau), v_K^1(\tau), v_K^2(\tau), v_K^3(\tau))$ where $K = 1, 2$ and $\mathbf{u}_K(\tau)$, $\mathbf{v}_K(\tau)$ are displacement and small rotation vectors, respectively, of the both end cross sections of the beam. Simultaneously, $\boldsymbol{\chi}^*(\xi, \tau) = (\chi_1^*(\xi, \tau), \chi_2^*(\xi, \tau), \chi_3^*(\xi, \tau))$, $\xi \in [0, l]$, $\tau \in [t_0, t_f]$ represent the displacements of the beam axis at the time instant τ, while $\boldsymbol{\chi}_0^*(\xi, \tau) \in \boldsymbol{R}^3$ describe the displacements of the beam axis under external loadings in the case in which $\mathbf{q}_1(\tau) = \mathbf{0}$, $\mathbf{q}_2(\tau) = \mathbf{0}$.

3.4.6 Special cases and matrix notation

In this section we shall consider problems in which the form of the constitutive equations is linear, and determined by eqs. (2.2). Neglecting the thermal effects we obtain from eq. $(2.2)_1$ the following constitutive law

$$\mathbf{S}_t(\mathbf{X}) = \mathfrak{C}(\mathbf{X})[\mathbf{E}_t(\mathbf{X})]$$

where the strain tensor $\mathbf{E}_t(\mathbf{X})$ is given by eq. (1.10).

The first goal in this section is to derive a special case of eqs. (3.55) for the linear stress-strain relation. By using eq. $(3.54)_1$ and denoting

$$Q^a(t) \equiv f^a(t) - M^{ab}\ddot{q}_b(t)$$

we get

$$\int_{\Omega_e} \mathrm{tr}\{\mathfrak{C}(\mathbf{X})[\mathbf{E}_t(\mathbf{X})]\nabla\boldsymbol{\chi}^T(\mathbf{X}, t)\nabla\mathbf{F}^a(\mathbf{X})\}\mathrm{d}V(\mathbf{X}) = Q^a(t), \qquad a = 1, \dots, m, \tag{3.73}$$

which holds under the discretization constraints (3.48). Setting $\boldsymbol{\chi}_0(\mathbf{X}) = \mathbf{X}$ in eq. (3.48) and introducing the displacement vector $\mathbf{u}_t(\mathbf{X})$ we have

$$\mathbf{u}_t(\mathbf{X}) = \boldsymbol{\chi}(\mathbf{X}, t) - \mathbf{X} = \mathbf{F}^a(\mathbf{X})q_a(t).$$

It is assumed throughout this section that $\mathbf{X} \in \Omega_e$, i.e., we confine ourselves to an arbitrary but fixed finite element. Substituting the displacement gradients $\nabla\mathbf{u}_t(\mathbf{X}) = \nabla\mathbf{F}^a(\mathbf{X})q_a(t)$ into eq. (1.10) we obtain the formula

$$\mathbf{E}_t(\mathbf{X}) = \tfrac{1}{2}[\nabla\mathbf{F}^a(\mathbf{X}) + (\nabla\mathbf{F}^a)^T(\mathbf{X})]q_a(t) + \tfrac{1}{2}(\nabla\mathbf{F}^a)^T(\mathbf{X})\nabla\mathbf{F}^b(\mathbf{X})q_a(t)q_b(t)$$

which with the notation

$$\mathbf{B}^a(\mathbf{X}) \equiv \tfrac{1}{2}[\nabla\mathbf{F}^a(\mathbf{X}) + (\nabla\mathbf{F}^a)^T(\mathbf{X})],$$

$$\overline{\mathbf{B}}^a(\mathbf{X}, \mathbf{q}) \equiv \tfrac{1}{2}(\nabla\mathbf{F}^a)^T(\mathbf{X})\nabla\mathbf{F}^b(\mathbf{X})q_b,$$

yields

$$\mathbf{E}_t(\mathbf{X}) = [\mathbf{B}^a(\mathbf{X}) + \bar{\mathbf{B}}^a(\mathbf{X}, \mathbf{q}(t))] q_a(t).$$

Substituting now $\nabla\boldsymbol{\chi}(\mathbf{X}, t) = \nabla\mathbf{F}^a(\mathbf{X}) q_a(t)$ and the right-hand side of the last equation into eq. (3.73), we obtain

$$\int_{\Omega_e} \{\mathrm{tr}\{\mathfrak{C}(\mathbf{X})[\mathbf{B}^b(\mathbf{X}) + \bar{\mathbf{B}}^b(\mathbf{X}, \mathbf{q})][\mathbf{B}^a(\mathbf{X}) + 2\bar{\mathbf{B}}^a(\mathbf{X}, \mathbf{q})]\}\mathrm{d}V(\mathbf{X}) q_b(t) = Q^a(t).$$

Hence, denoting

$$k_0^{ab} \equiv \int_{\Omega_e} \mathrm{tr}\{\mathfrak{C}(\mathbf{X})[\mathbf{B}^b(\mathbf{X})]\mathbf{B}^a(\mathbf{X})\}\mathrm{d}V(\mathbf{X}),$$

$$k_1^{ab}(\mathbf{q}) \equiv \int_{\Omega_e} \{2\,\mathrm{tr}\{\mathfrak{C}(\mathbf{X})[\mathbf{B}^b(\mathbf{X})]\bar{\mathbf{B}}^a(\mathbf{X}, \mathbf{q})\} + \\ + \mathrm{tr}\{\mathfrak{C}(\mathbf{X})[\bar{\mathbf{B}}^b(\mathbf{X}, \mathbf{q})]\mathbf{B}^a(\mathbf{X})\}\}\,\mathrm{d}V(\mathbf{X}),$$

$$k_2^{ab}(\mathbf{q}) \equiv 2\int_{\Omega_e} \mathrm{tr}\{\mathfrak{C}(\mathbf{X})[\bar{\mathbf{B}}^b(\mathbf{X}, \mathbf{q})]\bar{\mathbf{B}}^a(\mathbf{X}, \mathbf{q})\}\mathrm{d}V(\mathbf{X}),$$

we arrive at the final form of eq. (3.73) as

$$[k_0^{ab} + k_1^{ab}(\mathbf{q}(t)) + k_2^{ab}(\mathbf{q}(t))] q_b(t) = Q^a(t), \quad a = 1, \ldots, m. \tag{3.74}$$

We may observe that the matrix $\mathbf{k}_0 = [k_0^{ab}]$ is independent of the generalized coordinates $q_a(t)$ while the elements of the matrices $\mathbf{k}_1 = [k_1^{ab}(\mathbf{q})]$ and $\mathbf{k}_2 = [k_2^{ab}(\mathbf{q})]$ are linear and quadratic forms, respectively, in the variables $q_a(t)$, $a = 1, \ldots, m$.

Eq. (3.74) represents the discretized form of the equation of motion for materials with linear stress-strain constitutive law.

As the next step we shall derive now the incremental form of eq. (3.74). To this aim we shall use eqs. (3.54) and (3.55). By virtue of the relations

$$\boldsymbol{\chi}(\mathbf{X}, t) = \mathbf{X} + \mathbf{F}^b(\mathbf{X}) q_b(t)$$

and

$$\nabla\boldsymbol{\chi}^T(\mathbf{X}, t)\nabla\mathbf{F}^a(\mathbf{X}) = \nabla\mathbf{F}^a(\mathbf{X}) + (\nabla\mathbf{F}^a)^T(\mathbf{X})\nabla\mathbf{F}^b(\mathbf{X}) q_b(t),$$

we obtain

$$\int_{\Omega_e} \mathrm{tr}[\mathbf{S}_t(\mathbf{X})(\nabla\mathbf{F}^a(\mathbf{X}) + (\nabla\mathbf{F}^a)^T(\mathbf{X})\nabla\mathbf{F}^b(\mathbf{X}) q_b(t))]\mathrm{d}V(\mathbf{X}) = Q^a(t).$$

The incremental formulation follows from the foregoing equation by replacing $\mathbf{S}_t(\mathbf{X})$, $q_b(t)$, $Q^a(t)$ by $\mathbf{S}_t(\mathbf{X}) + \Delta\mathbf{S}_t(\mathbf{X})$, $q_b(t) + \Delta q_b(t)$, $Q^a(t) + \Delta Q^a(t)$, respectively, and by neglecting terms which are nonlinear with respect to the increments. Since

$$\operatorname{tr}[\Delta \mathbf{S}_t(\mathbf{X})(\nabla \mathbf{F}^a(\mathbf{X})+(\nabla \mathbf{F}^a)^T(\mathbf{X})\nabla \mathbf{F}^b(\mathbf{X})q_b(t))]$$
$$= \operatorname{tr}[\Delta \mathbf{S}_t(\mathbf{X})(\mathbf{B}^a(\mathbf{X})+2\bar{\mathbf{B}}^a(\mathbf{X},\mathbf{q}(t)))],$$
$$\Delta \mathbf{S}_t(\mathbf{X}) = \mathfrak{C}(\mathbf{X})[\Delta \mathbf{E}_t(\mathbf{X})],$$
$$\Delta \mathbf{E}_t(\mathbf{X}) = [\mathbf{B}^b(\mathbf{X})-2\bar{\mathbf{B}}^b(\mathbf{X},\mathbf{q}(t))]\Delta q_b(t),$$

where $\Delta \mathbf{E}_t(\mathbf{X})$ is the linear part of the strain increment with respect to $\Delta \mathbf{q}(t)$, then the incremental equation reads

$$\int_{\Omega_e} \operatorname{tr}\{\mathfrak{C}(\mathbf{X})[\mathbf{B}^b(\mathbf{X})+2\bar{\mathbf{B}}^b(\mathbf{X},\mathbf{q}(t))](\mathbf{B}^a(\mathbf{X})+2\bar{\mathbf{B}}^a(\mathbf{X},\mathbf{q}(t)))\}\,dV(\mathbf{X})\Delta q_b(t)+$$
$$+2\int_{\Omega_e} \operatorname{tr}\{\mathbf{S}_t(\mathbf{X})(\nabla \mathbf{F}^a)^T(\mathbf{X})\nabla \mathbf{F}^b(\mathbf{X})\}\,dV(\mathbf{X})\Delta q_b(t) = \Delta Q^a(t).$$

Hence, denoting

$$k_0^{ab} \equiv \int_{\Omega_e} \operatorname{tr}\{\mathfrak{C}(\mathbf{X})[\mathbf{B}^b(\mathbf{X})]\mathbf{B}^a(\mathbf{X})\}\,dV(\mathbf{X}),$$

$$\bar{k}^{ab}(\mathbf{q}) \equiv 2\int_{\Omega_e} \{\operatorname{tr}\{\mathfrak{C}(\mathbf{X})[\bar{\mathbf{B}}^b(\mathbf{X},\mathbf{q})]\mathbf{B}^a(\mathbf{X})\}+$$
$$+\operatorname{tr}\{\mathfrak{C}(\mathbf{X})[\mathbf{B}^b(\mathbf{X})]\bar{\mathbf{B}}^a(\mathbf{X},\mathbf{q})\}\}\,dV(\mathbf{X}), \tag{3.75}$$

$$\bar{\bar{k}}^{ab}(\mathbf{q}) \equiv 4\int_{\Omega_e} \operatorname{tr}\{\mathfrak{C}(\mathbf{X})[\bar{\mathbf{B}}^b(\mathbf{X},\mathbf{q})]\bar{\mathbf{B}}^a(\mathbf{X},\mathbf{q})\}\,dV(\mathbf{X}),$$

$$k^{ab}(\mathbf{S}_t) \equiv 2\int_{\Omega_e} \operatorname{tr}[\mathbf{S}_t(\mathbf{X})(\nabla \mathbf{F}^a)^T(\mathbf{X})\nabla \mathbf{F}^b(\mathbf{X})]\,dV(\mathbf{X}),$$

we finally obtain

$$[k_0^{ab}+k^{ab}(\mathbf{S}_t)+\bar{k}^{ab}(\mathbf{q}(t))+\bar{\bar{k}}^{ab}(\mathbf{q}(t))]\Delta q_b(t) = \Delta Q^a(t),$$
$$a = 1, \ldots, m. \tag{3.76}$$

The matrix $[k_0^{ab}]$ in eqs. (3.76) is called the constitutive matrix, the matrix $[k^{ab}(\mathbf{S}_t)]$ is called the initial stress matrix and the matrices $[\bar{k}^{ab}(\mathbf{q})]$ and $[\bar{\bar{k}}^{ab}(\mathbf{q})]$ are called initial displacement matrices, being linear and quadratic in the displacement vector $\mathbf{u}_t(\mathbf{X})$, respectively.

Eq. (3.76) can also be obtained directly from eq. (3.74) by substituting $q_b+\Delta q_b$, $Q^a+\Delta Q^a$ for q_b, Q^a and by retaining only the terms which are linear with respect to the increments Δq_b.

We also note that since in the definition of terms entering eq. (3.76) the constitutive law is used in its incremental form, the final incremental equation (3.76) applies in fact to the analysis of any material characterized by the law of such a type.

The resulting equation (3.76) represents the discretized incremental form of the equation of motion with the linear stress-strain constitutive law.

Let us now consider the linear heat-flux law given by eq. $(2.2)_2$. Taking into account the discretized heat conduction equation (3.57) and neglecting the coupling between the temperature and deformation fields (represented in eq. (3.57) by the term $L^{Aab}(t)$), we obtain

$$\int_{\Omega_e} [\mathbf{K}_0(\mathbf{X})\nabla\vartheta_t(\mathbf{X})]\cdot\nabla f^A(\mathbf{X})\,dV(\mathbf{X}) + \int_{\Omega_e} \varrho(\mathbf{X})\dot{\varepsilon}(\mathbf{X},t) f^A(\mathbf{X})\,dV(\mathbf{X}) = e^A(t),$$

$$A = 1, \ldots, n. \qquad (3.77)$$

This equation holds under the discretization constraints for temperature (3.51), which can be written as

$$\vartheta_t(\mathbf{X}) = \theta(\mathbf{X},t) - \theta_0 = f^A(\mathbf{X})p_A(\mathbf{X}).$$

The internal energy $\varepsilon(\mathbf{X},t) = \varphi(\mathbf{X},t) + \eta(\mathbf{X},t)\theta(\mathbf{X},t)$, by means of eqs. $(2.2)_{3,4}$, and with the terms dependent on deformation neglected, becomes

$$\varepsilon(\mathbf{X},t) = -\tfrac{1}{2}\beta(\mathbf{X})\vartheta_t(\mathbf{X})^2 - \beta(\mathbf{X})\vartheta_t(\mathbf{X})\theta_0.$$

Setting now

$$k^{AB} = \int_{\Omega_e} [\mathbf{K}_0(\mathbf{X})\nabla f^B(\mathbf{X})]\cdot\nabla f^A(\mathbf{X})\,dV(\mathbf{X}),$$

$$C_1^{AB} = -\int_{\Omega_e} \varrho(\mathbf{X})\beta(\mathbf{X})\theta_0 f^A(\mathbf{X})f^B(\mathbf{X})\,dV(\mathbf{X}), \qquad (3.78)$$

$$C_2^{AB} = -\int_{\Omega_e} \varrho(\mathbf{X})\beta(\mathbf{X})f^A(\mathbf{X})f^B(\mathbf{X})f^C(\mathbf{X})\,dV(\mathbf{X})\cdot p_C,$$

we obtain eq. (3.77) in the final form

$$k^{AB}p_B(t) + [C_1^{AB} + C_2^{AB}(\mathbf{p}(t))]\dot{p}_B(t) = e^A(t), \qquad A = 1, 2, \ldots, n. \qquad (3.79)$$

The matrix $\mathbf{C}_1$ is independent of the temperature while the elements of the matrix $\mathbf{C}_2 = [C_2^{AB}(\mathbf{p})]$ are linear forms in the variables $p_A(t)$, $A = 1, 2, \ldots, n$. Clearly, the matrix $\mathbf{k} = [k^{AB}]$ may be a function of the temperature if the thermal conductivities are temperature dependent. The matrices $\mathbf{C} = \mathbf{C}_1 + \mathbf{C}_2$ and $\mathbf{k}$ are called the elemental heat capacity and the elemental conductivity matrix, respectively—they should not be confused with the damping and stiffness matrices denoted before by the same symbols.

Replacing in eq. (3.79) $p_B(t)$ by $p_B(t) + \Delta p_B(t)$ and $e^A(t)$ by $e^A(t) + \Delta e^A(t)$ we obtain the following incremental form of the discretized heat conduction equation

$$[k^{AB}+C_2^{AB}(\dot{\mathbf{p}}(t))]\Delta p_B(t)+[C_1^{AB}+C_2^{AB}(\mathbf{p}(t))]\Delta\dot{p}_B(t) = \Delta e^A(t),$$
$$A = 1, 2, \ldots, n \qquad (3.80)$$

where the terms nonlinear in the increments $\Delta\mathbf{p}(t)$, $\Delta\dot{\mathbf{p}}(t)$ have been omitted.

We shall complete this section with transforming eqs. (3.76), (3.80) to the matrix form. To this aim we denote by $N_{(e)}$ the number of the degrees of freedom in the finite element under consideration. We further introduce the following matrices[1] where $k = 1, 2, 3$, $a = 1, \ldots, N_{(e)}$:

$$\mathbf{u}_{3\times 1} \equiv [u_k],$$
$$\mathbf{E}_{6\times 1} \equiv [E_{11}\ E_{22}\ E_{33}\ E_{12}\ E_{13}\ E_{23}]^T,$$
$$\mathbf{q}_{N_{(e)}\times 1} \equiv [q_a],$$
$$\boldsymbol{\varphi}_{3\times N_{(e)}} \equiv [F^{ak}],$$
$$\mathbf{B}_{6\times N_{(e)}} \equiv [B^{a11}\ B^{a22}\ B^{a33}\ B^{a12}\ B^{a13}\ B^{a23}],$$
$$\bar{\mathbf{B}}_{6\times N_{(e)}} \equiv [\bar{B}^{a11}\ \bar{B}^{a22}\ \bar{B}^{a33}\ \bar{B}^{a12}\ \bar{B}^{a13}\ \bar{B}^{a23}],$$
$$\mathbf{C}_{6\times 6} \equiv \begin{bmatrix} C^{1111} & C^{1122} & C^{1133} & 2C^{1112} & 2C^{1113} & 2C^{1123} \\ \cdot & C^{2222} & C^{2233} & 2C^{2212} & 2C^{2213} & 2C^{2223} \\ \cdot & \cdot & C^{3333} & 2C^{3312} & 2C^{3313} & 2C^{3323} \\ \cdot & \cdot & \cdot & 2C^{1212} & 2C^{1213} & 2C^{1223} \\ \cdot & \cdot & \cdot & \cdot & 2C^{1313} & 2C^{1323} \\ \cdot & \cdot & \cdot & \cdot & \cdot & 2C^{2323} \end{bmatrix},$$
$$\mathbf{k}^{(con)}_{N_{(e)}\times N_{(e)}} \equiv [k_0^{ab}],$$
$$\bar{\mathbf{k}}^{(u)}_{N_{(e)}\times N_{(e)}} \equiv [\bar{k}^{ab}(\mathbf{q})],$$
$$\bar{\bar{\mathbf{k}}}^{(u)}_{N_{(e)}\times N_{(e)}} \equiv [\bar{\bar{k}}^{ab}(\mathbf{q})],$$
$$\mathbf{k}^{(\sigma)}_{N_{(e)}\times N_{(e)}} \equiv [k^{ab}(\mathbf{S}_t)].$$

The equations

$$\mathbf{u}_t(\mathbf{X}) = \mathbf{F}^a(\mathbf{X})q_a(t)$$

and

$$\mathbf{E}_t(\mathbf{X}) = [\mathbf{B}^a(\mathbf{X})+\bar{\mathbf{B}}^a(\mathbf{X}, \mathbf{q}(t))]q_a(t)$$

---

[1] For the sake of notational simplicity we omit the arguments. The subindices R, C in any matrix $\mathbf{A}_{R\times C}$ are meant to indicate that the matrix has R rows and C columns (the so-called $R\times C$ matrix).

are given by

$$\mathbf{u}_{6\times 1} = \boldsymbol{\varphi}_{6\times N_{(e)}}\mathbf{q}_{N_{(e)}\times 1},$$
$$\mathbf{E}_{6\times 1} = (\mathbf{B}_{6\times N_{(e)}}+\overline{\mathbf{B}}_{6\times N_{(e)}})\mathbf{q}_{N_{(e)}\times 1}, \tag{3.81}$$

so that the incremental equations (3.76) assume the form

$$(\mathbf{k}^{(con)}_{N_{(e)}\times N_{(e)}}+\mathbf{k}^{(\sigma)}_{N_{(e)}\times N_{(e)}}+\mathbf{k}^{(u)}_{N_{(e)}\times N_{(e)}})\Delta\mathbf{q} = \Delta\mathbf{Q} \tag{3.82}$$

with

$$\mathbf{k}^{(u)}_{N_{(e)}\times N_{(e)}} \equiv \overline{\mathbf{k}}^{(u)}_{N_{(e)}\times N_{(e)}}+\overline{\overline{\mathbf{k}}}^{(u)}_{N_{(e)}\times N_{(e)}}, \tag{3.83}$$

and, by eqs. (3.75)

$$\mathbf{k}^{(con)}_{N_{(e)}\times N_{(e)}} = \int\limits_{\Omega_e} \mathbf{B}^T_{6\times N_{(e)}}\mathbf{C}_{6\times 6}\mathbf{B}_{6\times N_{(e)}}\,dV,$$

$$\overline{\mathbf{k}}^{(u)}_{N_{(e)}\times N_{(e)}} = 2\int\limits_{\Omega_e} (\overline{\mathbf{B}}^T_{6\times N_{(e)}}\mathbf{C}_{6\times 6}\mathbf{B}_{6\times N_{(e)}}+\mathbf{B}^T_{6\times N_{(e)}}\mathbf{C}_{6\times 6}\overline{\mathbf{B}}_{6\times N_{(e)}})\,dV, \tag{3.84}$$

$$\overline{\overline{\mathbf{k}}}^{(u)}_{N_{(e)}\times N_{(e)}} = 4\int\limits_{\Omega_e} \overline{\mathbf{B}}^T_{6\times N_{(e)}}\mathbf{C}_{6\times 6}\overline{\mathbf{B}}_{6\times N_{(e)}}\,dV.$$

Similarly, eq. (3.79) can be brought to the matrix form describing one element as

$$\mathbf{C}_{N_{(e)}\times N_{(e)}}\dot{\mathbf{p}}_{N_{(e)}\times 1}+\mathbf{k}_{N_{(e)}\times N_{(e)}}\mathbf{p}_{N_{(e)}\times 1} = \mathbf{e}_{N_{(e)}\times 1}. \tag{3.85}$$

3.4.7 *Interaction between elements*

The general idea of the interaction between adjacent elements of the structure at hand is very simple. From the physical point of view it is based on the condition that any two adjacent structural elements cannot deform independently of each other. In order to realize this condition we have to interrelate vector functions which determine deformations of the structural elements under consideration. This can be done by assuming that some of the vector components $q_1(\tau), \ldots, q_m(\tau)$ which by means of eq. (3.48) determine the deformation $\boldsymbol{\chi}(\mathbf{X}, \tau)$, $\mathbf{X}\in\overline{\Omega}_e$, of the structural element, are used for the description of the deformation in the adjacent elements. To outline this procedure let us consider two structural elements represented in the region Ω by its two subregions Ω_1, Ω_2 such that $\partial\Omega_1\cap\partial\Omega_2 \neq \emptyset$. Any two such elements are called the *adjacent elements*. Formulae (3.48) for the adjacent elements yields

$$\boldsymbol{\chi}(\mathbf{X}, \tau) = \mathbf{F}^a_1(\mathbf{X})q^1_a(\tau)+\boldsymbol{\chi}_0(\mathbf{X}), \quad \mathbf{X}\in\overline{\Omega}_1,$$
$$\boldsymbol{\chi}(\mathbf{X}, \tau) = \mathbf{F}^a_2(\mathbf{X})q^2_a(\tau)+\boldsymbol{\chi}_0(\mathbf{X}), \quad \mathbf{X}\in\overline{\Omega}_2,\ \tau\in[t_0, t_f]. \tag{3.86}$$

The values of the deformation $\boldsymbol{\chi}(\mathbf{X}, \tau)$ as well as the values of the deformation gradient $\nabla\boldsymbol{\chi}(\mathbf{X}, \tau)$ for every $\mathbf{X}\in\partial\Omega_1\cap\partial\Omega_2$ which are calculated from both the foregoing formulae have to be equal. This is due to the assumption that the deformation function together with its material and time derivatives cannot suffer discontinuities when passing from one structural element (say, that represented by the subregion Ω_1) to an adjacent one (say, that represented by Ω_2). However, in most cases the constraints imposed on the deformation of adjacent elements imply that such continuity conditions cannot be satisfied exactly and therefore must be replaced by some weaker conditions. For example, we can restrict ourselves to the continuity conditions for the deformation $\boldsymbol{\chi}(\mathbf{X}, t)$ imposed only at a finite number of points along the interelement boundary $\partial\Omega_1 \cap \partial\Omega_2$. If $\bar{\mathbf{X}}$ is one of such points then eqs. (3.86) yield the condition

$$\mathbf{F}_1^a(\bar{\mathbf{X}})q_a^1(\tau) = \mathbf{F}_2^a(\bar{\mathbf{X}})q_a^2(\tau), \qquad \tau \in [t_0, t_f],$$

which interrelates the components of the vectors $\mathbf{q}^1(\tau) \equiv (q_1^1(\tau), \dots, q_m^1(\tau))$ and $\mathbf{q}^2(\tau) \equiv (q_1^2(\tau), \dots, q_m^2(\tau))$. In the second part of the book we shall see that such limited continuity conditions may be fulfilled in practice by making some of the components of the vector $\mathbf{q}^1(\tau)$ equal to the pertinent components of the adjacent element vector $\mathbf{q}^2(\tau)$.

3.5 Evaluation and correction of solutions

3.5.1 Residual reactions

Using the term "problem" we shall mean now a problem of finding deformations, stresses, temperature distributions and heat fluxes in a certain structural element. It can be seen that the specification of constraints for deformations and temperature fields which is to be understood as the specification of the sets $\mathscr{D}_\tau(\Omega)$, $\mathscr{T}_\tau(\Omega)$, $\tau \in [t_0, t_f]$ (and hence the specification of the sets $\mathscr{U}_t = \mathscr{U}_t(\boldsymbol{\chi}(\cdot, t))$, $\mathscr{Z}_t = \mathscr{Z}_t(\theta(\cdot, t))$ in (2.14), (2.15)), plays an important part in the mathematical modelling of structural mechanics problems. Up to now we have always assumed that such specific forms of these sets are given. We have shown how the choice of constraints makes it possible to formulate various engineering theories of structures. In this section we shall come back to some general implications of internal constraints on the mathematical description of problems in thermomechanics.

We recall that the internal constraints for deformations $\boldsymbol{\chi}(\cdot, \tau)$ and for temperature fields $\theta(\cdot, \tau)$, $\tau \in [t_0, t_f]$, as determined by the sets $\mathscr{D}(\Omega)$, $\mathscr{T}(\Omega)$ are postulated mainly in order to simplify the mathematical description of various engineering problems and to facilitate their solution. For the sake of simplicity we put aside for the moment all the internal constraints which are due to the material properties of the body (such as incompressibility of the material). Thus the form of the internal constraints is not determined on the basis of physical informations about the structural element but refers rather to certain a priori hypotheses concerning the possible deformations and temperature distributions in this element. These hypotheses are based, as a rule, on rather general premises concerning the expected behaviour of the structural element in the given problem. Such an approach is well known in the field of structural mechanics constituting the basis for various engineering theories of beams, plates, shells etc. Thus in thermomechanics we deal with various mathematical descriptions of a certain fixed engineering problem, or of a certain class of such problems. The various mathematical descriptions correspond to different internal constraints. The natural question which arises immediately is how to compare the "quality" of these mathematical models constructed for the given problem by postulating different internal constraints. The answer to this question obtained on the grounds of the analysis of a physical sense of the reactions to the constraints will be outlined below.

To begin with let us observe that in every problem under consideration in which internal constraints are involved, we deal with two kinds of reactions. Firstly, there are reactions the existence of which can be explained in terms of the character of physical interactions between the structural element and its exterior. The reactions are due to the boundary constraints for deformations and temperature distributions. As a simple example of such reactions we can take the reaction forces in supports of the element which naturally have to be accounted for in constructing the mathematical description of this element. Such reactions are called the support reactions and are denoted by the symbols $\check{\mathbf{s}}_t(\mathbf{X})$, $\check{\pi}_t(\mathbf{X})$, $\mathbf{X} \in \partial\Omega$. Obviously, $\check{\mathbf{s}}_t(\mathbf{X}) = 0$ on the part of the boundary $\partial\Omega$ which is free from any supports (at the time instant t), whereas $\check{\pi}_t(\mathbf{X}) = 0$ on the part of the boundary across which a heat supply is known (at the time instant t). Under the assumption that there are no internal constraints due to the material properties of the body, all the internal reactions $\bar{\mathbf{b}}_t(\mathbf{X})$, $\bar{\alpha}_t(\mathbf{X})$, $\mathbf{X} \in \Omega$, together with the reactions $\bar{\mathbf{s}}_t(\mathbf{X}) - \check{\mathbf{s}}_t(\mathbf{X})$, $\bar{\pi}_t(\mathbf{X}) - \check{\pi}_t(\mathbf{X})$, $\mathbf{X} \in \partial\Omega$, can be roughly interpreted as reactions to constraints introduced to simplify the mathematical description of the given problem. Setting

$$\mathbf{r}_\tau(\mathbf{X}) = \begin{cases} (\bar{\mathbf{b}}_\tau(\mathbf{X}), \alpha_\tau(\mathbf{X})) & \text{for} \quad \mathbf{X} \in \Omega, \\ (\bar{\mathbf{s}}(\mathbf{X}) - \check{\mathbf{s}}(\mathbf{X}), \bar{\pi}(\mathbf{X}) - \check{\pi}(\mathbf{X})) & \text{for} \quad \mathbf{X} \in \partial\Omega, \end{cases}$$

the fields $\mathbf{r}_t(\mathbf{X})$, $\mathbf{X} \in \bar{\Omega}$, will be referred to as the residual reactions. If all the thermal effects can be neglected then

$$\mathbf{r}_\tau(\mathbf{X}) = \begin{cases} \bar{\mathbf{b}}_\tau(\mathbf{X}) & \text{for} \quad \mathbf{X} \in \Omega, \\ \bar{\mathbf{s}}_\tau(\mathbf{X}) - \check{\mathbf{s}}_\tau(\mathbf{X}) & \text{for} \quad \mathbf{X} \in \partial\Omega. \end{cases}$$

The residual reactions $\mathbf{r}_\tau(\mathbf{X})$, $\tau \in [t_0, t_f]$, may not be defined on some surfaces, lines or points in Ω as well as on certain lines or points on $\partial\Omega$.

It must be stressed that the decomposition of reactions into the support reactions and the residual reactions does not correspond to the decomposition of constraints into the boundary and internal ones, since the boundary constraints are superimposed on the internal constraints, cf. the definitions of sets $\mathscr{D}_\tau(\Omega)$, $\mathscr{T}_\tau(\Omega)$ in Sec. 2.3.1, and the support reactions are implied by both the boundary and internal constraints. It may happen, however, that for a certain problem in structural mechanics the constraints will be such that no residual reactions will be produced. In order to realize that such a situation is possible we can give the following example. Let us postulate in a given problem some internal and boundary constraints and assume that we are able to find the unique solution to the constrained problem. In this way we also find the reactions. Consider now a modified problem in which all the calculated residual reactions are treated as the known external agents. It is clear that in such a modified problem (which, as a rule, will appear very artificial from the engineering point of view) all the residual reaction will disappear. It follows from eqs. (2.13), (2.14) that problems in which the residual reactions are equal to zero can be treated as problems without any internal constraints, i.e. problems in which no simplifying hypotheses about the deformations $\boldsymbol{\chi}(\cdot, \tau)$ and the temperature fields $\theta(\cdot, \tau)$, $\tau \in [t_0, t_f]$, inside the body are postulated a priori. Hence, the residual reactions constitute a basis of informations on how the internal constraints influence the problem solution. It is evident that, roughly speaking, we can expect "better" solutions for "smaller" residual reactions.

3.5.2 *Evaluation of solutions*

In order to characterize the magnitude of the residual reactions we shall assign to every field $\mathbf{r}(\cdot)$ which can represent residual reactions, a nonnegative number $N(\mathbf{r}(\cdot))$ such that the following conditions hold:

1. $N(\lambda\mathbf{r}(\cdot)) = |\lambda| N(\mathbf{r}(\cdot))$ for every real number λ and every $\mathbf{r}(\cdot)$,

2. $N(\mathbf{r}_1(\cdot)) + N(\mathbf{r}_2(\cdot)) \geqslant N(\mathbf{r}_1(\cdot) + \mathbf{r}_2(\cdot))$ for every $\mathbf{r}_1(\cdot)$, $\mathbf{r}_2(\cdot)$. The number $N(\mathbf{r}(\cdot))$ is said to be a *seminorm* of $\mathbf{r}(\cdot)$[1]. It can be observed that the seminorm is not uniquely defined. As simple example of seminorms we can take the values of the resultant force (or of the resultant moment) of all the residual reactions acting on the whole body or its arbitrary, fixed part.

---

[1] We have here assumed that the set of all possible residual reactions constitute a linear space, cf. footnote on page 45 in Sec. 2.3.3. If 0 is a zero element of this space (i.e. if all the residual reactions can be neglected) then the condition 1 implies that $N(0) = 0$ but in general $N(\mathbf{r}(\cdot)) = 0$ may not imply $\mathbf{r}(\cdot) = \mathbf{0}$.

Assume now that we deal with a structural mechanics problem for which we decided to introduce two mathematical models. Let these models be based on two different internal constraints defined by two pairs of sets $(\mathscr{D}'(\Omega), \mathscr{T}'(\Omega))$ and $(\mathscr{D}''(\Omega), \mathscr{T}''(\Omega))$. Moreover, let all other data characterizing the problem, including boundary constraints given by $\mathscr{D}_\tau(\partial\Omega)$, $\mathscr{T}_\tau(\partial\Omega)$, $\tau \in [t_0, t_f]$, be identical in both cases. Assume that we have carried out with the same accuracy all the necessary calculations leading to two pertinent solutions. If the calculated residual reactions $r'_\tau(\cdot)$, $r''_\tau(\cdot)$, $\tau \in [t_0, t_f]$, satisfy the following condition for every τ:

$$N_\gamma(r'_\tau(\cdot)) \leqslant N_\gamma(r''_\tau(\cdot)), \quad \gamma = 1, \ldots, g, \quad g \geqslant 1, \quad \tau \in [t_0, t_f], \tag{3.87}$$

where every $N_\gamma(\cdot)$, $\gamma = 1, \ldots, g$, is an arbitrary but properly chosen and fixed seminorm, then the solution for the internal constraints $(\mathscr{D}'(\Omega), \mathscr{T}'(\Omega))$ can be treated as "better" than the solution for the internal constraints $(\mathscr{D}''(\Omega), \mathscr{T}''(\Omega))$. Also, the model of the problem considered based on the internal constraints $(\mathscr{D}'(\Omega), \mathscr{T}'(\Omega))$ can be justly called better than that based on the internal constraints $(\mathscr{D}''(\Omega), \mathscr{T}''(\Omega))$. It has to be remembered that:

1. The above mutual evaluation of different solutions and different internal constraints for the given structural mechanics problem depends heavily on the choice of the seminorm $N_\gamma(\cdot)$, $\gamma = 1, \ldots, g$,

2. The model which is better in one problem can be worse in another one even if we deal in both the cases with the same set of seminorms $N_\gamma(\cdot)$ for residual reactions,

3. We can compare various solutions to a certain problem, which are based on various internal constraints, only if all other data concerning this problem coincide (or the difference between them can be neglected from the physical or engineering point of view).

It has to be also emphasized that the mutual, or relative, evaluation of certain mathematical models for the given problem yields no information about the adequacy between the behaviour of the real structural element and its various mathematical representations within the framework of structural mechanics.

Apart from the relative criterion (3.87), in many special problems we can introduce an absolute criterion of the form

$$N_\gamma(r_\tau(\cdot)) \leqslant \lambda_\gamma, \quad \tau \in [t_0, t_f], \quad \gamma = 1, \ldots, g, \tag{3.88}$$

where $\lambda_1, \ldots, \lambda_g$ are postulated a priori positive numbers which can be called the absolute maximal admissible "errors" of the solution in the seminorms $N_\gamma(\cdot)$, $\gamma = 1, \ldots, g$, respectively. In engineering problems in which the deformations and temperature distributions can be treated as depending solely on the known external agents $\mathbf{b}_\tau(\cdot)$, $\alpha_\tau(\cdot)$, $\mathbf{s}_\tau(\cdot)$, $\pi_\tau(\cdot)$ for $\tau \in [t_0, t_f]$[1], by setting

$$\mathbf{a}_\tau(\mathbf{X}) = \begin{cases} (\mathbf{b}_\tau(\mathbf{X}), \alpha_\tau(\mathbf{X})) & \text{for} \quad \mathbf{X} \in \Omega, \\ (\mathbf{s}_\tau(\mathbf{X}) - \check{\mathbf{s}}_\tau(\mathbf{X}), \pi_\tau(\mathbf{X}) - \check{\pi}_\tau(\mathbf{X})) & \text{for} \quad \mathbf{X} \in \partial\Omega, \end{cases}$$

we can assume that

$$\lambda_\gamma = \alpha N_\gamma(\mathbf{a}_\tau(\cdot)), \qquad \gamma = 1, \ldots, g, \qquad \tau \in [t_0, t_f],$$

where α is the known positive number, $\alpha \ll 1$, which can be called the relative maximal admissible error of the solution obtained. The physical interpretation of condition (3.88) with $\lambda_\gamma = \alpha N_\gamma(\mathbf{a}_\tau(\cdot))$ is clear. Namely, we postulate that, roughly speaking, the residual reactions have to be sufficiently small with respect to the known external agents combined with support conditions.

3.5.3 *Correction of solutions*

Assume now that the solution to the structural mechanics problem under consideration can not be accepted in the light of the criterion (3.74). Such a situation will be typical rather than exceptional if we use internal constraints leading to the engineering theories discussed in Sec. 3.2 and 3.3. Let us observe that the values of stresses $\mathbf{S}_t(\mathbf{X})$ and heat fluxes $\mathbf{h}_t(\mathbf{X})$ have to satisfy by eqs. (3.16) the boundary conditions

$$\nabla\chi(\mathbf{X}, t)\mathbf{S}_t(\mathbf{X})\mathbf{n}(\mathbf{X}) = \mathbf{s}_t(\mathbf{X}) + \check{\mathbf{s}}_t(\mathbf{X}) + (\bar{\mathbf{s}}_t(\mathbf{X}) - \check{\mathbf{s}}_t(\mathbf{X})),$$
$$\mathbf{h}_t(\mathbf{X}) \cdot \mathbf{n}(\mathbf{X}) = \pi_t(\mathbf{X}) + \check{\pi}_t(\mathbf{X}) + (\bar{\pi}_t(\mathbf{X}) - \check{\pi}_t(\mathbf{X})), \qquad \mathbf{X} \in \partial\Omega.$$

Hence, for residual boundary reactions $(\bar{\mathbf{s}}_t(\mathbf{X}) - \check{\mathbf{s}}_t(\mathbf{X}), \bar{\pi}_t(\mathbf{X}) - \check{\pi}_t(\mathbf{X}))$, $\mathbf{X} \in \partial\Omega$, which are not sufficiently small, the values of $\mathbf{S}_t(\mathbf{X})\mathbf{n}(\mathbf{X})$, $\mathbf{h}_t(\mathbf{X})\mathbf{n}(\mathbf{X})$ can have no physical sense whatsoever. Thus we have to correct the solution or to construct another mathematical model of the given problem. Setting aside the second possibility we shall state here that the solution already obtained, as inaccurate as it may be, can still yield important informations about the problem. These informations can be treated as a basis for the introduction of certain additional constraints which have a special form given by $\mathbf{S}_\tau(\mathbf{X}) \in \{\tilde{\mathbf{S}}_\tau(\mathbf{X})\}$, $\mathbf{h}_\tau(\mathbf{X}) \in \{\tilde{\mathbf{h}}_\tau(\mathbf{X})\}$, $\mathbf{X} \in \Omega$, $\tau \in [t_0, t_f]$[2] where we deal with the known one element sets (singletons) of admissible values of stresses and heat fluxes.

---

[1] It means that deformations and temperature distributions for such problems would be constant in the time interval $[t_0, t_f]$ if the external agents were equal to zero.

[2] We assume for the time being that the deformations and temperature fields resulting from the solution need not be corrected.

At the same time the values of $\tilde{\mathbf{S}}_t(\mathbf{X})$ and $\tilde{\mathbf{h}}_t(\mathbf{X})$ are determined by the formulae of the form

$$\begin{aligned}\tilde{\mathbf{S}}_t(\mathbf{X}) &= \mathbf{S}_t^0(\mathbf{X}, \boldsymbol{\chi}^*(\cdot, t), \theta^*(\cdot, t)),\\ \tilde{\mathbf{h}}_t(\mathbf{X}) &= \mathbf{h}_t^0(\mathbf{X}, \boldsymbol{\chi}^*(\cdot, t), \theta^*(\cdot, t)),\end{aligned} \tag{3.89}$$

where $\mathbf{S}_t^0(\mathbf{X}, \cdot)$, $\mathbf{h}_t^0(\mathbf{X}, \cdot)$ are the postulated functionals with the values of their arguments obtained previously from the solution of the problem, cf. Sec. 3.2. The form of the RHS of eqs. (3.89) is based on additional physically motivated premises and it is assumed to be known in every engineering theory of thermomechanics. The additional constraints for stresses and heat fluxes introduced here and given by $\mathbf{S}_t(\mathbf{X}) = \tilde{\mathbf{S}}_t(\mathbf{X})$, $\mathbf{h}_t(\mathbf{X}) = \tilde{\mathbf{h}}_t(\mathbf{X})$, $\mathbf{X} \in \Omega$, $t \in [t_0, t_f]$, imply that $\mathscr{S}^0(\mathbf{S}(\cdot)) = \{0\}$, $\mathscr{H}^0(\mathbf{h}(\cdot)) = \{0\}$. Hence formulae (2.15) and (2.16) imply that reactions $L_t(\cdot)$, $l_t(\cdot)$ are arbitrary continuous functions which can be calculated from the pertinent constitutive relations of the form similar to that introduced in Sec. 2.3.5. For the incremental formulation we can easily go over from eqs. (3.89) to the formulae for $\Delta\tilde{\mathbf{S}}(\mathbf{X})$, $\Delta\tilde{\mathbf{h}}(\mathbf{X})$.

If additionally to stresses and heat fluxes also deformations and temperature distributions as derived from the solution to the given problem need to be corrected, then it is required to introduce additional constraints of the form $\boldsymbol{\chi}(\mathbf{X}, t) \in \{\tilde{\boldsymbol{\chi}}(\mathbf{X}, t)\}$, $\theta(\mathbf{X}, t) \in \{\tilde{\theta}(\mathbf{X}, t)\}$, $\mathbf{X} \in \Omega$, where $\tilde{\boldsymbol{\chi}}(\cdot, t)$, $\tilde{\theta}(\cdot, t)$ are known and given by

$$\begin{aligned}\tilde{\boldsymbol{\chi}}(\mathbf{X}, t) &= \boldsymbol{\chi}_t^0(\mathbf{X}, \boldsymbol{\chi}^*(\cdot, t)),\\ \tilde{\theta}(\mathbf{X}, t) &= \theta_t^0(\mathbf{X}, \theta^*(\cdot, t)),\end{aligned} \tag{3.90}$$

where the RHS of eqs. (3.90) represent certain postulated a priori functionals and the values of their arguments are determined by the solution of the given problem. The constraints here introduced imply that $\mathscr{U}_t(\tilde{\boldsymbol{\chi}}(\cdot, t)) = \{0\}$ and $\mathscr{T}_t(\tilde{\theta}(\cdot, t)) = \{0\}$ for every $t \in [t_0, t_f]$, cf. Secs. 2.3.2 and 2.3.3. Formulae (2.13), (2.14) are identically satisfied and reactions can be computed from eqs. (3.15), (3.16).

Examples of functionals appearing in eqs. (3.89), (3.90) will be given in the second part of the book. It has to be remembered that the arguments of these functionals are functions defined on hence, the aforementioned procedure is strictly related to the method of constraints and constitutes a very important supplement to this method. Discussion concerning the evaluation of solutions given in Sec. 3.4.2 remains valid also for the additional constraints introduced in this section. It is stressed in passing that eqs. (3.89), (3.90) can be used only after solving the given problem for $\boldsymbol{\chi}^*(\cdot, t)$ and $\theta^*(\cdot, t)$, $t \in [t_0, t_f]$ within a certain engineering theory, i.e., after employing the method of constraints.

3.5.4 *On adaptive solution refinements*

The evaluation of solution on the basis of residual reactions discussed in Sec. 3.5.2 and its correction mentioned in Sec. 3.5.3 can be effectively employed mainly in the case of the discrete description of structures that leads to using various numerical techniques. In this context so-called adaptive formulations of the finite element method have recently been developed [5]–[9]. Generally speaking, the idea of such adaptive techniques is based on the following computational steps:

1. Evaluation of solution via the assessment of the residual reactions. This kind of assessment is referred to as the a posteriori error estimate since a certain numerical solution to the problem has to be known. The error estimates are supposed to have a local character, being related independently to various parts of the structure.
2. Correction of solution indicated by the error estimate obtained. It has to be emphasized that because of the local character of the error estimates there exists a possibility of the local refinement of the discretized (numerical) model of the structure at hand. Thus, the local error estimates have to imply the local refinements of the model; such a feature of the procedure is referred to as the adaptivity of the model.

The main idea behind the adaptive solution refinement lies in the minimization of what can be called the engineering intuition in introducing new improved models (new internal constraints) on the basis of the known a posteriori error estimates. The intuition is replaced, at least partly, by certain formal procedures leading from the postulated a priori discretized (numerical) models of the structure to new discretized models.

So far, adaptive techniques have found numerous applications only in the framework of the finite element method. There, the adaptive techniques relate to the computation of local a posteriori errors that automatically imply a new discretization mesh or new interpolating functions. The list of references on this subject is rather extensive; for the foundations and examplary applications the reader is referred to [5]–[9].

Summary

The results of this chapter have been based on the physical notions and the fundamental principles of continuum mechanics (Chapter 1) as well as on the informations about the structural element to be analysed such as material properties, external agents and constraints (Chapter 2). In this way we have been able to formulate in Sec. 3.1 the general form of governing relations of structural and solid mechanics. Aiming at further applications we have used the variational formulation in its finite and incremental forms. After having postulated some special classes of internal constraints (motivated by the shape of the structural element, the character of external loadings, etc.) we could pass from a general variational formulation to various engineering theories of structural mechanics and thermomechanics. The approach proposed has been called the method of constraints; by using it consistently we have next obtained in succession the governing equations for shell and plate theories, rod and beams theories and plane problems. In Sec. 3.3 we have passed from finite to incremental formulation of the above engineering theories.

Solutions to various special problems obtained within such theories can be compared and evaluated; in many cases they also need to be corrected. In Sec. 3.4 we have used the concept of constraints and reactions to them in order to discuss and correct solutions to various special problem.

This chapter closes the theoretical foundation of nonlinear mechanics of structures. The general governing equations obtained in Secs. 3.2 and 3.3 constitute the point of departure for the formulations and solutions of various specific engineering problems to be discussed in subsequent parts of the book.

Problems

3.1 Consider eqs. (3.1)–(3.4) and derive the resulting form of the variational equation for linear thermo-elastic material under the assumption that the RHS of eqs. (3.4) are known functions of $\mathbf{X}$ and t.

3.2 Derive the linearized form of eqs. (3.6), (3.7) for the general case and for isotropic thermo-elastic material.

Hint: consider the constitutive equations introduced at the end of Sec. 2.1.2.

3.3 Derive the incremental variational formulation for the case of "small" (linear) increments, in which all nonlinear terms with respect to any increments may be neglected.

3.4 Consider eq. (3.13) in the special cases of uniaxial and plane state of stress.

3.5 Consider eqs. (3.15), (3.16) and specify them by assuming that internal constraints are absent and boundary constraints are given by $\mathbf{u}_\tau(\mathbf{X}) = \mathbf{0}$, $\vartheta_\tau(\mathbf{X}) = 0$, $\mathbf{X} \in \Gamma$, where Γ is a known part of the boundary $\partial\Omega$ and $\tau \in [t_0, t_f]$.

3.6 Using results of Sec. 3.1.4 find conditions which have to be satisfied by the reactions to constraints introduced in Problem 3.4.

3.7 Generalize the variational equation (3.13) to make it consistent with the internal constraints given by

$$\boldsymbol{\chi}(\mathbf{X}, t) = \boldsymbol{\Phi}(\mathbf{X}, \boldsymbol{\chi}^*(\mathbf{X}^*, t)),$$
$$\boldsymbol{\chi}^*(\mathbf{X}^*, t) \equiv (\chi_1^*(\mathbf{X}^*, t), \ldots, \chi_m^*(\mathbf{X}^*, t)),$$

where $\boldsymbol{\Phi}(\cdot)$ is a differentiable function of all its arguments.

3.8 Derive the variational equation for the internal temperature constraints given by

$$\theta(\mathbf{X}, t) = f(\mathbf{X}, \theta^*(\mathbf{X}^*, t)),$$
$$\theta^*(\mathbf{X}, t) \equiv (\theta_1^*(\mathbf{X}^*, t), \ldots, \theta_n^*(\mathbf{X}^*, t)),$$

where $f(\cdot)$ is a differentiable function of all its arguments.

3.9 Generalize the results of Problem 3.7 to the case in which $\mathbf{\Phi}(\cdot)$ depends also on derivatives of $\boldsymbol{\chi}^*(\mathbf{X}^*, t)$ with respect to $\mathbf{X}^*$.

3.10 Derive the general variational equations for shell and plate theories assuming that the internal constraints are given by

$$\boldsymbol{\chi}(\mathbf{X}^*, \eta, t) = \mathbf{\Phi}(\eta, \boldsymbol{\chi}^*(\mathbf{X}^*, t)),$$

where $\mathbf{\Phi}(\cdot)$ is a differentiable function.

3.11 Consider eqs. (3.32) and (3.34) and specify them to the case of linear thermo-elasticity; assume that body loadings can be neglected and boundary loadings are independent of deformations.

3.12 Formulate the governing relations of shell and plate theories under the assumption that the internal constraints are given by eqs. (3.41).

3.13 Solve Problem 3.12 under the additional condition that eqs. (3.42) hold.

3.14 Derive the linear equations of elastic shell and plate theories for the internal constraints given by eqs. (3.41).

Hint: assume $\mathbf{p}(\mathbf{X}^*, t) = \mathbf{p}_0(\mathbf{X}^*)+\Delta\mathbf{p}(\mathbf{X}^*, t)$, $\mathbf{d}(\mathbf{X}^*, t) = \mathbf{d}_0(\mathbf{X}^*)+\Delta\mathbf{d}(\mathbf{X}^*, t)$, where $\mathbf{p}_0(\cdot)$, $\mathbf{d}_0(\cdot)$ are known and then linearize the equations with respect to small increments $\Delta\mathbf{p}(\mathbf{X}^*, t)$, $\Delta\mathbf{d}(\mathbf{X}^*, t)$.

3.15 Solve Problem 3.14 under the condition that eqs. (3.42) hold and the shell material is isotropic.

3.16 Derive the governing relations for plane problems of linear elastic isotropic materials.

Hint: assume $\chi_K^*(\mathbf{X}^*, t) = X_K^* + u_K(\mathbf{X}^*, t)$ and linearize the equations with respect to the displacements $u_K(\mathbf{X}^*, t)$.

3.17 Consider eqs. (3.51)–(3.54) and go over to the case in which all the equations are linear with respect to the increments.

3.18 Derive the incremental formulation of shell and plate theories for the internal constraints given by eqs. (3.41).

3.19 Solve Problem 3.18 under the assumption that eqs. (3.42) hold.

4

Extensions and specifications of the general theory

Purpose of the chapter

As pointed out at the beginning of Section 3.4 our goal is to construct a discretized model of structure which on the basis of single element description will yield equations describing the whole system within the framework of a numerically-oriented formulation.

This book is not a finite element book. Therefore, no complete account of the finite element foundations and methodology will be given. The interested reader is referred to any of a great number of relevant books on this subject, [13], [39], [42], [44]–[46], [49], [79], [121], [136], [137], [183]. Instead, we describe the fundamentals of discretized formulation for structural mechanics problems by fully exploiting the concept of internal constraints. However, as we shall see in this chapter, equations will have the form to which the bulk of existing, highly sophisticated finite element techniques will directly apply. A general description of the relevant algebraic procedures follows in Chapter 5.

In this chapter we are going to explore in a more detailed way the possibilities of effective structural analysis opened up by having considered various discretizing constraints. Thus, we shall start with giving an outline of the approach typically employed to obtain the overall equations describing the discretized structural model on the basis of single element equations. We shall then consider a number of simple elements with the aim of illustrating the procedure rather than attempting to review existing element formulations. An unconventional approach to the analysis of lattice-type structures made of beams distributed regularly over given surfaces is discussed next. We shall also give in this chapter a more detailed description of thermo-elastic-plastic materials as well as discuss the fundamentals of elastic-plastic analysis for structures subjected to non-proportionally varying external loads and temperatures.

4.1 Domain discretization

To construct a discretized model of the structural mechanics problem we must first construct a discretized model of the domain $\bar{\Omega} = \Omega \cup \partial\Omega$, $\Omega \subset \boldsymbol{R}^n$, in which the problem is defined. To this aim we have first to divide the region $\bar{\Omega}$ into a number of elements and then to select some characteristic points on the interelement boundaries (and, sometimes, within the elements as well). The elements are referred to as "finite elements" and the points as nodal points or "nodes". The elements should have as simple a shape as possible. They should also be able to closely reproduce the possibly curvilinear boundaries of the domain $\bar{\Omega}$. We shall say that

> A *discretized model of the domain* $\bar{\Omega} \subset \boldsymbol{R}^n$ is a region $\tilde{\Omega} \subset \boldsymbol{R}^n$ which is the union of a finite number E of (closed, bounded) subregions Ω_e of $\boldsymbol{R}^n$, $e = 1, 2, \ldots, E$; $\bar{\Omega}_e = \Omega_e \cup \partial\Omega_e$, called *finite elements of* $\tilde{\Omega}$.

The region $\tilde{\Omega}$ should obviously be selected in such a way that it closely approximates $\bar{\Omega}$. The finite elements $\bar{\Omega}_e$ taken without boundaries (i.e. the open elements Ω_e) are pairwise disjoint. We have

$$\tilde{\Omega} = \bigcup_{e=1}^{E} \bar{\Omega}_e, \qquad \Omega_e \cap \Omega_f = \emptyset, \qquad e \neq f;\, e, f = 1, 2, \ldots, E.$$

In the discretized model of $\bar{\Omega}$ we select next a finite number N of points, called *global nodes*, and we label them as $X^1, X^2, \ldots, X^N$. Within each element we identify a number $N_{(e)}$ of points, called *local nodes*, and label them as $X^1_{(e)}, X^2_{(e)}, \ldots, X^{N(e)}_{(e)}$, $e = 1, 2, \ldots, E$.

A typical discretization of a two-dimensional problem with a curvilinear boundary is shown in Fig. 13. The region $\tilde{\Omega}$ is taken as the collection of three

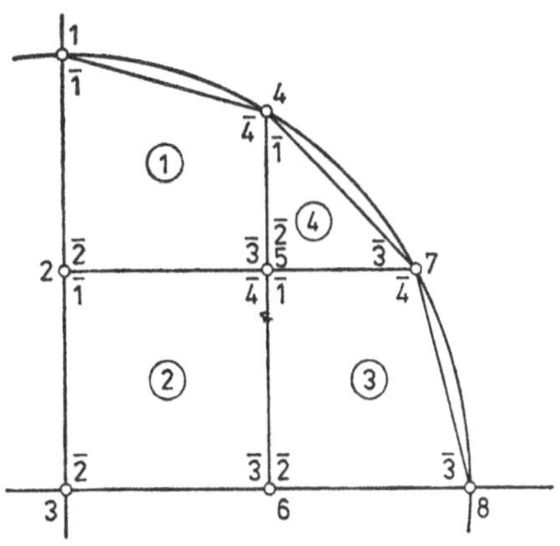

Fig. 13 Discretization of 2D problem

quadrilateral finite elements labelled 1, 2, 3 and of one triangular element labelled 4. They approximate the region $\bar{\Omega}$ assumed to be a sector of a circle. The numbers $1, 2, \ldots, 8$ stand for the global nodes, $N = 8$ whereas the sets of numbers $\bar{1}, \bar{2}, \bar{3}, \bar{4}$ or $\bar{1}, \bar{2}, \bar{3}$ stand for the local nodes in each element, $N_{(1)} = 4$, $N_{(2)} = 4$, $N_{(3)} = 4$, $N_{(4)} = 3$.

It is clear that a correspondence must exist between the global and local nodes if the elements are to fit together smoothly. This correspondence is established in the form of the following element connectivity table, Table 1.

Table 1 Element connectivity table

Element	Nodes locally labelled as $\bar{1}$	$\bar{2}$	$\bar{3}$	$\bar{4}$	Type
1	1	2	5	4	quadrilateral
2	2	3	6	5	quadrilateral
3	5	6	8	7	quadrilateral
4	4	5	7	—	triangular

For each element we construct now a Boolean (zero-one) connectivity matrix of the dimension $N_{(e)} \times N$ as

$$\Lambda^{(1)}_{4\times 8} = \begin{bmatrix} 1&0&0&0&0&0&0&0\\ 0&1&0&0&0&0&0&0\\ 0&0&0&0&1&0&0&0\\ 0&0&0&1&0&0&0&0 \end{bmatrix},$$

$$\Lambda^{(2)}_{4\times 8} = \begin{bmatrix} 0&1&0&0&0&0&0&0\\ 0&0&1&0&0&0&0&0\\ 0&0&0&0&0&1&0&0\\ 0&0&0&0&1&0&0&0 \end{bmatrix}, \tag{4.1}$$

$$\Lambda^{(3)}_{4\times 8} = \begin{bmatrix} 0&0&0&0&1&0&0&0\\ 0&0&0&0&0&1&0&0\\ 0&0&0&0&0&0&0&1\\ 0&0&0&0&0&0&1&0 \end{bmatrix},$$

$$\Lambda^{(4)}_{3\times 8} = \begin{bmatrix} 0&0&0&1&0&0&0&0\\ 0&0&0&0&1&0&0&0\\ 0&0&0&0&0&0&1&0 \end{bmatrix}.$$

The collection of all E matrices of this type makes it possible to put effectively all the elements into their proper positions in the discretized model.

We note in passing that the only approximation made so far was the replacement of the original region $\bar{\Omega}$ by its approximate $\tilde{\Omega}$. The discrete structure of $\tilde{\Omega}$ has been introduced to serve later as a convenient basis for constructing discretized models of the structural mechanics problems to be subsequently considered.

4.2 Solving differential equations by weighted residual methods

It is widely acknowledged that in order to solve approximately a differential equation it is usually advantageous to use for the unknown function an expansion in terms of the so-called trial functions. To be more explicit let us now consider any set of linear differential equations written in the symbolic form as

$$A_{ij}u_j(x_s)+B_i(x_s) = 0, \quad x_s \in \Omega, \qquad i,j = 1,2,\ldots,n, \quad s = 1,2,3 \tag{4.2}$$

and supplemented by boundary conditions

$$A_{rj}^{(b)}u_j(x_s)+B_r^{(b)}(x_s) = 0, \quad x_s \in \partial\Omega, \quad r = 1,2,\ldots,n'. \tag{4.3}$$

The symbols A_{ij} and $A_{rj}^{(b)}$ stand for linear differential operator matrices whereas $u_j(x_s)$ is an unknown function assumed sufficiently smooth. If we now postulate an expansion

$$u_i \cong \tilde{u}_i = \sum_{\alpha=1}^{N} \varphi_{i\alpha}(x_s)r_\alpha \tag{4.4}$$

where $\varphi_{11}, \varphi_{12}, \ldots, \varphi_{nN}$ is a set of independent, known trial functions (not necessarily satisfying the boundary conditions) then, by introducing (4.4) into (4.2), (4.3) we obtain

$$\begin{aligned} A_{ij}\tilde{u}_j+B_i &= S_i \quad \text{in } \Omega, \\ A_{rj}^{(b)}\tilde{u}_j+B_r^{(b)} &= S_r^{(b)} \quad \text{on } \partial\Omega. \end{aligned} \tag{4.5}$$

The domain residuals S_i and the boundary residuals $S_r^{(b)}$ are in general different from zero and are due to the approximate character of the anticipated solution (4.4).

We can now attempt to reduce the weighted "sum" of the residuals S_i and $S_r^{(b)}$ by writing

$$\int_{\Omega} W_{\alpha i}S_i\,\mathrm{d}\Omega + \int_{\partial\Omega} W_{\alpha r}^{(b)}S_r^{(b)}\,\mathrm{d}(\partial\Omega) = 0, \qquad \alpha = 1,2,\ldots,N, \quad i = 1,2,\ldots,n, \quad r = 1,2,\ldots,n' \tag{4.6}$$

where, in general, the weighting functions $W_{\alpha i}$ and $W_{\alpha r}^{(b)}$ can be chosen independently.

Clearly, if eq. (4.6) is satisfied for a very large number of arbitrary functions $W_{\alpha i}$ and $W_{\alpha r}^{(b)}$ then the approximation $\tilde{u}_i$ in eq. (4.4) must approach the exact solution u_i, provided the expansion (4.4) is capable of so doing. Introducing (4.4) into (4.5) and the resulting equations into (4.6) we obtain

$$\int_{\Omega} W_{\alpha i} A_{ij} \varphi_{j\beta} r_\beta \, \mathrm{d}\Omega + \int_{\partial\Omega} W_{\alpha r}^{(b)} A_{rj}^{(b)} \varphi_{j\beta} r_\beta \, \mathrm{d}(\partial\Omega) + \\ + \int_{\Omega} W_{\alpha i} B_i \, \mathrm{d}\Omega + \int_{\partial\Omega} W_{\alpha r}^{(b)} B_r^{(b)} \, \mathrm{d}(\partial\Omega) = 0 \qquad (4.7)$$

or, in a more compact form

$$K_{\alpha\beta} r_\beta = R_\alpha, \qquad \alpha, \beta = 1, 2, \ldots, N, \qquad (4.8)$$

where

$$K_{\alpha\beta} = \int_{\Omega} W_{\alpha i} A_{ij} \varphi_{j\beta} \, \mathrm{d}\Omega + \int_{\partial\Omega} W_{\alpha r}^{(b)} A_{rj}^{(b)} \varphi_{j\beta} \, \mathrm{d}(\partial\Omega), \\ R_\alpha = \int_{\Omega} W_{\alpha i} B_i \, \mathrm{d}\Omega + \int_{\partial\Omega} W_{\alpha r}^{(b)} B_r^{(b)} \, \mathrm{d}(\partial\Omega). \qquad (4.9)$$

The set of N linear algebraic equations (4.8) may now be solved for N unknowns r_α, $\alpha = 1, 2, \ldots, N$ and by using this solution in eq. (4.4) the approximation $\tilde{u}_i$ is obtained. We note that the most common choice of weighting functions $W_{\alpha i}$ is to take them as the trial functions themselves, $W_{\alpha i} = \varphi_{i\alpha}$. For $S_r^{(b)} = 0$ the name of Galerkin is usually attached to this version of the method.

It is possible to generalize the method of weighted residuals to cover nonlinear problems as well. To show this let us symbolically write the model nonlinear problem in the form

$$A_{ij}(u_k) u_j + B_i = 0 \quad \text{in } \Omega, \\ A_{rj}^{(b)}(u_k) u_j + B_r^{(b)} = 0 \quad \text{on } \partial\Omega, \qquad (4.10)$$

where $A_{ij}(u_k)$, $A_{rj}^{(b)}(u_k)$ are operator matrices dependent on the unknown vector function u_k. By following the same procedure as above we readily arrive at the set of nonlinear algebraic equations

$$K_{\alpha\beta}(r_\gamma) r_\beta = R_\alpha \qquad (4.11)$$

in which

$$K_{\alpha\beta}(r_\gamma) = \int_{\Omega} W_{\alpha i} A_{ij}(\varphi_{k\gamma} r_\gamma) \varphi_{j\beta} \, \mathrm{d}\Omega + \int_{\partial\Omega} W_{\alpha r}^{(b)} A_{rj}^{(b)}(\varphi_{k\gamma} r_\gamma) \varphi_{j\beta} \, \mathrm{d}(\partial\Omega). \qquad (4.12)$$

The nonlinear equations set (4.11) requires iterative solution, and any of the many standard methods can be applied. The simplest one (and often used in practical computations) consists in taking some initial guess $r_\alpha = r_\alpha^{(0)}$ and evaluating the matrix

$$K_{\alpha\beta}(r_\alpha^{(0)}) = K_{\alpha\beta}^{(0)}. \tag{4.13}$$

An improved approximation for r_α can be obtained as

$$r_\alpha^{(1)} = \overset{-1}{K}{}_{\alpha\beta}^{(0)} R_\beta, \tag{4.14}$$

provided $K_{\alpha\beta}^{(0)}$ is non-singular which is here assumed. The process can be continued in the form

$$r_\alpha^{(k)} = \overset{-1}{K}{}_{\alpha\beta}^{(k-1)} R_\beta, \quad k = 1, 2, \ldots \tag{4.15}$$

until the difference between $r_\alpha^{(k)}$ and $r_\alpha^{(k-1)}$ is within a suitable tolerance.

Another method of solving (4.11) is based on using a parametric representation of the right-hand side vector R_α (assumed known) and of the solution vector r_α in the form

$$\begin{aligned} R_\alpha &= R_\alpha(t), \\ r_\alpha &= r_\alpha(t). \end{aligned} \tag{4.16}$$

By taking the derivative with respect to t on both sides of eq. (4.11) we arrive at

$$\begin{aligned} &K_{\alpha\beta}(r_\gamma)\dot{r}_\beta + \frac{\partial K_{\alpha\beta}}{\partial r_\gamma}\dot{r}_\gamma r_\beta = \dot{R}_\alpha, \\ &\left[K_{\alpha\beta}(r_\gamma) + \frac{\partial K_{\alpha\gamma}}{\partial r_\beta} r_\gamma\right]\dot{r}_\beta = \dot{R}_\alpha, \\ &K_{\alpha\beta}^{(T)}(r_\gamma)\dot{r}_\beta = \dot{R}_\alpha, \end{aligned} \tag{4.17}$$

where $(\dot{\ldots}) = \dfrac{\mathrm{d}(\ldots)}{\mathrm{d}t}$ and $K_{\alpha\beta}^{(T)}$ is known as the "tangent" coefficient matrix. Eq. (4.17), when supplemented by a suitable initial condition

$$r_\alpha|_{t=0} = r_\alpha(0) = \hat{r}_\alpha, \tag{4.18}$$

forms a linear initial value problem which when solved repetitively for increasing t may be thought of as an equivalent to the set of nonlinear algebraic equations (4.11). The solution to eqs. $(4.17)_3$, (4.18) proceeds as follows:

$$\begin{aligned} &K_{\alpha\beta}^{(T)}(\hat{r}_\gamma) r_\beta(\Delta t) = R_\alpha(\Delta t) - R_\alpha(0) + K_{\alpha\beta}^{(T)}(\hat{r}_\gamma)\hat{r}_\beta, \\ &r_\alpha(\Delta t) = \overset{-1}{K}{}_{\alpha\beta}^{(T)}(\hat{r}_\gamma)[R_\beta(\Delta t) - R_\beta(0) + K_{\beta\delta}^{(T)}(\hat{r}_\gamma)\hat{r}_\delta], \end{aligned} \tag{4.19}$$

$$K_{\alpha\beta}^{(T)}[r_\gamma(\Delta t)]r_\beta(2\Delta t) = R_\alpha(2\Delta t)-R_\alpha(\Delta t)+K_{\alpha\beta}^{(T)}[r_\gamma(\Delta t)]r_\beta(\Delta t),$$

$$r_\alpha(2\Delta t) = \overset{-1}{K_{\alpha\beta}^{(T)}}[r_\gamma(\Delta t)]\{R_\beta(2\Delta t)-R_\beta(\Delta t)+K_{\beta\delta}^{(T)}[r_\gamma(\Delta t)]r_\delta(\Delta t)\}, \text{ etc.} \tag{4.19}$$

Using this algorithm the solution $r_\alpha(t)$ may be followed up to any required value of t (i.e. up to any preset value of R_α). Using the terminology to be introduced in Chapter 5 we easily identify the two nonlinear problem solution schemes described above as the global and incremental schemes, respectively.

The usefulness of the weighted residuals method is apparent. However, major difficulties still remain when attempting to construct "good" sets of trial functions. In particular, when Ω is a curvilinear, or other complex region the application of the method may become impossible. In addition, the matrix $K_{\alpha\beta}$ of the algebraic equation system produced by the method can sometimes become numerically ill-conditioned as the number of terms adopted in the approximation increases. Fortunately, there are ways to eliminate these deficiences. We have so far assumed that the trial functions $\varphi_{i\alpha}$ of the expansion (4.4) were defined by a single expression valid throughout the whole domain Ω. Therefore, the integrals in eqs. (4.9) were evaluated in one operation over the whole domain. An alternative is to use the discretized model of the region Ω following the lines of Sec. 4.1, and then to construct the approximation $\tilde{u}_i$ in a piecewise manner over each subdomain. The trial functions can then also be defined in a piecewise manner by using different expressions in the various subdomains Ω_e from which the total domain is developed. In such a case, neglecting for the time being the difference between Ω and $\tilde{\Omega}$, the definite integrals occurring in (4.9) can be obtained simply by summing the contributions from each subdomain as

$$K_{\alpha\beta} = \sum_{e=1}^{E}\Big[\int_{\Omega_e} W_{\alpha i}A_{ij}\varphi_{j\beta}\,\mathrm{d}\Omega + \int_{\partial\Omega_e} W_{\alpha r}^{(b)}A_{rj}^{(b)}\varphi_{j\beta}\,\mathrm{d}(\partial\Omega)\Big],$$

$$R_\alpha = \sum_{e=1}^{E}\Big[\int_{\Omega_e} W_{\alpha i}B_i\,\mathrm{d}\Omega + \int_{\partial\Omega_e} W_{\alpha r}^{(b)}B_r^{(b)}\,\mathrm{d}(\partial\Omega)\Big]. \tag{4.20}$$

Here, $\partial\Omega_e$, $e = 1, 2, \ldots, E$ is the boundary of the e-th element which also belongs to the boundary of the whole domain Ω and the summation involving $\partial\Omega_e$ is taken only over those elements which lie immediately adjacent to the boundary. If the elements are of a relatively simple shape we can always select the trial functions so as to make the computations involved in (4.20) suitable for applications of computer-oriented algorithms. It is exactly here that the essential idea of the finite element method lies.

4.3 Assembly of element matrices and vectors

In Section 3.4 we have derived the fundamental equations describing on the single element level the nonlinear problem of structural dynamics in the form

$$\mathbf{m}_{N_{(e)}\times N_{(e)}}\Delta\ddot{\mathbf{q}}_{N_{(e)}\times 1}+\mathbf{c}_{N_{(e)}\times N_{(e)}}\Delta\dot{\mathbf{q}}_{N_{(e)}\times 1}+\mathbf{k}_{N_{(e)}\times N_{(e)}}\Delta\mathbf{q}_{N_{(e)}\times 1} = \Delta\mathbf{Q}_{N_{(e)}\times 1} \tag{4.21}$$

where $N_{(e)}$ is the number of elemental degrees of freedom. In order to fulfill the main objective of this section, which is the construction of a discretized model for the whole structure at hand, we shall now consider ways to carry out the assembly process to derive global stiffness, damping and mass matrices. We emphasize again that the discrete nature of inter-element connectivity distinguishes our approach from methods of continuum mechanics and makes it possible to formulate problems of structural analysis in the form of matrix equations.

Numerous examples of realistic systems and their constituent finite elements may easily be found in every finite element textbook. Therefore, we shall rather concentrate on a brief review of major steps typical of assembly procedure.

The procedure of assembling the element matrices and vectors is based on the requirement of "compatibility" at the element nodes. This means that at the node where elements are connected the value of each unknown nodal degree of freedom (be that a translation, rotation, curvature or other quantity) is the same for all the elements joining at that node. When the degrees of freedom are matched at a common node, the appropriate stiffness (damping, mass) and load contributions from each element sharing that node are added to obtain the net stiffness (damping, mass) and the net load at that node.

Assume now that we analyze a structure for which we have already decided upon the discretization mesh and the global numbering of nodes. We have also locally numbered each element nodes, selected local coordinates and defined nodal degrees of freedom. Employing appropriate shape functions we are then able to arrive at the equation of the form (4.21) for each and every finite element.

The next task which has to be undertaken for each element is the transformation from local to global coordinates. The transformation is described below for a general 3D element; specifications for other element types and other space dimensions are similar and therefore are here not dealt with in detail.

Assume that the transformation of any vector **u** from the global coordinates z_1, z_2, z_3 taken as the reference coordinates at time t from which the incremental response is considered to the instantaneous local coordinates ξ, η, ζ is given by the relation

$$\begin{bmatrix} u_\xi \\ u_\eta \\ u_\zeta \end{bmatrix} = \mathbf{s}_{3\times 3} \begin{bmatrix} u_{z_1} \\ u_{z_2} \\ u_{z_3} \end{bmatrix} \tag{4.22}$$

where

$$\mathbf{s}_{3\times 3} = \begin{bmatrix} s_{\xi 1} & s_{\xi 2} & s_{\xi 3} \\ s_{\eta 1} & s_{\eta 2} & s_{\eta 3} \\ s_{\zeta 1} & s_{\zeta 2} & s_{\zeta 3} \end{bmatrix} \tag{4.23}$$

and $s_{\xi 1}$ is the direction cosine of the ξ-axis with respect to z_1-axis, similar interpretation holding for the remaining entries of $\mathbf{s}$.

Since the element generalized incremental displacement vector $\Delta\mathbf{q}$ usually consists of vectors representing generalized displacements at subsequent element nodes, we may write symbolically

$$\Delta\mathbf{q}_{N_{(e)}\times 1} = \begin{bmatrix} \mathbf{s}_{3\times 3} & & 0 \\ & \mathbf{s}_{3\times 3} & \\ & & \ddots & \\ 0 & & & \mathbf{s}_{3\times 3} \end{bmatrix} \Delta\mathbf{r}^{(e)} = \mathbf{S}^{(e)}_{N_{(e)}\times N_{(e)}} \Delta\mathbf{r}^{(e)}_{N_{(e)}\times 1} \tag{4.24}$$

where $\Delta\mathbf{r}^{(e)}$ is the element generalized incremental displacement vector with its components referred to global coordinates z_1, z_2, z_3. Clearly, if at any node not all the components of a generalized displacement vector are considered, the corresponding transformation matrix $\mathbf{s}_{3\times 3}$ must be appropriately reduced. Similarly

$$\Delta\mathbf{Q}_{N_{(e)}\times 1} = \mathbf{S}^{(e)}_{N_{(e)}\times N_{(e)}} \Delta\mathbf{R}^{(e)}_{N_{(e)}\times 1} \tag{4.25}$$

the vector $\Delta\mathbf{R}^{(e)}$ consisting of generalized incremental nodal force vectors with components taken with reference to the global coordinates. Using eqs. (4.24), (4.25) and left-multiplying eq. (4.21) by $\mathbf{S}^{(e)T}$ we readily arrive at

$$\mathbf{S}^{(e)T}\mathbf{m}\mathbf{S}^{(e)}\Delta\ddot{\mathbf{r}}^{(e)} + \mathbf{S}^{(e)T}\mathbf{c}\mathbf{S}^{(e)}\Delta\dot{\mathbf{r}}^{(e)} + \mathbf{S}^{(e)T}\mathbf{k}\mathbf{S}^{(e)}\Delta\mathbf{r}^{(e)} = \Delta\mathbf{R}^{(e)} \tag{4.26}$$

where the identity $\mathbf{S}^{(e)T}\mathbf{S}^{(e)}$ (orthogonality of the transformation matrix $\mathbf{S}^{(e)}$) was used to simplify the right-hand side expression. Denoting

$$\begin{aligned} \mathbf{S}^{(e)T}\mathbf{m}\mathbf{S}^{(e)} &= \mathbf{M}^{(e)}, \\ \mathbf{S}^{(e)T}\mathbf{c}\mathbf{S}^{(e)} &= \mathbf{C}^{(e)}, \\ \mathbf{S}^{(e)T}\mathbf{k}\mathbf{S}^{(e)} &= \mathbf{K}^{(e)}, \end{aligned} \tag{4.27}$$

we obtain from eq. (4.26)

$$\mathbf{M}^{(e)}_{N_{(e)}\times N_{(e)}}\Delta\ddot{\mathbf{r}}^{(e)}_{N_{(e)}\times 1}+\mathbf{C}^{(e)}_{N_{(e)}\times N_{(e)}}\Delta\dot{\mathbf{r}}^{(e)}_{N_{(e)}\times 1}+\mathbf{K}^{(e)}_{N_{(e)}\times N_{(e)}}\Delta\mathbf{r}^{(e)}_{N_{(e)}\times 1} = \Delta\mathbf{R}^{(e)}_{N_{(e)}\times 1}. \tag{4.28}$$

We emphasize that eqs. (4.21) and (4.28) are equivalent—they describe the dynamic continuing (or incremental) equilibrium of a given finite element in terms of components referred to the local and global coordinates, respectively.

The next step is to appropriately expand the elemental $N_{(e)}\times N_{(e)}$ matrices $\mathbf{M}^{(e)}$, $\mathbf{C}^{(e)}$, $\mathbf{K}^{(e)}$ and $N_{(e)}\times 1$ vector $\Delta\mathbf{R}^{(e)}$ to dimensions $N\times N$ and $N\times 1$, respectively, by inserting additional zeros at appropriate locations. To this aim we can use the Boolean connectivity matrix of Sec. 4.2 carrying out the transformation

$$\mathbf{\Lambda}^{(e)T}_{N\times N_{(e)}}\mathbf{K}^{(e)}_{N_{(e)}\times N_{(e)}}\mathbf{\Lambda}^{(e)}_{N_{(e)}\times N} = \mathbf{K}^{(e)}_{N\times N}. \tag{4.29}$$

The matrix $\mathbf{K}^{(e)}_{N\times N}$ has still an elemental character, it differs, however, from $\mathbf{K}^{(e)}_{N_{(e)}\times N_{(e)}}$ by having zeros everywhere but at row and column numbers which correspond to the global numbers of the elemental degrees of freedom describing the element at hand. The similar relations hold for the matrices $\mathbf{C}^{(e)}$ and $\mathbf{M}^{(e)}$. Noting that the global vector of generalized incremental displacements $\Delta\mathbf{r}$ may be related to $\Delta\mathbf{r}^{(e)}$ by means of the transformation

$$\Delta\mathbf{r}^{(e)}_{N_{(e)}\times 1} = \mathbf{\Lambda}^{(e)}_{N_{(e)}\times N}\Delta\mathbf{r}_{N\times 1} \tag{4.30}$$

and defining

$$\Delta\mathbf{R}^{(e)}_{N\times 1} = \mathbf{\Lambda}^{(e)T}_{N\times N_{(e)}}\Delta\mathbf{R}^{(e)}_{N_{(e)}\times 1}$$

we readily obtain from eq. (4.28)

$$\mathbf{M}^{(e)}_{N\times N}\Delta\ddot{\mathbf{r}}_{N\times 1}+\mathbf{C}^{(e)}_{N\times N}\Delta\dot{\mathbf{r}}_{N\times 1}+\mathbf{K}^{(e)}_{N\times N}\Delta\mathbf{r}_{N\times 1} = \Delta\mathbf{R}^{(e)}_{N\times 1}. \tag{4.31}$$

Since we now have the fully consistent numbering scheme for all the degrees of freedom in the structure analysed and one common coordinate system is used, the assembly procedure requires nothing more but taking the sum over all the elements thus yielding

$$\Big(\sum_{e=1}^{E}\mathbf{M}^{(e)}_{N\times N}\Big)\Delta\ddot{\mathbf{r}}_{N\times 1}+\Big(\sum_{e=1}^{E}\mathbf{C}^{(e)}_{N\times N}\Big)\Delta\dot{\mathbf{r}}_{N\times 1}+\Big(\sum_{e=1}^{E}\mathbf{K}^{(e)}_{N\times N}\Big)\Delta\mathbf{r}_{N\times 1} = \sum_{e=1}^{E}\Delta\mathbf{R}^{(e)}_{N\times 1}. \tag{4.32}$$

Noting that the vector $-\sum_{e=1}^{E} \Delta \mathbf{R}_{N\times 1}^{(e)}$ has as its entries the successive components of the resultant incremental generalized force acting from "inside the structure" on its nodes, we may simply write the incremental nodal equilibrium condition as

$$-\sum_{e=1}^{E} \Delta \mathbf{R}_{N\times 1}^{(e)} + \Delta \mathbf{R}_{N\times 1} = \mathbf{0}_{N\times 1} \tag{4.33}$$

or

$$\sum_{e=1}^{E} \Delta \mathbf{R}_{N\times 1}^{(e)} = \Delta \mathbf{R}_{N\times 1}$$

with $\mathbf{R}_{N\times 1}$ being the external load vector acting at nodes. Finally, defining the matrices

$$\begin{aligned} \mathbf{M}_{N\times N} &= \sum_{e=1}^{E} \mathbf{M}_{N\times N}^{(e)}, \\ \mathbf{C}_{N\times N} &= \sum_{e=1}^{E} \mathbf{C}_{N\times N}^{(e)}, \\ \mathbf{K}_{N\times N} &= \sum_{e=1}^{E} \mathbf{K}_{N\times N}^{(e)} \end{aligned} \tag{4.34}$$

we arrive at the ultimate form of the fundamental equation for the discretized nonlinear dynamics model in the form

$$\mathbf{M}\Delta\ddot{\mathbf{r}} + \mathbf{C}\Delta\dot{\mathbf{r}} + \mathbf{K}\Delta\mathbf{r} = \Delta\mathbf{R}. \tag{4.35}$$

Clearly, transformation (4.34) applies to any summond in the expression for the elemental stiffness matrix, cf. eq. (3.76). In this way we may define the global counterparts to the constitutive, initial stress and initial displacement matrices as

$$\mathbf{K}^{(con)} = \sum_{e=1}^{E} \mathbf{\Lambda}^{(e)T}\mathbf{S}^{(e)T}\mathbf{k}^{(con)}\mathbf{S}^{(e)}\mathbf{\Lambda}^{(e)},$$

$$\mathbf{K}^{(\sigma)} = \sum_{e=1}^{E} \mathbf{\Lambda}^{(e)T}\mathbf{S}^{(e)T}\mathbf{k}^{(\sigma)}\mathbf{S}^{(e)}\mathbf{\Lambda}^{(e)},$$

$$\mathbf{K}^{(u)} = \sum_{e=1}^{E} \mathbf{\Lambda}^{(e)T}\mathbf{S}^{(e)T}\mathbf{k}^{(u)}\mathbf{S}^{(e)}\mathbf{\Lambda}^{(e)},$$

with

$$\mathbf{K} = \mathbf{K}^{(con)}+\mathbf{K}^{(\sigma)}+\mathbf{K}^{(u)}. \tag{4.36}$$

Let us now define the internal force vector $\mathbf{F}$ corresponding to stresses $\boldsymbol{\sigma}$ and given by

$$\mathbf{F} = \sum_{e=1}^{E}\Big(\int_{\Omega_e} \mathbf{B}^T\boldsymbol{\sigma}\,\mathrm{d}\Omega\Big). \tag{4.37}$$

Introducing the simplified notation

$$\mathbf{r}_t = \mathbf{r}(t), \qquad \mathbf{R}_t = \mathbf{R}(t),$$
$$\Delta\mathbf{r} = \mathbf{r}_{t+\Delta t}-\mathbf{r}_t, \qquad \Delta\mathbf{R} = \mathbf{R}_{t+\Delta t}-\mathbf{R}_t, \qquad \text{etc.} \tag{4.38}$$

with t being the time instant at the beginning of the step considered, we rewrite eq. (4.35) as

$$\mathbf{M}\ddot{\mathbf{r}}_{t+\Delta t}+\mathbf{C}\dot{\mathbf{r}}_{t+\Delta t}+\mathbf{K}\Delta\mathbf{r} = \mathbf{R}_{t+\Delta t}-\mathbf{R}_t+\mathbf{M}\ddot{\mathbf{r}}_t+\mathbf{C}\dot{\mathbf{r}}_t. \tag{4.39}$$

By noting that after the previous time step there holds within the iteration accuracy the relation

$$\mathbf{M}\ddot{\mathbf{r}}_t+\mathbf{C}\dot{\mathbf{r}}_t+\mathbf{F}_t = \mathbf{R}_t \tag{4.40}$$

eq. (4.39) is transformed to the form

$$\mathbf{M}\ddot{\mathbf{r}}_{t+\Delta t}+\mathbf{C}\dot{\mathbf{r}}_{t+\Delta t}+\mathbf{K}\Delta\mathbf{r} = \mathbf{R}_{t+\Delta t}-\mathbf{F}_t \tag{4.41}$$

which, when using the expression (4.37) to calculate the vector $\mathbf{F}_t$, increases the accuracy of the incremental/iterative solution algorithm, cf. Chapter 5.

By neglecting in eq. (4.35) the mass and damping terms we obtain the fundamental equation of quasi-statics as

$$\mathbf{K}\Delta\mathbf{r} = \Delta\mathbf{R} \tag{4.42}$$

which we may prefer to rewrite in terms of rates rather than finite increments as

$$\mathbf{K}\dot{\mathbf{r}} = \dot{\mathbf{R}}. \tag{4.43}$$

Starting from the elemental heat conduction equation, cf. eq. (3.85)

$$\mathbf{c}_{N_{(e)}\times N_{(e)}}\dot{\mathbf{p}}_{N_{(e)}\times 1}+\mathbf{k}_{N_{(e)}\times N_{(e)}}\mathbf{p}_{N_{(e)}\times 1} = \mathbf{e}_{N_{(e)}\times 1} \tag{4.44}$$

and using the fully analogous approach, we can generate the fundamental equation governing the nonlinear, time-dependent heat transfer process in the form

$$\mathbf{C}_{N\times N}\dot{\boldsymbol{\theta}}_{N\times 1}+\mathbf{K}_{N\times N}\boldsymbol{\theta}_{N\times 1} = \mathbf{Q}_{N\times 1}. \tag{4.45}$$

The notation introduced in eqs. (4.44), (4.45) reads:

$\mathbf{p}$ — vector of elemental nodal temperatures,

θ — vector of nodal temperatures for the whole region,
c — elemental matrix of heat capacity (not to be confused with the damping matrix),
C — global matrix of heat capacity,
k — elemental matrix of conductivity (not to be confused with the stiffness matrix),
K — global matrix of conductivity,
e, **Q** — vectors of elemental and global nodal thermal "loads".

The matrices **K** and **C** and the vector **Q** are in general temperature-dependent.

4.4 Elastic pin-jointed element in space

To illustrate the procedure of obtaining the finite element stiffness matrices in specific nonlinear structural mechanics problems let us first consider the pin-jointed bar element shown in Fig. 14. The local ξ-axis is taken in the

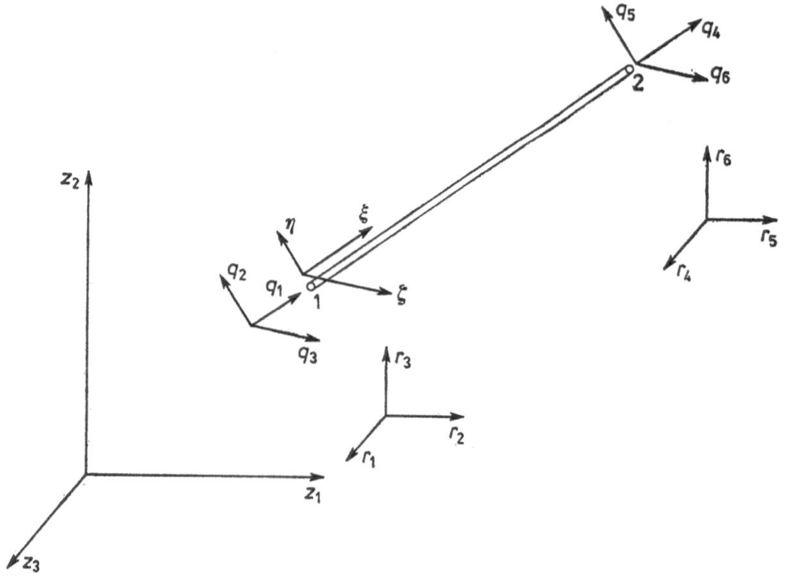

Fig. 14 3D truss element

axial direction of the element with origin at corner (or local node) 1. We assume linear fields $u_\xi(\xi)$, $u_\eta(\xi)$, $u_\zeta(\xi)$ to describe the displacement components in the directions of local coordinate axes ξ, η, ζ and express them as

$$u_\xi(\xi) = \alpha_1 + \alpha_2 \xi,$$
$$u_\eta(\xi) = \alpha_3 + \alpha_4 \xi, \tag{4.46}$$
$$u_\zeta(\xi) = \alpha_5 + \alpha_6 \xi.$$

The constants $\alpha_1, \alpha_2, \ldots, \alpha_6$ can be expressed in terms of the nodal displacement degrees of freedom by using the conditions

$$u_\xi(0) = q_1, \quad u_\xi(l) = q_4,$$
$$u_\eta(0) = q_2, \quad u_\eta(l) = q_5, \tag{4.47}$$
$$u_\zeta(0) = q_3, \quad u_\zeta(l) = q_6.$$

With the help of eqs. (4.47) we can present eqs. (4.46) in the form

$$\mathbf{u}_{3\times1}(\xi) = \{u_\xi(\xi) \;\; u_\eta(\xi) \;\; u_\zeta(\xi)\} = \mathbf{N}_{3\times6}(\xi)\mathbf{q}_{6\times1} \tag{4.48}$$

where

$$\mathbf{N}_{3\times6}(\xi) = \{\mathbf{N}^1_{1\times6}(\xi) \;\; \mathbf{N}^2_{1\times6}(\xi) \;\; \mathbf{N}^3_{1\times6}(\xi)\}$$
$$= \begin{bmatrix} 1-\xi/l & 0 & 0 & \xi/l & 0 & 0 \\ 0 & 1-\xi/l & 0 & 0 & \xi/l & 0 \\ 0 & 0 & 1-\xi/l & 0 & 0 & \xi/l \end{bmatrix}, \tag{4.49}$$

$$\mathbf{q}_{6\times1} = \{q_1 \;\; q_2 \;\; \cdots \;\; q_6\}.$$

Considering now the incremental updated Lagrangian description of the element, we shall use the relation

$$\Delta\mathbf{u}_{3\times1}(\xi) = \mathbf{N}_{3\times6}(\xi)\Delta\mathbf{q}_{6\times1}. \tag{4.50}$$

We further have the linear axial incremental strain in the element as

$$\Delta\varepsilon_{\xi\xi} = \Delta\boldsymbol{\epsilon}_{1\times1} = \mathbf{B}_{1\times6}\Delta\mathbf{q}_{6\times1} \tag{4.51}$$

where

$$\mathbf{B}_{1\times6} = \left[-\frac{1}{l} \;\; 0 \;\; 0 \;\; \frac{1}{l} \;\; 0 \;\; 0\right]. \tag{4.52}$$

The expression $\sigma_{ij}\Delta u_{k,i}\Delta u_{k,j}$ needed to construct the initial stress stiffness matrix takes for the one-dimensional problem the form $\sigma_{\xi\xi}(\Delta u^2_{\xi,\xi} + \Delta u^2_{\eta,\xi} + \Delta u^2_{\zeta,\xi})$ which can be expressed in terms of nodal incremental displacements $\Delta\mathbf{q}$ as

$$\sigma_{ij}\Delta u_{k,i}\Delta u_{k,j} = \boldsymbol{\sigma}_{\xi\xi}\Delta\mathbf{q}^T_{1\times6}\left[\left(\frac{\partial\mathbf{N}^1}{\partial\xi}\right)^T_{6\times1}\left(\frac{\partial\mathbf{N}^1}{\partial\xi}\right)_{1\times6} + \left(\frac{\partial\mathbf{N}^2}{\partial\xi}\right)^T_{6\times1}\left(\frac{\partial\mathbf{N}^2}{\partial\xi}\right)_{1\times6} + \right.$$
$$\left. + \left(\frac{\partial\mathbf{N}^3}{\partial\xi}\right)^T_{6\times1}\left(\frac{\partial\mathbf{N}^3}{\partial\xi}\right)_{1\times6}\right]\Delta\mathbf{q}_{6\times1}. \tag{4.53}$$

The constitutive law of one-dimensional linear elasticity is given by

$$\sigma_{\xi\xi} = E\varepsilon_{\xi\xi}$$

or, in the matrix notation, by

$$\boldsymbol{\sigma}_{1\times 1} = \mathbf{C}_{1\times 1}\boldsymbol{\epsilon}_{1\times 1} \tag{4.54}$$

with

$$\mathbf{C}_{1\times 1} = \mathbf{E}, \quad \boldsymbol{\sigma}_{1\times 1} = \sigma_{\xi\xi} = \sigma, \quad \boldsymbol{\epsilon}_{1\times 1} = \varepsilon_{\xi\xi} = \varepsilon.$$

The element constitutive (elastic) stiffness matrix in the local coordinate system can now be obtained as, cf. eq. $(3.75)_1$

$$\mathbf{k}^{(el)}_{6\times 6} = \int_{\Omega_e} \mathbf{B}^T_{6\times 1}\mathbf{C}_{1\times 1}\mathbf{B}_{1\times 6}\,\mathrm{d}\Omega = \frac{AE}{l}\begin{bmatrix} 1 & 0 & 0 & -1 & 0 & 0 \\ 0 & 0 & 0 & 0 & 0 & 0 \\ 0 & 0 & 0 & 0 & 0 & 0 \\ -1 & 0 & 0 & 1 & 0 & 0 \\ 0 & 0 & 0 & 0 & 0 & 0 \\ 0 & 0 & 0 & 0 & 0 & 0 \end{bmatrix} \tag{4.55}$$

where A is the constant cross-sectional area of the bar element. The element initial stress stiffness matrix is obtained from the definition $(3.75)_4$ by using the relation (4.53) as

$$\mathbf{k}^{(\sigma)}_{6\times 6} = \int_{\Omega_e} \sigma\left[\left(\frac{\partial \mathbf{N}^1}{\partial \xi}\right)^T_{6\times 1}\left(\frac{\partial \mathbf{N}^1}{\partial \xi}\right)_{1\times 6} + \left(\frac{\partial \mathbf{N}^2}{\partial \xi}\right)^T_{6\times 1}\left(\frac{\partial \mathbf{N}^2}{\partial \xi}\right)_{1\times 6} + \right.$$

$$\left. + \left(\frac{\partial \mathbf{N}^3}{\partial \xi}\right)^T_{6\times 1}\left(\frac{\partial \mathbf{N}^3}{\partial \xi}\right)_{1\times 6}\right]\mathrm{d}\Omega = \frac{Q}{l}\begin{bmatrix} 1 & 0 & 0 & -1 & 0 & 0 \\ & 1 & 0 & 0 & -1 & 0 \\ & & 1 & 0 & 0 & -1 \\ & & & 1 & 0 & 0 \\ & \text{sym.} & & & 1 & 0 \\ & & & & & 1 \end{bmatrix} \tag{4.56}$$

where $Q = A\sigma$ is the axial force in the element. If we note that the matrices $\mathbf{k}^{(el)}$ and $\mathbf{k}^{(\sigma)}$ enter the elemental relation additively, cf. eq. (4.36), and that normally in the elastic range $Q \ll EA$, eq. (4.56) simplifies to the commonly used form given by

$$\mathbf{k}^{(\sigma)}_{6\times 6} = \frac{Q}{l}\begin{bmatrix} 0 & 0 & 0 & 0 & 0 & 0 \\ & 1 & 0 & 0 & -1 & 0 \\ & & 1 & 0 & 0 & -1 \\ & & & 0 & 0 & 0 \\ & \text{sym.} & & & 1 & 0 \\ & & & & & 1 \end{bmatrix}. \tag{4.57}$$

The stiffness relation for the space truss in the local coordinates may thus be written within the framework of updated Lagrangian description as

$$[\mathbf{k}^{(el)} + \mathbf{k}^{(\sigma)}]_{6\times 6} \Delta \mathbf{q}_{6\times 1} = \Delta \mathbf{Q}_{6\times 1} \tag{4.58}$$

with the stiffness matrices given by eqs. (4.55) and (4.57), respectively, and with the nodal force vector $\Delta \mathbf{Q}_{6\times 1}$ formed in full analogy to its displacement counterpart $\Delta \mathbf{q}_{6\times 1}$, cf. Fig. 14.

The derivation of the initial displacement matrix $\mathbf{k}^{(u)}_{6\times 6}$ required for the total Lagrangian approach and linearized buckling analysis will be commented upon in Sec. 4.5 below when describing the 3D frame element—dropping the appropriate rows and columns in the frame matrix $\mathbf{k}^{(u)}_{12\times 12}$ will result in the initial displacement matrix for the truss element.

The derivation of the mass matrix for the pin-jointed element follows directly from the definition as, cf. eqs. $(3.54)_3$, (4.48)

$$\mathbf{m}_{6\times 6} = \int_{\Omega_e} \varrho \mathbf{N}^T_{6\times 3} \mathbf{N}_{3\times 6} d\Omega \tag{4.59}$$

where the matrix $\mathbf{N}_{3\times 6}$ is defined in eq. (4.49). Carrying out the necessary integrations leads to the final form of the mass matrix in the local coordinates as

$$\mathbf{m}_{6\times 6} = \frac{\varrho A l}{6} \begin{bmatrix} 2 & 0 & 0 & 1 & 0 & 0 \\ & 2 & 0 & 0 & 1 & 0 \\ & & 2 & 0 & 0 & 1 \\ & & & 2 & 0 & 0 \\ & \text{sym.} & & & 2 & 0 \\ & & & & & 2 \end{bmatrix}. \tag{4.60}$$

The dynamic equilibrium equations in the local coordinates for the space truss element takes the form

$$\mathbf{m}_{6\times 6} \Delta \ddot{\mathbf{q}}_{6\times 1} + [\mathbf{k}^{(el)}_{6\times 6} + \mathbf{k}^{(\sigma)}_{6\times 6}] \Delta \mathbf{q}_{6\times 1} = \Delta \mathbf{Q}_{6\times 1} \tag{4.61}$$

in which a damping matrix may be additionally included if required.

The specifications of the matrices to simpler cases like plane truss elements is trivial and need not be explicitly described.

As emphasized a few times in this section, the general stiffness relation for a single space truss element (4.61) is referred to local coordinates ξ, η, ζ. Remembering the detailed discussion of the transformation rules given in Sec. 4.1 we shall now proceed by showing how such a transformation is accomplished in the particular problem under consideration.

The first step in the transformation procedure is to express the nodal incremental displacement and force vectors $\Delta \mathbf{q}, \Delta \mathbf{Q}$ in terms of their components referred to the global coordinates z_1, z_2, z_3. Let us denote the transformed vectors as

$$\Delta \mathbf{r}^{(e)}_{6\times 1} = \{\Delta r^{(e)}_1\ \Delta r^{(e)}_2\ \dots\ \Delta r^{(e)}_6\},$$
$$\Delta \mathbf{R}^{(e)}_{6\times 1} = \{\Delta R^{(e)}_1\ \Delta R^{(e)}_2\ \dots\ \Delta R^{(e)}_6\}, \tag{4.62}$$

so that we have the relations, cf. eqs. (4.25)

$$\Delta \mathbf{q}_{6\times 1} = \mathbf{S}^{(e)}_{6\times 6}\Delta \mathbf{r}^{(e)}_{6\times 1}, \quad \Delta \mathbf{Q} = \mathbf{S}^{(e)}_{6\times 6}\Delta \mathbf{R}^{(e)}_{6\times 1}. \tag{4.63}$$

By looking at the ordering of entries in the vectors $\Delta\mathbf{q}$, $\Delta\mathbf{r}^{(e)}$ and $\Delta\mathbf{Q}$, $\Delta\mathbf{R}^{(e)}$ we readily establish the transformation matrix $\mathbf{S}^{(e)}_{6\times 6}$ as

$$\mathbf{S}^{(e)}_{6\times 6} = \begin{bmatrix} s_{\xi 1} & s_{\xi 2} & s_{\xi 3} & & & \\ s_{\eta 1} & s_{\eta 2} & s_{\eta 3} & & \mathbf{0}_{3\times 3} & \\ s_{\zeta 1} & s_{\zeta 2} & s_{\zeta 3} & & & \\ & & & s_{\xi 1} & s_{\xi 2} & s_{\xi 3} \\ & \mathbf{0}_{3\times 3} & & s_{\eta 1} & s_{\eta 2} & s_{\eta 3} \\ & & & s_{\zeta 1} & s_{\zeta 2} & s_{\zeta 3} \end{bmatrix} \tag{4.64}$$

where $s_{\xi 1}$ stands again, cf. Sec. 4.3, for the direction cosine of the ξ-axis with respect to the z_1-axis and the rest of the direction cosines is defined similarly. By transforming the matrices defined in eq. (4.55), (4.57) and (4.60) as

$$\mathbf{K}^{(e)(el)}_{6\times 6} = \mathbf{S}^{(e)T}_{6\times 6}\mathbf{k}^{(el)}_{6\times 6}\mathbf{S}^{(e)}_{6\times 6},$$
$$\mathbf{K}^{(e)(\sigma)} = \mathbf{S}^{(e)T}_{6\times 6}\mathbf{k}^{(\sigma)}_{6\times 6}\mathbf{S}^{(e)}_{6\times 6}, \tag{4.65}$$
$$\mathbf{M}^{(e)} = \mathbf{S}^{(e)T}_{6\times 6}\mathbf{m}_{6\times 6}\mathbf{S}^{(e)}_{6\times 6}$$

we arrive at the elemental matrices referred to global coordinates so that now the fundamental relation (4.61) takes the form

$$\mathbf{M}^{(e)}_{6\times 6}\Delta\ddot{\mathbf{r}}^{(e)}_{6\times 1} + [\mathbf{K}^{(e)(el)}_{6\times 6} + \mathbf{K}^{(e)(\sigma)}_{6\times 6}]\Delta\mathbf{r}^{(e)}_{6\times 1} = \Delta\mathbf{R}^{(e)}_{6\times 1}. \tag{4.66}$$

The assembly of the global stiffness matrix follows by identifying the numbering of global and local degrees of freedom as described in Sec. 4.3.

We note in passing that it is obviously enough to consider only three (out of nine in eq. (4.64)) independent cosines (say, $s_{\xi 1}, s_{\xi 2}, s_{\xi 3}$) to uniquely fix the direction of the truss member in the space $z_1 z_2 z_3$. Using the simple relations between different cosines we may carry out the transformations (4.65) in a rather compact way. For the constitutive stiffness matrix we obtain, for instance

$$\mathbf{K}^{(e)(el)}_{6\times 6} = \frac{AE}{l}\begin{bmatrix} s^2_{\xi 1} & s_{\xi 1}s_{\xi 2} & s_{\xi 1}s_{\xi 3} & -s^2_{\xi 1} & -s_{\xi 1}s_{\xi 2} & -s_{\xi 1}s_{\xi 3} \\ & s^2_{\xi 2} & s_{\xi 2}s_{\xi 3} & -s_{\xi 1}s_{\xi 2} & -s^2_{\xi 2} & -s_{\xi 2}s_{\xi 3} \\ & & s^2_{\xi 3} & -s_{\xi 1}s_{\xi 3} & -s_{\xi 2}s_{\xi 3} & -s^2_{\xi 3} \\ & \text{sym.} & & s^2_{\xi 1} & s_{\xi 1}s_{\xi 2} & s_{\xi 1}s_{\xi 3} \\ & & & & s^2_{\xi 2} & s_{\xi 2}s_{\xi 3} \\ & & & & & s^2_{\xi 3} \end{bmatrix} \tag{4.67}$$

The similar calculations for the remaining matrices are left to the reader as exercises.

4.5 Elastic frame element in space

A space frame element considered in this section is a straight bar of uniform cross-section, which is capable of resisting axial force, shear forces acting along and bending moments acting about the two principal axes in the plane of the element cross-section, and twisting moment acting about its centroidal

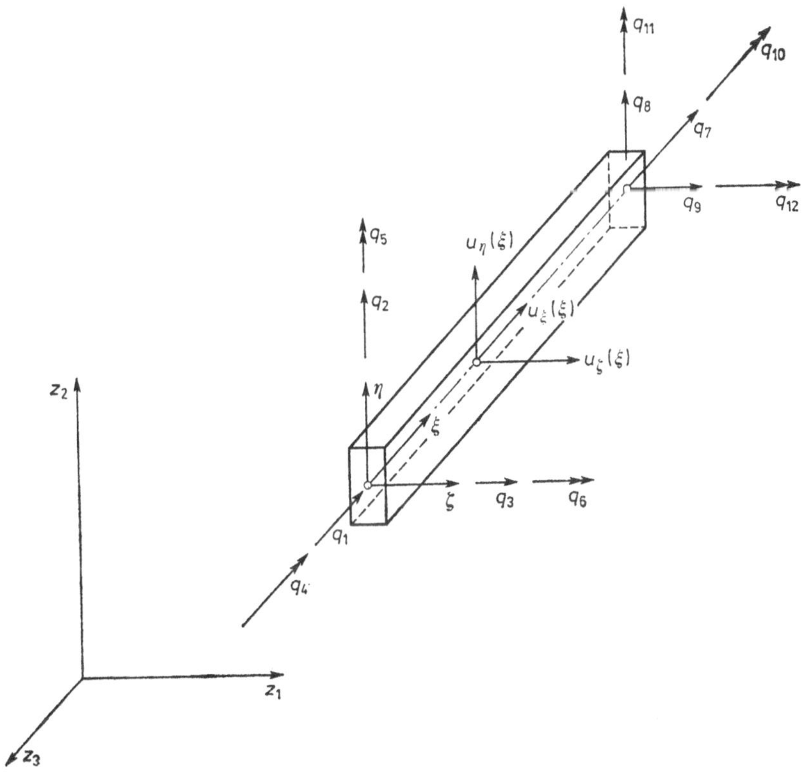

Fig. 15 3D beam element

axis. The corresponding displacement-type degrees of freedom are shown in Fig. 15 with reference to the local coordinates ξ, η, ζ.

The material of the beam is assumed to be linear elastic—the way the inelastic properties may be incorporated will be discussed in Chapter 6.

In full accordance with our incremental approach we consider now the beam at a certain time t assuming that the response up to this time instant has been solved for and that the position of the beam at time $t+\Delta t$ is to be determined. To simplify the description we assume that the local deformations of the beam have been small enough to make it possible to consider the beam as the straight one at the typical time instant t.

Assuming the local coordinate axes ξ, η, ζ to coincide with the beam centroidal axis and the principal axes of the cross-section, respectively, the elemental vector of generalized incremental displacements has the form

$$\Delta \mathbf{q}_{12\times1} = \{\Delta q_1 \; \Delta q_2 \; \dots \; \Delta q_{12}\}. \tag{4.68}$$

Using the kinematic constraint of plane cross-sections during the whole deformation process we may decompose the incremental displacement vector

$$\Delta \mathbf{u}(\xi, \eta, \zeta) = \{\Delta u_\xi(\xi, \eta, \zeta) \; \Delta u_\eta(\xi, \eta, \zeta) \; \Delta u_\zeta(\xi, \eta, \zeta)\} \tag{4.69}$$

into a 3×6 matrix $\mathbf{H}(\eta, \zeta)$ and another generalized incremental displacement vector $\Delta \mathbf{q}_{0_{6\times1}}$ as

$$\Delta \mathbf{u}_{3\times1}(\xi, \eta, \zeta) = \mathbf{H}_{3\times6}(\eta, \zeta)\Delta \mathbf{q}_0(\xi) \tag{4.70}$$

where

$$\mathbf{H}_{3\times6}(\eta, \zeta) = \begin{bmatrix} 1 & 0 & 0 & 0 & \zeta & -\eta \\ 0 & 1 & 0 & -\zeta & 0 & 0 \\ 0 & 0 & 1 & \eta & 0 & 0 \end{bmatrix}$$

and

$$\Delta \mathbf{q}_{0_{6\times1}}(\xi) = \Big\{\Delta u_\xi(\xi, 0, 0) \; \Delta u_\eta(\xi, 0, 0) \; \Delta u_\zeta(\xi, 0, 0) \; \Delta\varphi(\xi, 0, 0) \; -\frac{\partial \Delta u_\zeta}{\partial \xi}(\xi, 0, 0) \; \frac{\partial \Delta u_\eta}{\partial \xi}(\xi, 0, 0)\Big\}$$

is defined for points along the centroidal axis. The incremental rotation $\Delta\varphi$ corresponds to the twist angle whereas the next two entries of the vector $\Delta \mathbf{q}_0$ describe the material fiber rotations about η- and ζ-axes, respectively.

For the vector $\Delta \mathbf{q}_0$ we adopt the following expansion in the ξ-direction

$$\Delta \mathbf{q}_{0_{6\times1}}(\xi) = \bar{\mathbf{N}}_{6\times12}(\xi)\Delta \mathbf{q}_{12\times1} \tag{4.71}$$

where

$$\bar{\mathbf{N}}_{6\times12} = \begin{bmatrix} \bar{N}_1 & 0 & 0 & 0 & 0 & 0 & \bar{N}_7 & 0 & 0 & 0 & 0 & 0 \\ 0 & \bar{N}_2 & 0 & 0 & 0 & \bar{N}_6 & 0 & \bar{N}_8 & 0 & 0 & 0 & \bar{N}_{12} \\ 0 & 0 & \bar{N}_3 & 0 & \bar{N}_5 & 0 & 0 & 0 & \bar{N}_9 & 0 & \bar{N}_{11} & 0 \\ 0 & 0 & 0 & \bar{N}_4 & 0 & 0 & 0 & 0 & 0 & \bar{N}_{10} & 0 & 0 \\ 0 & 0 & -\bar{N}_{3,\xi} & 0 & -\bar{N}_{5,\xi} & 0 & 0 & 0 & -\bar{N}_{9,\xi} & 0 & -\bar{N}_{11,\xi} & 0 \\ 0 & \bar{N}_{2,\xi} & 0 & 0 & 0 & \bar{N}_{6,\xi} & 0 & \bar{N}_{8,\xi} & 0 & 0 & 0 & \bar{N}_{12,\xi} \end{bmatrix} \tag{4.72}$$

$$\begin{aligned} &\bar{N}_1 = 1-\xi/l, \quad \bar{N}_2 = 1-3(\xi/l)^2+2(\xi/l)^3, \quad \bar{N}_3 = \bar{N}_2, \quad \bar{N}_4 = \bar{N}_1, \\ &\bar{N}_5 = l[-\xi/l+2(\xi/l)^2-(\xi/l)^3], \quad \bar{N}_6 = -\bar{N}_5, \quad \bar{N}_7 = \xi/l, \\ &\bar{N}_8 = 3(\xi/l)^2-2(\xi/l)^3, \quad \bar{N}_9 = \bar{N}_8, \quad \bar{N}_{10} = \bar{N}_7, \\ &\bar{N}_{11} = l[(\xi/l)^2-(\xi/l)^3], \quad \bar{N}_{12} = -\bar{N}_{11}. \end{aligned} \tag{4.73}$$

Using eqs. (4.70) and (4.72) we arrive at

$$\Delta \mathbf{u}_{3\times 1} = \mathbf{H}_{3\times 6}\bar{\mathbf{N}}_{6\times 12}\Delta \mathbf{q}_{12\times 1} = \mathbf{N}_{3\times 12}\Delta \mathbf{q}_{12\times 1}. \tag{4.74}$$

The expressions (4.70), (4.72) have been derived by assuming

$$\begin{aligned}
\Delta u_\xi(\xi) &= \alpha_1 + \alpha_2 \xi, \\
\Delta \varphi(\xi) &= \alpha_3 + \alpha_4 \xi, \\
\Delta u_\eta(\xi) &= \alpha_5 + \alpha_6 \xi + \alpha_7 \xi^2 + \alpha_8 \xi^3, \\
\Delta u_\zeta(\xi) &= \alpha_9 + \alpha_{10} \xi + \alpha_{11} \xi^2 + \alpha_{12} \xi^3
\end{aligned} \tag{4.75}$$

and calculating the constants α_1, α_2 from the conditions

$$\Delta u_\xi(0) = \Delta q_1, \quad \Delta u_\xi(l) = \Delta q_7,$$

the constants α_3, α_4 from the conditions

$$\Delta \varphi(0) = \Delta q_4, \quad \Delta \varphi(l) = \Delta q_{10},$$

the constants $\alpha_5, \ldots, \alpha_8$ from the conditions

$$\Delta u_\eta(0) = \Delta q_2, \quad \Delta u_\eta(l) = \Delta q_8,$$

$$\frac{\partial \Delta u_\eta(0)}{\partial \xi} = \Delta q_6, \quad \frac{\partial \Delta u_\eta(l)}{\partial \xi} = \Delta q_{12},$$

and the constants $\alpha_9, \ldots, \alpha_{12}$ from the conditions

$$\Delta u_\zeta(0) = \Delta q_3, \quad \Delta u_\zeta(l) = \Delta q_9,$$

$$\frac{\partial \Delta u_\zeta(0)}{\partial \xi} = \Delta q_5, \quad \frac{\partial \Delta u_\zeta(l)}{\partial \xi} = \Delta q_{11}.$$

The linear incremental strain vector is conveniently defined as

$$\Delta \boldsymbol{\varepsilon}_{3\times 1}(\xi, \eta, \zeta) = \left\{ \frac{\partial \Delta u_\xi}{\partial \xi} \quad \frac{\partial \Delta u_\xi}{\partial \eta} + \frac{\partial \Delta u_\eta}{\partial \xi} \quad \frac{\partial \Delta u_\xi}{\partial \zeta} + \frac{\partial \Delta u_\zeta}{\partial \xi} \right\} \tag{4.76}$$

which by eq. (4.74) leads to

$$\Delta \boldsymbol{\varepsilon}_{3\times 1}(\xi, \eta, \zeta) = \mathbf{B}_{3\times 12}(\xi, \eta, \zeta)\Delta \mathbf{q}_{12\times 1} \tag{4.77}$$

where

$$\mathbf{B}_{3\times 12} = \begin{bmatrix}
\bar{N}_{1,\xi} & -\eta\bar{N}_{2,\xi\xi} & -\zeta\bar{N}_{3,\xi\xi} & 0 & -\zeta\bar{N}_{5,\xi\xi} & -\eta\bar{N}_{6,\xi\xi} \\
0 & 0 & 0 & -\zeta\bar{N}_{4,\xi} & 0 & 0 \\
0 & 0 & 0 & \eta\bar{N}_{4,\xi} & 0 & 0
\end{bmatrix.$$
$$\left.\begin{matrix}
\bar{N}_{7,\xi} & -\eta\bar{N}_{8,\xi\xi} & -\zeta\bar{N}_{9,\xi\xi} & 0 & -\zeta\bar{N}_{11,\xi\xi} & -\eta\bar{N}_{12,\xi\xi} \\
0 & 0 & 0 & -\zeta\bar{N}_{10,\xi} & 0 & 0 \\
0 & 0 & 0 & \eta\bar{N}_{10,\xi} & 0 & 0
\end{matrix}\right]. \tag{4.78}$$

Assuming the constitutive relation in the form

$$\Delta\boldsymbol{\sigma}_{3\times1} = \{\sigma_{\xi\xi} \quad \sigma_{\xi\eta} \quad \sigma_{\xi\zeta}\} = \mathbf{C}_{3\times3}\Delta\boldsymbol{\varepsilon}_{3\times1} \tag{4.79}$$

with

$$\mathbf{C}_{3\times3} = E\begin{bmatrix} 1 & 0 & 0 \\ 0 & \dfrac{1}{2(1+\nu)} & 0 \\ 0 & 0 & \dfrac{1}{2(1+\nu)} \end{bmatrix} = \begin{bmatrix} E & 0 & 0 \\ 0 & G & 0 \\ 0 & 0 & G \end{bmatrix} \tag{4.80}$$

we may find the elastic stiffness matrix for the frame element in space from the definition, cf. eq. $(3.75)_1$

$$\mathbf{k}^{(el)}_{12\times12} = \int_{\Omega_e} \mathbf{B}^T_{12\times3}\mathbf{C}_{3\times3}\mathbf{B}_{3\times12}\,\mathrm{d}\Omega. \tag{4.81}$$

By substituting into eq. (4.81) the matrices **B** and **C** given in eqs. (4.78) and (4.80), respectively, and carrying out the integration we end up with the following matrix

$$\mathbf{k}^{(el)}_{12\times12} = \begin{bmatrix}
\frac{EA}{l} & 0 & 0 & 0 & 0 & 0 & -\frac{EA}{l} & 0 & 0 & 0 & 0 & 0 \\
 & \frac{12EJ_\zeta}{l^3} & 0 & 0 & 0 & \frac{6EJ_\zeta}{l^2} & 0 & -\frac{12EJ_\zeta}{l^3} & 0 & 0 & 0 & \frac{6EJ_\zeta}{l^2} \\
 & & \frac{12EJ_\eta}{l^3} & 0 & \frac{-6EJ_\eta}{l^2} & 0 & 0 & 0 & -\frac{12EJ_\eta}{l^3} & 0 & -\frac{6EJ_\eta}{l^2} & 0 \\
 & & & \frac{GJ}{l} & 0 & 0 & 0 & 0 & 0 & -\frac{GJ}{l} & 0 & 0 \\
 & & & & \frac{4EJ_\eta}{l^2} & 0 & 0 & 0 & \frac{6EJ_\eta}{l^3} & 0 & \frac{2EJ_\eta}{l} & 0 \\
 & & & & & \frac{4EJ_\zeta}{l} & 0 & -\frac{6EJ_\zeta}{l^2} & 0 & 0 & 0 & \frac{2EJ_\zeta}{l} \\
 & & & & & & \frac{EA}{l} & 0 & 0 & 0 & 0 & 0 \\
 & & & & & & & \frac{12EJ_\zeta}{l^3} & 0 & 0 & 0 & -\frac{6EJ_\zeta}{l^2} \\
 & & \text{sym.} & & & & & & \frac{12EJ_\eta}{l^3} & 0 & \frac{6EJ_\eta}{l^2} & 0 \\
 & & & & & & & & & \frac{GJ}{l} & 0 & 0 \\
 & & & & & & & & & & \frac{4EJ_\eta}{l} & 0 \\
 & & & & & & & & & & & \frac{4EJ_\zeta}{l}
\end{bmatrix} \tag{4.82}$$

where we have additionally introduced the following notation:

J_η, J_ζ — area moments of inertia of the cross-section about η–η, and ζ–ζ axes, respectively,

J — polar moment of inertia of the cross-section.

In order to derive the initial stress matrix for the element at hand we first note that the term $\int\limits_{\Omega_e} \sigma_{ij} \Delta u_{k,i} \Delta u_{k,j} \mathrm{d}\Omega$, cf. eq. (3.75)$_4$, may be written in matrix notation as, cf. eq. (4.53)

$$\int\limits_{\Omega_e} \boldsymbol{\sigma}_{1\times 1} \Delta \mathbf{q}^T_{1\times 12} \left[\left(\frac{\partial \mathbf{N}^1}{\partial \xi} \right)^T_{12\times 1} \left(\frac{\partial \mathbf{N}^1}{\partial \xi} \right)_{1\times 12} + \left(\frac{\partial \mathbf{N}^2}{\partial \xi} \right)^T_{12\times 1} \left(\frac{\partial \mathbf{N}^2}{\partial \xi} \right)_{1\times 12} + \right.$$

$$\left. + \left(\frac{\partial \mathbf{N}^3}{\partial \xi} \right)^T_{12\times 1} \left(\frac{\partial \mathbf{N}^3}{\partial \xi} \right)_{1\times 12} \right] \Delta \mathbf{q}_{12\times 1} \mathrm{d}\Omega = \Delta \mathbf{q}^T_{1\times 12} \mathbf{k}^{(\sigma)}_{12\times 12} \Delta \mathbf{q}_{12\times 1} \tag{4.83}$$

where it was assumed that the only stress component contributing to the initial stress stiffness of the element is the axial stress $\sigma_{\xi\xi}$ and

$$\begin{aligned} \mathbf{N}^1_{1\times 12} &= [\bar{N}_1 \;\; -\eta \bar{N}_{2,\xi} \;\; -\zeta \bar{N}_{3,\xi} \;\; 0 \;\; -\zeta \bar{N}_{5,\xi} \;\; -\eta \bar{N}_{6,\xi} \\ &\qquad \bar{N}_7 \;\; -\eta \bar{N}_{8,\xi} \;\; -\zeta \bar{N}_{8,\xi} \;\; 0 \;\; -\zeta \bar{N}_{11,\xi} \;\; -\eta \bar{N}_{12,\xi}], \\ \mathbf{N}^2_{1\times 12} &= [0 \;\; \bar{N}_2 \;\; 0 \;\; -\zeta \bar{N}_4 \;\; 0 \;\; \bar{N}_6 \;\; 0 \;\; \bar{N}_8 \;\; 0 \;\; -\zeta \bar{N}_{10} \;\; 0 \;\; \bar{N}_{12}], \\ \mathbf{N}^3_{1\times 12} &= [0 \;\; 0 \;\; \bar{N}_3 \;\; \eta \bar{N}_4 \;\; \bar{N}_5 \;\; 0 \;\; 0 \;\; 0 \;\; \bar{N}_9 \;\; \eta \bar{N}_{10} \;\; \bar{N}_{11} \;\; 0] \end{aligned} \tag{4.84}$$

are rows of the matrix $\mathbf{N}_{3\times 12}$, cf. eq. (4.74). Using eqs. (4.83), (4.84) we arrive at the final form of the initial stress matrix for the space frame element as

$$\mathbf{k}^{(\sigma)}_{12\times 12} = \begin{bmatrix} 0 & 0 & 0 & 0 & 0 & 0 & 0 & 0 & 0 & 0 & 0 & 0 \\ & \frac{6Q}{5l} & 0 & 0 & 0 & \frac{Q}{10} & 0 & -\frac{6Q}{5l} & 0 & 0 & 0 & \frac{Q}{10} \\ & & \frac{6Q}{5l} & 0 & -\frac{Q}{10} & 0 & 0 & 0 & -\frac{6Q}{5l} & 0 & -\frac{Q}{10} & 0 \\ & & & 0 & 0 & 0 & 0 & 0 & 0 & 0 & 0 & 0 \\ & & & & \frac{2Ql}{15} & 0 & 0 & 0 & \frac{Q}{10} & 0 & -\frac{Ql}{30} & 0 \\ & & & & & \frac{2Ql}{15} & 0 & -\frac{Q}{10} & 0 & 0 & 0 & -\frac{Ql}{30} \\ & & & & & & 0 & 0 & 0 & 0 & 0 & 0 \\ & & & & & & & \frac{6Q}{5l} & 0 & 0 & 0 & -\frac{Q}{10} \\ & & & & & & & & \frac{6Q}{5l} & 0 & \frac{Q}{10} & 0 \\ & & & & & & & & & 0 & 0 & 0 \\ & & \text{sym.} & & & & & & & & \frac{2Ql}{15} & 0 \\ & & & & & & & & & & & \frac{2Ql}{15} \end{bmatrix} \tag{4.85}$$

where we have employed the definition of the axial force as

$$Q = \int_A \sigma_{\xi\xi} \mathrm{d}A .$$

We note that neglecting the rows and columns in $\mathbf{k}^{(\sigma)}_{12\times 12}$ corresponding to the degrees of freedom q_4, q_5, q_6 and q_{10}, q_{11}, q_{12} leads to the matrix which can be used for the analysis of truss-like elements. Note, however, the slight difference in entries of such a matrix when compared against the one previously obtained using a different approach, cf. eq. (4.56).

The linear part of the initial displacement matrix is defined in local coordinates as, cf. eq. $(3.75)_2$

$$\bar{\mathbf{k}}^{(u)}_{12\times 12} = 2 \int_{\Omega_e} [\mathbf{B}^T_{12\times 3}\mathbf{C}_{3\times 3}\bar{\mathbf{B}}_{3\times 12} + \bar{\mathbf{B}}^T_{12\times 3}\mathbf{C}_{3\times 3}\mathbf{B}_{3\times 12}]\mathrm{d}\Omega \tag{4.86}$$

where $\mathbf{B}$ is given by eq. (4.52) and the matrix $\bar{\mathbf{B}}$ may be derived by considering the relations

$$\boldsymbol{\epsilon}^{NL}_{3\times 1} = \left\{ \frac{1}{2}\left[\left(\frac{\partial u_\eta}{\partial \xi}\right)^2 + \left(\frac{\partial u_\zeta}{\partial \xi}\right)^2\right] \quad 0 \quad 0 \right\} = \{\overset{*}{\mathbf{B}}_{1\times 12} \quad \mathbf{O}_{1\times 12} \quad \mathbf{O}_{1\times 12}\}\mathbf{q}_{12\times 1}$$

$$= \bar{\mathbf{B}}_{3\times 12}(\xi;\mathbf{q})\mathbf{q}_{12\times 1},$$

$$\overset{*}{\mathbf{B}}_{1\times 12} = \frac{1}{2}\mathbf{q}^T_{1\times 12}\mathbf{A}_{12\times 12},$$

$$\mathbf{A} = \begin{bmatrix}
0 & 0 & 0 & 0 & 0 & 0 & 0 & 0 & 0 & 0 & 0 & 0 \\
 & \bar{N}_{2,\xi}\bar{N}_{2,\xi} & 0 & 0 & 0 & \bar{N}_{2,\xi}\bar{N}_{6,\xi} & 0 & \bar{N}_{2,\xi}\bar{N}_{8,\xi} & 0 & 0 & 0 & \bar{N}_{2,\xi}\bar{N}_{12,\xi} \\
 & & \bar{N}_{3,\xi}\bar{N}_{3,\xi} & 0 & \bar{N}_{3,\xi}\bar{N}_{5,\xi} & 0 & 0 & 0 & \bar{N}_{3,\xi}\bar{N}_{9,\xi} & 0 & \bar{N}_{3,\xi}\bar{N}_{11,\xi} & 0 \\
 & & & 0 & 0 & 0 & 0 & 0 & 0 & 0 & 0 & 0 \\
 & & & & \bar{N}_{5,\xi}\bar{N}_{5,\xi} & 0 & 0 & 0 & \bar{N}_{5,\xi}\bar{N}_{9,\xi} & 0 & \bar{N}_{5,\xi}\bar{N}_{11,\xi} & 0 \\
 & & & & & \bar{N}_{6,\xi}\bar{N}_{6,\xi} & 0 & \bar{N}_{6,\xi}\bar{N}_{8,\xi} & 0 & 0 & 0 & \bar{N}_{6,\xi}\bar{N}_{12,\xi} \\
 & & & & & & 0 & 0 & 0 & 0 & 0 & 0 \\
 & & & & & & & \bar{N}_{8,\xi}\bar{N}_{8,\xi} & 0 & 0 & 0 & \bar{N}_{8,\xi}\bar{N}_{12,\xi} \\
 & & & & & & & & \bar{N}_{9,\xi}\bar{N}_{9,\xi} & 0 & \bar{N}_{9,\xi}\bar{N}_{11,\xi} & 0 \\
 & & \text{sym.} & & & & & & & 0 & 0 & 0 \\
 & & & & & & & & & & \bar{N}_{11,\xi}\bar{N}_{12,\xi} & 0 \\
 & & & & & & & & & & & \bar{N}_{12,\xi}\bar{N}_{12,\xi}
\end{bmatrix} \tag{4.87}$$

We note that in the nonlinear expression for the axial strain the contribution $\left(\frac{\partial u_\xi}{\partial \xi}\right)^2$ has been neglected. Introducing the matrices $\mathbf{B}$ and $\bar{\mathbf{B}}$ into the definition (4.86) results in the following form of the linear part of the initial displacement matrix

$$\bar{\mathbf{k}}^{(u)}_{12\times 12} = \begin{bmatrix} 0 & a & b & 0 & d & c & 0 & -a & -b & 0 & f & e \\ & 0 & 0 & 0 & 0 & 0 & -a & 0 & 0 & 0 & 0 & 0 \\ & & 0 & 0 & 0 & 0 & -b & 0 & 0 & 0 & 0 & 0 \\ & & & 0 & 0 & 0 & 0 & 0 & 0 & 0 & 0 & 0 \\ & & & & 0 & 0 & -d & 0 & 0 & 0 & 0 & 0 \\ & & & & & 0 & -c & 0 & 0 & 0 & 0 & 0 \\ & & & & & & 0 & u & b & 0 & -f & -e \\ & & & & & & & 0 & 0 & 0 & 0 & 0 \\ & & & & & & & & 0 & 0 & 0 & 0 \\ & & \text{sym.} & & & & & & & 0 & 0 & 0 \\ & & & & & & & & & & 0 & 0 \\ & & & & & & & & & & & 0 \end{bmatrix} \tag{4.88}$$

where

$$a = [72(q_8-q_2)-6l(q_{12}-q_6)]\,\alpha,$$
$$b = [72(q_9-q_3)+6l(q_5+q_{11})]\,\alpha,$$
$$c = [6(q_8-q_2)+2l^2(q_{12}-4q_6)]\,\alpha,$$
$$d = [-6l(q_9-q_3)+2l^2(q_{11}-q_5)]\,\alpha,$$
$$e = [6l(q_8-q_2)-2l^2(4q_{12}-q_6)]\,\alpha,$$
$$f = [-6l(q_9-q_3)-2l^2(4q_{11}-q_5)]\,\alpha,$$
$$\alpha = QA/60l^2.$$

The matrix $\bar{\bar{\mathbf{k}}}^{(u)}$ (nonlinear in $\mathbf{q}$ contribution to $\mathbf{k}^{(u)}$) may be obtained as, cf. eq. (3.75)$_3$

$$\bar{\bar{\mathbf{k}}}^{(u)}_{12\times 12} = \int_{\Omega_e} \mathbf{A}^T_{12\times 12}\mathbf{q}_{12\times 1}\mathbf{q}^T_{1\times 12}\mathbf{A}_{12\times 12}E\,\mathrm{d}\Omega \tag{4.89}$$

with $\mathbf{A}$ given in eq. (4.87); derivation of the explicit form of this matrix is left as an excercise to the reader. By neglecting in eq. (4.88) the rows and columns corresponding to the rotational degrees of freedom (i.e. those numbered 3, 4, 5, 10, 11, 12) we may obtain the linear part of the initial displacement matrix for the truss element.

The consistent mass matrix of the space frame element in local coordinates is derived from the definition

$$\mathbf{m}_{12\times 12} = \int_{\Omega_e} \varrho \mathbf{N}^T_{12\times 3}\mathbf{N}_{3\times 12}\,\mathrm{d}\Omega \tag{4.90}$$

which after using eq. (4.84) becomes

$$\mathbf{m}_{12\times 12} = \varrho A l \begin{bmatrix} \frac{1}{3} & 0 & 0 & 0 & 0 & 0 & \frac{1}{6} & 0 & 0 & 0 & 0 & 0 \\ & \frac{13}{35} & 0 & 0 & 0 & \frac{11l}{210} & 0 & \frac{9}{70} & 0 & 0 & 0 & -\frac{13l}{420} \\ & & \frac{13}{35} & 0 & -\frac{11l}{210} & 0 & 0 & 0 & \frac{9}{70} & 0 & \frac{13l}{420} & 0 \\ & & & \frac{J}{3A} & 0 & 0 & 0 & 0 & 0 & \frac{J}{6A} & 0 & 0 \\ & & & & \frac{l^2}{105} & 0 & 0 & 0 & -\frac{13l}{420} & 0 & -\frac{l^2}{140} & 0 \\ & & & & & \frac{l^2}{105} & 0 & \frac{13l}{420} & 0 & 0 & 0 & -\frac{l^2}{140} \\ & & & & & & \frac{1}{3} & 0 & 0 & 0 & 0 & 0 \\ & & & & & & & \frac{13}{35} & 0 & 0 & 0 & -\frac{11l}{210} \\ & & & \text{sym.} & & & & & \frac{13}{35} & 0 & \frac{11l}{210} & 0 \\ & & & & & & & & & \frac{J}{3A} & 0 & 0 \\ & & & & & & & & & & \frac{l^2}{105} & 0 \\ & & & & & & & & & & & \frac{l^2}{105} \end{bmatrix}. \tag{4.91}$$

The stiffness and mass matrices given in eqs. (4.82), (4.85), (4.88) and (4.91) are expressed in the local coordinate system $\xi\eta\zeta$. The transformation to the global system $z_1 z_2 z_3$ may be accomplished by using the transformation

$$\Delta\mathbf{q}_{12\times 1} = \begin{bmatrix} \mathbf{s} & & & 0 \\ & \mathbf{s} & & \\ & & \mathbf{s} & \\ 0 & & & \mathbf{s} \end{bmatrix} \Delta\mathbf{r}^{(e)}_{12\times 1} \tag{4.92}$$

where $\mathbf{s}$ is the matrix of eq. (4.23) and $\Delta\mathbf{r}^{(e)}$ lists the components of the elemental generalized incremental displacement vector expressed in global coordinates.

4.6 Elastic lattice-type structures

4.6.1 Discrete models

By a *lattice-type structure* we shall understand here a system of a rather large number of rod or beam elements interconnected by deformable or rigid nodes. The axes of the elements in the undeformed state are assumed to form a certain regular mesh on a plane or on a surface. We shall also assume that: the number of the degrees of freedom is finite and the same at each node, the mass and load distributions can be approximated by masses and loads assigned to the nodes of the structure only, and the motion of every rod or beam element is uniquely determined (in terms of some formulas known a priori) by the motion of the two nodes that bound this element. If the regularity of the mesh, together with the remaining assumptions listed above, allow to represent the governing relations for the lattice-type structure in the form of a system of finite difference equations then the model of the structure will be referred to as the discrete model.

We shall confine ourselves in this section to elastic lattice-type structures with rigid nodes (the joints in two- or three-dimensional frames are typical examples of rigid nodes). In order to construct the discrete model for a lattice-type structure we introduce the reference configuration in which all the nodes are represented by points in the plane parametrized by Cartesian orthogonal coordinates with the vector basis $\mathbf{a}_1, \mathbf{a}_2$. The set of all these points is denoted by $\mathscr{X}$.

We further assume that the structure at hand is composed of three families of rods at the most—this implies that any point $\mathbf{X} \in \mathscr{X}$ can be connected, via the pertinent rod element, with not more than six other points denoted by, cf. Fig. 16, $\mathbf{X}+\Delta_A\mathbf{X}$, $\mathbf{X}-\bar{\Delta}_A\mathbf{X}$, provided that $\mathbf{X}+\Delta_A\mathbf{X}$, $\mathbf{X}-\bar{\Delta}_A\mathbf{X} \in \mathscr{X}$, $A = \mathrm{I, II, III}$, for the fixed $\Delta_A\mathbf{X}$ or $\bar{\Delta}_A\mathbf{X}$. The current and reference configurations for the structure are shown in Fig. 16 ($\boldsymbol{\chi}(\mathbf{X}, \tau)$ is the place occupied by the mass center of the node $\mathbf{X}$ at time τ). In order to simplify further analysis we explicitly distinguish two kinds of nodes, setting

$$\mathscr{X}^0 \equiv \{\mathbf{X} \in \mathscr{X}; \mathbf{X}+\Delta_A\mathbf{X} \in \mathscr{X}, \mathbf{X}-\Delta_A\mathbf{X} \in \mathscr{X},\ A = \mathrm{I, II, III}\},$$

$$\partial\mathscr{X} \equiv \mathscr{X} \setminus \mathscr{X}^0$$

The nodes which belong to the set $\mathscr{X}^0$ are called *internal* and those in the set $\partial\mathscr{X}$ are referred to as the *boundary nodes.*

It has to be emphasized that nodes of the real structure analysed need not be located in the plane.

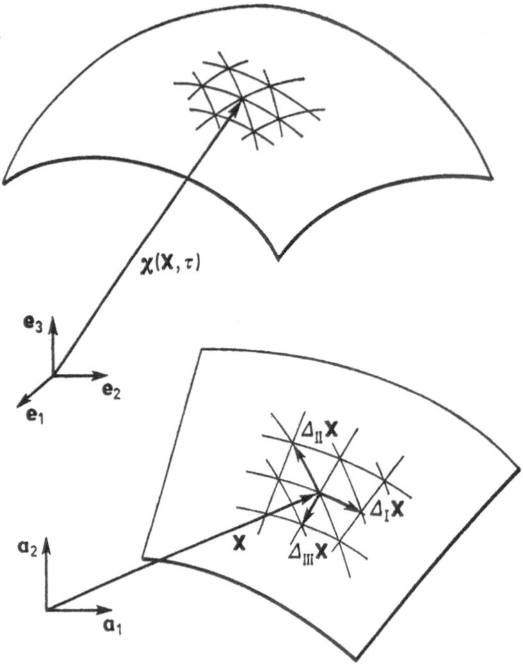

Fig. 16 Current and reference configuration of lattice structure

In a special case we can introduce such a reference configuration that the condition $\Delta_A \mathbf{X} = \bar{\Delta}_A \mathbf{X}$ holds for every $\mathbf{X} \in \mathscr{X}^0$ and for $A = \mathrm{I}, \mathrm{II}, \mathrm{III}$. In this case vectors $\Delta_A \mathbf{X}$ are constant (independent of $\mathbf{X}$) for every A.

Since the vector basis $\mathbf{a}_1, \mathbf{a}_2$ and the set $\mathscr{X}$ are assumed to be known a priori and $\mathbf{X} = X^\alpha \mathbf{a}_\alpha$ ($\alpha = 1, 2$; summation convention holds[1]), we may identify, slightly abusing the notation, every $\mathbf{X} \in \mathscr{X}$ with the pair (X^1, X^2). We shall also treat the points of $\mathscr{X}$ as the material coordinates of the nodes. For the sake of simplicity every $\mathbf{X} \in \mathscr{X}$ will be referred to as the node $\mathbf{X}$, instead of "the node with the material coordinates $\mathbf{X} = (X^1, X^2)$".

Let $\mathbf{e}_k$, $k = 1, 2, 3$, be the orthonormal vector basis of the spatial coordinate system in the physical space in which all "current" configurations of the structure are embedded. Thus

$$\boldsymbol{\chi}(\mathbf{X}, \tau) = \chi^k(\mathbf{X}, \tau)\mathbf{e}_k, \quad \mathbf{X} \in \mathscr{X}, \quad \tau \in [t_0, t_f].$$

Moreover, let $\mathbf{W}(\mathbf{X}) = W^k(\mathbf{X})\mathbf{e}_k$, $\mathbf{X} \in \mathscr{X}$, be a certain vector field defined on $\mathscr{X}$. We shall use the following notation

---

[1] Summation convention holds throughout the whole section while the indices α, β run over the sequence 1, 2.

$$\Delta_A \mathbf{W}(\mathbf{X}) \equiv \mathbf{W}(\mathbf{X}+\Delta_A \mathbf{X})-\mathbf{W}(\mathbf{X}),$$
$$\bar{\Delta}_A \mathbf{W}(\mathbf{X}) \equiv \mathbf{W}(\mathbf{X})-\mathbf{W}(\mathbf{X}-\bar{\Delta}_A \mathbf{X}),$$

provided $\mathbf{X}+\Delta_A\mathbf{X} \in \mathscr{X}$, $\mathbf{X} \in \mathscr{X}$ and $\mathbf{X}-\bar{\Delta}_A\mathbf{X} \in \mathscr{X}$. In this way we have defined the right- and left-hand side finite differences, respectively, of the vector field $\mathbf{W}(\mathbf{X}) = W^k(\mathbf{X})\mathbf{e}_k$, $\mathbf{X} \in \mathscr{X}$. Similarly, we define the finite differences of any scalar field φ: $\mathscr{X} \to \boldsymbol{R}$ as

$$\Delta_A \varphi(\mathbf{X}) \equiv \varphi(\mathbf{X}+\Delta_A \mathbf{X})-\varphi(\mathbf{X}),$$
$$\bar{\Delta}_A \varphi(\mathbf{X}) \equiv \varphi(\mathbf{X})-\varphi(\mathbf{X}-\bar{\Delta}_A \mathbf{X}).$$

Additionally, we shall find useful to denote

$$\{\mathbf{W}(\mathbf{X})\}_A \equiv \tfrac{1}{2}[\mathbf{W}(\mathbf{X})+\mathbf{W}(\mathbf{X}+\Delta_A \mathbf{X})],$$
$$\{\varphi(\mathbf{X})\}_A \equiv \tfrac{1}{2}[\varphi(\mathbf{X})+\varphi(\mathbf{X}+\Delta_A \mathbf{X})],$$

provided both $\mathbf{X}$ and $\mathbf{X}+\Delta_A\mathbf{X}$ belong to the set $\mathscr{X}$.

If a certain scalar or vector quantity is assigned not to a fixed node $\mathbf{X} \in \mathscr{X}$ but to a rod connecting, say, the nodes $\mathbf{X}$ and $\mathbf{X}+\Delta_A\mathbf{X}$, then it will be denoted by $\psi_A(\mathbf{X})$ and $\mathbf{V}_A(\mathbf{X})$, respectively (or by $\psi^A(\mathbf{X})$, $\mathbf{V}^A(\mathbf{X})$; the position of the index A, $A = \mathrm{I, II, III}$, is immaterial). We also define

$$\bar{\psi}_A(\mathbf{X}+\Delta_A \mathbf{X}) \equiv \psi_A(\mathbf{X}),$$
$$\bar{\mathbf{V}}_A(\mathbf{X}+\Delta_A \mathbf{X}) \equiv \mathbf{V}_A(\mathbf{X}),$$

and hence

$$\bar{\psi}_A(\mathbf{X}) = \psi_A(\mathbf{X}-\bar{\Delta}_A \mathbf{X}),$$
$$\bar{\mathbf{V}}_A(\mathbf{X}) = \mathbf{V}_A(\mathbf{X}-\bar{\Delta}_A \mathbf{X}).$$

The physical meaning of the fields $\psi_A(\cdot)$, $\mathbf{V}_A(\cdot)$ as well as the fields $\bar{\psi}_A(\cdot)$, $\bar{\mathbf{V}}_A(\cdot)$ clearly suggests that they should be defined not on the set $\mathscr{X}$ but on the sets

$$\mathscr{X}^A \equiv \{\mathbf{X} \in \mathscr{X}; \mathbf{X}+\Delta_A \mathbf{X} \in \mathscr{X}\},$$
$$\bar{\mathscr{X}}^A \equiv \{\mathbf{X} \in \mathscr{X}; \mathbf{X}-\bar{\Delta}_A \mathbf{X} \in \mathscr{X}\}, \qquad A = \mathrm{I, II, III},$$

respectively. It can be further observed that if $\varphi(\cdot)$, $\mathbf{W}(\cdot)$ are functions defined on $\mathscr{X}$ then $\Delta_A\varphi(\cdot)$, $\Delta_A\mathbf{W}(\cdot)$ are defined on $\mathscr{X}^A$ while $\bar{\Delta}_A\varphi(\cdot)$, $\bar{\Delta}_A\mathbf{W}(\cdot)$ are defined on $\bar{\mathscr{X}}^A$. At the same time, by simple calculations, we obtain the identities[1]

$$\bar{\Delta}_A \psi_A(\mathbf{X}) = \Delta_A \bar{\psi}_A(\mathbf{X}),$$
$$\bar{\Delta}_A \mathbf{V}_A(\mathbf{X}) = \Delta_A \bar{\mathbf{V}}_A(\mathbf{X}), \qquad A = \mathrm{I, II, III},$$

[1] Summation convention with respect to $A = \mathrm{I, II, III}$ does not hold unless otherwise stated. If the lattice structure consists of two families of rods only then $A = \mathrm{I, II}$.

for every $\mathbf{X} \in \mathscr{X}^A \cap \bar{\mathscr{X}}^A$. The foregoing formulas will make it possible to symmetrize the form of finite difference equations we are going to obtain below.

In some special problems we shall assign to every $\mathbf{X} \in \mathscr{X}$ a separate vector basis $\mathbf{g}_a(\mathbf{X})$ in the physical space ($a = 1, 2, 3$ in general) which is interrelated with the basis $\mathbf{e}_k$, $k = 1, 2, 3$, by means of the transformation laws

$$\mathbf{g}_a(\mathbf{X}) = A_a^{\cdot k}(\mathbf{X})\mathbf{e}_k, \qquad \mathbf{e}_k = A_k^{\cdot a}(\mathbf{X})\mathbf{g}_a(\mathbf{X})$$

where $A_a^{\cdot k}(\mathbf{X})$ and $A_k^{\cdot a}(\mathbf{X})$ are elements of the known non-singular 3×3 matrix such that

$$A_a^{\cdot k}(\mathbf{X})A_k^{\cdot b}(\mathbf{X}) = \delta_a^b,$$

where a, b run over 1, 2, 3. Hence

$$\begin{aligned}\Delta_A\mathbf{W}(\mathbf{X}) &= \Delta_A[W^a(\mathbf{X})\mathbf{g}_a(\mathbf{X})] \\ &= W^a(\mathbf{X}+\Delta_A\mathbf{X})\mathbf{g}_a(\mathbf{X}+\Delta_A\mathbf{X}) - W^a(\mathbf{X})\mathbf{g}_a(\mathbf{X}) \\ &= W^a(\mathbf{X}+\Delta_A\mathbf{X})[\mathbf{g}_a(\mathbf{X})+\Delta_A\mathbf{g}_a(\mathbf{X})] - W^a(\mathbf{X})\mathbf{g}_a(\mathbf{X}),\end{aligned}$$

$$A = \mathrm{I}, \mathrm{II}, \mathrm{III}.$$

Introducing matrices $G^b_{Aa}(\mathbf{X})$ such that

$$\Delta_A\mathbf{g}_a(\mathbf{X}) = G^b_{Aa}(\mathbf{X})\mathbf{g}_b(\mathbf{X}),$$

we obtain

$$\Delta_A\mathbf{W}(\mathbf{X}) = \delta_A W^a(\mathbf{X})\mathbf{g}_a(\mathbf{X}),$$

where we have denoted

$$\delta_A W^a(\mathbf{X}) \equiv \Delta_A W^a(\mathbf{X}) + G^a_{Ab}(\mathbf{X})W^b(\mathbf{X}+\Delta_A\mathbf{X}).$$

We may similarly write

$$\bar{\Delta}_A\mathbf{W}(\mathbf{X}) = \bar{\delta}_A W^a(\mathbf{X})\mathbf{g}_a(\mathbf{X})$$

where

$$\bar{\delta}_A W^a(\mathbf{X}) \equiv \bar{\Delta}_A W^a(\mathbf{X}) + G^a_{Ab}(\mathbf{X})W^b(\mathbf{X}-\bar{\Delta}_A\mathbf{X}).$$

The finite difference $\Delta_A\mathbf{W}(\mathbf{X})$ and $\bar{\Delta}_A\mathbf{W}(\mathbf{X})$ of the vector valued function $\mathbf{W}(\cdot)$ defined on $\mathscr{X}$ are expressed here in the vector basis $\mathbf{g}_a(\mathbf{X})$, $a = 1, 2, 3$, which depends on $\mathbf{X} \in \mathscr{X}$. It can be easily observed that

$$G^b_{Aa}(\mathbf{X}) = A_k^{\cdot b}(\mathbf{X})\Delta_A A_a^{\cdot k}(\mathbf{X}).$$

For further particulars concerning the formalism introduced above the reader is referred to [173].

It has been so far assumed that every node $\mathbf{X} \in \mathscr{X}$ of the lattice-type structure can be treated as rigid (hence, it has not more than six degrees of freedom). Moreover, let every internal node $\mathbf{X} \in \mathscr{X}^o$ be loaded by the resultant force $\mathbf{F}(\mathbf{X}, \tau)$ acting at its mass center $\boldsymbol{\chi}(\mathbf{X}, \tau)$ and the resultant moment $\mathbf{H}(\mathbf{X}, \tau)$; no other loads are applied to the structure. Let $m(\mathbf{X})$, $\mathbf{X} \in \mathscr{X}^0$, stands for the mass distribution in the lattice-type structure under consideration; this is in accordance with out initial assumption that the mass of the structure is approximated by the system of concentrated masses assigned exclusively to the internal nodes. Such an approximation is plausible for the number of internal nodes (which belong to $\mathscr{X}^0$) that is large and much bigger than the number of boundary nodes (which are elements of $\partial\mathscr{X}$); in the sequel only such lattice-type structures will be considered.

Let finally $\mathbf{P}^A(\mathbf{X}, \tau)$, $\mathbf{M}^A(\mathbf{X}, \tau)$ be the vectors of the resultant force and of the resultant moment, respectively, which act from the node $\mathbf{X}+\Delta_A\mathbf{X}$ on the node $\mathbf{X}$ and are applied at the point $\boldsymbol{\chi}(\mathbf{X}, \tau)+\eta\Delta_A\boldsymbol{\chi}(\mathbf{X}, \tau)$, $\eta \in [0, 1]$, in the physical space, cf. Fig. 17. Using all the aforementioned assumptions

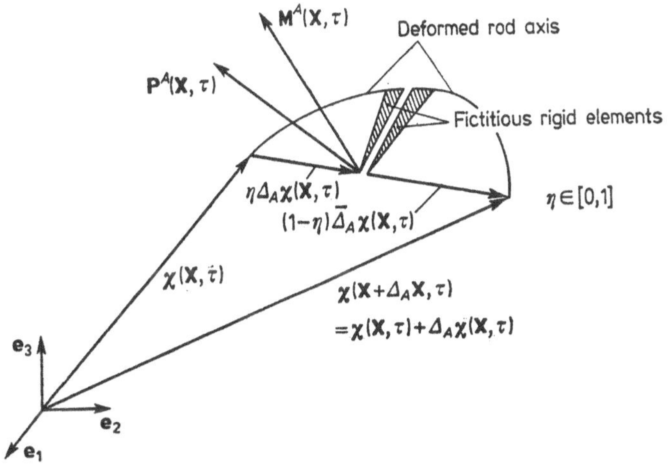

Fig. 17 Internal forces in rod element

and notations we may now present the equations of motion for an arbitrary node $\mathbf{X} \in \mathscr{X}^0$ as (summation convention with respect to A holds!)

$$\bar{\Delta}_A\mathbf{P}^A(\mathbf{X}, \tau)+\mathbf{F}(\mathbf{X}, \tau) = m(\mathbf{X})\ddot{\boldsymbol{\chi}}(\mathbf{X}, \tau),$$
$$\bar{\Delta}_A\mathbf{M}^A(\mathbf{X}, \tau)+\eta\Delta_A\boldsymbol{\chi}(\mathbf{X}, \tau)\times\mathbf{P}^A(\mathbf{X}, \tau)+ +(1-\eta)\bar{\Delta}_A\boldsymbol{\chi}(\mathbf{X}, \tau)\times\bar{\mathbf{P}}^A(\mathbf{X}, \tau)+\mathbf{H}(\mathbf{X}, \tau) = 0, \quad (4.93)$$

where η is an arbitrary but fixed parameter, $\eta \in [0, 1]$, cf. Fig. 17. Setting

$$\mathbf{P}^A(\mathbf{X}, \tau) = P^{Aa}(\mathbf{X}, \tau)\mathbf{g}_a(\mathbf{X}),$$
$$\mathbf{F}(\mathbf{X}, \tau) = F^a(\mathbf{X}, \tau)\mathbf{g}_a(\mathbf{X}),$$
$$\mathbf{M}^A(\mathbf{X}, \tau) = M^{Aa}(\mathbf{X}, \tau)\mathbf{g}_a(\mathbf{X}),$$
$$\mathbf{H}(\mathbf{X}, \tau) = H^a(\mathbf{X}, \tau)\mathbf{g}_a(\mathbf{X}),$$

we obtain the coordinate form of the equations of motion (indices a, b, c run over 1, 2, 3)

$$\begin{aligned}
&\bar{\delta}_A P^{Aa}(\mathbf{X}, \tau)+F^a(\mathbf{X}, \tau) = m(\mathbf{X})\ddot{\chi}^a(\mathbf{X}, \tau),\\
&\bar{\delta}_A M^{Aa}(\mathbf{X}, \tau)+[\eta\delta_A\chi^b(\mathbf{X}, \tau)P^{Ac}(\mathbf{X}, \tau)+\\
&\qquad +(1-\eta)\bar{\delta}_A\chi^b(\mathbf{X}, \tau)\overline{P^{Ac}}(\mathbf{X}, \tau)]\,\varepsilon^{a}_{\cdot bc}(\mathbf{X})+H^a(\mathbf{X}, \tau) = 0,
\end{aligned}$$

where we have introduced the symbol $\varepsilon^{a}_{\cdot bc}(\mathbf{X})$ as

$$\varepsilon^{a}_{\cdot bc}(\mathbf{X})\mathbf{g}_a(\mathbf{X}) = \mathbf{g}_b(\mathbf{X})\times\mathbf{g}_c(\mathbf{X}).$$

Introducing the vector basis $\mathbf{g}^a(\mathbf{X})$, $a = 1, 2, 3$, defined by

$$\mathbf{g}^a(\mathbf{X})\cdot\mathbf{g}_b(\mathbf{X}) = \delta^a_b,$$

we conclude that

$$\varepsilon^{a}_{\cdot bc}(\mathbf{X}) = [\mathbf{g}_b(\mathbf{X})\times\mathbf{g}_c(\mathbf{X})]\cdot\mathbf{g}^a(\mathbf{X}),$$

i.e. $\varepsilon^{a}_{\cdot bc}(\mathbf{X})$ is the mixed product of the vectors $\mathbf{g}^a(\mathbf{X})$, $\mathbf{g}_b(\mathbf{X})$, $\mathbf{g}_c(\mathbf{X})$.

The coordinate form of the equations of motion introduced above takes into account the fact that to every node $\mathbf{X}\in\mathscr{X}$ there is assigned a separate vector basis $\mathbf{g}_a(\mathbf{X})$, $a = 1, 2, 3$. If all such bases are made to coincide with the (orthonormal) vector basis $\mathbf{e}_i$, $i = 1, 2, 3$, of the spatial coordinate system, then the equations of motion will assume the form (summation with respect to A holds)

$$\begin{aligned}
&\bar{\Delta}_A P^{Ai}(\mathbf{X}, \tau)+F^i(\mathbf{X}, \tau) = m(\mathbf{X})\ddot{\chi}^i(\mathbf{X}, \tau),\\
&\bar{\Delta}_A M^{Ai}(\mathbf{X}, \tau)+[\eta\Delta_A\chi^j(\mathbf{X}, \tau)P^{Ak}(\mathbf{X}, \tau)+\\
&\qquad +(1-\eta)\bar{\Delta}_A\chi^j(\mathbf{X}, \tau)\bar{P}^{Ak}(\mathbf{X}, \tau)]\,\varepsilon^{i}_{\cdot jk}+H^i(\mathbf{X}, \tau) = 0,
\end{aligned}$$

where $\varepsilon^{i}_{\cdot jk}$ stands for the Ricci symbol.

According to the notation introduced at the beginning of the section the form of the equations of motion can be symmetrized by replacing the terms

$$\bar{\Delta}_A\mathbf{P}^A(\mathbf{X}, t),\qquad \bar{\Delta}_A\mathbf{M}^A(\mathbf{X}, t)$$

by the terms

$$\tfrac{1}{2}[\bar{\Delta}_A\mathbf{P}^A(\mathbf{X}, t)+\Delta_A\bar{\mathbf{P}}^A(\mathbf{X}, t)],\qquad \tfrac{1}{2}[\bar{\Delta}_A\mathbf{M}^A(\mathbf{X}, t)+\Delta_A\bar{\mathbf{M}}^A(\mathbf{X}, t)],$$

respectively.

From the assumption that the nodes can be treated as rigid we obtain the obvious conclusion that the deformation of the structure can be determined by the position $\boldsymbol{\chi}(\mathbf{X}, t)$ of the nodes as well as by their rotations. The rotations will be described by the rotation matrices $\mathbf{Q}(\mathbf{X}, t)$ with the components $Q_a^{\cdot b}(\mathbf{X}, t)$ related to the basis $\mathbf{g}_a(\mathbf{X})$, $a = 1, 2, 3$, or the components $Q_i^{\cdot j}(\mathbf{X}, t)$ which are related to the orthonormal vector basis $\mathbf{e}_i$, $i = 1, 2, 3$.

Let us now assume that $\varkappa(\mathbf{X})$, $\mathbf{X} \in \mathscr{X}$, determine a certain known (undeformed) configuration of the lattice-type structure. Every vector basis $\mathbf{g}_a(\mathbf{X})$, $\mathbf{X} \in \mathscr{X}$, can then be interpreted as the basis assigned in the undeformed configuration to the node $\mathbf{X} \in \mathscr{X}$ treated as the rigid body. Setting

$$\mathbf{G}_a(\mathbf{X}, t) = Q_a^{\cdot b}(\mathbf{X}, t)\mathbf{g}_b(\mathbf{X}),$$

we shall interpret $G_a(\mathbf{X}, t)$ as the vector basis which rotates with the node treated as the rigid body. Hence, $G_a(\mathbf{X}, t)$ is assigned in the actual configuration to the node $\mathbf{X} \in \mathscr{X}$. The foregoing formula yields the interpretation of the rotation matrices $\mathbf{Q}(\mathbf{X}, t)$ as well as the interpretation of the vector basis $\mathbf{g}_a(\mathbf{X})$, $a = 1, 2, 3$, which up to now has been introduced in a quite formal manner. The functions $\boldsymbol{\chi}(\mathbf{X}, t)$, $\mathbf{Q}(\mathbf{X}, t)$, $\mathbf{X} \in \mathscr{X}$, $t \in [t_0, t_f]$, describe the motion of the structure under consideration.

We shall now pass to considering the constitutive equations assuming the elastic properties of the material. Under the assumptions introduced at the beginning of the section we shall postulate for every rod element (connecting nodes $\mathbf{X}$ and $\mathbf{X}+\Delta_A\mathbf{X}$), the relations

$$\begin{aligned}\mathbf{P}^A(\mathbf{X}, t) &= \mathscr{P}^A(\mathbf{X}; \Delta_A\boldsymbol{\chi}(\mathbf{X}, t), \mathbf{Q}(\mathbf{X}, t), \Delta_A\mathbf{Q}(\mathbf{X}, t)),\\ \mathbf{M}^A(\mathbf{X}, t) &= \mathscr{M}^A(\mathbf{X}; \Delta_A\boldsymbol{\chi}(\mathbf{X}, t), \mathbf{Q}(\mathbf{X}, t), \Delta_A\mathbf{Q}(\mathbf{X}, t)),\end{aligned} \tag{4.94}$$

where $\mathscr{P}^A(\mathbf{X};\ \cdot)$, $\mathscr{M}^A(\mathbf{X};\ \cdot)$ are assumed to be known continuous functions. Eqs. (4.94) have to hold for $\mathbf{X} \in \mathscr{X}^A$, provided that $\mathbf{P}^A(\mathbf{X}, t)$ or $\mathbf{M}^A(\mathbf{X}, t)$ are not known a priori. Such a situation can take place if either $\mathbf{X} \in \partial\mathscr{X}$ or $\mathbf{X}+\Delta_A\mathbf{X} \in \partial\mathscr{X}$ (cf. the boundary conditions specified below). Since the functions $\mathscr{P}^A(\mathbf{X};\ \cdot)$, $\mathscr{M}^A(\mathbf{X};\ \cdot)$ have to satisfy the principle of the material frame indifference (i.e. they must be invariant under any rigid body motion of the rod), they can be represented as functions of certain strain measures. As the strain measures for the rod we shall take vectors $\boldsymbol{\gamma}_A(\mathbf{X}, t)$, $\boldsymbol{\varkappa}_A(\mathbf{X}, t)$ with components given by

$$\begin{aligned}\gamma_{Ai}(\mathbf{X}, t) &= \{Q_i^{\cdot j}(\mathbf{X}, t)\}_A \Delta_A \chi_j(\mathbf{X}, t) - \Delta_A \varkappa_i(\mathbf{X}, t),\\ \varkappa_{Ai}(\mathbf{X}, t) &= \tfrac{1}{2}\{Q_j^{\cdot l}(\mathbf{X}, t)\}_A \Delta_A Q_{lk}(\mathbf{X}, t)\, \varepsilon_i^{\cdot jk},\end{aligned} \tag{4.95}$$

which can be checked to attain zero values for rigid body motions of the rod, cf. [167], p. 68. Thus, we arrive at the constitutive equations for the elastic rods in the form

$$\begin{aligned}\mathbf{P}^A(\mathbf{X}, t) &= \mathscr{P}^A(\mathbf{X}; \boldsymbol{\gamma}_A(\mathbf{X}, t), \boldsymbol{\varkappa}_A(\mathbf{X}, t)),\\ \mathbf{M}^A(\mathbf{X}, t) &= \mathscr{M}^A(\mathbf{X}; \boldsymbol{\gamma}_A(\mathbf{X}, t), \boldsymbol{\varkappa}_A(\mathbf{X}, t)),\end{aligned} \tag{4.96}$$

$\mathscr{P}^A(\mathbf{X};\ \cdot)$, $\mathscr{M}^A(\mathbf{X};\ \cdot)$ being the known functions (for the sake of notational simplicity we use here the same symbols as before). Equations of motion

(4.93), constitutive equations (4.96) and equations (4.95) defining the strain measures have to be considered together with the initial conditions

$$\boldsymbol{\chi}(\mathbf{X}, t_0) = \boldsymbol{\varkappa}_0(\mathbf{X}), \quad \dot{\boldsymbol{\chi}}(\mathbf{X}, t_0) = \mathbf{v}_0(\mathbf{X}), \quad \mathbf{X} \in \mathscr{X}^0, \tag{4.97}$$

where $\boldsymbol{\varkappa}_0(\cdot)$ and $\mathbf{v}_0(\cdot)$ are known, as well as with equations involving the boundary nodes which will be refered to as the boundary conditions. We shall restrict ourselves to two special kinds of the boundary conditions by setting $\partial\mathscr{X} = \partial_1\mathscr{X} \cup \partial_2\mathscr{X}$ where $\partial_1\mathscr{X}$ and $\partial_2\mathscr{X}$ are two disjointed subsets of the boundary nodes. Firstly, we shall assume that $\mathbf{P}^A(\mathbf{X}, t)$, $\mathbf{M}^A(\mathbf{X}, t)$ are known for $t \in [t_0, t_f]$ either if $\mathbf{X} \in \partial_1\mathscr{X}$ and $\mathbf{X}+\Delta_A\mathbf{X} \in \mathscr{X}^0$ or if $\mathbf{X} \in \mathscr{X}^0$ and $\mathbf{X}+$ $+\Delta_A\mathbf{X} \in \partial_1\mathscr{X}$. Secondly, we shall assume that $\boldsymbol{\chi}(\mathbf{X}, t)$, $\mathbf{Q}(\mathbf{X}, t)$ are known for $t \in [t_0, t_f]$ either if $\mathbf{X} \in \partial_2\mathscr{X}$ and $\mathbf{X}+\Delta_A\mathbf{X} \in \mathscr{X}^0$ or if $\mathbf{X}-\bar{\Delta}_K\mathbf{X} \in \mathscr{X}^0$ and $\mathbf{X} \in \partial_2\mathscr{X}$.

It can be easily verified that the equations of motion combined with the constitutive equations and the strain measure definitions lead, after taking into account the boundary conditions, to a system of the second order ordinary differential equations for $\boldsymbol{\chi}(\mathbf{X}, t)$ and to a system of algebraic equations for $\mathbf{Q}(\mathbf{X}, t)$ where $\mathbf{X} \in \mathscr{X}^0$, $t \in [t_0, t_f]$. Because every rotation matrix $\mathbf{Q}(\mathbf{X}, t)$ is uniquely determined by three independent functions we can also note that the total number of the resulting equations is equal to the total number of unknown functions. Namely, if the number of internal nodes (number of elements in the set $\mathscr{X}^0$) is equal to n then the total number of resulting equations as well as the total number of unknown functions are both equal to $6n$. The ordinary differential equations for $\boldsymbol{\chi}(\mathbf{X}, t)$, $t \in [t_0, t_f]$, have to be considered together with the initial conditions (4.97).

Let us now define the displacement fields $\mathbf{u}(\cdot, t)\colon \mathscr{X} \to \boldsymbol{R}^3$, $t \in [t_0, t_f]$, putting

$$\mathbf{u}(\mathbf{X}, t) \equiv \boldsymbol{\chi}(\mathbf{X}, t) - \boldsymbol{\varkappa}(\mathbf{X}).$$

Assume further that the rotations $\mathbf{Q}(\mathbf{X}, t)$ of all the nodes are small. Such being the case the rotation matrix can be approximated by the use of the known formula

$$Q_i^{\cdot j} \simeq \delta_i^j + \varepsilon_i^{\cdot jk} v_k,$$

where $\mathbf{v}(\mathbf{X}, t)$ is the small rotation vector. This leads to the following strain measures

$$\begin{aligned} \gamma_{Ai}(\mathbf{X}, t) &= \Delta_A u_i(\mathbf{X}, t) + \varepsilon_{ij}^{\cdot\cdot k} \Delta_A \varkappa^j(\mathbf{X}) \{v_k(\mathbf{X}, t)\}_A, \\ \varkappa_{Ai}(\mathbf{X}, t) &= \Delta_A v_i(\mathbf{X}, t). \end{aligned} \tag{4.98}$$

Independently of the aforementioned approximation we can also assume that the strain measures $\boldsymbol{\gamma}_A$, $\boldsymbol{\varkappa}_A$ are small. The constitutive equations can then be assumed in the linear form as

$$
\begin{aligned}
P^{Ai}(\mathbf{X}, t) &= \mathbf{A}^{Aij}(\mathbf{X})\gamma_{Aj}(\mathbf{X}, t)+\mathbf{B}^{Aij}(\mathbf{X})\varkappa_{Aj}(\mathbf{X}, t),\\
M^{Ai}(\mathbf{X}, t) &= \mathbf{B}^{Aji}(\mathbf{X})\gamma_{Aj}(\mathbf{X}, t)+\mathbf{C}^{Aij}(\mathbf{X})\varkappa_{Aj}(\mathbf{X}, t),
\end{aligned}
\tag{4.99}
$$

where it is assumed that the 6×6 matrices

$$
\begin{bmatrix} \mathbf{A}^A(\mathbf{X}) & \mathbf{B}^A(\mathbf{X}) \\ (\mathbf{B}^A(\mathbf{X}))^T & \mathbf{C}^A(\mathbf{X}) \end{bmatrix}, \quad \mathbf{X}\in\mathscr{X}^A,
$$

are symmetric and positive definite for every $A = \mathrm{I}, \mathrm{II}, \mathrm{III}$. The latter assumption is implied by the postulate of the existence of the elastic potential for every rod which is given by

$$
\tfrac{1}{2}\mathbf{A}^{Aij}\gamma_{Ai}\gamma_{Aj}+\mathbf{B}^{Aij}\gamma_{Ai}\varkappa_{Aj}+\tfrac{1}{2}\mathbf{C}^{Aij}\varkappa_{Ai}\varkappa_{Aj}.
$$

Therein, our initial assumption that the rod deformation is uniquely determined by the displacements and rotations of the nodes that bound the rod is effectively utilized.

Finally, we can also analyse problems in which the displacement increments $\Delta_A\mathbf{u}(\mathbf{X}, t)$ can be neglected with respect to $\Delta_A\boldsymbol{\varkappa}(\mathbf{X})$. The equations of motion can then be postulated in the following form (summation convention with respect to A holds):

$$
\begin{aligned}
&\bar{\Delta}_A\mathbf{P}^A(\mathbf{X}, t)+\mathbf{F}(\mathbf{X}, t) = m(\mathbf{X})\ddot{u}(\mathbf{X}, t),\\
&\bar{\Delta}_A\mathbf{M}^A(\mathbf{X}, t)+\eta\Delta_A\boldsymbol{\varkappa}(\mathbf{X})\times\mathbf{P}^A(\mathbf{X}, t)+\\
&\qquad\qquad+(1-\eta)\bar{\Delta}_A\boldsymbol{\varkappa}(\mathbf{X})\times\bar{\mathbf{P}}^A(\mathbf{X}, t)+\mathbf{H}(\mathbf{X}, t) = 0.
\end{aligned}
\tag{4.100}
$$

Eqs. (4.98), (4.99), (4.100) together with the appropriate boundary and initial conditions constitute the linearized discrete model of the elastic lattice-type structures. As an example let us consider equations for a thin prismatic rod made of an isotropic homogeneous linear-elastic material. Let $E_A(\mathbf{X})$, $A_A(\mathbf{X})$ stand for the Young modulus and the area of the cross section, respectively, $C_A(\mathbf{X})$ be the torsional rigidity and $I_{A\eta}(\mathbf{X})$, $I_{A\zeta}(\mathbf{X})$ denote the main central inertia moments of the cross-section. Denote also $l_A(\mathbf{X}) \equiv |\Delta_A\boldsymbol{\varkappa}(\mathbf{X})|$. Further, let $d_\xi^A(\mathbf{X})$, $d_\eta^A(\mathbf{X})$, $d_\zeta^A(\mathbf{X})$ be the triple of mutually orthogonal unit vectors with $d_\xi^A(\mathbf{X})$ directed along the axis of the undeformed rod and the remaining two coinciding with the principal axes of the cross-section. Setting

$$
\begin{aligned}
P^{Ai} &= d_\xi^{Ai}S_\xi^A+d_\eta^{Ai}S_\eta^A+d_\zeta^{Ai}S_\zeta^A,\\
M^{Ai} &= d_\xi^{Ai}G_\xi^A+d_\eta^{Ai}G_\eta^A+d_\zeta^{Ai}G_\zeta^A,
\end{aligned}
$$

assuming $\eta = \frac{1}{2}$ and denoting

$$
\begin{aligned}
&\varepsilon_{A\xi} \equiv d_\xi^{Ai}\gamma_{Ai}, \quad &&\varepsilon_{A\eta} \equiv d_\eta^{Ai}\gamma_{Ai}, \quad &&\varepsilon_{A\zeta} \equiv d_\zeta^{Ai}\gamma_{Ai},\\
&\mu_{A\xi} \equiv d_\xi^{Ai}\varkappa_{Ai}, \quad &&\mu_{A\eta} \equiv d_\eta^{Ai}\varkappa_{Ai}, \quad &&\mu_{A\zeta} \equiv d_\zeta^{Ai}\varkappa_{Ai}
\end{aligned}
$$

we may write down the following formulae:

$$S_\xi^A = \frac{E_A A_A}{l_A} \varepsilon_{A\xi},$$

$$S_\eta^A = \frac{12 E_A J_{A\zeta}}{l_A^2} \varepsilon_{A\eta},$$

$$S_\zeta^A = \frac{12 E_A J_{A\eta}}{l_A^2} \varepsilon_{A\zeta},$$

$$G_\xi^A = \frac{C_A}{l_A} \mu_{A\xi},$$

$$G_\eta^A = \frac{E_A J_{A\eta}}{l_A} \mu_{A\eta},$$

$$G_\zeta^A = \frac{E_A J_{A\zeta}}{l_A} \mu_{A\zeta},$$

which are implied by the well-known constitutive relations of the theory of rods under consideration. Simple algebra leads to the following results

$$A^{Aij} = \frac{E_A A_A}{l_A} d_\xi^{Ai} d_\xi^{Aj} + \frac{12 E_A J_{A\zeta}}{l_A^3} d_\eta^{Ai} d_\eta^{Aj} + \frac{12 E_A J_{A\eta}}{l_A^3} d_\zeta^{Ai} d_\zeta^{Aj},$$

$$B^{Aij} = 0, \tag{4.101}$$

$$C^{Aij} = \frac{C_A}{l_A} d_\xi^{Ai} d_\xi^{Aj} + \frac{E_A J_{A\eta}}{l_A} d_\eta^{Ai} d_\eta^{Aj} + \frac{E_A J_{A\zeta}}{l_A} d_\zeta^{Ai} d_\zeta^{Aj}$$

which are the specific form of the general constitutive equations (4.99).

We may note that an alternative definition of strain measures given by

$$\eta_{Ai} = \Delta_A u_i + \varepsilon_i^{\ jk} \Delta_A \varkappa_j v_k,$$

$$\varkappa_{Ai} = \Delta_A v_i$$

leads to the following expression for elastic strain energy

$$\varepsilon = \tfrac{1}{2} A^{Aij} \eta_{Ai} \eta_{Aj} + H^{Aij} \eta_{Ai} \varkappa_{Aj} + \tfrac{1}{2} F^{Aij} \varkappa_{Ai} \varkappa_{Aj}$$

in which the rigidities H^{Aij} and F^{Aij} are given by (A^{Aij} is the same as in eq. (4.101))

$$H^{Aij} = \frac{6 E_A J_{A\zeta}}{l_A^2} d_\zeta^{Ai} d_\eta^{Aj} - \frac{6 E_A J_{A\zeta}}{l_A^2} d_\eta^{Ai} d_\zeta^{Aj},$$

$$F^{Aij} = \frac{C_A}{l_A} d_\xi^{Ai} d_\xi^{Aj} + \frac{4 E_A J_{A\eta}}{l_A} d_\eta^{Ai} d_\eta^{Aj} + \frac{4 E_A J_{A\zeta}}{l_A} d_\zeta^{Ai} d_\zeta^{Aj}.$$

We may also observe that the constitutive equations expressed in terms of the new strain measures read

$$P^{Ai}(\mathbf{X}, t) = A^{Aij}(\mathbf{X})\eta_{Aj}(\mathbf{X}, t)+H^{Aij}(\mathbf{X})\varkappa_{Aj}(\mathbf{X}, t),$$
$$M^{Ai}(\mathbf{X}, t) = H^{Aji}(\mathbf{X})\eta_{Aj}(\mathbf{X}, t)+F^{Aij}(\mathbf{X})\varkappa_{Aj}(\mathbf{X}, t),$$

where now the generalized forces refer to the nodal cross-sections of the beams ($\eta = 1$).

This section is completed with the following summarizing remarks. The lattice-type structures investigated above may be composed of two (in which case $A = \text{I}, \text{II}$) or three families of rods ($A = \text{I}, \text{II}, \text{III}$). For the equations concerning more complex structures the reader is referred to [98], [99]. We have confined ourselves to elastic structures only—the way inelastic structures can be handled is outlined in Sec. 6.3. The finite difference formalism results in the concise form of the governing relations which can be used to advantage particularly for the analysis of regular lattice-type structures with a rather large number of the internal nodes. The formulation proposed in this section may be taken as the starting point to derive equations of another useful model of lattice-type structures which may be called the continuous model. This we shall undertake below.

4.6.2 *Continuous models*

Sometimes we have to deal with very dense lattice-type structures for which all the geometric and physical properties, and also the character of external load, can, roughly speaking, be treated as regularly distributed over the whole structure. In such cases it seems natural to try to approximate the functions defined on the discrete set $\mathscr{X}$ by continuous functions defined on the plane region Ω which, in a certain sense, approximates the set $\mathscr{X}$, Fig. 18.

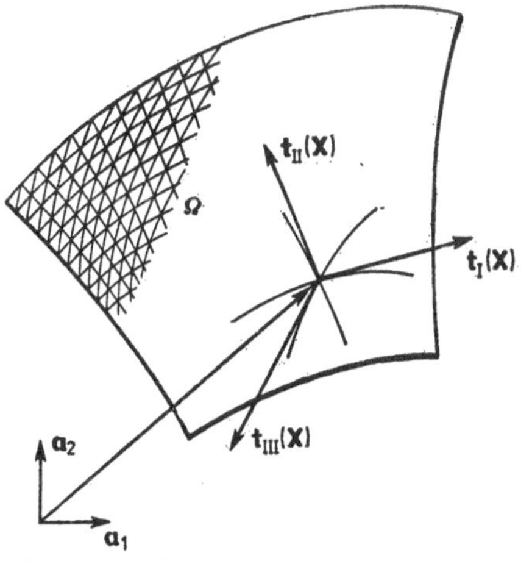

Fig. 18 Unit tangent vectors

Using this philosophy we shall show below how to pass from the discretized model to what is called a continuous model of the lattice-type structures. Such an approach has been thoroughly investigated in [164]–[166] and applied to many problems of engineering significance in [167].

Let is define

$$\Pi(\mathbf{X}) \equiv |\Delta_{\mathrm{I}}\mathbf{X}\times\Delta_{\mathrm{II}}\mathbf{X}|, \quad l_A(\mathbf{X}) \equiv |\Delta_A\mathbf{X}|, \quad \tilde{l}_A(\mathbf{X}) = \Pi(\mathbf{X})/l_A(\mathbf{X})$$

and introduce the following densities defined, for every t, on the finite systems of points belonging to $\mathscr{X}$:

$$\mathbf{p}^A(\mathbf{X},t) \equiv \frac{\mathbf{P}^A(\mathbf{X},t)}{\tilde{l}_A(\mathbf{X})}, \quad \mathbf{m}^A(\mathbf{X},t) \equiv \frac{\mathbf{M}^A(\mathbf{X},t)}{\tilde{l}_A(\mathbf{X})}, \quad \mathbf{X}\in\mathscr{X}^A,$$

$$\mathbf{f}(\mathbf{X},t) \equiv \frac{\mathbf{F}(\mathbf{X},t)}{\Pi(\mathbf{X})}, \quad \varrho(\mathbf{X}) \equiv \frac{m(\mathbf{X})}{\Pi(\mathbf{X})}, \quad \mathbf{h}(\mathbf{X},t) \equiv \frac{\mathbf{H}(\mathbf{X},t)}{\Pi(\mathbf{X})}, \quad \mathbf{X}\in\mathscr{X}^0.$$

Let us also introduce for an arbitrary scalar or vector field $\psi(\cdot)$ defined on $\mathscr{X}$ (or on $\mathscr{X}^A$, $A = \mathrm{I}, \mathrm{II}, \mathrm{III}$) the following finite difference operators

$$\nabla_A\psi(\mathbf{X}) \equiv \frac{\Delta_A\psi(\mathbf{X})}{l_A(\mathbf{X})}, \quad \bar{\nabla}_A\psi(\mathbf{X}) \equiv \frac{\bar{\Delta}_A\psi(\mathbf{X})}{l_A(\mathbf{X})}.$$

The equation of motion (4.93) may then be rewritten in the form (summation convention holds with respect to A)

$$\begin{aligned} &\frac{1}{\tilde{l}_A}\nabla_A(\mathbf{p}^A\tilde{l}_A)+\mathbf{f} = \varrho\ddot{\boldsymbol{\chi}}, \\ &\frac{1}{\tilde{l}_A}\nabla_A(\mathbf{m}^A\tilde{l}_A)+\eta\nabla_A\boldsymbol{\chi}\times\mathbf{p}^A+(1-\eta)\bar{\nabla}_A\boldsymbol{\chi}\times\bar{\mathbf{p}}^A+\mathbf{h} = 0. \end{aligned} \tag{4.102}$$

We introduce the new strain measures (instead of (4.95))

$$\begin{aligned} &\tilde{\gamma}_{Ai} \equiv \frac{\gamma_{Ai}}{l_A} = \{Q_i^{\cdot j}\}_A\nabla_A\chi_j-\nabla_A\varkappa_i, \\ &\tilde{\varkappa}_{Ai} \equiv \frac{\varkappa_{Ai}}{l_A} = \tfrac{1}{2}\{Q_j^{\cdot l}\}_A\nabla_A Q_{lk}\,\varepsilon_i^{\cdot jk}. \end{aligned} \tag{4.103}$$

Hence, we can postulate in the case of the elastic rod that

$$\begin{aligned} &\mathbf{p}^A(\mathbf{X},t) = \mathscr{P}^A\big(\mathbf{X};\tilde{\boldsymbol{\gamma}}_A(\mathbf{X},t)l_A(\mathbf{X}),\tilde{\boldsymbol{\varkappa}}_A(\mathbf{X},t)l_A(\mathbf{X})\big)/\tilde{l}_A(\mathbf{X}), \\ &\mathbf{m}^A(\mathbf{X},t) = \mathscr{M}^A\big(\mathbf{X};\tilde{\boldsymbol{\gamma}}_A(\mathbf{X},t)l_A(\mathbf{X}),\tilde{\boldsymbol{\varkappa}}_A(\mathbf{X},t)l_A(\mathbf{X})\big)/\tilde{l}_A(\mathbf{X}). \end{aligned} \tag{4.104}$$

Functions in eqs. (4.102)–(4.104), for every time instant $t\in[t_0,t_f]$, are defined either on the set $\mathscr{X}^0$ or on the sets $\mathscr{X}^A$. As we have stated before, the basic idea of passing from the discrete to a continuous model of the lattice-type structure consists in the extrapolation of the discrete functions by certain sufficiently regular functions defined on the region Ω of the plane. Under such assumption we shall treat all the functions in eqs. (4.102)–(4.104) as defined, for every $t \in [t_0, t_f]$, on the region Ω. As the second fundamental postulate we assume the possibility of approximation of the finite difference operators $\bar{\nabla}_A$, ∇_A by the derivatives in the direction of the vector $\mathbf{t}_A(\mathbf{X}) \equiv \Delta_A\mathbf{X}/|\Delta_A\mathbf{X}|$:

$$\nabla_A \psi(\mathbf{X}) \simeq \bar{\nabla}_A \psi(\mathbf{X}) \simeq \psi_{,\alpha}(\mathbf{X}) t^\alpha_A(\mathbf{X}).$$

Thus, from now on we may assume that $\mathbf{X} \in \Omega$. Taking into account the equations of motion (4.102) modified to the form in which $\mathbf{X} \in \Omega$, we can see that

$$\frac{1}{\tilde{l}_A} \nabla_A(\mathbf{p}^A \tilde{l}_A) \simeq \frac{1}{\tilde{l}_A} (\mathbf{p}^A \tilde{l}_A)_{,\alpha} t^\alpha_A = (\mathbf{p}^A t^\alpha_A)_{,\alpha} - \mathbf{p}^A \tilde{l}_A \left(\frac{t^\alpha_A}{\tilde{l}_A}\right)_{,\alpha}.$$

It can also be proved that

$$\left(\frac{t^\alpha_A}{\tilde{l}_A}\right)_{,\alpha} = 0.$$

Introducing the notation

$$T^{k\alpha}(\mathbf{X}, t) \equiv \sum_{A=\mathrm{I}}^{\mathrm{III}} p^{Ak}(\mathbf{X}, t) t^\alpha_A(\mathbf{X}),$$

$$M^{k\alpha}(\mathbf{X}, t) \equiv \sum_{A=\mathrm{I}}^{\mathrm{III}} m^{Ak}(\mathbf{X}, t) t^\alpha_A(\mathbf{X}),$$

we obtain from eq. (4.102) after simple calculations the following differential equations:

$$\begin{aligned} &T^{k\alpha}{}_{,\alpha}(\mathbf{X}, t) + f^k(\mathbf{X}, t) = \varrho(\mathbf{X}) \ddot{\chi}^k(\mathbf{X}, t), \\ &M^{k\alpha}{}_{,\alpha}(\mathbf{X}, t) + \varepsilon^{k}_{\cdot ij} \chi^i{}_{,\alpha}(\mathbf{X}, t) T^{j\alpha}(\mathbf{X}, t) + h^k(\mathbf{X}, t) = 0. \end{aligned} \tag{4.105}$$

Eqs. (4.105) represent the equations of motion for the continuous model of the lattice-type structures.

Performing further simple calculations we can show that after defining the symbols

$$\begin{aligned} \gamma_{i\alpha} &\equiv Q_i^{\cdot j} \chi_{j,\alpha} - \varkappa_{i,\alpha}, \\ \varkappa_{i\alpha} &\equiv \tfrac{1}{2} Q_j^{\cdot l} Q_{lk,\alpha} \varepsilon_i^{\cdot jk}, \end{aligned} \tag{4.106}$$

we arrive at

$$\begin{aligned} \tilde{\gamma}_{Ai} &= \gamma_{i\alpha} t^\alpha_A, \\ \tilde{\varkappa}_{Ai} &= \varkappa_{i\alpha} t^\alpha_A. \end{aligned}$$

Thus, $\boldsymbol{\gamma}(\mathbf{X}, t) = (\gamma_{i\alpha}(\mathbf{X}, t))$, $\boldsymbol{\varkappa}(\mathbf{X}, t) = (\varkappa_{i\alpha}(\mathbf{X}, t))$ can be taken as the strain measures in the continuous model at hand. Now, by the constitutive equations (4.94) and using the notation

$$\tilde{T}^{k\alpha}(\mathbf{X}; \boldsymbol{\gamma}, \boldsymbol{\varkappa}) \equiv \sum_{A=\mathrm{I}}^{\mathrm{III}} \mathscr{T}^{Ak}(\mathbf{X}; \boldsymbol{\gamma}\mathbf{t}_A l_A, \boldsymbol{\varkappa} t_A l_A) t_A^{\alpha}/\tilde{l}_A,$$

$$\tilde{M}^{k\alpha}(\mathbf{X}; \boldsymbol{\gamma}, \boldsymbol{\varkappa}) \equiv \sum_{A=\mathrm{I}}^{\mathrm{III}} \mathscr{M}^{Ak}(\mathbf{X}; \boldsymbol{\gamma}\mathbf{t}_A l_A, \boldsymbol{\varkappa} t_A l_A) t_A^{\alpha}/\tilde{l}_A,$$

we arrive at the following general form of the constitutive equations for the continuous model of the elastic lattice-type structures:

$$\begin{aligned} T^{k\alpha}(\mathbf{X}, t) &= \tilde{\mathscr{T}}^{k\alpha}(\mathbf{X}; \boldsymbol{\gamma}(\mathbf{X}, t), \boldsymbol{\varkappa}(\mathbf{X}, t)), \\ M^{k\alpha}(\mathbf{X}, t) &= \tilde{\mathscr{M}}^{k\alpha}(\mathbf{X}; \boldsymbol{\gamma}(\mathbf{X}, t), \boldsymbol{\varkappa}(\mathbf{X}, t)). \end{aligned} \tag{4.107}$$

Equations of motion (4.105), geometric equations (4.106) and constitutive equations (4.107) are the governing relation of the problem treated within the assumptions of the continuous approach. The basic unknowns are: the deformation vector $\boldsymbol{\chi}(\mathbf{X}, t)$ and the rotation matrix $\mathbf{Q}(\mathbf{X}, t)$, the latter determined by the three independent real valued functions. The unknowns are defined, for every $t \in [t_0, t_f]$, on the region Ω of the plane.

Let us now introduce the displacement field

$$\mathbf{u}(\mathbf{X}, t) \equiv \boldsymbol{\chi}(\mathbf{X}, t) - \boldsymbol{\varkappa}(\mathbf{X}), \quad \mathbf{X} \in \Omega,\ t \in [t_0, t_f],$$

where $\boldsymbol{\varkappa}\colon \Omega \to \boldsymbol{R}^3$ is the known smooth mapping. Assuming that the rotations are small we obtain

$$Q_i^{\cdot j} \simeq \delta_i^j + \varepsilon_i^{\cdot jk} v_k.$$

Thus, the strain measures can now be postulated in the form

$$\begin{aligned} \gamma_{i\alpha}(\mathbf{X}, t) &= u_{i,\alpha}(\mathbf{X}, t) + \varepsilon_i^{\cdot jk} v_k(\mathbf{X}, t) \varkappa_{j,\alpha}(\mathbf{X}), \\ \varkappa_{i\alpha}(\mathbf{X}, t) &= v_{i,\alpha}(\mathbf{X}, t). \end{aligned} \tag{4.108}$$

In many problems the displacement gradients $\nabla \mathbf{u}(\mathbf{X}, t)$ can be treated as small. Then, by means of the approximation

$$\chi^i_{,\alpha} T^{j\alpha} = (\varkappa^i_{,\alpha} + u^i_{,\alpha}) T^{j\alpha} \simeq \varkappa^i_{,\alpha} T^{j\alpha}$$

instead of the general form (4.105) of the equation of motion we postulate the following simplified form of them

$$\begin{aligned} T^{k\alpha}_{,\alpha}(\mathbf{X}, t) + f^k(\mathbf{X}, t) &= \varrho(\mathbf{X}) \ddot{u}^k(\mathbf{X}, t), \\ M^{k\alpha}_{,\alpha}(\mathbf{X}, t) + \varepsilon^{k}_{\cdot ij} \varkappa^j_{,\alpha}(\mathbf{X}) T^{j\alpha}(\mathbf{X}, t) + h^k(\mathbf{X}, t) &= 0. \end{aligned} \tag{4.109}$$

It may also be observed that from the purely formal point of view the equations of motion obtained for the continuous model of lattice-type structures have a similar form to those of the Cosserat media, [166].

The linearized form of the constitutive equations (4.107) expressed in terms of the small strain measures is written as

$$\begin{aligned} T^{k\alpha}(\mathbf{X}, t) &= \tilde{\mathbf{A}}^{k\alpha l\beta}(\mathbf{X})\gamma_{l\beta}(\mathbf{X}, t)+\tilde{\mathbf{B}}^{k\alpha l\beta}(\mathbf{X})\varkappa_{l\beta}(\mathbf{X}, t), \\ M^{k\alpha}(\mathbf{X}, t) &= \tilde{\mathbf{B}}^{l\beta k\alpha}(\mathbf{X})\gamma_{l\beta}(\mathbf{X}, t)+\tilde{\mathbf{C}}^{k\alpha l\beta}(\mathbf{X})\varkappa_{l\beta}(\mathbf{X}, t), \end{aligned} \tag{4.110}$$

where by eqs. (4.101) we have

$$\tilde{\mathbf{A}}^{k\alpha l\beta} = \sum_{A=\mathrm{I}}^{\mathrm{III}} A^{Aij} t_A^\alpha t_A^\beta \frac{l_A}{\tilde{l}_A},$$

$$\tilde{\mathbf{B}}^{k\alpha l\beta} = \sum_{A=\mathrm{I}}^{\mathrm{III}} B^{Aij} t_A^\alpha t_A^\beta \frac{l_A}{\tilde{l}_A},$$

$$\tilde{\mathbf{C}}^{k\alpha l\beta} = \sum_{A=\mathrm{I}}^{\mathrm{III}} C^{Aij} t_A^\alpha t_A^\beta \frac{l_A}{\tilde{l}_A}.$$

The moduli $A^{Aij}(\mathbf{X})$, $B^{Aij}(\mathbf{X})$, $C^{Aij}(\mathbf{X})$ defined on the set Ω, for $\mathbf{X} \in \mathscr{X}^0$ have the same meaning as those appearing in eqs. (4.101) for the discrete model of the structure.

Considering now the lattices of this prismatic rods made of an isotropic linear-elastic homogeneous material discussed at the end of Sec. 4.6.1, we may obtain

$$\begin{aligned} \tilde{\mathbf{A}}^{k\alpha l\beta} &= \sum_{A=\mathrm{I}}^{\mathrm{III}} (\alpha_A d_{A\xi}^k d_{A\xi}^l + \alpha_{A\eta} d_{A\eta}^k d_{A\eta}^l + \alpha_{A\zeta} d_{A\zeta}^k d_{A\zeta}^l) t_A^\alpha t_A^\beta, \\ \tilde{\mathbf{B}}^{k\alpha l\beta} &= 0, \\ \tilde{\mathbf{C}}^{k\alpha l\beta} &= \sum_{A=\mathrm{I}}^{\mathrm{III}} (\delta_A d_{A\xi}^k d_{A\xi}^l + \delta_{A\eta} d_{A\eta}^k d_{A\eta}^l + \delta_{A\zeta} d_{A\zeta}^k d_{A\zeta}^l) t_A^\alpha t_A^\beta, \end{aligned} \tag{4.111}$$

where the following additional notation have been introduced

$$\begin{aligned} \alpha_A &\equiv \frac{E_A A_A}{\tilde{l}_A}, \quad \alpha_{A\eta} \equiv \frac{12 E_A J_{A\zeta}}{\tilde{l}_A l_A^2}, \quad \alpha_{A\zeta} \equiv \frac{12 E_A J_{A\eta}}{\tilde{l}_A l_A^2}, \\ \delta_A &\equiv \frac{C_A}{\tilde{l}_A}, \quad \delta_{A\eta} \equiv \frac{E_A J_{A\eta}}{\tilde{l}_A}, \quad \delta_{A\zeta} \equiv \frac{E_A J_{A\zeta}}{\tilde{l}_A}. \end{aligned} \tag{4.112}$$

It has to be emphasized again that for the discrete model the functions $\alpha_A(\mathbf{X}), \ldots, \delta_{A\zeta}(\mathbf{X})$ are defined on $\mathscr{X}^A$ so that passing to the continuous model they have to be extrapolated on the whole region Ω as certain sufficiently regular functions. The similar extrapolation must be applied to all other functions present in the equations describing the continuous model.

Equations of motion (4.109), geometric equations (4.108) and constitutive equations (4.110) represent the system of governing equations for the continuous model of linear-elastic lattice-type structures. The basic unknowns are: the displacement vector field $\mathbf{u}(\cdot, t)$ and the small rotation vector field $\mathbf{v}(\cdot, t)$ defined on Ω for every $t \in [t_0, t_f]$.

The governing relations, both in the general and the linearized cases, have to be considered together with the pertinent boundary and initial conditions. Initial conditions have the form similar to that of the discrete model

$$\boldsymbol{\chi}(\mathbf{X}, t_0) = \boldsymbol{\varkappa}_0(\mathbf{X}), \quad \dot{\boldsymbol{\chi}}(\mathbf{X}, t_0) = \mathbf{v}_0(\mathbf{X}); \quad \mathbf{X} \in \Omega,$$

with $\boldsymbol{\varkappa}_0(\cdot)$, $\mathbf{v}_0(\cdot)$ being known. To give an example of such boundary conditions let us assume that there is given a decomposition $\partial\Omega = \partial_1\Omega \cup \partial_2\Omega$ of the boundary of the region Ω into two disjointed parts. We assume the following conditions to hold on $\partial_1\Omega$:

$$T^{k\alpha}(\mathbf{X}, t)n_\alpha(\mathbf{X}) = t^k(\mathbf{X}, t),$$

$$M^{k\alpha}(\mathbf{X}, t)n_\alpha(\mathbf{X}) = m^k(\mathbf{X}, t), \quad t \in [t_0, t_f],$$

in which $n_\alpha(\mathbf{X})$ are the components of the unit outward normal vector to $\partial_1\Omega$, and $t^k(\cdot)$, $m^k(\cdot)$ are the known functions. At the same time we assume that the values of $\boldsymbol{\chi}(\mathbf{X}, t)$, $\mathbf{Q}(\mathbf{X}, t)$ are known on $\partial_2\Omega$ for every $t \in [t_0, t_f]$:

$$\boldsymbol{\chi}(\mathbf{X}, t) = \boldsymbol{\chi}_0(\mathbf{X}, t),$$

$$\mathbf{Q}(\mathbf{X}, t) = \mathbf{Q}_0(\mathbf{X}, t).$$

By means of $\boldsymbol{\chi}(\mathbf{X}, t) = \boldsymbol{\varkappa}(\mathbf{X}) + \mathbf{u}(\mathbf{X}, t)$, we can also formulate the initial and boundary conditions (on $\partial_2\Omega$) in terms of the displacements $\mathbf{u}(\mathbf{X}, t)$. If the rotations are small then instead of the condition $\mathbf{Q}(\mathbf{X}, t) = \mathbf{Q}_0(\mathbf{X}, t)$ on $\partial_2\Omega$, we postulate the condition $\mathbf{v}(\mathbf{X}, t) = \mathbf{v}_0(\mathbf{X}, t)$, $\mathbf{X} \in \partial_2\Omega$, $t \in [t_0, t_f]$, where $\mathbf{v}_0(\mathbf{X}, t)$ is known. The number of the boundary conditions (for an arbitrary but fixed $\mathbf{X} \in \partial\Omega$) is equal to six; this fact is implied by the order of partial differential equations describing the elastic lattice-type structure within its continuous model.

The thorough discussion of various boundary-value problems in the field of continuous modelling of lattice-type structures have been given in [72], [73], [87]–[89] and summarized in the textbook [167]. Some special examples of nonlinear solutions will be presented in this book, cf. Sec. 6.3.

4.7 Engineering shell theories

4.7.1 Foundations

In Sections 3.2 and 3.3 there have been described general procedures leading from the governing relations of solid mechanics to theories of shells, plates and rods. The results obtained in Sec. 3.2.2 will now be used as the basis for the derivation of some special shell theories which can be referred to as the engineering theories. They are the theories in which the constraints imposed on the motion and the temperature field have to satisfy conditions of the form (3.31). In order to formally simplify the analysis we shall restrict ourselves to pure mechanical theories, i.e. we shall neglect any thermal effects.

As a starting point we assume eqs. (3.22), (3.23) and (3.24) to hold. To simplify notation we replace symbols Ω^*, $\mathbf{X}^*$ by Ω, $\mathbf{X}$ respectively. We also keep in mind that small Latin as well as small Greek indices ($k, l, \ldots$ and $\alpha, \beta, \ldots$) run over the sequence 1, 2, 3 and are related to the spatial and material coordinate systems, respectively. Moreover, indices $K, L, \ldots$ run over the sequence 1, 2 and are related to the first and second material coordinates while the third material coordinate X^3 is denoted by η, $\eta \equiv X^3$. Finally, indices a, b run over the sequence $1, 2, \ldots, m$. Summation convention holds with respect to the all aforementioned sub- and superscripts.

Under this slightly changed notation, eq. (3.24) now reads

$$\int_{\Omega} H^{aK}(\mathbf{X}, t) v^*_{a,K}(\mathbf{X}) \mathrm{d}V(\mathbf{X}) = \oint_{\partial\Omega} s^a(\mathbf{X}, t) v^*_a(\mathbf{X}) \mathrm{d}S(\mathbf{X}) +$$

$$+ \int_{\Omega} [f^a(\mathbf{X}, t) + h^a(\mathbf{X}, t) - M^{ab}(\mathbf{X}) \ddot{\chi}^*_b(\mathbf{X}, t)] v^*_a(\mathbf{X}) \mathrm{d}V(\mathbf{X}) \tag{4.113}$$

and has to hold for every $\mathbf{v}^*(\cdot) = (v^*_1(\cdot), \ldots, v^*_m(\cdot)) \in \mathscr{U}^*$, where $\mathscr{U}^*$ is a set of generalized virtual displacement (we tacitly assume now that $\mathscr{U}^*$ is independent of the time coordinate). Here, Ω is a region on the plane, $\mathrm{d}V(\mathbf{X}) \equiv \mathrm{d}X^1 \mathrm{d}X^2$ and $\mathrm{d}S(\mathbf{X})$ is a line segment of $\partial\Omega$. Eq. (4.113) has to be considered together with eqs. (3.22) and (3.23) which are now rewritten in the coordinate form as

$$H^{aK}(\mathbf{X}, t) = \int_{-\delta}^{\delta} S^{\alpha\beta}(\mathbf{X}, \eta, t) \chi_{k,\beta}(\mathbf{X}, \eta, t) F^{ak}(\eta) \delta^K_\alpha \mathrm{d}\eta,$$

$$h^a(\mathbf{X}, t) = -\int_{-\delta}^{\delta} S^{\alpha\beta}(\mathbf{X}, \eta, t) \chi_{k,\beta}(\mathbf{X}, \eta, t) F^{ak}{}_{,\alpha}(\eta) \mathrm{d}\eta, \tag{4.114}$$

$$f^a(\mathbf{X}, t) = \int_{-\delta}^{\delta} b^k(\mathbf{X}, \eta, t) F^{a}_{\cdot k}(\eta) d\eta + s^k(\mathbf{X}, \delta, t) F^{a}_{\cdot k}(\delta) +$$

$$+ s^k(\mathbf{X}, -\delta, t) F^{a}_{\cdot k}(-\delta),$$

$$M^{ab}(\mathbf{X}) = \int_{-\delta}^{\delta} \varrho(\mathbf{X}, \eta) F^{ak}(\eta) F^{b}_{\cdot k}(\eta) d\eta; \quad \mathbf{X} \in \Omega,$$

$$s^a(\mathbf{X}, t) = \int_{-\delta}^{\delta} s^k(\mathbf{X}, \eta, t) F^{a}_{\cdot k}(\eta) d\eta; \quad \mathbf{X} \in \partial\Omega.$$

It has been shown in Sec. 3.2.2 that the variational equation (4.113) with the denotations (4.114) describes the situation in which on the motion $\mathbf{x} = \boldsymbol{\chi}(\mathbf{X}, \eta, t)$, $\mathbf{X} \in \Omega \subset \boldsymbol{R}^2$, $\eta \in (-\delta, \delta)$, $t \in [t_0, t_f]$ are imposed constraints of the form

$$\chi^k(\mathbf{X}, \eta, t) = F^{ak}(\eta) \chi^*_a(\mathbf{X}, t) + \chi^k_0(\mathbf{X}, \eta, t),$$

where $F^{ak}(\cdot)$, $\chi^k_0(\cdot)$ are assumed to be known and $\chi^*_a(\cdot, t)$ are, for every $t \in [t_0, t_f]$, the unknown functions defined on Ω such that

$$\boldsymbol{\chi}^*(\cdot, t) = (\chi^*_1(\cdot, t), \ldots, \chi^*_m(\cdot, t)) \in \mathscr{D}^*(\Omega).$$

Here, $\mathscr{D}^*(\Omega)$ is the known set of (sufficiently regular) functions which are called the generalized deformations.

By the engineering shell theories we shall mean the theories in which the constraints imposed on the motion can be expressed in the form

$$\chi^k(\mathbf{X}, \eta, t) = p^k(\mathbf{X}, t) + \eta d^k(\mathbf{X}, t). \tag{4.115}$$

This means that when passing to the engineering shell theories we have

$$\chi_0 \equiv 0,$$

$$\chi^*_a(\mathbf{X}, t) = \delta^k_a p_k(\mathbf{X}, t) + \delta^{k+3}_a d_k(\mathbf{X}, t), \quad a = 1, \ldots, 6,$$

and $F^{ak}(\eta) = \delta^{kl}$ for $a = l$ as well as $F^{ak}(\eta) = \eta\delta^{kl}$ for $a = l+3$. At the same time

$$(\mathbf{p}(\cdot, t), \mathbf{d}(\cdot, t)) \in \mathscr{D}^*(\Omega), \quad t \in [t_0, t_f], \tag{4.116}$$

which means that the pairs of new unknown vector fields $\mathbf{p}(\cdot, t)$, $\mathbf{d}(\cdot, t)$ (which now determine the motion of the shell) have to fulfill certain additional conditions. These conditions have to be specified for every special engineering shell theory and can be termed the shell internal constraints. The condition (4.115) implies that $z^k = p^k(\mathbf{X}, t)$, $\mathbf{X} \in \Omega$, for every $t \in [t_0, t_f]$ represents a certain regular (smooth) surface in the physical space which is said to be the midsurface of the shell (at the time instant t). The meaning of $\mathbf{p}(\mathbf{X}, t)$ and $\mathbf{d}(\mathbf{X}, t)$ is shown in Fig. 19.

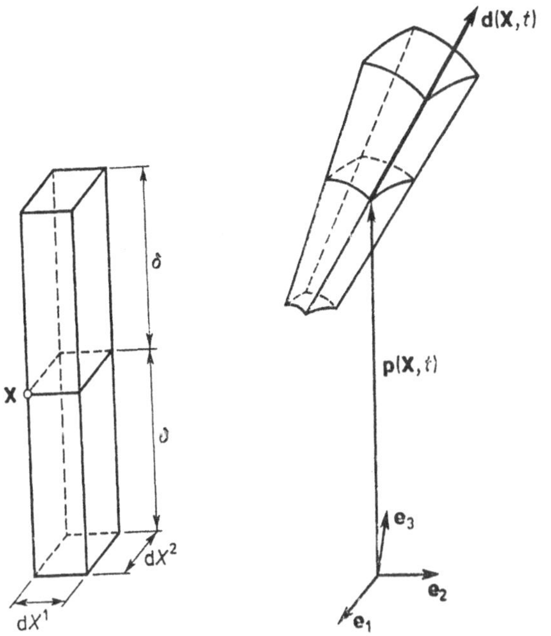

Fig. 19 Shell element before and after deformation

Taking into account the constraints (4.115) as a special case of eq. (3.18) and setting $v_a(\cdot) = \delta_a^k v_k(\cdot) + \delta_a^{k+3}\tilde{v}_k(\cdot)$, we obtain the following form of the variational equation (4.113)

$$\int_{\Omega} (H^{lK} v_{l,K} + \tilde{H}^{lK}\tilde{v}_{l,K})\,\mathrm{d}V$$

$$= \oint_{\partial\Omega} (t^l v_l + \tilde{t}^l \tilde{v}_l)\,\mathrm{d}S + \int_{\Omega} [(f^l - i^l)v_l + (\tilde{f}^l + \tilde{h}^l - \tilde{i}^l)\tilde{v}_l]\,\mathrm{d}V \qquad (4.117)$$

in which we have denoted

$$H^{lK} \equiv \int_{-\delta}^{\delta} S^{K\beta}(\mathbf{X}, \eta, t)\chi^l{}_{,\beta}(\mathbf{X}, \eta, t)\,\mathrm{d}\eta,$$

$$\tilde{H}^{lK} \equiv \int_{-\delta}^{\delta} S^{K\beta}(\mathbf{X}, \eta, t)\chi^l{}_{,\beta}(\mathbf{X}, \eta, t)\eta\,\mathrm{d}\eta,$$

$$\tilde{h}^l \equiv -\int_{-\delta}^{\delta} S^{3\beta}(\mathbf{X}, \eta, t)\chi^l{}_{,\beta}(\mathbf{X}, \eta, t)\,\mathrm{d}\eta,$$

$$f^l \equiv \int_{-\delta}^{\delta} b^l(\mathbf{X}, \eta, t)\mathrm{d}\eta + s^l(\mathbf{X}, \delta, t) + s^l(\mathbf{X}, -\delta, t), \tag{4.118}$$

$$\tilde{f}^l \equiv \int_{-\delta}^{\delta} b^l(\mathbf{X}, \eta, t)\eta\mathrm{d}\eta + \delta s^l(\mathbf{X}, \delta, t) - \delta s^l(\mathbf{X}, -\delta, t),$$

$$i^l \equiv \int_{-\delta}^{\delta} \varrho(\mathbf{X}, \eta)\mathrm{d}\eta \cdot \ddot{p}^l(\mathbf{X}, t) + \int_{-\delta}^{\delta} \varrho(\mathbf{X}, \eta)\eta\mathrm{d}\eta \cdot \ddot{d}^l(\mathbf{X}, t),$$

$$\tilde{i}^l \equiv \int_{-\delta}^{\delta} \varrho(\mathbf{X}, \eta)\eta\mathrm{d}\eta \cdot \ddot{p}^l(\mathbf{X}, t) + \int_{-\delta}^{\delta} \varrho(\mathbf{X}, \eta)\eta^2\mathrm{d}\eta \cdot \ddot{d}^l(\mathbf{X}, t),$$

for $\mathbf{X} \in \Omega$, $t \in [t_0, t_f]$ and

$$t^l \equiv \int_{-\delta}^{\delta} s^l(\mathbf{X}, \eta, t)\mathrm{d}\eta,$$

$$\tilde{t}^l \equiv \int_{-\delta}^{\delta} s^l(\mathbf{X}, \eta, t)\eta\mathrm{d}\eta, \tag{4.119}$$

for $\mathbf{X} \in \partial\Omega$, $t \in [t_0, t_f]$. The variational equation (4.117) has to hold for every

$$(\mathbf{v}(\cdot), \tilde{\mathbf{v}}(\cdot)) \in \mathscr{U}^*$$

where $\mathscr{U}^*$ is a set of the generalized virtual displacements. The set $\mathscr{U}^*$ is uniquely determined by $\mathscr{D}^*(\Omega)$ and depends on the generalized deformations $\mathbf{p}(\cdot, t)$, $\mathbf{d}(\cdot, t)$, i.e. in general $\mathscr{U}^* = \mathscr{U}^*(\mathbf{p}(\cdot, t), \mathbf{d}(\cdot, t))$. The interrelation between the sets $\mathscr{D}^*(\Omega)$ and $\mathscr{U}^*$ was explained in Sec. 2.3.2, where the sets $\mathscr{D}_t(\Omega)$ and $\mathscr{U}_t(\boldsymbol{\varkappa}(\cdot), \mathbf{v}(\cdot))$ were considered). The special cases of this interrelation will be detailed below. Eqs. (4.117)–(4.119) and the specification of $\mathscr{D}^*(\Omega)$ constitute the foundations for formulating various engineering shell theories. It has to be emphasized that the fields $S^{\alpha\beta}(\cdot, t)$, $b^l(\cdot, t)$, $s^l(\cdot, t)$, $\varrho(\cdot)$ in the integrands of eqs. (4.118) and (4.119) are densities related to the space of the material coordinates $(\mathbf{X}, \eta) \in \Omega \times (-\delta, \delta)$. Hence, for every $t \in [t_0, t_f]$, the fields of densities in the variational equation (4.117) are related to the region Ω on the plane or to its boundary $\partial\Omega$. The region $\Omega \times (-\delta, \delta)$ may not coincide with any region of the space occupied either by undeformed or deformed shell. Therefore, we shall now transform the shell equations to the form in which the basic fields are related to a certain surface $z^k = \varkappa^k(\mathbf{X})$, $\mathbf{X} \in \Omega$, which coincides with the midsurface of the undeformed (or deformed, at a fixed time instant) shell.

4.7.2 General form of equations of motion

It can be easily verified that the variational equation (4.117) can be expressed in the form[1]

$$\int_{\Omega} (r^k v_k + \tilde{r}^k \tilde{v}_k) \mathrm{d}V + \oint_{\partial\Omega} (p^k v_k + \tilde{p}^k \tilde{v}_k) \mathrm{d}S = 0 \tag{4.120}$$

which holds for every $(\mathbf{v}(\cdot), \tilde{\mathbf{v}}(\cdot)) \in \mathscr{U}^*$ and where r^k, $\tilde{r}^k$, p^k, $\tilde{p}^k$ satisfy the conditions

$$\begin{aligned} H^{lK}{}_{,K} + f^l + r^l &= i^l, \\ \tilde{H}^{lK}{}_{,K} + \tilde{f}^l + \tilde{h}^l + \tilde{r}^l &= \tilde{i}^l, \end{aligned} \tag{4.121}$$

in $\Omega \times (t_0, t_f)$ and the conditions

$$\begin{aligned} H^{lK} n_K &= t^l + p^l, \\ \tilde{H}^{lK} n_K &= \tilde{t}^l + \tilde{p}^l, \end{aligned} \tag{4.122}$$

on $\partial\Omega \times (t_0, t_f)$, where n_K are the components of the unit normal outward to $\partial\Omega$. The fields $r^l(\cdot, t)$, $\tilde{r}^l(\cdot, t)$ defined for every $t \in [t_0, t_f]$ on Ω and the fields $p^l(\cdot, t)$, $\tilde{p}^l(\cdot, t)$ defined for every $t \in [t_0, t_f]$ on $\partial\Omega$, have the very simple physical interpretation; they constitute the reactions to the shell internal constraints (4.116).

The procedure leading from eqs. (4.120)–(4.122) to the equations of motion will now consist of two steps. In the first, we shall introduce the alternative form for the second of eqs. (4.121). To this aim we define the new fields

$$\begin{aligned} \bar{H}^{kK} &\equiv \varepsilon^{klm} d_l \tilde{H}_m^{\cdot K}, & \bar{f}^k &\equiv \varepsilon^{klm} d_l \tilde{f}_m, \\ \bar{r}^k &\equiv \varepsilon^{klm} d_l \tilde{r}_m, & \bar{i}^k &\equiv \varepsilon^{klm} d_l \tilde{i}_m \end{aligned}$$

which represent the moments of pertinent forces taken with respect to the axes tangent to the midsurface of the deformed shell. Hence, the second of eqs. (4.121) yields

$$\bar{H}^{kK}{}_{,K} - \varepsilon^{klm} d_{l,K} \tilde{H}_m{}^K + \varepsilon^{klm} d_l \tilde{h}_m + \bar{f}^k + \bar{r}^k = \bar{i}^k.$$

From the symmetry of the second Piola–Kirchhoff stress tensor $S^{\alpha\beta} = S^{\beta\alpha}$ we obtain

$$\varepsilon_{klm} \int_{-\delta}^{\delta} \chi^m{}_{,\alpha} S^{\alpha\beta} \chi^l{}_{,\beta} \, \mathrm{d}\eta = 0$$

[1] For the sake of simplicity we shall from now on drop the arguments of all the functions, bearing in mind that $\mathbf{v}(\cdot)$, $\tilde{\mathbf{v}}(\cdot)$ are defined on $\bar{\Omega}$ and all other fields are defined either on $\Omega \times (t_0, t_f)$ or on $\partial\Omega \times (t_0, t_f)$.

and thus

$$\varepsilon_{klm}\Big[\int_{-\delta}^{\delta} \chi^m{}_{,\alpha} S^{\alpha K} d\eta \cdot p^l{}_{,K} + \int_{-\delta}^{\delta} \chi^m{}_{,\alpha} S^{\alpha K} \eta d\eta \cdot d^l{}_{,K} + \int_{-\delta}^{\delta} \chi^m{}_{,\alpha} S^{\alpha 3} d\eta \cdot d^l\Big]$$

$$= \varepsilon_{klm}[H^{mK} p^l{}_{,K} + \tilde{H}^{mK} d^l{}_{,K} - \tilde{h}^m d^l] = 0.$$

It follows that

$$-\varepsilon^{klm} d_{l,K} \tilde{H}_m{}^K + \varepsilon^{klm} d_l \tilde{h}_m = \varepsilon^{klm} \tilde{H}_m{}^K p_{l,K}.$$

Thus, instead of eqs. (4.121) we get

$$\begin{aligned} &H^{lK}{}_{,K} + f^l + r^l = i^l, \\ &\bar{H}^{lK}{}_{,K} + \varepsilon^{lmn} H_m{}^K p_{n,K} + \bar{f}^l + \bar{r}^l = \bar{i}^l. \end{aligned} \tag{4.123}$$

In the second step, we transform eqs. (4.123) to the form in which all terms will be related as densities not to the region Ω on the plane but to the curved midsurface of the deformed shell. We assume that t is an arbitrary but fixed time instant. The reference midsurface is defined by means of the mapping $\mathbf{p}(\cdot, t)\colon \Omega \to R^3$ which can also be expressed in the form

$$z^k = p^k(\mathbf{X}, t), \quad \mathbf{X} \in \Omega.$$

The deformed midsurface of the shell will be denoted by $\mathbf{p}(\Omega, t)$. For each $\mathbf{X} \in \Omega$ we introduce the vector basis $\mathbf{p}_{,K}(\mathbf{X}, t)$ tangent at the point $\mathbf{z} = \mathbf{p}(\mathbf{X}, t)$ to the midsurface $\mathbf{p}(\Omega, t)$ and we define the unit vector $\mathbf{N}(\mathbf{X}, t)$ normal to this surface, setting

$$\mathbf{N}(\mathbf{X}, t) = \frac{\mathbf{p}_{,1}(\mathbf{X}, t) \times \mathbf{p}_{,2}(\mathbf{X}, t)}{|\mathbf{p}_{,1}(\mathbf{X}, t) \times \mathbf{p}_{,2}(\mathbf{X}, t)|}.$$

The first and the second fundamental (metric) tensors of the surface $\mathbf{p}(\Omega, t)$ are given by the known formulas

$$\begin{aligned} a_{KL}(\mathbf{X}, t) &= p^k{}_{,K}(\mathbf{X}, t) p^l{}_{,L}(\mathbf{X}, t) \delta_{kl}, \\ b_{KL}(\mathbf{X}, t) &= N^k(\mathbf{X}, t) p^l{}_{,KL}(\mathbf{X}, t) \delta_{kl}. \end{aligned}$$

We also define $a \equiv \det a_{KL}$; by a^{KL} we denote the components of the matrix inverse to the matrix with components a_{KL} and we introduce the Christoffel symbols on $\mathbf{p}(\Omega, t)$ as

$$\begin{Bmatrix} K \\ LM \end{Bmatrix} = a^{KN} p_{k,LM} p^k{}_{,N} = \tfrac{1}{2} a^{KN} (a_{NL,M} + a_{NM,L} - a_{LM,N}).$$

We tacitly assume here that the reader is familiar with the basic notions of tensor analysis on two-dimensional differentiable manifolds.

Let $F^l = F^l(\mathbf{X}, t)$ stand for the components of an arbitrary vector field $F^l \mathbf{e}_l$ in eqs. (4.123). Every such a field is a vector density related to the plane element dX^1dX^2. Since an arbitrary element $dA(\mathbf{X}, t)$ of the surface $\mathbf{p}(\Omega, t)$ is given by $dA(\mathbf{X}, t) = \sqrt{a(\mathbf{X}, t)}dX^1dX^2$ then the pertinent vector density related to $\mathbf{p}(\Omega, t)$ has the components $F^l(\mathbf{X}, t)/\sqrt{a(\mathbf{X}, t)}$. They are the components in the vector basis $\mathbf{e}_l$, $l=1, 2, 3$, of the spatial coordinate system. Passing to the components of $F^l\mathbf{e}_l$ in the vector basis $\mathbf{p}_{,M}(\mathbf{X}, t)$, $\mathbf{N}(\mathbf{X}, t)$, $M = 1, 2$, we obtain

$$\frac{1}{\sqrt{a}} F^l = F^M p^l{}_{,M} + FN^l,$$
$$F^M = \frac{1}{\sqrt{a}} F^l p_{l,N} a^{NM}, \qquad F = \frac{1}{\sqrt{a}} F^l N_l. \tag{4.124}$$

Moreover, if $F^l - S^{lK}{}_{,K}$ then

$$\frac{1}{\sqrt{a}} F^l = \frac{1}{\sqrt{a}} S^{lK}{}_{,K} = \left(\frac{1}{\sqrt{a}} S^{lK}\right)_{,K} + \frac{(\sqrt{a})_{,K}}{\sqrt{a}} \frac{1}{\sqrt{a}} S^{lK}$$
$$= (S^{MK} p^l{}_{,M} + S^K N^l)_{,K} + \frac{(\sqrt{a})_{,K}}{\sqrt{a}} (S^{MK} p^l{}_{,M} + S^K N^l).$$

Now, taking into account that

$$\frac{(\sqrt{a})_{,K}}{\sqrt{a}} = \begin{Bmatrix} L \\ LK \end{Bmatrix}$$

and using the known Gauss–Weingarten formulas

$$p^l{}_{,KL} = \begin{Bmatrix} M \\ KL \end{Bmatrix} p^l{}_{,M} + b_{KL} N^l,$$
$$N^k{}_{,L} = -b^N_L p^k{}_{,N}, \qquad b^N_L = a^{NM} b_{ML}, \tag{4.125}$$

we obtain

$$\frac{1}{\sqrt{a}} F^l = \frac{1}{\sqrt{a}} S^{lK}{}_{,K} = (S^{KL}|_K - b^L_K S^K) p^l{}_{,L} + (S^K|_K + b_{KL} S^{KL}) N^l \tag{4.126}$$

where

$$S^K|_L = S^K{}_{,L} + \begin{Bmatrix} K \\ LM \end{Bmatrix} S^M,$$
$$S^{KL}|_M = S^{KL}{}_{,M} + \begin{Bmatrix} K \\ NM \end{Bmatrix} S^{NL} + \begin{Bmatrix} L \\ NM \end{Bmatrix} S^{KN},$$

are the covariant derivatives of the vector and tensor fields.

Formulas (4.124) and (4.126) make it possible to represent $f^l, r^l, \ldots, \bar{i}^l$ and $H^{lK}{}_{,K}$, $\bar{H}^{lK}{}_{,K}$ in the form of the densities related to the surface $\mathbf{p}(\Omega, t)$

and expressed in the basis $p^k{}_{,K}$, N^k. Using eq. (4.124) we also obtain

$$\frac{1}{\sqrt{a}}\,\varepsilon^{lmn}H_m{}^K p_{n,K} = \varepsilon^{lmn}(H^{MK}p_{m,M}+H^K N_m)p_{n,K}$$
$$= (\varepsilon^{Nmn}p^l{}_{,N}+\varepsilon^{mn}N^l)(H^{MK}p_{m,M}+H^K N_m)p_{n,K},$$

where

$$\varepsilon^{mn} \equiv \varepsilon^{lmn}N_l, \qquad \varepsilon^{Nmn} \equiv \varepsilon^{lmn}p_{l,M}a^{MN}.$$

Since

$$\varepsilon^{Nmn}p_{m,M}p_{n,K} = \varepsilon^{lmn}p_{l,P}p_{m,M}p_{n,K}a^{PN} = 0,$$

then denoting

$$e_{KL} \equiv \varepsilon^{mn}p_{m,K}p_{n,L} = \varepsilon^{lmn}p_{m,K}p_{n,L}N_l,$$
$$e^{L}_{\cdot K} \equiv a^{LM}e_{MK},$$

we obtain

$$\frac{1}{\sqrt{a}}\,\varepsilon^{lmn}H_m{}^K p_{n,K} = e^{L}_{\cdot K}H^K p^l{}_{,L}+e_{KL}H^{KL}N^l.$$

Applying now formulas (4.124) and (4.126) to the equations of motion (4.123) we arrive after simple calculations at the following alternative form of these equations

$$\begin{aligned}
&H^{KL}|_K - b^L_K H^K + f^L + r^L = i^L,\\
&H^K|_K + b_{KL}H^{KL} + f + r = i,\\
&\bar{H}^{KL}|_K - b^L_K \bar{H}^K + e^{L}_{\cdot K}H^K + \bar{f}^L + \bar{r}^L = \bar{i}^L,\\
&\bar{H}^K|_K + b_{KL}\bar{H}^{KL} + e_{KL}H^{KL} + \bar{f} + \bar{r} = \bar{i},
\end{aligned} \tag{4.127}$$

where by means of eqs. (4.118) we have

$$\begin{aligned}
H^{KL} &= \frac{1}{\sqrt{a}}\int_{-\delta}^{\delta} S^{K\beta}\chi^l{}_{,\beta}\,\mathrm{d}\eta\cdot p_{l,N}a^{NL},\\
H^{K} &= \frac{1}{\sqrt{a}}\int_{-\delta}^{\delta} S^{K\beta}\chi^l{}_{,\beta}\,\mathrm{d}\eta\cdot N_l,\\
\bar{H}^{KL} &= \varepsilon_k{}^{\cdot lm}\frac{1}{\sqrt{a}}\int_{-\delta}^{\delta} S^{K\beta}\chi_{m,\beta}\eta\,\mathrm{d}\eta\cdot d_l p^k{}_{,M}a^{ML},\\
\bar{H}^{K} &= \varepsilon_k{}^{\cdot lm}\frac{1}{\sqrt{a}}\int_{-\delta}^{\delta} S^{K\beta}\chi_{m,\beta}\eta\,\mathrm{d}\eta\cdot d_l N^k,
\end{aligned} \tag{4.128}$$

with

$$S^{K\beta}\chi_{m,\beta} = S^{KM}(p_{m,M}+\eta d_{m,M})+S^{K3}d_m$$

and

$$\begin{aligned}
\Phi^L &= \frac{1}{\sqrt{a}}\Phi^k p_{k,M}a^{LM}, \quad \Phi = \frac{1}{\sqrt{a}}\Phi^k N_k,\\
\bar{\Phi}^L &= \frac{1}{\sqrt{a}}\varepsilon_{klm}\tilde{\Phi}^m d^l p^k{}_{,M}a^{LM}, \quad \bar{\Phi} = \frac{1}{\sqrt{a}}\varepsilon_{klm}\tilde{\Phi}^m d^l N^k,
\end{aligned} \tag{4.129}$$

where the symbol Φ stands for f, r and i. Eqs. (4.128) define the shell stresses. From eqs. (4.129) it follows that the reactions r^L, r, $\bar{r}^L$, $\bar{r}$ are expressed by the reactions r^k, $\tilde{r}^k$ which in turn are constrained by the variational condition (4.120). Eqs. (4.127) with the shell stresses defined by eqs. (4.128) and with the notation of eqs. (4.129) for the external forces (loadings) f^k, $\tilde{f}^k$, the reactions r^k, $\tilde{r}^k$ and the inertia forces $-i^k$, $-\tilde{i}^k$, represent the general form of the equations of motion for the engineering shell theories. It has to be emphasized that all entities as well as the covariant derivatives in eqs. (4.127) are referred to the midsurface $\mathbf{p}(\Omega, t)$ of the deformed shell (at an arbitrary but fixed time instant $t \in [t_0, t_f]$, which is not known a priori.)

So far the set $\mathscr{D}^*(\Omega)$ of the generalized deformations $\big(\mathbf{p}(\cdot, t). \ \mathbf{d}(\cdot, t)\big)$ and hence the set $\mathscr{U}^* = \mathscr{U}^*\big(\mathbf{p}(\cdot, t), \mathbf{d}(\cdot, t)\big)$ of the generalized virtual displacements $\big(\mathbf{v}(\cdot), \tilde{\mathbf{v}}(\cdot)\big)$ have been left unspecified. Putting aside the restrictions imposed on $\mathbf{p}(\cdot, t)$, $\mathbf{d}(\cdot, t)$, $t \in [t_0, t_f]$, by the kinematic boundary conditions (for $\mathbf{X} \in \partial\Omega$) we shall analyse only the restrictions imposed on $\mathbf{p}(\cdot, t)$, $\mathbf{d}(\cdot, t)$ for $\mathbf{X} \in \Omega$. We confine ourselves to the case in which all these restrictions have the form

$$\varphi_A(\mathbf{X}, \mathbf{p}, \mathbf{d}, \nabla\mathbf{p}, \nabla\mathbf{d}) = 0, \quad A = 1, \ldots, M, \mathbf{X} \in \Omega, \tag{4.130}$$

where $\varphi_A(\cdot)$ are assumed to be the known real valued differentiable functions. This means that the condition (4.116) holds if and only if eqs. (4.130) are fulfilled. Let $\mathscr{U}_0^*\big(\mathbf{p}(\cdot, t), \mathbf{d}(\cdot, t)\big)$ be a subset of $\mathscr{U}^*\big(\mathbf{p}(\cdot, t), \mathbf{d}(\cdot, t)\big)$ that contains all the generalized virtual displacements $\big(\mathbf{v}(\cdot), \tilde{\mathbf{v}}(\cdot)\big)$ which on the boundary $\partial\Omega$ attain the values equal to zero: $\mathbf{v}(\mathbf{X}) = \tilde{\mathbf{v}}(\mathbf{X}) = 0$ for $\mathbf{X} \in \partial\Omega$[1]. Then

$$\big(\mathbf{v}(\cdot), \tilde{\mathbf{v}}(\cdot)\big) \in \mathscr{U}_0^*\big(\mathbf{p}(\cdot, t), \mathbf{d}(\cdot, t)\big)$$

---

[1] We tacitly assume that the generalized deformations $\big(\mathbf{p}(\cdot, t), \mathbf{d}(\cdot, t)\big)$ and the generalized virtual displacements $\big(\mathbf{v}(\cdot), \tilde{\mathbf{v}}(\cdot)\big)$ are defined on Ω but have the well defined traces on the boundary $\partial\Omega$.

holds if and only if for every $\mathbf{X} \in \Omega$

$$\frac{\partial \varphi_A}{\partial p^k} v^k + \frac{\partial \varphi_A}{\partial p^k{}_{,K}} v^k{}_{,K} + \frac{\partial \varphi_A}{\partial d^k} \tilde{v}^k + \frac{\partial \varphi_A}{\partial d^k{}_{,K}} \tilde{v}^k{}_{,K} = 0, \qquad A = 1, \ldots, M,$$

and (4.131)

$$v^k(\mathbf{X}) = \tilde{v}^k(\mathbf{X}) = 0 \quad \text{for every } \mathbf{X} \in \partial\Omega .$$

The variational condition (4.120) has to hold for every $(\mathbf{v}, \tilde{\mathbf{v}}) \in \mathscr{U}^*$ and hence it has also to hold for every $(\mathbf{v}, \tilde{\mathbf{v}}) \in \mathscr{U}_0^*$ since $\mathscr{U}_0^* \subset \mathscr{U}^*$. Thus, from the condition (4.120) we obtain that

$$\int_\Omega (r^l v_l + \tilde{r}^l \tilde{v}_l) \mathrm{d}V = 0 \tag{4.132}$$

for every $(\mathbf{v}(\cdot), \tilde{\mathbf{v}}(\cdot)) \in \mathscr{U}_0^*$.

Let $\lambda^A(\mathbf{X}, t)$, $\mathbf{X} \in \Omega$, $A = 1, \ldots, M$, be arbitrary differentiable real valued functions. Using the divergence theorem and taking into account eq. (4.131) it can be easily verified that eq. (4.132) is satisfied if[1]

$$\begin{aligned} r^k &= -\lambda^A \frac{\partial \varphi_A}{\partial p^k} + \left(\lambda^A \frac{\partial \varphi_A}{\partial p^k{}_{,K}}\right)_{,K}, \\ \tilde{r}^k &= -\lambda^A \frac{\partial \varphi_A}{\partial d^k} + \left(\lambda^A \frac{\partial \varphi_A}{\partial d^k{}_{,K}}\right)_{,K}. \end{aligned} \tag{4.133}$$

The notation of eqs. (4.129) yields at the same time

$$\begin{aligned} r^L &= \frac{1}{\sqrt{a}} r^k p_{k,M} a^{LM}, \qquad r = \frac{1}{\sqrt{a}} r^k N_k, \\ \tilde{r}^L &= \frac{1}{\sqrt{a}} \varepsilon^{klm} p_{l,M} d_m \tilde{r}_k a^{LM}, \qquad \tilde{r} = \frac{1}{\sqrt{a}} \varepsilon^{klm} N_l d_m \tilde{r}_k . \end{aligned} \tag{4.134}$$

Combining eqs. (4.133) and (4.134) we conclude that the reactions $r^L, r, \tilde{r}^L, \tilde{r}$ in the equations of motion (4.127) can be expressed in terms of the generalized deformations $\mathbf{p}$, $\mathbf{d}$ and of the functions $\lambda^A(\mathbf{X}, t)$, $\mathbf{X} \in \Omega$, $t \in [t_0, t_f]$, $A = 1, \ldots, M$. The functions $\lambda^A(\cdot)$ are called the constraint functions and together with generalized deformations $p^k(\cdot, t)$, $d^k(\cdot, t)$, $t \in [t_0, t_f]$, constitute the system of the basic unknowns in the equations of motion (4.127).

At the end of this section we shall introduce the strain measures for the engineering shell theories. To this aim let $\varkappa\colon \Omega \to \boldsymbol{R}^3$ represents the smooth surface $\boldsymbol{\varkappa}(\Omega)$ and let

$$\zeta^k(\mathbf{X}, \eta) = \varkappa^k(\mathbf{X}) + \eta \nu^k(\mathbf{X}), \qquad \boldsymbol{\nu}(\mathbf{X}) = \frac{\boldsymbol{\varkappa}_{,1}(\mathbf{X}) \times \boldsymbol{\varkappa}_{,2}(\mathbf{X})}{|\boldsymbol{\varkappa}_{,1}(\mathbf{X}) \times \boldsymbol{\varkappa}_{,2}(\mathbf{X})|} .$$

[1] Summation convention with respect to $A = 1, \ldots, M$ holds.

The vectors $\zeta^k(\mathbf{X}, \eta)\mathbf{e}_k$ will be treated as the radius vectors of the region occupied by the shell in a certain known configuration. For any $t \in (t_0, t_f)$ the vectors $\chi^k(\mathbf{X}, \eta, t)\mathbf{e}_k$ with the components

$$\chi^k(\mathbf{X}, \eta, t) = p^k(\mathbf{X}, t) + \eta d^k(\mathbf{X}, t)$$

are the radius vectors of the shell which undergoes the deformation. Thus the Lagrange strain measures are given by

$$\begin{aligned}
E_{KL} &= \tfrac{1}{2}(\chi^k{}_{,K}\chi_{k,L} - \zeta^k{}_{,K}\zeta_{k,L}) = e_{KL} + 2\eta e'_{KL} + \eta^2 e''_{KL},\\
E_{K3} &= \tfrac{1}{2}(\chi^k{}_{,K}\chi_{k,3} - \zeta^k{}_{,K}\zeta_{k,3}) = e_K + \eta e'_K,\\
E_{33} &= \tfrac{1}{2}(\chi^k{}_{,3}\chi_{k,3} - \zeta^k{}_{,3}\zeta_{k,3}) = e,
\end{aligned} \tag{4.135}$$

where the following shell strain measures have been introduced

$$\begin{aligned}
e_{KL} &\equiv \tfrac{1}{2}(p^k{}_{,K}p_{k,L} - \varkappa^k{}_{,K}\varkappa_{k,L}),\\
e'_{KL} &\equiv \tfrac{1}{4}[(p^k{}_{,K}d_{k,L} + p^k{}_{,L}d_{k,K}) - (\varkappa^k{}_{,K}\nu_{k,L} + \varkappa^k{}_{,L}\nu_{k,K})],\\
e''_{KL} &\equiv \tfrac{1}{2}(d^k{}_{,K}d_{k,L} - \nu^k{}_{,K}\nu_{k,L}),\\
e_K &\equiv \tfrac{1}{2}p^k{}_{,K}d_k,\\
e'_K &\equiv \tfrac{1}{2}d^k{}_{,K}d_k,\\
e &\equiv \tfrac{1}{2}(d^k d_k - 1).
\end{aligned} \tag{4.136}$$

Eqs. (4.136) represent the interrelation between the generalized deformations and the shell strain measures in the engineering shell theories.

Summing up the derivations carried out in this part of the book we state that the following equations have been obtained for the class of shell theories considered above:

1. Equations of motion (4.127) with the shell stresses (4.128) and $f^L, r^L, \ldots, \bar{r}, \bar{i}$ given in eqs. (4.129),
2. Equations of internal constraints (4.130) with $\varphi_A(\cdot)$ being the known real valued differentiable functions,
3. Equations for reactions (4.134) with r^k, $\tilde{r}^k$ given by (4.133), which define the reactions $r^L, r, \bar{r}^L, \bar{r}$ in terms of the constraint functions $\lambda^A(\cdot)$, $A = 1, \ldots$ $\ldots, M$,
4. Equations for strain measures (4.136) which combine the shell strain measures and the shell generalized deformations.

The aforementioned equations hold for $\mathbf{X} \in \Omega$, $t \in [t_0, t_f]$. They have to be considered together with the constitutive relations which interrelate the shell stresses and the shell strain measures given by the right-hand sides of eqs. (4.128) and (4.136), respectively. The constitutive relations depend also on the specific form of the shell internal constraints (4.130). Two examples of such a specification will be given below.

4.7.3 Special theories

In this section we shall consider two special cases of the engineering shell theories obtained by specifying the form of the functions $\varphi^A(\mathbf{X}, \cdot)$, $\mathbf{X} \in \Omega$, $A = 1, \ldots, M$, in eqs. (4.130).

In the first case we assume that the generalized shell deformations $p^k(\cdot, t)$, $d^k(\cdot, t)$ for every $t \in [t_0, t_f]$ are functions which are sufficiently regular, independent of each other and restricted only by the boundary conditions. It means that the functions $\varphi_A(\cdot)$ in eqs. (4.130) can be treated as identically equal to zero (the number M of these functions is arbitrary). Hence, we conclude by eq. (4.131) that the generalized virtual displacements $v^k(\cdot)$, $\tilde{v}^k(\cdot)$ are independent functions which may be restricted by the boundary conditions only. The equations for reactions (4.133) yield $r^l = \bar{r}^l = 0$ and hence $r^L = \bar{r}^L = 0$, $r = \bar{r} = 0$. Thus, in the equations of motion (4.127) there are no reaction forces. In this case we deal with the six basic unknowns $p^k(\cdot)$, $d^k(\cdot)$, $k = 1, 2, 3$, and this is why such shell theories are referred to as the six-parameter shell theories. For the elastic materials, remembering that the second Piola–Kirchhoff stress tensor $S^{\alpha\beta}(\mathbf{X}, \eta, t)$ is uniquely determined by the Lagrange strain $E_{\alpha\beta}(\mathbf{X}, \eta, t)$, the use of the formulas (4.135), (4.136) leads to the elastic shell constitutive equations in the form

$$\begin{aligned}
H^{KL} &= \mathscr{H}^{KL}(\mathbf{X}; \hat{\mathrm{e}}, \hat{\mathrm{e}}', \hat{\mathrm{e}}''),\\
H^{K} &= \mathscr{H}^{K}(\mathbf{X}; \hat{\mathrm{e}}, \hat{\mathrm{e}}', \hat{\mathrm{e}}''),\\
\bar{H}^{KL} &= \bar{\mathscr{H}}^{KL}(\mathbf{X}; \hat{\mathrm{e}}, \hat{\mathrm{e}}', \hat{\mathrm{e}}''),\\
\bar{H}^{K} &= \bar{\mathscr{H}}^{K}(\mathbf{X}; \hat{\mathrm{e}}, \hat{\mathrm{e}}', \hat{\mathrm{e}}''),
\end{aligned} \tag{4.137}$$

where[1]

$$\begin{aligned}
\hat{\mathrm{e}} &\equiv ([e_{MN}], [e_M], e),\\
\hat{\mathrm{e}}' &\equiv ([e'_{MN}], [e'_M]),\\
\hat{\mathrm{e}}'' &\equiv [e''_{MN}],
\end{aligned}$$

and where $\mathscr{H}^{KL}(\mathbf{X}; \cdot)$, $\mathscr{H}^{K}(\mathbf{X}; \cdot)$, $\bar{\mathscr{H}}^{KL}(\mathbf{X}; \cdot)$, $\bar{\mathscr{H}}^{K}(\mathbf{X}; \cdot)$ are the known functions. The form of these functions is implied by eqs. (4.128), by the constitutitve equations for the second Piola–Kirchhoff stress tensor $S^{\alpha\beta}$ and by eqs. (4.135).

The equations of motion (4.127) with $r^L = r = \bar{r}^L = \bar{r} = 0$, the constitutive equations (4.137) and the geometric relations (4.136) lead to the system of six resulting equations for six unknown functions.

---

[1] Symbol $[A_{MN}]$ stands for the 2×2 matrix with the components A_{MN} and $[A_M] \equiv (A_1, A_2)$.

In the second special case we assume that the generalized shell deformations $(\mathbf{p}(\cdot, t), \mathbf{d}(\cdot, t))$, $t \in [t_0, t_f]$, are restricted by three equations of internal constraints (4.130) which have the form

$$d_k p^k{}_{,K} = 0, \quad d^k d_k - 1 = 0, \quad K = 1, 2. \tag{4.138}$$

It means that $d^k(\mathbf{X}, t)$ are components of the unit vector normal to the midsurface $z^k = p^k(\mathbf{X}, t)$, $\mathbf{X} \in \Omega$, of the deformed shell, i.e. that $d^k(\mathbf{X}, t) = N^k(\mathbf{X}, t)$. Bearing in mind that $\mathbf{p}(\cdot, t)$ has to represent for every fixed $t \in [t_0, t_f]$, the smooth surface (midsurface of the shell), we obtain now

$$\mathbf{d}(\mathbf{X}, t) = \frac{\mathbf{p}_{,1}(\mathbf{X}, t) \times \mathbf{p}_{,2}(\mathbf{X}, t)}{|\mathbf{p}_{,1}(\mathbf{X}, t) \times \mathbf{p}_{,2}(\mathbf{X}, t)|}.$$

It follows that as the basic kinematic unknowns we can take the three functions $p^k(\mathbf{X}, t)$, $\mathbf{X} \in \Omega$, $t \in [t_0, t_f]$, which for every fixed $t \in [t_0, t_f]$ determine the midsurface of the deformed shell. The theories under consideration are called the three-parameter shell theories of the Love–Kirchhoff shell theories. Since the last of the conditions (4.138) is too restrictive in the case of large strains (note that it implies the thickness to be constant during deformation) the Love–Kirchhoff shell theories find applications for problems in which the strains are small.

Equations for reactions (4.133) now yield

$$r^k = (\lambda^K d^k)_{,K},$$
$$\bar{r}^k = -\lambda^K p^k{}_{,K} - \lambda^3 d^k,$$

where $\lambda^K = \lambda^K(\mathbf{X}, t)$, $\lambda^3 = \lambda^3(\mathbf{X}, t)$ are arbitrary, sufficiently regular real valued functions defined on $\Omega \times [t_0, t_f]$. Denoting $\bar{\lambda}^K \equiv \lambda^K/\sqrt{a}$, from eqs. (4.134) we may obtain after some calculation

$$r^L = \frac{1}{\sqrt{a}} (\lambda^K d^k)_{,K} p_{k,M} a^{ML} = -b^L_K \bar{\lambda}^K,$$

$$r = \frac{1}{\sqrt{a}} (\lambda^K d^k)_{,K} N_k = \bar{\lambda}^K|_K,$$

$$\bar{r}^L = -\frac{1}{\sqrt{a}} \varepsilon^{klm} p_{l,M} d_m (\lambda^K p_{k,K} + \lambda^3 d_k) a^{LM} = -e^L_{\cdot K} \bar{\lambda}^K,$$

$$\bar{r} = -\frac{1}{\sqrt{a}} \varepsilon^{klm} N_l d_m (\lambda^K p_{k,K} + \lambda^3 d_k) = 0.$$

Since now $N_k = d_k$, then from eqs. (4.129) we get $\bar{f} = 0$, $\bar{i} = 0$ and from eqs. (4.128) the condition $\bar{H}^K = 0$. Defining

$$Q^K \equiv H^K + \bar{\lambda}^K,$$

we shall treat $Q^K = Q^K(\mathbf{X}, t)$ as an arbitrary function. The equation of motion (4.127) may be reduced to the form

$$\begin{aligned} &H^{KL}|_K - b^L_K Q^K + f^L = i^L, \\ &Q^K|_K + b_{KL} H^{KL} + f = i, \\ &\bar{H}^{KL}|_K + e^{L}_{\cdot K} Q^K + \bar{f}^L = \bar{i}, \\ &b_{KL}\bar{H}^{KL} + e_{KL} H^{KL} = 0. \end{aligned} \tag{4.139}$$

In the above equations only H^{KL} and $\bar{H}^{KL}$ are determined by the stresses within the shell. Moreover, after some calculations we arrive by eqs. (4.128) at

$$\begin{aligned} H^{KL} &= \frac{1}{\sqrt{a}} \int_{-\delta}^{\delta} (\delta^L_N - \eta b^L_N) S^{NK} \mathrm{d}\eta, \\ \bar{H}^{KL} &= \frac{1}{\sqrt{a}} e^{\cdot L}_{M} \int_{-\delta}^{\delta} (\delta^M_N - \eta b^M_N) \eta S^{NK} \mathrm{d}\eta. \end{aligned} \tag{4.140}$$

Eqs. (4.138) now imply that $e_K = e'_K = 0$, $e = 0$. Hence, for the elastic material the constitutive equations are given by

$$\begin{aligned} H^{KL} &= \mathscr{H}^{KL}(\mathbf{X}; [e_{MN}], [e'_{MN}], [e''_{MN}]), \\ \bar{H}^{KL} &= \bar{\mathscr{H}}^{KL}(\mathbf{X}; [e_{MN}], [e'_{MN}], [e''_{MN}]), \end{aligned} \tag{4.141}$$

where $\mathscr{H}^{KL}(\mathbf{X}; \cdot)$, $\bar{\mathscr{H}}^{KL}(\mathbf{X}; \cdot)$ are the known functions the form of which is implied by eqs. (4.140) and by the constitutive equations for the components S^{KL} of the second Piola–Kirchhoff stress tensor, cf. Sec. 2.1. By virtue of $d_k = N_k$ we have for the Love–Kirchhoff constraints $e_K = e'_K = 0$, $e = 0$ and eqs. (4.135) reduce to the form

$$E_{KL} = e_{KL} + 2\eta e'_{KL} + \eta^2 e''_{KL} \tag{4.142}$$

where

$$\begin{aligned} e_{KL} &\equiv \tfrac{1}{2}(p^k{}_{,K} p_{k,L} - \varkappa^k{}_{,K} \varkappa_{k,L}), \\ e'_{KL} &\equiv \tfrac{1}{4}[(p^k{}_{,K} N_{k,L} + p^k{}_{,L} N_{k,K}) - (\varkappa^k{}_{,K} \nu_{k,L} + \varkappa^k{}_{,L} \nu_{k,K})], \\ e''_{KL} &\equiv \tfrac{1}{2}(N^k{}_{,K} N_{k,L} - \nu^k{}_{,K} \nu_{k,L}). \end{aligned}$$

On the other hand, using the Gauss–Weingarten formulas (4.125) we get

$$\begin{aligned} e_{KL} &= \tfrac{1}{2}(a_{KL} - \mathring{a}_{KL}), \\ e'_{KL} &= -\tfrac{1}{2}(b_{KL} - \mathring{b}_{KL}), \\ e''_{KL} &= \tfrac{1}{2}(c_{KL} - \mathring{c}_{KL}), \end{aligned} \tag{4.143}$$

where a_{KL}, b_{KL}, c_{KL} ($\mathring{a}_{KL}$, $\mathring{b}_{KL}$, $\mathring{c}_{KL}$) are the first, second and third fundamental tensors, respectively, of the surface $\mathbf{p}(\Omega, t)$ (surface $\boldsymbol{\varkappa}(\Omega)$):

$$\begin{aligned} a_{KL} &= p^k{}_{,K} p_{k,L}, & \mathring{a}_{KL} &= \varkappa^k{}_{,K} \varkappa_{k,L}, \\ b_{KL} &= -N_k p^k{}_{,KL}, & \mathring{b}_{KL} &= -\nu_k \varkappa^k{}_{,KL}, \\ c_{KL} &= N_{k,K} N^k{}_{,L} = b_{KM} b^M_L, & \mathring{c}_{KL} &= \nu_{k,K} \nu^k{}_{,L} = \mathring{b}_{KM} \mathring{b}^M_L. \end{aligned} \tag{4.144}$$

Substituting the right-hand sides of eqs. (4.140) into the last of eqs. (4.139) we obtain the identity. It means that for the Love–Kirchhoff shell theory we have obtained the system of five equations of motion. These equations for elastic shells (i.e. after taking into account eqs. (4.141)) lead to the system of five differential equations for five unknown functions $p^k(\cdot)$, $k = 1, 2, 3$, $Q^K(\cdot)$, $K = 1, 2$, defined on $\Omega \times (t_0, t_f)$.

Remember that $\boldsymbol{\varkappa}: \Omega \to \boldsymbol{R}^3$ is a mapping which represents a certain smooth surface $\boldsymbol{\varkappa}(\Omega)$. Setting

$$\mathbf{u}(\mathbf{X}, t) \equiv \mathbf{p}(\mathbf{X}, t) - \boldsymbol{\varkappa}(\mathbf{X}), \tag{4.145}$$

the vector $\mathbf{u}(\mathbf{X}, t)$ will be referred to as the displacement vector of the shell midsurface. For the known $\boldsymbol{\varkappa}(\cdot)$ the functions $u^k(\cdot)$, $k = 1, 2, 3$, can be taken, instead of $p^k(\cdot)$ as the new unknowns. If the gradients $\nabla \mathbf{u}(\mathbf{X}, t)$ are sufficiently small (as related in some norm to $\nabla \boldsymbol{\varkappa}(\mathbf{X})$), then we can linearize the strain measures e_{MN}, e'_{MN}, e''_{MN} in eqs. (4.135) with respect to $\nabla \mathbf{u}$ and then to linearize also the constitutive equations (4.141), provided the material of the shell is elastic. In this way we can also approximate the equations of (4.139) of the Love–Kirchhoff shell theory by assuming that all the geometric entities in eqs. (4.139) (i.e. b_{KL}, $b^L_K = a^{LM} b_{MK}$, e_{KL}, $e^L_{\cdot K} = e_{MK} a^{ML}$ as well as the Christoffel symbols in the covariant derivatives) are related not to the unknown deformed midsurface $\mathbf{p}(\Omega, t)$ but to the known surface $\boldsymbol{\varkappa}(\Omega)$. This kind of approximation leads to what is called the small deformation theory and can also be applied to any other engineering shell theory. To this aim we have to define the increments

$$\begin{aligned} \mathbf{u}(\mathbf{X}, t) &\equiv \mathbf{p}(\mathbf{X}, t) - \boldsymbol{\varkappa}(\mathbf{X}), \\ \mathbf{w}(\mathbf{X}, t) &\equiv \mathbf{d}(\mathbf{X}, t) - \boldsymbol{\nu}(\mathbf{X}), \end{aligned} \tag{4.146}$$

where $\boldsymbol{\nu}(\mathbf{X})$ is the unit normal to the surface $\boldsymbol{\varkappa}(\Omega)$ at $\mathbf{X} \in \Omega$. Then, we linearize the strain measures with respect to $\nabla \mathbf{u}$, $\nabla \mathbf{w}$ and $\mathbf{w}$, we linearize the constitutive equations (4.141) with respect to the strain measures (provided the material of the shell is elastic) and we consider all the geometric entities in the equations of motion (4.139) as referred to the known surface $\boldsymbol{\varkappa}(\Omega)$.

The governing relations of various engineering shell theories (such as the six-parameter theory and the Love–Kirchhoff theory) constitute the basis for the formulation of different boundary-value shell problems. Formulating such problems we have to take into account how the shell is loaded and supported, what material it is made of and what are the initial conditions (in the problems of dynamics) for the generalized shell deformations. The formulation of special problems often admits some further simplifications in the form of the governing equations. The particulars can easily be found in many textbooks and monograph books on shell theory and applications, [112], [152], [153].

4.7.4 Example: axisymmetric deformations of shells

As an example of application of the general considerations given in Sec. 4.7.3 we shall now consider axisymmetric deformations of shells of revolution. The midsurface of such shells represents at each time instant $t \in [t_0, t_f]$ a surface of revolution. Every such surface may be determined by a curve c_t in the plane $Oz_1 z_3$ with no double points that is rotated by the angle 2π around the z_3-axis of the spatial coordinate system $Oz_1 z_2 z_3$. The curve c_t cannot intersect the z_3-axis except possibly at its end points. The z_3-axis and the curve c_t are referred to as the axis of revolution and the meridian curve, respectively, of the shell midsurface, cf. Fig. 20. Introducing the

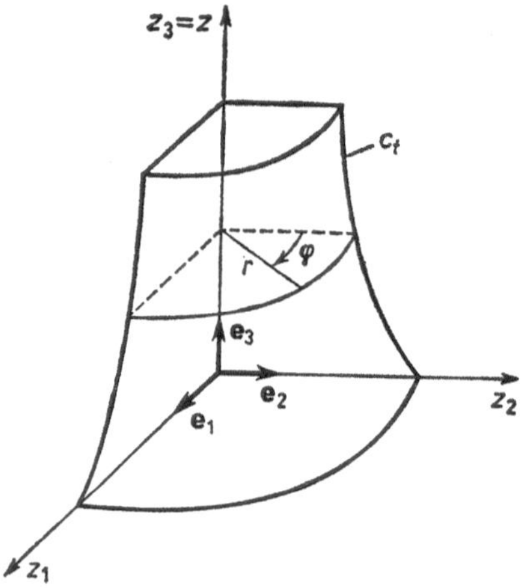

Fig. 20 Geometry of shell of revolution

cylindrical coordinates $r = \sqrt{(z_1)^2 + (z_2)^2}$, $\varphi = \arctan(z_1/z_2)$, $z = z_3$ in the physical space we assume the meridian curve c_t in the form (t is again an arbitrary but fixed time instant, $t \in [t_0, t_f]$)

$$r = r_t(\xi), \quad z = z_t(\xi), \quad \xi \in [\xi_0, \xi_f],$$

where $r_t(\cdot), z_t(\cdot)$ are differentiable functions and ξ is the parameter of the curve. Thus the material coordinates on the midsurface can be given by $X^1 = \xi$, $X^2 = \varphi$ with $\xi \in (\xi_0, \xi_f)$, $\varphi \in (0, 2\pi)$, i.e. $\Omega = (\xi_0, \xi_f) \times (0, 2\pi)$. It has to be emphasized that the midsurface of the deformed shell is not known a priori and hence the functions $r_t(\cdot)$, $z_t(\cdot)$ are not known for any $t \in (t_0, t_f]$. Since now the vector $\mathbf{p}(\mathbf{X}, t)$, $\mathbf{X} = (\xi, \varphi)$, has the components

$$p_1(\mathbf{X}, t) = r_t(\xi)\cos\varphi, \quad p_2(\mathbf{X}, t) = r_t(\xi)\sin\varphi, \quad p_3(\mathbf{X}, t) = z_t(\xi),$$

the functions $z_t(\cdot), r_t(\cdot)$, $t \in [t_0, t_f]$ for the axisymmetric deformations analysed within the Love–Kirchhoff shell theory can be taken as the basic unknowns. However, in the formulation of shell problems we introduce as a rule the concept of the displacement vector. Therefore, we have to introduce a surface $\boldsymbol{\varkappa}(\Omega)$ which can be interpreted as the known midsurface of a certain "undeformed" shell. The surface $\boldsymbol{\varkappa}(\Omega)$ is clearly the surface of revolution and is assumed to be determined by the meridian curve γ given by

$$z = z(\xi), \quad r = r(\xi), \quad \xi \in [\xi_0, \xi_f],$$

where $z(\cdot)$, $r(\cdot)$ are the known functions. In the case of the axisymmetric deformations of shells of revolution the displacement vector has only two non-zero components, namely

$$u_r(\xi, t) = r_t(\xi) - r(\xi),$$
$$u_z(\xi, t) = z_t(\xi) - z(\xi), \quad \xi \in [\xi_0, \xi_f].$$

Introducing the orthonormal vector basis $\mathbf{e}_k$ of the spatial coordinate system we obtain the following representation

$$\begin{aligned} \mathbf{x} &= \mathbf{p}(\xi, \varphi, t) = r_t(\xi)\cos\varphi\,\mathbf{e}_1 + r_t(\xi)\sin\varphi\,\mathbf{e}_2 + z_t(\xi)\mathbf{e}_3, \\ \mathbf{z} &= \boldsymbol{\varkappa}(\xi, \varphi) = r(\xi)\cos\varphi\,\mathbf{e}_1 + r(\xi)\sin\varphi\,\mathbf{e}_2 + z(\xi)\mathbf{e}_3, \end{aligned} \tag{4.147}$$

for the mappings $\mathbf{p}(\cdot, t)$, $\boldsymbol{\varkappa}(\cdot)$, respectively in which

$$r_t(\xi) = r(\xi) + u_r(\xi, t), \quad z_t(\xi) = z(\xi) + u_z(\xi).$$

Taking into account the above formulas we can calculate all the geometric objects which occur in the Love–Kirchhoff theory of shells at hand. Remembering that $X^1 = \xi$, $X^2 = \varphi$, we obtain the vectors $\boldsymbol{\varkappa}_{,1}(\mathbf{X})$, $\boldsymbol{\varkappa}_{,2}(\mathbf{X})$ tangent to $\boldsymbol{\varkappa}(\Omega)$ at $\mathbf{X} \in \Omega = (\xi_0, \xi_f) \times (0, 2\pi)$ and then the unit vector $\boldsymbol{\nu}(\mathbf{X})$ normal to $\boldsymbol{\varkappa}(\Omega)$ at $\mathbf{X}$:

$$\boldsymbol{\nu}(\mathbf{X}) = \frac{\boldsymbol{\varkappa}_{,1}(\mathbf{X}) \times \boldsymbol{\varkappa}_{,2}(\mathbf{X})}{|\boldsymbol{\varkappa}_{,1}(\mathbf{X}) \times \boldsymbol{\varkappa}_{,2}(\mathbf{X})|}.$$

Denoting by "prime" the derivatives with respect to the parameter ξ of the meridian curve we obtain the following components of the first and second fundamental tensors of the surface $\boldsymbol{\varkappa}(\Omega)$:

$$\mathring{a}_{11} = (r'(\xi))^2 + (z'(\xi))^2, \quad \mathring{a}_{12} = \mathring{a}_{21} = 0, \quad \mathring{a}_{22} = (r(\xi))^2,$$
$$\mathring{b}_{11} = -\mathring{a}_{11}/R_s, \quad \mathring{b}_{12} = \mathring{b}_{21} = 0, \quad \mathring{b}_{22} = -\mathring{a}_{22}/R_\varphi,$$

where R_s and R_φ are the principal radii of curvature. The similar formulas hold for the midsurface $\mathbf{p}(\Omega, t)$. After calculating the Christoffel symbols and postulating the axisymmetric state of stress (the isotropy of the material is also assumed) we may obtain the explicit form of the equations of motion (4.139). Similarly, for the axisymmetric strain deformations the only non-zero strain components of those given by eqs. (4.135) are the meridional E_{11} and the circumferential E_{22} components

$$\begin{aligned} E_{11} &= e_{11} + 2\eta e'_{11} + \eta^2 e''_{11}, \\ E_{22} &= e_{22} + 2\eta e'_{22} + \eta^2 e''_{22}, \end{aligned} \tag{4.148}$$

where $e_{11}, \ldots, e''_{22}$ have to be determined from eqs. (4.143), (4.144) accounting for eqs. (4.146).

In the numerical analysis of shell problems vector and tensor fields are represented in the unit basis $\mathbf{a}_K(\mathbf{X}) \equiv \boldsymbol{\varkappa}_{,K}(\mathbf{X})/|\boldsymbol{\varkappa}_{,K}(\mathbf{X})|$, $\boldsymbol{\nu}(\mathbf{X})$, $K = 1, 2$. The components in this basis (for every $\mathbf{X} \in \Omega$) are called the *physical components*. Since $|\boldsymbol{\varkappa}_{,K}| = \sqrt{\mathring{a}_{KK}}$ (no summation) then the physical meridional and circumferential components are given by

$$\varepsilon_{11} = \frac{E_{11}}{\mathring{a}_{11}}, \quad \varepsilon_{22} = \frac{E_{22}}{\mathring{a}_{22}}. \tag{4.149}$$

The physical components of the displacement vector are assumed as

$$\begin{aligned} w(\xi, t) &\equiv [\mathbf{p}(\mathbf{X}, t) - \boldsymbol{\varkappa}(\mathbf{X})] \cdot \boldsymbol{\nu}(\mathbf{X}), \\ u(\xi, t) &\equiv [\mathbf{p}(\mathbf{X}, t) - \boldsymbol{\varkappa}(\mathbf{X})] \cdot \frac{\boldsymbol{\varkappa}_{,1}(\mathbf{X})}{|\boldsymbol{\varkappa}_{,1}(\mathbf{X})|}. \end{aligned} \tag{4.150}$$

Taking into account eqs. (4.149), (4.142) and (4.143), (4.144), introducing the components (4.150) of the displacement vector and performing some rather tedious calculations we may arrive at the relations between the physical strain components ε_{11}, ε_{22} and the physical components w, u of the displacement vector describing the axisymmetric deformations of the shell of revolution. A simplified form of these relationships will be the subject of our considerations in the next section.

4.8 Thin shell finite element for nonlinear axisymmetric analysis

Using the results concerning the derivation of finite element matrices, Sec. 3.4, and the description of the axisymmetric shell deformations, Sec. 4.7, in the current section the stiffness and mass matrices are developed for large deformation analysis of thin elastic or inelastic shells of revolution under axisymmetric loading. The distribution of the load along the shell meridian as well as the axisymmetric shell geometry and support conditions are arbitrary. The load can vary either very slowly with time (static or quasi-static problems) or have a dynamic character.

The finite element shell model is based upon the division of the continuous shell structure into a number of short frusta (to be referred to as the shell elements) connected at their edges called nodal circles, Fig. 21. The earliest

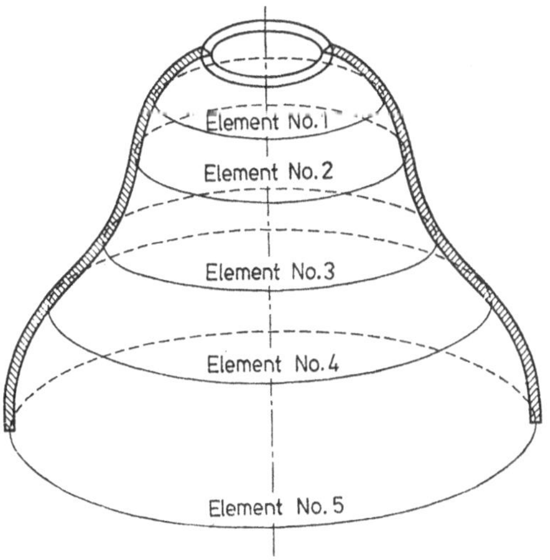

Fig. 21 Axisymmetric shell elements

and even today most popular elements have been conical frusta. However, in such elements meridional membrane and bending action are artificially decoupled within each element. In reality these actions interact with one another throughout a doubly curved shell. This is why the doubly curved axisymmetric elements have been developed and implemented into computer programs, [29], [69], [74], [181].

The element described in this section has 10 degrees of freedom four of which are eliminated at the element level in the process of the so-called static condensation. The element makes it possible to represent conveniently the shape of meridional curve so that the continuity of slopes and curvatures at the nodal circles is assured.

The meridional profile of the middle surface of the curved element connecting the nodal circles (1) and (2) is shown in Fig. 22. To describe the kinematics of the deformation process for the element we use the local co-rotational Cartesian frame of reference $O\xi\eta$. The ξ-coordinate, when normalized with

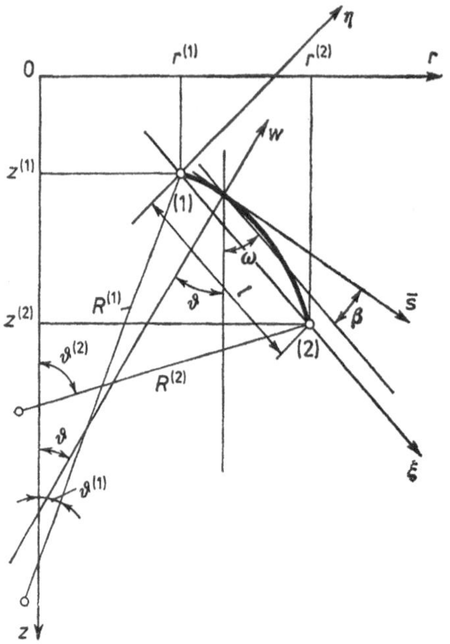

Fig. 22 Geometry of axisymmetric element

values 0 at (1) and 1 at (2) node, is denoted by $\bar{\xi}$. The axisymmetric state of strain of the element is referred to the orthogonal curvilinear coordinates s, φ coinciding with the principal curvature lines of the shell. The incremental strains of the shell can be expressed as (the physical components are used throughout)

$$\Delta\boldsymbol{\epsilon}_{2\times1}(s,\chi) = \mathbf{W}_{2\times4}(\chi)\Delta\boldsymbol{\epsilon}^0_{4\times1} \tag{4.151}$$

where

$$\Delta\boldsymbol{\epsilon} = \{\Delta\varepsilon_s \ \Delta\varepsilon_\varphi\}, \quad \Delta\boldsymbol{\epsilon}^0 = \{\Delta\varepsilon_s^0 \ \Delta\varepsilon_\varphi^0 \ \Delta\varkappa_s^0 \ \Delta\varkappa_\varphi^0\},$$

$$\mathbf{W} = \begin{bmatrix} 1 & 0 & \chi & 0 \\ 0 & 1 & 0 & \chi \end{bmatrix},$$

$\Delta\varepsilon_s, \Delta\varepsilon_\varphi$ are meridional and circumferential incremental strains based upon current configuration, $\Delta\varepsilon_s^0$, $\Delta\varepsilon_\varphi^0$, $\Delta\varkappa_s^0$, $\Delta\varkappa_\varphi^0$ are meridional and circumferential incremental strains and curvatures of the reference (middle) surface of the shell and χ is the coordinate normal to this surface. It is assumed that the deformation of the shell follows Love–Kirchhoff's hypothesis which asserts that the unit normal vector to the middle surface remains normal in any configuration and that its length does not change. Then the convected form of the coordinate system $0s\varphi\tau$ remains orthogonal and coincides with the principal curvatures in the new configuration. The expression defining the strain increment can be divided into a linear and a nonlinear part as

$$\Delta\boldsymbol{\epsilon} = \Delta\bar{\boldsymbol{\epsilon}} + \Delta\bar{\bar{\boldsymbol{\epsilon}}} = \mathbf{W}(\Delta\bar{\boldsymbol{\epsilon}}^0 + \Delta\bar{\bar{\boldsymbol{\epsilon}}}^0)$$

where the linear part of the middle surface generalized strain increment is defined by

$$\Delta\bar{\boldsymbol{\epsilon}}^0 = \{\Delta\bar{\varepsilon}_s^0 \;\; \Delta\bar{\varepsilon}_\varphi^0 \;\; \Delta\bar{\varkappa}_s^0 \;\; \Delta\bar{\varkappa}_\varphi^0\},$$

$$\Delta\bar{\varepsilon}_s^0 = \Delta u_{,s} + \Delta w/R_s, \quad \Delta\bar{\varepsilon}_\varphi^0 = (\Delta u \cos\vartheta + \Delta w \sin\vartheta)\frac{1}{r}, \tag{4.152}$$

$$\Delta\bar{\varkappa}_s^0 = \Delta\psi_{,s} + \frac{\Delta\bar{\varepsilon}_s^0}{R_s}, \quad \Delta\bar{\varkappa}_\varphi^0 = \frac{\cos\vartheta}{r}\Delta\psi + \frac{\Delta\bar{\varepsilon}_\varphi^0}{R_\varphi},$$

while the nonlinear one by

$$\Delta\bar{\bar{\boldsymbol{\epsilon}}}^0 = \{\Delta\bar{\bar{\varepsilon}}_s^0 \;\; \Delta\bar{\bar{\varepsilon}}_\varphi^0 \;\; \Delta\bar{\bar{\varkappa}}_s^0 \;\; \Delta\bar{\bar{\varkappa}}_\varphi^0\},$$

$$\Delta\bar{\bar{\varepsilon}}_s^0 = (\Delta\psi)^2/2, \quad \Delta\bar{\bar{\varepsilon}}_\varphi^0 = 0, \tag{4.153}$$

$$\Delta\bar{\bar{\varkappa}}_s^0 = \frac{(\Delta\psi)^2}{2R_s}, \quad \Delta\bar{\bar{\varkappa}}_\varphi^0 = -\frac{(\Delta\psi)^2 \sin\vartheta}{2r}, \quad \Delta\psi = \frac{\Delta u}{R_s} - \Delta w_{,s}.$$

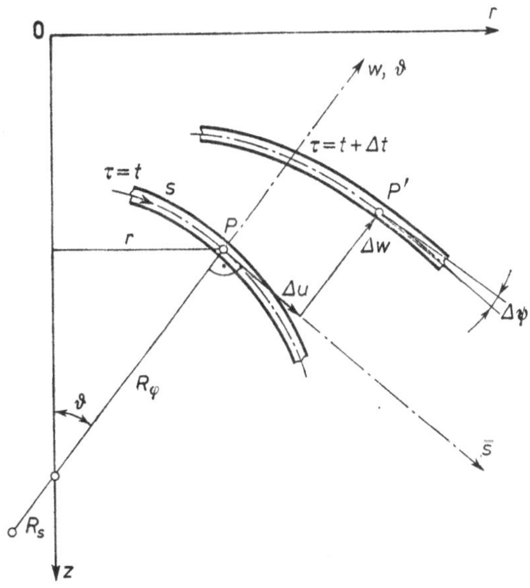

Fig. 23 Increments of generalized displacements

In the above expressions Δu, Δw, $\Delta\psi$ stand for the generalized displacement increments of the shell (the normal and tangent displacements to the shell middle surface and the rotation, respectively, cf. Fig. 23), R_s, R_φ are the meridional and circumferential curvature radii while r and ϑ are defined in Fig. 22.

The constitutive law describing the shell material behaviour is written in the incremental form as

$$\Delta\boldsymbol{\sigma}_{2\times1} = \mathbf{C}_{2\times2}\Delta\boldsymbol{\epsilon}_{2\times1} \tag{4.154}$$

where $\Delta\boldsymbol{\sigma} = \{\Delta\sigma_s\ \Delta\sigma_\varphi\}$ is the incremental stress vector formed from the incremental Piola–Kirchhoff stress components based on the shell configuration at the beginning of the time step considered, $\Delta\boldsymbol{\epsilon} = \{\Delta\varepsilon_s\ \Delta\varepsilon_\varphi\}$ is the incremental strain vector formed from the incremental Green strain components based on the same configuration and

$$\mathbf{C} = \begin{bmatrix} C_{ss} & C_{s\varphi} \\ C_{s\varphi} & C_{\varphi\varphi} \end{bmatrix}$$

is the constitutive matrix defined by, cf. Secs. 2.1.4 and 4.10,

$$C_{ss} = \frac{E}{\Omega}[e+(1-e)S_2^2], \quad C_{\varphi\varphi} = \frac{E}{\Omega}[e+(1-e)S_1^2],$$

$$C_{s\varphi} = \frac{E}{\Omega}[\nu e-(1-e)S_1S_2], \quad e = \frac{E_T}{E},$$

$$\Omega = (1-\nu^2)e+(1-e)(S_1^2+2\nu S_1S_2+S_2^2),$$

$$S_1 = \frac{1}{\bar{\sigma}}(\sigma_s-\tfrac{1}{2}\sigma_\varphi), \quad S_2 = \frac{1}{\bar{\sigma}}(\sigma_\varphi-\tfrac{1}{2}\sigma_s),$$

$$\bar{\sigma} = (\sigma_s^2-\sigma_s\sigma_\varphi+\sigma_\varphi^2)^{1/2}.$$

In the above expressions the following notation is used, cf. Sec. 2.1.4:

σ_s, σ_φ—meridional and circumferential stresses (note that at the beginning of the time step Cauchy and Piola–Kirchhoff stresses are equal),

ν —Poisson's ratio,

E —Young modulus,

E_T—tangent modulus,

$\bar{\sigma}$ —equivalent stress (corresponding to the Huber–Mises yield condition).

We shall also employ the notational convention that the value of any function, say σ_s, at the beginning and end of the current time step is denoted as ${}^-\sigma_s$ and ${}^+\sigma_s$, respectively.

Eq. (4.154) reads in the elastic range as

$$\Delta\boldsymbol{\sigma} = \mathbf{C}^{(e)}\Delta\boldsymbol{\epsilon}^{(el)}$$

where

$$\Delta \boldsymbol{\epsilon}^{(el)} = \{\Delta \varepsilon_s^{(el)} \;\; \Delta \varepsilon_\varphi^{(el)}\},$$

$$\mathbf{C}^{(e)} = \frac{E}{1-\nu^2}\begin{bmatrix} 1 & \nu \\ \nu & 1 \end{bmatrix}.$$

In deriving eq. (4.154) the additivity of the incremental elastic and plastic strain components was assumed with the latter part defined according to the associated flow rule as, cf. Secs. 2.1.5, 4.10, eq. (4.205)

$$\Delta \boldsymbol{\epsilon}^{(pl)} = \frac{3}{2\bar{\sigma}} \Delta \bar{\varepsilon}^{(pl)} \boldsymbol{\sigma}^D$$

where $\Delta \bar{\varepsilon}^{(pl)}$ is the equivalent incremental plastic strain while $\boldsymbol{\sigma}^D$ stands for the deviatoric stress vector

$$\boldsymbol{\sigma}^D = \left\{\sigma_s - \frac{\sigma_s+\sigma_\varphi}{3} \qquad \sigma_\varphi - \frac{\sigma_s+\sigma_\varphi}{3}\right\}.$$

In order to include additionally some reological effects the incremental strain decomposition may be generalized to read

$$\Delta \boldsymbol{\epsilon} = \Delta \boldsymbol{\epsilon}^{(el)} + \Delta \boldsymbol{\epsilon}^{(pl)} + \Delta \boldsymbol{\epsilon}^{(c)} \tag{4.155}$$

with $\Delta \boldsymbol{\epsilon}^{(c)}$ being the creep contribution to the total incremental strain. Similarly as in the case of a purely inviscid plastic response it is assumed that the creep strain increment has the direction of the deviatoric stress:

$$\Delta \boldsymbol{\epsilon}^{(c)} = \frac{3}{2\bar{\sigma}} \Delta \bar{\varepsilon}^{(c)} \boldsymbol{\sigma}^D = \frac{3}{2\bar{\sigma}} \dot{\bar{\varepsilon}}^{(c)} \Delta t \boldsymbol{\sigma}^D$$

where Δt is the time increment while $\Delta \bar{\varepsilon}^{(c)}$ the equivalent creep strain increment. In the general case $\dot{\bar{\varepsilon}}^{(c)}$ should be assumed to be a function of the stresses, time, temperature and total creep strain:

$$\dot{\bar{\varepsilon}}^{(c)} = \dot{\bar{\varepsilon}}^{(c)}(\bar{\sigma}, \bar{\varepsilon}^{(c)}, \vartheta, t).$$

The specific form of this equation adopted here reads

$$\dot{\bar{\varepsilon}}^{(c)} = A \bar{\sigma}^m t^n \tag{4.156}$$

where A, m and n are material constants.

Another generalization of the constitutive equations discussed so far is the elastic-viscoplastic law useful in the analysis of behaviour of certain metallic materials under dynamic load conditions. It is based on the incremental strain decomposition

$$\Delta \boldsymbol{\epsilon} = \Delta \boldsymbol{\epsilon}^{(el)} + \Delta \boldsymbol{\epsilon}^{(vp)}$$

in which the incremental viscoplastic strain contribution is given by

$$\Delta \boldsymbol{\epsilon}^{(vp)} = \gamma \Delta t \left\langle \Phi\left(\frac{\bar{\sigma}-\sigma_y}{\sigma_y}\right)\right\rangle \frac{\partial f}{\partial \boldsymbol{\sigma}^T},$$

γ being the viscosity, f the yield function such that $f = \bar{\sigma}-\sigma_y$, and $\Phi(\cdot)$ a nonlinear function of its argument. The symbol $\langle \Phi(\cdot)\rangle$ is defined as

$$\langle \Phi(\cdot)\rangle = \begin{cases} 0 & \text{for} \quad f \leqslant 0, \\ \Phi(\cdot) & \text{for} \quad f > 0. \end{cases}$$

Here, we shall assume specifically

$$\Delta \boldsymbol{\epsilon}^{(vp)} = \gamma \Delta t \left(\frac{\bar{\sigma}-\sigma_y}{\sigma_y}\right)^k \frac{3}{2\bar{\sigma}} \boldsymbol{\sigma}^D \quad \text{for} \quad f > 0, \tag{4.157}$$

where k is treated as a material constant. A more thorough discussion of eq. (4.157) will be given later in Sec. 8.2 when describing its particular application.

In order to assure the necessary versatility of the finite element at hand we first need to work out a method of replacing the given meridional curve (defining the geometry of the real shell to be analysed) by a substitute curve which matches with the original curve at selected points. The substitute curve is best described in local coordinates, Fig. 22. The curve is assumed to pass through the given end (nodal) points and to have at these points the same slopes and curvatures as those of the specified curve itself. If the normalized abscissa $\bar{\xi}$ is used—i.e. the abscissa is divided by the cord length l—the equation for the substitute curve may be represented as, [69], [181]

$$\bar{\eta} = \bar{\xi}(1-\bar{\xi})(a_1+a_2\bar{\xi}+a_3\bar{\xi}^2+a_4\bar{\xi}^3), \qquad \bar{\xi} = \frac{\xi}{l}, \bar{\eta} = \frac{\eta}{l} \tag{4.158}$$

where

$$\begin{aligned} a_1 &= \tan\beta^{(1)}, \\ a_2 &= \tan\beta^{(1)}+\tfrac{1}{2}\bar{\eta}''^{(1)}, \\ a_3 &= -(5\tan\beta^{(1)}+4\tan\beta^{(2)})+\tfrac{1}{2}\bar{\eta}''^{(2)}-\bar{\eta}''^{(1)}, \\ a_4 &= 3(\tan\beta^{(1)}+\tan\beta^{(2)})+\tfrac{1}{2}(\bar{\eta}''^{(2)}-\bar{\eta}''^{(1)}), \\ \bar{\eta}'' &= \frac{d^2\bar{\eta}}{d\bar{\xi}^2} = -\frac{l}{R_s\cos^3\beta} \end{aligned} \tag{4.159}$$

and R_s is the meridional curvature radius for the element described by eq. (4.158). We shall assume $R_s > 0$ for meridional curves which are convex with respect to the positive η-axis. The parameters in eq. (4.158) can be determined from input data (i.e. nodal coordinates) as follows

$$\Delta r = r^{(2)}-r^{(1)}, \quad \Delta z = z^{(2)}-z^{(1)}, \quad \Delta s = (\Delta r^2+\Delta z^2)^{1/2},$$

$$\sin\omega = \frac{\Delta r}{\Delta s}, \quad \cos\omega = \frac{\Delta z}{\Delta s}, \tag{4.160}$$

$$\sin\beta^{(n)} = \cos\vartheta^{(n)}\cos\omega - \sin\vartheta^{(n)}\sin\omega,$$

$$\cos\beta^{(n)} = \sin\vartheta^{(n)}\cos\omega - \cos\vartheta^{(n)}\sin\omega.$$

Having established the substitute curve, we can express all the geometric quantities that enter eqs. (4.152), (4.153) as

$$\tan\beta = \bar{\eta}',$$

$$r = r^{(1)} + l\xi\left(\sin\omega + \frac{\eta}{\xi}\cos\omega\right),$$

$$\frac{\mathrm{d}}{\mathrm{d}s} = \frac{\cos\beta}{l}\frac{\mathrm{d}}{\mathrm{d}\xi}, \quad \frac{1}{R_s} = -\frac{1}{l}\frac{\mathrm{d}^2\bar{\eta}}{\mathrm{d}\xi^2}\cos^3\beta,$$

$$\cos\vartheta = \sin\beta\cos\omega + \cos\beta\sin\omega = (\bar{\eta}'\cos\omega + \sin\omega)\cos\beta,$$

$$\sin\vartheta = \cos\beta\cos\omega - \sin\beta\sin\omega = (\cos\omega - \bar{\eta}'\sin\omega)\cos\beta,$$

where

$$\bar{\eta}' = \frac{\mathrm{d}\bar{\eta}}{\mathrm{d}\xi}.$$

The most essential feature of the discretized analysis advocated in this book is clearly the selection of constraints imposed upon the structural deformations. In the finite element terminology this corresponds to the problem of expanding the displacements in terms of a relatively complete set of approximating functions. Although these functions can be chosen arbitrarily, polynomials are the most suitable forms for numerical work. Here, we adopt the following expansions for the displacement components expressed in terms of the local coordinates, cf. Fig. 24:

$$\begin{aligned} \Delta u_\xi(\xi) &= \alpha_1 + \alpha_2\xi + \alpha_3\xi^2 + \alpha_4\xi^3 + \alpha_5\xi^4, \\ \Delta u_\eta(\xi) &= \alpha_6 + \alpha_7\xi + \alpha_8\xi^2 + \alpha_9\xi^3 + \alpha_{10}\xi^4. \end{aligned} \tag{4.161}$$

We note that the finite element defined by eqs. (4.158), (4.161) has three desirable properties that, to a certain degree, determine the quality of any shell element. These are: (a) use of doubly curved elements for doubly curved shells, (b) freedom from strain under rigid body motion, (c) the same competence for all assumed displacement fields.

The above derivations are meant to describe any frustum element—to adequately represent a central cap element the displacement patterns (4.161) must be specialized by assuming for the point located on the axis of symmetry

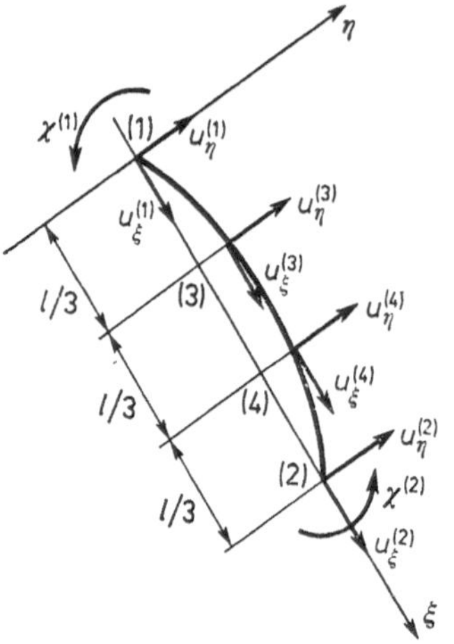

Fig. 24 Axisymmetric shell element: degrees of freedom and displacement expansions

that both the displacement component normal to this axis and meridional rotation vanish, cf. Fig. 25. We have

$$\begin{bmatrix} \Delta u \\ \Delta w \end{bmatrix} = \begin{bmatrix} \cos\beta & \sin\beta \\ -\sin\beta & \cos\beta \end{bmatrix} \begin{bmatrix} \Delta u_\xi \\ \Delta u_\eta \end{bmatrix} \tag{4.162}$$

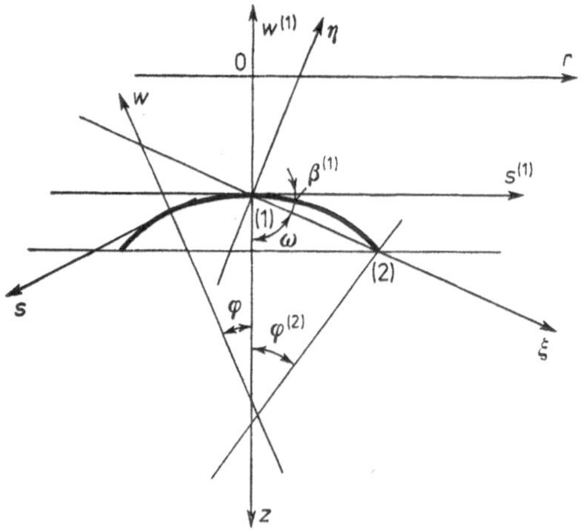

Fig. 25 Central cap element

and for the point (1) at which

$$\Delta u = \Delta u^{(1)}, \quad \Delta w = \Delta w^{(1)}, \quad \beta^{(1)} = \frac{\pi}{2} - \omega$$

eq. (4.162) takes the form

$$\begin{bmatrix} \Delta u^{(1)} \\ \Delta w^{(1)} \end{bmatrix} = \begin{bmatrix} \sin\omega & \cos\omega \\ -\cos\omega & \sin\omega \end{bmatrix} \begin{bmatrix} \Delta u_\xi^{(1)} \\ \Delta u_\eta^{(1)} \end{bmatrix}.$$

Further

$$\Delta u^{(1)} = \Delta u_\xi^{(1)} \sin\omega + \Delta u_\eta^{(1)} \cos\omega = 0,$$

$$\Delta \psi^{(1)} = \left.\frac{\mathrm{d}\Delta w}{\mathrm{d}s}\right|_{(1)} = \frac{1}{l}\left[-\left.\frac{\mathrm{d}\Delta u_\xi}{\mathrm{d}\xi}\right|_{(1)} \tan\beta^{(1)} + \left.\frac{\mathrm{d}\Delta u_\eta}{\mathrm{d}\xi}\right|_{(1)} \right] \cos^2\beta^{(1)}$$

which leads to

$$\alpha_1 \sin\omega + \alpha_6 \cos\omega = 0,$$
$$\alpha_2 \tan\beta^{(1)} - \alpha_7 = 0.$$

Thus, eq. (4.161) takes for the cap element the form

$$\begin{aligned} \Delta u_\xi &= -\bar{\alpha}_1 \cos\omega + \bar{\alpha}_2 \xi + \bar{\alpha}_3 \xi^2 + \bar{\alpha}_4 \xi^3 + \bar{\alpha}_5 \xi^4, \\ \Delta u_\eta &= \bar{\alpha}_1 \sin\omega + \bar{\alpha}_2 \tan\beta^{(1)} \xi + \bar{\alpha}_6 \xi^2 + \bar{\alpha}_7 \xi^3 + \bar{\alpha}_8 \xi^4 \end{aligned} \tag{4.163}$$

with $\bar{\alpha}_1, \bar{\alpha}_2, \ldots, \bar{\alpha}_8$ being the new set of generalized coordinates.

In view of the relationships

$$\mathrm{d}s = \frac{l\,\mathrm{d}\bar{\xi}}{\cos\beta}, \qquad \frac{\mathrm{d}\beta}{\mathrm{d}\bar{\xi}} = \frac{\mathrm{d}^2\eta}{\mathrm{d}\xi^2}\cos^2\beta$$

the geometric equations (4.152), (4.153) may be transformed to the form

$$\Delta\varepsilon_s^0 = \frac{\cos^2\beta}{l}\left(\frac{\mathrm{d}\Delta u_\xi}{\mathrm{d}\xi} + \tan\beta \frac{\mathrm{d}\Delta u_\eta}{\mathrm{d}\xi}\right) + \frac{(\Delta\psi)^2}{2},$$

$$\Delta\varepsilon_\varphi^0 = \frac{1}{r}(\sin\omega \Delta u_\xi + \cos\omega \Delta u_\eta),$$

$$\begin{aligned} \Delta\varkappa_s^0 = \frac{1}{l^2}\Bigg[& \frac{\mathrm{d}\Delta u_\xi}{\mathrm{d}\xi}\eta''(1 - \tan^2\beta)\cos^2\beta + \frac{\mathrm{d}^2\Delta u_\xi}{\mathrm{d}\xi^2}\tan\beta + \\ & + 2\frac{\mathrm{d}\Delta u_\eta}{\mathrm{d}\xi}\eta''\tan\beta\cos^2\beta - \frac{\mathrm{d}^2\Delta u_\eta}{\mathrm{d}\xi^2}\Bigg]\cos^3\beta + \frac{(\Delta\psi)^2}{2R_s}, \end{aligned}$$

$$\Delta\varkappa_\varphi^0 = \frac{1}{lr}\left(\frac{\mathrm{d}\Delta u_\xi}{\mathrm{d}\xi}\tan\beta - \frac{\mathrm{d}\Delta u_\eta}{\mathrm{d}\xi}\right)(\sin\omega + \cos\omega\tan\beta)\cos^3\beta - \frac{\sin\vartheta}{2r}(\Delta\psi)^2,$$

$$\Delta\psi = -\frac{\cos^2\beta}{l}\left(\tan\frac{\mathrm{d}\Delta u_\xi}{\mathrm{d}\xi} - \frac{\mathrm{d}\Delta u_\eta}{\mathrm{d}\xi}\right)$$

which, after introducing the displacement expansions (4.161), (4.163) become

$$\Delta\boldsymbol{\varepsilon}_{4\times1} = \mathbf{B}^{(1)}_{4\times10}\boldsymbol{\alpha}_{10\times1}, \qquad \Delta\bar{\boldsymbol{\varepsilon}}_{4\times1} = \mathbf{B}^{(2)}_{4\times8}\bar{\boldsymbol{\alpha}}_{8\times1} \tag{4.164}$$

where

$$\boldsymbol{\alpha}_{10\times 1}=\{\alpha_1\ \ \alpha_2\ \ \ldots\ \ \alpha_{10}\},\qquad \bar{\boldsymbol{\alpha}}_{8\times 1}=\{\bar{\alpha}_1\ \ \bar{\alpha}_2\ \ \ldots\ \ \bar{\alpha}_8\}.$$

The matrices $\mathbf{B}^{(1)}$ (for the frustum element) and $\mathbf{B}^{(2)}$ (for the cap element) are defined as

$$\mathbf{B}^{(1)}_{4\times 10}=[{}'\mathbf{B}^{(1)}_{4\times 5}\ \ {}''B^{(1)}_{4\times 5}],\qquad \mathbf{B}^{(2)}_{4\times 8}=[{}'\mathbf{B}^{(2)}_{4\times 4}\ \ {}''\mathbf{B}^{(2)}_{4\times 4}],$$

$${}'\mathbf{B}^{(1)}_{4\times 5}=\begin{bmatrix} 0 & \varrho & 2\xi\varrho & 3\xi^2\varrho & 4\xi^3\varrho \\ \delta_c & \xi\delta_c & \xi^2\delta_c & \xi^3\delta_c & \xi^4\delta_c \\ 0 & (1-\eta'^2)\,\delta_1 & 2\xi(1-\eta'^2)\,\delta_1+\eta'\mu & 3\xi[\xi(1-\eta'^2)\,\delta_1+\eta'\mu] & 4\xi^2[\xi(1-\eta'^2)\,\delta_1+\tfrac{3}{2}\eta'\mu] \\ 0 & \eta'\delta_2 & 2\xi\eta'\delta_2 & 3\xi^2\eta'\delta_2 & 4\xi^3\eta'\delta_2 \end{bmatrix},$$

$${}''\mathbf{B}^{(1)}_{4\times 5}=\begin{bmatrix} 0 & \eta'\varrho & 2\xi\eta'\varrho & 3\xi^2\eta'\varrho & 4\xi^3\eta'\varrho \\ \delta_s & \xi\delta_s & \xi^2\delta_s & \xi^3\delta_s & \xi^4\delta_s \\ 0 & 2\eta'\delta_1 & 4\xi\eta'\delta_1-\mu & 3\xi(2\xi\eta'\delta_1-\mu) & 4\xi^2(2\xi\eta'\delta_1-3\mu) \\ 0 & -\delta_2 & -2\xi\delta_2 & -3\xi^2\delta_2 & -4\xi^3\delta_2 \end{bmatrix},$$

$${}'\mathbf{B}^{(2)}_{4\times 4}=\begin{bmatrix} 0 & \varrho(1+\eta'\tan\beta^{(1)}) & 2\xi\varrho & 3\xi^2\varrho \\ 0 & \dfrac{\cos v^{(1)}}{r\cos\beta^{(1)}} & \xi\delta_s & \xi^2\delta_s \\ 0 & \delta_1(1+2\eta'\tan\beta^{(1)}-\eta'^2) & \delta_3+2(1-\eta'^2)\xi\delta_1 & 3\xi[\delta_3+\xi(1-\eta'^2)\delta_1] \\ 0 & \dfrac{\eta'-\tan\beta^{(1)}}{\xi}\,\delta_2 & 2\eta'\delta_2 & 3\xi\eta'\delta_2 \end{bmatrix},$$

$${}''\mathbf{B}^{(2)}_{4\times 4}$$

$$=\begin{bmatrix} 4\xi^3\varrho & 2\xi\eta'\varrho & 3\xi^2\eta'\varrho & 4\xi^3\eta'\varrho \\ \xi^3\delta_s & \xi\delta_c & \xi^2\delta_c & \xi^3\delta_c \\ 4\xi^2[3\delta_3+\xi^2(1-\eta'^2)\,\delta_1] & 4\xi\eta'\delta_1-\mu & 3\xi(2\xi\eta'\delta_1-\mu) & 2\xi^2(4\xi\eta'\delta_1-3\mu) \\ 4\xi^2\eta'\delta_2 & -2\delta_2 & -3\xi\delta_2 & -4\xi^2\delta_2 \end{bmatrix},$$

$$\varrho=\frac{1}{l(1+\eta'^2)},\qquad \mu=\frac{2}{l^2(1+\eta'^2)^{3/2}},$$

$$\delta_c=\frac{\cos\omega}{r},\qquad \delta_s=\frac{\sin\omega}{r},$$

$$\delta_1=\frac{\eta''}{l^2(1+\eta'^2)^{5/2}},\qquad \delta_2=\frac{\sin\omega+\eta'\cos\omega}{l^2(1+\eta'^2)^{3/2}},\qquad \delta_3=\frac{2\eta'}{l^2(1+\eta'^2)^{3/2}}.$$

Now, the constitutive stiffness matrix $\bar{\mathbf{k}}^{(con)}$ related to the generalized coordinates $\boldsymbol{\alpha} = \{\alpha_1 \ \alpha_2 \ \dots \ \alpha_{10}\}$ (or $\bar{\boldsymbol{\alpha}} = \{\bar{\alpha}_1 \ \bar{\alpha}_2 \ \dots \ \bar{\alpha}_8\}$) may be expressed as (frustum element)

$$^{(1)}\bar{k}^{(con)}_{\mu\nu} = \int_0^1 l B^{(1)}_{a\mu}(\xi) D_{ab}(\xi) B^{(1)}_{b\nu}(\xi) r(\xi) (1+\eta'^2)^{1/2} d\xi,$$

$$\mu, \nu = 1, 2, \dots, 10, \quad a, b = 1, 2, 3, 4 \qquad (4.165)$$

or as (cap element)

$$^{(2)}\bar{k}^{(con)}_{\bar{\mu}\bar{\nu}} = \int_0^1 l B^{(2)}_{a\bar{\mu}}(\xi) D_{ab}(\xi) B^{(2)}_{b\bar{\nu}}(\xi) r(\xi) (1+\eta'^2)^{1/2} d\xi,$$

$$\mu, \nu = 1, 2, \dots, 8, \quad a, b = 1, 2, 3, 4, \qquad (4.166)$$

where the matrix $\mathbf{D}_{4\times4}(\xi) = \tilde{\mathbf{D}}_{4\times4}(s)$ is given by

$$\tilde{\mathbf{D}}_{4\times4}(s) = \begin{bmatrix} \underset{2\times2}{\tilde{\mathbf{D}}_{11}(s)} & \underset{2\times2}{\tilde{\mathbf{D}}_{12}(s)} \\ \underset{2\times2}{\tilde{\mathbf{D}}_{12}(s)} & \underset{2\times2}{\tilde{\mathbf{D}}_{22}(s)} \end{bmatrix}, \qquad (4.167)$$

$$\tilde{\mathbf{D}}_{11}(s) = \int_{-h/2}^{h/2} \mathbf{C}_{2\times2}(s, \chi) d\chi,$$

$$\tilde{\mathbf{D}}_{12}(s) = \int_{-h/2}^{h/2} \mathbf{C}_{2\times2}(s, \chi) \chi d\chi,$$

$$\tilde{\mathbf{D}}_{22}(s) = \int_{-h/2}^{h/2} \mathbf{C}_{2\times2}(s, \chi) \chi^2 d\chi, \quad \chi \in [-h/2, h/2],$$

the matrix $\mathbf{C}_{2\times2}$ being given in the incremental constitutive law (4.154). Since the matrix $\mathbf{C}$ is a complex function of the coordinate χ (it depends on stresses which vary across the shell thickness), the numerical integration must be employed. A simple method has been used in this study, which can be given a mechanical interpretation by considering the shell divided into L layers of equal thickness h/L, Fig. 26. The integration in eqs. (4.167) is replaced by the finite sum over the layers

$$\mathbf{D}^{*}_{2\times2}(s) = \frac{h(s)}{L} \sum_{k=1}^{L} \mathbf{C}_{2\times2}(s, \bar{h}_k),$$

$$\mathbf{D}^{**}_{2\times2}(s) = \frac{h^2(s)}{L^2} \sum_{k=1}^{L} \mathbf{C}_{2\times2}(s, \bar{h}_k) (k - \tfrac{1}{2}),$$

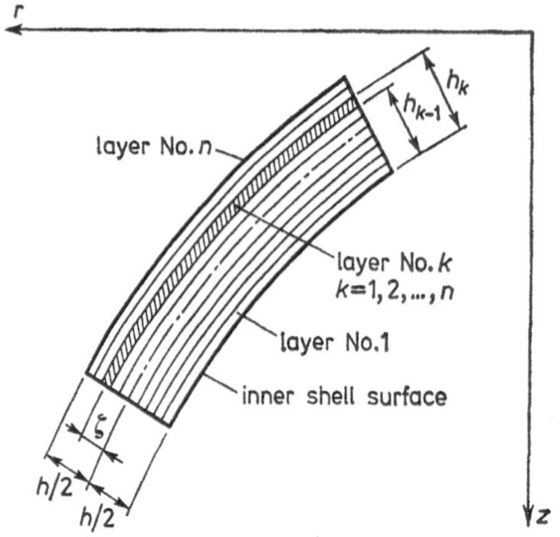

Fig. 26 Layered model of axisymmetric shell element

$$\mathbf{D}^{***}_{2\times 2}(s) = \frac{h^3(s)}{L^3} \sum_{k=1}^{L} \mathbf{C}_{2\times 2}(s, \bar{h}_k)\,(k^2-k+\tfrac{1}{2})$$

whereby

$$\tilde{\mathbf{D}}_{11}(s) = \mathbf{D}^{*}(s),$$

$$\tilde{\mathbf{D}}_{12}(s) = \mathbf{D}^{**}(s)-\tfrac{1}{2}h(s)\mathbf{D}^{*}(s),$$

$$\tilde{\mathbf{D}}_{22}(s) = \mathbf{D}^{***}(s)-h(s)\mathbf{D}^{**}(s)+\tfrac{1}{2}h^2(s)\mathbf{D}^{*}(s).$$

The initial stress matrix is obtained analogously as (frustum element)

$$^{(1)}\bar{\mathbf{k}}^{(\sigma)}_{10\times 10} = \int_0^1 l\mathbf{G}^{(1)T}_{10\times 4}(\xi)\mathbf{F}_{4\times 4}(\xi)\mathbf{N}_{4\times 4}(\xi)\mathbf{G}^{(1)}_{4\times 10}(\xi)r(\xi)\,(1+\eta'^2)^{1/2}\mathrm{d}\xi \qquad (4.168)$$

and (cap element)

$$^{(2)}\bar{\mathbf{k}}^{(\sigma)}_{8\times 8} = \int_0^1 l\mathbf{G}^{(2)T}_{8\times 4}(\xi)\mathbf{F}_{4\times 4}(\xi)\mathbf{N}_{4\times 4}(\xi)\mathbf{G}^{(2)}_{4\times 8}(\xi)r(\xi)\,(1+\eta'^2)^{1/2}\mathrm{d}\xi \qquad (4.169)$$

where the matrices $\mathbf{G}^{(1)}_{4\times 10}$ and $\mathbf{G}^{(2)}_{4\times 8}$ are given as

$$\mathbf{G}^{(1)} = \begin{bmatrix} 0 & -\eta'\varrho & -2\xi\eta'\varrho & -3\xi^2\eta'\varrho & -4\xi^3\eta'\varrho & 0 & \varrho & 2\xi\varrho & 3\xi^2\varrho & 4\xi^3\varrho \\ 0 & 0 & 0 & 0 & 0 & 0 & 0 & 0 & 0 & 0 \\ 0 & -\eta'\varrho & -2\xi\eta'\varrho & -3\xi^2\eta'\varrho & -4\xi^3\eta'\varrho & 0 & \varrho & 2\xi\varrho & 3\xi^2\varrho & 4\xi^3\varrho \\ 0 & -\eta'\varrho & -2\xi\eta'\varrho & -3\xi^2\eta'\varrho & -4\xi^3\eta'\varrho & 0 & \varrho & 2\xi\varrho & 3\xi^2\varrho & 4\xi^3\varrho \end{bmatrix},$$

$$\mathbf{G}^{(2)} = \begin{bmatrix} 0 & \varrho(\tan\beta^{(1)}-\eta') & -2\xi\eta'\varrho & -3\xi^2\eta'\varrho & -4\xi^3\eta'\varrho & 2\xi\varrho & 3\xi^2\varrho & 4\xi^3\varrho \\ 0 & 0 & 0 & 0 & 0 & 0 & 0 & 0 \\ 0 & \varrho(\tan\beta^{(1)}-\eta') & -2\xi\eta'\varrho & -3\xi^2\eta'\varrho & -4\xi^3\eta'\varrho & 2\xi\varrho & 3\xi^2\varrho & 4\xi^3\varrho \\ 0 & \varrho(\tan\beta^{(1)}-\eta') & -2\xi\eta'\varrho & -3\xi^2\eta'\varrho & -4\xi^3\eta'\varrho & 2\xi\varrho & 3\xi^2\varrho & 4\xi^3\varrho \end{bmatrix}$$

the matrix $\mathbf{F}_{4\times4}$ has the form

$$\mathbf{F}(\xi) = \begin{bmatrix} 1 & 0 & 0 & 0 \\ 0 & 0 & 0 & 0 \\ 0 & 0 & \dfrac{1}{R_s} & 0 \\ 0 & 0 & 0 & -\dfrac{\sin\vartheta}{2} \end{bmatrix}$$

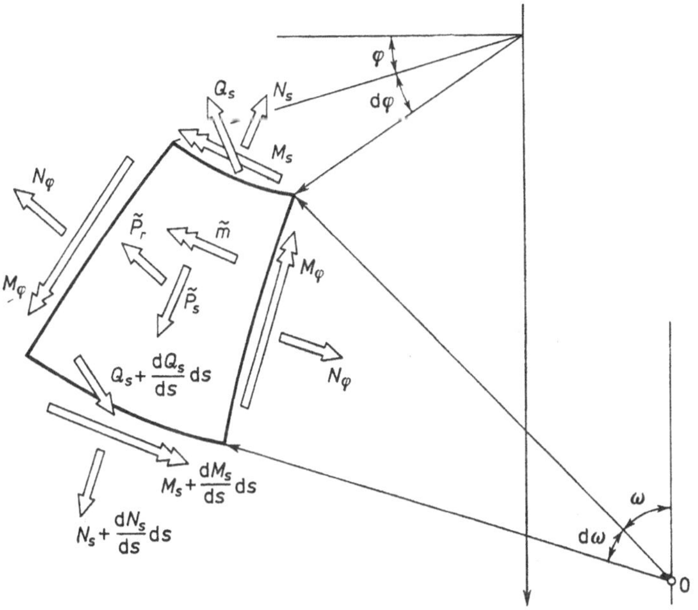

Fig. 27 Sign convention for internal forces and moments

and the matrix $\mathbf{N}_{4\times4}$ is defined in terms of internal forces N_s, N_φ and internal moments M_s, M_φ, cf. Fig. 27, as

$$\mathbf{N}(\xi) = \begin{bmatrix} N_s & 0 & 0 & 0 \\ 0 & N_\varphi & 0 & 0 \\ 0 & 0 & M_s & 0 \\ 0 & 0 & 0 & M_\varphi \end{bmatrix}.$$

Let now denote $\Delta\mathbf{u}(\xi) = \{\Delta u_\xi \;\; \Delta u_\eta \;\; \Delta\psi\}$. We may then write

$$\begin{aligned} \Delta\mathbf{u}_{3\times1}(\xi) &= \boldsymbol{\varphi}^{(1)}_{3\times10}(\xi)\boldsymbol{\alpha}_{10\times1}, \\ \Delta\mathbf{u}_{3\times1}(\xi) &= \boldsymbol{\varphi}^{(2)}_{3\times8}(\xi)\bar{\boldsymbol{\alpha}}_{8\times1}, \end{aligned} \tag{4.170}$$

where (frustum element)

$$\boldsymbol{\varphi}^{(1)} = \begin{bmatrix} 1 & \xi & \xi^2 & \xi^3 & \xi^4 & 0 & 0 & 0 & 0 & 0 \\ 0 & 0 & 0 & 0 & 0 & 1 & \xi & \xi^2 & \xi^3 & \xi^4 \\ 0 & -\eta'\varrho & -2\xi\eta'\varrho & -3\xi^2\eta'\varrho & -4\xi^3\eta'\varrho & 0 & \varrho & 2\xi\varrho & 3\xi^2\varrho & 4\xi^3\varrho \end{bmatrix} \tag{4.171}$$

and (cap element)

$$\boldsymbol{\varphi}^{(2)} = \begin{bmatrix} \cos\omega & \xi & \xi^2 & \xi^3 & \xi^4 & 0 & 0 & 0 \\ \sin\omega & \xi\tan\beta^{(1)} & 0 & 0 & 0 & \xi^2 & \xi^3 & \xi^4 \\ 0 & \varrho(\tan\beta^{(1)}-\eta') & -2\xi\eta'\varrho & -3\xi^2\eta'\varrho & -4\xi^3\eta'\varrho & 2\xi\varrho & 3\xi^2\varrho & 4\xi^3\varrho \end{bmatrix}. \tag{4.172}$$

Defining the elemental generalized displacement vector for the frustum element

$$\overset{(1)}{\Delta\mathbf{q}}_{10\times 1} = \{\Delta u_\xi^{(1)}\ \Delta u_\eta^{(1)}\ \Delta\psi^{(1)}\ \Delta u_\xi^{(2)}\ \Delta u_\eta^{(2)}\ \Delta\psi^{(2)}\ \Delta u_\xi^{(3)}\ \Delta u_\eta^{(3)}\ \Delta u_\xi^{(4)}\ \Delta u_\eta^{(4)}\} \tag{4.173}$$

and for the cap element[1]

$$\overset{(2)}{\Delta\mathbf{q}}_{8\times 1} = \{\Delta w^{(1)}\ \Delta u^{(2)}\ \Delta w^{(2)}\ \Delta\psi^{(2)}\ \Delta u^{(3)}\ \Delta w^{(3)}\ \Delta u^{(4)}\ \Delta w^{(4)}\} \tag{4.174}$$

and using eqs. (4.170), (4.171), (4.172) we get

$$\overset{(1)}{\Delta\mathbf{q}}_{10\times 1} = \overset{(1)}{\mathbf{A}}_{10\times 10}\boldsymbol{\alpha}_{10\times 1}, \quad \overset{(2)}{\Delta\mathbf{q}}_{8\times 1} = \overset{(2)}{\mathbf{A}}_{8\times 8}\bar{\boldsymbol{\alpha}}_{8\times 1} \tag{4.175}$$

where, denoting for simplicity

$$s = \sin\beta^{(1)}, \quad c = \cos\beta^{(1)}, \quad t = \tan\beta^{(1)},$$
$$\bar{s} = \sin\beta^{(2)}, \quad \bar{c} = \cos\beta^{(2)}, \quad \bar{t} = \tan\beta^{(2)}$$

---

[1] Note the different coordinate systems used for representing the vector components of both the vectors $\overset{(1)}{\Delta\mathbf{q}}$ and $\overset{(2)}{\Delta\mathbf{q}}$.

the matrices $\overset{(1)}{\mathbf{A}}$ and $\overset{(2)}{\mathbf{A}}$ are given by

$$\overset{(1)}{\mathbf{A}}_{10\times 10}=\begin{bmatrix} 1 & 0 & 0 & 0 & 0 & 0 & 0 & 0 & 0 & 0 \\ 0 & 0 & 0 & 0 & 0 & 1 & 0 & 0 & 0 & 0 \\ 0 & -\frac{sc}{l} & 0 & 0 & 0 & 0 & \frac{c^2}{l} & 0 & 0 & 0 \\ 1 & 1 & 1 & 1 & 1 & 0 & 0 & 0 & 0 & 0 \\ 0 & 0 & 0 & 0 & 0 & 1 & 1 & 1 & 1 & 1 \\ 0 & -\frac{\bar{s}\bar{c}}{l} & -\frac{2\bar{s}\bar{c}}{l} & -\frac{3\bar{s}\bar{c}}{l} & -\frac{4\bar{s}\bar{c}}{l} & 0 & \frac{\bar{c}^2}{l} & \frac{2\bar{c}^2}{l} & \frac{3c^2}{l} & \frac{4c^2}{l} \\ 1 & \frac{1}{3} & \frac{1}{9} & \frac{1}{27} & \frac{1}{81} & 0 & 0 & 0 & 0 & 0 \\ 0 & 0 & 0 & 0 & 0 & 1 & \frac{1}{3} & \frac{1}{9} & \frac{1}{27} & \frac{1}{81} \\ 1 & \frac{2}{3} & \frac{4}{9} & \frac{8}{27} & \frac{16}{81} & 0 & 0 & 0 & 0 & 0 \\ 0 & 0 & 0 & 0 & 0 & 1 & \frac{2}{3} & \frac{4}{9} & \frac{8}{27} & \frac{16}{81} \end{bmatrix},$$

$$\overset{(2)}{\mathbf{A}}_{8\times 8}=\begin{bmatrix} 1 & 0 & 0 & 0 & 0 & 0 & 0 & 0 \\ -\cos\omega & 1 & 1 & 1 & 1 & 0 & 0 & 0 \\ \sin\omega & t & 0 & 0 & 0 & 1 & 1 & 1 \\ 0 & \frac{t-\bar{t}}{l}\bar{c}^2 & -\frac{2\bar{s}\bar{c}}{l} & -\frac{3\bar{s}\bar{c}}{l} & -\frac{4\bar{s}\bar{c}}{l} & \frac{2\bar{c}^2}{l} & \frac{3\bar{c}^2}{l} & \frac{4\bar{c}^2}{l} \\ \cos\omega & \frac{1}{3} & \frac{1}{9} & \frac{1}{27} & \frac{1}{81} & 0 & 0 & 0 \\ \sin\omega & t\cdot\frac{1}{3} & 0 & 0 & 0 & \frac{1}{9} & \frac{1}{27} & \frac{1}{81} \\ -\cos\omega & \frac{2}{3} & \frac{4}{9} & \frac{8}{27} & \frac{16}{81} & 0 & 0 & 0 \\ \sin\omega & t\cdot\frac{2}{3} & 0 & 0 & 0 & \frac{4}{9} & \frac{8}{27} & \frac{16}{81} \end{bmatrix}.$$

The stiffness matrices $\bar{\mathbf{k}}^{(con)}$ and $\bar{\mathbf{k}}^{(\sigma)}$ referred to the generalized coordinates $\boldsymbol{\alpha}=\{\alpha_1\ \alpha_2\ \dots\ \alpha_{10}\}$ and $\bar{\boldsymbol{\alpha}}=\{\bar{\alpha}_1\ \bar{\alpha}_2\ \dots\ \bar{\alpha}_8\}$ may now be presented in the form referred to the generalized displacements as

$${}^{(1)}\mathbf{k}^{(con)}_{10\times 10}=\overset{(1)}{\mathbf{A}}{}^{-1,T}_{10\times 10}\,{}^{(1)}\bar{\mathbf{k}}^{(con)}_{10\times 10}\overset{(1)}{\mathbf{A}}{}^{-1}_{10\times 10},$$

$${}^{(2)}\mathbf{k}^{(con)}_{8\times 8}=\overset{(2)}{\mathbf{A}}{}^{-1,T}_{8\times 8}\,{}^{(2)}\bar{\mathbf{k}}^{(con)}_{8\times 8}\overset{(2)}{\mathbf{A}}{}^{-1}_{8\times 8},$$

$$
\begin{aligned}
{}^{(1)}\mathbf{k}^{(\sigma)}_{10\times 10} &= \overset{(1)}{\mathbf{A}}{}^{-1,T}_{10\times 10}\,{}^{(1)}\bar{\mathbf{k}}^{(\sigma)}_{10\times 10}\,\overset{(1)}{\mathbf{A}}{}^{-1}_{10\times 10},\\
{}^{(2)}\mathbf{k}^{(\sigma)}_{8\times 8} &= \overset{(2)}{\mathbf{A}}{}^{-1,T}_{8\times 8}\,{}^{(2)}\bar{\mathbf{k}}^{(\sigma)}_{8\times 8}\,\overset{(2)}{\mathbf{A}}{}^{-1}_{8\times 8}
\end{aligned}
$$

where ${}^{(1)}\mathbf{k}^{(con)}_{10\times 10}$, ${}^{(1)}\mathbf{k}^{(\sigma)}_{10\times 10}$ are the matrices (constitutive and initial stress, respectively) for the frustum element expressed in the local coordinates ξ, η whereas ${}^{(2)}\mathbf{k}^{(con)}_{8\times 8}$, ${}^{(2)}\mathbf{k}^{(\sigma)}_{8\times 8}$ are the matrices for the cap element in the global coordinates u, w, Fig. 25.

Using the general definition of the elemental mass matrix, cf. eq. (3.54)$_3$, we may easily obtain the mass matrix for the axisymmetric shell element as (frustum element)

$$
\begin{aligned}
{}^{(1)}\mathbf{m}_{10\times 10} = \overset{(1)}{\mathbf{A}}{}^{-1,T}_{10\times 10}\Bigg[\int_0^1 l\,\overset{(1)}{\boldsymbol{\varphi}}{}^{T}_{10\times 3}(\xi)\mathbf{W}^{T}_{3\times 3}(\xi)\hat{\mathbf{D}}_{3\times 3}(\xi)\mathbf{W}_{3\times 3}(\xi)\,\times\\
\times\,\overset{(1)}{\boldsymbol{\varphi}}_{3\times 10}(\xi)r(\xi)\,(1+\eta'^2)^{1/2}\mathrm{d}\xi\Bigg]\overset{(1)}{\mathbf{A}}{}^{-1}_{10\times 10}
\end{aligned}
\tag{4.176}
$$

and (cap element)

$$
\begin{aligned}
{}^{(2)}\mathbf{m}_{8\times 8} = \overset{(1)}{\mathbf{A}}{}^{-1,T}_{8\times 8}\Bigg[\int_0^1 l\,\overset{(2)}{\boldsymbol{\varphi}}{}^{T}_{8\times 3}(\xi)\mathbf{W}^{T}_{3\times 3}(\xi)\hat{\mathbf{D}}_{3\times 3}(\xi)\mathbf{W}_{3\times 3}(\xi)\,\times\\
\times\,\overset{(2)}{\boldsymbol{\varphi}}_{3\times 8}(\xi)r(\xi)\,(1+\eta'^2)^{1/2}\mathrm{d}\xi\Bigg]\overset{(2)}{\mathbf{A}}_{8\times 8}
\end{aligned}
\tag{4.177}
$$

where

$$
\mathbf{W}_{3\times 3} = \begin{bmatrix} \sin\omega & -\cos\omega & 0\\ \cos\omega & \sin\omega & 0\\ 0 & 0 & 1\end{bmatrix}, \qquad \hat{\mathbf{D}}_{3\times 3} = \varrho\begin{bmatrix} h & 0 & 0\\ 0 & h & 0\\ 0 & 0 & \dfrac{h^3}{12}\end{bmatrix}.
$$

In order to form the stiffness and mass matrices for the whole shell it is necessary to transform the elemental contributions to a common coordinate system, cf. Sec. 4.3. The global coordinates are employed here as the common system. We have, cf. eq. (4.24)

$$
\begin{aligned}
\overset{(1)}{\Delta\mathbf{q}}_{10\times 1} &= \begin{bmatrix} \mathbf{W}_{3\times 3} & \mathbf{0}_{3\times 3} & \mathbf{0}_{3\times 4}\\ & \mathbf{W}_{3\times 3} & \mathbf{0}_{3\times 4}\\ \text{sym.} & & \mathbf{I}_{4\times 4}\end{bmatrix}\overset{(1)}{\Delta\mathbf{r}}_{10\times 1} = \overset{(1)}{\mathbf{L}}_{10\times 10}\overset{(1)}{\Delta\mathbf{r}}_{10\times 1},\\
\overset{(2)}{\Delta\mathbf{q}}_{8\times 1} &= \begin{bmatrix} 1 & \mathbf{0}_{1\times 3} & \mathbf{0}_{1\times 4}\\ & \mathbf{W}_{3\times 3} & \mathbf{0}_{3\times 4}\\ \text{sym.} & & \mathbf{I}_{4\times 4}\end{bmatrix}\overset{(2)}{\Delta\mathbf{r}}_{8\times 1} = \overset{(2)}{\mathbf{L}}_{8\times 8}\overset{(2)}{\Delta\mathbf{r}}_{8\times 1}
\end{aligned}
$$

where the vectors of elemental generalized displacements referred to the system rz have been defined as

$$\overset{(1)}{\Delta \mathbf{r}}_{10\times1} = \{\Delta u_r^{(1)}\ \Delta u_z^{(1)}\ \Delta\psi^{(1)}\ \Delta u_r^{(2)}\ \Delta u_z^{(2)}\ \Delta\psi^{(2)}\ \Delta u_\xi^{(3)}\ \Delta u_\eta^{(3)}\ \Delta u_\xi^{(4)}\ \Delta u_\eta^{(4)}\},$$
$$\overset{(2)}{\Delta \mathbf{r}}_{8\times1} = \{\Delta u_z^{(1)}\ \Delta u_r^{(2)}\ \Delta u_z^{(2)}\ \Delta\psi^{(2)}\ \Delta u_\xi^{(3)}\ \Delta u_\eta^{(3)}\ \Delta u_\xi^{(4)}\ \Delta u_\eta^{(4)}\} \qquad (4.178)$$

and

$$\mathbf{I}_{4\times4} = \begin{bmatrix} 1 & 0 & 0 & 0 \\ & 1 & 0 & 0 \\ \text{sym.} & & 1 & 0 \\ & & & 1 \end{bmatrix},$$

$$\overset{(1)}{\mathbf{K}}{}^{(e)}_{10\times10} = \overset{(1)}{\mathbf{L}}{}^{T}_{10\times10}\, {}^{(1)}\mathbf{k}_{10\times10}\overset{(1)}{\mathbf{L}}_{10\times10}, \text{ etc.}$$

$$\overset{(2)}{\mathbf{K}}{}^{(e)}_{8\times8} = \overset{(2)}{\mathbf{L}}_{8\times8}\, {}^{(2)}\mathbf{k}_{8\times8}\overset{(2)}{\mathbf{L}}_{8\times8}, \text{ etc.}$$

Thus, we finally have the elemental (frustum and cap) equations as, cf. eq. (4.28)

$$\begin{aligned} &\overset{(1)}{\mathbf{M}}{}^{(e)}_{10\times10}\Delta\ddot{\mathbf{r}}^{(1)}_{10\times1} + (\overset{(1)}{\mathbf{K}}{}^{(e)(con)}_{10\times10} + \overset{(1)}{\mathbf{K}}{}^{(e)(\sigma)}_{10\times10})\Delta\mathbf{r}^{(1)}_{10\times1} = \Delta\mathbf{R}^{(e)}_{10\times1}, \\ &\overset{(2)}{\mathbf{M}}{}^{(e)}_{8\times8}\Delta\ddot{\mathbf{r}}^{(2)}_{8\times1} + (\overset{(2)}{\mathbf{K}}{}^{(e)(con)}_{8\times8} + \overset{(2)}{\mathbf{K}}{}^{(e)(\sigma)}_{8\times8})\Delta\mathbf{r}^{(2)}_{8\times1} = \Delta\mathbf{R}^{(e)}_{8\times1}. \end{aligned} \qquad (4.179)$$

The further steps in the assembly procedure are those described in Sec. 4.3. However, prior to this we may eliminate the internal degrees of freedom (i.e. those located at the internal nodes in the element) in the process of static condensation. The procedure may significantly reduce the dimension of the global problem—for E elements the discretized shell model has $10\times E - 3(E-1)$ degrees of freedom (boundary conditions not taken into account), out of which $4\times E$ degrees of freedom are eliminated. For large discretized systems this may reduce the size of the problem by some 57%

$$\left(\frac{4\times E}{7\times E+3} \xrightarrow[E\to\infty]{} 0.57\right).$$

To perform the static condensation we partition the (frustum) element stiffness equation as

$$\begin{bmatrix} \mathbf{K}_{\mathrm{I\,I}6\times6} & \mathbf{K}_{\mathrm{I\,II}6\times4} \\ \mathbf{K}_{\mathrm{II\,I}4\times6} & \mathbf{K}_{\mathrm{II\,II}4\times4} \end{bmatrix} \begin{bmatrix} \Delta\mathbf{r}_{\mathrm{I}6\times1} \\ \Delta\mathbf{r}_{\mathrm{II}4\times1} \end{bmatrix} = \begin{bmatrix} \Delta\mathbf{R}_{\mathrm{I}6\times1} \\ \Delta\mathbf{R}_{\mathrm{II}4\times1} \end{bmatrix} \qquad (4.180)$$

where the vectors corresponding to the boundary and internal nodes in the particular element under considerations are defined as

$$\Delta\mathbf{r}_{\mathrm{I}6\times1} = \{\Delta u_r^{(1)}\ \Delta u_z^{(1)}\ \Delta\psi^{(1)}\ \Delta u_r^{(2)}\ \Delta u_z^{(2)}\ \Delta\psi^{(2)}\},$$
$$\Delta\mathbf{r}_{\mathrm{II}4\times1} = \{\Delta u_\xi^{(3)}\ \Delta u_\eta^{(3)}\ \Delta u_\xi^{(4)}\ \Delta u_\eta^{(4)}\}$$

and $\Delta\mathbf{R}_{\mathrm{I}}$ and $\Delta\mathbf{R}_{\mathrm{II}}$ are the generalized forces acting on the boundary and internal nodes of the element, respectively. Solving eq. (4.180) for $\Delta\mathbf{r}_{\mathrm{II}}$ we obtain

$$\Delta\mathbf{r}_{\mathrm{II}} = \mathbf{K}_{\mathrm{II\,II}}^{-1}\Delta\mathbf{R}_{\mathrm{II}} - \mathbf{K}_{\mathrm{II\,II}}^{-1}\mathbf{K}_{\mathrm{II\,I}}\Delta\mathbf{r}_{\mathrm{I}}$$

which, when introduced into eq. (4.180), gives

$$[\mathbf{K}_{\mathrm{I\,I}} - \mathbf{K}_{\mathrm{I\,II}}\mathbf{K}_{\mathrm{II\,II}}^{-1}\mathbf{K}_{\mathrm{II\,I}}]\Delta\mathbf{r}_{\mathrm{I}} = \Delta\mathbf{R}_{\mathrm{I}} - \mathbf{K}_{\mathrm{I\,II}}\mathbf{K}_{\mathrm{II\,II}}^{-1}\Delta\mathbf{R}_{\mathrm{II}}$$

or

$$\hat{\mathbf{K}}_{6\times 6}\Delta\mathbf{r}_{\mathrm{I}6\times 1} = \Delta\hat{\mathbf{R}}_{6\times 1}$$

where

$$\hat{\mathbf{K}}_{6\times 6} = \mathbf{K}_{\mathrm{I\,I}6\times 6} - \mathbf{K}_{\mathrm{I\,II}6\times 4}\mathbf{K}_{\mathrm{II\,II}4\times 4}^{-1}\mathbf{K}_{\mathrm{II\,I}4\times 6},$$

$$\Delta\hat{\mathbf{R}}_{6\times 1} = \Delta\mathbf{R}_{\mathrm{I}6\times 1} - \mathbf{K}_{\mathrm{I\,II}6\times 4}\mathbf{K}_{\mathrm{II\,II}4\times 4}^{-1}\Delta\mathbf{R}_{\mathrm{II}4\times 1}.$$

The so-obtained elemental matrices $\hat{\mathbf{K}}_{6\times 6}$ may now be assembled to yield

$$\mathbf{K}_{\bar{N}\times\bar{N}}\Delta\mathbf{r}_{\bar{N}\times 1} = \Delta\mathbf{R}_{\bar{N}\times 1} \tag{4.181}$$

with $\bar{N}$ equal to $3\times$ (number of nodes) and $\mathbf{K}_{\bar{N}\times\bar{N}}$ being the band matrix with the band width equal to 6.

In the case of dealing with creep or viscoplastic problems we may want to use the initial load technique which is based on replacing the initial strains $\Delta\boldsymbol{\epsilon}^{(in)}$ due to these inelastic effects by equivalent nodal loads to be used in the iterative procedure of solving the system of algebraic equations. The expression for the initial load may be shown to have the form (frustum element)

$$\mathbf{J}_{10\times 1}^{(e)} = \mathbf{A}_{10\times 10}^{-1,T}\int_0^1 l\mathbf{B}_{10\times 4}^{(1)T}\left(\int_{-h/2}^{h/2}\mathbf{D}_{4\times 4}\Delta\boldsymbol{\epsilon}_{4\times 1}^{(in)}\,\mathrm{d}\chi\right)r(1+\eta'^2)^{1/2}\mathrm{d}\xi,$$

the fundamental system of equations assuming the form

$$\mathbf{M}\Delta\ddot{\mathbf{r}} + \mathbf{K}\Delta\mathbf{r} = \Delta\mathbf{R} + \sum_{e=1}^{E}\mathbf{J}^{(e)}. \tag{4.182}$$

This completes the derivation of all the ingredients necessary to effectively perform the numerical analysis of thin axisymmetric shells within the large displacement Love–Kirchhoff approximation. Numerous examples will be given in Sec. 8.1.

4.9 Linear strain triangular element for nonlinear plane stress analysis

In order to further illustrate the general discretized approach discussed in Sec. 3.4, we shall now consider the nonlinear formulation for a plane element, known as the linear strain triangular element for plane stress. For reasons which will become clear at the end of this section, the element may be considered to be a good illustration for the general finite element approach, preserving at the same time the simplicity necessary in this book. Thus, even if it is now generally believed that some numerically integrated, isoparametric finite elements are the most effective elements in practical computations, an analytical derivation is presented below with the aim to display the details of a typical stiffness matrix built-up, cf. [48], [75].

We shall begin our discussion with some general remarks on polynomial interpolation of the displacement field in an arbitrary element. Let $\boldsymbol{\xi} = (\xi^1, \xi^2, \ldots, \xi^s)$ be an arbitrary point of a bounded region Ω in the s-dimensional Euclidean space $\boldsymbol{E}^s$, $\boldsymbol{\xi} \in \Omega \subset \boldsymbol{E}^s$ and V^m be a m-dimensional vector space of functions $f(\boldsymbol{\xi})$ which are differentiable in Ω. The space V^m is assumed to be spanned by the basis

$$\boldsymbol{\varphi}(\boldsymbol{\xi}) = \{\varphi_1(\boldsymbol{\xi}) \;\; \varphi_2(\boldsymbol{\xi}) \;\; \ldots \;\; \varphi_m(\boldsymbol{\xi})\} \tag{4.183}$$

which implies that for any function $f(\boldsymbol{\xi}) \in V^m$ there exists a set of m numbers $\mathbf{a} = \{a_1 \;\; a_2 \;\; \ldots \;\; a_m\}$ such that

$$\begin{aligned} f(\boldsymbol{\xi}) &= \mathbf{a}^T \boldsymbol{\varphi}(\boldsymbol{\xi}) \\ &= a_1\varphi_1(\boldsymbol{\xi}) + a_2\varphi_2(\boldsymbol{\xi}) + \ldots + a_m\varphi_m(\boldsymbol{\xi}). \end{aligned} \tag{4.184}$$

The numbers $a_1, a_2, \ldots, a_m$ are referred to as the generalized coordinates of the function $f(\boldsymbol{\xi}) \in V^m$ in the basis $\boldsymbol{\varphi}(\boldsymbol{\xi})$.

Let us select m different points in the region Ω, $\boldsymbol{\xi}_i = (\xi_i^1, \ldots, \xi_i^s) \in \Omega$, $i = 1, 2, \ldots, m$, at which the function $f(\boldsymbol{\xi})$ takes on the values

$$f(\boldsymbol{\xi}_i) = f_i, \qquad i = 1, 2, \ldots, m.$$

Denote

$$\mathbf{f} = \{f_1 \;\; f_2 \;\; \ldots \;\; f_m\}$$

and introduce the $m \times m$ matrix given by

$$\mathbf{A} = [\varphi_{ij}]_{m\times m} = [\varphi_i(\boldsymbol{\xi}_j)].$$

The following relation holds

$$\mathbf{f} = \mathbf{A}\mathbf{a}, \quad f_i = \varphi_{ij}a_j; \quad i, j = 1, 2, \ldots, m$$

and, under the condition that $\det \mathbf{A} \neq 0$, we have also

$$\mathbf{a} = \mathbf{A}^{-1}\mathbf{f}.$$

Let us now construct a set of m functions

$$\Phi_i(\boldsymbol{\xi}) = \frac{1}{f_i}\mathbf{a}^T\boldsymbol{\varphi}(\boldsymbol{\xi}) = \frac{1}{f_i}\mathbf{f}^T\mathbf{A}^{-T}\boldsymbol{\varphi}(\boldsymbol{\xi}) \tag{4.185}$$

which satisfy the conditions

$$\Phi_i(\xi_j) = \delta_{ij}. \tag{4.186}$$

The set $\mathbf{\Phi} = \{\Phi_1 \xi () \ \Phi_2(\xi) \ \dots \ \Phi_m(\xi)\}$ forms the vector basis in the space V^m. If $\det \mathbf{A} \neq 0$ then the basis $\mathbf{\Phi}$ satisfying conditions (4.186) is unique. An arbitrary element $f \in V^m$ may be now presented as

$$f(\xi) = f_1 \Phi_1(\xi) + f_2 \Phi_2(\xi) + \dots + f_m \Phi_m(\xi) = \mathbf{f}^T \mathbf{\Phi}(\xi). \tag{4.187}$$

The function $\mathbf{\Phi}(\xi)$ is referred to as the interpolation vector with respect to the set of points ξ_i, $i = 1, 2, \dots, m$ in the region Ω.

The interpolation formula (4.187) may be extended to apply to any vector or matrix function of the variable ξ. Let

$$\mathbf{g}(\xi) = \{g^1(\xi) \ g^2(\xi) \ \dots \ g^q(\xi)\}, \quad g^n \in V^m, \quad n = 1, 2, \dots, q;$$

$$\mathbf{h}(\xi) = \begin{bmatrix} h^{11}(\xi) & \dots & h^{1p}(\xi) \\ \dots\dots & \dots & \dots\dots \\ h^{r1}(\xi) & \dots & h^{rp}(\xi) \end{bmatrix}, \quad h^{kl} \in V^m,$$

$$k = 1, 2, \dots, r, \quad l = 1, 2, \dots, p.$$

Values of these functions taken at nodal points ξ_i, $i = 1, 2, \dots, m$ will be denoted as

$$g_i^n = g^n(\xi_i), \quad i = 1, 2, \dots, m; \quad n = 1, 2, \dots, q,$$
$$h_i^{kl} = h^{kl}(\xi_i), \quad k = 1, 2, \dots, r; \quad l = 1, 2, \dots, p$$

and thus

$$\mathbf{g}^n(\xi) = g_1^n \Phi_1(\xi) + g_2^n \Phi_2(\xi) + \dots + g_m^n \Phi_m(\xi) = \mathbf{g}^{nT} \mathbf{\Phi}(\xi),$$
$$\mathbf{h}^{kl}(\xi) = h_1^{kl} \Phi_1(\xi) + h_2^{kl} \Phi_2(\xi) + \dots + h_m^{kl} \Phi_m(\xi) = \mathbf{h}^{klT} \mathbf{\Phi}(\xi)$$

where

$$\mathbf{g}^n = \{g_1^n \ g_2^n \ \dots \ g_m^n\}, \quad \mathbf{h}^{kl} = \{h_1^{kl} \ h_2^{kl} \ \dots \ h_m^{kl}\}.$$

Let us consider two simple examples to show the explicit forms of all the functions which have appear so far in our formulation.

Example I. Let $\Omega = [\xi_1, \xi_2] \subset \boldsymbol{E}^1$ and $V^m = V^2$ be the vector space of linear functions defined on Ω with basis $\varphi_1(\xi) = 1, \varphi_2(\xi) = \xi$. Basis $\mathbf{\Phi}$ will take the form, cf. Fig. 28

$$\Phi_1(\xi) = \frac{\xi_2 - \xi}{\xi_2 - \xi_1}, \quad \Phi_2(\xi) = \frac{\xi - \xi_1}{\xi_2 - \xi_1}$$

whereas the application of the space V^2 to interpolate an arbitrary function $u(\xi)$, $\xi \in \Omega$ is shown in Fig. 28.

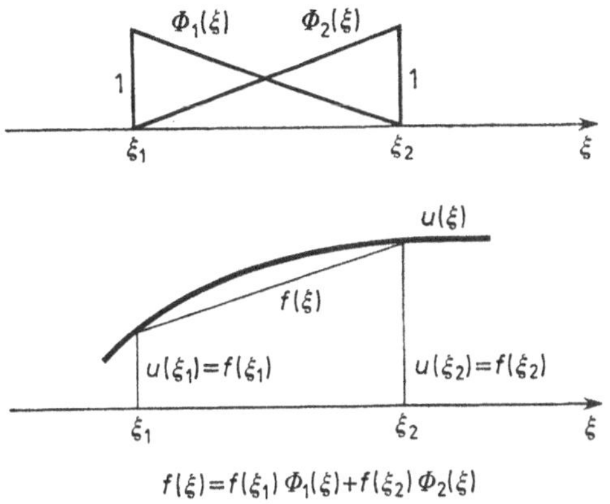

Fig. 28 Example of interpolating functions

Example II. Let Ω be a two-dimensional region shown in Fig. 29 and $V^m = V^3$ be the vector space of linear functions of two arguments ξ^1, ξ^2 defined on Ω with basis $\varphi_1(\xi^1, \xi^2) = 1, \varphi_2(\xi^1, \xi^2) = \xi^1, \ \varphi_3(\xi^1, \xi^2) = \xi^2$. Basis $\mathbf{\Phi}$ will take the form

$$\Phi_i(\xi^1, \xi^2) = \Psi_i + \sum_{t=1}^{2} \chi_i^t \xi^t, \qquad i = 1, 2, 3$$

where

$$\mathbf{\Psi} = \{\Psi_1 \ \ \Psi_2 \ \ \Psi_3\} = \frac{1}{2A}\begin{bmatrix} \xi_2^1\xi_3^2 - \xi_3^1\xi_2^2 \\ \xi_3^1\xi_1^2 - \xi_1^1\xi_3^2 \\ \xi_1^1\xi_2^2 - \xi_2^1\xi_1^2 \end{bmatrix},$$

$$2A = \det\begin{bmatrix} 1 & \xi_1^1 & \xi_1^2 \\ 1 & \xi_2^1 & \xi_2^2 \\ 1 & \xi_3^1 & \xi_3^2 \end{bmatrix}$$

and

$$\boldsymbol{\chi}^1 = \{\chi_1^1 \ \ \chi_2^1 \ \ \chi_3^1\} = \begin{bmatrix} \xi_2^2 - \xi_3^2 \\ \xi_3^2 - \xi_1^2 \\ \xi_1^2 - \xi_2^2 \end{bmatrix},$$

$$\boldsymbol{\chi}^2 = \{\chi_1^2 \ \ \chi_2^2 \ \ \chi_3^2\} = \begin{bmatrix} \xi_3^1 - \xi_2^1 \\ \xi_1^1 - \xi_3^1 \\ \xi_2^1 - \xi_1^1 \end{bmatrix}.$$

$2A$ is the doubled area of the triangle Ω. Functions Φ_i and the interpolation for an arbitrary function $\mathbf{u}(\boldsymbol{\xi})$, $\boldsymbol{\xi} \in \Omega$ are shown in Fig. 29.

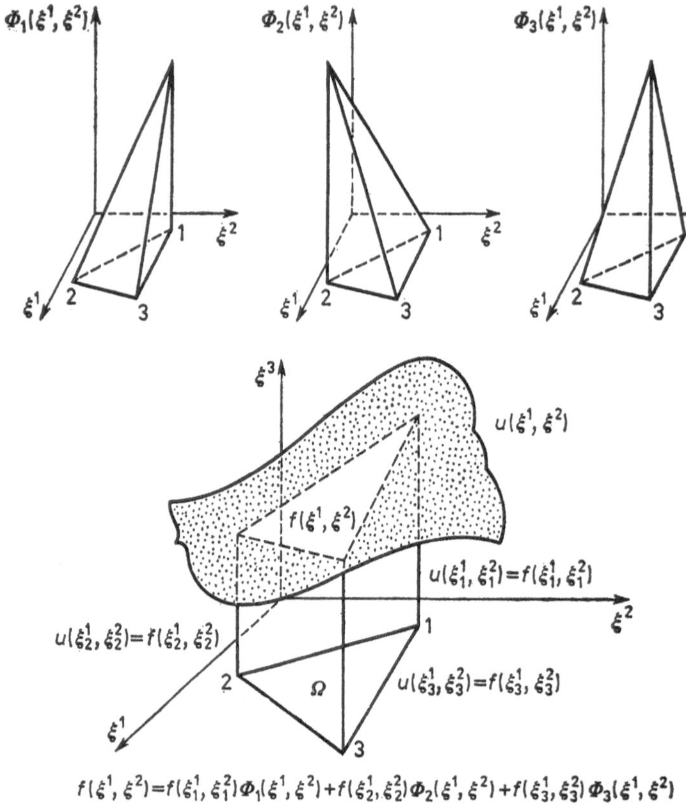

Fig. 29 Example of interpolating functions

In order to facilitate the discussion of more complex interpolation cases it turns out useful to introduce the so-called natural coordinates. Consider the triangle 1–2–3 lying in the plane $\xi\eta$ of the Cartesian coordinate system, Fig. 30. The vertex coordinates are denoted as (ξ_i, η_i), $i = 1, 2, 3$; the remaining geometric parameters are defined in Fig. 30. In particular, let A_1, A_2 and A_3 stand for the areas of triangles shown for any point P in Fig. 30. The natural coordinates of the point P are defined as the triple $\boldsymbol{\gamma} = (\gamma_1, \gamma_2, \gamma_3)$ given by

$$\gamma_1 = \frac{A_1}{A}, \quad \gamma_2 = \frac{A_2}{A}, \quad \gamma_3 = \frac{A_3}{A}.$$

Since $A_1 + A_2 + A_3 = A$ and consequently $\gamma_1 + \gamma_2 + \gamma_3 = 1$, the natural coordinates $\gamma_1, \gamma_2, \gamma_3$ are not independent.

The natural coordinates of any point P are invariants under arbitrary linear transformations of the triangle. The equation $\gamma_i = \text{const}$ represents

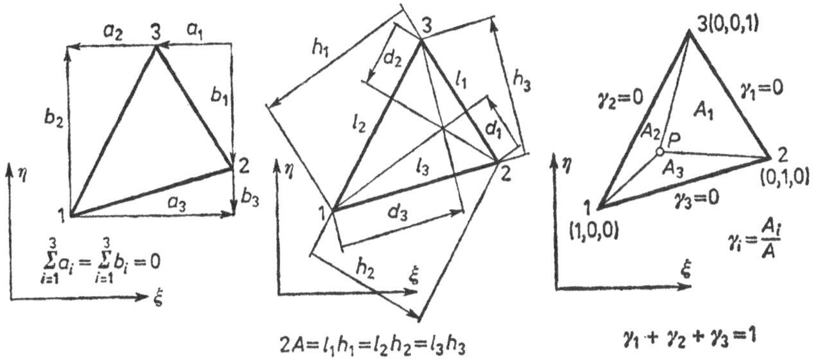

Fig. 30 Natural coordinates

the straight line parallel to the i-th side of the triangle with $\gamma_i = 0$ being just the equation of this side. Three vertices of the triangle are given by $(1, 0, 0)$, $(0, 1, 0)$ and $(0, 0, 1)$, three middle points of triangle sides are given by $(\frac{1}{2}, \frac{1}{2}, 0)$, $(0, \frac{1}{2}, \frac{1}{2})$ and $(\frac{1}{2}, 0, \frac{1}{2})$, the center of the inscribed circle is given by $(\frac{1}{3}, \frac{1}{3}, \frac{1}{3})$.

The use of natural coordinates may make it significantly easier to construct elemental stiffness matrices. In order to exploit this concept we have to express all the functions dependent upon ξ, η in terms of the variables γ_i, $i = 1, 2, 3$. The transformation is based on the following relations

$$\begin{bmatrix} 1 \\ \xi \\ \eta \end{bmatrix} = \begin{bmatrix} 1 & 1 & 1 \\ \xi_1 & \xi_2 & \xi_3 \\ \eta_1 & \eta_2 & \eta_3 \end{bmatrix} \begin{bmatrix} \gamma_1 \\ \gamma_2 \\ \gamma_3 \end{bmatrix},$$

$$\begin{bmatrix} \gamma_1 \\ \gamma_2 \\ \gamma_3 \end{bmatrix} = \frac{1}{2A} \begin{bmatrix} \xi_2\eta_3 - \xi_3\eta_2 & \eta_2 - \eta_3 & \xi_3 - \xi_2 \\ \xi_3\eta_1 - \xi_1\eta_3 & \eta_3 - \eta_1 & \xi_1 - \xi_3 \\ \xi_1\eta_2 - \xi_2\eta_1 & \eta_1 - \eta_2 & \xi_2 - \xi_1 \end{bmatrix} \begin{bmatrix} 1 \\ \xi \\ \eta \end{bmatrix},$$

$$\frac{\partial \gamma_i}{\partial \xi} = \frac{b_i}{2A}, \quad \frac{\partial \gamma_i}{\partial \eta} = \frac{a_i}{2A}, \quad \frac{\partial \xi}{\partial \gamma_i} = \xi_i, \quad \frac{\partial \eta}{\partial \gamma_i} = \eta_i,$$

$$a_1 = \xi_3 - \xi_2, \quad a_2 = \xi_1 - \xi_3, \quad a_3 = \xi_2 - \xi_1,$$

$$b_1 = \eta_2 - \eta_3, \quad b_2 = \eta_3 - \eta_1, \quad b_3 = \eta_1 - \eta_2,$$

$$\frac{\partial f}{\partial \xi} = \frac{b_i}{2A} \frac{\partial f}{\partial \gamma_i}, \quad \frac{\partial f}{\partial \eta} = \frac{a_i}{2A} \frac{\partial f}{\partial \gamma_i},$$

$$\frac{\partial^2 f}{\partial \xi \, \partial \eta} = \frac{a_i b_j}{4A^2} \frac{\partial^2 f}{\partial \gamma_i \, \partial \gamma_j}, \quad \text{etc.}$$

Let us now use the above concept to carry out interpolation of any function defined on the triangle in which we select six nodal points as shown in Fig. 31.

The natural coordinates make it possible to straightforwardly construct the interpolation functions $\mathbf{\Phi}$ thus avoiding the tedious process of inverting the matrix $\mathbf{A}$.

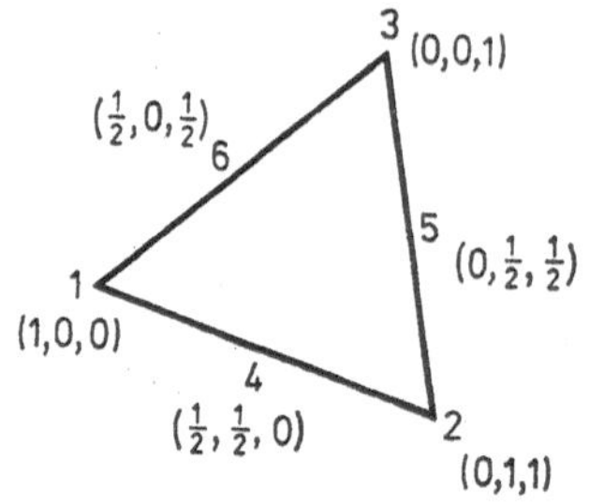

Fig. 31 Interpolation of functions defined on the triangular region

By adopting the interpolation functions in the form

$$\Phi_i^{(1)} = \gamma_i, \quad i = 1, 2, 3,$$

i.e. by putting for any function f

$$f(\gamma_1, \gamma_2, \gamma_3) = \mathbf{f}^T \mathbf{\Phi}^{(1)} = f_1 \gamma_1 + f_2 \gamma_2 + f_3 \gamma_3,$$

we easily find out that the interpolation is identical to that considered above in Example II. By adopting the interpolation functions in the form

$$\Phi_i^{(2)} = \gamma_i (2\gamma_i - 1), \quad i = 1, 2, 3, \tag{4.188}$$

$$\Phi_4^{(2)} = 4\gamma_1 \gamma_2, \quad \Phi_5^{(2)} = 4\gamma_2 \gamma_3, \quad \Phi_6^{(2)} = 4\gamma_1 \gamma_3,$$

i.e. by putting for any function f

$$f(\gamma_1, \gamma_2, \gamma_3) = \mathbf{f}^T \mathbf{\Phi}^{(2)} = f_1 \Phi_1^{(2)} + f_2 \Phi_2^{(2)} + \ldots + f_6 \Phi_6^{(2)}, \tag{4.189}$$

we obtain the quadratic interpolation corresponding to the choice of six nodal points at triangle vertices and mid-side points, Fig. 32. The further advantage of using the natural coordinates concept results from the fact

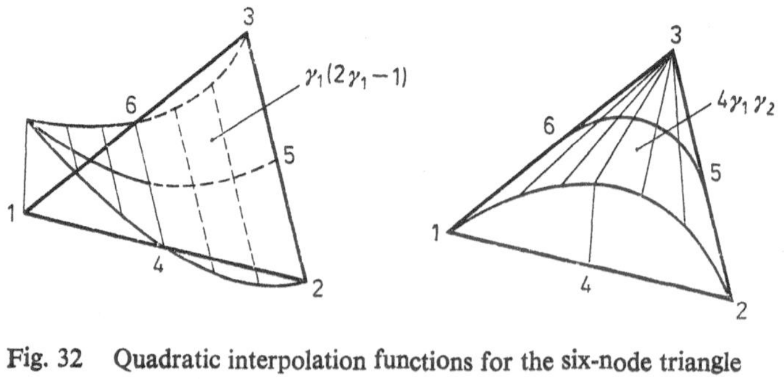

Fig. 32 Quadratic interpolation functions for the six-node triangle

that the integral over the triangle area of any polynomial of variables $\gamma_1, \gamma_2, \gamma_3$ is independent of the triangle shape and can be expressed as a fraction of the triangle area by means of the formula

$$I(m_i, m_j, m_k) = \int_{\Omega} \gamma_i^{m_i} \gamma_j^{m_j} \gamma_k^{m_k} \mathrm{d}\Omega = \alpha\Omega \tag{4.190}$$

where the triple (i, j, k) is an arbitrary permutation of 1, 2, 3 whereas m_i, m_j and m_k are arbitrary natural numbers. The values of the coefficients α are given in Table 2 for different values of the parameter $n = m_i + m_j + m_k$.

Table 2 Coefficient α for integration formula (4.190)

$n = m_i+m_j+m_k$	m_i	m_j	m_k	α
1	1	0	0	1/3
2	2	0	0	1/6
	1	1	0	1/12
3	3	0	0	1/10
	2	1	0	1/30
	1	1	1	1/60
4	4	0	0	1/15
	3	1	0	1/60
	2	2	0	1/90
	2	1	1	1/180
5	5	0	0	1/21
	4	1	0	1/105
	3	2	0	1/210
	3	1	1	1/420
	2	2	1	1/630

Formula (4.190) allows to avoid the necessity of using costly procedures of numerical integration while developing stiffness matrices.

The remainder of this section will be devoted to the derivation of explicit forms of stiffness matrices for the plane stress triangular element with six nodes and, consequently, twelve kinematic degrees of freedom. According to eq. (4.189) the explicit form of the kinematic constraints imposed on the incremental displacement field within the element is given by

$$\begin{aligned} \Delta u_\xi(\gamma_1, \gamma_2, \gamma_3) &= \mathbf{\Phi}^{(2)T}_{1\times 6} \Delta \mathbf{q}_{\xi_{6\times 1}}, \\ \Delta u_\eta(\gamma_1, \gamma_2, \gamma_3) &= \mathbf{\Phi}^{(2)T}_{1\times 6} \Delta \mathbf{q}_{\eta_{6\times 1}}, \end{aligned} \tag{4.191}$$

where

$$\begin{aligned} \{\Delta \mathbf{q}_{\xi_{6\times 1}} \;\; \Delta \mathbf{q}_{\eta_{6\times 1}}\} &= \Delta \mathbf{q}_{12\times 1}, \\ \Delta \mathbf{q}_{\xi_{6\times 1}} &= \{\Delta u_\xi^{(1)} \;\; \Delta u_\xi^{(2)} \;\; \dots \;\; \Delta u_\xi^{(6)}\}, \\ \Delta \mathbf{q}_{\eta_{6\times 1}} &= \{\Delta u_\eta^{(1)} \;\; \Delta u_\eta^{(2)} \;\; \dots \;\; \Delta u_\eta^{(6)}\}, \end{aligned}$$

and the vector function $\mathbf{\Phi}^{(2)}$ is defined in eq. (4.188). Introducing the elemental nodal displacement vector $\Delta \mathbf{q}_{12\times 1}$ we may compactly write

$$\Delta \mathbf{u}_{2\times 1}(\gamma_1, \gamma_2, \gamma_3) = \{\Delta u_\xi \;\; \Delta u_\eta\} = \begin{bmatrix} \mathbf{\Phi}^{(2)T}_{1\times 6} & \mathbf{0}_{1\times 6} \\ \mathbf{0}_{1\times 6} & \mathbf{\Phi}^{(2)T}_{1\times 6} \end{bmatrix}_{2\times 12} \Delta \mathbf{q}_{12\times 1}. \tag{4.192}$$

According to our basic philosophy the element is considered at a typical time instant t during the deformation process, and the components of vectors in eqs. (4.192) are referred to the local Cartesian coordinates $\xi\eta$. Defining the incremental strain vector at every point within the element as

$$\Delta \mathbf{\epsilon}(\gamma_1, \gamma_2, \gamma_3) = \{\Delta \varepsilon_{\xi\xi} \;\; \Delta \varepsilon_{\eta\eta} \;\; \Delta \gamma_{\xi\eta}\}_{3\times 1} \tag{4.193}$$

we arrive at

$$\Delta \mathbf{\epsilon}_{3\times 1} = \begin{bmatrix} \mathbf{\Phi}^{(2)T}_{\xi_{1\times 6}} & \mathbf{0}_{1\times 6} \\ \mathbf{0}_{1\times 6} & \mathbf{\Phi}^{(2)T}_{\eta_{1\times 6}} \\ \mathbf{\Phi}^{(2)T}_{\eta_{1\times 6}} & \mathbf{\Phi}^{(2)T}_{\xi_{1\times 6}} \end{bmatrix}_{3\times 12} \Delta \mathbf{q}_{12\times 1} = \mathbf{B}_{3\times 12} \Delta \mathbf{q}_{12\times 1} \tag{4.194}$$

where

$$\mathbf{\Phi}^{(2)}_{\xi_{6\times 1}} = \frac{\partial \mathbf{\Phi}^{(2)}}{\partial \xi} = \frac{1}{2A} \begin{bmatrix} (4\gamma_1 - 1) b_1 \\ (4\gamma_2 - 1) b_2 \\ (4\gamma_3 - 1) b_3 \\ 4(\gamma_2 b_1 + \gamma_1 b_2) \\ 4(\gamma_3 b_2 + \gamma_2 b_3) \\ 4(\gamma_1 b_3 + \gamma_3 b_1) \end{bmatrix}_{6\times 1};$$

$$\mathbf{\Phi}^{(2)}_{\eta_{6\times 1}} = \frac{\partial \mathbf{\Phi}^{(2)}}{\partial \eta} = \frac{1}{2A} \begin{bmatrix} (4\gamma_1 - 1) a_1 \\ (4\gamma_2 - 1) a_2 \\ (4\gamma_3 - 1) a_3 \\ 4(\gamma_2 a_1 + \gamma_1 a_2) \\ 4(\gamma_3 a_2 + \gamma_2 a_3) \\ 4(\gamma_1 a_3 + \gamma_3 a_1) \end{bmatrix}_{6\times 1}$$

are expressions for the derivative of the interpolation function obtained by using transformations given before. The elemental constitutive matrix is defined as usual by means of the relationship

$$\mathbf{k}_{12\times12} = \int_{\Omega} \mathbf{B}^T_{12\times3}\mathbf{C}_{3\times3}\mathbf{B}_{3\times12}\,d\Omega \tag{4.195}$$

in which the constitutive matrix $\mathbf{C}$ derives from the incremental material law $\Delta\boldsymbol{\sigma}_{3\times1} = \mathbf{C}_{3\times3}\Delta\boldsymbol{\epsilon}_{3\times1}$; explicit form of $\mathbf{C}$ for elastic-plastic material under plane stress will be given in Sec. 8.3. Generally, matrix $\mathbf{C}$ has the components

$$\mathbf{C}_{3\times3} = \begin{bmatrix} C_{11} & C_{12} & C_{13} \\ & C_{22} & C_{23} \\ \text{sym.} & & C_{33} \end{bmatrix}$$

which can be computed at every nodal point in the triangle to give

$$\mathbf{C}_{11\,6\times1} = \{C^{(1)}_{11}\ C^{(2)}_{11}\ \dots\ C^{(6)}_{11}\},\quad \dots,\quad \mathbf{C}_{33\,6\times1} = \{C^{(1)}_{33}\ C^{(2)}_{33}\ \dots\ C^{(6)}_{33}\}.$$

Assuming the quadratic variations of stiffness within the element we may express the matrix $\mathbf{C}$ at any point in terms of its values at the nodes as

$$\mathbf{C}_{3\times3} = \begin{bmatrix} \boldsymbol{\Phi}^{(2)T}_{1\times6}\mathbf{C}_{11\,6\times1} & \boldsymbol{\Phi}^{(2)T}_{1\times6}\mathbf{C}_{12\,6\times1} & \boldsymbol{\Phi}^{(2)T}_{1\times6}\mathbf{C}_{13\,6\times1} \\ & \boldsymbol{\Phi}^{(2)T}_{1\times6}\mathbf{C}_{22\,6\times1} & \boldsymbol{\Phi}^{(2)T}_{1\times6}\mathbf{C}_{23\,6\times1} \\ \text{sym.} & & \boldsymbol{\Phi}^{(2)T}_{1\times6}\mathbf{C}_{33\,6\times1} \end{bmatrix}$$

$$= \begin{bmatrix} \boldsymbol{\Phi}^{(2)T}_{1\times6} & \mathbf{0}_{1\times6} & \mathbf{0}_{1\times6} \\ \mathbf{0}_{1\times6} & \boldsymbol{\Phi}^{(2)T}_{1\times6} & \mathbf{0}_{1\times6} \\ \mathbf{0}_{1\times6} & \mathbf{0}_{1\times6} & \boldsymbol{\Phi}^{(2)T}_{1\times6} \end{bmatrix}_{3\times18} \begin{bmatrix} \mathbf{C}_{11\,6\times1} & \mathbf{C}_{12\,6\times1} & \mathbf{C}_{13\,6\times1} \\ & \mathbf{C}_{22\,6\times1} & \mathbf{C}_{23\,6\times1} \\ \text{sym.} & & \mathbf{C}_{33\,6\times1} \end{bmatrix}_{18\times3}$$

$$= \hat{\boldsymbol{\Phi}}_{3\times18}(\gamma_1, \gamma_2, \gamma_3)\hat{\mathbf{C}}_{18\times3} \tag{4.196}$$

which allows to rewrite the element stiffness matrix $\mathbf{k}_{12\times12}$ as

$$\mathbf{k}_{12\times12} = \int_{\Omega} \mathbf{B}^T_{12\times3}\hat{\boldsymbol{\Phi}}_{3\times18}\hat{\mathbf{C}}_{18\times3}\mathbf{B}_{3\times12}\,d\Omega. \tag{4.197}$$

Clearly, for any given nodal values of the constitutive matrix $\mathbf{C}$ (i.e. for any given matrix (4.196), the stiffness (4.197) may be readily computed without any reference to numerical integration schemes.

The explicit form of the initial stress stiffness matrix can be obtained from definition $(3.75)_4$ which leads in the present case to the relationship

$$\tfrac{1}{2}\Delta\mathbf{q}^T_{1\times12}\mathbf{k}^{(\sigma)}_{12\times12}\Delta\mathbf{q}_{12\times1}$$

$$= \tfrac{1}{2}\int_{\Omega} T^{ij}\Delta u^k_{,i}\Delta u_{k,j}\,d\Omega, \quad i,j = \xi, \eta. \tag{4.198}$$

The right-hand side can be rewritten as

$$\frac{1}{2}\Delta\mathbf{q}^T \int_\Omega \left\{ T^{\xi\xi} \begin{bmatrix} \mathbf{\Phi}_\xi^{(2)}\mathbf{\Phi}_\xi^{(2)T} & \mathbf{0} \\ \mathbf{0} & \mathbf{\Phi}_\xi^{(2)}\mathbf{\Phi}_\xi^{(2)T} \end{bmatrix}_{12\times 12} + \right.$$

$$+T^{\eta\eta} \begin{bmatrix} \mathbf{\Phi}_\eta^{(2)}\mathbf{\Phi}_\eta^{(2)T} & \mathbf{0} \\ \mathbf{0} & \mathbf{\Phi}_\eta^{(2)}\mathbf{\Phi}_\eta^{(2)T} \end{bmatrix}_{12\times 12} +$$

$$\left. +T^{\xi\eta} \begin{bmatrix} \mathbf{\Phi}_\xi^{(2)}\mathbf{\Phi}_\eta^{(2)T} & \mathbf{0} \\ \mathbf{0} & \mathbf{\Phi}_\xi^{(2)}\mathbf{\Phi}_\eta^{(2)T} \end{bmatrix}_{12\times 12} \right\} \mathrm{d}\Omega\Delta\mathbf{q}$$

since clearly

$$\Delta u^k_{,i}\Delta u_{k,j} = \Delta\mathbf{q}^T \begin{bmatrix} \mathbf{\Phi}_i^{(2)}\mathbf{\Phi}_j^{(2)T} & \mathbf{0} \\ \mathbf{0} & \mathbf{\Phi}_i^{(2)}\mathbf{\Phi}_j^{(2)T} \end{bmatrix} \Delta\mathbf{q}; \qquad i, j = \xi, \eta,$$

the derivatives $\mathbf{\Phi}_\xi^{(2)}$ and $\mathbf{\Phi}_\eta^{(2)}$ being defined in eq. (4.194). Expressing the elemental stresses $T^{\xi\xi}$, $T^{\eta\eta}$, $T^{\xi\eta}$ at a point in terms of the stresses at each of the six nodal points we have

$$T^{\xi\xi} = \mathbf{\Phi}_{1\times 6}^{(2)T}\boldsymbol{\sigma}_{\xi\xi_{6\times 1}},$$

$$T^{\eta\eta} = \mathbf{\Phi}_{1\times 6}^{(2)T}\boldsymbol{\sigma}_{\eta\eta_{6\times 1}},$$

$$T^{\xi\eta} = \mathbf{\Phi}_{1\times 6}^{(2)T}\boldsymbol{\sigma}_{\xi\eta_{6\times 1}},$$

$$\boldsymbol{\sigma}_{\xi\xi_{6\times 1}} = \{\sigma_{\xi\xi}^{(1)} \ \sigma_{\xi\xi}^{(2)} \ \dots \ \sigma_{\xi\xi}^{(6)}\},$$

$$\boldsymbol{\sigma}_{\eta\eta_{6\times 1}} = \{\sigma_{\eta\eta}^{(1)} \ \sigma_{\eta\eta}^{(2)} \ \dots \ \sigma_{\eta\eta}^{(6)}\},$$

$$\boldsymbol{\sigma}_{\xi\eta_{6\times 1}} = \{\sigma_{\xi\eta}^{(1)} \ \sigma_{\xi\eta}^{(2)} \ \dots \ \sigma_{\xi\eta}^{(6)}\}$$

and eq. (4.198) may be transformed to yield

$$\mathbf{k}_{12\times 12}^{(\sigma)} = \int_\Omega \left\{ \mathbf{\Phi}_{1\times 6}^{(2)T}\boldsymbol{\sigma}_{\xi\xi 6\times 1} \begin{bmatrix} \mathbf{\Phi}_\xi^{(2)}\mathbf{\Phi}_\xi^{(2)T} & \mathbf{0} \\ \mathbf{0} & \mathbf{\Phi}_\xi^{(2)}\mathbf{\Phi}_\xi^{(2)T} \end{bmatrix}_{12\times 12} + \right.$$

$$+\mathbf{\Phi}_{1\times 6}^{(2)T}\boldsymbol{\sigma}_{\eta\eta 6\times 1} \begin{bmatrix} \mathbf{\Phi}_\eta^{(2)}\mathbf{\Phi}_\eta^{(2)T} & \mathbf{0} \\ \mathbf{0} & \mathbf{\Phi}_\eta^{(2)}\mathbf{\Phi}_\eta^{(2)T} \end{bmatrix}_{12\times 12} +$$

$$\left. +\mathbf{\Phi}_{1\times 6}^{(2)T}\boldsymbol{\sigma}_{\xi\eta 6\times 1} \begin{bmatrix} \mathbf{\Phi}_\xi^{(2)}\mathbf{\Phi}_\eta^{(2)T} & \mathbf{0} \\ \mathbf{0} & \mathbf{\Phi}_\xi^{(2)}\mathbf{\Phi}_\eta^{(2)T} \end{bmatrix}_{12\times 12} \right\} \mathrm{d}\Omega. \tag{4.199}$$

The last relation can further be transformed to finally give

$$\mathbf{k}_{12\times 12}^{(\sigma)} = \boldsymbol{\sigma}_{\xi\xi}\mathbf{k}_{\xi\xi}^{(\sigma)} + \boldsymbol{\sigma}_{\eta\eta}\mathbf{k}_{\eta\eta}^{(\sigma)} + \boldsymbol{\sigma}_{\xi\eta}\mathbf{k}_{\xi\eta}^{(\sigma)} \tag{4.200}$$

with the matrices $\mathbf{k}_{\xi\xi}^{(\sigma)}$, $\mathbf{k}_{\eta\eta}^{(\sigma)}$ and $\mathbf{k}_{\xi\eta}^{(\sigma)}$ following directly by comparing eqs. (4.199) and (4.200). These matrices can easily be computed using Table 2. The sum of the so-obtained matrices $\mathbf{k}_{12\times 12}$ and $\mathbf{k}_{12\times 12}^{(\sigma)}$ allows to determine the total stiffness of the element at the given time instant—examples of application of such a formulation will be given in Chapter 7.

4.10 Constitutive matrices for thermo-elastic-plastic materials

In Section 2.1.4 we presented the general theory of thermo-elastic-plastic materials as formally defined within the framework of nonlinear solid mechanics. Such a general form will now be briefly reconsidered and specified to facilitate its use for solving boundary-value problems by the discretized approach. For simplicity of notation small deformation are considered only. The way the kinematically nonlinear effects may be additionally incorporated is indicated at the end of this section. The dependence of all functions to be considered on the variable $\mathbf{z}$ (i.e. points of the region Ω) is omitted for the sake of compactness.

The constitutive equation of linear elasticity is recalled as

$$\boldsymbol{\sigma}_{6\times1} = \mathbf{C}_{6\times6}(\boldsymbol{\epsilon}_{6\times1} - \boldsymbol{\epsilon}^{(\vartheta)}_{6\times1}) \tag{4.201}$$

in which the 6×6 constitutive matrix $\mathbf{C}$ was specified in Sec. 2.1.4 for the case of isotropic elasticity. The thermal strain $\boldsymbol{\epsilon}^{(\vartheta)}$ is defined as, cf. eq. (2.4)

$$\boldsymbol{\epsilon}^{(\vartheta)} = \mathbf{a}(\vartheta)\vartheta$$

so that

$$\dot{\boldsymbol{\epsilon}}^{(\vartheta)} = \left[\mathbf{a}(\vartheta) + \frac{\mathrm{d}\mathbf{a}}{\mathrm{d}\vartheta}\,\vartheta\right]\dot{\vartheta} = \bar{\mathbf{a}}(\vartheta)\,\dot{\vartheta}$$

where $\bar{\mathbf{a}}(\vartheta)$ is the instantaneous thermal expansion vector. Denoting, as before, the plastic strain vector by $\boldsymbol{\epsilon}^{(pl)}$, introducing additionally the vector of creep strain $\boldsymbol{\epsilon}^{(c)}$ (to be specified later in Chapter 7) and postulating additivity of all specific strain rate contributions we get from eq. (4.201)

$$\dot{\boldsymbol{\sigma}} = \mathbf{C}(\dot{\boldsymbol{\epsilon}} - \dot{\boldsymbol{\epsilon}}^{(pl)} - \dot{\boldsymbol{\epsilon}}^{(c)} - \bar{\mathbf{a}}\dot{\vartheta}) + \overset{*}{\boldsymbol{\sigma}} \tag{4.202}$$

in which

$$\overset{*}{\boldsymbol{\sigma}} = \dot{\mathbf{C}}(\boldsymbol{\epsilon} - \boldsymbol{\epsilon}^{(pl)} - \boldsymbol{\epsilon}^{(c)} - \mathbf{a}\vartheta) \tag{4.203}$$

is the contribution to the stress rate due to the change in temperature dependent elastic moduli.

To start with we shall consider the von Mises yield condition and the isotropic hardening concept which result in postulating that

$$f = F(\boldsymbol{\sigma}) - \sigma_y(\bar{\epsilon}^{(pl)}, \vartheta) = \sqrt{\tfrac{3}{2}\boldsymbol{\sigma}^{D^T}\boldsymbol{\sigma}^D} - \sigma_y = \bar{\sigma} - \sigma_y = 0 \tag{4.204}$$

where the tensile yield stress σ_y is a known function of its two arguments, $\bar{\sigma}$ is the stress intensity taken as the positive square root of $\sqrt{\tfrac{3}{2}\boldsymbol{\sigma}^{D^T}\boldsymbol{\sigma}^D}$, $\boldsymbol{\sigma}^D$ is the deviatoric stress vector given by

$$\boldsymbol{\sigma}^D = \boldsymbol{\sigma} - \tfrac{1}{3}(\sigma_{\xi\xi} + \sigma_{\eta\eta} + \sigma_{\zeta\zeta})\{1\ \ 1\ \ 1\ \ 0\ \ 0\ \ 0\}$$

and $\bar{\boldsymbol{\varepsilon}}^{(pl)}$ is the effective plastic strain. The associated plastic flow rule is postulated as

$$\dot{\boldsymbol{\varepsilon}}^{(pl)} = \dot{\lambda}\frac{\partial f}{\partial \boldsymbol{\sigma}^T} = \dot{\lambda}\frac{3}{2}\frac{\boldsymbol{\sigma}^D}{\sigma_y} = \dot{\lambda}\sqrt{\tfrac{3}{2}}\mathbf{n} = \dot{\bar{\varepsilon}}^{(pl)}\sqrt{\tfrac{3}{2}}\mathbf{n} \tag{4.205}$$

where we accept the convention that the derivative of a function with respect to a column vector is a row vector and $\mathbf{n} = \sqrt{\tfrac{3}{2}}\dfrac{\boldsymbol{\sigma}^D}{\sigma_y}$ is the unit normal to the yield surface (4.204) in the stress space while $\dot{\bar{\varepsilon}}^{(pl)} = \sqrt{\tfrac{2}{3}\dot{\varepsilon}^{(pl)T}\dot{\varepsilon}^{(pl)}} = \dot{\lambda}$ is the plastic strain rate intensity (or rate of effective plastic strain). The consistency condition which assures that the stress point remains on the yield surface during plastic flow reads

$$\dot{f} = \frac{\partial F}{\partial \boldsymbol{\sigma}}\dot{\boldsymbol{\sigma}} - \left.\frac{\partial \sigma_y}{\partial \bar{\varepsilon}^{(pl)}}\right|_{\vartheta=\text{const}}\dot{\bar{\varepsilon}}^{(pl)} - \left.\frac{\partial \sigma_y}{\partial \vartheta}\right|_{\bar{\varepsilon}^{(pl)}=\text{const}}\dot{\vartheta} = 0. \tag{4.206}$$

The isothermic hardening parameter $\zeta = \zeta(\bar{\varepsilon}^{(pl)}, \vartheta)$ is defined as

$$\zeta = \left.\frac{\partial \bar{\sigma}}{\partial \bar{\varepsilon}^{(pl)}}\right|_{\vartheta=\text{const}}. \tag{4.207}$$

By using the identity $\dfrac{\partial F}{\partial \boldsymbol{\sigma}^T} = \dfrac{3}{2}\dfrac{\boldsymbol{\sigma}^D}{\sigma_y}$ eq. (4.206) may be transformed to give

$$\dot{\bar{\sigma}} - \zeta\dot{\bar{\varepsilon}}^{(pl)} - \frac{\partial \bar{\sigma}}{\partial \vartheta}\dot{\vartheta} = 0$$

or

$$\zeta = \frac{\dot{\bar{\sigma}}}{\dot{\bar{\varepsilon}}^{(pl)}} - \frac{\partial \bar{\sigma}}{\partial \vartheta}\frac{\dot{\vartheta}}{\dot{\bar{\varepsilon}}^{(pl)}} = \frac{d\bar{\sigma}}{d\bar{\varepsilon}^{(pl)}} - \frac{\partial \bar{\sigma}}{\partial \vartheta}\frac{d\vartheta}{d\bar{\varepsilon}^{(pl)}}.$$

Defining another parameter

$$\zeta^* = \frac{d\bar{\sigma}}{d\bar{\varepsilon}^{(pl)}} \tag{4.208}$$

we arrive at

$$\zeta^* = \zeta + \frac{\partial \bar{\sigma}}{\partial \vartheta}\frac{d\vartheta}{d\bar{\varepsilon}^{(pl)}}. \tag{4.209}$$

It is often convenient to use the hardening moduli defined as

$$h = \tfrac{2}{3}\zeta, \quad h^* = \tfrac{2}{3}\zeta^* \tag{4.210}$$

and then the flow rule (4.205) may be transformed to the form

$$\dot{\boldsymbol{\varepsilon}}^{(pl)} = \frac{1}{h^*}(\mathbf{n}^T\dot{\boldsymbol{\sigma}})\mathbf{n}. \tag{4.211}$$

Noting that

$$h^* = h + \frac{2}{3}\frac{\partial\bar{\sigma}}{\partial\vartheta}\frac{d\theta}{d\bar{\varepsilon}^{(pl)}}$$

eq. (4.211) becomes

$$^{(pl)} = \frac{1}{h}(\mathbf{n}^T\dot{\boldsymbol{\sigma}})\mathbf{n} - \sqrt{\tfrac{2}{3}}\frac{1}{h}\frac{\partial\bar{\sigma}}{\partial\vartheta}\dot{\vartheta}\mathbf{n} \tag{4.212}$$

which may turn out more useful in computational practice than eq. (4.211). The first part of eq. (4.212) corresponds to isothermic process ($\dot{\vartheta} = 0$) whereas the second part is the correction due to the thermal effects. Eq. (4.212) is here considered as the final form of the plastic flow rule. By substituting eq. (4.212) into eq. (4.202) we obtain

$$\dot{\boldsymbol{\sigma}} = \mathbf{C}\left[\dot{\boldsymbol{\varepsilon}} - \frac{1}{h}(\mathbf{n}^T\dot{\boldsymbol{\sigma}})\mathbf{n} + \sqrt{\tfrac{2}{3}}\frac{1}{h}\frac{\partial\bar{\sigma}}{\partial\vartheta}\dot{\vartheta}\mathbf{n} - \dot{\boldsymbol{\varepsilon}}^{(c)} - \bar{\mathbf{a}}\dot{\vartheta}\right] + \overset{*}{\boldsymbol{\sigma}}. \tag{4.213}$$

Solving eq. (4.213) for the stress rate $\dot{\boldsymbol{\sigma}}$ results in the relationship

$$\dot{\boldsymbol{\sigma}} = \left[\mathbf{C} - \frac{\mathbf{C}\mathbf{n}\mathbf{n}^T\mathbf{C}}{h + \mathbf{n}^T\mathbf{C}\mathbf{n}}\right](\dot{\boldsymbol{\varepsilon}} - \dot{\boldsymbol{\varepsilon}}^{(c)} - \bar{\mathbf{a}}\dot{\vartheta}) + \overset{*}{\boldsymbol{\sigma}} + \overset{**}{\boldsymbol{\sigma}} \tag{4.214}$$

where the vector $\overset{*}{\boldsymbol{\sigma}}$ has been defined in eq. (4.203) while

$$\overset{**}{\boldsymbol{\sigma}} = \mathbf{C}\mathbf{n}\left[\frac{-\mathbf{n}^T\mathbf{C}\mathbf{n}\sqrt{\tfrac{2}{3}}\frac{1}{h}\frac{\partial\bar{\sigma}}{\partial\vartheta}\dot{\vartheta} - \overset{*}{\boldsymbol{\sigma}}{}^T\mathbf{n}}{h + \mathbf{n}^T\mathbf{C}\mathbf{n}} + \sqrt{\tfrac{2}{3}}\frac{1}{h}\frac{\partial\bar{\sigma}}{\partial\vartheta}\dot{\vartheta}\right]. \tag{4.215}$$

Using the explicit form of the elastic constitutive matrix eqs. (4.213), (4.214) and (4.215) simplify to the form

$$\dot{\boldsymbol{\sigma}} = \left[\mathbf{C} - \delta\frac{4G^2\mathbf{n}\mathbf{n}^T}{h + 2G}\right](\dot{\boldsymbol{\varepsilon}} - \dot{\boldsymbol{\varepsilon}}^{(c)} - \bar{\mathbf{a}}\dot{\vartheta}) + \overset{*}{\boldsymbol{\sigma}} + \overset{**}{\boldsymbol{\sigma}},$$

$$\overset{*}{\boldsymbol{\sigma}} = \dot{\mathbf{C}}(\boldsymbol{\varepsilon} - \boldsymbol{\varepsilon}^{(pl)} - \boldsymbol{\varepsilon}^{(c)} - \mathbf{a}\vartheta), \tag{4.216}$$

$$\overset{**}{\boldsymbol{\sigma}} = 2G\mathbf{n}\frac{\sqrt{\tfrac{2}{3}}\frac{\partial\bar{\sigma}}{\partial\vartheta}\dot{\vartheta} - \overset{*}{\boldsymbol{\sigma}}{}^T\mathbf{n}}{h + 2G}$$

where δ is the loading/unloading parameter to be defined below, cf. also Sec. 2.1.4.

Now, in order to generalize the above results to the case of a mixed isotropic/kinematic hardening we consider the following yield condition

$$\mathscr{F} = F(\boldsymbol{\sigma}, \boldsymbol{\alpha}) - \sigma_y(\bar{\boldsymbol{\varepsilon}}^{(pl)}, \vartheta) = \sqrt{\tfrac{3}{2}\bar{\boldsymbol{\sigma}}^{D^T}\bar{\boldsymbol{\sigma}}^D} - \sigma_y = \bar{\sigma} - \sigma_r = 0 \tag{4.217}$$

where now the stress intensity $\bar{\sigma}$ is formed from the components of the vector $\bar{\boldsymbol{\sigma}}^D$ which is the deviatoric vector corresponding to the "shifted" stress vector

$$\bar{\boldsymbol{\sigma}} = \boldsymbol{\sigma} - \boldsymbol{\alpha}, \tag{4.218}$$

by the so-called back stress $\boldsymbol{\alpha}$ representing the translation of the yield surface in stress space. Introducing the concept of a mixed hardening by considering the stress rate as being composed of two parts corresponding to the isotropic and kinematic hardening, respectively

$$\begin{aligned} &\dot{\boldsymbol{\sigma}} = \dot{\boldsymbol{\sigma}}^{(i)} + \dot{\boldsymbol{\sigma}}^{(k)}, \\ &\dot{\boldsymbol{\sigma}}^{(i)} = \omega\dot{\boldsymbol{\sigma}}, \quad \dot{\boldsymbol{\sigma}}^{(k)} = (1-\omega)\dot{\boldsymbol{\sigma}}, \quad \omega \in [0, 1], \\ &\sigma_r = \omega\sigma_y + (1-\omega)\sigma_y^0, \end{aligned} \tag{4.219}$$

ω being the material parameter which determines the proportion of each particular hardening type, we employ the consistency condition $\dot{f} = 0$ in the form of the two following equations

$$\begin{aligned} &\frac{\partial f}{\partial \boldsymbol{\sigma}} \dot{\boldsymbol{\sigma}}^{(k)} - \frac{\partial f}{\partial \boldsymbol{\alpha}} \dot{\boldsymbol{\alpha}} = 0, \\ &\frac{\partial f}{\partial \boldsymbol{\sigma}} \dot{\boldsymbol{\sigma}}^{(i)} - \frac{\partial \bar{\sigma}}{\partial \bar{\varepsilon}^{(pl)}} \dot{\bar{\varepsilon}}^{(pl)} - \frac{\partial \bar{\sigma}}{\partial \vartheta} \dot{\vartheta} = 0. \end{aligned}$$

Using eq. (4.219) we have

$$\begin{aligned} &\frac{\partial f}{\partial \boldsymbol{\sigma}} \dot{\boldsymbol{\sigma}}(1-\omega) - \frac{\partial f}{\partial \boldsymbol{\alpha}} \dot{\boldsymbol{\alpha}} = 0, \\ &\frac{\partial f}{\partial \boldsymbol{\sigma}} \dot{\boldsymbol{\sigma}}\omega - \frac{\partial \bar{\sigma}}{\partial \bar{\varepsilon}^{(pl)}} \dot{\bar{\varepsilon}}^{(pl)} - \frac{\partial \bar{\sigma}}{\partial \vartheta} \dot{\vartheta} = 0. \end{aligned}$$

The second of the above equations leads to the evolution equation for the yield limit as

$$\dot{\sigma}_r = \omega\zeta^* \dot{\bar{\varepsilon}}^{(pl)} \tag{4.220}$$

whereas the first equation leads similarly to the evolution for the back stress $\boldsymbol{\alpha}$ as

$$\dot{\boldsymbol{\alpha}} = (1-\omega)h^* \dot{\boldsymbol{\varepsilon}}^{(pl)}. \tag{4.221}$$

Replacing in eqs. (4.216) the stresses $\boldsymbol{\sigma}$ by $\boldsymbol{\sigma} - \boldsymbol{\alpha}$, together with eqs. (4.220), (4.221) completes the simple theory of associated non-isothermal plastic flow with isotropic/kinematic hardening provided appropriate loading/unloading conditions are employed. These may be taken as

$$\begin{aligned} &\text{elastic process:} \quad \bar{\sigma} < \sigma_r \text{ or } \bar{\sigma} = \sigma_r \text{ and } \mathbf{n}^T\dot{\boldsymbol{\sigma}} - \sqrt{\tfrac{2}{3}}\frac{\partial \sigma_r}{\partial \vartheta}\dot{\vartheta} \leqslant 0, \\ &\text{inelastic process:} \quad \bar{\sigma} = \sigma_r \text{ and } \mathbf{n}^T\dot{\boldsymbol{\sigma}} - \sqrt{\tfrac{2}{3}}\frac{\partial \sigma_r}{\partial \vartheta}\dot{\vartheta} > 0. \end{aligned} \tag{4.222}$$

The expression $\mathbf{n}^T\dot{\boldsymbol{\sigma}} - \sqrt{\frac{2}{3}}\frac{\partial \sigma_r}{\partial \vartheta}\dot{\vartheta}$ may be conveniently redefined by employing the so-called trial rate of stress given by

$$\dot{\boldsymbol{\sigma}}^{(tr)} = \mathbf{C}(\dot{\boldsymbol{\epsilon}} - \bar{\mathbf{a}}\dot{\vartheta}). \tag{4.223}$$

Using the relation

$$\mathbf{C}(\dot{\boldsymbol{\epsilon}} - \bar{\mathbf{a}}\dot{\vartheta}) = \dot{\boldsymbol{\sigma}} + \frac{\mathbf{Cn}}{h}(\mathbf{n}^T\dot{\boldsymbol{\sigma}}) - \sqrt{\frac{2}{3}}\frac{\mathbf{Cn}}{h}\frac{\partial \sigma_r}{\partial \vartheta}\dot{\vartheta}$$

which, when left-multiplied by $\mathbf{C}^{-1}$, expresses simply the additivity of strain rate contributions, we obtain

$$\mathbf{n}^T\dot{\boldsymbol{\sigma}} = \mathbf{n}^T\dot{\boldsymbol{\sigma}}^{(tr)}\frac{h}{h+\mathbf{n}^T\mathbf{Cn}} + \sqrt{\frac{2}{3}}\frac{\partial \sigma_r}{\partial \vartheta}\dot{\vartheta}\,\frac{\mathbf{n}^T\mathbf{Cn}}{h+\mathbf{n}^T\mathbf{Cn}}$$

or

$$\mathbf{n}^T\dot{\boldsymbol{\sigma}} - \sqrt{\frac{2}{3}}\frac{\partial \sigma_r}{\partial \vartheta}\dot{\vartheta} = \frac{h}{h+\mathbf{n}^T\mathbf{Cn}}\left[\mathbf{n}^T\dot{\boldsymbol{\sigma}}^{(tr)} - \sqrt{\frac{2}{3}}\frac{\partial \sigma_r}{\partial \vartheta}\dot{\vartheta}\right].$$

It turns out algorithmically more consistent to define the local plastic loading as, cf. eg. $(4.222)_2$

$$\mathbf{n}^T\dot{\boldsymbol{\sigma}}^{(tr)} - \sqrt{\frac{2}{3}}\frac{\partial \sigma_r}{\partial \vartheta}\dot{\vartheta} > 0 \quad \text{for} \quad \bar{\sigma} = \sigma_r. \tag{4.224}$$

The reason for taking the definition of plastic process in the form (4.224) is that the theory becomes directly applicable to softening and no attention needs to be paid to the sign of the actual hardening parameters. We also note that for $h^* < 0$ and $\omega > 0$ a stopping criterion needs to be introduced to ensure that the actual yield limit does not become negative.

The specific features of the numerical algorithm for dealing with inelastic problems are discussed in Chapter 5 and later when presenting exemplary solutions to specific boundary-value problems.

4.11 On inelastic analysis under non-proportionally varying loads

4.11.1 *Elastic-plastic structures subjected to mechanical loads*

Usually, some components of external agents acting upon engineering structures do not increase proportionally to each other in a monotone way. In the majority of cases they vary between certain given limits rather than increase proportionally to one loading parameter. This clearly makes the conventional elastic-plastic one-parameter-load analysis incomplete and definitely calls for additional considerations regarding the low-cycle fatigue due to alternating plastic strains and the possibility of an unlimited growth of permanent deformations leading to the so-called incremental collapse. A given structure is safe against both these effects if a condition known as shakedown, or adaptation condition is satisfied, [94], [96]. This requirement states that for structural shakedown the plastic energy dissipated in an arbitrary process contained within the preset limits should be bounded. The detailed presentation of the shakedown theory is clearly beyond the scope of this book. The more so that from the algorithmic viewpoint the conventional methods of shakedown analysis have not much in common with the incremental philosophy consequently followed in the book. However, the progress in developing the step-by-step numerical algorithms has prompted researchers to seek a reformulation of the fundamental shakedown theorems in such a way that the existing elastic-plastic computer codes turned useful. The first look at the problem seems to be extremely gloomy—to investigate the structural shakedown by means of incremental calculations seems to require checking an infinite number of load paths. The way out of this problem is offered by the following considerations. Therein, we shall consider the kinematically linearized problems only.

In most practical cases loads acting upon a structure can be described with a sufficient accuracy by assuming that they depend linearly on a finite number, say r, of load factors β_p such that

$$\mathbf{s}(\mathbf{z}, t) = \sum_{p=1}^{r} \beta_p(t)\mathbf{s}^{(p)}(\mathbf{z}), \quad \mathbf{z} \in \Omega. \tag{4.225}$$

Here, $\mathbf{s}$ stands for the total instantaneous load, $\mathbf{s}^{(p)}$ is a reference, constant (in time) load in the p-th load mode and β_p denotes the p-th load factor (or load intensity). Body forces are exluded from the considerations for simplicity but may also be presented in the form (4.225).

The external loading defined by eq. (4.225) is referred to as the r-parameter loading with $\beta_p(t)$, $p = 1, 2, \ldots, r$ being the independently varied parameters. For $r = 1$ we have clearly

$$\mathbf{s}(\mathbf{z}, t) = \beta(t)\mathbf{s}^{(1)}(\mathbf{z})$$

which is a typical one-parameter load representation.

Limits of the load variations are defined by a certain domain Γ in the r-dimensional space B of parameters β_p, $p = 1, 2, \ldots, r$. If there is no correlation between different load modes, the domain Γ becomes a r-dimensional parallelepiped

$$\beta_p^- \leqslant \beta_p \leqslant \beta_p^+, \quad p = 1, 2, \ldots, r, \tag{4.226}$$

β_p^-, β_p^+ being given constants defined by design codes or by specific structural service conditions.

We assume without any loss of generality that the domain Γ is convex and can be approximated with any desired accuracy by an r-dimensional polyhedron with the vertices given as radius vectors $\boldsymbol{\gamma}_j$, $j = 1, 2, \ldots, t$; cf. Fig. 33 for $r = 2$ and $t = 4$.

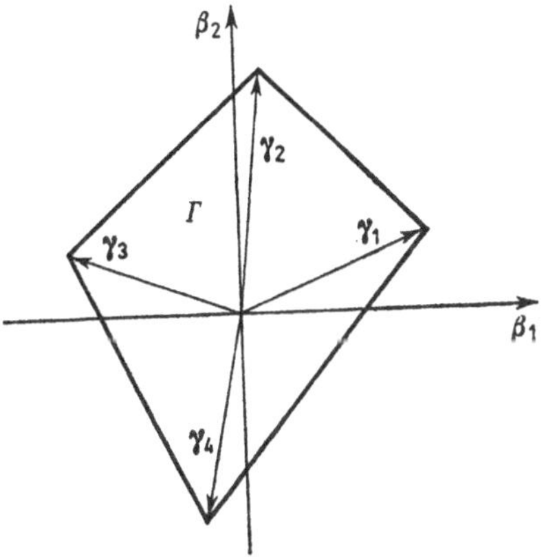

Fig. 33 Load domain

Assume that the structure is elastic-perfectly plastic and the loadings are such that the linearity of kinematic relations is assured. We have then in accordance with the derivation of Chapter 2 the following set of relations to hold:

$\boldsymbol{\epsilon} = \boldsymbol{\epsilon}^{(el)} + \boldsymbol{\epsilon}^{(pl)}$ — decomposition of total strain into elastic and plastic parts,

$f = F(\boldsymbol{\sigma}) - \sigma_y = 0$ — yield condition,

$\boldsymbol{\epsilon}^{(el)} = \overset{-1}{\mathbf{C}}\boldsymbol{\sigma}$ — Hooke's law,

$\dot{\boldsymbol{\epsilon}}^{(pl)} = \dot{\lambda}\dfrac{\partial f}{\partial \boldsymbol{\sigma}^T}$ — associated flow rule,

$$\left.\begin{aligned} &\dot{\lambda} \geqslant 0 \quad \text{if} \quad f = 0 \text{ and } \frac{\partial f}{\partial \boldsymbol{\sigma}}\dot{\boldsymbol{\sigma}} = 0 \\ &\dot{\lambda} = 0 \quad \text{if} \quad f < 0 \text{ or } \frac{\partial f}{\partial \boldsymbol{\sigma}}\dot{\boldsymbol{\sigma}} < 0 \end{aligned}\right\} \text{— loading/unloading conditions.}$$

If the domain $F(\boldsymbol{\sigma}) - \sigma_y \leqslant 0$ is convex (which is here assumed), then the yield function f may be defined as to satisfy the following relations:

$$\begin{aligned} &f(\boldsymbol{\sigma}) \geqslant 0, \quad f(0) = 0, \\ &f(\boldsymbol{\sigma}_1 + \boldsymbol{\sigma}_2) \leqslant f(\boldsymbol{\sigma}_1) + f(\boldsymbol{\sigma}_2), \\ &f(\mu\boldsymbol{\sigma}) > f(\boldsymbol{\sigma}), \quad f(\mu\boldsymbol{\sigma}) < |\mu| f(\boldsymbol{\sigma}) \quad \text{if} \quad |\mu| > 1, \end{aligned} \tag{4.227}$$

$$f[\mu\boldsymbol{\sigma}_1+(1-\mu)\boldsymbol{\sigma}_2] \leqslant \sigma_y \quad \text{if} \quad 0 \leqslant \mu \leqslant 1 \text{ provided}$$
$$f(\boldsymbol{\sigma}_1) \leqslant \sigma_y \text{ and} f(\boldsymbol{\sigma}_2) \leqslant \sigma_y . \tag{4.227}$$

Any stress state $\boldsymbol{\sigma}$ appearing in elastic-plastic structure may always be uniquely presented as the sum of stresses which would occur if the structure was perfectly elastic and of residual stresses $\boldsymbol{\rho}$:

$$\boldsymbol{\sigma} = \boldsymbol{\sigma}^{(el)}+\boldsymbol{\rho}. \tag{4.228}$$

We assume that the elastic stress field $\boldsymbol{\sigma}^{(el)}$ is not singular, i.e. all its stress components are finite within the region considered. It follows from theorems of linear elasticity that the stress fields $\boldsymbol{\sigma}^{(el)}$ and $\boldsymbol{\rho}$ are linear functionals of the actual load parameters β_p, $p = 1, 2, \ldots, r$ and of the actual plastic strain field, respectively

$$\boldsymbol{\sigma}^{(el)} = \sum_{p=1}^{r} \beta_p \boldsymbol{\sigma}^{(el)(p)},$$
$$\boldsymbol{\rho} = \mathbf{L}[\boldsymbol{\epsilon}^{(pl)}], \tag{4.229}$$

$\boldsymbol{\sigma}^{(el)(p)}$ denoting the elastic stress due to the reference load in its p-th mode, cf. eq. (4.225).

All the elastic stress fields for a given structure subjected to the r-parameter loading (4.225) constitute a r-dimensional linear vector space M. Those of them which belong to the domain defined by eqs. (4.226) constitute a certain domain Σ contained in M. According to eq. (4.229) and due to the elastic stress boundness assumed, the transformation φ: $B \to M$ is linear although, in general, not reversible. Thus, as any point $\boldsymbol{\gamma} \in \Gamma$ can be presented (not uniquely) as the linear combination of the vertices $\boldsymbol{\gamma}_1, \boldsymbol{\gamma}_2, \ldots, \boldsymbol{\gamma}_t$ of the domain Γ:

$$\boldsymbol{\gamma} = \lambda_1\boldsymbol{\gamma}_1+\lambda_2\boldsymbol{\gamma}_2+ \ldots +\lambda_t\boldsymbol{\gamma}_t,$$
$$\lambda_j \geqslant 0, \quad j = 1, 2, \ldots, t; \quad \lambda_1+\lambda_2+ \ldots +\lambda_t = 1$$

then also every element $\boldsymbol{\xi} \in \Sigma$ can be presented as follows:

$$\boldsymbol{\xi} = \lambda_1\boldsymbol{\xi}_1+\lambda_2\boldsymbol{\xi}_2+ \ldots +\lambda_t\boldsymbol{\xi}_t$$

where

$$\boldsymbol{\xi}_j = \varphi(\boldsymbol{\gamma}_j), \quad j = 1, 2, \ldots, t.$$

The linearity of the transformation φ implies also that the domain $\Sigma \subset M$ is convex and finite and that the boundary $\partial\Gamma$ of the load domain Γ transforms onto the boundary $\partial\Sigma$ of the domain Σ. It is also easy to show that for every $\boldsymbol{\xi} \in \Sigma$ there exists at least one pair of boundary elements $\boldsymbol{\xi}', \boldsymbol{\xi}'' \in \partial\Sigma$ and a number λ such that

$$0 \leqslant \lambda \leqslant 1, \quad \boldsymbol{\xi} = \lambda\boldsymbol{\xi}' + (1-\lambda)\boldsymbol{\xi}''. \tag{4.230}$$

Within the limits of validity of the assumptions presented above the following fundamental shakedown theorem holds, [94], [96]:

A given structure shakes down in an arbitrary loading path contained in a given load domain Γ if and only if there exists a time independent residual stress field $\boldsymbol{\rho}(\mathbf{z})$ such that the condition

$$f[\boldsymbol{\sigma}^{(el)}(\mathbf{z}, t) + \boldsymbol{\rho}(\mathbf{z})] \leqslant \sigma_y \tag{4.231}$$

holds for all the fields $\boldsymbol{\sigma}^{(el)}(\mathbf{z}, t) \in \Sigma$.

The linearity of the transformation φ, the convexity of the domain Σ and formulae $(4.227)_{1,4}$ imply that if condition (4.231) is satisfied for all the stress fields $\boldsymbol{\sigma}^{(el)} \in \partial\Sigma$ then it also holds for all the fields $\boldsymbol{\sigma}^{(el)} \in \Sigma$. For if condition (4.231) holds for the pair $\boldsymbol{\xi}', \boldsymbol{\xi}'' \in \partial\Sigma$ of a stress field $\boldsymbol{\sigma}^{(el)} \in \Sigma$ considered, then according to (4.230) and $(4.227)_4$ it holds also for $\boldsymbol{\sigma}^{(el)}$. In other words:

THEOREM A. *If a given structure shakes down in any load path which is on the boundary of a given load domain, then it also shakes down in an arbitrary load path contained within that domain.*

Let us now introduce the notion of a cyclic loading process. By such a process we mean a time and space distribution of the external load given by eq. (4.225) in which all the load factors $\beta_p(t)$, $p = 1, 2, \ldots, r$, are explicitly given periodic functions of time with a common period T:

$$\beta_p(t+T) = \beta_p(t), \quad p = 1, 2, \ldots, r. \tag{4.232}$$

For a given cyclic load one can construct in the r-dimensional stress field space M an envelope Σ' of all the elastic stress fields associated with this load. (The domain Σ' will not in general be convex.) The natural question to be asked now is: Does shakedown in an explicitly given cyclic loading process (4.232) imply shakedown in the general case (i.e. for arbitrary loading varying within the domain) as well, provided the envelope Σ' contains the domain Σ associated with the load domain Γ. The answer to this question is of great practical importance as all experimental investigations are performed for cyclic loadings; it is thus important to know whether the results are applicable for other load histories as well. The affirmative answer would also increase the significance of numerical computations performed for cyclic loads. The answer is offered by the following three theorems:

THEOREM B. *If a given structure shakes down in a prescribed cyclic loading process then it also shakes down in any process for which the envelope Σ of elastic stress fields is contained within (or coincides with) the envelope Σ' relative to the given cyclic process.*

Proof: Due to condition (4.231), if the structure has shaken down there exists a time-independent residual stress field satisfying (4.231) for all $\boldsymbol{\sigma}^{(el)} \in \Sigma'$ and thus for all $\boldsymbol{\sigma}^{(el)} \in \Sigma$.

THEOREM C. *If a structure shakes down in a cyclic load process which covers the whole boundary $\partial\Gamma$ of a given load domain Γ, then it shakes down in any load path contained within the domain Γ.*

The proof follows trivially from Theorems A and B.

THEOREM D. *If a given structure shakes down in a cyclic loading process which contains all the vertices $\boldsymbol{\gamma}_j$, $j = 1, 2, \ldots, t$, of a given load domain Γ then it shakes down in an arbitrary loading path contained within the domain Γ.*

Proof: If the structure has shaken down in the given cyclic load path containing the vertices $\boldsymbol{\gamma}_j$ of Γ then there exists a residual stress field $\boldsymbol{\rho}'(\mathbf{x})$ which satisfies condition (4.231) for all the elastic stress fields $\boldsymbol{\xi}_1, \ldots, \boldsymbol{\xi}_t$ associated with the respective vertices $\boldsymbol{\gamma}_j$. Then, according to eq. $(4.227)_{3,4}$ for an arbitrary elastic stress field $\boldsymbol{\sigma}^{(el)} \in \Sigma$ the following relation holds provided $\lambda_1 + \lambda_2 + \ldots + \lambda_t = 1$, $\lambda_1 \geqslant 0, \ldots, \lambda_t \geqslant 0$:

$$\begin{aligned} F(\boldsymbol{\sigma}^{(el)} + \boldsymbol{\rho}') &= F(\lambda_1 \boldsymbol{\xi}_1 + \ldots + \lambda_t \boldsymbol{\xi}_t + \boldsymbol{\rho}') \\ &= F[\lambda_1 \boldsymbol{\xi}_1 + \ldots + \lambda_t \boldsymbol{\xi}_t + (\lambda_1 + \ldots + \lambda_t)\boldsymbol{\rho}'] \\ &= F[\lambda_1(\boldsymbol{\xi}_1 + \boldsymbol{\rho}') + \ldots + \lambda_t(\boldsymbol{\xi}_t + \boldsymbol{\rho}')] \\ &\leqslant \lambda_1 F(\boldsymbol{\xi}_1 + \boldsymbol{\rho}') + \lambda_2 F(\boldsymbol{\xi}_2 + \boldsymbol{\rho}') + \ldots + \lambda_t F(\boldsymbol{\xi}_t + \boldsymbol{\rho}') \\ &\leqslant (\lambda_1 + \lambda_2 + \ldots + \lambda_t)\sigma_y = \sigma_y . \end{aligned}$$

The inequality $F(\boldsymbol{\sigma}^{(el)} + \boldsymbol{\rho}') \leqslant \sigma_y$ implies shake down in arbitrary loading path contained within the domain Γ.

The above theorem provides a simple way to check whether a given structure is able to shakedown under an arbitrary loading path contained in the hyper-polyhedral load domain Γ. All that is needed is to reproduce incrementally a critical cycle of loading which connects the origin of the load space with all the subsequent vertices of Γ. In Fig. 34 the critical load path is defined as *O–A–O–B–O–C–O–D–O*. Several such cycles may turn out necessary to run on the computer with the plastic strain increments and the dissipated energy monitored. If the latter ceases to grow after a finite number of cycles then the structure has adapted itself to the critical cyclic load and hence it is able to shake down under any variable loading contained in Γ.

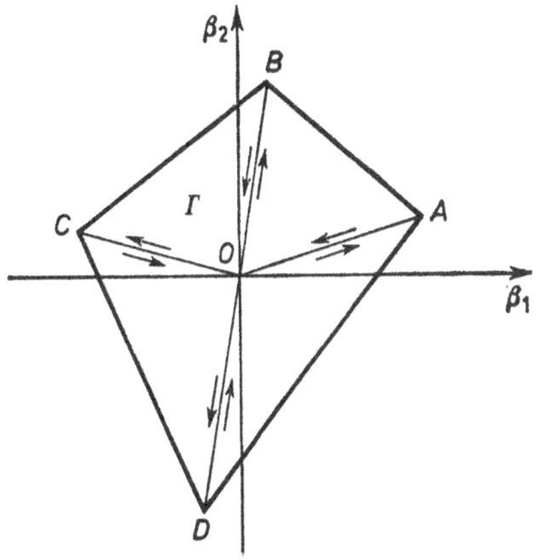

Fig. 34 Critical load path

On the contrary, if the dissipated energy grows steadily then the adaptation is impossible. Further inspection of the signs and magnitudes of plastic strain increments shows whether this is due to incremental collapse or alternating plasticity.

Details of numerical realization of the above shakedown algorithm will be given in Chapters 6 and 7 while discussing specific examples of its application.

4.11.2 *Thermo-elastic-plastic structures subjected to mechanical and thermal loads*

Let us assume the yield condition to vary with temperature in the following way:

$$f = F(\boldsymbol{\sigma}) - \sigma_y(\vartheta) = 0, \tag{4.233}$$

ϑ denoting the relative temperature. We assume the conditions (4.227) to hold also in the present case. The temperature field $\vartheta(\mathbf{z}, t)$ within a region considered may be assumed without any loss of generality as a linear functional of the arguments β_p, $p = 1, 2, \ldots, r$, cf. eq. (4.225):

$$\vartheta(\mathbf{z}, t) = \sum_{p=1}^{r} \beta_p(t)\,\vartheta^{(p)}(\mathbf{z}), \quad \mathbf{z} \in \Omega$$

where $\vartheta^{(p)}$ is a reference constant (in time) temperature in the p-th temperature mode. If external mechanical load and temperature are independent then obviously some factors β_p will correspond only to s, cf. eq. (4.225) while the remaining ones only to temperature ϑ. The limits of load and temperature variations may be set either by means of the inequalities (4.226) or by the coordinates of the r-dimensional polyhedron vertices $\boldsymbol{\gamma}_j$, $j = 1, 2, \ldots, t$, cf. Fig. 33.

The (small) total strain $\boldsymbol{\epsilon}$ is the sum of elastic, thermal and plastic parts

$$\boldsymbol{\epsilon} = \boldsymbol{\epsilon}^{(el)} + \boldsymbol{\epsilon}^{(\vartheta)} + \boldsymbol{\epsilon}^{(pl)}$$

where

$$\boldsymbol{\epsilon}^{(\vartheta)} = \mathbf{a}\vartheta,$$

$\mathbf{a}$ being the vector of thermal expansion coefficients. The stress vector is again decomposed as

$$\boldsymbol{\sigma} = \boldsymbol{\sigma}^{(el)} + \boldsymbol{\rho}. \tag{4.234}$$

Here, similarly to eq. (4.228), $\boldsymbol{\sigma}^{(el)}$ is the thermoelastic stress computed under the assumption of unlimited elastic response and for the same loads as $\boldsymbol{\sigma}$, $\boldsymbol{\rho}$ being thus a residual stress field (in equilibrium with vanishing loads and under zero relative temperature). Within the framework of linear thermoelasticity

$$\boldsymbol{\sigma}^{(el)}(\mathbf{z}, t) = \sum_{p=1}^{r} \beta_p(t)\boldsymbol{\sigma}^{(el)(p)}(\mathbf{z}), \tag{4.235}$$

$\boldsymbol{\sigma}^{(el)(p)}$ being the thermoelastic stress due to the reference load in its p-th mode. The fundamental shakedown theorem extended to include thermal actions, [96], reads:

A given structure will shake down in an arbitrary loading path contained in a given load/temperature domain if and only if there exists a time-independent residual stress field $\boldsymbol{\rho}(\mathbf{x})$ *such that*

$$F[\boldsymbol{\sigma}^{(el)}(\mathbf{z}, t) + \boldsymbol{\rho}(\mathbf{z})] \leqslant \sigma_y[\mathbf{z}, \vartheta^{(el)}(\mathbf{z}, t)] \tag{4.236}$$

at each point $\mathbf{z}$ of the structure and for all the thermoelastic stresses $\boldsymbol{\sigma}^{(el)}$ and temperatures $\vartheta^{(el)}$ which may appear ($\vartheta^{(el)}$ is the temperature field associated with the stress state $\boldsymbol{\sigma}^{(el)}$).

Thus, if the given structure shakes down in a certain cyclic load/temperature path, each cycle of which contains all the vertices $\boldsymbol{\gamma}_1, \ldots, \boldsymbol{\gamma}_t$ then there must exist a certain time-independent residual stress field $\boldsymbol{\rho}'(\mathbf{z})$ such that

$$F(\mathbf{z}, \boldsymbol{\xi}_j + \boldsymbol{\rho}') \leqslant \sigma_y(\mathbf{z}, \eta_j) \tag{4.237}$$

at every point $\mathbf{z}$ and for every $j = 1, 2, \ldots, t$ where $\boldsymbol{\xi}_j$, η_j, $j = 1, 2, \ldots, t$ are the thermoelastic stress fields and temperatures corresponding to the respective vertices $\boldsymbol{\gamma}_j$, $j = 1, 2, \ldots, t$.

In the case of metals, the yield condition (4.233) at a given constant (in time) temperature is convex with respect to the stress component $\boldsymbol{\sigma}$. Therefore, the domain defined by eq. (4.233) in the 7-dimensional space of the stress components and temperature will be convex for $\vartheta^- \leqslant \vartheta \leqslant \vartheta^+$ provided the function $\sigma_y(\mathbf{z}, \vartheta)$ is a concave function of ϑ within the segment $[\vartheta^-, \vartheta^+]$, i.e. provided

$$\lambda\sigma_y(\mathbf{z}, \vartheta_1)+(1-\lambda)\sigma_y(\mathbf{z}, \vartheta_2) \leqslant \sigma_y[\mathbf{z}, \lambda\vartheta_1+(1-\lambda)\vartheta_2] \tag{4.238}$$

for any $0 \leqslant \lambda \leqslant 1$, $\vartheta^- \leqslant \vartheta_1, \vartheta_2 \leqslant \vartheta^+$.

It can be seen that if the condition (4.238) holds then eq. (4.237) implies condition (4.236) to hold for all the admissible load/temperature states as defined by the appropriate polyhedron in the r-dimensional space of parameters β_p, $p = 1, 2, \ldots, r$. The proof duplicates the considerations presented in Sec. 4.9.1 and goes as follows ($\lambda_1+\lambda_2+ \ldots +\lambda_t = 1$; $\lambda_j \geqslant 0$, $j = 1, 2, \ldots, t$):

$$\begin{aligned} F(\boldsymbol{\sigma}^{(el)}+\boldsymbol{\rho}') &= F[\lambda_1\boldsymbol{\xi}_1+ \ldots +\lambda_t\boldsymbol{\xi}_t+(\lambda_1+ \ldots +\lambda_t)\boldsymbol{\rho}'] \\ &= F[\lambda_1(\boldsymbol{\xi}_1+\boldsymbol{\rho}')+ \ldots +\lambda_t(\boldsymbol{\xi}_t+\boldsymbol{\rho}')] \\ &\leqslant \lambda_1 F(\boldsymbol{\xi}_1+\boldsymbol{\rho}')+ \ldots +\lambda_t F(\boldsymbol{\xi}_t+\boldsymbol{\rho}') \\ &\leqslant \lambda_1 \sigma_y(\eta_1)+ \ldots +\lambda_t \sigma_y(\eta_t) \end{aligned} \tag{4.239}$$

where the last inequality results from (4.237). In view of eqs. (4.238) we have

$$\begin{aligned} \sigma_y(\vartheta^{(el)}) &= \sigma_y(\lambda_1\eta_1+ \ldots +\lambda_t\eta_t) \\ &\geqslant \lambda_1\sigma_y(\eta_1)+ \ldots +\lambda_t\sigma_y(\eta_t). \end{aligned} \tag{4.240}$$

Thus, relations (4.239), (4.240) finally imply that

$$F(\boldsymbol{\sigma}^{(el)}+\boldsymbol{\rho}') \leqslant \sigma_y(\vartheta^{(el)})$$

which completes the proof. It is clear now that the method of shakedown analysis suggested in Sec. 4.9.1 can also be used in the case of thermal effects, provided condition (4.238) holds (i.e. provided the yield-point stress $\sigma_y(\vartheta)$ is a non-convex function of temperature).

Summary

A fairly broad range of problems has been discussed in this chapter. The main goal in it was to specify some of the previous results in order to make them more easily applicable to numerical analyses which follow in Part II of the book. In particular, we have discussed the general philosophy of a discretized approach to structural mechanics problems showing how closely it is related to a purely mathematical concept of weighted residual methods used for solving sets of differential equations. We have then passed on to discuss, at different places within the chapter, some specific finite elements to be later used in the analysis of specific boundary-value problems. We have thus considered both the truss and frame elements located in 3D space, the linear strain triangular element for plane stress analysis and the thin shell element for axisymmetric analysis.

Besides, some extensions to the general formulation of Chapter 3 have been given. In particular, a major part of the chapter has been devoted to the derivation of governing relations for lattice-type structures. A difference description has been proposed which allows to automatically generate equations for structures composed of up to three families of spatially distributed rods. A continuous model applicable to dense lattices has also been developed.

Next, we have thoroughly discussed different engineering theories of shells and exemplified it by considering thin shell axisymmetric deformations. This has been followed by the section on certain algorithmic aspects of using constitutive equations for thermo-elastic-plastic materials. The chapter closes with the presentation of some theoretical background for the shakedown analysis of elastic-plastic structures.

Problems

4.1 Using eq. (4.55) derive the elemental stiffness for the linear 2D analysis of trusses. Form the global stiffness matrix for the two-member truss in Fig. 35 and find the displacement u_2 under the force P.

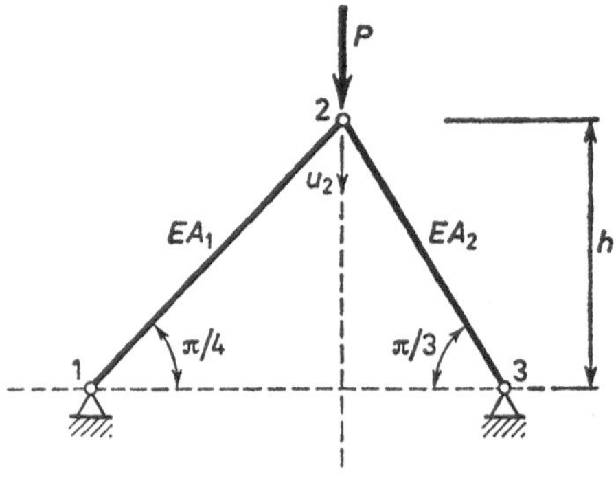

Fig. 35 Two-member truss

4.2 For the plane mesh shown in Fig. 36 propose the global and local numbering of nodes and form the element connectivity table.

4.3 Specify the weighting functions in eq. (4.6) in such a way as to obtain in turn the point collocation method, the subdomain collocation method, the Galerkin method, the method of moments, method of least squares, etc., cf. [38], [49], [134], [184], for instance.

4.4 Compare the stiffness matrices (4.55) and (4.82). Specify the 3D beam stiffness (4.82) to plane problem. Solve using the obtained matrix the problem of plane cantilever bending

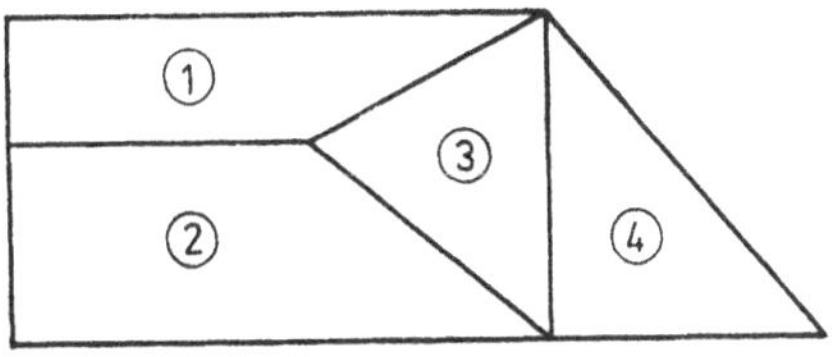

Fig. 36 Discretization mesh

under end transverse concentrated force employing one- and two-element discretization for the beam.

4.5 Derive the lumped elemental mass matrix for the beam element, cf. eq. (4.91).

4.6 Explain the difference between the truss initial stiffness matrix (4.56) and the analogous matrix derived from the beam matrix (4.85), cf. remark below eq. (4.85).

Hint: In eq. (4.83), instead of $\dfrac{\partial \Delta u_\eta}{\partial \xi}$ and $\dfrac{\partial \Delta u_\zeta}{\partial \xi}$ given in terms of the vectors **N**, cf. eq. (4.84), consider the approximate relations

$$\frac{\partial \Delta u_\eta}{\partial \xi} \approx \frac{1}{l}(\Delta q_{12} - \Delta q_6), \qquad \frac{\partial \Delta u_\zeta}{\partial \xi} \approx \frac{1}{l}(\Delta q_{11} - \Delta q_5).$$

Note that such an assumption implies the use of an average constant slope over the whole length of the element and can be justified only if the size of the element is small in relation to the overall length of the actual structure. Use the same approximation as above in deriving the linear initial displacement matrix (4.88).

4.7 Derive explicit forms of global elastic and initial stress stiffness matrices for the structure shown in Fig. 37.

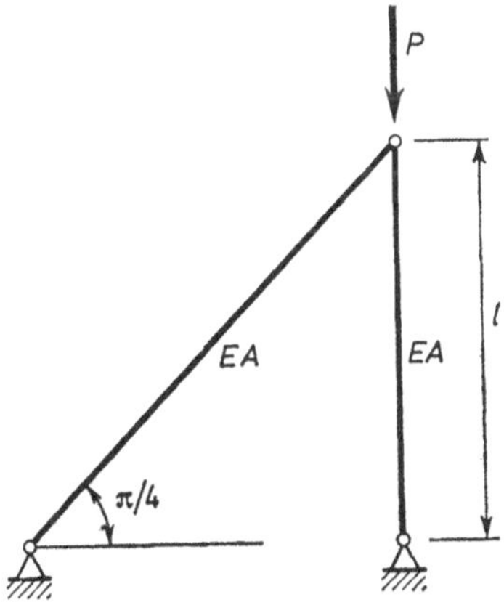

Fig. 37 Two-member truss

4.8 Perform calculations leading to eqs. (4.101).

4.9 Using eqs. (4.99), (4.100), (4.101) derive for the grid in Fig. 38 the explicit form of the equations of motion expressed in terms of the generalized displacements.

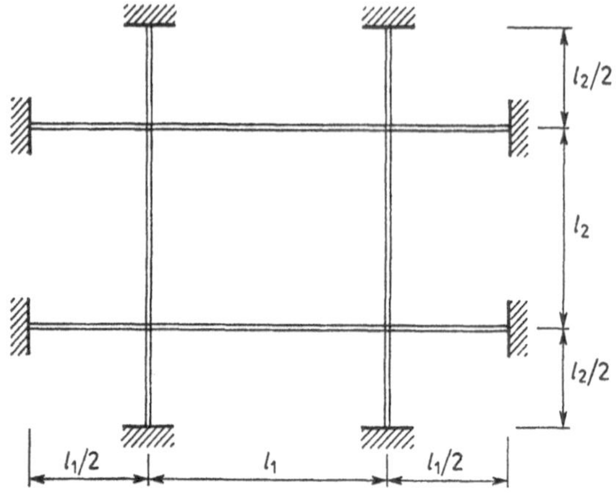

Fig. 38 Grid

4.10 Generalize eqs. (4.93) to the case in which the lattice structure consists of four families of rods. Such structure has two kinds of nodes: the one at which eight rods intersect and the other at which only four rods meet.

Hint: Confine yourself to the case in which the external forces and masses are assigned to the first kind of nodes only. Remember that eqs. (4.93) have to be a special case of the new equations.

4.11 Eqs. (4.93) have been obtained under the condition that the mass of the structure is approximated by the system of masspoints assigned to the internal nodes (or, to be more exact, to the points at which the axes of rods intersect). Hence, the right-hand sides in eqs. (4.93) are equal to zero. Consider the situation in which the inertia terms are present in both the equations of motion.

4.12 The vector bases $\mathbf{g}_a(\mathbf{X})$, $a = 1, 2, 3, \mathbf{X} \in \mathscr{X}$, in the equations of the discrete models of lattice-type structures in Sec. 4.6.1 can be introduced independently of each other. However, the choice of $\mathbf{g}_a(\mathbf{X})$ and hence the coordinate form of the pertinent governing equations depend, as a rule, on the character of the undeformed configuration of the structure, $\varkappa(\mathbf{X})$, $\mathbf{X} \in \mathscr{X}$. In many problems points $\varkappa(\mathbf{X})$, $\mathbf{X} \in \mathscr{X}$, can be treated as located on a certain smooth surface with the minimum radius of curvature much bigger than the maximum member length $|\Delta_A \varkappa(\mathbf{X})|$, $\mathbf{X} \in \mathscr{X}$, $A = \mathrm{I, II, III}$. Then it may turn out useful to select the vectors $\mathbf{g}_K(\mathbf{X})$, $K = 1, 2$, as tangent to the surface S with $\mathbf{g}_3(\mathbf{X}) = \mathbf{N}(\mathbf{X})$, $\mathbf{N}(\mathbf{X})$ being the unit normal to S at $\mathbf{X}$.

Introducing the notation

$$b_{AK}(\mathbf{X}) \equiv G^3_{AK}(\mathbf{X}), \qquad b_A^{\cdot L}(\mathbf{X}) \equiv G^L_{A3}(\mathbf{X}), \qquad b_A(\mathbf{X}) \equiv G^3_{A3}(\mathbf{X}),$$

$$\bar{\delta}_A W^K(\mathbf{X}) \equiv \bar{\Delta}_A W^K(\mathbf{X}) + G^K_{AL}(\mathbf{X}) W^L(\mathbf{X} - \bar{\Delta}_A \mathbf{X}); \qquad K, L = 1, 2,$$

transform eqs. (4.98)–(4.100) to a new form which can be referred to as describing the discrete models of linear-elastic surface lattice-type shells, cf. also Sec. 6.3.1.

4.13 Eqs. (4.101) hold under the assumption that $\eta = 1/2$. Formulate the alternative forms of eqs. (4.101) to hold for $\eta = 1$ and $\eta = 0$, cf. Sec. 6.3.1.

4.14 Scalar and vector densities in eqs. (4.105)–(4.111) are related to the plane region Ω and the vector components are related to the basis $\mathbf{e}_k$, $k = 1, 2, 3$, of the spatial Cartesian coordinate system. Transform these equations to the form in which all mathematical entities are related to the surface $\chi(\Omega, t)$ and to the vector bases $g_\alpha(X, t) \equiv \chi_{,\alpha}(X, t)$, $\alpha = 1, 2$, $N(X, t)$ where $N(X, t)$ is the unit normal to $\chi(\Omega, t)$ at X.

Hint: To solve this problem one should take into account the fundamental equations of tensor calculus on the two-dimensional differentiable manifolds. The basic transformation formulae will have the form (α, β run over 1, 2)

$$\tilde{\varrho} = \frac{1}{\sqrt{a}} \varrho,$$

$$\tilde{f}^\alpha = \frac{1}{\sqrt{a}} f^k \chi_{k,\beta} a^{\alpha\beta}, \qquad \tilde{f} = \frac{1}{\sqrt{a}} f^k N_k,$$

$$\tilde{T}^{\beta\alpha} = \frac{1}{\sqrt{a}} T^{k\alpha} \chi_{k,\gamma} a^{\gamma\beta}, \qquad \tilde{T}^\beta = \frac{1}{\sqrt{a}} T^{k\beta} N_k,$$

where $a \equiv \det[a_{\alpha\beta}]$, $a_{\alpha\beta} = \chi^k{}_{,\alpha} \chi_{k,\beta}$ are components of the first fundamental (metric) tensor of the surface $\chi(\Omega, t)$ and $a^{\gamma\beta} a_{\gamma\alpha} = \delta^\beta_\alpha$ (similar formulae hold for h^k, $M^{k\alpha}$). The resulting transformed form of the equations of motion (4.109) will be given by (inertia terms being neglected)

$$\tilde{T}^{\alpha\beta}|_\beta - b^\alpha_\beta \tilde{T}^\beta + \tilde{f}^\alpha = 0,$$

$$\tilde{T}^\beta|_\beta + b_{\alpha\beta} \tilde{T}^{\alpha\beta} + \tilde{f} = 0,$$

$$\tilde{M}^{\alpha\beta}|_\beta - b^\alpha_\beta \tilde{M}^\beta + e^{\alpha}_{\cdot\beta} \tilde{T}^\beta + \tilde{h}^\alpha = 0,$$

$$\tilde{M}^\beta|_\beta + b_{\alpha\beta} \tilde{M}^{\alpha\beta} + e_{\alpha\beta} \tilde{T}^{\alpha\beta} + h = 0,$$

where the vertical lines stand for the covariant derivative operators on the surface $\chi(\Omega, t)$, $b_{\alpha\beta}$ are components of the second fundamental tensor of this surface, $b^\alpha_\beta = a^{\gamma\alpha} b_{\gamma\beta}$, $e_{\alpha\beta}$ are components of the Ricci bivector and $e^{\alpha}_{\cdot\beta} = a^{\gamma\alpha} e_{\gamma\beta}$.

4.15 Show that equations of the continuous model for lattice-type shells may be reduced within the linear approximation to the equations of motion

$$T^{\alpha\beta}|_\beta - b^\alpha_\beta T^\beta + f^\alpha = 0, \qquad M^{\alpha\beta}|_\beta - b^\alpha_\beta M^\beta + e^{\alpha}_{\cdot\beta} T^\beta + h^\alpha = 0,$$

$$T^\beta|_\beta + b_{\alpha\beta} T^{\alpha\beta} + f = 0, \qquad M^\beta|_\beta + e_{\alpha\beta} T^{\alpha\beta} + b_{\alpha\beta} M^{\alpha\beta} + h = 0,$$

equations of compatibility

$$\gamma_{\alpha\beta} = u_\alpha|_\beta - b_{\alpha\beta} u + e_{\alpha\beta} v, \qquad \varkappa_{\alpha\beta} = v_\alpha|_\beta - b_{\alpha\beta} v,$$

$$\gamma_\alpha = u|_\alpha + b^\beta_\alpha u_\beta + e_{\alpha\beta} v^\beta, \qquad \varkappa_\alpha = v|_\alpha + b^\beta_\alpha v_\beta,$$

constitutive equations

$$T^{\alpha\beta} = A^{\alpha\beta\gamma\delta} \gamma_{\gamma\delta}, \qquad M^{\alpha\beta} = C^{\alpha\beta\gamma\delta} \varkappa_{\gamma\delta},$$

$$T^\alpha = A^{\alpha\beta} \gamma_\beta, \qquad M^\alpha = C^{\alpha\beta} \varkappa_\beta.$$

Using eqs. (4.111) derive explicit forms for $A^{\alpha\beta\gamma\delta}$, $A^{\alpha\beta}$, $C^{\alpha\beta\gamma\delta}$, $C^{\alpha\beta}$.

4.16 Specify the equilibrium equations formulated in the previous problem to obtain equations of the linear theory of lattice-type plates.

4.17 Consider a grid composed of a polar net of bars the central, principal axes of inertia of which are tangent and normal to the plane generated by the axes of bars, cf. also Fig. 118. The grid is subjected to a transverse, uniformly distributed load of intensity q. Using the polar set of coordinates the lines of which coincide with the axes of the actual ribs show that the equilibrium equations take the form

$$T^{\varrho}{}_{,\varrho}+\overset{-1}{\varrho}\,T^{\varrho}+q=0,$$

$$M^{\varrho\varphi}{}_{,\varrho}+2\overset{-1}{\varrho}\,M^{\varrho\varphi}+\overset{-1}{\varrho}\,M^{\varphi\varrho}-\overset{-1}{\varrho}\,P^{\varrho}=0,$$

where comma denotes ordinary differentiation with respect to ϱ. Show that the bending moments $M_{(\varrho)}$, $M_{(\varphi)}$ and the shear forces $\check{P}_{(\varrho)}$ defined at the central cross-sections of bars are given by

$$M_{(\varrho)}=\frac{\tilde{d}_{(\varphi)\alpha}\tilde{d}_{(\varrho)\beta}}{\tilde{d}_{(\varphi)\gamma}d^{\gamma}_{(\varrho)}}\,\tilde{l}_{\varphi}M^{\alpha\beta},$$

$$M_{(\varphi)}=\frac{\tilde{d}_{(\varrho)\alpha}\tilde{d}_{(\varphi)\beta}}{\tilde{d}_{(\varrho)\gamma}d^{\gamma}_{(\varphi)}}\,l_{\varrho}M^{\alpha\beta},$$

$$\check{P}_{(\varrho)}=\frac{\tilde{d}_{(\varphi)\alpha}}{\tilde{d}_{(\varphi)\beta}d^{\beta}_{(\varrho)}}\,\tilde{l}_{\varrho}P^{(\alpha)}$$

where $d_{(\alpha)}=\{d^{\varrho}_{(\alpha)}\ d^{\varphi}_{(\alpha)}\}$ and $\tilde{d}_{(\alpha)}=\{\tilde{d}^{\varrho}_{(\alpha)}\ \tilde{d}^{\varphi}_{(\alpha)}\}$, $\alpha=\varrho,\varphi$, are unit vectors tangent and normal to the directions ϱ and φ of the mesh, respectively, so that

$$d^{\varrho}_{(\varrho)}=1,\quad d^{\varrho}_{(\varphi)}=0,\quad \tilde{d}^{\varrho}_{(\varrho)}=0,\quad \tilde{d}^{\varrho}_{(\varphi)}=-1,$$
$$d_{(\varrho)\varrho}=1,\quad d_{(\varphi)\varrho}=0,\quad \tilde{d}_{(\varrho)\varrho}=0,\quad \tilde{d}_{(\varphi)\varrho}=-1,$$

$$d^{\varphi}_{(\varrho)}=0,\quad d^{\varphi}_{(\varphi)}=\overset{-1}{\varrho},\quad \tilde{d}^{\varphi}_{(\varrho)}=\overset{-1}{\varrho},\quad \tilde{d}^{\varphi}_{(\varphi)}=0,$$
$$d_{(\varrho)\varphi}=0,\quad d_{(\varphi)\varphi}=\varrho,\quad \tilde{d}_{(\varrho)\varphi}=\varrho,\quad \tilde{d}_{(\varphi)\varphi}=0.$$

4.18 Applying eqs. (4.124), (4.126) to eqs. (4.123) show by direct calculation that eqs. (4.127) with the notation (4.128), (4.129) hold true.

4.19 Using eqs. (4.135) and (4.136) derive the form of the Lagrange strain measures in terms of the displacements of the shell midsurface and the increments of the vector field.

Hint: Use the definitions (4.146).

4.20 Taking into account the procedure leading to eqs. (4.139), derive the equations of motion introducing:

(a) internal constraints of the form $d_k p^k{}_{,K}=0$, $K=1,2$,

(b) internal constraints of the form $d^k d_k-1=0$.

4.21 For the constraints (a) and (b) in the previous problem derive the form of the pertinent strain measures.

4.22 Derive the exact form of the inertia terms on the right-hand sides of the equations of motion (4.139) for the Love–Kirchhoff shell theory.

4.23 Using eqs. (4.146) and applying the linearization procedure with respect to the gradients $\nabla\mathbf{u}(\mathbf{X},t)$, $\nabla\mathbf{w}(\mathbf{X},t)$ and to the stress components $S^{\alpha\beta}(\mathbf{X},\eta,t)$ derive the equations of motion in both the special cases of the shell theories detailed in Sec. 4.7.3.

4.24 Under the assumptions made in the previous problem derive the form of the linearized strain measures.

4.25 On the basis of considerations in Sec. 4.7.4 derive eqs. (4.152).

4.26 Perform explicitly calculations necessary to specify the meridional substitute curve (4.158) for spherical and conical shell elements.

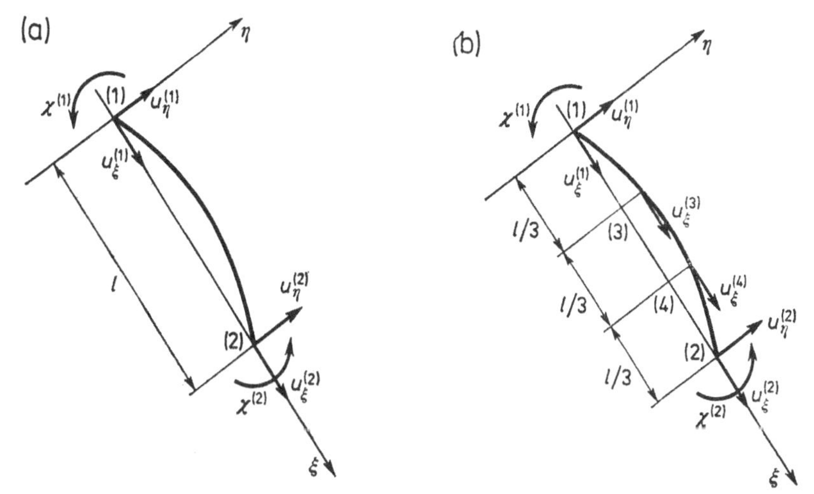

Fig. 39 Displacement expansions for axisymmetric element

4.27 Develop equations for thin axisymmetric element based on the following displacement field expansions, cf. eqs. (4.161), Fig. 39:

(a) $\Delta u_\xi(\xi) = \alpha_1 + \alpha_2 \xi,$

$\Delta u_\eta(\xi) = \alpha_3 + \alpha_4 \xi + \alpha_5 \xi^2 + \alpha_6 \xi^3,$

(b) $\Delta u_\xi(\xi) = \alpha_1 + \alpha_2 \xi + \alpha_3 \xi^2 + \alpha_4 \xi^3,$

$\Delta u_\eta(\xi) = \alpha_5 + \alpha_6 \xi + \alpha_7 \xi^2 + \alpha_8 \xi^3.$

4.28 Justify the integration formulae (4.167).

4.29 Following the derivation of Sec. 4.9 develop elastic, elastic-plastic and initial stress stiffness matrices for the constant strain triangular element under plane stress.

4.30 Specify final results of Sec. 4.10 to obtain equations describing elastic-plastic material with isotropic hardening and elastic-plastic material with kinematic hardening, respectively. Identify in each case all the material parameters and suggest experiments which can be used to determine their values.

4.31 Formulate two fundamental shake down theorems and discuss their relation to two fundamental theorems of limit analysis, [94], [96].

5

Numerical algorithms and software concepts

Purpose of the chapter

The objective in this chapter is to give a general outline of numerical algorithms for solving nonlinear equations typical of structural mechanics discretized formulations discussed in Chapter 3 (Secs. 3.4–3.6) and Chapter 4. The algorithms to be reviewed are believed to belong to those most commonly used in modern computer programs in the field of structural mechanics applications.

5.1 Introductory comments

This book is devoted to the fundamental aspects of structural mechanics formulations and is by no means intended to give a full account of the numerical procedures necessary for a successful computer implementation. Nevertheless, it seems desirable to include in the book some general comments on numerical aspects of modern structural analysis with the intention to set up a more realistic picture of practical usefulness of the formulations discussed and, on the other hand, to call the reader's attention to this fascinating branch of the contemporary nonlinear mechanics.

In this chapter we assume on the part of the reader his familiarity with the fundamental algorithmic aspects of the finite element method as applied to linear problems. Those of our readers who are also familiar with the nonlinear finite element applications may skip this chapter and pass directly to the next one.

Solution procedures for non-linear problems have been discussed by a great number of authors, see [13], [31], [35], [41], [51], [108], [121], [138], [140], [161] for representative reviews and references. In contrary to linear problems it is extremely difficult, if not impossible at all, to develop one single method of general validity that can be used in a routine manner. The procedures described below are believed to at least illustrate the current directions of research effort and to be general enough to apply to a relatively wide class of structural mechanics problems. Unavoidably, certain important topics had to be omitted to keep the chapter within reasonable length. To help the reader in pursuing further studies an abundant referencing is employed throughout.

5.2 Nonlinear quasi-statics

To be general enough we shall now consider the fundamental equations describing the continuing equilibrium of a discretized structural system under slowly varying load as derived in Chapter 4 in the form, cf. eq. (4.43)

$$\mathbf{K}(\mathbf{r})\dot{\mathbf{r}} = \dot{\mathbf{R}}. \tag{5.1}$$

This is a system of N simultaneous ordinary differential equations of the first order with N unknown functions $\mathbf{r} = \mathbf{r}(t)$. The system has to be solved with appropriate initial conditions of the form

$$\mathbf{r}(0) = \hat{\mathbf{r}}_0. \tag{5.2}$$

Any kind of a numerical integration scheme known in the vast numerical analysis literature can be employed to solve eqs. (5.1), (5.2). Below we find it more instructive to give the reader a more physically based description of some of the algorithms. There are two fundamental types of numerical schemes routinely used to integrate eqs. (5.1): explicit and implicit ones. The typical representative of the explicit algorithm group is the one-step Euler forward method which is based on the relationship, cf. eq. (4.42)

$$\mathbf{K}(\mathbf{r}_t)\Delta\mathbf{r} = \Delta\mathbf{R} \tag{5.3}$$

where the notation $\mathbf{r}_t = \mathbf{r}(t)$ is the same as that introduced to simplify the presentation in Chapter 4, cf. eq. (4.38), and t is the time instant at the beginning of the step considered. Physically, eq. (5.3) corresponds to applying the load in several (or more) small steps and to assuming the structure to respond linearly within each step, the configuration at the end of the step being obtained without iteration. In describing the algorithm by eq. (5.3) we assume that the solution for the kinematic and static variables for all time steps from time 0 to time t, inclusive, have been solved for, and that the solution for time $t+\Delta t$ is required next. It is noted that the solution process for the next required equilibrium position is typical and has to be applied repetitively until the complete solution path has been solved for.

Since the solution of such a problem requires the repetitive solution of the set of linear simultaneous equations (5.3), the total numerical effort corresponds roughly to the effort necessary to solve a corresponding linear problem times the number of incremental steps we have chosen to apply.

The procedure implied by eq. (5.3) is simple to use and has been widely employed, particularly in the early literature on elasto-plastic finite element analysis. However, errors are likely to accumulate after several steps unless very fine steps are adopted. The solution may, therefore, diverge considerably from the true response.

In order to avoid large integration errors we may choose to iterate at each load level until, within the assumptions of the integration scheme employed, eq. (5.1) is satisfied within a required tolerance. This is typical of the implicit algorithms among which the Euler one-step backward method is based on the relationship

$$\mathbf{K}(\mathbf{r}_{t+\Delta t})\Delta\mathbf{r} = \Delta\mathbf{R}. \tag{5.4}$$

Since the value of $\mathbf{r}_{t+\Delta t}$ (and consequently the matrix $\mathbf{K}_{t+\Delta t} = \mathbf{K}(\mathbf{r}_{t+\Delta t})$) is unknown at the time instant t considered, it is necessary to iterate in the solution of eq. (5.4). The simplest iteration scheme is based on the following sequence of relations, $i = 1, 2, 3, \ldots$

$$\begin{aligned}
\delta\tilde{\mathbf{R}}^{(i-1)} &= \mathbf{R}_{t+\Delta t} - \mathbf{F}^{(i-1)}_{t+\Delta t},\\
\mathbf{K}(\mathbf{r}^{(0)}_{t+\Delta t})\,\delta\mathbf{r}^{(i)} &= \delta\tilde{\mathbf{R}}^{(i-1)},\\
\mathbf{r}^{(i)}_{t+\Delta t} &= \mathbf{r}^{(i-1)}_{t+\Delta t} + \delta\mathbf{r}^{(i)},\\
\mathbf{F}^{(i)}_{t+\Delta t} &= \mathbf{F}^{(i-1)}_{t+\Delta t} + \mathbf{K}(\mathbf{r}^{(i)}_{t+\Delta t})\delta\mathbf{r}^{(i)}
\end{aligned} \tag{5.5}$$

where $\mathbf{F}^{(0)}_{t+\Delta t} = \mathbf{F}_t$, $\mathbf{r}^{(0)}_{t+\Delta t} = \mathbf{r}_t$ and the vector $\mathbf{F}_t$ has, as before, the mechanical interpretation as the nodal vector of internal forces, i.e. the resultant forces acting on each node from the surrounding elements. In the computational practice the vector $\mathbf{F}^{(i)}_{t+\Delta t}$ is often calculated from actual stresses $\boldsymbol{\sigma}^{(i)}_{t+\Delta t}$ rather than by using eq. $(5.5)_4$, i.e., cf. eq. (4.37)

$$\mathbf{F}^{(i)}_{t+\Delta t} = \sum_{e=1}^{E}\Big(\int_{\Omega_e} \mathbf{B}^T\boldsymbol{\sigma}^{(i)}_{t+\Delta t}\,\mathrm{d}\Omega\Big). \tag{5.6}$$

In eqs. (5.5) we calculate in each iteration an out-of-balance load vector which yields a correction to the displacement increment, and we continue the iteration until the out-of-balance load vector $\delta\mathbf{R}^{(i)}$ (or the displacement correction $\delta\mathbf{r}^{(i)}$) becomes sufficiently small.

It is very easy to recognize that eqs. (5.6) form the basis for the iterative solution of the nodal equilibrium equation

$$\mathbf{F}_{t+\Delta t} = \mathbf{R}_{t+\Delta t} \tag{5.7}$$

by means of the well-known modified Newton–Raphson (or constant stiffness) iteration method, Fig. 40. The main advantage of this approach is that the stiffness matrix is formed (and decomposed) only once at each step. On the

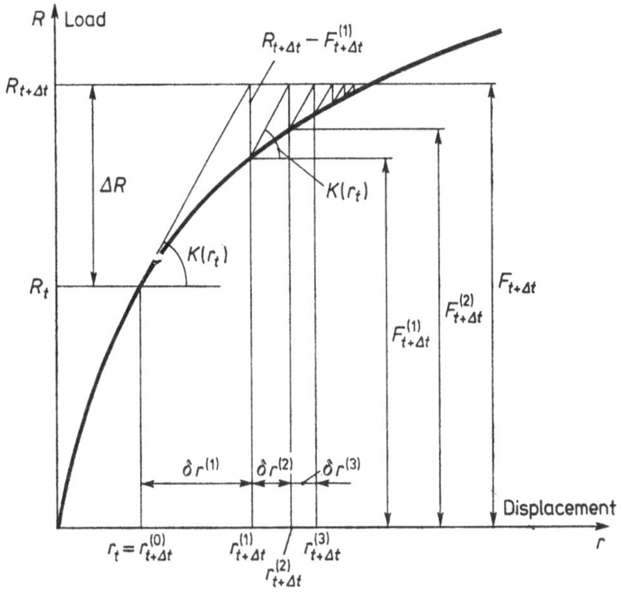

Fig. 40 Modified Newton–Raphson iteration

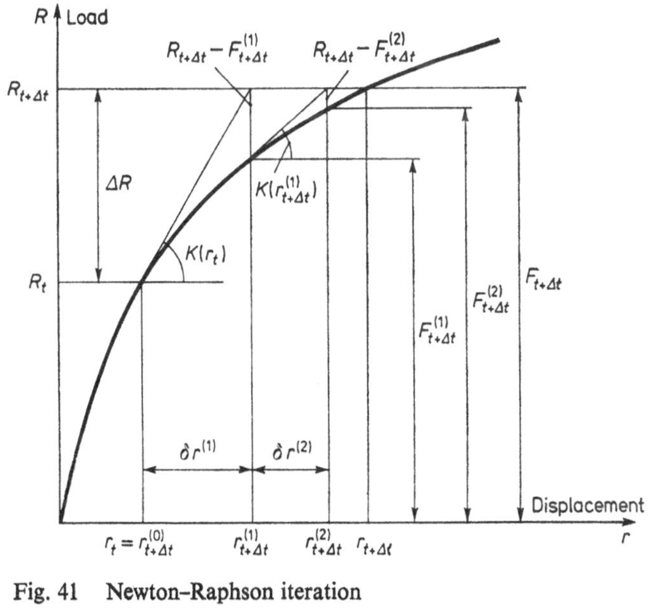

Fig. 41 Newton–Raphson iteration

contrary, the (full) Newton–Raphson iteration requires the stiffness matrix to be reformulated at every iteration so that eq. $(5.5)_2$ is replaced by, cf. Fig. 41

$$\mathbf{K}(\mathbf{r}_{t+\Delta t}^{(i-1)})\,\delta\mathbf{r}^{(i)} = \delta\mathbf{R}^{(i-1)}. \tag{5.8}$$

A disadvantage of this procedure is that a large amount of computational effort may be required to repetitively form and decompose the matrix $\mathbf{K}_{t+\Delta t}^{(i-1)}$. However, constant stiffness iteration will typically converge more slowly than Newton–Raphson iteration, and schemes to accelerate convergence may be desirable. It may also be advantageous to devise mixed iteration schemes combining the features of both procedures.

As an alternative to forms of Newton–Raphson iteration a class of methods known as matrix update methods or quasi-Newton methods has been developed and successfully tested on different nonlinear structural mechanics problems. These methods represent a reasonable compromise between updating the matrix at each iteration as in the full Newton–Raphson approach and using the previous tangent as in the modified Newton–Raphson approach. The typical, very effective and broadly used algorithm in this group of iterative methods is the one called BFGS (Broyden–Fletcher–Goldfarb–Shanno), [13], [108]. Without going into theoretical details concerning the derivation of the BFGS update equations let us just briefly review the basic steps of this algorithm with reference to the formulation (5.5) for a given load level:

— solve the equation

$$\tilde{\mathbf{K}}_{t+\Delta t}^{(0)}\,\delta\mathbf{r}^{(1)} = \delta\mathbf{R}^{(0)}, \qquad \tilde{\mathbf{K}}_{t+\Delta t}^{(0)} = \mathbf{K}_t, \qquad \delta\mathbf{R}^{(0)} = \mathbf{R}_{t+\Delta t} - \mathbf{F}_t,$$

— compute the internal force vector $\mathbf{F}_{t+\Delta t}^{(1)}$ corresponding to the displacement vector $\mathbf{r}_{t+\Delta t}^{(1)} = \mathbf{r}_t + \delta\mathbf{r}^{(1)}$,

— form the auxiliary vector

$$\delta\mathbf{F}^{(1)} = \mathbf{F}_{t+\Delta t}^{(1)} - \mathbf{F}_t,$$

— compute the vector of unequilibrated nodal forces

$$\delta\mathbf{R}^{(1)} = \mathbf{R}_{t+\Delta t} - \mathbf{F}_{t+\Delta t}^{(1)},$$

— form an approximate inverse to the secant matrix as

$$\tilde{\mathbf{K}}_{t+\Delta t}^{-1(1)} = \left[\mathbf{I} - \frac{\delta\mathbf{r}^{(1)}\delta\mathbf{F}^{(1)T}}{\delta\mathbf{r}^{(1)T}\delta\mathbf{F}^{(1)}}\right]\mathbf{K}_t^{-1}\left[\mathbf{I} - \frac{\delta\mathbf{F}^{(1)}\delta\mathbf{r}^{(1)T}}{\delta\mathbf{r}^{(1)T}\delta\mathbf{F}^{(1)}}\right] + \frac{\delta\mathbf{r}^{(1)}\delta\mathbf{r}^{(1)T}}{\delta\mathbf{r}^{(1)T}\delta\mathbf{F}^{(1)}},$$

— find the next displacement correction by using the relation (the matrix $\tilde{\mathbf{K}}_{t+\Delta t}^{-1(1)}$ is given!)

$$\delta\mathbf{r}^{(2)} = \tilde{\mathbf{K}}_{t+\Delta t}^{-1(1)}\,\delta\mathbf{R}^{(1)},$$

— compute the internal force vector $\mathbf{F}^{(2)}_{t+\Delta t}$ corresponding to the displacement vector $\mathbf{r}^{(2)}_{t+\Delta t} = \mathbf{r}^{(1)}_{t+\Delta t} + \delta\mathbf{r}^{(2)}$,

— form the auxiliary vector

$$\delta\mathbf{F}^{(2)} = \mathbf{F}^{(2)}_{t+\Delta t} - \mathbf{F}^{(1)}_{t+\Delta t},$$

— compute the vector of unequilibrated nodal forces

$$\delta\mathbf{R}^{(2)} = \mathbf{R}_{t+\Delta t} - \mathbf{F}^{(2)}_{t+\Delta t},$$

— form an approximate inverse to the secant matrix as

$$\tilde{\mathbf{K}}^{-1(2)}_{t+\Delta t} = \left[\mathbf{I} - \frac{\delta\mathbf{r}^{(2)}\delta\mathbf{F}^{(2)T}}{\delta\mathbf{r}^{(2)T}\delta\mathbf{F}^{(2)}}\right]\tilde{\mathbf{K}}^{-1(1)}_{t+\Delta t}\left[\mathbf{I} - \frac{\delta\mathbf{F}^{(2)}\delta\mathbf{r}^{(2)T}}{\delta\mathbf{r}^{(2)T}\delta\mathbf{F}^{(2)}}\right] + \frac{\delta\mathbf{r}^{(2)}\delta\mathbf{r}^{(2)T}}{\delta\mathbf{r}^{(2)T}\delta\mathbf{F}^{(2)}},$$

— find the next displacement correction

$$\delta\mathbf{r}^{(3)} = \tilde{\mathbf{K}}^{-1(2)}_{t+\Delta t}\,\delta\mathbf{R}^{(2)}.$$

Further iterations follow in the exactly similar fashion. We note that in the computational practice slightly different realization of the BFGS algorithm may turn out more useful, cf. [13].

If an incremental solution strategy is to be effective, realistic criteria should be used for the termination of the iterative process. At the end of every iteration there should be a convergence check to see whether the iterative solution has converged within preset tolerances. The most commonly used convergence criteria are:

(a) *Displacement criterion*

$$\frac{\|\delta\mathbf{r}^{(i)}\|}{\|\mathbf{r}_{t+\Delta t}\|} \leqslant \varepsilon_D \tag{5.9}$$

where $\|\cdot\|$ is a vector norm, ε_D is a preset displacement convergence tolerance and since $\mathbf{r}_{t+\Delta t}$ is yet unknown it must be approximated, by its last available value $\mathbf{r}^{(i)}_{t+\Delta t} = \mathbf{r}^{(i-1)}_{t+\Delta t} + \delta\mathbf{r}^{(i)}$, for instance.

(b) *Force criterion*

$$\frac{\|\delta\mathbf{R}^{(i)}\|}{\|\Delta\mathbf{R}\|} \leqslant \varepsilon_F \tag{5.10}$$

where ε_F is a preset out-of-balance force tolerance.

(c) *Energy criterion*

$$\frac{\delta\mathbf{r}^{(i)T}(\mathbf{R}_{t+\Delta t} - \mathbf{F}^{(i-1)}_{t+\Delta t})}{\delta\mathbf{r}^{(1)T}(\mathbf{R}_{t+\Delta t} - \mathbf{F}_t)} \leqslant \varepsilon_E \tag{5.11}$$

in which the amount of work done by the residual out-of-balance loads on the displacement increment correction is compared against this value at the first iteration and ε_E is a preset quantity.

The standard and most effective methods of solving sets of linear algebraic equations require positive-definiteness of the coefficients matrix. This is assured in the conforming displacement finite element model at the so-called regular points along the equilibrium path. The detailed discussion of solution difficulties that arise in passing over singular (limit or bifurcation) points is beyond the scope of this presentation. We shall only give below some general comments on basic difficulties which are encountered in the analysis of such problems, referring the interested readers to specific presentations such as [22], [31], [41].

Fig. 42 illustrates a typical nonlinear behaviour of a statically loaded structure characterized in terms of an external load measure λ (say, a multiplier

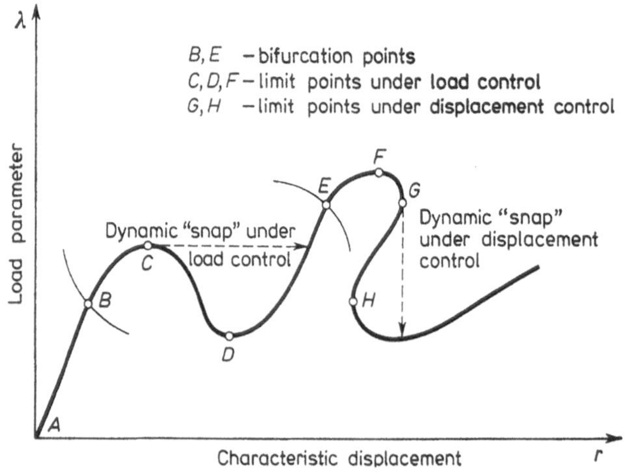

Fig. 42 Typical nonlinear behaviour of statically loaded structure

defining the proportional load increase) versus a displacement state measure r (say, a typical structure deflection). The points B, E are referred to as the bifurcation (or equilibrium branching) points, the points C, D, F as the limit points under load control whereas the points G, H as the limit points under displacement control. Clearly, as for the load levels at C, D, F (or B, E) there is no (or non-unique) displacement solution for any incremental load, the total stiffness matrix must be singular at these points. The hypothetical solution depicted in Fig. 42 suggest that we have a means to overcome numerical difficulties at the singular points to continue the analysis and that we follow the fundamental equilibrium path rather than the secondary ones discovered at the bifurcation points (the opposite may happen to be required in a practical situation).

For most problems of engineering importance it is quite unnecessary to trace such a convoluted load/displacement path as shown in Fig. 42. In many situations all that may appear to be needed is the load level at the first singular point, be that a bifurcation point B or a limit point C. However, without analysis techniques that allow the limit points to be passed, even this information may be unavailable or unreliable. For instance, identification of a singular point as the point of the iterative solution collapse may be totally false as the divergence may be due to, say, computer round-off errors rather than to the singularity of the stiffness matrix. For other problems, the analysis may be performed on an individual component of a complete structure. In such a situation, it may be important to obtain information on the nature of the load shedding, following the singular points, in order to assess the performance of the complete structure.

In this section we shall briefly discuss strategies for tracing nonlinear response near limit points. A simple method for dealing with the bifurcation points will be suggested in Sec. 5.4.

The discussion is based on the incremental/iterative procedure for solving nonlinear problems of statics as described above. We assume for simplicity in notation that the solid is subjected to a proportionally varying load. Thus, the solution algorithm can be presented as

$$\mathbf{K}_{t+\Delta t}^{(i-1)}\delta\mathbf{r}^{(i)} = \lambda_{t+\Delta t}\mathbf{R}^* - \mathbf{F}_{t+\Delta t}^{(i-1)} \tag{5.12}$$

or

$$\mathbf{K}_{t+\Delta t}^{(i-1)}\delta\mathbf{r}^{(i)} = \Delta\lambda\mathbf{R}^* - \mathbf{J}_{t+\Delta t}^{(i-1)} \tag{5.13}$$

where

$$\mathbf{J}_{t+\Delta t}^{(i-1)} = \mathbf{F}_{t+\Delta t}^{(i-1)} - \lambda_t\mathbf{R}^* \tag{5.14}$$

and λ_t is the load factor that describes the intensity of the reference load $\mathbf{R}^*$ to be applied at time t. The iteration in eq. (5.12) is performed under constant external load $\lambda_{t+\Delta t}\mathbf{R}^*$, i.e. in the N-dimensional space of displacements $\mathbf{r}$.

As we already pointed out the equilibrium iterations usually break down near the limit point (in load controlled problems). Many procedures have been devised in the literature to circumvent this problem. The most obvious one consists in switching to the displacement control. The procedure is based on selecting a displacement component as a controlling parameter with the corresponding load taken as unknown and calculated as a reaction. The method, often excellent in performing academic studies, has inherent shortcomings in dealing with more realistic structural configurations and loadings.

Another simple method to avoid difficulties with convergence of the iterative approach is to suppress the iterations in the critical zone. This procedure is particularly useful when it is combined with the so-called current stiffness parameter to guide the algorithm, [22], Fig. 43. At a point A close to the maximum load point F the incremental/iterative algorithm is discontinued and the pure incremental scheme is used instead. If a norm of the incremental

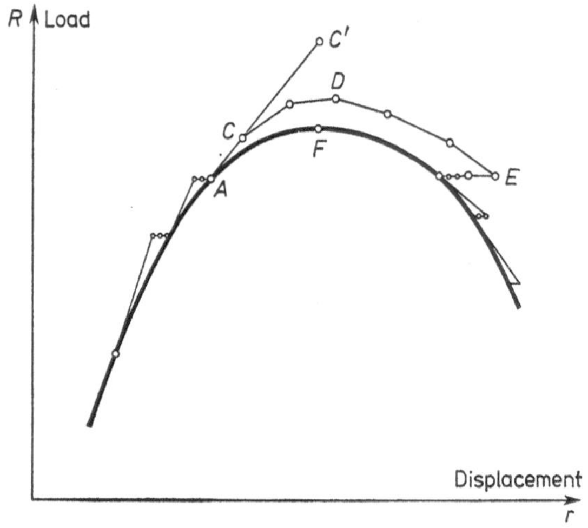

Fig. 43 Suppressing iterations in the critical zone

displacement vector exceeds a preset value, point C' in Fig. 43, the load and displacement vector are linearly scaled back, point C. Starting the next step calculations at C a negative diagonal elements in the stiffness matrix may be detected in which case negative load increments are applied, point D. The iteration procedure is resumed when the stiffness parameter reaches a point E far enough from the critical zone. The technique requires very small steps to avoid drifting away from the equilibrium path.

The currently most promising area of research in regard to the problem under consideration lies in the automatic selection of the incremental load levels and the iteration with the load parameter and the displacement components varying simultaneously. When the iteration is performed in the load-displacement space, eqs. (5.12) and (5.13) are used in the form

$$\mathbf{K}_{t+\Delta t}^{(j)} \delta \mathbf{r}^{(i)} = \lambda_{t+\Delta t}^{(i)} \mathbf{R}^* - \mathbf{F}_{t+\Delta t}^{(i-1)} \tag{5.15}$$

or

$$\mathbf{K}_{t+\Delta t}^{(j)} \delta \mathbf{r}^{(i)} = \delta \lambda_{t+\Delta t}^{(i)} \mathbf{R}^* - {}'\mathbf{J}_{t+\Delta t}^{(i-1)}$$

where

$$\lambda_{t+\Delta t}^{(i)} = \lambda_{t+\Delta t}^{(i-1)} + \delta \lambda^{(i)} = \lambda_t + \Delta \lambda^{(i)} \tag{5.16}$$

and

$$'\mathbf{J}^{(i-1)}_{t+\Delta t} = \mathbf{F}^{(i-1)}_{t+\Delta t} - \lambda^{(i-1)}_{t+\Delta t}\mathbf{R}^* \tag{5.17}$$

is the difference between the internal (i.e. calculated from stresses) and external nodal load vectors after $(i-1)$ iteration at the step $(t, t+\Delta t)$. The symbol $\mathbf{K}^{(j)}_{t+\Delta t}$ represents the tangent matrix at stage j in the time interval $[t, t+\Delta t]$ with $j = i-1$ in the standard Newton–Raphson solution procedure and $j = 0$ ($\mathbf{K}^{(0)}_{t+\Delta t} = \mathbf{K}_t$) in the typical modified Newton–Raphson iteration. As before, $\mathbf{r}^{(i)}_{t+\Delta t}$ stands for the vector of total displacement at iteration (i), $\Delta\mathbf{r}^{(i)}$ is the incremental displacement vector after iteration (i) and $\delta\mathbf{r}^{(i)}$ designates the i-th iterative displacement. The similar notation applies to the proportional load factor $\lambda^{(i)}_{t+\Delta t}$.

For the purpose of reviewing some widely used arc-length strategies to control the incremental iterative procedure for the solution of eq. (5.15), it proves convenient to denote

$$\begin{aligned}
\delta\mathbf{t}^{(1)} &= [\delta\mathbf{r}^{(1)T} \;\; \delta\lambda^{(1)}]^T,\\
\delta\mathbf{t}^{(i)} &= [\delta\mathbf{r}^{(i)T} \;\; \delta\lambda^{(i)}]^T,\\
\Delta\mathbf{t}^{(i)} &= [\Delta\mathbf{r}^{(i)T} \;\; \Delta\lambda^{(i)}]^T,\\
\mathbf{r}^*_{t+\Delta t} &= \overset{-1}{\mathbf{K}^{(j)}_{t+\Delta t}}\mathbf{R}^*,\\
\delta\mathbf{r}^{(1)} &= \delta\lambda^{(1)}\mathbf{r}^*_{t+\Delta t}.
\end{aligned} \tag{5.18}$$

The original form of the arc-length procedure, sometimes also referred to as the bordering algorithm, requires that the scalar products of the tangent vector $\delta\mathbf{t}^{(1)}$ and the vectors $\Delta\mathbf{t}^{(i)}$, $i = 1, 2, \dots$ take a prescribed value s_0^2, cf. Fig. 44

$$\delta\mathbf{t}^{(1)T}\Delta\mathbf{t}^{(i)} = s_0^2, \qquad i = 1, 2, 3, \dots \tag{5.19}$$

From an algorithmic point of view the scalar quantity s_0, referred to as the arc-length, is used to control the solution procedure by suitably adjusting its value, if necessary, according to experimentally observed convergence behaviour. From a geometrical standpoint this procedure may be viewed as an iteration on hyperplanes normal to the tangent line spanned by $\delta\mathbf{t}^{(1)}$.

In a computational context it is advantageous to enforce the constraint (5.19) in the course of the following two-step procedure:

Step 1: enforce the constraint (5.19) for $i = 1$, i.e.

$$\delta\mathbf{t}^{(1)T}\delta\mathbf{t}^{(1)} = s_0^2.$$

This is accomplished by setting

$$\delta\lambda^{(1)^2}\mathbf{r}^{*T}_{t+\Delta t}\mathbf{r}^*_{t+\Delta t} + \delta\lambda^{(1)^2} = s_0^2$$

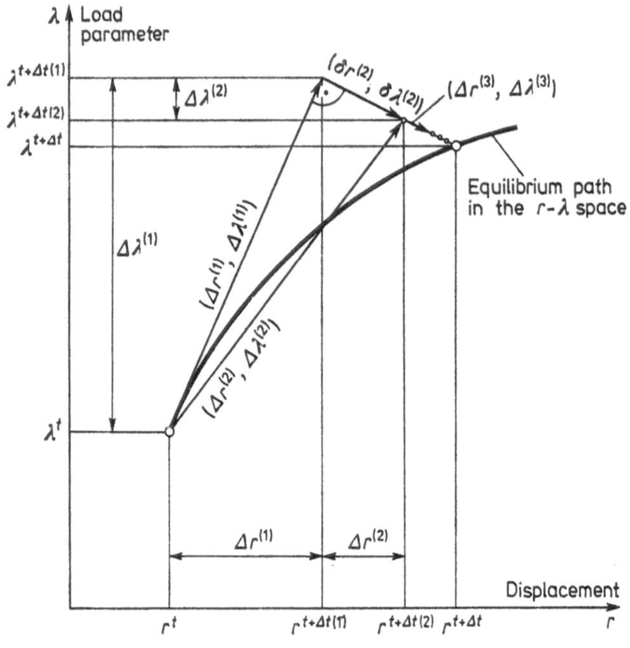

Fig. 44 Iterations in the (r, λ)-space

where $\mathbf{r}^*_{t+\Delta t}$ is determined by solving $\mathbf{K}^{(j)}_{t+\Delta t}\,\mathbf{r}^*_{t+\Delta t} = \mathbf{R}^*$. Thus, the load increment for the first step becomes

$$\Delta\lambda^{(1)} = \left[\frac{s_0^2}{\mathbf{r}^{*T}_{t+\Delta t}\mathbf{r}^*_{t+\Delta t}+1}\right]^{1/2}.$$

Step 2: for $i = 2, 3, \ldots$ enforce the normal hyperplane constraint defined as

$$\delta\mathbf{t}^{(1)T}\delta\mathbf{t}^{(i)} = 0, \quad i = 2, 3, \ldots$$

Using eqs. (5.18) we have equivalently

$$\delta\mathbf{r}^{(1)T}\delta\mathbf{r}^{(i)} + \delta\lambda^{(1)}\delta\lambda^{(i)} = 0, \quad i = 2, 3, \ldots \tag{5.20}$$

It is clear that the introduction of the constraint (5.19) directly into the original system of equations results in a non-symmetric matrix of coefficients. To overcome this drawback, a system of equations with matrix coefficient $\mathbf{K}^{(j)}_{t+\Delta t}$ is solved twice as follows. Noting that (5.15) may be rewritten as

$$\mathbf{K}^{(j)}_{t+\Delta t}\,\delta\mathbf{r}^{(i)} = \delta\lambda^{(i)}\mathbf{R}^* + (\lambda_t + \Delta\lambda^{(i-1)})\,\mathbf{R}^* - \mathbf{F}^{(i-1)}_{t+\Delta t} \tag{5.21}$$

since $\lambda^{(i)}_{t+\Delta t} = \lambda_t + \Delta\lambda^{(i-1)} + \delta\lambda^{(i)} = \lambda_t + \Delta\lambda^{(i)}$, we may solve the following two systems of equations with $\mathbf{K}^{(j)}_{t+\Delta t}$ as the coefficient matrix

$$\mathbf{K}^{(j)}_{t+\Delta t}\mathbf{r}^*_{t+\Delta t} = \mathbf{R}^*,$$

$$\mathbf{K}^{(j)}_{t+\Delta t}\,\delta\hat{\mathbf{r}}^{(i)} = (\lambda_t + \Delta\lambda^{(i-1)})\,\mathbf{R}^* - \mathbf{F}^{(i-1)}_{t+\Delta t}. \tag{5.22}$$

Since the vector $\delta\mathbf{r}^{(i)}$ may be expressed as

$$\delta\mathbf{r}^{(i)} = \mathbf{r}^*_{t+\Delta t}\,\delta\lambda^{(i)} + \delta\hat{\mathbf{r}}^{(i)} \tag{5.23}$$

substitution of eq. (5.23) into eq. (5.22) yields

$$\delta\lambda^{(i)} = -\frac{\delta\mathbf{r}^{(1)T}\delta\hat{\mathbf{r}}^{(i)}}{\delta\mathbf{r}^{(1)T}\mathbf{r}^*_{t+\Delta t} + \delta\lambda^{(1)}}. \tag{5.24}$$

It is possible to modify the normal hyperplane constraint in several ways. One of the alternative procedures is to replace the tangent vector $\delta\mathbf{t}^{(1)}$ in Step 2 by the updated tangent vector defined as

$$\Delta\mathbf{t}^{(i-1)} = [\Delta\mathbf{r}^{(i-1)T} \;\; \Delta\lambda^{(i-1)}]^T.$$

This results in a slightly different equation for the load increment

$$\delta\lambda^{(i)} = -\frac{\Delta\mathbf{r}^{(i-1)T}\delta\hat{\mathbf{r}}^{(i)}}{\Delta\mathbf{r}^{(i-1)T}\mathbf{r}^*_{t+\Delta t} + \Delta\lambda^{(i-1)}}.$$

A different alternative form of the constant arc-length method (known as the Crisfield's procedure) is achieved by enforcing throughout all the iterations corresponding to a given load step the spherical constraint

$$\Delta\mathbf{t}^{(i)T}\Delta\mathbf{t}^{(i)} = s_0^2, \quad i = 1, 2, \ldots$$

Equivalently

$$\Delta\mathbf{r}^{(i)T}\Delta\mathbf{r}^{(i)} + \Delta\lambda^{(i)^2} = s_0^2. \tag{5.25}$$

Observing that now

$$\Delta\mathbf{r}^{(i)} = \Delta\mathbf{r}^{(i-1)} + \delta\mathbf{r}^{(i)} = \Delta\mathbf{r}^{(i-1)} + \delta\hat{\mathbf{r}}^{(i)} + \delta\lambda^{(i)}\mathbf{r}^*_{t+\Delta t},$$

$$\Delta\lambda^{(i)} = \Delta\lambda^{(i-1)} + \delta\lambda^{(i)}$$

we obtain the constraint (5.25) in the form of a quadratic equation for the change in load parameter $\delta\lambda^{(i)}$ as

$$a_2\,\delta\lambda^{(i)^2} + a_1\,\delta\lambda^{(i)} + a_0 = 0 \tag{5.26}$$

where the coefficients a_0, a_1, a_2 are known at the i-th iteration and are given by

$$a_0 = (\Delta\mathbf{r}^{(i-1)} + \delta\hat{\mathbf{r}}^{(i)})^T(\Delta\mathbf{r}^{(i-1)} + \delta\hat{\mathbf{r}}^{(i)}) + \Delta\lambda^{(i-1)^2} - s_0^2,$$

$$a_1 = 2(\Delta\mathbf{r}^{(i-1)} + \delta\hat{\mathbf{r}}^{(i)})^T\mathbf{r}^*_{t+\Delta t} + 2\Delta\lambda^{(i-1)},$$

$$a_2 = \mathbf{r}^{*T}_{t+\Delta t}\mathbf{r}^*_{t+\Delta t} + 1.$$

The solution of the proper root for $\delta\lambda^{(i)}$ may be accomplished by choosing (for softening systems) a positive angle between the old incremental vector $\Delta\mathbf{t}^{(i-1)}$ and the new one $\Delta\mathbf{t}^{(i)}$. For stiffening systems one has to choose the corresponding negative angle. If both angles exhibit the same sign, the solution closest to the linear "solution" $\delta\lambda^{(i)} = -a_0/a_1$ of eq. (5.26) has to be taken.

Numerical experience has shown, [131], that instead of applying the constraint (5.25) it is preferable to fix the "incremental" length s_0 in N dimensional space, i.e. eq. (5.25) is replaced by

$$\Delta \mathbf{r}^{(i)T} \Delta \mathbf{r}^{(i)} = s_0^2. \tag{5.27}$$

Accordingly, the coefficients a_0, a_1, a_2 simplify to

$$\begin{aligned} a_0 &= (\Delta \mathbf{r}^{(i-1)} + \delta \hat{\mathbf{r}}^{(i)})^T (\Delta \mathbf{r}^{(i-1)} + \delta \hat{\mathbf{r}}^{(i)}) - s_0^2, \\ a_1 &= 2(\Delta \mathbf{r}^{(i-1)} + \delta \hat{\mathbf{r}}^{(i)}) \mathbf{r}^*_{t+\Delta t}, \\ a_2 &= \mathbf{r}^{*T}_{t+\Delta t} \mathbf{r}^*_{t+\Delta t}. \end{aligned} \tag{5.28}$$

All the forms of the constraint conditions used in arc-length control procedures involve expressions where sums over displacement quantities, rotational quantities and load parameters occur in mixed form. This lack of homogeneity raises the natural objection that one could in principle change the character of the method simply by changing the scale of the quantities defining the constraints. To circumvent this problem we may introduce a "scaling vector" for the purpose of homogenizing the quantities that define the constraint condition. In this way a non-dimensional form of the constraint condition may be obtained. Let us define the scaling vector as

$$\hat{\mathbf{t}} = \{\hat{\mathbf{r}}^T \;\; \hat{\lambda}\}$$

the components of which include scaling factors $\hat{r}_1, \hat{r}_2, \ldots, \hat{r}_N$ for the displacement components and a scaling factor $\hat{\lambda}$ for the load parameter. Any vector $\delta \mathbf{t} = [\delta \mathbf{r}^T \;\; \delta \lambda]$ is now rewritten as

$$\left\{ \frac{\delta r_1}{\hat{r}_1} \;\; \frac{\delta r_2}{\hat{r}_2} \;\; \cdots \;\; \frac{\delta r_N}{\hat{r}_N} \;\; \frac{\delta \lambda}{\hat{\lambda}} \right\}.$$

As the scaling factor $\hat{\mathbf{t}}$ we may take, for instance, the displacement vector resulting from linear analysis under the load $\mathbf{R}^*$ together with a proportional load factor. An alternative scaling may be obtained by using averages of the absolute rotations and displacements of linear analysis as scaling parameters for all rotational and displacement nodal unknowns, respectively. (In particular, by letting all the displacement scaling factors go to infinity, $r_\alpha \rightarrow \infty$, $\alpha = 1, 2, \ldots, N$ and by putting $\hat{\lambda}$ = const we recover the classical load control procedure.)

By letting $\hat{\lambda} \rightarrow \infty$ we may formally justify the procedure indicated in eq. (5.27).

Numerical effort involved in all currently available arc-length methods is similar. We note that normal hyperplane constraint method may fail if no intersection between the load path and hyperplane is found. In the spherical constraint condition this drawback is no longer present at the expense of additional effort involved in the selection of the correct root in the equation (5.26).

5.3 Integration of elastic-plastic constitutive law

In the previous section we have discussed the numerical procedures for solving nonlinear problems of statics, described by the sets of ordinary differential equations of the form (5.1). Strictly speaking, this form applies to nonlinear elastic problems only and inelastic analysis is in fact outside the scope of such a description. This is so because in the latter case the total stresses can not be resolved in terms of total displacements so that eq. (5.1) assumes the form

$$\mathbf{K}(\boldsymbol{\sigma}, \mathbf{r})\dot{\mathbf{r}} = \dot{\mathbf{R}} \tag{5.29}$$

where the notation $\mathbf{K}(\boldsymbol{\sigma}, \mathbf{r})$ is meant to emphasize that the stiffness matrix in the current configuration (dependence on $\mathbf{r}$!) is also a given function, cf. eq. (3.82), of the stress components at every integration point.

In this section we would like to complement the discussion of Sec. 5.2 by considering some related issues pertaining to inelastic material properties. In accordance with eq. (5.29) the accuracy of the inelastic analysis relies heavily on the way the inelastic constitutive law of the form $\dot{\boldsymbol{\sigma}} = \mathbf{C}\dot{\boldsymbol{\epsilon}}$, cf. eq. $(2.3)_1$, is integrated to yield both the incremental stresses $\Delta\boldsymbol{\sigma}$ and, by using a stress accumulation procedure, the total stresses $\boldsymbol{\sigma}$ at the end of the step. For simplicity, the assumption of small deformations is employed throughout this section.

To discuss the algorithm let us first recall the fundamental equations describing isothermal theory of elastic-plastic materials with the von Mises yield surface, associated flow rule and linear combination of isotropic and kinematic hardening (or softening), cf. Sec. 4.8:

Stress rate—strain rate equation:

$$\dot{\boldsymbol{\sigma}} = \mathbf{C}(\dot{\boldsymbol{\epsilon}} - \dot{\boldsymbol{\epsilon}}^{(pl)}). \tag{5.30}$$

Evolution equation for the yield limit:

$$\dot{\sigma}_y = \omega\zeta\dot{\bar{\varepsilon}}^{(pl)}. \tag{5.31}$$

Evolution equation for the back stress:

$$\dot{\boldsymbol{\alpha}} = (1-\omega)\tfrac{2}{3}\zeta\dot{\boldsymbol{\epsilon}}^{(pl)}. \tag{5.32}$$

Flow rule:

$$\dot{\boldsymbol{\epsilon}}^{(pl)} = \begin{cases} 0 & \text{if } \; f(\sigma) < \sigma_y \quad \text{or} \quad f(\sigma) = \sigma_y \\ & \text{and} \quad \mathbf{n}^T\dot{\boldsymbol{\sigma}}^{(tr)} \leqslant 0 \text{ (elastic process)}, \\ \dot{\bar{\varepsilon}}^{(pl)}\mathbf{n} & \text{if } f(\sigma) = \sigma_y \text{ and } \mathbf{n}^T\dot{\boldsymbol{\sigma}}^{(tr)} > 0 \text{ (plastic process)}. \end{cases} \tag{5.33}$$

Evolution equation for the equivalent (effective) plastic strain:

$$\dot{\bar{\varepsilon}}^{(pl)} = \sqrt{\tfrac{2}{3}}\,\frac{1}{2G}\,\frac{\mathbf{n}^T\dot{\boldsymbol{\sigma}}^{(tr)}}{1+(\zeta/3G)}\,. \tag{5.34}$$

Definition of the so-called trial stress rate:

$$\dot{\boldsymbol{\sigma}}^{(tr)} = \mathbf{C}\dot{\boldsymbol{\epsilon}}\,. \tag{5.35}$$

Definition of the unit normal to the yield surface:

$$\mathbf{n} = \sqrt{\tfrac{3}{2}}\,\frac{\bar{\boldsymbol{\sigma}}^D}{\sigma_y}\,. \tag{5.36}$$

As before, we use here the notation

$$\boldsymbol{\sigma}^D = \boldsymbol{\sigma}-\tfrac{1}{3}(\sigma_1+\sigma_2+\sigma_3)\{1 \;\; 1 \;\; 1 \;\; 0 \;\; 0 \;\; 0\},$$

$$\bar{\boldsymbol{\sigma}} = \boldsymbol{\sigma}-\boldsymbol{\alpha},$$

$$\dot{\bar{\varepsilon}}^{(pl)} = \sqrt{\tfrac{2}{3}\dot{\boldsymbol{\epsilon}}^{(pl)T}\dot{\boldsymbol{\epsilon}}^{(pl)}}\,.$$

ω is the material parameter which determines the proportion of isotropic and kinematic hardening or softening ($\omega = 0$—kinematic hardening, $\omega = 1$—isotropic hardening, intermediate values involve both mechanisms), ζ is the plastic modulus describing the slope of the effective stress vs effective plastic strain curve so that $\zeta > 0$ corresponds to strain hardening, $\zeta = 0$ to perfect plasticity and $\zeta < 0$ to strain softening.

We recall that the conditions of the RHS of eqs. (5.33), as well as the equation (5.34) follow from the following simple derivation:

$$\dot{\boldsymbol{\sigma}} = \mathbf{C}\dot{\boldsymbol{\epsilon}}-\frac{1}{h}\,\mathbf{C}\mathbf{n}(\mathbf{n}^T\dot{\boldsymbol{\sigma}}),$$

$$\dot{\boldsymbol{\sigma}} = \dot{\boldsymbol{\sigma}}^{(tr)}-\frac{1}{h}\,\mathbf{C}\mathbf{n}(\mathbf{n}^T\dot{\boldsymbol{\sigma}}),$$

$$\mathbf{n}^T\dot{\boldsymbol{\sigma}} = \mathbf{n}^T\dot{\boldsymbol{\sigma}}^{(tr)}-\frac{1}{h}\,\mathbf{n}^T\mathbf{C}\mathbf{n}(\mathbf{n}^T\dot{\boldsymbol{\sigma}}),$$

$$\mathbf{n}^T\dot{\boldsymbol{\sigma}} = \frac{h}{h+\mathbf{n}^T\mathbf{C}\mathbf{n}}\,\mathbf{n}^T\dot{\boldsymbol{\sigma}}^{(tr)},$$

$$h = \tfrac{2}{3}\zeta,$$

$\mathbf{Cn} = 2G\mathbf{n}$ (for elastically isotropic materials only),

$$\frac{\mathbf{n}^T\dot{\boldsymbol{\sigma}}}{h} = \frac{1}{2G}\,\frac{1}{1+(\zeta/3G)}\,\mathbf{n}^T\dot{\boldsymbol{\sigma}}^{(tr)},$$

$$\dot{\bar{\varepsilon}}^{(pl)} = \sqrt{\tfrac{2}{3}\dot{\boldsymbol{\epsilon}}^{(pl)T}\dot{\boldsymbol{\epsilon}}^{(pl)}} = \sqrt{\tfrac{2}{3}}\,\frac{\mathbf{n}^T\dot{\boldsymbol{\sigma}}}{h} = \sqrt{\tfrac{2}{3}}\,\frac{1}{2G}\,\frac{\mathbf{n}^T\dot{\boldsymbol{\sigma}}^{(tr)}}{1+(\zeta/3G)}\,.$$

We further note that the definitions of elastic and plastic processes given in eqs. (5.33) are more consistent from the computational standpoint than those employed previously, cf. Sec. 2.1.4. This is so because the conditions (5.33) invoke the elastic constitutive law whenever it makes sense to do so, that is, whenever the elastically-induced stress is not attempting to get outside of the yield surface. Otherwise, a plastic process is invoked and the consistency condition is used to define $\dot{\bar{\varepsilon}}^{(pl)}$, cf. Sec. 2.1.4, so that the stress remains on the yield surface. We also note that the use of the loading—unloading conditions in the form (5.33) makes it possible to deal directly with softening materials.

Eqs. (5.30)–(5.36) can be used to construct integration algorithms for elastic-plastic constitutive law. The widely accepted and effective algorithm is known as the radial return algorithm. It is based on the idea that the normal vector at the subsequent iteration is approximated in terms of the stress $\bar{\boldsymbol{\sigma}}^{(tr)D}$, cf. eq. (5.35), where

$$\bar{\boldsymbol{\sigma}}^{(tr)D} = \boldsymbol{\sigma}^{(tr)D} - \boldsymbol{\alpha}.$$

The iterative algorithm can be summarized as follows, $i = 1, 2, \ldots$:

Step 1. Obtain the i-th approximation to incremental strain $\Delta\boldsymbol{\epsilon}$ by solving eq. $(5.5)_2$ for $\delta\mathbf{r}^{(i)}$ and by computing $\mathbf{r}^{(i)}_{t+\Delta t}$ from eq. $(5.5)_3$ and $\Delta\boldsymbol{\epsilon}^{(i)}$ from the strain-displacement relation.

Step 2. Calculate the trial stress increment:

$$\Delta\boldsymbol{\sigma}^{(tr)(i+1)} = \mathbf{C}\Delta\boldsymbol{\epsilon}^{(i)}.$$

Step 3. Calculate the trial stresses:

$$\boldsymbol{\sigma}^{(tr)(i+1)} = \boldsymbol{\sigma}^{(tr)(i)} + \Delta\boldsymbol{\sigma}^{(tr)(i+1)},$$

$$\bar{\boldsymbol{\sigma}}^{(tr)(i+1)} = \boldsymbol{\sigma}^{(tr)(i+1)} - \boldsymbol{\alpha}^{(i)}.$$

Step 4. Calculate the deviatoric part of $\bar{\boldsymbol{\sigma}}^{(tr)(i+1)}$:

$$\bar{\boldsymbol{\sigma}}^{(tr)D(i+1)} = \bar{\boldsymbol{\sigma}}^{(tr)(i+1)} - \tfrac{1}{3}\left(\bar{\sigma}_1^{(tr)(i+1)} + \bar{\sigma}_2^{(tr)(i+1)} + \bar{\sigma}_3^{(tr)(i+1)}\right)\{1\ 1\ 1\ 0\ 0\ 0\}.$$

Step 5. Calculate the product of $\bar{\boldsymbol{\sigma}}^{(tr)D(i+1)}$ by itself:

$$A = \bar{\boldsymbol{\sigma}}^{(tr)D(i+1)T}\bar{\boldsymbol{\sigma}}^{(tr)D(i+1)}.$$

Step 6. Check the yield condition

$$\text{if } \tfrac{3}{2}A \leqslant \sigma_y^{(i)2} \text{ then } \boldsymbol{\sigma}^{(i+1)} = \boldsymbol{\sigma}^{(tr)(i+1)},$$
$$\sigma_y^{(i+1)} = \sigma_y^{(i)},$$
$$\boldsymbol{\alpha}^{(i+1)} = \boldsymbol{\alpha}^{(i)} \text{ (elastic process)}$$
go to Step 1,

$$\text{if } \tfrac{3}{2}A > \sigma_y^{(i)^2} \text{ then go to Step 7 (plastic process).}$$

Step 7. Calculate the approximation to the unit normal

$$\mathbf{n} = \sqrt{\frac{3}{2}}\,\frac{\bar{\boldsymbol{\sigma}}^{(tr)D(i+1)}}{\sigma_y^{(i)}}.$$

Step 8. Calculate the effective plastic strain increment

$$\Delta\bar{\varepsilon}^{(pl)} = \frac{1}{3G+\zeta}\left(\sqrt{\frac{3}{2}}\,||\bar{\boldsymbol{\sigma}}^{(tr)D(i+1)}|| - \sigma_y^{(i)}\right).$$

Step 9. Update

$$\boldsymbol{\sigma}^{(i+1)} = \boldsymbol{\sigma}^{(tr)(i+1)} - 2G\Delta\bar{\varepsilon}^{(pl)}\mathbf{n},$$
$$\sigma_y^{(i+1)} = \sigma_y^{(i)} + \omega\zeta\Delta\bar{\varepsilon}^{(pl)},$$
$$\boldsymbol{\alpha}^{(i+1)} = \boldsymbol{\alpha}^{(i)} + \tfrac{2}{3}(1-\omega)\zeta\Delta\bar{\varepsilon}^{(pl)}\mathbf{n},$$
$$\boldsymbol{\epsilon}^{(pl)(i+1)} = \boldsymbol{\epsilon}^{(pl)(i)} + \Delta\bar{\varepsilon}^{(pl)}\mathbf{n},$$
$$\bar{\varepsilon}^{(pl)(i+1)} = \bar{\varepsilon}^{(pl)(i)} + \Delta\bar{\varepsilon}^{(pl)}.$$

For the finite deformation case the constitutive equation (5.30) is usually modified to read, cf. the last equation of Sec. 1.3.3,

$$\mathring{\boldsymbol{\sigma}} = \mathbf{C}(\dot{\boldsymbol{\epsilon}} - \dot{\boldsymbol{\epsilon}}^{(pl)})$$

where $\mathring{\boldsymbol{\sigma}}$ is the rate of stress vector formed from the components of the so-called Truesdell rate of the Cauchy stress vector $\boldsymbol{\sigma}$. In the context of the updated Lagrangian description we note that the components of $\mathring{\boldsymbol{\sigma}}$ are equal to the components of the usual time derivative of the second Piola–Kirchhoff stress based on the current configuration—this property clearly results from the last equation in Sec. 1.3.3 by identifying the reference and current configurations of the body. Thus, the algorithm described in this section applies also to finite deformation problems provided the stresses are correctly interpreted and appropriate transformations are performed at the end of the step to update the reference configuration, [79].

Sometimes other objective stress rates are taken to appear in the rate-type constitutive equations describing finite deformation problems. An appropriate re-definition of the constitutive moduli **C** (addition of some terms which are all bi-linear in stress and velocity gradient components) is then needed.

The above simple generalization of the kinematically linearized theory to cover finite deformation problems neglects the numerical errors related to the violation of the objectivity conditions within each time step—details of an approach circumventing this problem are beyond the scope of this book.

5.4 Initial and linearized buckling

The objective of any nonlinear analysis is basically to estimate the maximum load that a structure can carry prior to its collapse. However, due to the high computational cost involved in solving problems that are posed incrementally we are sometimes satisfied with a partial information on the critical behaviour of the solid. To get such an information we try to find an answer to the following question:

Assume we know the fundamental elastic solution to a given solid mechanics problem. Is there any possibility, and for which value of load, for another solution into which the solid could bifurcate if it were slightly perturbed from its primary equilibrium path?

The problem is illustrated in Fig. 45. The first load level at which there is a solution nonuniqueness is called the buckling load. Let us recall again

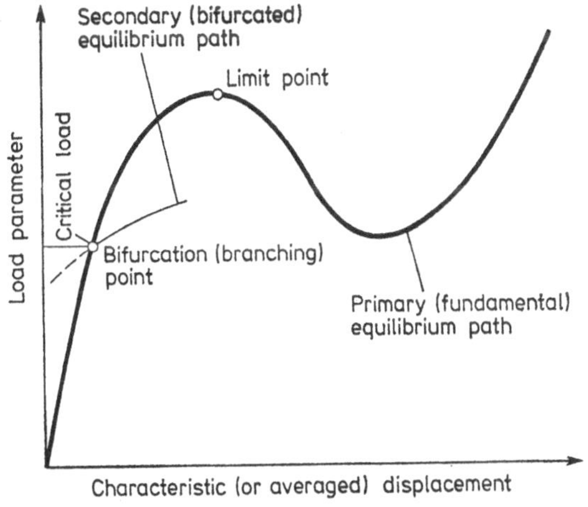

Fig. 45 Linearized buckling

the fundamental set of equations describing the nonlinear static response of an elastic solid within the explicit finite difference approximation and updated Lagrangian description

$$[\mathbf{K}^{(el)}(\mathbf{r}_t)+\mathbf{K}^{(\sigma)}(\mathbf{r}_t, \boldsymbol{\sigma}_t)]\Delta\mathbf{r} = \Delta\mathbf{R}. \tag{5.37}$$

It is here assumed that the system is characterized by the existing nodal displacements $\mathbf{r}_t$ and elemental (monitored at the integration points) stresses $\boldsymbol{\sigma}_t$ all being the result of the previous load history up to its value $\mathbf{R}_t$. The argument $\mathbf{r}_t$ in $\mathbf{K}^{(el)}$ and $\mathbf{K}^{(\sigma)}$ indicates explicitly that these matrices are formed in the current configuration of the structure. Let us now consider a given load increment $\Delta\mathbf{R}$ and put

$$\Delta\mathbf{R}^{(\mu)} = \mu\Delta\mathbf{R} \tag{5.38}$$

where μ is a scalar parameter ranging from zero (beginning of the incremental step) to one (end of the step). Since the incremental problem as described by eq. (5.37) is linear, the load $\mu\Delta\mathbf{R}$ will give rise to the displacement increment $\mu\Delta\mathbf{r}$ and to the corresponding stress increment $\mu\Delta\boldsymbol{\sigma}$. Having solved the problem for any fixed μ we have for the next step the equation

$$[\mathbf{K}^{(el)}(\mathbf{r}_t+\mu\Delta\mathbf{r})+\mathbf{K}^{(\sigma)}(\mathbf{r}_t+\mu\Delta\mathbf{r},\ \boldsymbol{\sigma}_t+\mu\Delta\boldsymbol{\sigma})]\Delta\mathbf{r}' = \Delta\mathbf{R}' \tag{5.39}$$

where $\Delta\mathbf{r}' = \mathbf{r}_{t+2\Delta t}-\mathbf{r}_{t+\Delta t}$, $\Delta\mathbf{R}' = \mathbf{R}_{t+2\Delta t}-\mathbf{R}_{t+\Delta t}$. Our objective now is to find such a value of the parameter $\mu = \mu_{cr}$ that eq. (5.39) admits more than one solution $\Delta\mathbf{r}'$. This will clearly mean that the load value $\mathbf{R}_t+\mu_{cr}\Delta\mathbf{R}$ corresponds to the buckling behaviour of the solid. To simplify the analytic description of the problem we assume now that (the so-called initial buckling):

(a) the prebuckling deformations are so small that the stiffness matrices can be formed on the initial undeformed configuration so that $\mathbf{K}^{(el)} = \text{const}$, $\mathbf{K}^{(\sigma)} = \mathbf{K}^{(\sigma)}(\boldsymbol{\sigma}_t)$,

(b) the whole value of load $\mathbf{R}$ has been applied proportionally to a single parameter λ such that

$$\lambda\mathbf{R}^* = \mathbf{R}_t+\mu\Delta\mathbf{R} \tag{5.40}$$

where $\mathbf{R}^*$ is a reference load vector,

(c) the solid is linearly elastic up to buckling, i.e.

$$\begin{aligned}\lambda\mathbf{r}^* &= \mathbf{r}_t+\mu\Delta\mathbf{r},\\ \lambda\boldsymbol{\sigma}^* &= \boldsymbol{\sigma}_t+\mu\Delta\boldsymbol{\sigma},\end{aligned} \tag{5.41}$$

where $\mathbf{r}^*$ and $\boldsymbol{\sigma}^*$ correspond to the load $\mathbf{R}^*$.

Using now the linearity of the initial stress matrix with respect to $\boldsymbol{\sigma}$ we arrive at

$$[\mathbf{K}^{(el)}+\lambda\mathbf{K}^{(\sigma)}(\boldsymbol{\sigma}^*)]\Delta\mathbf{r}' = \Delta\mathbf{R}'. \tag{5.42}$$

If now for a certain value of $\lambda = \lambda_{cr}$ there are two different solutions $\Delta\mathbf{r}_1'$, $\Delta\mathbf{r}_2'$ to eq. (5.42) for the same load $\Delta\mathbf{R}'$ then by subtracting sidewise

$$[\mathbf{K}^{(el)}+\lambda_{cr}\mathbf{K}^{(\sigma)}(\boldsymbol{\sigma}^*)]\mathbf{v} = \mathbf{0} \tag{5.43}$$

with $\mathbf{v} = \Delta\mathbf{r}_1' - \Delta\mathbf{r}_2'$. Eq. (5.43) is clearly the condition for the load $\lambda_{cr}\mathbf{R}^*$ to be the buckling load. In terms of linear algebra, it represents the generalized eigenvalue problem. The solution to eq. (5.43) consists of N nontrivial eigenpairs $(\lambda_1, \mathbf{v}_1), \ldots, (\lambda_N, \mathbf{v}_N)$ with the eigenvalues $\lambda_1, \lambda_2, \ldots, \lambda_N$ and the eigenvectors $\mathbf{v}_1, \mathbf{v}_2, \ldots, \mathbf{v}_N$. Usually, in buckling analysis we are interested in the smallest eigenvalue λ_{cr} (and possibly the corresponding buckling mode $\mathbf{v}_{cr}$) only.

To sum up the considerations of this section we characterize the initial buckling behaviour as follows:

(a) the solid behaves linearly under proportionally increasing load $\lambda\mathbf{R}^*$ up to the value of the load parameter $\lambda = \lambda_{cr}$,

(b) configuration changes up to the buckling load are negligible,

(c) for $\lambda = \lambda_{cr}$ the solid can buckle into the secondary equilibrium form with the difference between the two solutions described by the buckling mode vector $\mathbf{v}_{cr}$,

(d) the postbuckling behaviour ($\lambda > \lambda_{cr}$) can not be found by using the simplified analysis presented in this section—the full step-by-step analysis would be required to solve such a problem.

Alternatively, the Koiter's postbuckling theory may be used.

A more precise estimate of the buckling load may be obtained by dropping the condition (b) and assuming that configuration changes can approximately be represented by means of the linear part of the initial stiffness matrix, cf. eq. (3.83). The buckling problem takes then the form

$$\{\mathbf{K}^{(el)} + \lambda_{cr}[\mathbf{K}^{(\sigma)}(\boldsymbol{\sigma}^*) + \bar{\mathbf{K}}^{(u)}(\mathbf{r}^*)]\}\mathbf{v} = \mathbf{0} \tag{5.44}$$

which is often referred to as the linearized buckling. Clearly, formulations (5.43) and (5.44) will lead to differences in critical load assessment only if the matrix $\bar{\mathbf{K}}^{(u)}$ is different from zero up to buckling—detailed comparisons are discussed in [31], for instance.

It should be emphasized that an extremely simplified picture was here discussed only. In a more general analysis the incremental solution is carried out including any geometric and material nonlinearities, and at every load step it is checked whether a bifurcation is possible. Once buckling has been found we may wish to continue the analysis beyond it. Without going into details of this highly complex problem of structural stability we just suggest here a simplified procedure which has been applied with some success in [77]. Having found the bifurcation point and the corresponding buckling mode the next load step may be performed by constraining the response in such a way that only one of the two (or more) possible postbifurcation paths can be followed. Thus the algebraic problem becomes unique again. After a few steps the original incremental/iterative scheme can be resumed and the analysis continued along the selected postbifurcation path up to the next singular point.

The reader is referred to [31], [49], [98], [136] for further discussion of stability problems in the context of the finite element analysis.

5.5 Nonlinear dynamics

In this section we attempt to describe some numerical procedures for the solution of sets of ordinary differential equations of the second order. In the discussion of numerical algorithms of temporal integration particular attention should as a rule be focused on the concept of numerical stability, which decisively influences performance of each and every algorithm. Some notion of stability is required to prove convergence whereas some, usually stronger, stability definitions are desirable for a qualitative assessment of the response with respect to properties of the governing differential equations. For linear problems there seems to be little dispute as to how to proceed with such an analysis. However, for nonlinear problems the situation is much more complex and its description, if possible at all, would require another book. Therefore, keeping with the character of this presentation, no attempts will be made to precisely characterize the properties of algorithms to be described. Instead, rather practical indications will be given with reference to most often used procedures.

Let us recall the fundamental set of equations describing the fundamental problem of nonlinear dynamics in the finite element setting

$$\mathbf{M}\Delta\ddot{\mathbf{r}}+\mathbf{C}\Delta\dot{\mathbf{r}}+\mathbf{K}\Delta\mathbf{r} = \Delta\mathbf{R} \tag{5.45}$$

which can also be presented as, cf. eq. (4.41)

$$\mathbf{M}\ddot{\mathbf{r}}_{t+\Delta t}+\mathbf{C}\dot{\mathbf{r}}_{t+\Delta t}+\mathbf{K}_{t+\Delta t}\Delta\mathbf{r} = \mathbf{R}_{t+\Delta t}-\mathbf{F}_t. \tag{5.46}$$

As a matter of fact, eq. (5.46) represents the dynamic equilibrium condition for the discretized structure at time $t+\Delta t$. By using some finite difference expressions (an example will be shown below) to approximate the acceleration and velocity vectors $\ddot{\mathbf{r}}_{t+\Delta t}$, $\dot{\mathbf{r}}_{t+\Delta t}$ in terms of the unknown incremental displacement vector (and some other quantities known at time t) this equation can be solved for $\Delta\mathbf{r}$. It should also be again emphasized that eq. (5.46) holds for the static analysis provided the mass and damping effects are neglected. For we simply have in such a case

$$\mathbf{K}_{t+\Delta t}\Delta\mathbf{r} = \mathbf{R}_{t+\Delta t}-\mathbf{F}_t. \tag{5.47}$$

Consider now the dynamic equilibrium of the system at time t. This can be described by eq. (5.46) specified for the time interval $[t-\Delta t, t]$ thus reading

$$\mathbf{M}\ddot{\mathbf{r}}_t+\mathbf{C}\dot{\mathbf{r}}_t+\mathbf{K}_t\Delta\mathbf{r} = \mathbf{R}_t-\mathbf{F}_{t-\Delta t}. \tag{5.48}$$

Since, within the incremental linearization error

$$\mathbf{F}_{t-\Delta t}+\mathbf{K}_t\Delta\mathbf{r} = \mathbf{F}_t$$

eq. (5.48) can be rewritten as

$$\mathbf{M}\ddot{\mathbf{r}}_t+\mathbf{C}\dot{\mathbf{r}}_t = \mathbf{R}_t-\mathbf{F}_t. \qquad (5.49)$$

By using some finite difference expressions to approximate the acceleration and velocity vectors $\ddot{\mathbf{r}}_t$ and $\dot{\mathbf{r}}_t$ in terms of the unknown incremental displacement vector $\Delta\mathbf{r}_{(t,\,t+\Delta t)}$ (and other quantities known up to time t), eq. (5.49) can be solved for $\Delta\mathbf{r}$. Such an algorithm is referred to as the *explicit integration scheme* while the previous approach defined by eq. (5.46) is called the *implicit integration scheme*. A noteworthy point is that the step-by-step solution scheme based on the explicit approach does not reduce to the incremental static analysis, if mass and damping effects are neglected.

Both the methods are illustrated below. We begin with a version of the explicit scheme and assume

$$\begin{aligned}\dot{\mathbf{r}}_t &= \frac{1}{2\Delta t}(-\mathbf{r}_{t-\Delta t}+\mathbf{r}_{t+\Delta t}) = \frac{1}{2\Delta t}(-\mathbf{r}_{t-\Delta t}+\mathbf{r}_t+\Delta\mathbf{r}),\\ \ddot{\mathbf{r}}_t &= \frac{1}{(\Delta t)^2}(\mathbf{r}_{t-\Delta t}-2\mathbf{r}_t+\mathbf{r}_{t+\Delta t}) = \frac{1}{(\Delta t)^2}(\mathbf{r}_{t-\Delta t}-\mathbf{r}_t+\Delta\mathbf{r}).\end{aligned} \qquad (5.50)$$

According to the form of these equations the method is referred to as the central difference scheme. Substituting the relations for $\dot{\mathbf{r}}_t$ and $\ddot{\mathbf{r}}_t$ into eq. (5.49) we obtain

$$\left[\frac{1}{(\Delta t)^2}\mathbf{M}+\frac{1}{2\Delta t}\mathbf{C}\right]\Delta\mathbf{r} = \mathbf{R}_t-\frac{1}{(\Delta t)^2}\mathbf{M}(\mathbf{r}_{t-\Delta t}-\mathbf{r}_t)-\frac{1}{2\Delta t}\mathbf{C}(-\mathbf{r}_{t-\Delta t}+\mathbf{r}_t)-\mathbf{F}_t \qquad (5.51)$$

from which we can solve for $\Delta\mathbf{r}$. Having calculated $\Delta\mathbf{r}$ we may proceed by finding the vector $\mathbf{r}_{t+\Delta t}$ as $\mathbf{r}_{t+\Delta t} = \mathbf{r}_t+\Delta\mathbf{r}$ which opens the way to solve eq. (5.51) for the next displacement increment. It is noted that such an integration scheme does not require a factorization of the effective stiffness matrix, provided the mass and damping matrices are diagonal. Moreover, none of the global matrices needs in fact to be assembled. The advantages of using the central difference (or any other explicit) method becomes apparent. Since no matrices of the complete structure need to be calculated, the solution can essentially be carried out on the element level and relatively little high-speed storage is required. Considering the shortcomings of the central difference method, it must be first recognized that the effectiveness of the procedure depends on the use of the diagonal mass and damping matrices. However, this is usually not a very serious disadvantage because a possible loss of accuracy can be compensated by using a fine-enough finite element discretization.

A further observation is that using the central difference method the calculation of $\Delta\mathbf{r}$ involves $\mathbf{r}_t$ and $\mathbf{r}_{t-\Delta t}$. Therefore, to calculate the solution at time Δt (the first step), a special starting procedure must be used. Since $\hat{\mathbf{r}}_0$, $\hat{\dot{\mathbf{r}}}_0$ and $\ddot{\mathbf{r}}_0$ are known (note that with $\hat{\mathbf{r}}_0$ and $\hat{\dot{\mathbf{r}}}_0$ known, $\ddot{\mathbf{r}}_0$ can be computed using the dynamic equilibrium condition at time 0), the relations (5.50) may be used to get $\mathbf{r}_{-\Delta t}$ as

$$\mathbf{r}_{-\Delta t} = \hat{\mathbf{r}}_0 - \Delta t\,\hat{\dot{\mathbf{r}}}_0 + \frac{(\Delta t)^2}{2}\ddot{\mathbf{r}}_0. \tag{5.52}$$

The last very important consideration in the use of the central difference scheme (and any other explicit scheme) is that for algorithm stability reasons this integration method requires the time step Δt to be smaller than a critical value Δt_{cr} which can be calculated from the mass, damping and stiffness properties of the complete element assemblage. Therefore, the central difference scheme is said to be conditionally stable. If a time step is used larger than Δt_{cr} the integration is unstable which means that any errors resulting from the numerical integration or round-off in the computer grow and make the solution worthless in many situations.

The second direct integration method to be described in this chapter is a typical implicit integration scheme known as the Newmark method. The following finite difference expansions are employed in this method:

$$\begin{aligned} \mathbf{r}_{t+\Delta t} &= \mathbf{r}_t + \dot{\mathbf{r}}_t\Delta t + [(\tfrac{1}{2}-\alpha)\ddot{\mathbf{r}}_t + \alpha\ddot{\mathbf{r}}_{t+\Delta t}](\Delta t)^2, \\ \dot{\mathbf{r}}_{t+\Delta t} &= \dot{\mathbf{r}}_t + [(1-\delta)\ddot{\mathbf{r}}_t + \delta\ddot{\mathbf{r}}_{t+\Delta t}]\Delta t \end{aligned} \tag{5.53}$$

where α and δ are parameters that are selected to obtain best stability and accuracy characteristics of the integration scheme. Solving from eq. $(5.53)_1$ for $\ddot{\mathbf{r}}_{t+\Delta t}$ in terms of $\mathbf{r}_{t+\Delta t}$ and then substituting for $\ddot{\mathbf{r}}_{t+\Delta t}$ into $(5.53)_2$ we obtain equations for $\ddot{\mathbf{r}}_{t+\Delta t}$ and $\dot{\mathbf{r}}_{t+\Delta t}$, each in terms of the unknown displacements $\mathbf{r}_{t+\Delta t}$ (or displacement increments $\Delta\mathbf{r}$) only. These two relations for $\dot{\mathbf{r}}_{t+\Delta t}$ and $\ddot{\mathbf{r}}_{t+\Delta t}$ are substituted into eq. (5.46) yielding

$$\begin{aligned} &\left[\frac{1}{(\alpha\Delta t)^2}\mathbf{M} + \frac{\delta}{\alpha\Delta t}\mathbf{C} + \mathbf{K}_{t+\Delta t}\right]\Delta\mathbf{r} \\ &= \mathbf{R}_{t+\Delta t} + \mathbf{M}\left[\frac{1}{\alpha\Delta t}\dot{\mathbf{r}}_t + \left(\frac{1}{2\alpha}-1\right)\ddot{\mathbf{r}}_t\right] + \\ &\qquad + \mathbf{C}\left[\left(\frac{\delta}{\alpha}-1\right)\dot{\mathbf{r}}_t + \frac{\Delta t}{2}\left(\frac{\delta}{\alpha}-2\right)\ddot{\mathbf{r}}_t\right] - \mathbf{F}_t, \end{aligned} \tag{5.54}$$

from which one can solve for $\Delta\mathbf{r}$. A basic difference between the Newmark method and the central difference method is the appearance of the stiffness matrix $\mathbf{K}_{t+\Delta t}$ as a factor to the required incremental displacement $\Delta\mathbf{r}$. The term $\mathbf{K}\Delta\mathbf{r}$ appears because in eq. (5.46) equilibrium is considered at time $t+\Delta t$ and not at time t as in the central difference method. However, for some values of parameters α and δ there is no critical time-step limit in this case, and Δt can be selected much larger. The Newmark algorithm can also be used for static problems provided the mass and damping effects are neglected.

One additional remark must be made at this place with reference to the methods discussed so far. The direct solution of the incremental finite element equations (5.51) may be considered entirely consistent with the dynamic equilibrium conditions (5.49) at time t and no further improvements are necessary. This is so because the configuration at time t is known. On the contrary, in the implicit approach the configuration at time $t+\Delta t$ for which the equilibrium conditions are established is unknown. This makes it in general necessary to perform additional iterations to find a more accurate solution within each interval of time, and such a solution can be obtained effectively using a Newton-type iteration. This amounts to seeking the solution of the equations

$$\mathbf{M}\ddot{\mathbf{r}}^{(i)}_{t+\Delta t}+\mathbf{C}\dot{\mathbf{r}}^{(i)}_{t+\Delta t}+\mathbf{K}^{(i-1)}_{t+\Delta t}\,\delta\mathbf{r}^{(i)} = \mathbf{R}_{t+\Delta t}-\mathbf{F}^{(i-1)}_{t+\Delta t}, \qquad i = 1, 2, \ldots, \tag{5.55}$$

in which the vectors $\ddot{\mathbf{r}}_{t+\Delta t}$ and $\dot{\mathbf{r}}_{t+\Delta t}$ are given by eqs. (5.50), the internal force vector $\mathbf{F}^{(i-1)}_{t+\Delta t}$ corresponds to the state of stress in the configuration with the displacements

$$\mathbf{r}^{(i-1)}_{t+\Delta t} = \mathbf{r}^{(i-2)}_{t+\Delta t}+\delta\mathbf{r}^{(i-1)}. \tag{5.56}$$

$\delta\mathbf{r}^{(i)}$ is the i-th correction to the incremental displacement vector $\Delta\mathbf{r}$ and i stands for the iteration number. Eq. (5.55) defines the full Newton–Raphson iteration scheme which must be considered together with initial conditions

$$\begin{aligned} \mathbf{K}^{(0)}_{t+\Delta t} &= \mathbf{K}_t, \\ \mathbf{F}^{(0)}_{t+\Delta t} &= \mathbf{F}_t, \\ \mathbf{r}^{(0)}_{t+\Delta t} &= \mathbf{r}_t. \end{aligned} \tag{5.57}$$

The iteration continues until appropriate termination criteria are met. Although the use of the full Newton–Raphson iteration, in which the updating and factorizing of the effective stiffness matrix takes place anew in each iteration, can be efficient in some specific nonlinear analyses, the use of the method is usually not very efficient in general geometric and material nonlinear response calculations. We can then better use a modified iteration scheme given by

$$\mathbf{M}\ddot{\mathbf{r}}^{(i)}_{t+\Delta t}+\mathbf{C}\dot{\mathbf{r}}^{(i)}_{t+\Delta t}+\mathbf{K}_t\,\delta\mathbf{r}^{(i)} = \mathbf{R}_{t+\Delta t}-\mathbf{F}^{(i-1)}_{t+\Delta t}, \qquad i = 1, 2, \ldots, \tag{5.58}$$

in which the effective stiffness matrix need be factorized only once at each time step. It proves sometimes advantageous to choose another form of the modified Newton–Raphson approach by using the relation

$$\mathbf{M}\ddot{\mathbf{r}}^{(i)}_{t+\Delta t}+\mathbf{C}\dot{\mathbf{r}}^{(i)}_{t+\Delta t}+\mathbf{K}_{\bar{t}}\,\delta\mathbf{r}^{(i)} = \mathbf{R}_{t+\Delta t}-\mathbf{F}^{(i-1)}_{t+\Delta t}, \qquad i = 1, 2, \ldots, \tag{5.59}$$

in which the index $\bar{t}$ takes on some a priori selected values from the set $\{0, \Delta t, 2\Delta t, \ldots\}$ thus yielding a significant flexibility in controlling the errors and efficiency of the approach. The possibility of using the quasi-Newton iterations should also be indicated in this context.

It is emphasized that the use of any formulation (5.55), (5.58) or (5.59) should lead to the same results provided all the algorithms are convergent. In particular, all nonlinearities are taken fully into account in the evaluation of the vector $\mathbf{F}^{(i)}_{t+\Delta t}$. Clearly, the number of algebraic operations required in each case to reach the convergence may differ considerably.

It should also be pointed out that a number of so-called mixed (implicit/explicit) algorithms have recently been developed, [61], [63], based essentially on breaking down the region Ω into two subregions in which explicit and implicit time integrations are carried out, respectively. Such methods have undoubtedly great potential but their description is beyond the scope of this book.

It may generally be said that the number of operations required in the direct integration methods is directly proportional to the number of time steps used in the analysis. Therefore, the use of direct integration can be expected to be effective when the response for a relatively short duration is required. However, if the integration must be carried out for many time steps, it may be more effective to first transform the equilibrium equation (5.46) into a form in which the step-by-step solution is less costly. The basic operation in this solution is a change of basis from the N nodal point generalized displacements to P modal generalized displacements, $P \ll N$ prior to the step-by-step solution. In this transformation we use

$$\mathbf{r}_{t+\Delta t} = \mathbf{\Phi}\boldsymbol{\rho}_{t+\Delta t} \tag{5.60}$$

where $\boldsymbol{\rho}_{t+\Delta t}$ are the P modal generalized displacements at time $t+\Delta t$ and $\mathbf{\Phi}$ is a $N\times P$ matrix

$$\mathbf{\Phi}_{N\times P} = [\mathbf{\Phi}_1 \;\; \mathbf{\Phi}_2 \;\; \ldots \;\; \mathbf{\Phi}_P] \tag{5.61}$$

consisting of the eigenvectors $\mathbf{\Phi}_{1_{N\times 1}}, \ldots, \mathbf{\Phi}_{P_{N\times 1}}$ of the linearized eigenproblem at time 0

$$[\mathbf{K}_0-\omega^2_{0,\alpha}\mathbf{M}]\mathbf{\Phi}_\alpha = 0, \qquad \alpha = 1, 2, \ldots, P, \tag{5.62}$$

$\omega_{0,\alpha}$, $\alpha = 1, 2, \ldots, P$, being the natural frequencies of the system. Because of the condition $P \ll N$ eq. (5.60) holds in an approximate manner only.

However, by selecting P sufficiently large the error in the solution can be made arbitrarily small.

Substituting from eq. (5.60) into eq. (5.59) written for $\bar{t} = 0$ yields

$$\ddot{\boldsymbol{\rho}}^{(i)}_{t+\Delta t} + \boldsymbol{\Lambda}\dot{\boldsymbol{\rho}}^{(i)}_{t+\Delta t} + \boldsymbol{\Omega}^2 \Delta\boldsymbol{\rho}^{(i)} = \boldsymbol{\Phi}^T(\mathbf{R}_{t+\Delta t} - \mathbf{F}^{(i-1)}_{t+\Delta t}) \tag{5.63}$$

where $\boldsymbol{\Omega}^2$ is a $P \times P$ diagonal matrix listing the frequencies squared on its diagonal,

$$\boldsymbol{\Omega}^2 = \begin{bmatrix} \omega_{0,1}^2 & & & \\ & \omega_{0,2}^2 & & 0 \\ & & \ddots & \\ & 0 & & \omega_{0,P}^2 \end{bmatrix} \tag{5.64}$$

and, assuming the so-called proportional damping,

$$\boldsymbol{\Lambda} = \begin{bmatrix} 2\xi_1\omega_{0,1} & & & \\ & 2\xi_2\omega_{0,2} & & 0 \\ & & \ddots & \\ & 0 & & 2\xi_P\omega_{0,P} \end{bmatrix} \tag{5.65}$$

where the ξ_α, $\alpha = 1, 2, \ldots, P$, are the modal damping ratios corresponding to $\omega_{0,\mathbf{P}}$, respectively.

It must be pointed out that different from linear analysis, the incremental equilibrium equations in the new basis, eq. (5.63), are still coupled because the nodal point vector $\mathbf{F}^{(i-1)}_{t+\Delta t}$ can only be evaluated once all displacements are known

$$\mathbf{r}^{(i-1)}_{t+\Delta t} = \boldsymbol{\Phi}\boldsymbol{\rho}^{(i-1)}_{t+\Delta t}. \tag{5.66}$$

Therefore, the solution of the P equations (5.63) must be performed simultaneously. However, since $\boldsymbol{\Lambda}$ and $\boldsymbol{\Omega}$ are diagonal matrices and only P equations are considered, $P \ll N$, the solution of the eigenproblem (5.62) plus the step-by-step solution of eq. (5.63) using, for instance, the Newmark method, may turn out significantly more cost-effective than the direct step-by-step solution of eq. (5.59). The numerical gain depends on how significant are the nonlinearities in the system, though. The reader interested in using modal superposition methods in nonlinear dynamics should consult [16], [79], [11], for instance.

5.6 Dynamic stability under non-periodic loads

The analysis of structural buckling as illustrated in Sec. 5.4 may now be considered a classical subject with the tremendous practical importance in all attempts to realistically assess the structural safety. In the last few decades the subject has attracted literally thousands of researchers which resulted in the development of refined and firm foundations for many branches of structural stability under static loads. On the contrary, the development of both theoretical and practical aspects of stability under dynamic loadings has not yet reached a comparable level of maturity. In particular, practically oriented treatment of dynamic structural buckling has undoubtedly been hampered in the past by the lack of a simple and appealing algebraic representation in terms of the tangent stiffness matrix singularity so useful in static analysis. What has often been only pursued is for instance the repetitive solution of the structural response for increasing load parameters with the attempt to identify the so-called dynamic snap-through behaviour, [77], [82]. The amount of computer time necessary, the lack of precisely defined stability criterion and the negligence of any solution path branching and related problems of imperfection sensitivity are clearly the major deficiences of such an approach. The classical papers on dynamic structural buckling, [32], are of limited value since they are based on a number of simplifying assumptions (inertial forces in the pre-buckling state are neglected, some nonlinear terms are omitted in the equations of motion, dynamic and static buckling shapes are assumed identical) which are not applicable in the analysis to follow in this section.

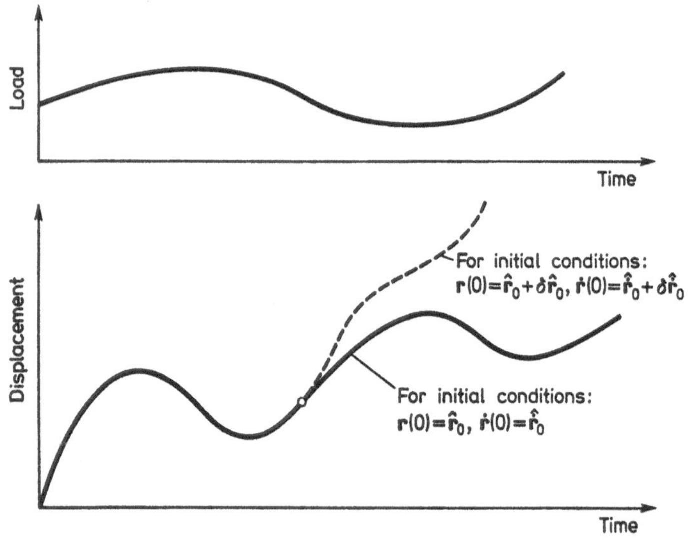

Fig. 46 Dynamic "quasi-bifurcation"

It has been observed, [81], [103], [104] that there might exist some "bifurcated" solutions in structural dynamics. A hypothetical situation which illustrates such a phenomenon is shown in Fig. 46—specific structural mechanics examples related to such a behaviour have been published in the recent literature, [77], [82]. Here, we shall present a theoretical substantiation of the dynamic buckling phenomenon and give (in Chapter 6) a number of simple illustrations stressing the practical significance of the approach.

Let us consider the (discrete in space) representation of the nonlinear dynamic behaviour of structures without damping in the form, cf. eq. (5.49)

$$\mathbf{M}\ddot{\mathbf{r}}(t)+\mathbf{F}[\mathbf{r}(t)] = \mathbf{R}(t) \tag{5.67}$$

where, as before, **F** is the internal nodal load vector calculated from stresses. The notation **F**[**r**] is used here for the sake of simplicity only—the dependence of **F** on the process history is not excluded. The initial conditions are

$$\mathbf{r}(0) = \hat{\mathbf{r}}_0, \quad \dot{\mathbf{r}}(0) = \hat{\dot{\mathbf{r}}}_0 \tag{5.68}$$

$\hat{\mathbf{r}}_0, \hat{\dot{\mathbf{r}}}_0$ being given vectors. Let us assume for the future reference that the mass matrix is positive-definite, which is true for both the consistent and diagonal mass matrices provided the massless degrees of freedom have been condensed in the latter approach. We rewrite eq. (5.67) as

$$\ddot{\mathbf{r}}(t) = \overset{-1}{\mathbf{M}}\{\mathbf{R}(t)-\mathbf{F}[\mathbf{r}(t)]\} \tag{5.69}$$

or, simpler still, as

$$\ddot{\mathbf{r}}(t) = \mathbf{P}[t; \mathbf{r}(t)] \tag{5.70}$$

$\mathbf{P} = \overset{-1}{\mathbf{M}}(\mathbf{R}-\mathbf{F})$ being the generalized "force" vector. The solution to eq. (5.70) may generally be written as

$$\mathbf{r}(t) = \mathbf{p}(t; \hat{\mathbf{r}}_0, \hat{\dot{\mathbf{r}}}_0). \tag{5.71}$$

Changing now the initial values $\hat{\mathbf{r}}_0$, $\hat{\dot{\mathbf{r}}}_0$ to $\hat{\mathbf{r}}_0+\delta\hat{\mathbf{r}}_0$, $\hat{\dot{\mathbf{r}}}_0+\delta\hat{\dot{\mathbf{r}}}_0$, respectively, we have the resulting motion given by

$$\mathbf{r}(t)+\delta\mathbf{r}(t) = \mathbf{p}(t; \hat{\mathbf{r}}_0+\delta\hat{\mathbf{r}}_0, \hat{\dot{\mathbf{r}}}_0+\delta\hat{\dot{\mathbf{r}}}_0) \tag{5.72}$$

with $\delta\mathbf{r}(t)$ being the perturbation of motion. According to the Liapunov stability concept we say that the motion $\mathbf{p}(t; \hat{\mathbf{r}}_0, \hat{\dot{\mathbf{r}}}_0)$ is *stable* if a small distance between the points $(\hat{\mathbf{r}}_0, \hat{\dot{\mathbf{r}}}_0)$ and $(\hat{\mathbf{r}}_0+\delta\hat{\mathbf{r}}_0, \hat{\dot{\mathbf{r}}}_0+\delta\hat{\dot{\mathbf{r}}}_0)$ in the phase space implies the distance between the points $(\mathbf{r}(t), \dot{\mathbf{r}}(t))$, $(\mathbf{r}(t)+\delta\mathbf{r}(t), \dot{\mathbf{r}}(t)+\delta\dot{\mathbf{r}}(t))$ be small for all $t > 0$. Knowing the explicit dependence of $\mathbf{r}(t)$ on $\hat{\mathbf{r}}_0$, $\hat{\dot{\mathbf{r}}}_0$, cf. eq. (5.71), it is an easy matter to check the above criterion directly. Our goal is, however, to formulate a criterion which will make it possible to analyze the stability on the basis of the fundamental solution only, that is the solution obtained for one particular set of the initial values $\hat{\mathbf{r}}_0$, $\hat{\dot{\mathbf{r}}}_0$. Using eqs. (5.70) and (5.72) we obtain the relationship

$$\delta\ddot{\mathbf{r}}(t) = \mathbf{P}(t; \mathbf{r}+\delta\mathbf{r}) - \mathbf{P}(t; \mathbf{r}) = \delta\mathbf{P}(t; \mathbf{r}, \delta\mathbf{r}) \tag{5.73}$$

or, by eq. (5.69) under the assumption that the load history is the same for both the fundamental and perturbed processes, the relationship

$$\delta\ddot{\mathbf{r}}(t) = \overset{-1}{\mathbf{M}}[-\mathbf{F}(\mathbf{r}+\delta\mathbf{r}) + \mathbf{F}(\mathbf{r})]. \tag{5.74}$$

Linearizing the right-hand side of eq. (5.74) with respect to $\delta\mathbf{r}$ we arrive at

$$\delta\ddot{\mathbf{r}}(t) = \overset{-1}{\mathbf{M}}\left[-\frac{\partial\mathbf{F}}{\partial\mathbf{r}}\,\delta\mathbf{r}(t)\right] \tag{5.75}$$

or, introducing the concept of the tangent stiffness matrix $\mathbf{K}^{(T)} = \dfrac{\partial\mathbf{F}}{\partial\mathbf{r}}$, at

$$\delta\ddot{\mathbf{r}}(t) = -\overset{-1}{\mathbf{M}}\mathbf{K}^{(T)}[\mathbf{r}(t)]\,\delta\mathbf{r}(t). \tag{5.76}$$

It can now be easily shown that for any deviated motion, stable or not, there is the following identity

$$\tfrac{1}{2}\delta\mathbf{r}(t)^T\delta\mathbf{r}(t) = \tfrac{1}{2}\delta\hat{\mathbf{r}}_0^T\delta\hat{\mathbf{r}}_0 + \int_0^t \Big[\delta\hat{\dot{\mathbf{r}}}_0^T\delta\hat{\mathbf{r}}_0 + \int_0^{t'} \delta\ddot{\mathbf{r}}(t'')^T\delta\mathbf{r}(t'')\,\mathrm{d}t'' + \int_0^{t'} \delta\dot{\mathbf{r}}(t'')^T\delta\dot{\mathbf{r}}(t'')\,\mathrm{d}t''\Big]\,\mathrm{d}t' \tag{5.77}$$

which results directly from the trivial relationships

$$\begin{aligned}
&\int_0^{t'} \delta\ddot{\mathbf{r}}(t'')^T\delta\mathbf{r}(t'')\,\mathrm{d}t'' \\
&\qquad = \int_0^{t'} \left\{\frac{\mathrm{d}}{\mathrm{d}t''}[\delta\dot{\mathbf{r}}(t'')^T\delta\mathbf{r}(t'')] - \delta\dot{\mathbf{r}}(t'')^T\delta\dot{\mathbf{r}}(t'')\right\}\mathrm{d}t'', \\
&\int_0^t \left[\int_0^{t'} \delta\ddot{\mathbf{r}}(t'')^T\delta\mathbf{r}(t'')\,\mathrm{d}t'' + \int_0^{t'} \delta\dot{\mathbf{r}}(t'')^T\delta\dot{\mathbf{r}}(t'')\,\mathrm{d}t''\right]\mathrm{d}t' \\
&\qquad = \int_0^t \delta\dot{\mathbf{r}}(t')^T\delta\mathbf{r}(t')\,\mathrm{d}t' - \int_0^t \delta\hat{\dot{\mathbf{r}}}_0^T\delta\hat{\mathbf{r}}_0\,\mathrm{d}t'
\end{aligned} \tag{5.78}$$

because in general

$$\delta\dot{\mathbf{r}}(t')^T\delta\mathbf{r}(t') = \frac{\mathrm{d}}{\mathrm{d}t'}[\delta\mathbf{r}(t')^T\delta\mathbf{r}(t')] - \delta\mathbf{r}(t')^T\delta\dot{\mathbf{r}}(t'), \tag{5.79}$$

so that

$$\delta\dot{\mathbf{r}}(t')^T\delta\mathbf{r}(t') = \frac{1}{2}\frac{\mathrm{d}}{\mathrm{d}t'}[\delta\mathbf{r}(t')^T\delta\mathbf{r}(t')]. \tag{5.80}$$

Introducing eq. (5.80) into eq. (5.78) we obtain identity (5.77).

Assume for simplicity that $\delta\hat{\dot{\mathbf{r}}}_0 = 0$. Since the term $\delta\dot{\mathbf{r}}^T\delta\dot{\mathbf{r}}$ is positive for any t, eq. (5.77) implies that the boundedness of the deviated motion $\delta\mathbf{r}(t)$ (or the stability of the motion $\mathbf{p}$, eq. (5.71)) is governed by the history of the term $\delta\ddot{\mathbf{r}}^T\delta\mathbf{r}$. This term can be presented as, cf. eq. (5.76)

$$\delta\ddot{\mathbf{r}}(t)^T\delta\mathbf{r}(t) = -\delta\mathbf{r}(t)^T\overset{-1}{\mathbf{M}}\mathbf{K}^{(T)}[\mathbf{r}(t)]\,\delta\mathbf{r}(t). \tag{5.81}$$

Now, if there exists a $\delta\mathbf{r}(t)$ such that for a certain time t_{cr}

$$\delta\ddot{\mathbf{r}}(t)^T\delta\mathbf{r}(t) > 0 \quad \text{for} \quad t > t_{cr} \tag{5.82}$$

then $\delta\mathbf{r}$ grows without bound and the fundamental motion is unstable. In other words, by using eq. (5.81), if for any $\delta\mathbf{r}(t)$

$$\delta\mathbf{r}(t)^T\overset{-1}{\mathbf{M}}\mathbf{K}^{(T)}[\mathbf{r}(t)]\,\delta\mathbf{r}(t) < 0, \tag{5.83}$$

then one (or more) of the vibration modes begins to grow in an unlimited manner. Clearly, the tangent stiffness matrix $\mathbf{K}^{(T)}$ for any realistic structural configuration is positive-definite at $t = 0$ and at a number of subsequent time instants $t < t_{cr}$. We have then at each instant the eigenproblem

$$[\overset{-1}{\mathbf{M}}\mathbf{K}^{(T)} - \lambda\mathbf{I}]\mathbf{v} = \mathbf{0}$$

or

$$[\mathbf{K}^{(T)} - \lambda\mathbf{M}]\mathbf{v} = \mathbf{0} \tag{5.84}$$

which yields the positive eigenvalues

$$\lambda_1(t) \leqslant \lambda_2(t) \leqslant \lambda_3(t) \leqslant \ldots, \qquad t < t_{cr} \tag{5.85}$$

with the corresponding normalized eigenvectors

$$\mathbf{v}_1(t),\ \mathbf{v}_2(t),\ \mathbf{v}_3(t),\ \ldots \tag{5.86}$$

At $t = t_{cr}$ the smallest eigenvalue $\lambda_1(t_{cr})$ changes its sign (becomes negative) and the deviated motion mode $\mathbf{v}_1(t)$ begins its unlimited growth. The foregoing discussion indicates that the condition $\lambda_1 = 0$ corresponding to the singularity of the tangent matrix, plays a crucial role in the dynamic stability analysis of structures. Since the criterion bears striking resemblance to the static branching condition, two important remarks have to be additionally made:

(i) To discover singularity of $\mathbf{K}^{(T)}[\mathbf{r}(t)]$ the dynamic solution path $\mathbf{r} = \mathbf{r}(t)$, must be followed so that generally the condition $\det \mathbf{K}^{(T)} = 0$ corresponds to different displacements and stresses than those rendering the singularity of $\mathbf{K}^{(T)}$ in the static process. The dynamic analysis is clearly more general and may be considered to contain the static uniqueness criterion due to Hill, [56], as a special case.

(ii) Contrary to statics, the dynamic branching behaviour should be understood as the existence of at least one perturbed motion, which is initially in the small neighbourhood of the fundamental undisturbed motion and which starts at $t = t_{cr}$ to depart monotonically from the neighbourhood. This phenomenon is proposed to be referred to as quasi-bifurcation.

The computational procedure based on the above development can be applied to any nonlinear structural dynamics problem in the following way:

— The fundamental solution (5.71) is obtained by using any implicit method of direct integration and the smallest eigenvalue of (5.84) is monitored at each step to discover at time $t = t_{cr}$ the critical point at which $\mathbf{K}^{(T)}$ ceases to be strictly positive-definite. Clearly, the effective stiffness matrix of the integration scheme is still non-singular.

— The step-by-step analysis can easily be continued beyond this point but it must be remembered that only a slight modification in the initial conditions in accordance with the discovered mode of unbounded growth will amplify the solution leading to instability. This can be checked numerically by repeating the step-by-step computations with the only difference being the appropriately perturbed initial conditions.

Numerous examples of the application of the above approach will be given in Chapter 6.

5.7 Nonlinear heat transfer

It is important to realize that many of the basic concepts of the numerical analysis of structures can directly be used in the analysis of heat transfer problems. To illustrate this let us first recall the governing finite element equations for the nonlinear heat transfer, cf. eq. (4.45)

$$\mathbf{C}\dot{\boldsymbol{\theta}}+\mathbf{K}\boldsymbol{\theta} = \mathbf{Q} \tag{5.87}$$

in which $\mathbf{C}$ is the heat capacity matrix, $\mathbf{Q}$ is a vector of thermal nodal loads and $\mathbf{K}$ is the conductivity matrix. The matrices $\mathbf{C}$ and $\mathbf{K}$ and the vector $\mathbf{Q}$ are in general temperature dependent.

Eq. (5.87) is a system of nonlinear ordinary differential equations which must be solved for the consecutive values of the nodal temperatures.

As an example of temporal integration algorithms let us consider a family of one-step methods with the following assumptions:

$$\begin{aligned} \dot{\boldsymbol{\theta}}_{t+\alpha\Delta t} &= (\boldsymbol{\theta}_{t+\Delta t}-\boldsymbol{\theta}_t)\frac{1}{\Delta t}, \\ \boldsymbol{\theta}_{t+\alpha\Delta t} &= (1-\alpha)\boldsymbol{\theta}_t+\alpha\boldsymbol{\theta}_{t+\Delta t} \end{aligned} \tag{5.88}$$

where $\alpha \in [0, 1]$. The fundamental heat transfer equation written for the time instant $t+\alpha\Delta t$ takes the form

$$\mathbf{C}_{t+\alpha\Delta t}\dot{\boldsymbol{\theta}}_{t+\alpha\Delta t}+\mathbf{K}_{t+\alpha\Delta t}\boldsymbol{\theta}_{t+\alpha\Delta t} = \mathbf{Q}_{t+\alpha\Delta t} \tag{5.89}$$

which, upon substituting expansions (5.88), becomes

$$\begin{aligned} &\left[\frac{1}{\Delta t}\mathbf{C}_{t+\alpha\Delta t}+\alpha\mathbf{K}_{t+\alpha\Delta t}\right]\boldsymbol{\theta}_{t+\Delta t} \\ &\qquad = \mathbf{Q}_{t+\alpha\Delta t}+\frac{1}{\Delta t}\mathbf{C}_{t+\alpha\Delta t}\boldsymbol{\theta}_t-(1-\alpha)\mathbf{K}_{t+\alpha\Delta t}\boldsymbol{\theta}_t. \end{aligned} \tag{5.90}$$

Since the solution of eq. (5.90) for the "next" nodal temperature vector at time $t+\Delta t$ requires in general the determination of the matrices **C**, **K** and of the vector **Q** at time $t+\Delta t$ for which the temperature is yet to be calculated, the procedure clearly needs iteration. The exception to this rule takes place for $\alpha = 0$ (the explicit Euler forward method), for which we have

$$\frac{1}{\Delta t}\mathbf{C}_t\boldsymbol{\theta}_{t+\Delta t} = \mathbf{Q}_t+\frac{1}{\Delta t}\mathbf{C}_t\boldsymbol{\theta}_t-\mathbf{K}_t\boldsymbol{\theta}_t. \tag{5.91}$$

However, since this scheme is only conditionally stable, and it possesses unattractive accuracy characteristics, it is usually replaced by one of the implicit algorithms obtained for $\alpha \neq 0$. In particular, the Euler backward method, which corresponds to $\alpha = 1$ is described by the equation

$$\left[\frac{1}{\Delta t}\mathbf{C}_{t+\Delta t}+\mathbf{K}_{t+\Delta t}\right]\boldsymbol{\theta}_{t+\Delta t} = \mathbf{Q}_{t+\Delta t}+\frac{1}{\Delta t}\mathbf{C}_{t+\Delta t}\boldsymbol{\theta}_t \tag{5.92}$$

supplemented by the iteration scheme of the form

$$\left[\frac{1}{\Delta t}\mathbf{C}^{(i-1)}_{t+\Delta t}+\mathbf{K}^{(i-1)}_{t+\Delta t}\right]\boldsymbol{\theta}^{(i)}_{t+\Delta t} = \mathbf{Q}^{(i-1)}_{t+\Delta t}+\frac{1}{\Delta t}\mathbf{C}^{(i-1)}_{t+\Delta t}\boldsymbol{\theta}_t. \tag{5.93}$$

The counterpart of eq. (5.93) valid for any $\alpha \neq 0$ reads

$$\left[\frac{1}{\Delta t}\,\mathbf{C}^{(i-1)}_{t+\alpha\Delta t}+\alpha\mathbf{K}^{(i-1)}_{t+\alpha\Delta t}\right]\boldsymbol{\theta}^{(i)}_{t+\alpha\Delta t}$$

$$=\mathbf{Q}^{(i-1)}_{t+\alpha\Delta t}+\frac{1}{\Delta t}\,\mathbf{C}^{(i-1)}_{t+\alpha\Delta t}\boldsymbol{\theta}_t-(1-\alpha)\mathbf{K}^{(i-1)}_{t+\alpha\Delta t}\boldsymbol{\theta}_t \tag{5.94}$$

with

$$\boldsymbol{\theta}^{(i)}_{t+\alpha\Delta t}=(1-\alpha)\boldsymbol{\theta}_t+\alpha\boldsymbol{\theta}^{(i)}_{t+\Delta t}.$$

5.8 Development of software

Every computer must be able to perform three basic functions: (i) to receive data which define the particular problem to be analyzed, (ii) to process the data in accordance with some preset rules, (iii) to produce the results. The structure of each general-purpose finite element computer code can accordingly be broken down into three corresponding functional units which are usually called directly from the main program as shown in Fig. 47. The term "general-purpose" refers to the ability of the program to specify any boundary-value problem in a certain class (frames, shells, etc.) by means of the input data alone, i.e. without any changes in the program itself.

The preprocessor consists of subroutines which read input data and define problem-dependent parameters and data arrays. Perhaps the most important among the tasks performed by every efficient preprocessor is the ability to generate an "optimal" finite element mesh based on a minimum of input data prepared by the analyst.

Clearly, the processor is the most important part of the computer program. It is a decisive factor as to the efficiency of the entire code. The major tasks performed by the processor are shown in Fig. 47. In particular, in this part of the program element matrices and vectors are computed and assembled, boundary conditions are imposed and the fundamental set of equations is solved for the nodal point values of the nonlinear solution. The time spent on the repetitive solution of the stiffness equations of the structure represents a large percentage of the total computation cost. Thus the method of solving these equations is critical to an efficient solution in terms of both the computer core storage requirements and the solution time.

It should be stressed at this point that no unique optimum program organization currently exist, and various different and effective strategies are followed at different research centers. In addition, new ideas are being developed particularly as new computer equipment becomes available. Never-

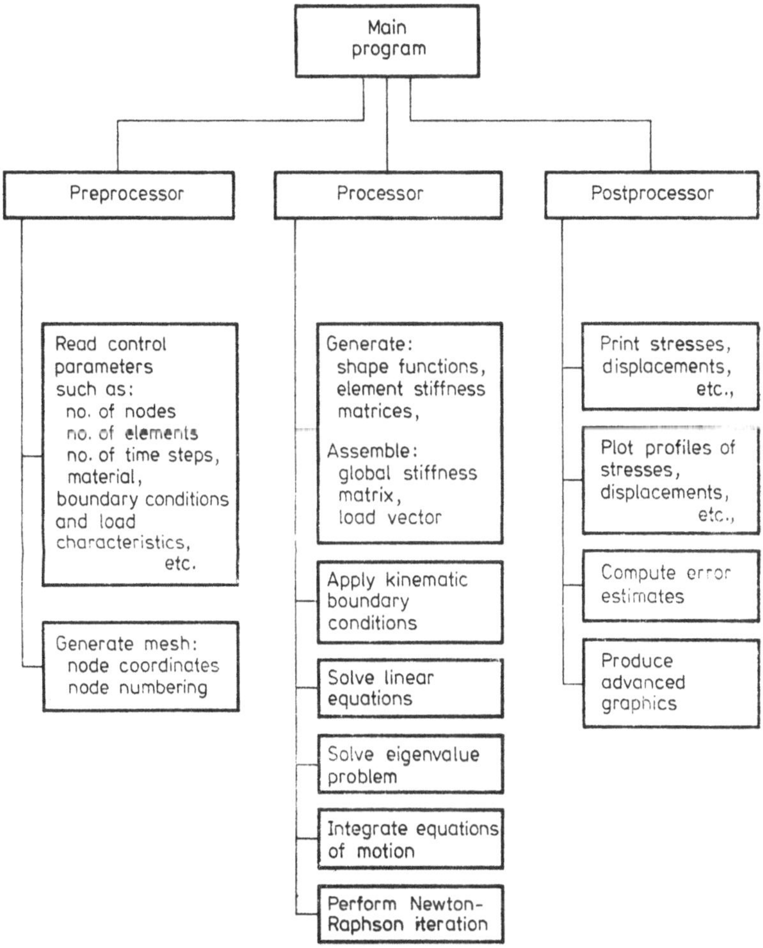

Fig. 47 Functional units in typical finite element programs

theless, even if program design might at first sight look quite differently, in effect the same basic steps are essentially followed as illustrated in Fig. 47.

The postprocessor can vary in sophistication from simply printing the solution values at nodes, via calculating the values of secondary variables (stresses in the displacements-type models, for instance) at selected points and plotting the results up to the color graphics enhancement of output, films, etc...

In currently written programs many useful self-adaptive features are incorporated with appropriate feedback routes leading from error estimate subroutines to preprocessor and processor routines. The reaction force concept presented in Sec. 3.5 is obviously an example of such an approach.

Finite element programs are called self-adaptive when they possess a local a posteriori error estimation capability and a capability to assign automatically (or with a minimal user interaction) additional degrees of freedom to regions where greater accuracy is required. This automated process reduces the discretization error until the desired, preset level of accuracy is attained, thus establishing confidence in the solution.

It should be noted that until very recently the implementation of adaptive programs for the finite element analysis has been prevented by the lack of a formulation allowing for efficient reanalysis. This is clearly a crucial point because the cost of successive analyses increases rapidly unless the results of an available solution can be fully utilized in obtaining succeeding solutions.

In finite element analysis convergence can be achieved in two ways. First, the type of elements (i.e. their constraint equation) can be fixed and the maximum dimension of elements, say h, made to approach zero. This process is often called the h-convergence, in contrast to the so-called p-convergence that will occur in meshes of fixed form in which the degree of the interpolating polynomial p is successively increased. At this stage of the finite element method development it is not at all clear which kind of the convergence study is more effective—even if some numerical experiences tend to suggest that the p-convergence rate is higher, in practice a compromise is always necessary as small elements may be needed to model the boundaries or inhomogeneities of the problem at hand. It may be indicated as a rule-of-thumb that second- or third-degree polynomials often represent the optimal choice, [184].

Summary

The main objective in this chapter was to present some general background concerning numerical algorithms commonly employed for solving nonlinear problems of structural mechanics. Without going into details of computer implementation we have consecutively discussed theoretical aspects of the solution procedures for nonlinear statics (including integration of inelastic constitutive laws), linearized buckling, nonlinear dynamics and nonlinear transient heat flow. The chapter closes with comments on current problems in software development.

Problems

5.1 Characterize basic algorithms commonly employed for solving large systems of linear simultaneous algebraic equations by both direct and iterative methods, cf. [13], [42], [46], for instance.

5.2 Generalize the considerations of Sec. 5.3 to cover thermo-elastic-plastic materials.

5.3 Explain why the loading/unloading conditions given as (4.224) appear to be more convenient in the numerical analysis of inelastic structures.

5.4 Using the literature on solving eigenvalue problems, cf. [13], discuss methods best suited for the buckling analysis.

5.5 On the basis of Secs. 5.2 and 5.4 discuss the difficulties encountered when looking for bifurcation points in problems with a nonlinear prebuckling equilibrium path.

5.6 Develop a numerical algorithm for passing through the bifurcation points by constraining the postbuckling path using the eigenmode obtained from the solution of the eigenvalue problem at the bifurcation point.

5.7 Characterize the stability properties of the Newmark method as applied to linear-problems; indicate difficulties in generalizing them to nonlinear situations.

5.8 Write out basic relationships of the ϑ-Wilson and Houboldt implicit integration methods for nonlinear dynamics problems, [13].

5.9 In Sec. 5.6 only one aspect of dynamic structural instabilities has been discussed. Try to discuss other phenomena related to the subject of dynamic stability such as:
— flutter stability due to nonconservative quasi-static follower loads,
— dynamic snap-through,
— parametric resonance.

5.10 Develop an algorithm for integrating linear heat transfer equations by means of modal superposition technique, cf. [13], [46], for instance.

PART II

SELECTED APPLICATIONS

6

Trusses, frames, lattice-type shells

Purpose of the chapter

It is natural to start the discussion of some exemplary solutions to nonlinear structural mechanics problems with structures made of bars. This task is undertaken in the present chapter in which we aim at showing the potential and effectiveness of the formulations and algorithms described earlier. Within this chapter we shall move forward by gradually increasing the degree of complexity of structural configurations considered; thus we begin with trusses in which only axial forces are transmitted by their particular members. Then we pass to frames made of beams which may effectively respond to axial forces as well as to bending and twisting moments. The last part of the chapter is devoted to the discussion of a highly complex problem of nonlinear static analysis of structures made of bars forming a repetitive pattern on given surfaces and thus behaving essentially in a shell-type manner.

We note that even if the book attempts to give a unifying basis for any nonlinear structural analysis, the subsequent chapters will also emphasize the variety of techniques necessary for specific applications.

6.1 Trusses

6.1.1 Nonlinear effects in truss analysis—a model problem

The whole variety of possible nonlinear effects can be illustrated with reference to a simple 3D truss configuration shown in Fig. 48. The truss is assumed to deform symmetrically with respect to the z_3-axis (i.e. $u_1^{(2)}=0$ at node 2) and to have an initial imperfection $u_0^{(2)}$; displacements $u_2^{(2)}$, $u_3^{(2)}$ determine the actual configuration at time t from which the next configuration at $t+\Delta t$ is sought within the framework of updated Lagrangian description. The 6×6 elastic and initial stress matrices for the element 1 at time t, given by eqs. (4.55), (4.56), may be transformed to global coordinates z_1, z_2, z_3 indicated in Fig. 48 yielding, cf. eq. (4.67)

$$\mathbf{k}_{6\times 6}^{(el)(1)} = \frac{AE}{l}\begin{bmatrix} \hat{\mathbf{c}}_{3\times 1}^{(1)}\hat{\mathbf{c}}_{1\times 3}^{(1)T} & -\hat{\mathbf{c}}_{3\times 1}^{(1)}\hat{\mathbf{c}}_{1\times 3}^{(1)T} \\ -\hat{\mathbf{c}}_{3\times 1}^{(1)}\hat{\mathbf{c}}_{1\times 3}^{(1)T} & \hat{\mathbf{c}}_{3\times 1}^{(1)}\hat{\mathbf{c}}_{1\times 3}^{(1)T} \end{bmatrix}$$

$$= \frac{AE}{l} \begin{bmatrix} 1 & 0 & 0 & -1 & 0 & 0 \\ & 0 & 0 & 0 & 0 & 0 \\ & & 0 & 0 & 0 & 0 \\ & \text{sym.} & & 1 & 0 & 0 \\ & & & & 0 & 0 \\ & & & & & 0 \end{bmatrix}, \tag{6.1}$$

$$\mathbf{k}^{(\sigma)(1)}_{6\times 6} = \frac{Q}{l} \begin{bmatrix} \mathbf{I}_{3\times 3} - \hat{\mathbf{c}}^{(1)}_{3\times 1}\hat{\mathbf{c}}^{(1)T}_{1\times 3} & -(\mathbf{I}_{3\times 3} - \hat{\mathbf{c}}^{(1)}_{3\times 1}\hat{\mathbf{c}}^{(1)T}_{1\times 3}) \\ -(\mathbf{I}_{3\times 3} - \hat{\mathbf{c}}^{(1)}_{3\times 1}\hat{\mathbf{c}}^{(1)T}_{1\times 3}) & \mathbf{I}_{3\times 3} - \hat{\mathbf{c}}^{(1)}_{3\times 1}\hat{\mathbf{c}}^{(1)T}_{1\times 3} \end{bmatrix}$$

$$= \frac{Q}{l} \begin{bmatrix} 0 & 0 & 0 & 0 & 0 & 0 \\ & 1 & 0 & 0 & -1 & 0 \\ & & 1 & 0 & 0 & -1 \\ & & & 0 & 0 & 0 \\ & \text{sym.} & & & 1 & 0 \\ & & & & & 1 \end{bmatrix} \tag{6.2}$$

where

$$\hat{\mathbf{c}}^{(1)}_{3\times 1} = \{c^{(1)}_{\xi 1} \;\; c^{(1)}_{\xi 2} \;\; c^{(1)}_{\xi 3}\} = \left\{\frac{l'}{l} \;\; \frac{u^{(2)}_0 + u^{(2)}_2}{l} \;\; \frac{h + u^{(2)}_3}{l}\right\}, \tag{6.3}$$

Q is the axial elemental force and the vector of incremental elemental degrees of freedom has the form, cf. eqs. (4.47)–(4.49)

$$\Delta\mathbf{r}^{(1)}_{6\times 1} = \{\Delta u^{(1)}_1 \;\; \Delta u^{(1)}_2 \;\; \Delta u^{(1)}_3 \;\; \Delta u^{(2)}_1 \;\; \Delta u^{(2)}_2 \;\; \Delta u^{(2)}_3\} = \{\Delta\mathbf{u}^{(1)} \;\; \Delta\mathbf{u}^{(2)}\}, \tag{6.4}$$

$\mathbf{u}^{(1)}$, $\mathbf{u}^{(2)}$ being the displacement vectors of nodes 1 and 2, Fig. 48. Introducing the incremental displacement vectors $\Delta\mathbf{r}^{(2)}_{6\times 1}$, $\Delta\mathbf{r}^{(3)}_{6\times 1}$ for the remaining two

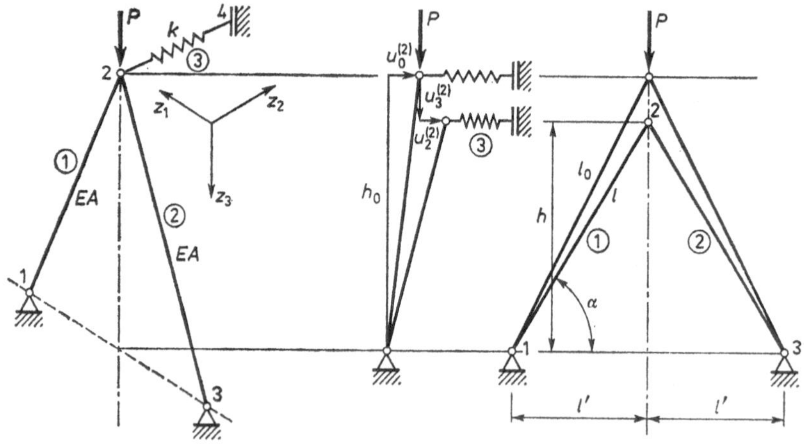

Fig. 48 3D truss — a model problem; $E = 2.1\times 10^{11}\,\frac{\text{N}}{\text{m}^2}$, $k = 4\times 10^4\,\frac{\text{N}}{\text{m}}$, $A = 6\times 10^{-4}\text{ m}^2$, $h = 0.22$ m, $l = 2.1888618$ m

truss elements at hand (the spring 3 can clearly be treated as a truss element as well) we note that the general expressions (6.1), (6.2) still hold for them whereas the vectors of directional cosines take the form

$$\hat{\mathbf{c}}^{(2)}_{3\times 1} = \left\{ -\frac{l'}{l} \quad \frac{u_0^{(2)}+u_2^{(2)}}{l} \quad \frac{h+u_3^{(2)}}{l} \right\}, \tag{6.5}$$

$$\hat{\mathbf{c}}^{(3)}_{3\times 1} = \{0 \;\; 1 \;\; 0\}. \tag{6.6}$$

The assembly procedure follows the lines thoroughly described in Sec. 4.3; after adjusting local and global numbering schemes and introducing the boundary and symmetry conditions we end up with the elastic and initial stress matrices of the form

$$\mathbf{K}^{(el)}_{2\times 2} = \begin{bmatrix} \dfrac{2EA}{l^3}(u_0^{(2)}+u_2^{(2)})+k & \dfrac{2EA}{l^3}(h_0+u_3^{(2)})(u_0^{(2)}+u_2^{(2)}) \\ \text{sym.} & \dfrac{2EA}{l^3}(h_0+u_3^{(2)})^2 \end{bmatrix}, \tag{6.7}$$

$$\mathbf{K}^{(\sigma)}_{2\times 2} = P\begin{bmatrix} \dfrac{1}{h_0+u_3^{(2)}}\left(\dfrac{u_0^{(2)}+u_2^{(2)}}{(h_0+u_3^{(2)})^2}-1\right) & \dfrac{u_0^{(2)}+u_2^{(2)}}{l^2} \\ \text{sym.} & \dfrac{h_0+u_3^{(2)}}{l^2}-\dfrac{1}{h_0+u_3^{(2)}} \end{bmatrix} \tag{6.8}$$

corresponding to the incremental displacement and load vectors for the whole structure of the form

$$\Delta\mathbf{r}_{2\times 1} = \{\Delta r_1 \;\; \Delta r_2\}, \qquad \Delta\mathbf{R} = \{\Delta R_1 \;\; \Delta R_2\}$$

with

$$\Delta r_1 = \Delta u_2^{(2)}, \qquad \Delta r_2 = \Delta u_3^{(2)},$$
$$\Delta R_1 = 0, \qquad \Delta R_2 = \Delta P.$$

The incremental/iterative algorithm of Sec. 4.2 employed for the solution of the equation

$$[\mathbf{K}^{(el)}+\mathbf{K}^{(\sigma)}]_{2\times 2}\Delta\mathbf{r} = \Delta\mathbf{R} \tag{6.9}$$

in which $u_0^{(2)} = 0$ (perfect truss) yields the solution shown by solid line in Fig. 49. The limit point (point of maximum load) L can clearly be identified along the primary equilibrium path so obtained. The possibility of any buckling (out-of-plane for the whole structure, or local by exceeding the Euler buckling forces in the compressed elements) has been so far excluded for simplicity. Now, the initial buckling problem is posed by introducing

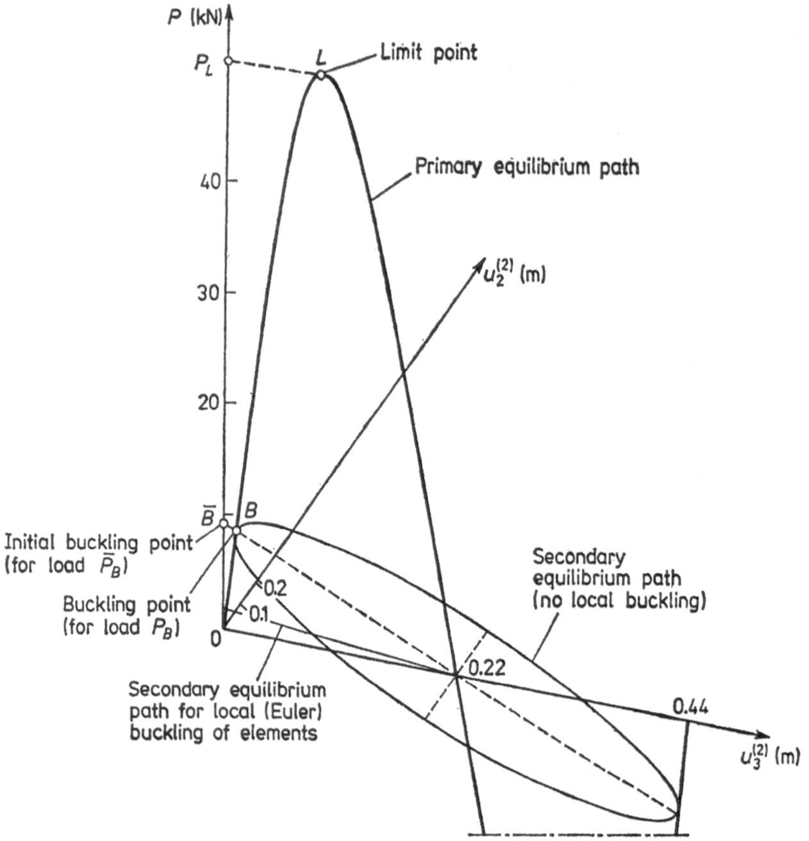

Fig. 49 Equilibrium paths for perfect truss

into eq. (6.9) the zero current displacements ($u_2^{(2)} = u_3^{(2)} = 0$) and the no-imperfection condition which yields

$$\begin{bmatrix} k - \dfrac{P}{h_0} & 0 \\ 0 & \dfrac{2h_0^2 AE}{l^3} - \dfrac{Pl'^2}{h_0 l^2} \end{bmatrix} \begin{bmatrix} v_1 \\ v_2 \end{bmatrix} = \begin{bmatrix} 0 \\ 0 \end{bmatrix}. \tag{6.10}$$

It is seen that for bifurcation it is necessary that the external load be equal $P = kh_0$; this value is shown in Fig. 49 as $\bar{P}_B = 8.4873$ kN. A more accurate estimate of the bifurcation load may be obtained by accounting for deformations which has taken place prior to buckling (linearized buckling instead of initial buckling, cf. Sec. 5.4). This may be done by including in the buckling criterion the linear part of the initial displacement matrix, cf. eq. (5.44), or by performing the incremental analysis with checking for bifurcation at each step using the "initial" buckling formulation with matrices $\mathbf{K}^{(el)}$, $\mathbf{K}^{(\sigma)}$ based each time on the constantly updated geometry. In the case considered both approaches yield the first buckling load reduced down to the point B for which $P_B = 8.4713$ kN. The point B obviously lies on the solution curve.

Assuming that from this point onwards the structure follows the "out-of-plane" secondary (rather than primary) equilibrium path we obtain the solution in the form of an ellipsoidal curve, Fig. 49. The solution branching at the point B is circumvented by appropriately constraining the governing equilibrium equation at that point, cf. Sec. 5.4.

The buckling phenomena just described may be termed "global" in the sense that they refer to the overall critical behaviour of the truss. To indicate the possibility of a local buckling let us assume that the elements 1 and 2 may buckle according to the classical Euler formula $Q_{cr} = \pi^2 EJ/l^2$. Assuming for simplicity that in the post-buckling range the elemental axial forces Q remain constant and equal to Q_{cr} we observe that simple equilibrium of the node 2 gives for $u_0^{(2)} = u_2^{(2)} = 0$

$$P_{cr} = 2Q_{cr}\sin\alpha \approx \frac{2\pi^2 EJ}{l^3}(h_0 - u_3^{(2)}). \tag{6.11}$$

Eq. (6.11) represents approximately the straight line in the plane $P - u_3^{(2)}$ as shown in Fig. 49 and may be given the interpretation of the secondary equilibrium path after local buckling. In general, the local buckling may precede the global one, or the converse situation may take place. In the case both bifurcation points are close to each other the danger of the so-called interactive global-local buckling exists, which often implies very unstable post-critical behaviour.

Let us now consider the truss with a geometry imperfection in the form of the initial displacement u_0 at the node 2. The thick solid line in Fig. 50 corresponds to the incremental/iterative solution obtained for such a situation with $u_0 = 0.022$ m. The bifurcation point B is apparently "replaced" in the case of the real (i.e. imperfect) structure by the limit load point $L^{(r)}$ (for which $P = P_L^{(r)}$); after some time the primary equilibrium path for the real structure gets very close to the secondary equilibrium path for the ideal structure.

The ratio $P_L^{(r)}/P_B$ may be used to assess the so-called structural sensitivity to imperfections; this ratio indicates the reduction of the critical load due to the existence of imperfections. The ratio $P_L^{(r)}/P_B$ calculated for different imperfections and different spring constants is shown in Table 3 (note that $P_L^{(r)}/P_B$ is practically independent of k!). Considering the so-obtained points $A, B, C, \ldots, I$, we may attempt to construct a curve $u_0^{(2)} = f(P_L^{(r)}/P_B)$ passing

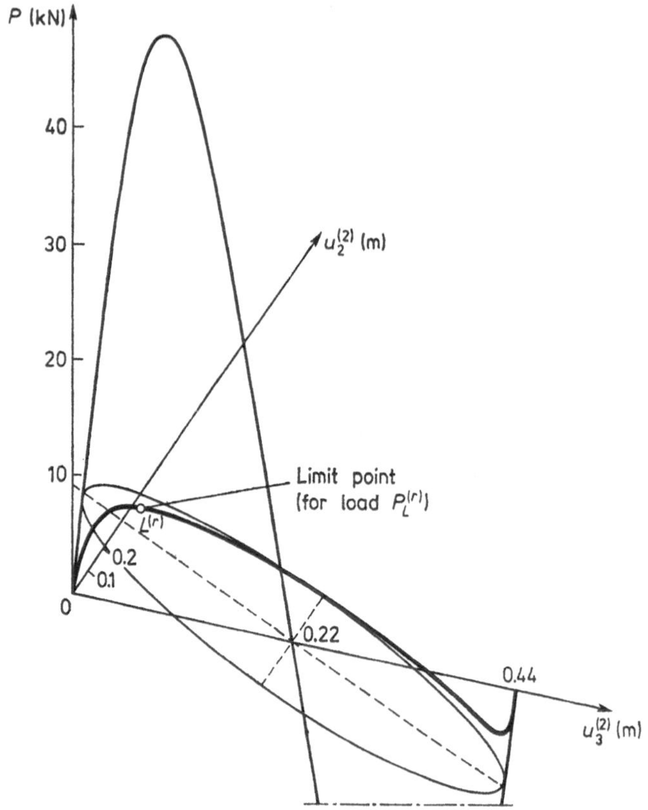

Fig. 50 Equilibrium paths for imperfect truss

through them; the quadratic interpolation (passing through A, B, C) results in the equation

$$u_0^{(2)} = a\left(1-\frac{P_L^{(r)}}{P_B}\right)^2 + b\left(1-\frac{P_L^{(r)}}{P_B}\right) \tag{6.12}$$

with $a = 0.2077$ m, $b = 0.0140$ m, whereas the cubic interpolation (passing through A, B, D, G) leads to the relationship

$$u_0^{(2)} = a\left(1-\frac{P_L^{(r)}}{P_B}\right)^4 + b\left(1-\frac{P_L^{(r)}}{P_B}\right)^3 + \\ + c\left(1-\frac{P_L^{(r)}}{P_B}\right)^2 + d\left(1-\frac{P_L^{(r)}}{P_B}\right), \tag{6.13}$$

$a = 4.1015$ m, $b = 2.0610$ m, $c = 0.4643$ m, $d = 0.0087$ m.

Table 3 Sensitivity to imperfections of the simple truss

Point	$u_0^{(2)}$ (m)	$k = 40$ (kN/m)		$k = 100$ (kN/m)	
		$P_L^{(r)} \times 10^4$ N	$P_L^{(r)}/P_B$	$P_L^{(r)} \times 10^4$ N	$P_L^{(r)}/P_B$
A	0.0000	0.8487308	1	1.999032	1
B	0.0005	0.8268549	0.9742252	1.947726	0.9743346
C	0.0010	0.8141142	0.9592136	1.9180347	0.9594814
D	0.0030	0.7773137	0.9158542	1.8321431	0.9165151
E	0.0080	0.7136581	0.8408533	1.6834360	0.8421255
F	0.0150	0.6475286	0.7629375	1.5288387	0.7647895
G	0.0220	0.5940361	0.6999110	1.4037789	0.7022293
H	0.0300	0.5422822	0.6389330	1.2828163	0.6417187
I	0.0400	0.4872590	0.5741030	1.1543240	0.5774414

Using these curves one can assess the reduction in critical load for any value of imperfection $u_0^{(2)}$. We note that according to the so-called Koiter's theory of imperfection sensitivity the general expression for the function f considered above should for this type of structural behaviour have the form

$$\left(1-\frac{P_L^{(r)}}{P_B}\right)^2 + 4au_0^{(2)}\frac{P_L^{(r)}}{P_B} = 0. \tag{6.14}$$

The value of an undetermined parameter a may here be established by imposing, for instance, the condition that the initial curvatures of the functions (6.14) and (6.12) (or (6.13)) are equal. The imperfection sensitivity described by eq. (6.14) with a corresponding to the form (6.12) or (6.13) are displayed in Fig. 51 as the curves I and II, respectively. An approximate curve passing through the points $A, B, \ldots, I$ is also shown in this figure.

The same truss problem was considered next under the assumption of inelastic material properties in the elements 1 and 2 and the elastic properties of the spring k. The inelastic material model was taken as elastic-plastic with kinematic hardening, Fig. 52; the ratio of the constant tangent modulus E_T to the Young modulus of elasticity was assumed as $E_T/E = 0.01$ and the yield limit was $\sigma_y = 3.5 \cdot 10^7$ N/m^2. Because of the incremental solution philosophy

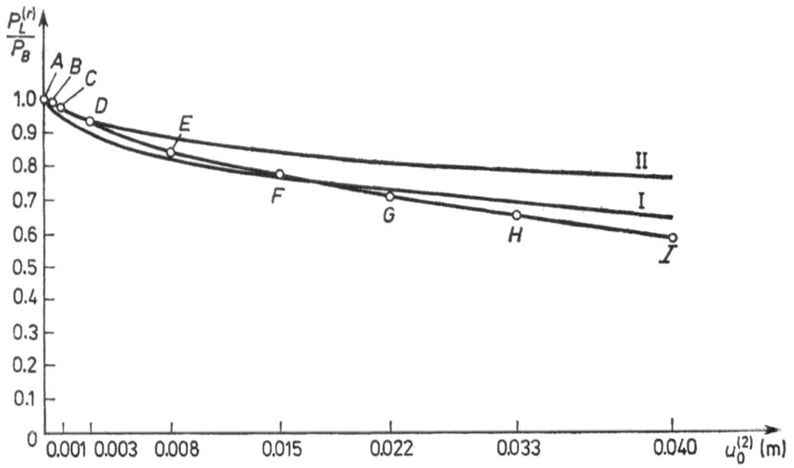

Fig. 51 Sensitivity to imperfections of the simple truss

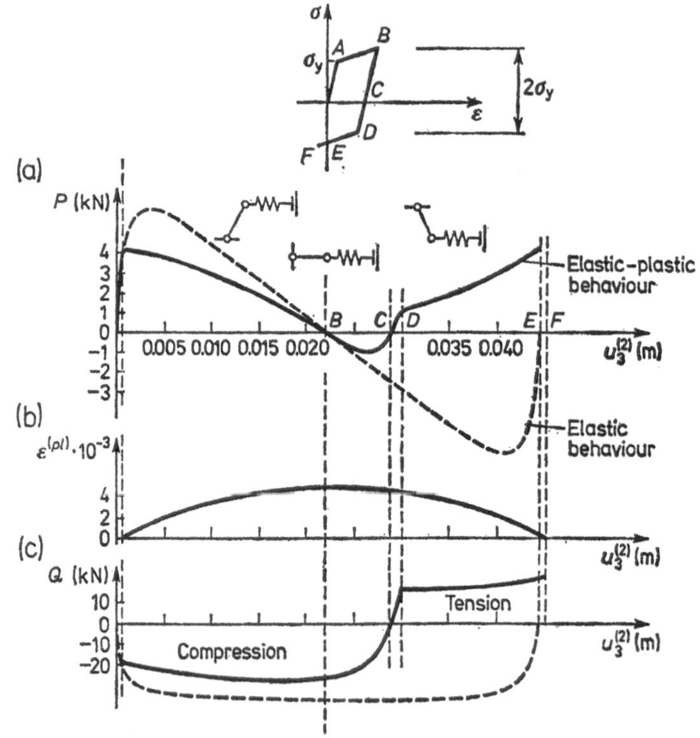

Fig. 52 Behaviour of elastic-plastic truss

and of the one-dimensional character of elemental deformations, the generalization of the constitutive matrix (4.55) is very simple in this case—we merely need to replace the Young modulus E in it by E_T for all steps which refer to the plastic states in the element considered. The results of the analysis for the perfect truss ($u_0^{(2)} = u_2^{(2)} = 0$) are shown in Fig. 52: Fig. 52(a) displays the $P - u_3^{(2)}$ relationship, Fig. 52(b) the value of the varying plastic strain in each inelastic element, and Fig. 52(c) illustrates changes in the axial force Q in elements 1 and 2 during the deformation process.

To illustrate the theoretical considerations on dynamical stability presented in Sec. 5.6 let us now consider the same 3D truss loaded dynamically by a step-type loading shown in Fig. 53. For a given load magnitude the step-by-step

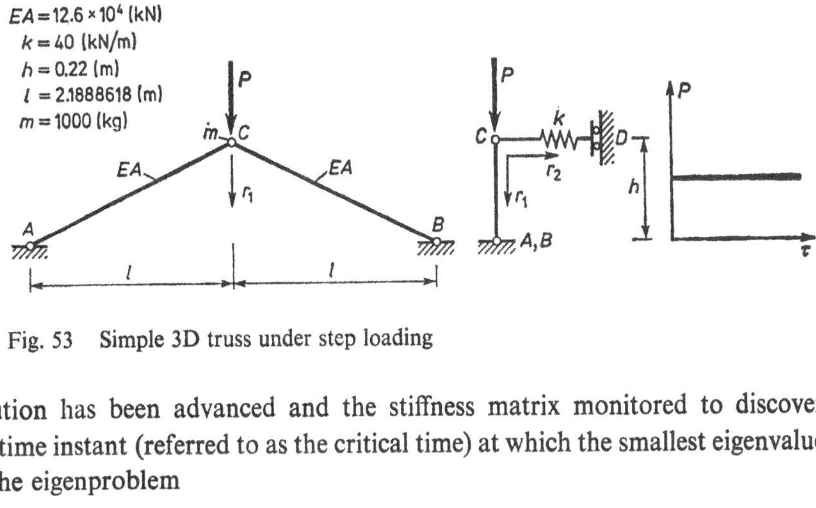

Fig. 53 Simple 3D truss under step loading

solution has been advanced and the stiffness matrix monitored to discover the time instant (referred to as the critical time) at which the smallest eigenvalue of the eigenproblem

$$[\mathbf{K}^{(T)} - \lambda \mathbf{M}]\mathbf{v} = 0$$

becomes zero. The curve of critical time vs load magnitude is plotted in Fig 54. For the loads leading to quasi-bifurcations the eigenvector corresponding to the zero eigenvalue is $\{0 \;\; a\}$ where a is any real number. By calculating the nonlinear response for any no-critical-time case, say for $P = 4.3$ kN, it has been shown that both the curves corresponding to the ideal and perturbed (by the above eigenvector with $a = 0.0001$ m) geometries remain close to each other in the whole time interval considered. On the contrary, and in full accordance with our theoretical result of Sec. 5.6, the perturbation in the initial geometry by the eigenvector $\{0 \;\; 0.0001\}$ for the loads $P = 10$ kN and $P = 15$ kN (which imply the existence of the critical time) lead to the response curves departing in a distinct way from the fundamental solutions for the unperturbed configurations, Fig. 55. Also, the points of departure along the

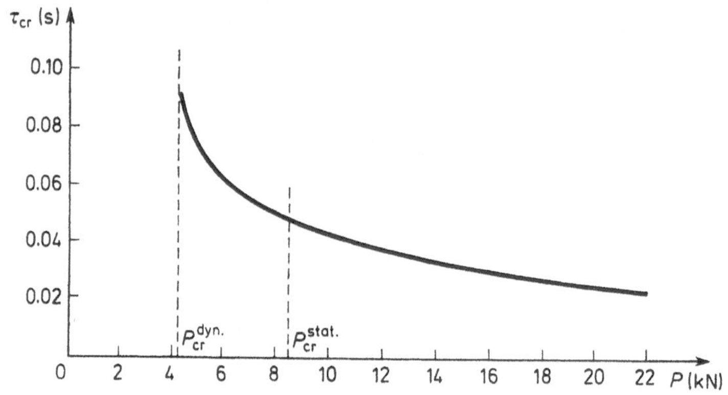

Fig. 54 Critical time vs load magnitude curve

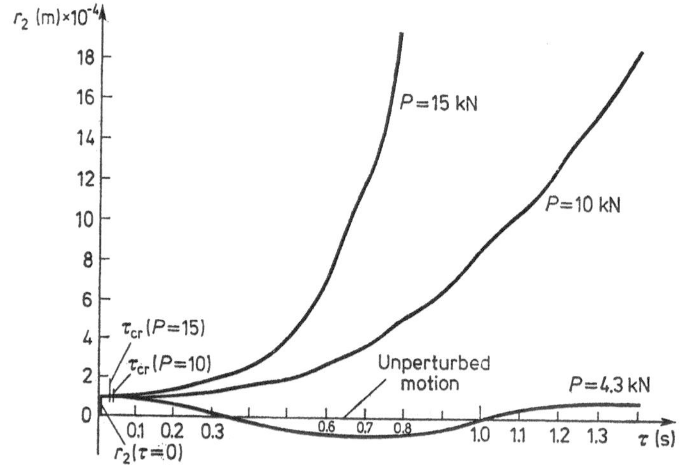

Fig. 55 Perturbed and unperturbed response curves for the 3D truss

time axis may be identified as the critical times for these particular load magnitudes.

In the requirement that a perturbed motion becomes unbounded it is clearly essential that the condition $\lambda_{min} < 0$ related to the fundamental motion holds for all time instants such that $\tau > \tau_{cr}$. If this is not the case, i.e. if after some time $\tau_1 > \tau_{cr}$ the value of λ_{min} becomes positive again, the perturbed motion may or may not be bounded (although in the former case both the fundamental and perturbed motions may not be close to each other any more). As a matter of fact, the appropriately perturbed motion remains bounded in the above example provided $P < 8.47$ kN even if there exist critical times for such load amplitudes, cf. Fig. 54. The fact that $P = 8.47$ kN is also the static buckling load in this case results from the simplicity of the configuration at hand and is by no means a general property of the dynamic stability solutions.

The time variation of λ_{min} for different values of the load P are shown in Fig. 56. It may be said that for $P < 8.47$ kN the perturbed motion is bounded

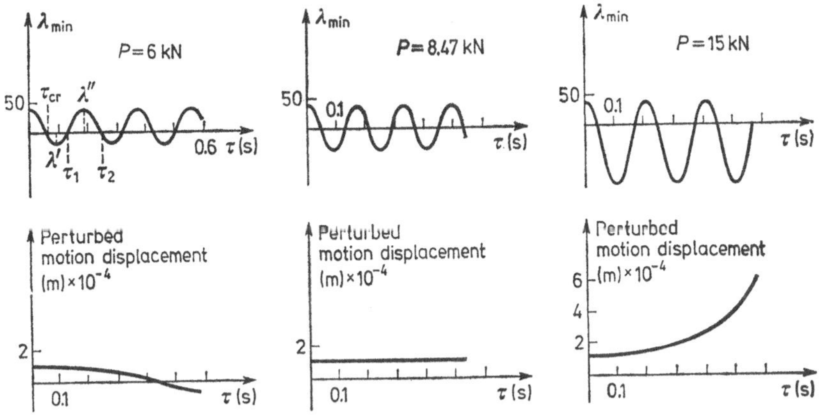

Fig. 56 Time variation of λ_{min} for different values of the load P

since the negative value of λ_{min} has not lasted "long enough" to enforce the solution to diverge. On the contrary, for $P > 8.47$ kN we can clearly observe the unboundedness of the perturbed motion since the time interval $[\tau_{cr}, \tau_1]$ is long enough to induce the snap-through.

From the practical point of view it is certainly important to be able to recognize the boundedness or unboundedness of the perturbed solution without having to trace the λ_{min} behaviour in the infinite time interval $[\tau_{cr}, \infty)$. As a rule of thumb resulting from some limited experience we may suggest comparing the minimum (in time) value of $\lambda_{min} = \lambda'$, Fig. 56, against the maximum (in time) value of $\lambda_{min} = \lambda''$ in the time interval $[\tau_1, \tau_2]$. If $\lambda' < \lambda''$ then instability is likely to occur whereas $\lambda' > \lambda''$ suggests the boundedness of the motion.

6.1.2 Accuracy of computations and further examples

In order to approximately assess the accuracy of the solution algorithms used for solving nonlinear structural mechanics problems let us first consider the simplified 2D version of the test truss example analyzed in Sec. 6.1.1. This example is useful in this respect since the closed-form static solution is readily available. In Fig. 57 the load vs vertical displacement curves are shown corresponding to the analytical solution (curve A), numerical solution by the incremental algorithm with 6 steps and no iterative corrections (curve C),

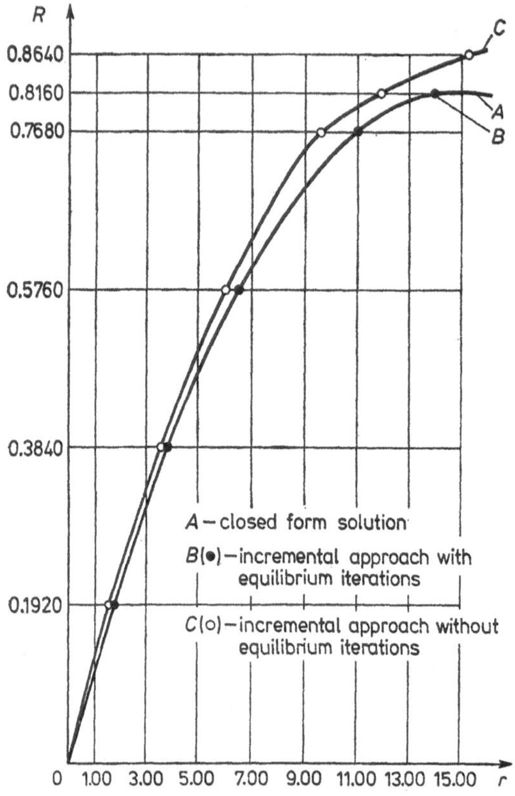

Fig. 57 Nonlinear test computations

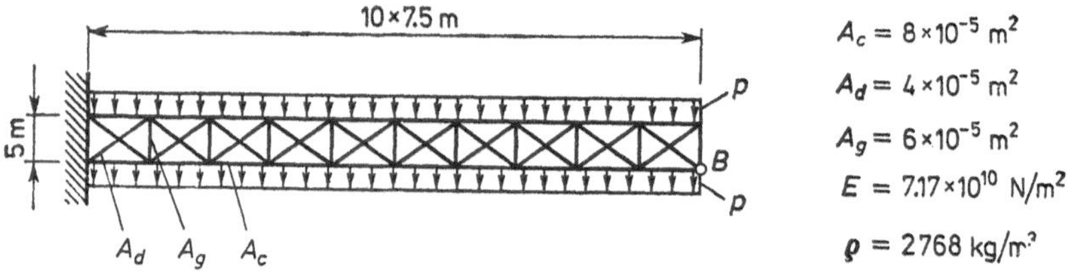

Fig. 58 Truss cantilever under uniformly distributed suddenly applied load

and the numerical solution with the same steps length and equilibrium corrections (points B along the curve A). The effect of incremental error accumulation is clearly seen documenting the need for iterative algorithms.

The next example shown in Fig. 58 is a truss cantilever under uniformly distributed suddenly applied load $p = 1.362 \cdot 10^5$ N/m modelled by concentrated forces acting at nodes. The bar characteristics are given in the figure. The linear and nonlinear dynamic responses are shown in Fig. 59, the linear

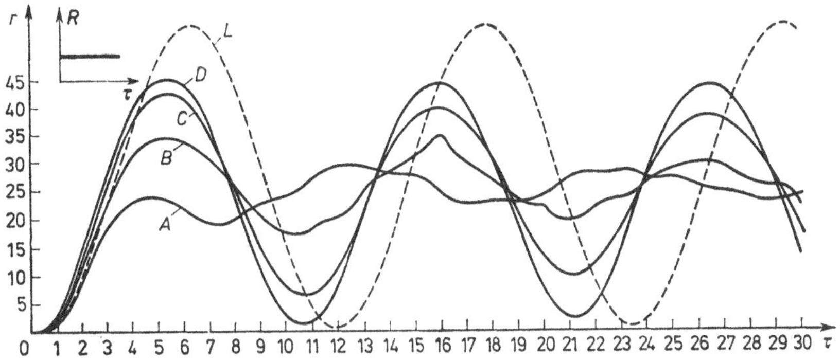

Fig. 59 Dynamic truss response with and without large displacement effects — displacement at point *B*

response being described by the curve *L*. In order to assess the step length sensitivity of the nonlinear solution four different cases were considered with $\Delta t = 0.1$, $\Delta t = 0.05$, $\Delta t = 0.025$, $\Delta t = 0.0125$. The resulting solutions are shown as curves *A*, *B*, *C* and *D*, respectively. The fast convergence of the solution with decreasing step length is clearly documented. All the curves were obtained without equilibrium iteration. Employing the Newton–Raphson iteration scheme the solutions for $\Delta t = 0.1$ and $\Delta t = 0.05$ were divergent whereas for $\Delta t = 0.025$ and $\Delta t = 0.0125$ the curve *D* was essentially rediscovered.

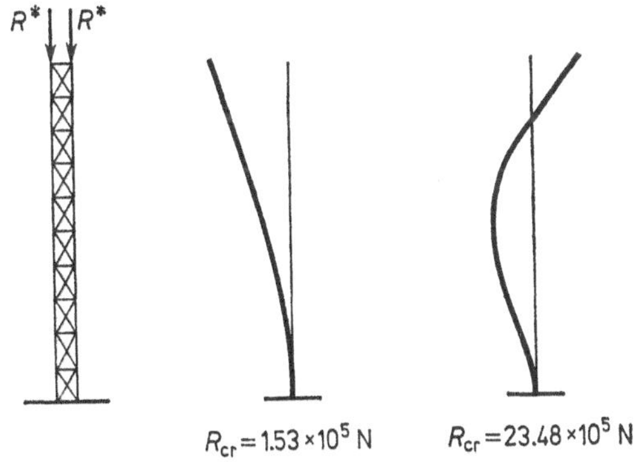

Fig. 60 Initial in-plane buckling of the compressed truss cantilever

Fig. 60 illustrates the initial in-plane buckling solution for the same structure loaded by static compressive forces; the first two buckling loads together with the associated buckling modes are shown.

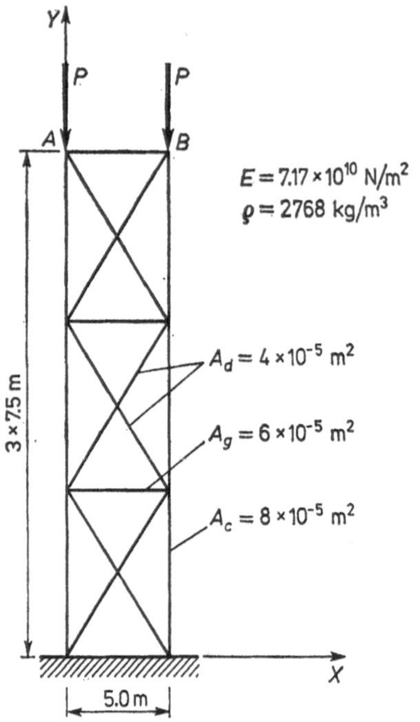

Fig. 61 In-plane deformed truss under step-type dynamic loading

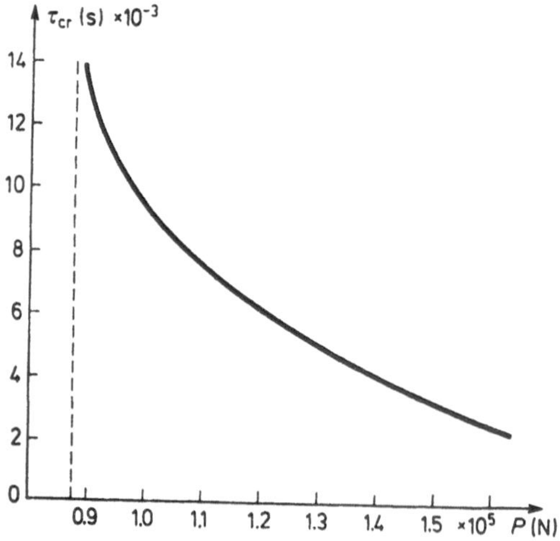

Fig. 62 Critical time vs load magnitude

A similar structure was next analyzed with respect to its dynamic characteristics, Fig. 61. The truss is again constrained to deform in-plane only. The plot of the critical time vs load magnitude is shown in Fig. 62. The perturbation to the initial geometry is introduced by means of an eigenvector corresponding for the given load to the zero eigenvalue. The solid and broken lines in Fig. 63

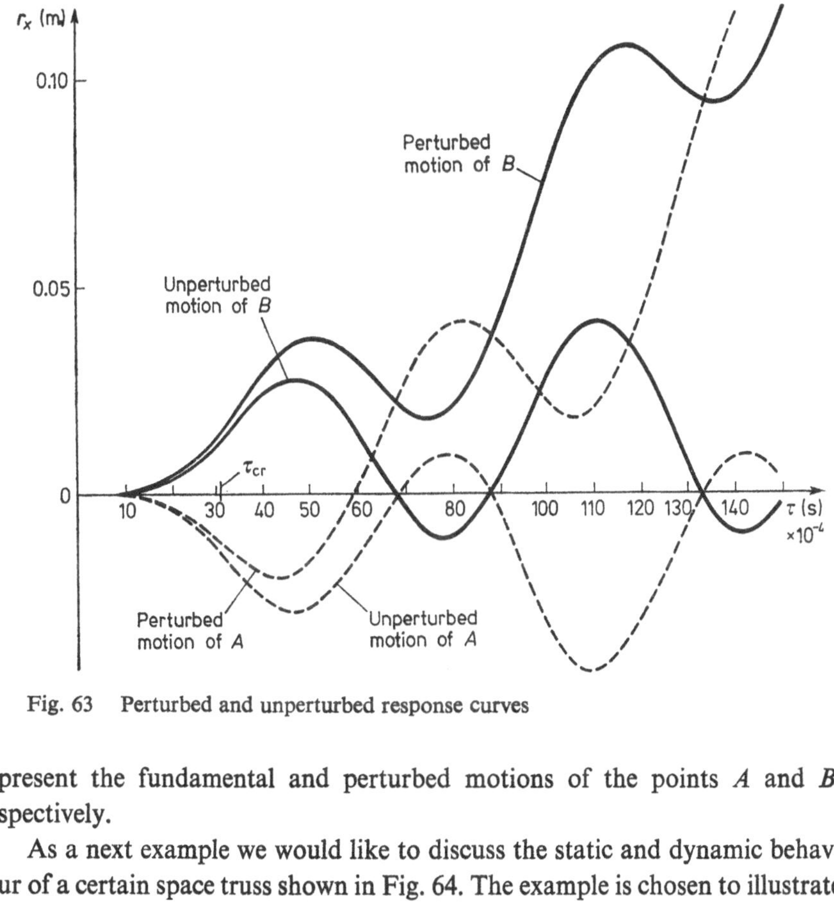

Fig. 63 Perturbed and unperturbed response curves

represent the fundamental and perturbed motions of the points A and B, respectively.

As a next example we would like to discuss the static and dynamic behaviour of a certain space truss shown in Fig. 64. The example is chosen to illustrate the possible complexity of the structural behaviour even in the case of relatively simple geometry and load. For simplicity we exclude in what follows the possibility of local buckling of the truss elements. Let this shell-like structure be loaded by a concentrated force applied at the apex (or, when more convenient, by the prescribed displacement at this point). After first passing through the limit point L_1, cf. Fig. 65, the analysis was continued until the first bifurcation point B_1 was encountered. At this point two further equilibrium paths are possible—the first one fully symmetric (fundamental) and the other with certain asymmetry in the deformation pattern. The analysis was continued

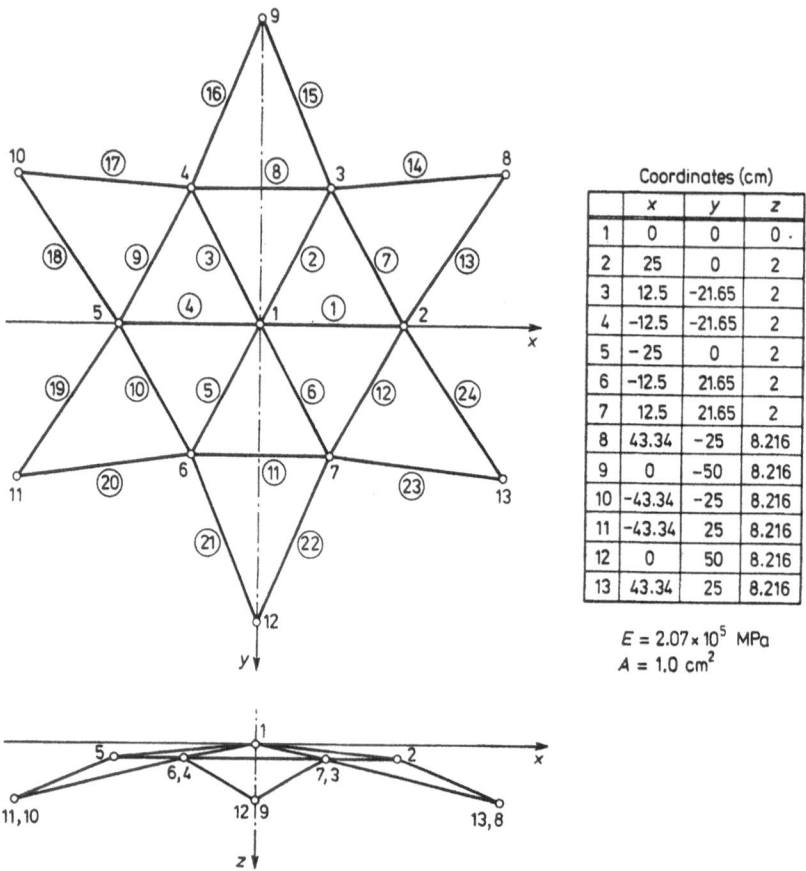

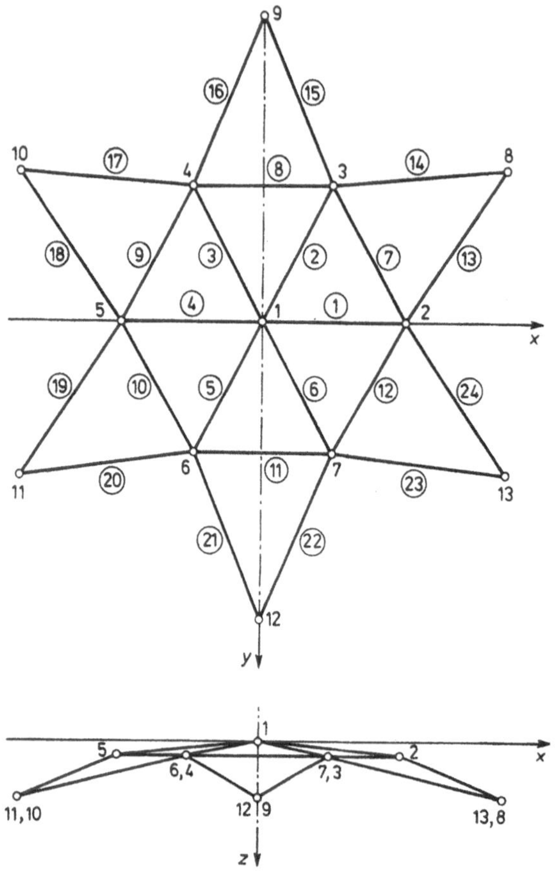

Coordinates (cm)

	x	y	z
1	0	0	0
2	25	0	2
3	12.5	-21.65	2
4	-12.5	-21.65	2
5	-25	0	2
6	-12.5	21.65	2
7	12.5	21.65	2
8	43.34	-25	8.216
9	0	-50	8.216
10	-43.34	-25	8.216
11	-43.34	25	8.216
12	0	50	8.216
13	43.34	25	8.216

$E = 2.07 \times 10^5$ MPa
$A = 1.0$ cm^2

Fig. 64 Geometry of space truss

beyond the point B_1 by imposing upon the truss vertical displacements at nodes 2, 3, ..., 7 the conditions of strict response symmetry. By using such constraints we were able to proceed with the incremental/iterative analysis up to the point E_1 at which the displacement-controlled problem broke down due to its non-uniqueness. Highly unstable post-buckling behaviour led to immediate snap-through after passing the bifurcation points B_2 and B_3; the characteristics of the post-buckling behaviour are given in Fig. 65 together with the strengthening curves obtained in each case from further incremental/iterative analysis. The results could be completed and discussed in a much more detailed way but even only those displayed in Fig. 65 make it possible to reiterate our initial observation that even very simple elastic structures can be characterized by extremely complicated load-displacement diagrams when the full range nonlinear behaviour is of interest.

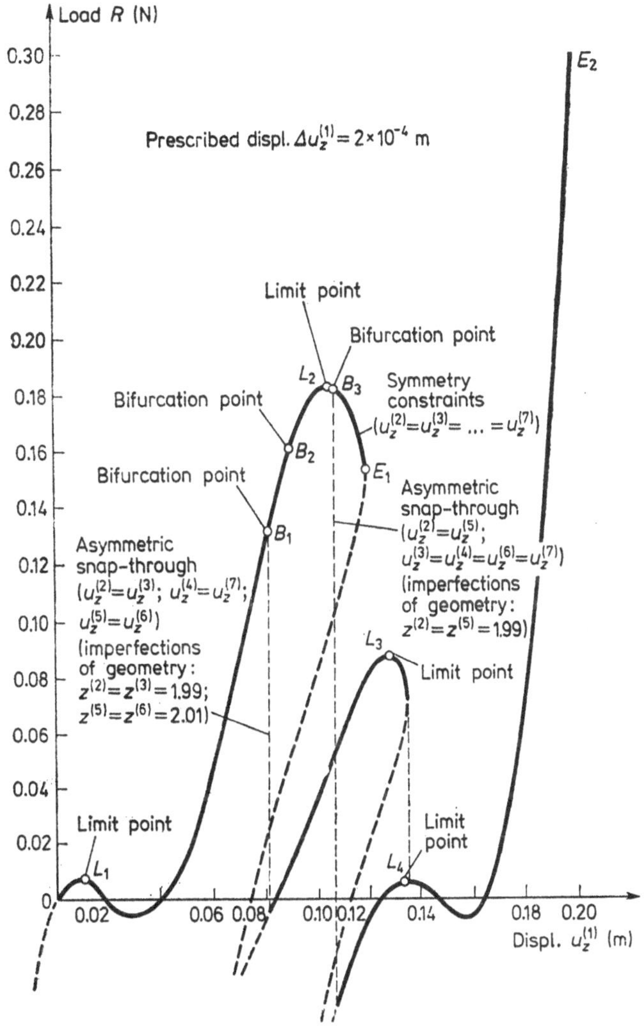

Fig. 65 Statics of the elastic 3D truss

The similar analysis was carried out assuming elastic-plastic material for the truss elements. Because the element buckling is again neglected, the analysis has clearly only an academic value. The results are shown in Fig. 66

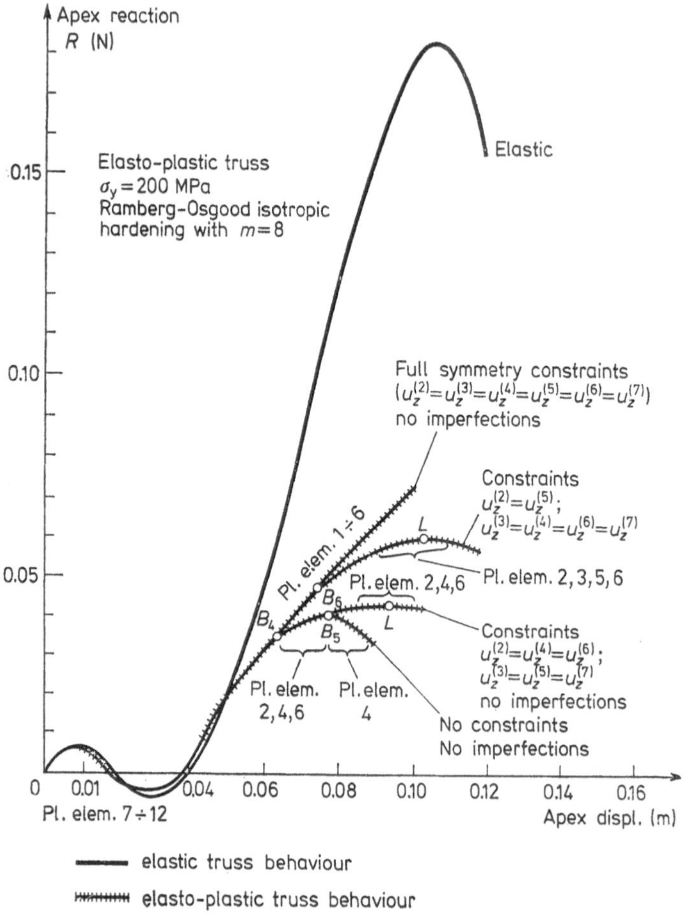

Fig. 66 Statics of the elastic-plastic 3D truss

and compared with the elastic solution. The limit and bifurcation points as well as plastified elements at different deformation stages are indicated. The imposed-symmetry approach allows again to characterize the post-critical behaviour. Asymmetries of the secondary equilibrium paths were always very distinct and resulted in the plastic unloading of some truss elements.

The elastic truss of the same geometry was next analyzed under the set of concentrated forces described in Table 4. Different small geometry and load imperfections patterns were considered, Table 4. The results given in

Table 4 **Geometry and load imperfections of the simple truss**

Node	Coordinates (cm)				Vertical loading ratios		
	Perfect geometry			Imperfect	Perfect	Imperfect	
	x	y	z	z–I	z	z–II	z–III
1	0.00	0.00	0.00	0.00	1.0	1.0	1.0
2	25.00	0.00	2.00	1.80	2.0	2.0	2.0
3	12.50	−21.65	2.00	1.80	2.0	2.0	2.0
4	12.50	21.65	2.00	2.00	2.0	1.5	1.5
5	−25.00	0.00	2.00	2.00	2.0	2.0	2.0
6	−12.50	21.65	2.00	2.00	2.0	2.0	2.0
7	12.50	21.65	2.00	1.80	2.0	1.5	2.5
8	43.30	−25.00	8.22	8.22	0.0	0.0	0.0
9	0.00	−50.00	8.22	8.22	0.0	0.0	0.0
10	−43.30	−25.00	8.22	8.22	0.0	0.0	0.0
11	−43.30	25.00	8.22	8.22	0.0	0.0	0.0
12	0.00	50.00	8.22	8.22	0.0	0.0	0.0
13	43.30	25.00	8.22	8.22	0.0	0.0	0.0

Figs. 67, 68 show the essential influence of the imperfections upon the response curve characterizing the ideal problem. The truss under such a loading may certainly be termed imperfection sensitive.

As a final illustration in this section we shall briefly discuss behaviour of the same elastic truss under the step load in the form of a concentrated force applied at the apex, Fig. 64. For the value of $P = 5$ kN the critical time was obtained as $\tau_{cr} = 0.065$ s. Introducing into the structure small imperfections in accordance with the corresponding eigenvector (the z_3 coordinate of the point 1 is perturbed by -0.01 with the rest of nodal coordinates changed accordingly to preserve the overall shape of the eigenmode) we obtain the

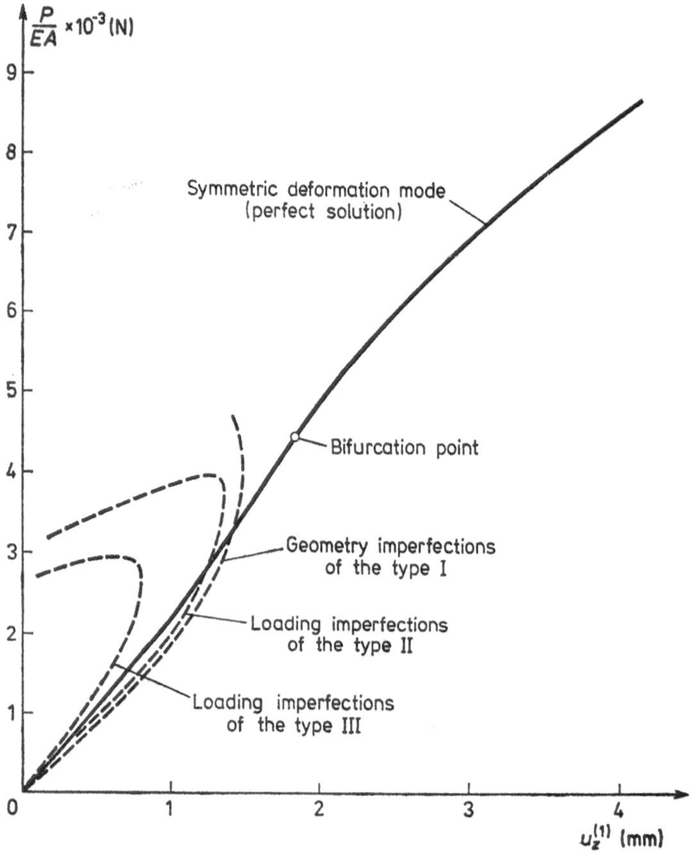

Fig. 67 Truss with imperfections: vertical displacement of the node 1 vs load parameter

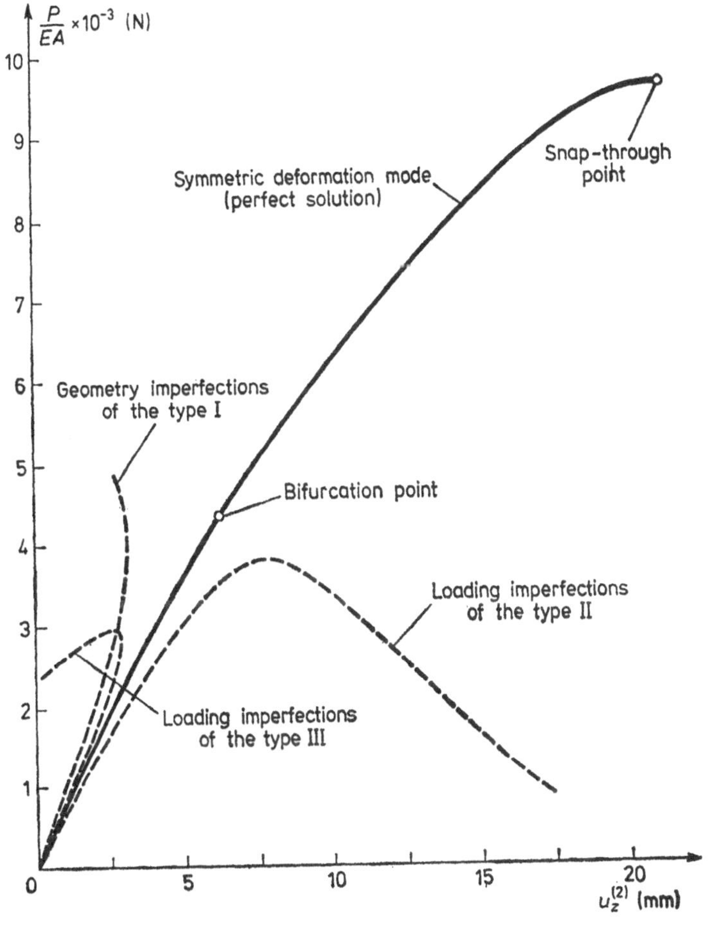

Fig. 68 Truss with imperfections: vertical displacement of the node 2 vs load parameter

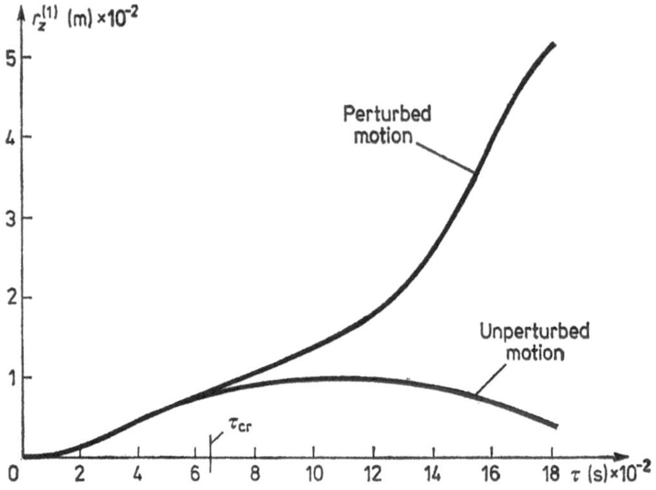

Fig. 69 Perturbed and unperturbed response curves for dynamically loaded 3D truss

response curve for the same load shown in Fig. 69. The branching of the perfect and imperfect structure responses indicates again very clearly the significance of the critical time concept.

6.1.3 Elastic-plastic truss under non-proportionally varying loads

In order to investigate the performance of the algorithms described in Sec. 4.9 a series of numerical tests were performed for a hyperstatic truss shown in Fig. 70. The truss was loaded by the repeated-variable forces H

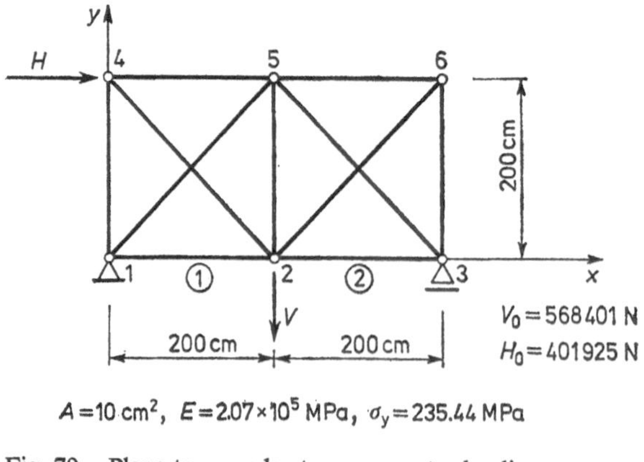

Fig. 70 Plane truss under two-parameter loading

and V. Introducing the non-dimensional parameters

$$\beta_1 = \frac{H}{H_0}, \quad \beta_2 = \frac{V}{V_0}$$

where H_0 and V_0 are the ultimate values of H and V when acting separately, some characteristic regions in the plane $\beta_1 \beta_2$ can be found by either linear programming or, in the case of limit analysis, by simple inspection of possible collapse mechanism, [28]. A region of the initial elastic response is bounded by the line ABC in Fig. 71. Its segments AB and BC correspond to the first

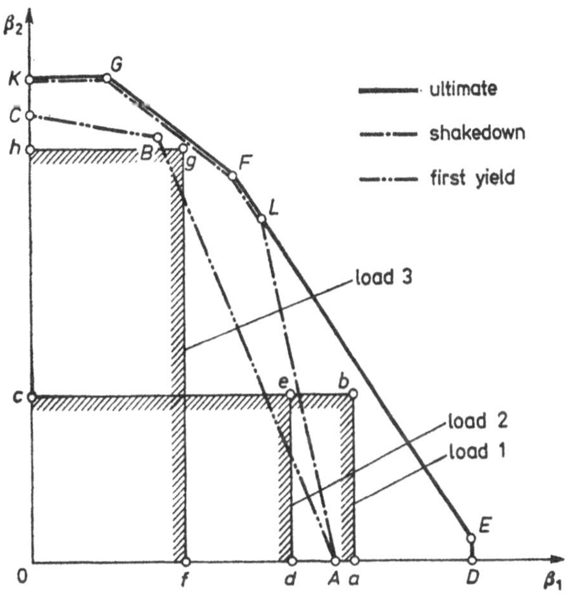

Fig. 71 Characteristic load domains for the truss

yield of bars 4–5 and 2–5, respectively. The boundary $DEFGK$ represents the ultimate load. The collapse modes corresponding to the segments DE, EF, FG and GK are shown in Figs. 72(a), 72(b), 72(c) and 72(d) respectively. Note that the ultimate load does not depend upon the loading history. Therefore, any load path (proportional or nonproportional) must terminate at the intersection with the boundary $DEFGK$ because of the plastic collapse. A shakedown domain for the load

$$-\xi\bar{\beta}_1 \leqslant \beta_1 \leqslant \xi\bar{\beta}_1, \quad 0 \leqslant \beta_2 \leqslant \xi\bar{\beta}_2$$

is bounded by the line $ALFGK$. The values $\bar{\beta}_1$, $\bar{\beta}_2$ correspond to a given reference load. If we expand the domain by increasing ξ monotonically, then at some ultimate value $\xi = \xi_0$ the shakedown possibilities will be exhausted.

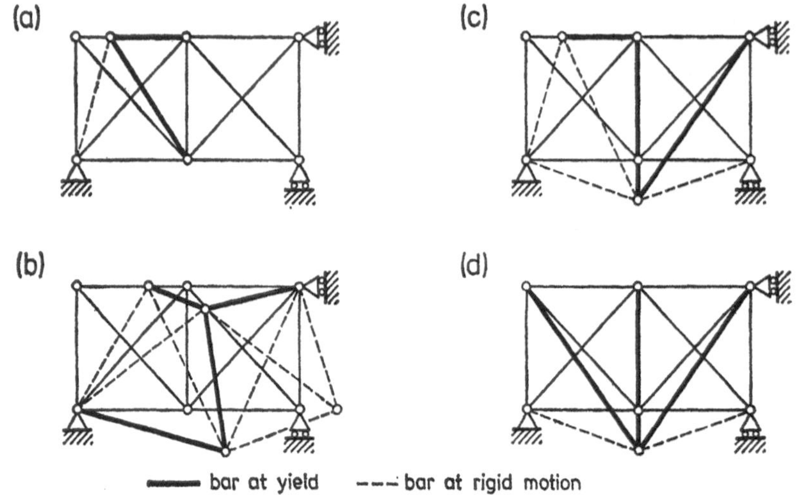

Fig. 72 Collapse mechanisms for the truss

Having a clear picture of the truss behaviour we choose three typical loads to be checked by incremental calculations. They are referred to as loads 1, 2 and 3 in Table 5 and Fig. 71. Load 1 is expected to produce the so-called alternating plasticity because its domain extends beyond the segment *AL*. Load 2 should lead to adaptation as well as load 3, the latter in the vicinity

Table 5 Test loads for incremental procedure

Load number	Load bounds $\bar{\beta}_1$	Load bounds $\bar{\beta}_2$
1	0.732	0.345
2	0.586	0.345
3	0.350	0.850

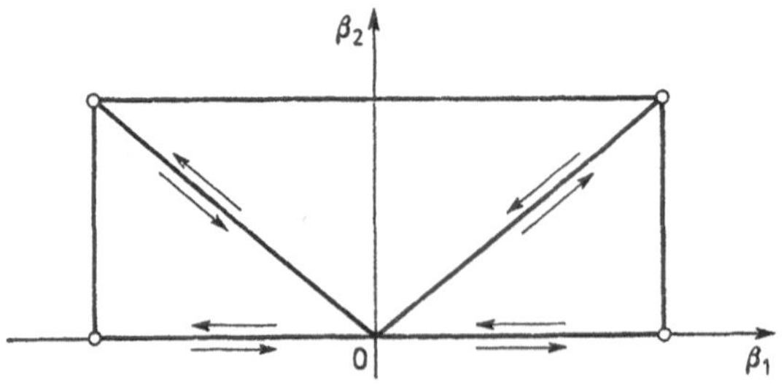

Fig. 73 Critical load cycles

Table 6 **Plastic strain increments in bar 4–5 of the truss under test loads**

Cycle number	Segment number	$\Delta\varepsilon_{4-5}^{(pl)} \times 10^3$		
		Load 1	Load 2	Load 3
1	1	−0.645	0.000	0.000
	2	−2.659	−0.771	−0.794
	3	1.285	0.000	0.000
	4	2.272	0.000	0.000
2	1	−1.291	0.000	0.000
	2	−2.270	0.000	0.000
	3	1.291	0.000	0.000
	4	2.270	0.000	0.000
3	1	−1.291	0.000	0.000
	2	−2.270	0.000	0.000
	3	1.291	0.000	0.000
	4	2.270	0.000	0.000

Table 7 **Total accumulated plastic strain for the truss under test loads**

Cycle number	Segment number	$\bar{\varepsilon}^{(pl)} \times 10^3$		
		Load 1	Load 2	Load 3
1	1	0.645	0.000	0.000
	2	3.304	0.771	1.013
	3	4.589	0.771	1.013
	4	6.861	0.771	1.013
2	1	8.152	0.771	1.013
	2	10.432	0.771	1.023
	3	11.723	0.771	1.023
	4	13.993	0.771	1.023
3	1	15.284	0.771	1.023
	2	17.554	0.771	1.023
	3	18.845	0.771	1.023
	4	21.115	0.771	1.023

of the so-called incremental collapse. The results are given in Tables 6, 7 and Figs. 74, 75. Figs. 74(a), (b) show the history of the critical cyclic load. The load increments are adjusted so as to avoid unnecessary waste of time in elastic regime but to maintain accuracy after yielding has taken place. From Table 6 and Fig. 74(c) it is seen that the alternating plastic strain

$$\varepsilon_{4-5}^{(pl)} = \pm 3.561 \cdot 10^{-3}$$

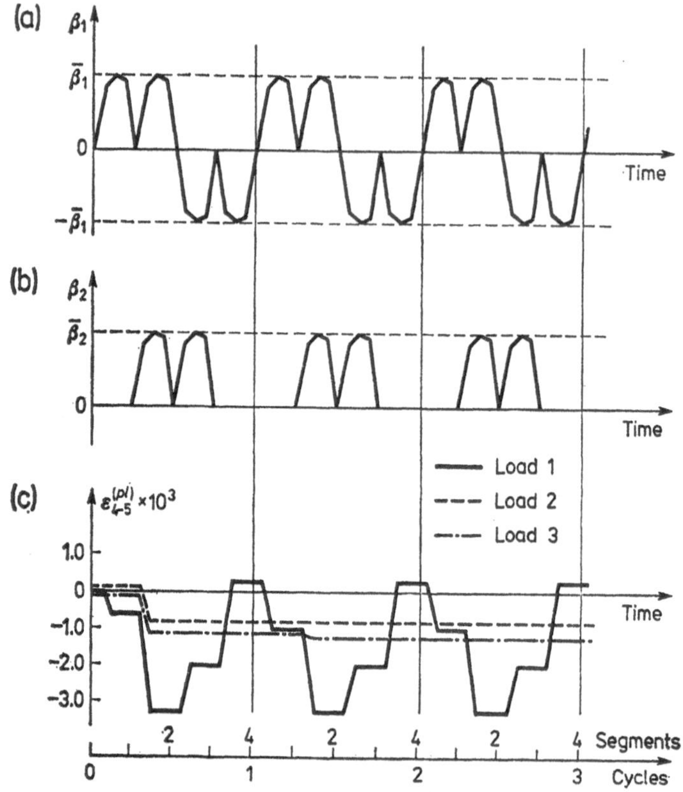

Fig. 74 Evolution of the load and plastic strain for truss under critical cyclic load

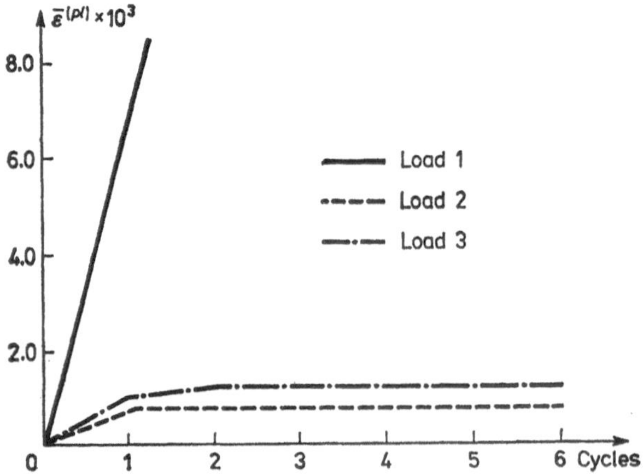

Fig. 75 Evolution of the total accumulated plastic strain under test loads 1, 2, 3

develops in bar 4–5 after the second load cycle. This leads to the steady growth of the dissipated energy which is proportional to the total accumulated absolute value of the plastic strain $\varepsilon^{(pl)}$ (see Table 7 and Fig. 75). After the bound $\bar{\beta}_1$ was made smaller by 20% which corresponds to load 2, the plastic strain of bar 4–5 has stabilized at the level

$$\varepsilon_{4\text{-}5}^{(pl)} = -0.771 \cdot 10^{-3}$$

achieved during the second segment of the first cycle. The response of the truss to subsequent cycles was purely elastic.

The incremental approach to shakedown provides not only residual stresses (as the linear programming approach, for instance) but residual displacements as well. The latter may turn out important for an engineer because it happens that shakedown is achieved at the expense of unacceptably large structural distortion. In this example, the vertical residual displacement of node 2

$$u_{2,\mathrm{res}}^{(2)} = -0.417 \cdot 10^{-3}\ \mathrm{cm}$$

after the adaptation to load 2 is very small compared to the truss dimensions.

The upper bound $\bar{\beta}_1$ for load 3 is only $\sim$ 3% less than the ultimate value

$$\bar{\beta}_{2,0} = 0.874$$

corresponding to the intersection of the line *fg* with the boundary segment *FG*. Despite this small safety margin the incremental procedure firmly confirmed the adaptation. The total accumulated plastic strain ceased to grow after the second load cycle.

6.2 Frames

6.2.1 *Elastic-plastic constitutive matrices for beam elements*

As we have already emphasized throughout the book, an important cause of nonlinearity in the behaviour of structures is the development of material plasticity. In the case of frame structures the overall nonlinearity of the load-deflection relationship may often be adequately modelled by considering plastic effects concentrated at discrete sections in the structure.

The frames considered below consist, as in Sec. 4.5, of slender, prismatic and straight (or at least sectionally prismatic and straight) members with rigid connections at the joints. A typical joint has six degrees of freedom consisting of the usual three rotations about and three translations along the local coordinate axes. The longitudinal axes of the members may be in any position relative to the global coordinate system.

Plastic deformations are assumed to be restricted to the cross-sections at the ends of the members. This implies that external loads act only upon joints of the frames. The inelastic effects together with large deformations which are admitted in the formulation call for an incremental approach. This may directly be accepted in the form previously considered in Sec. 4.5, provided the constitutive matrix $\mathbf{k}^{(el)}$ is modified to account for inelastic effects at the beam ends. Even if the formulation to follow is based on a number of highly idealized assumptions we shall document its validity on many representative tests of nonlinear frame behaviour.

We shall first recall and slightly extend the definitions of kinematic and static quantities related to the problem of 3D beam analysis as discussed in Sec. 4.5. The vector of the generalized nodal incremental displacements is defined as

$$\Delta \mathbf{q}_{12\times 1} = \{\Delta q_1 \ \Delta q_2 \ \dots \ \Delta q_{12}\} \tag{6.15}$$

or as

$$\Delta \mathbf{q}_{12\times 1} = \{\Delta q_1^{(1)} \ \Delta q_2^{(1)} \ \Delta q_3^{(1)} \ \Delta q_4^{(1)} \ \Delta q_5^{(1)} \ \Delta q_6^{(1)} \ \Delta q_1^{(2)} \ \Delta q_2^{(2)} \ \Delta q_3^{(2)} \ \Delta q_4^{(2)} \ \Delta q_5^{(2)} \ \Delta q_6^{(2)}\}$$

where the vector $\Delta \mathbf{q}^{(\alpha)} = \{\Delta q_1^{(\alpha)} \ \Delta q_2^{(\alpha)} \ \dots \ \Delta q_6^{(\alpha)}\}$, $\alpha = 1, 2$, collects the generalized incremental displacements at the node α and $\Delta q_1^{(\alpha)}$, $\Delta q_2^{(\alpha)}$, $\Delta q_3^{(\alpha)}$ are three components of the nodal incremental displacement vector at this node whereas $\Delta q_4^{(\alpha)}$, $\Delta q_5^{(\alpha)}$, $\Delta q_6^{(\alpha)}$ represent the components of the incremental rotation vector. The analogous notation applies to the corresponding static variables so that $Q_1^{(\alpha)}$, $Q_2^{(\alpha)}$, $Q_3^{(\alpha)}$ stand for the force components (axial force and two shear forces) whereas $Q_4^{(\alpha)}$, $Q_5^{(\alpha)}$, $Q_6^{(\alpha)}$ represent the moments (twisting moment and two bending moments), all at the node α, $\alpha = 1, 2$[1].

The first assumption we make concerns the additivity of the elastic and plastic parts of the generalized incremental displacement vector and has the form

$$\Delta \mathbf{q} = \Delta \mathbf{q}^{(el)} + \Delta \mathbf{q}^{(pl)}. \tag{6.16}$$

The plastic deformations at an end cross-section are generated if a condition known as the limit state condition is satisfied at this point having the form

$$\bar{F}^{(\alpha)} = \bar{f}^{(\alpha)}(\mathbf{Q}^{(\alpha)}, \mathbf{Q}_0^{(\alpha)}) - 1 = 0 \tag{6.17}$$

the index $\alpha = 1, 2$ referring again to the node 1 and node 2 of the element on

[1] An alternative notation $q_1^{(\alpha)} = u_\xi^{(\alpha)}$, $q_2^{(\alpha)} = u_\eta^{(\alpha)}$, $q_3^{(\alpha)} = u_\zeta^{(\alpha)}$, $q_4^{(\alpha)} = \varphi_\xi^{(\alpha)}$, $q_5^{(\alpha)} = \varphi_\eta^{(\alpha)}$, $q_6^{(\alpha)} = \varphi_\zeta^{(\alpha)}$; $Q_1^{(\alpha)} = N^{(\alpha)}$, $Q_2^{(\alpha)} = T_\eta^{(\alpha)}$, $Q_3^{(\alpha)} = T_\zeta^{(\alpha)}$, $Q_4^{(\alpha)} = M_\xi^{(\alpha)}$, $Q_5^{(\alpha)} = M_\eta^{(\alpha)}$, $Q_6^{(\alpha)} = M_\zeta^{(\alpha)}$ will also be used if more convenient.

hand. According to our assumptions, these cross-sections are the only ones likely to go plastic during the deformation process. The vector $\mathbf{Q}_0$ has the form, cf. Fig. 15

$$\mathbf{Q}_0 = \{\mathbf{Q}_0^{(1)}\ \mathbf{Q}_0^{(2)}\} = \{Q_{0,1}^{(1)}\ \dots\ Q_{0,6}^{(1)}\ Q_{0,1}^{(2)}\ \dots\ Q_{0,6}^{(2)}\}$$

where the Q_0's are the ultimate values established for the generalized forces. The next assumption is that the functions appearing in the limit state conditions may be taken to serve as potentials to determine the generalized plastic displacements, i.e.

$$\Delta\mathbf{q}^{(1)(pl)} = \Delta\lambda_1 \frac{\partial \bar{f}^{(1)}}{\partial \mathbf{Q}^{(1)T}}, \quad \Delta\mathbf{q}^{(2)(pl)} = \Delta\lambda_2 \frac{\partial \bar{f}^{(2)}}{\partial \mathbf{Q}^{(2)T}} \tag{6.18}$$

for the plastic zones at the ends 1 and 2 of the beam. $\Delta\lambda_1$ and $\Delta\lambda_2$ are non-negative proportionality constants, cf. eq. (4.205).

Note that inelastic effects in the cross-section are ignored unless its load capacity has been reached.

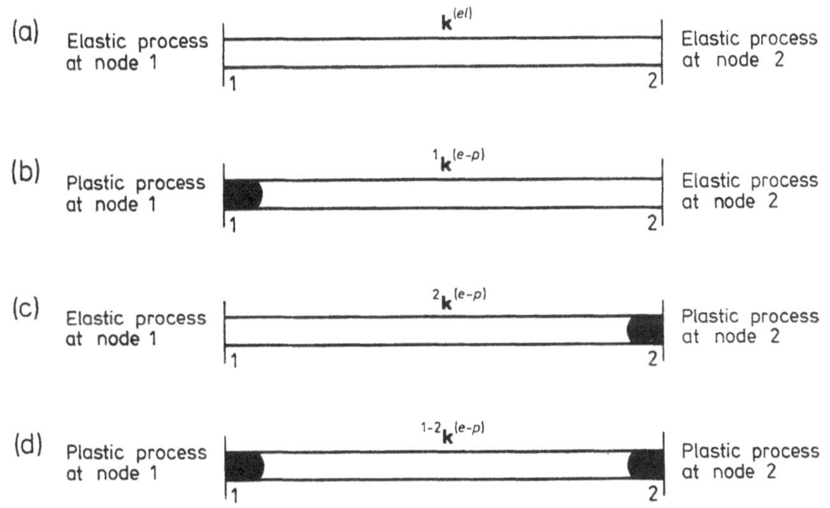

Fig. 76 Four cases for beam element stiffness (elastic and elastic-plastic), cf. eq. (6.22);

elastic process at node α:

$$F^{(\alpha)} < 0 \text{ or } F^{(\alpha)} = 0 \text{ and } \left(\frac{\partial f^{(\alpha)}}{\partial \mathbf{Q}_{6\times 1}^{(\alpha)}}\right)_{1\times 6} [\mathbf{k}_{1\alpha}^{(el)}\ \mathbf{k}_{\alpha 2}^{(el)}]_{6\times 12}\Delta\mathbf{q}_{12\times 1} \leqslant 0$$

plastic process at node α:

$$F^{(\alpha)} = 0 \text{ and } \left(\frac{\partial f^{(\alpha)}}{\partial \mathbf{Q}_{6\times 1}^{(\alpha)}}\right)_{1\times 6} [\mathbf{k}_{1\alpha}^{(el)}\ \mathbf{k}_{\alpha 2}^{(el)}]_{6\times 12}\Delta\mathbf{q}_{12\times 1} > 0$$

For the beam element considered we distinguish four different forms of the stiffness matrix in question:

Case 1: The elastic stiffness matrix $\mathbf{k}^{(el)}_{12\times 12}$ is used when no limit state condition has been met at the beam ends, Fig. 76(a).

Case 2: The elastic-plastic stiffness matrix ${}^1\mathbf{k}^{(e-p)}$ (to be determined below) is used when a plastic zone has developed at end 1 of the beam, Fig. 76(b).

Case 3: The elastic-plastic stiffness matrix ${}^2\mathbf{k}^{(e-p)}$ is used when a plastic zone has developed at end 2 of the beam, Fig. 76(c).

Case 4: The elastic-plastic stiffness matrix ${}^{1-2}\mathbf{k}^{(e-p)}$ is used when both ends have developed plastic deformations, Fig. 76(d).

It should be emphasized that the cases 2 and 3 have been distinguished from each other merely to simplify the programming.

We shall now derive explicit expressions for the stiffness matrices ${}^1\mathbf{k}^{(e-p)}$, ${}^2\mathbf{k}^{(e-p)}$ and ${}^{1-2}\mathbf{k}^{(e-p)}$. We proceed by forming the gradient of the yield function in the form of a column vector as

$$\mathbf{V}^{(\alpha)}_{6\times 1} = \frac{\partial \bar{f}^{(\alpha)}}{\partial \mathbf{Q}^{(\alpha)T}_{1\times 6}} = \left\{ \frac{\partial \bar{f}^{(\alpha)}}{\partial Q_1^{(\alpha)}} \quad \frac{\partial \bar{f}^{(\alpha)}}{\partial Q_2^{(\alpha)}} \quad \cdots \quad \frac{\partial \bar{f}^{(\alpha)}}{\partial Q_6^{(\alpha)}} \right\} \tag{6.19}$$

so that

$$\Delta \mathbf{q}^{(\alpha)(pl)}_{6\times 1} = \Delta \lambda_{(\alpha)} \mathbf{V}^{(\alpha)}_{6\times 1}, \qquad \alpha = 1, 2. \tag{6.20}$$

Assuming that the cross-section on hand is characterized by the elastic-plastic constitutive law which bears resemblance to the classical law of elastic-ideally plastic material we postulate the normality condition as

$$\Delta \mathbf{q}^{(\alpha)(pl)T}_{1\times 6} \Delta \mathbf{Q}^{(\alpha)}_{6\times 1} = 0 \tag{6.21}$$

to hold at each element nodal cross-section with limit state condition (6.17) satisfied. Using the elastic constitutive law

$$\begin{bmatrix} \Delta \mathbf{Q}^{(1)}_{6\times 1} \\ \Delta \mathbf{Q}^{(2)}_{6\times 1} \end{bmatrix} = \begin{bmatrix} \mathbf{k}^{(el)}_{11_{6\times 6}} & \mathbf{k}^{(el)}_{12_{6\times 6}} \\ \mathbf{k}^{(el)}_{12_{6\times 6}} & \mathbf{k}^{(el)}_{22_{6\times 6}} \end{bmatrix} \begin{bmatrix} \Delta \mathbf{q}^{(1)}_{6\times 1} \\ \Delta \mathbf{q}^{(2)}_{6\times 1} \end{bmatrix} \tag{6.22}$$

and the additivity of the generalized displacement vectors (6.16) we arrive at

$$\begin{bmatrix} \Delta \mathbf{Q}^{(1)} \\ \Delta \mathbf{Q}^{(2)} \end{bmatrix}_{12\times 1} = \begin{bmatrix} \mathbf{k}^{(el)}_{11} & \mathbf{k}^{(el)}_{12} \\ \mathbf{k}^{(el)}_{12} & \mathbf{k}^{(el)}_{22} \end{bmatrix}_{12\times 12} \begin{bmatrix} \Delta \mathbf{q}^{(1)} - \Delta \mathbf{q}^{(1)(pl)} \\ \Delta \mathbf{q}^{(2)} - \Delta \mathbf{q}^{(2)(pl)} \end{bmatrix}_{12\times 1}; \tag{6.23}$$

hence by eq. (6.20)

$$\begin{bmatrix} \Delta \mathbf{Q}^{(1)} \\ \Delta \mathbf{Q}^{(2)} \end{bmatrix} = \begin{bmatrix} \mathbf{k}^{(el)}_{11} & \mathbf{k}^{(el)}_{12} \\ \mathbf{k}^{(el)}_{12} & \mathbf{k}^{(el)}_{22} \end{bmatrix} \begin{bmatrix} \Delta \mathbf{q}^{(1)} - \Delta \lambda_1 \mathbf{V}^{(1)} \\ \Delta \mathbf{q}^{(2)} - \Delta \lambda_2 \mathbf{V}^{(2)} \end{bmatrix}. \tag{6.24}$$

An arbitrary change in generalized forces cannot violate the limit state condition, we have therefore the condition

$$\begin{bmatrix} \Delta f^{(1)} \\ \Delta f^{(2)} \end{bmatrix}_{2\times 1} = \begin{bmatrix} \dfrac{\partial f^{(1)}}{\partial \mathbf{Q}^{(1)}} \Delta \mathbf{Q}^{(1)} \\ \dfrac{\partial f^{(2)}}{\partial \mathbf{Q}^{(2)}} \Delta \mathbf{Q}^{(2)} \end{bmatrix}_{2\times 1} = \begin{bmatrix} \mathbf{V}^{(1)T}_{1\times 6} & \mathbf{0}_{1\times 6} \\ \mathbf{0}_{1\times 6} & \mathbf{V}^{(2)T}_{1\times 6} \end{bmatrix}_{2\times 12} \begin{bmatrix} \Delta \mathbf{Q}^{(1)}_{6\times 1} \\ \Delta \mathbf{Q}^{(2)}_{6\times 1} \end{bmatrix}_{12\times 1}. \tag{6.25}$$

Substituting eq. (6.24) into eq. (6.25) leads for $\Delta\lambda_1 \neq 0, \Delta\lambda_2 = 0$ (Fig. 76(b)) to the following self-explanatory derivation

$$\Delta \mathbf{Q}_{12\times 1} = \begin{bmatrix} \mathbf{k}_{11}^{(el)} & \mathbf{k}_{12}^{(el)} \\ \mathbf{k}_{12}^{(el)} & \mathbf{k}_{22}^{(el)} \end{bmatrix} \begin{bmatrix} \Delta \mathbf{q}_1 - \Delta\lambda_1 \mathbf{V}^{(1)} \\ \Delta \mathbf{q}_2 \end{bmatrix},$$

$$\begin{bmatrix} 0 \\ \cdot \end{bmatrix}_{2\times 1} = \begin{bmatrix} \mathbf{V}^{(1)T} & 0 \\ 0 & \mathbf{V}^{(2)T} \end{bmatrix}_{2\times 12} \begin{bmatrix} \mathbf{k}_{11}^{(el)} & \mathbf{k}_{12}^{(el)} \\ \mathbf{k}_{12}^{(el)} & \mathbf{k}_{22}^{(el)} \end{bmatrix}_{12\times 12} \begin{bmatrix} \Delta \mathbf{q}_1 - \Delta\lambda_1 \mathbf{V}^{(1)} \\ \Delta \mathbf{q}_2 \end{bmatrix}_{12\times 1}, \tag{6.26}$$

$$\Delta\lambda_1 = \frac{\mathbf{V}^{(1)T}_{1\times 6}[\mathbf{k}_{11}^{(el)} \;\; \mathbf{k}_{12}^{(el)}]_{6\times 12}}{\mathbf{V}^{(1)T}_{1\times 6}\mathbf{k}^{(el)}_{11_{6\times 6}}\mathbf{V}^{(1)}_{6\times 1}} \begin{bmatrix} \Delta \mathbf{q}_1 \\ \Delta \mathbf{q}_2 \end{bmatrix}_{12\times 1}.$$

Similarly for the case $\Delta\lambda_1 = 0, \Delta\lambda_2 \neq 0$ (Fig. 76(c)) we obtain

$$\Delta\lambda_2 = \frac{\mathbf{V}^{(2)T}[\mathbf{k}_{12}^{(el)} \;\; \mathbf{k}_{22}^{(el)}]}{\mathbf{V}^{(2)T}\mathbf{k}_{22}^{(el)}\mathbf{V}^{(2)}} \begin{bmatrix} \Delta \mathbf{q}_1 \\ \Delta \mathbf{q}_2 \end{bmatrix}. \tag{6.27}$$

In the case of plasticity at both ends we have $\Delta\lambda_1 \neq 0$, $\Delta\lambda_2 \neq 0$ and thus

$$\begin{bmatrix} \mathbf{V}^{(1)T}\mathbf{k}_{11}^{(el)} & \mathbf{V}^{(1)T}\mathbf{k}_{12}^{(el)} \\ \mathbf{V}^{(2)T}\mathbf{k}_{12}^{(el)} & \mathbf{V}^{(2)T}\mathbf{k}_{22}^{(el)} \end{bmatrix}_{2\times 12} \begin{bmatrix} \Delta \mathbf{q}_1 \\ \Delta \mathbf{q}_2 \end{bmatrix}_{12\times 1}$$
$$= \begin{bmatrix} \mathbf{V}^{(1)T}\mathbf{k}_{11}^{(el)}\Delta\lambda_1\mathbf{V}^{(1)} + \mathbf{V}^{(1)T}\mathbf{k}_{12}^{(el)}\Delta\lambda_2\mathbf{V}^{(2)} \\ \mathbf{V}^{(2)T}\mathbf{k}_{12}^{(el)}\Delta\lambda_1\mathbf{V}^{(1)} + \mathbf{V}^{(2)T}\mathbf{k}_{22}^{(el)}\Delta\lambda_2\mathbf{V}^{(2)} \end{bmatrix}_{2\times 1}. \tag{6.28}$$

This set of equations may be solved for $\Delta\lambda_1$, $\Delta\lambda_2$ yielding

$$\Delta\lambda_1 = \frac{(\mathbf{V}^{(1)T}[\mathbf{k}_{11}^{(el)} \;\; \mathbf{k}_{12}^{(el)}]\Delta\mathbf{q})(\mathbf{V}^{(2)T}\mathbf{k}_{22}^{(el)}\mathbf{V}^{(2)}) - (\mathbf{V}^{(2)T}[\mathbf{k}_{12}^{(el)} \;\; \mathbf{k}_{22}^{(el)}]\Delta\mathbf{q})(\mathbf{V}^{(1)T}\mathbf{k}_{12}^{(el)}\mathbf{V}^{(2)})}{(\mathbf{V}^{(1)T}\mathbf{k}_{11}^{(el)}\mathbf{V}^{(1)})(\mathbf{V}^{(2)T}\mathbf{k}_{22}^{(el)}\mathbf{V}^{(2)}) - (\mathbf{V}^{(1)T}\mathbf{k}_{12}^{(el)}\mathbf{V}^{(2)})^2},$$

$$\Delta\lambda_2 = \frac{(\mathbf{V}^{(2)T}[\mathbf{k}_{12}^{(el)} \;\; \mathbf{k}_{22}^{(el)}]\Delta\mathbf{q})(\mathbf{V}^{(1)T}\mathbf{k}_{11}^{(el)}\mathbf{V}^{(1)}) - (\mathbf{V}^{(1)T}[\mathbf{k}_{11}^{(el)} \;\; \mathbf{k}_{12}^{(el)}]\Delta\mathbf{q})(\mathbf{V}^{(2)T}\mathbf{k}_{12}^{(el)}\mathbf{V}^{(1)})}{(\mathbf{V}^{(1)T}\mathbf{k}_{11}^{(el)}\mathbf{V}^{(1)})(\mathbf{V}^{(2)T}\mathbf{k}_{22}^{(el)}\mathbf{V}^{(2)}) - (\mathbf{V}^{(1)T}\mathbf{k}_{12}^{(el)}\mathbf{V}^{(2)})^2}. \tag{6.29}$$

Substituting now the computed values of plastic multipliers $\Delta\lambda_1$, $\Delta\lambda_2$ into eq. (6.24) we get the following set of constitutive relations describing the behaviour of the beam element considered:

(i) for the case of $\Delta\lambda_1 \neq 0$, $\Delta\lambda_2 = 0$ (Fig. 76(b))

$$\Delta\mathbf{Q} = \begin{bmatrix} \mathbf{k}_{11}^{(el)} - \dfrac{\mathbf{k}_{11}^{(el)}\mathbf{V}^{(1)}\mathbf{V}^{(1)T}\mathbf{k}_{11}^{(el)}}{C_1} & \mathbf{k}_{12}^{(el)} - \dfrac{\mathbf{k}_{11}^{(el)}\mathbf{V}^{(1)}\mathbf{V}^{(1)T}\mathbf{k}_{12}^{(el)}}{C_1} \\ \mathbf{k}_{12}^{(el)} - \dfrac{\mathbf{k}_{12}^{(el)}\mathbf{V}^{(1)}\mathbf{V}^{(1)T}\mathbf{k}_{11}^{(el)}}{C_1} & \mathbf{k}_{22}^{(el)} - \dfrac{\mathbf{k}_{12}^{(el)}\mathbf{V}^{(1)}\mathbf{V}^{(1)T}\mathbf{k}_{12}^{(el)}}{C_1} \end{bmatrix}_{12\times 12} \times$$

$$\times\, \Delta\mathbf{q}_{12\times 1} = {}^{1}\mathbf{k}^{(e-p)}\Delta\mathbf{q} \tag{6.30}$$

where

$$C_1 = \mathbf{V}^{(1)T}\mathbf{k}_{11}^{(el)}\mathbf{V}^{(1)},$$

(ii) for the case of $\Delta\lambda_1 = 0$, $\Delta\lambda_2 \neq 0$ (Fig. 76(c))

$$\Delta\mathbf{Q} = \begin{bmatrix} \mathbf{k}_{11}^{(el)} - \dfrac{\mathbf{k}_{12}^{(el)}\mathbf{V}^{(2)}\mathbf{V}^{(2)T}\mathbf{k}_{12}^{(el)}}{C_2} & \mathbf{k}_{12}^{(el)} - \dfrac{\mathbf{k}_{12}^{(el)}\mathbf{V}^{(2)}\mathbf{V}^{(2)T}\mathbf{k}_{22}^{(el)}}{C_2} \\ \mathbf{k}_{12}^{(el)} - \dfrac{\mathbf{k}_{22}^{(el)}\mathbf{V}^{(2)}\mathbf{V}^{(2)T}\mathbf{k}_{12}^{(el)}}{C_2} & \mathbf{k}_{22}^{(el)} - \dfrac{\mathbf{k}_{22}^{(el)}\mathbf{V}^{(2)}\mathbf{V}^{(2)T}\mathbf{k}_{22}^{(el)}}{C_2} \end{bmatrix} \Delta\mathbf{q}$$

$$= {}^{2}\mathbf{k}^{(e-p)}\Delta\mathbf{q} \tag{6.31}$$

where

$$C_2 = \mathbf{V}^{(2)T}\mathbf{k}_{22}^{(el)}\mathbf{V}^{(2)},$$

(iii) for the case of $\Delta\lambda_1 \neq 0$, $\Delta\lambda_2 \neq 0$ (Fig. 76(d))

$$\Delta\mathbf{Q} = {}^{1-2}\mathbf{k}^{(e-p)}\Delta\mathbf{q}, \tag{6.32}$$

where

$${}^{1-2}\mathbf{k}^{(e-p)} = \begin{bmatrix} \mathbf{k}_{11}^{(e-p)} & \mathbf{k}_{12}^{(e-p)} \\ \mathbf{k}_{12}^{(e-p)} & \mathbf{k}_{22}^{(e-p)} \end{bmatrix},$$

$$\mathbf{k}_{11}^{(e-p)} = \mathbf{k}_{11}^{(el)} - \frac{1}{C_{1-2}} [\mathbf{k}_{11}^{(el)}\mathbf{V}^{(1)}(\mathbf{V}^{(2)T}\mathbf{k}_{22}^{(el)}\mathbf{V}^{(2)}\mathbf{V}^{(1)T}\mathbf{k}_{11}^{(el)} -$$
$$- \mathbf{V}^{(1)T}\mathbf{k}_{12}^{(el)}\mathbf{V}^{(2)}\mathbf{V}^{(2)T}\mathbf{k}_{21}^{(el)}) + \mathbf{k}_{12}^{(el)}\mathbf{V}^{(2)}(\mathbf{V}^{(1)T}\mathbf{k}_{11}^{(el)}\mathbf{V}^{(1)}\mathbf{V}^{(2)T}\mathbf{k}_{12}^{(el)} -$$
$$- \mathbf{V}^{(2)T}\mathbf{k}_{12}^{(el)}\mathbf{V}^{(1)}\mathbf{V}^{(1)T}\mathbf{k}_{11}^{(el)})],$$

$$\mathbf{k}_{12}^{(e-p)} = \mathbf{k}_{12}^{(el)} - \frac{1}{C_{1-2}} [\mathbf{k}_{11}^{(el)}\mathbf{V}^{(1)}(\mathbf{V}^{(2)T}\mathbf{k}_{22}^{(el)}\mathbf{V}^{(2)}\mathbf{V}^{(1)T}\mathbf{k}_{12}^{(el)} - \tag{6.33}$$
$$- \mathbf{V}^{(1)T}\mathbf{k}_{12}^{(el)}\mathbf{V}^{(2)}\mathbf{V}^{(2)T}\mathbf{k}_{22}^{(el)}) + \mathbf{k}_{12}^{(el)}\mathbf{V}^{(2)}(\mathbf{V}^{(1)T}\mathbf{k}_{11}^{(el)}\mathbf{V}^{(1)}\mathbf{V}^{(2)T}\mathbf{k}_{22}^{(el)} -$$
$$- \mathbf{V}^{(2)T}\mathbf{k}_{12}^{(el)}\mathbf{V}^{(1)}\mathbf{V}^{(1)T}\mathbf{k}_{12}^{(el)})],$$

$$\mathbf{k}_{22}^{(e-p)} = \mathbf{k}_{22}^{(el)} - \frac{1}{C_{1-2}} [\mathbf{k}_{12}^{(el)}\mathbf{V}^{(1)}(\mathbf{V}^{(2)T}\mathbf{k}_{22}^{(el)}\mathbf{V}^{(2)}\mathbf{V}^{(1)T}\mathbf{k}_{12}^{(el)} -$$
$$- \mathbf{V}^{(1)T}\mathbf{k}_{12}^{(el)}\mathbf{V}^{(2)}\mathbf{V}^{(2)T}\mathbf{k}_{22}^{(el)}) + \mathbf{k}_{22}^{(el)}\mathbf{V}^{(2)}(\mathbf{V}^{(1)T}\mathbf{k}_{11}^{(el)}\mathbf{V}^{(1)}\mathbf{V}^{(2)T}\mathbf{k}_{22}^{(el)} -$$
$$- \mathbf{V}^{(2)T}\mathbf{k}_{12}^{(el)}\mathbf{V}^{(1)}\mathbf{V}^{(1)T}\mathbf{k}_{12}^{(el)})], \tag{6.33}$$
$$C_{1-2} = C_1 \cdot C_2 - (\mathbf{V}^{(1)T}\mathbf{k}_{12}^{(el)}\mathbf{V}^{(2)})^2.$$

Note that the above derivation is to a great extent just a formal repetition of the procedure we used in Sec. 4.8 to derive the constitutive law for the elastic-plastic material.

For any given limit state functions $\bar{f}(\mathbf{Q}, \mathbf{Q}_0)$ we may readily set up the detailed forms of stiffness matrices ${}^1\mathbf{k}^{(e-p)}$, ${}^2\mathbf{k}^{(e-p)}$, ${}^{1-2}\mathbf{k}^{(e-p)}$ by explicitly calculating the current values of the vectors $\mathbf{V}^{(1)}$ and $\mathbf{V}^{(2)}$.

As has already been indicated, for finite deformation analysis the constitutive matrices obtained above must be supplemented with the initial generalized stress matrices as derived in Sec. 4.8.

6.2.2 *Limit state conditions*

In order to solve the limit load problem for an elastic-ideally plastic beam or frame it is necessary to locate the zones of extensive plastic flow which by taking ultimately the form of generalized plastic "hinges" turn the structure into mechanism. For the member cross-section of a two-dimensional beam structure loaded in its plane with the bending moment M, shear force T and longitudinal force N taken as the internal stress resultants the problem of finding the cross-sectional limit state conditions is essentially that of transforming the yield condition described in terms of normal and shear stresses $\sigma_{\xi\xi}$ and $\sigma_{\xi\eta}$ into the limit state condition expressed in terms of the stress resultants as

$$f(\sigma_{\xi\xi}, \sigma_{\xi\eta}; \sigma_y) = 0 \rightarrow \bar{F}(M, N, T; M_0, N_0, T_0) = 0.$$

The equation $\bar{F} = 0$ is sometimes also referred to as the plastic interaction surface in the space of generalized stresses. In more complex situations, typical for 3D frames, for instance, the problem is of similar nature although by far more complex analytically.

It can be said in general that the derivation of limit state conditions constitutes in all non-trivial situations an extremely difficult problem. There exists a vast literature on the subject. The variety of methods used and their only approximate character is reflected in the existence of a large number of different explicit expressions describing limit state conditions for specific situations. Clearly, the expressions depend on a number of factors such as the set of generalized forces assumed relevant in the given problem, specific shape of the beam cross-section, the beam theory employed and the material yield condition $f = 0$. Among the methods most commonly used those based on static and kinematic theorems of limit analysis as well as those derivable from the Prandtl stress functions are certainly among the leaders.

Plastic interaction surfaces depend strongly on the cross-sectional geometric properties. With this in mind, beams may be considered to be of two basic types, namely those with a solid and those with a thin-walled cross-section. The latter types have usually been solved on the basis of thin-walled beam theory and consequently have an extremely complicated arithmetical structure which often renders them useless in engineering applications.

In what follows we shall present two simple examples of an effective approach to the problem of finding limit state conditions and quote some of other expressions for plastic interaction surfaces which have proved to be effective in the numerical analysis of frames. Our approach will be essentially based on static theorem of limit analysis. However, since the stress fields we are going to postulate will not be strictly statically admissible (violation of stress smoothness conditions at some discontinuity lines), the surfaces will not necessarily be lower bounds to the real plastic interaction surfaces.

Consider a beam with a bi-symmetric solid cross-section and assume that the limit state condition depends on the shear force and bending moment only. Assume further that a part of the cross-section supports the normal stresses corresponding to bending action of the beam whereas the remaining

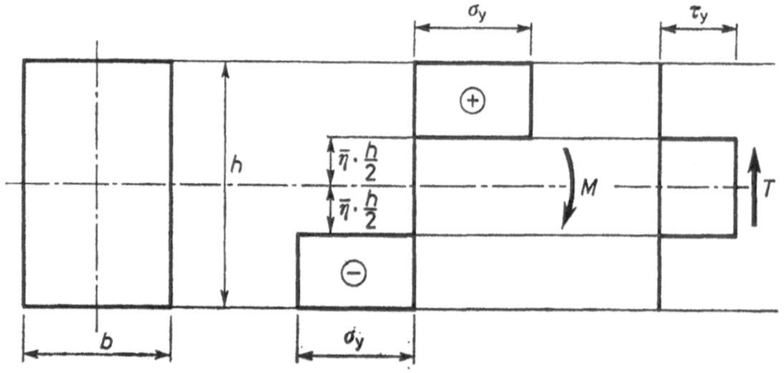

Fig. 77 Approximate stress distribution in bi-symmetric cross-section under bending and shear

part carries the shear, Fig. 77. Denoting the ultimate values of the stress resultants by

$$T_0 = \tau_y bh = k\sigma_y bh,$$
$$M_0 = \sigma_y \frac{bh^2}{4} \tag{6.34}$$

we have for either Tresca ($k = 0.5$) or Huber–Mises $\left(k = 1/\sqrt{3}\right)$ yield condition

$$M = \sigma_y b(1-\bar{\eta})\frac{h}{2}\cdot\frac{h}{2}\left(\bar{\eta}+\frac{1-\bar{\eta}}{2}\right)\cdot 2 = M_0(1-\bar{\eta}^2),$$

$$T = k\sigma_y\bar{\eta}hb = T_0$$

which upon eliminating $\bar{\eta}$ becomes

$$\frac{M}{M_0}+\left(\frac{T}{T_0}\right)^2 = 1. \tag{6.35}$$

The plastic interaction curve (6.35) is clearly an approximation to the exact relationship, Fig. 78. We note that the state of stress in the beam as assumed

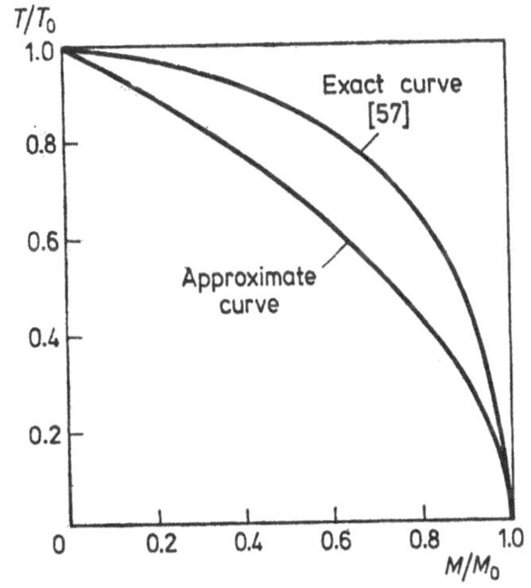

Fig. 78 Approximate and exact limit state curves for rectangular cross-section under bending and shear

above is deficient in that it does not satisfy internal equilibrium for $\eta = \pm\bar{\eta}h/2$. Thus, even if in this case the approximate limit state condition is a lower bound to the real one, this may not be so in other problems analyzed similarly.

Consider now the bi-symmetric I-cross-section and assume, cf. Fig. 79, that the shear force T is carried by the web only, the axial force N is carried by the whole cross-section whereas the bending moment M is carried either by the whole cross-section or merely by the flanges, depending on the value

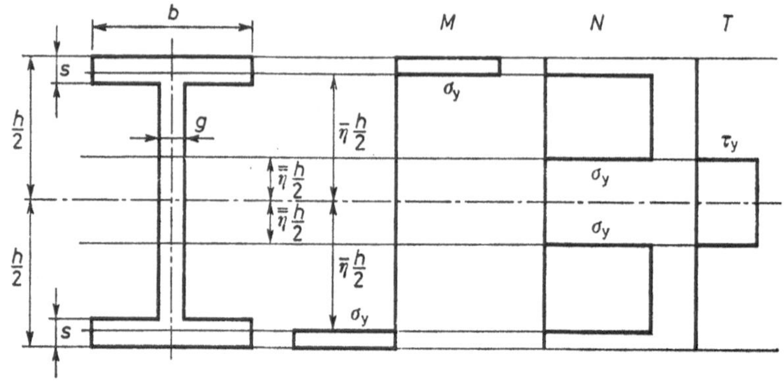

Fig. 79 Stress distribution in I-cross-section subjected to bending, shear and tension/compression

of $\bar{\eta}$. In accordance with the notation introduced in Fig. 79 the limit values of stress resultants considered separately are given by

$$\begin{aligned} T_0 &= (h-2s)g\tau_y, \\ N_0 &= [2bs+(h-2s)g]\sigma_y, \\ M_0^f &= bs(h-s)\sigma_y, \\ M_0^{f,w} &= \left[bh(h-s)+g\left(\frac{h}{2}-s\right)^2\right]\sigma_y \end{aligned} \tag{6.36}$$

where T_0, N_0, M_0^f, $M_0^{f,w}$ are the limit values of the shear force, axial force and the bending moments carried by the flanges and the whole cross-section, respectively, all the geometric parameters are defined in Fig. 79 and τ_y, σ_y are the material yield stresses in pure shear and simple tension, respectively. Fig. 79 displays also the stress distributions accepted for this problem in accordance with the previous discussion for $M \leqslant M_0^f$. They can be analytically written as

$$\begin{aligned} T &= hg\tau_y\bar{\bar{\eta}}, \\ N &= \sigma_y h[(2\alpha-1)(b-g)-g\bar{\bar{\eta}}+b\bar{\eta}], \\ M &= \frac{bh^2}{4}\sigma_y(1-\bar{\eta}^2), \end{aligned} \tag{6.37}$$

where $\alpha = \dfrac{s}{h}$, $\bar{\eta} \in [1-2\alpha, 1]$, $\bar{\bar{\eta}} \in [0, 1-2\alpha]$.

Eliminating the geometric parameters $\bar{\eta}$ and $\bar{\bar{\eta}}$ we obtain the equation of the plastic interaction surface in the space of the generalized forces. Thus, the derivation corresponds to the transformation indicated symbolically at the beginning of this section.

As previously noted, the stress distribution defined by eq. (6.37) is not statically admissible as it violates along some lines the differential equations of equilibrium.

Let us introduce the nondimensional generalized forces as

$$\begin{aligned} t &= T/T_0 = c_1\bar{\bar{\eta}}, \\ n &= N/N_0 = c_2+c_3\bar{\eta}+c_4\bar{\bar{\eta}}, \\ m &= M/M_0^f = c_5(1-\bar{\eta}^2) \end{aligned} \tag{6.38}$$

where

$$1-2\alpha \leqslant \bar{\eta} \leqslant 1, \quad 0 \leqslant \bar{\bar{\eta}} \leqslant 1-2\alpha,$$

$$c_1 = \frac{1}{1-2\alpha}, \quad c_2 = \frac{\sigma_y h(2\alpha-1)(b-g)}{N_0},$$

$$c_3 = \frac{\sigma_y hb}{N_0}, \quad c_4 = -\frac{\sigma_y hg}{N_0}, \quad c_5 = \frac{bh^2\sigma_y}{4M_0^f}.$$

Elimination of $\bar{\eta}$ and $\bar{\bar{\eta}}$ from eqs. (6.38) results in the equation

$$\bar{F} = \frac{4m}{c_5}+\frac{4}{c_4^2}\left[n-\left(\frac{c_3}{c_1}t+c_2\right)\right]^2-1 = 0 \tag{6.39}$$

which explicitly defines the plastic interaction surface. For the positive values of the internal forces M, N and T, the limit value of the bending moment

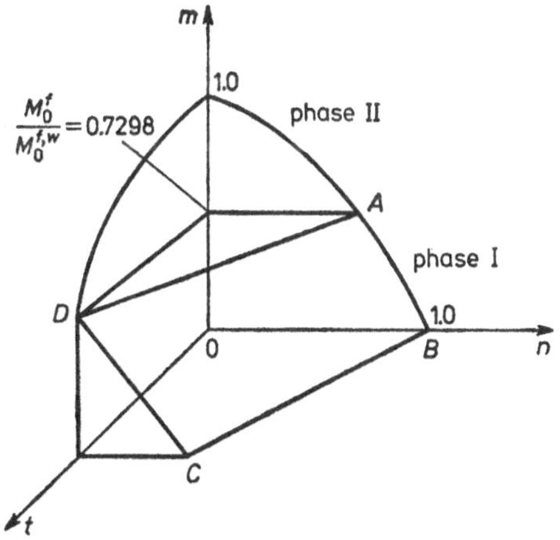

Fig. 80 Limit state surface for bi-symmetric I-cross-section

$M_0 = M_0^f$ and the following geometric characteristics of the cross-section: $h = 40$ cm, $b = 25$ cm, $s = 1$ cm the limit state condition is shown in Fig. 80 as the surface $ABCD$.

In order to consider the interaction surface for the case in which bending moment may be carried by the flanges as well as the web we have to analyse the stress distribution displayed in Fig. 81; its analytical description is formulated as

$$t = c_1 \bar{\bar{\eta}}, \quad n = -c_3(\bar{\eta} - \bar{\bar{\eta}}), \quad m = 1 + c_6 \bar{\eta}^2 \tag{6.40}$$

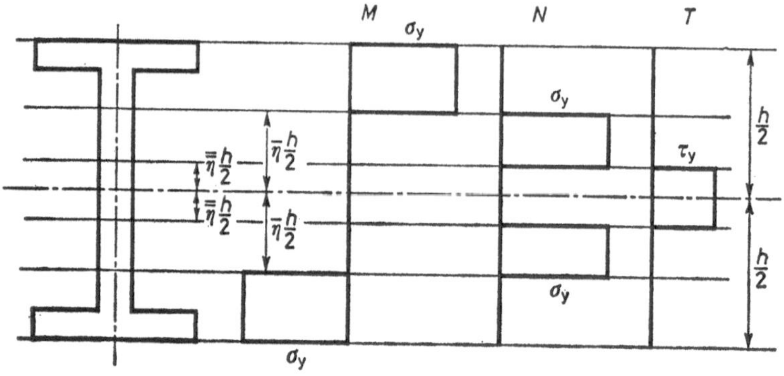

Fig. 81 Stress distribution in I-cross-section subjected to bending, shear and tension/compression

where

$$c_6 = -\frac{gh^2\sigma_y}{4M_0^{f,w}}, \qquad 0 \leqslant \bar{\bar{\eta}} \leqslant \bar{\eta} \leqslant 1-2\alpha.$$

This leads to the interaction surface in the form

$$\bar{F} = m - c_6\left(\frac{n}{c_3} + \frac{t}{c_1}\right)^2 - 1 = 0. \tag{6.41}$$

For the positive values of the generalized forces the surface is shown in Fig. 80. The parametric representation of the curve AD which separates phase I (Fig. 79) and phase II (Fig. 81) is given by

$$t = \frac{hg\tau_y}{T_0}\bar{\bar{\eta}},$$

$$n = \frac{hg\sigma_y}{N_0}(1-2\alpha) - \frac{hg\sigma_y}{N_0}\bar{\bar{\eta}}$$

which upon eliminating the parameter $\bar{\bar{\eta}}$ becomes the straight line given by

$$n = \frac{hg\sigma_y}{N_0}(1-2\alpha) - \frac{T_0\sigma_y}{N_0\tau_y}t.$$

The plastic interaction curves for other cross-sections and different combinations of the generalized forces can be derived by employing the similar approach. We recall below some commonly used expressions:
— uniaxial bending with axial force — rectangular cross-section

$$|m|+n^2-1 = 0,$$

— uniaxial bending with axial force — thin-walled I-cross-section

$$|m|+|n|-1 = 0,$$

— uniaxial bending with shear force — rectangular cross-section

$$m^2+t^2-1 = 0$$

or

$$|m|+\tfrac{3}{4}t^2-1 = 0 \quad \text{for} \quad |t| \leqslant \tfrac{2}{3}$$

or

$$|m|-0.98(1-t^4) = 0,$$

— uniaxial bending with shear force — thin-walled I-cross-section

$$|m|+0.44\frac{A-A_f}{A+A_f}t^2-1 = 0 \quad \text{for} \quad |t| \leqslant 0.79$$

or

$$|m|+\frac{A-A_f}{A+A_f}[1-(1-t^2)^{1/2}]-1 = 0$$

or

$$m^2+t^2-\alpha m^2t^2-1 = 0$$

(the symbols A, A_f, α stand for cross-section area, flange area and a coefficient depending on the profile of the cross section, [187]),
— uniaxial bending and torsion — rectangular cross-section

$$m^2+m_\xi^2-1 = 0,$$

— uniaxial bending with axial force (tension) and torsion — rectangular cross-section

$$|m|(1-m_\xi^2)^{1/2}+m_\xi^2+n^2-1 = 0,$$

— uniaxial bending with axial force (tension) and shear — rectangular cross-section

$$|m|+\tfrac{3}{4}t^2+n^2-1 = 0 \quad \text{for} \quad |t| \leqslant \tfrac{2}{3}$$

or

$$|m|(1-t^2)^{1/2}+n^2+t^2-1 = 0,$$

— uniaxial bending with axial force (tension) and shear — thin-walled I-cross-section

$$\tfrac{1}{2}[m^2+m(m^2+\alpha n^2)^{1/2}]+\tfrac{1}{4}\alpha n^2+t^2-1 = 0$$

where

$$\alpha = (2A-A_f)^2/A^2-A_f^2,$$

— biaxial bending — rectangular cross-section

$$|m_\eta|+\tfrac{3}{4}m_\xi^2-1 = 0,$$

— biaxial bending and torsion — square cross-section

$$|m_\eta|(1-m_\xi^2)^{1/2}+\tfrac{3}{4}m_\xi^2+m_\xi^2-1 = 0.$$

6.2.3 *Nonlinear analysis of curved beams*

It is quite obvious that by neglecting all the terms corresponding to the circumferential direction in equations of Sec. 4.8 describing the axisymmetric shell element we may obtain equations of the corresponding curved beam theory deformed in-plane. Adopting such an approach in this section, with all the inelastic effects accounted for by means of the layered description, we are able to describe any elastic-plastic structure made of arbitrarily curved beams in the range of large deformations. This theoretical framework of elastic-plastic beam analysis, different from the one based on the generalized plastic hinges as presented in Sec. 6.2.2, will thus create opportunity to perform comparisons of both the formalisms as to their effectiveness and accuracy.

Using the theoretical background, notation and the equations thoroughly discussed in Sec. 4.8 the computer program for axisymmetric thin shell analysis has been modified and used for the analysis of the following examples (only elastic structures are considered in this section, examples concerning inelastic structures being described in Sec. 6.2.4).

Example 1. Initial buckling of elastic circular arch, Fig. 82.

The arch is supported on the two fixed hinges and loaded by the constant pressure p. The discretization by means of 16 equal curve beam finite elements is employed. The geometric data are given in Fig. 82. The solution of the buckling equation (5.43) yields the first buckling load, associated with the asymmetric buckling shape, as $p_{cr} = 26.3025$ MPa, which is different from the one given in [146] ($p_{cr} = 3EJ/R^3$) by 0.2%.

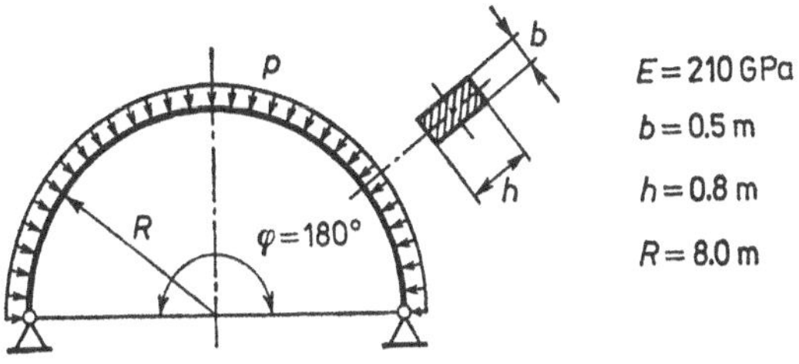

Fig. 82 Circular hinged arch under pressure

Example 2. Shallow elastic arch clamped at the ends.

16 finite elements are taken again for the analysis of geometrically nonlinear problem shown in Fig. 83. Sensitivity of the computed response to the size of the load step is clearly visible. The results are compared with those given in [139].

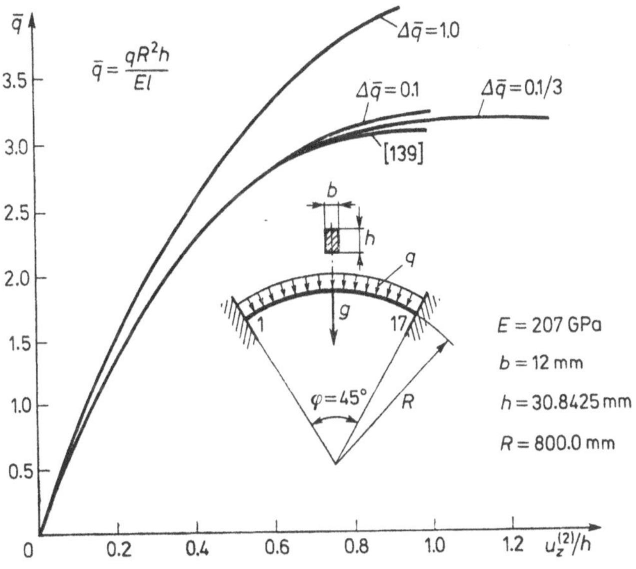

Fig. 83 Shallow clamped arch under pressure

6.2.4 *Examples of numerical frame analysis up to collapse*

The effectiveness of the finite elements proposed in Secs. 6.2.1 (plastic "hinge" element, referred below as *PH*-element) and 6.2.3 (layered model, referred below as *L*-element) was verified by a number of test examples. The first one was a beam of a rectangular cross-section, clamped at both ends and subjected to a concentrated load, cf. Fig. 84, in which the geometrical and

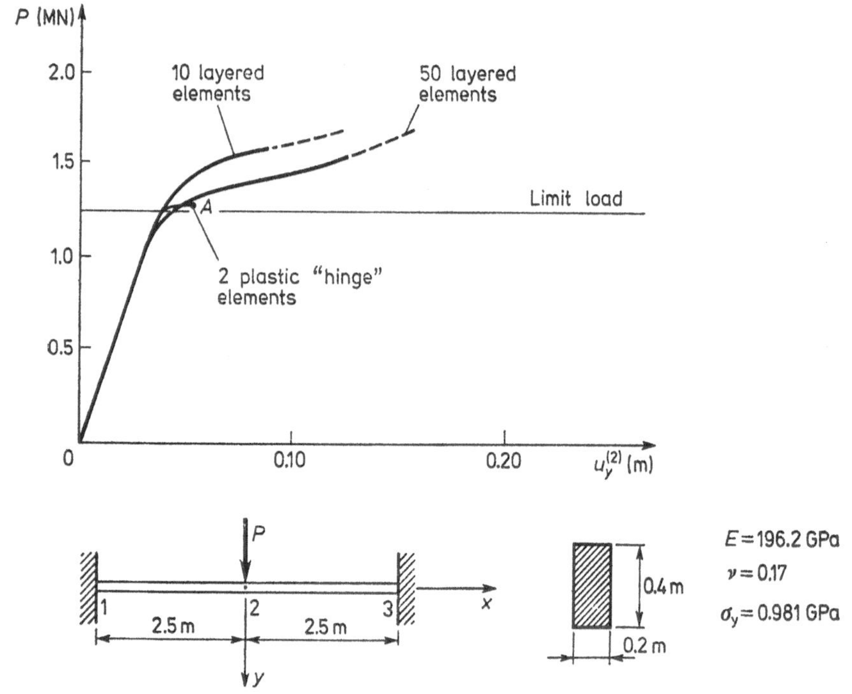

Fig. 84 Elastic-plastic analysis of clamped beam

material data are also shown. The solution obtained by using two *PH*-elements is compared against results obtained by using 10 and 50 *L*-elements. The curves shown in Fig. 84 illustrate the dependence of the vertical displacement at the central point of the beam upon the load applied. The limit load solution is also shown. The analysis with *PH*-elements was terminated at the point *A* due to the singularity of the global stiffness matrix. On the other hand, the analysis with *L*-elements was continued with no singularity at all. This can be attributed to the incompressibility of advanced plastic flow which can not be captured correctly by the *L*-elements. The distribution of plastic zones at the theoretical limit load level for both *L*-element analyses is shown in Fig. 85. The asymmetry of the results with respect to the horizontal mid-axis is due to the large deformation effects included in the analysis. The above example may also serve well to illustrate the significant reduction in computation time for the *PH*-element analysis which needed only a tenth of the CPU time necessary for the computation with 50 *L*-elements.

The second test example was a one-bay frame with a geometry and loading as shown in Fig. 86. This frame problem was solved using three *PH*-elements and results were compared with the limit load solution. Both large displacement

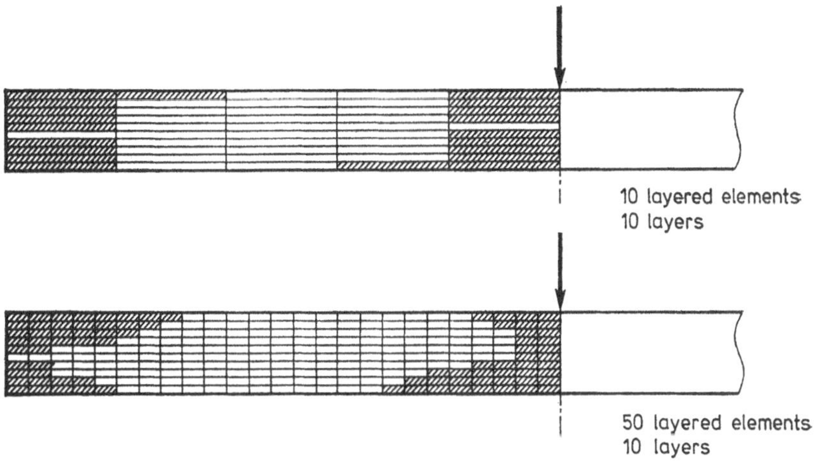

Fig. 85 Distribution of plastic zones at theoretical limit load

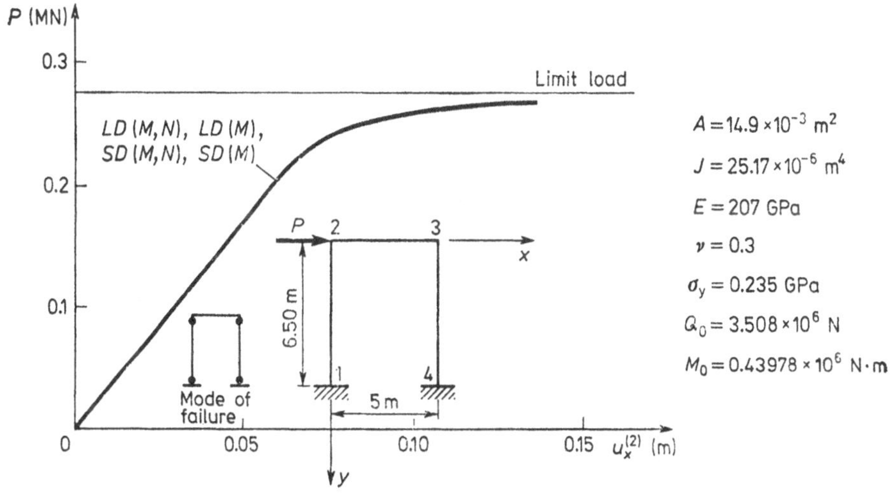

Fig. 86 Elastic-plastic frame analysis: geometric and material data, load-displacement plot, mode of failure

(LD) and small displacement (SD) analyses gave practically the same results. Also, the analysis was insensitive to the number of generalized forces in the limit state condition. In Fig. 86 the symbols (*M*) and (*M*, *N*) indicate that the bending moment only and both the bending moment and the axial force entered the yield conditions, respectively. The kinematic mechanism (mode of failure) is also shown in Fig. 86.

The third example consisted of the same frame as in the second example under different loading conditions. The problem is shown in Fig. 87. Again

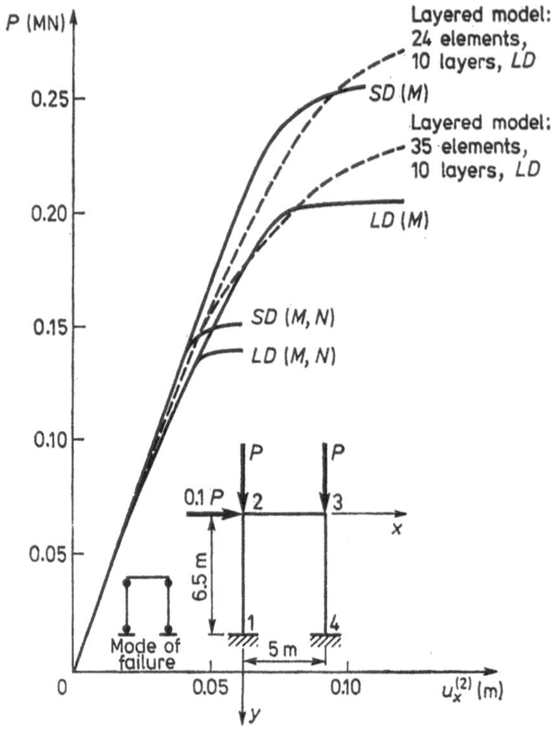

Fig. 87 Elastic-plastic frame analysis: significance of the limit yield condition form

only three *PH*-elements were used. In Fig. 87 the results are compared with those obtained with the *L*-element idealization. The same notation as in the previous example was used, and will be further used in the following examples. The significance of including the axial force in the limit state condition is clearly seen in this case.

In Fig. 88 the geometrical and material data are given for the next example in the form of a two bay asymmetric frame. Fig. 89 shows a synopsis of the different solutions. The exact solution of the limit analysis given in [100] is compared with the different solutions obtained with the *PH*-elements. The inclusion of the axial force in the limit state condition did not influence the small displacement results at all, and the large displacement results only marginally.

The two bay two storey frame of Fig. 90 was analysed for the small and large displacement cases. The final mode of failure obtained with the large displacement analysis is the same as the one with small displacements, while the limit load is slightly lower. Fig. 90 also shows the order of the subsequent plastic zone development for both calculations.

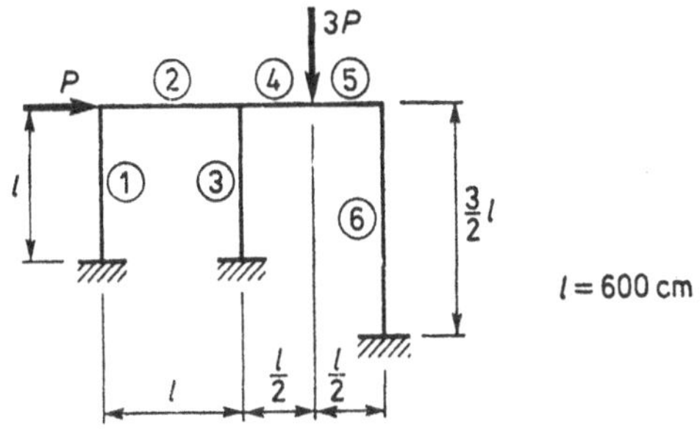

Element No.	Cross-section characteristic	Material data
1, 2, 3, 6	20 × 40; $A = 800\ \text{cm}^2$ $J_\eta = 106\,667\ \text{cm}^4$ $Q_0 = 62.784 \times 10^5$ N $M_0 = 62.784 \times 10^6$ N·cm	$E = 1.962 \times 10^4\ \text{N/mm}^2$ $\nu = 0.3$ $\sigma_y = 0.981 \times 10^2\ \text{N/mm}^2$
4, 5	30.25 × 46; $A = 1391.5\ \text{cm}^2$ $J_\eta = 245\,368\ \text{cm}^4$ $Q_0 = 109.185 \times 10^5$ N $M_0 = 125.568 \times 10^6$ N·cm	

Fig. 88 Geometric and material data; sketch of the frame with external loading

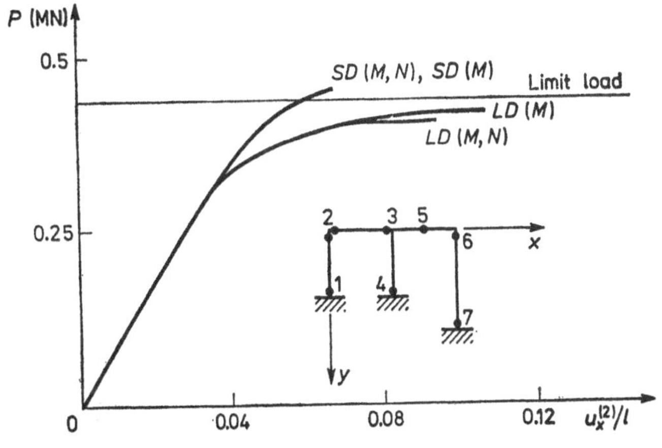

Fig. 89 Load-displacement plot and mode of failure

	Material data	A (m^2)	J (m^4)	M_0 (N·m)	N_0 (N)
Columns	$E = 207$ GPa	0.0192	0.511×10^{-3}	718.1×10^3	470.9×10^4
Girders	$\nu = 0.3$	0.0118	0.292×10^{-3}	412.0×10^3	290.4×10^4

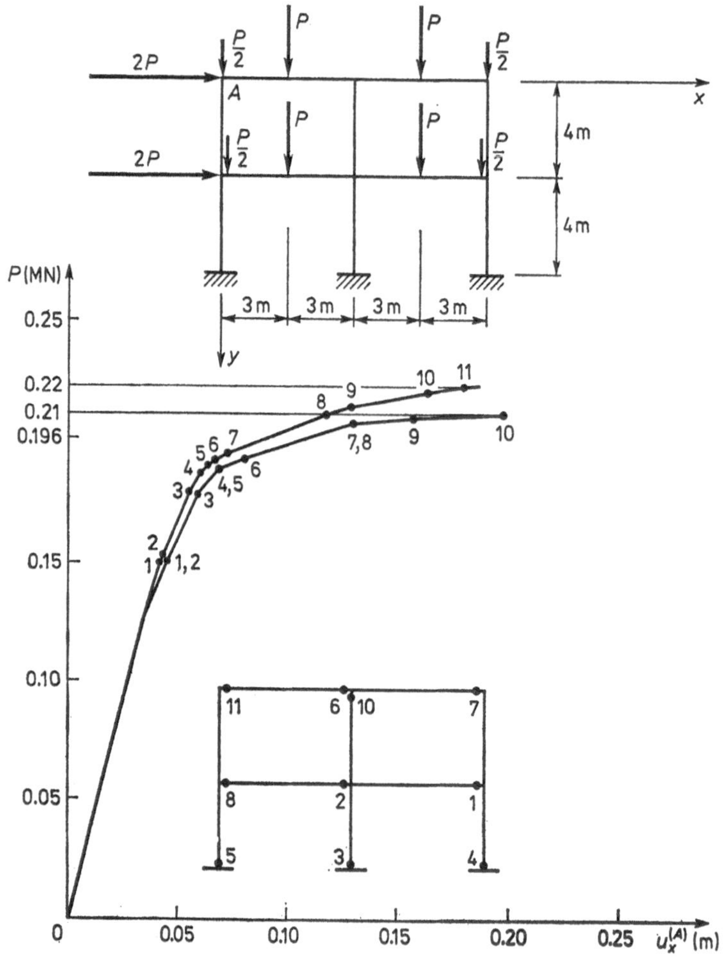

Fig. 90 Frame data, loading, load-displacement plot and successive plastic zone development

In Fig. 91 the geometrical and material data are shown for the example of a two beam structure. Also shown are different idealizations by means of 2, 6 and 30 *PH*-elements; the 30 *L*-element idealization was also used. The solution for the same configuration was reported in [132] where the

Element	Material data	A (cm^2)	J (cm^4)	M_0 (N·cm)	N_0 (N)
①	$E = 0.6867 \times 10^7$ N/cm^2 $\sigma_y = 0.2612 \times 10^5$ N/cm^2 $\nu = 0.3$	4.85	1.475	60724	12.7167
②	$E = 0.6867 \times 10^7$ N/cm^2 $\sigma_y = 0.33109 \times 10^5$ N/cm^2 $\nu = 0.3$			76695	16.0619

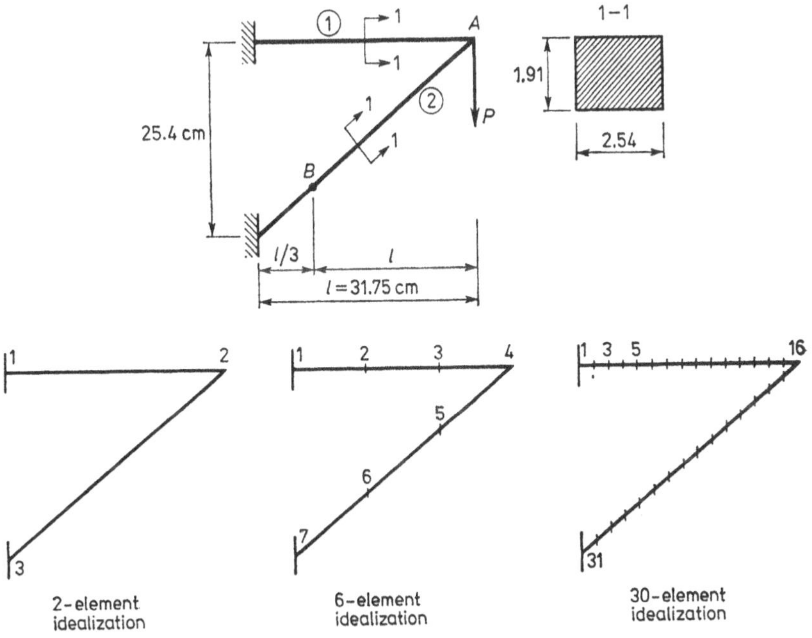

Fig. 91 Frame data, loading, idealizations

emphasis was on its stability characteristics using both analytical and experimental results. It can be seen from the plots given in Fig. 92 that at a certain load level the instability of the structure is initialized by the rapid increase of the inward displacement of the point *B*. The final instability is detected by the singularity of the global stiffness matrix.

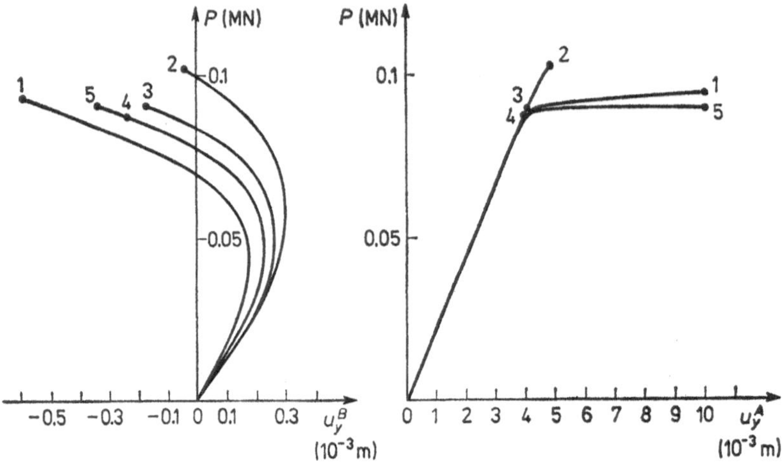

Idealization	Curves	Critical load (N)
30 *L*-elements	0−1	90742
2 *PH*-elements	0−2	99081
6 *PH*-elements	0−3	89271
30 *PH*-elements	0−4−5	88290

Fig. 92 Results obtained with different idealizations

As the next example the three-dimensional structure shown in Fig. 93 was analysed. The limit state condition was taken in the form given as the last equation in Sec. 6.2.2. According to the response curves shown in Fig. 93, the inclusion of the shear force in the plastic interaction curve appears to have a negligible effect.

As a final example, the space frame shown in Fig. 94 was analysed under the prescribed downward displacement at the top point. One sixth of the frame was considered so that only the fully symmetric deformation mode could be followed. The characteristic curves of the reaction force R_A at the point A versus the vertical displacement $u_z^{(A)}$ of the point A and $u_z^{(C)}$ of the point C during the subsequent plastifications are shown in Fig. 95.

6.2.5 Elastic-plastic beam under non-proportionally varying loads

In this section we shall briefly consider an inelastic cantilever beam subjected to varying loads so that the shakedown approach is again necessary. We shall employ the method described in Sec. 4.11.1 and exemplified in Sec.

Elements	Geometrical properties		Material properties
Columns	$A = 0.08\,m^2$	$J_y = 2.667 \times 10^{-4}\ m^4$ $J_z = 10.667 \times 10^{-4}\ m^4$ $J_0 = 7.3 \times 10^{-4}\ m^4$	$E = 19.62$ GPa $\sigma_y = 235.4$ MPa $\nu = 0.17$
Beams	$A = 0.08\,m^2$	$J_y = 10.667 \times 10^{-4}\ m^4$ $J_z = 2.667 \times 10^{-4}\ m^4$ $J_0 = 7.3 \times 10^{-4}\ m^4$	

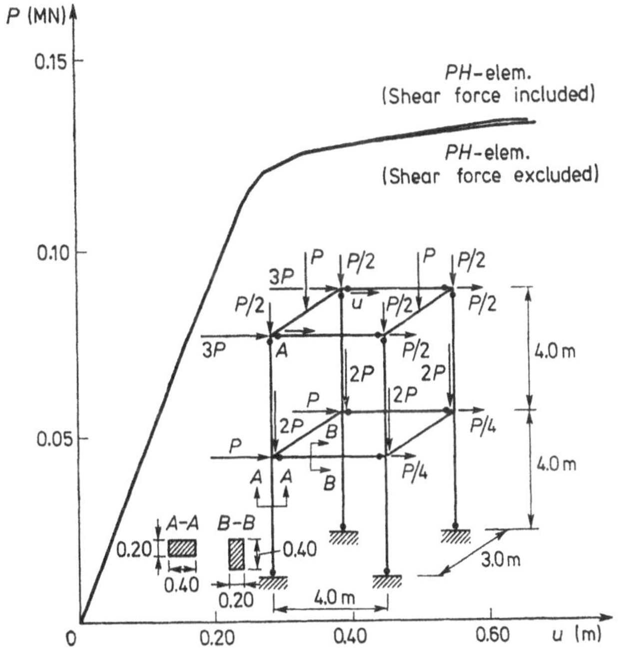

Fig. 93 Data for the space frame; loading, mode of failure and load-displacement plot

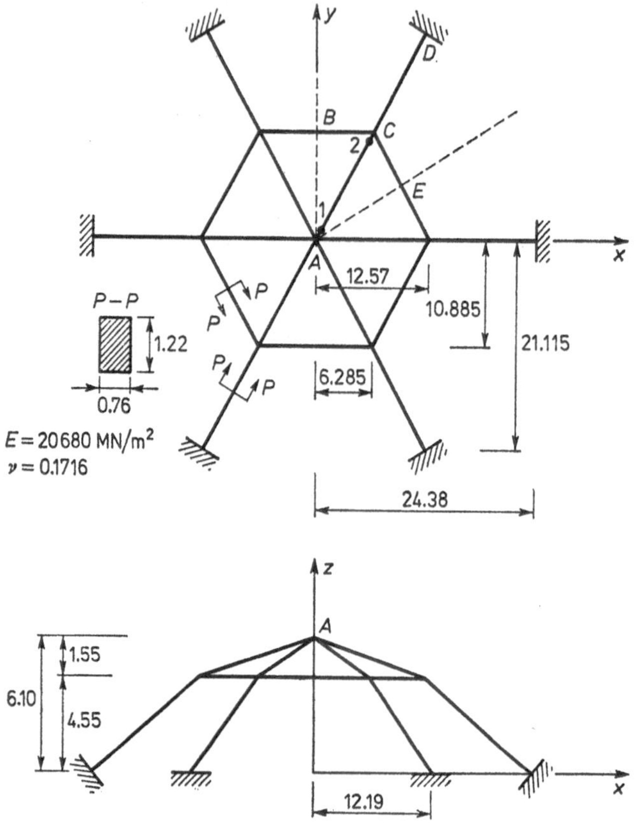

Fig. 94 Sketch of the frame dome

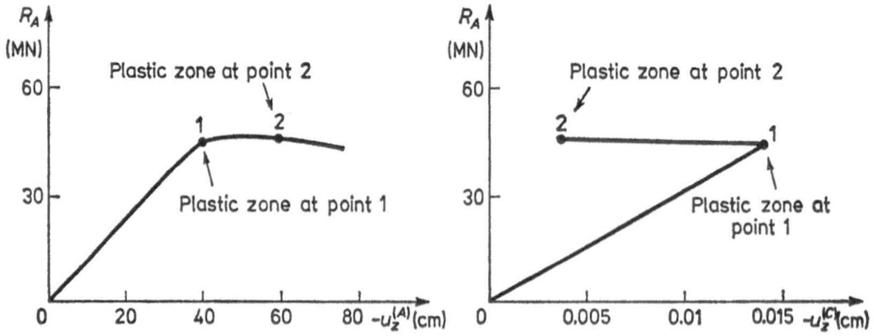

Fig. 95 Load-displacement plots

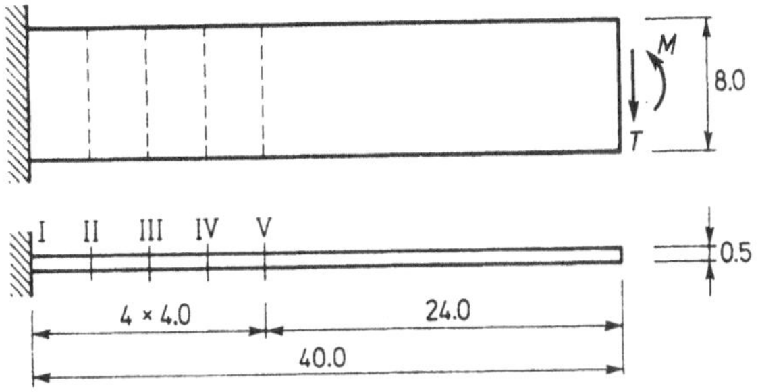

Fig. 96 Cantilever under repeated-variable bending moment and shear force

6.1.3 on the analysis of a 2D truss. The beam geometry and material properties are defined in Fig. 96. The example was chosen because of the possibility of comparing analytical and numerical results. Taking the moment M and the shear force T as independent loading modes, we introduce the nondimensional parameters

$$\beta_1 = \frac{M}{M_0}, \quad \beta_2 = \frac{T}{T_0} \tag{6.42}$$

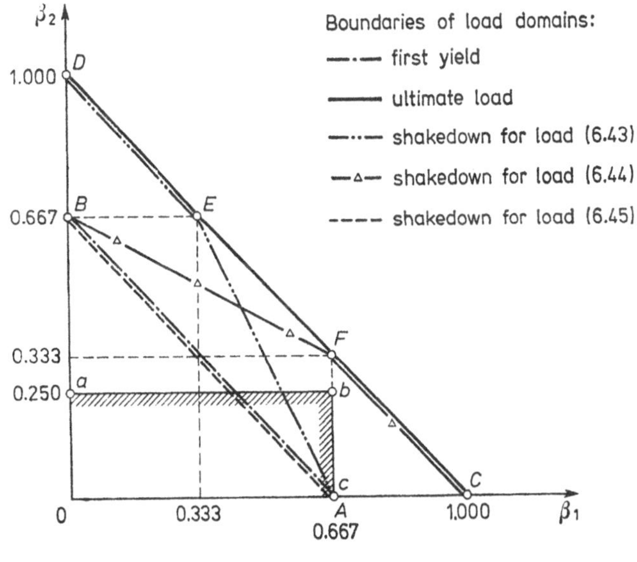

Fig. 97 Characteristic load domains for the cantilever

where M_0, T_0 are the ultimate values of M and T when acting alone on the beam. Using a simple beam theory it is easy to find analytically the characteristic domains in the plane $\beta_1 - \beta_2$ as shown in Fig. 97. An initial elastic response is bounded by the line AB whereas the ultimate load corresponds to the parallel line CD. The boundary AED defines the shakedown domain for the load case I

$$-\xi\bar{\beta}_1 \leqslant \beta_1 \leqslant \xi\bar{\beta}_1, \quad 0 \leqslant \beta_2 \leqslant \xi\bar{\beta}_2 \tag{6.43}$$

where the values $\bar{\beta}_1$, $\bar{\beta}_2$ correspond again, cf. Sec. 6.1.3, to a given reference (service) load. If we expand the domain by increasing ξ monotonically, then at some ultimate value $\xi = \xi_0$ the shakedown possibilities will be exhausted. The value ξ_0 is exactly a safety factor for overloading.

For the loading case II

$$0 \leqslant \beta_1 \leqslant \xi\bar{\beta}_1, \quad -\xi\bar{\beta}_2 \leqslant \beta_2 \leqslant \xi\bar{\beta}_2 \tag{6.44}$$

the corresponding boundary is BFC. Finally, when both M and T are allowed to take positive and negative values, i.e. for the load case III

$$-\xi\bar{\beta}_1 \leqslant \beta_1 \leqslant \xi\bar{\beta}_1, \quad -\xi\bar{\beta}_2 \leqslant \beta_2 \leqslant \xi\bar{\beta}_2 \tag{6.45}$$

the shakedown domain coincides with the initial elastic one. The segments AB, AE and BF correspond to the alternating plasticity at the outer fibres of the clamped end of the bar. The segments DE and FC imply the incremental collapse caused by the plastic hinge that develops at the same cross-section.

Table 8 **Test loads**

Load number	Load bounds	
	$\bar{\beta}_1$	$\bar{\beta}_2$
1	0.667	0.250
2	0.667	0.250

Two test loads defined in Table 8 and Fig. 98 were taken for incremental calculations. Despite identical bounds $\bar{\beta}_1$, $\bar{\beta}_2$ they should lead to opposite results as far as shakedown is concerned. Load 1 is of type (6.43) and its domain $Oabc$ violates the boundary segment AE. Therefore, the alternating plasticity is to be expected. The same rectangle, when regarded as the loading domain of type (6.44), is situated inside the relevant shakedown region. Hence, load 2 of Table 8 results in plastic adaptation.

Incremental calculations were performed for two different discrete models of the cantilever. In the first model reported in this section the beam was

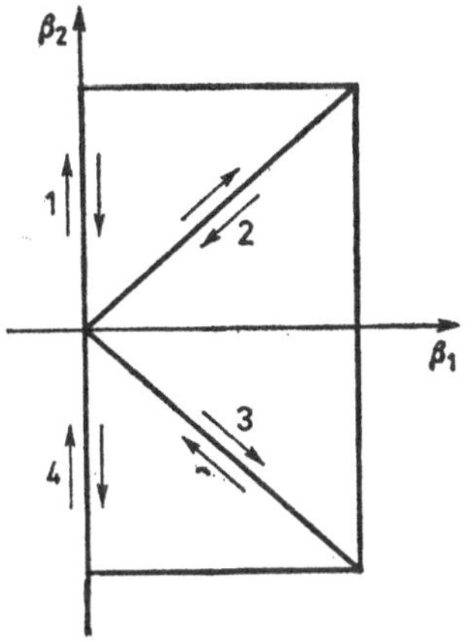

Fig. 98 Critical load cycles

represented by 6 layered elements decribed in Sec. 6.2.3. Eight layers of equal depth were chosen for the present study. The discretization mesh shown in Fig. 99 resulted in a small problem (12 degrees of freedom after static con-

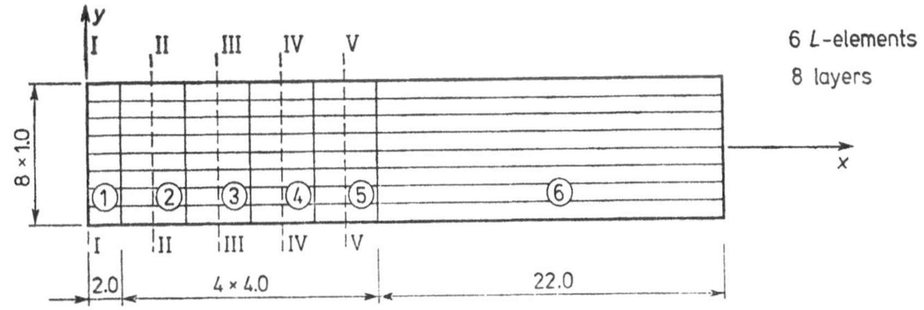

Fig. 99 Discrete model of the cantilever using L-elements

densation, cf. Sec. 4.7), but allowed one to obtain a detailed enough picture of the stress-strain state near the clamped end, where yielding was expected.

The results are given in Table 9 and Fig. 100. The adaptation to load 2 was achieved in the first critical cycle which is clearly demonstrated in Fig. 100. The residual stresses observed in the cross-sections I–I to V–V after adaptation to load 2 are shown in Table 10 and Fig. 101. The residual vertical displacement of the free end of the bar was

$$u_y^{(7)} = 0.0636 \text{ cm}.$$

It must be noted that the elements employed in this study are based upon the Kirchhoff–Bernoulli hypothesis and thus any influence of the shear stress

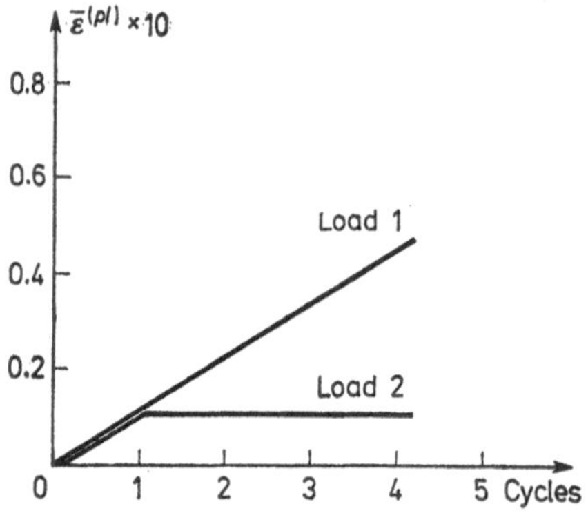

Fig. 100 Evolution of the total accumulated equivalent plastic strain

Table 9 Accumulated effective plastic strains

Cycle number	Load 1	Load 2
1	0.1096	0.1096
2	0.221	0.1096
3	0.332	0.1096
4	0.443	0.1096

Table 10 Residual stresses

Cross-section	I–I	II–II	III–III	IV–IV	V–V
y \ x	0.0	4.0	8.0	12.0	16.0
−4.0	878.7	811.1	720.8	630.4	540.0
−2.0	−715.2	−604.8	−457.3	−309.7	−162.0
0.0	0.000	0.000	0.000	0.000	0.000
2.0	715.2	604.8	457.3	309.7	162.0
4.0	−878.7	−811.1	−720.8	−630.4	−540.0

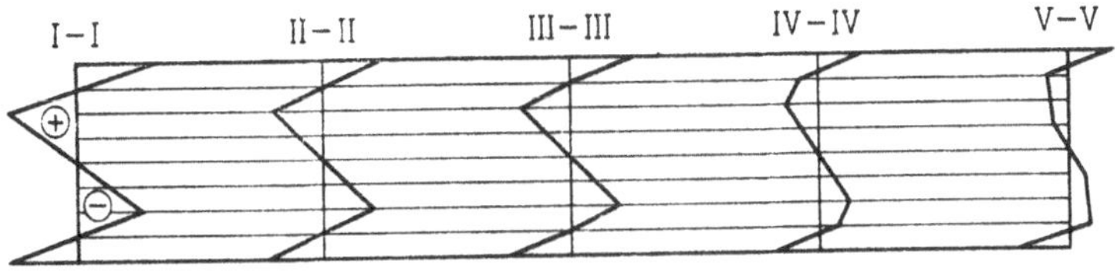

Fig. 101 Residual stresses σ_{xx} in selected cross-sections of the cantilever after adaptation to load 2

upon yielding is neglected. In order to evaluate an error introduced by this assumption an alternative discretization was carried out by means of some plane stress elements and will be presented in Sec. 7.4. Comparisons against the present results as well as some overall conclusions will also be given there.

6.2.6 *Elastic buckling of plane grid*

In this section merely a simple illustration on the analysis of initial structural buckling will be given with reference to a plane grid loaded in-plane

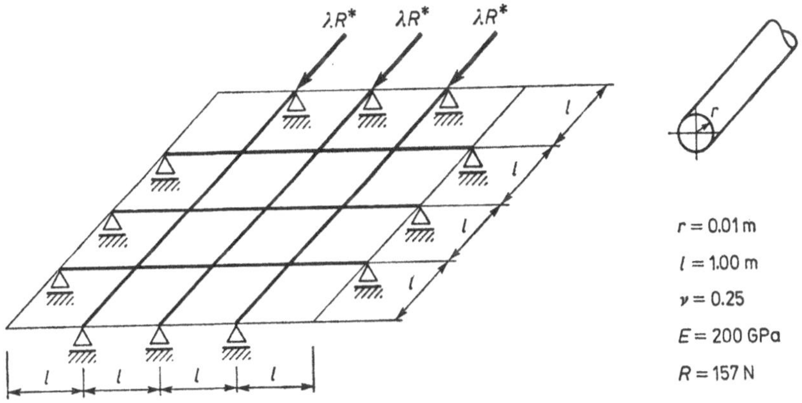

Fig. 102 Elastic buckling analysis of plane grid

as shown in Fig. 102, [162]. The solution of the buckling equation of the form, cf. eq. (5.43)

$$[\mathbf{K}^{(el)} + \lambda \mathbf{K}^{(\sigma)}(\boldsymbol{\sigma}^*)]\mathbf{v} = \mathbf{0}$$

has yielded a set of pairs (eigenvalue, eigenvector), six of which are illustrated in Fig. 103.

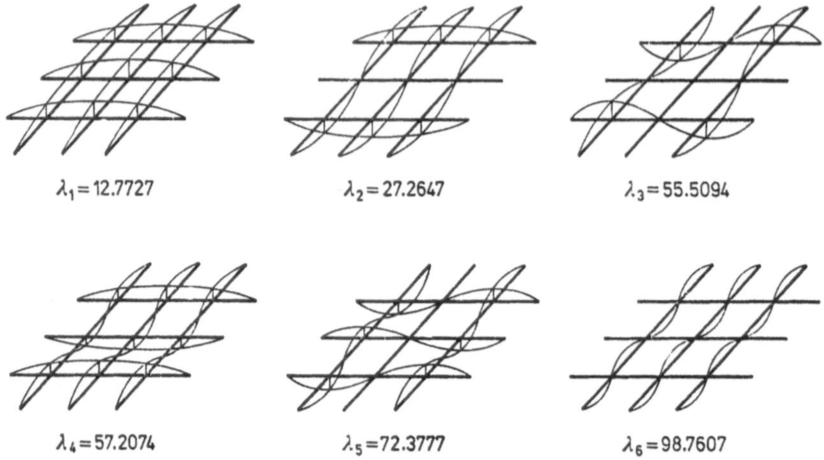

Fig. 103 Critical loads and buckling modes for elastic grid loaded in-plane

6.2.7 *Approximate large displacement analysis of frames using buckling mode superposition*

Let us consider a class of elastic structural configurations for which a "mildly" nonlinear deformations can approximately be described by means of the following nonlinear equation

$$[\mathbf{K}^{(el)} + \mathbf{K}^{(\sigma)}(\boldsymbol{\sigma})]\mathbf{r} = \mathbf{R}. \tag{6.46}$$

This equation describes the nonlinear problem in terms of the secant (or global) rather than tangent (or incremental) approach. The approximations employed in eq. (6.46) consist in assuming that the matrices $\mathbf{K}^{(el)}$ and $\mathbf{K}^{(\sigma)}$ are built upon the initial geometry, the initial displacement matrix $\mathbf{K}^{(u)}$ is negligible and the Piola–Kirchhoff stress components may be replaced by the Cauchy stress components. Thus, the elastic stiffness matrix $\mathbf{K}^{(el)}$ is assumed constant whereas the initial stress matrix $\mathbf{K}^{(\sigma)}$ depends linearly in the whole range of structure deformations on the "true" stress components.

Consider a one-parameter family of external loadings

$$\mathbf{R} = \lambda \mathbf{R}^* \tag{6.47}$$

where $\mathbf{R}^*$ stands for a reference load and λ is a scalar multiplier. We further assume that the linear increase of the external load implies a linear change in the state of stress, i.e.

$$\boldsymbol{\sigma} = \lambda \boldsymbol{\sigma}^*$$

where $\boldsymbol{\sigma}^*$ is the stress at the load $\mathbf{R}^*$. The bifurcation (or initial buckling) problem is described by the equation, cf. Sec. 5.4,

$$[\mathbf{K}^{(el)} + \lambda \mathbf{K}^{(\sigma)}(\boldsymbol{\sigma}^*)]\mathbf{v} = \mathbf{0}. \tag{6.48}$$

This equation defines the so-called generalized eigenproblem, the solution to which consist of N pairs $(\lambda_1, \mathbf{v}_1)$, $(\lambda_2, \mathbf{v}_2), \ldots, (\lambda_N, \mathbf{v}_N)$; $\lambda_1, \lambda_2, \ldots, \lambda_N$ being the eigenvalues and $\mathbf{v}_1, \mathbf{v}_2, \ldots, \mathbf{v}_N$ the eigenvectors. Let us form the $N \times N$ matrices

$$\begin{aligned} \boldsymbol{\Lambda}_{N\times N} &= \begin{bmatrix} \lambda_1 & & & \\ & \lambda_2 & 0 & \\ & & \ddots & \\ & 0 & & \lambda_N \end{bmatrix}, \\ \mathbf{X}_{N\times N} &= \begin{bmatrix} v_{11} & v_{21} & \cdots & v_{N1} \\ v_{12} & v_{22} & \cdots & v_{N2} \\ \cdot & \cdot & \cdot & \cdot \\ v_{1N} & v_{2N} & \cdots & v_{NN} \end{bmatrix} = [\mathbf{v}_1 \ \ \mathbf{v}_2 \ \ \ldots \ \ \mathbf{v}_N] \end{aligned} \tag{6.49}$$

which allow to rewrite eigenproblem (6.48) in the form

$$\mathbf{K}^{(el)}\mathbf{X} = -\mathbf{K}^{(\sigma)}(\boldsymbol{\sigma}^*)\mathbf{X}\boldsymbol{\Lambda}. \tag{6.50}$$

Note that the eigenvectors $\mathbf{v}_1, \mathbf{v}_2, \ldots, \mathbf{v}_N$ may be computed from eq. (6.48) only up to arbitrary scalar multipliers. It is also easy to show that these vectors satisfy the so-called $\mathbf{K}^{(el)}$- and $\mathbf{K}^{(\sigma)}$-orthogonality conditions which after appropriate scaling may be presented as

$$\mathbf{X}^T\mathbf{K}^{(el)}\mathbf{X} = \mathbf{I}, \quad \mathbf{X}^T\mathbf{K}^{(\sigma)}\mathbf{X} = -\boldsymbol{\Lambda}^{-1}.$$

Hence

$$\mathbf{K}^{(el)} = \mathbf{X}^{-T}\mathbf{X}^{-1}, \quad \mathbf{K}^{(\sigma)} = -\mathbf{X}^{-T}\boldsymbol{\Lambda}^{-1}\mathbf{X}^{-1}. \tag{6.51}$$

Eq. (6.46) can be presented in the form which in the pre-critical range (for which $\det[\mathbf{K}^{(el)} + \lambda\mathbf{K}^{(\sigma)}(\boldsymbol{\sigma}^*)] \neq 0$) may be solved for $\mathbf{r}$ yielding

$$\mathbf{r}(\lambda) = [\mathbf{K}^{(el)} + \lambda\mathbf{K}^{(\sigma)}(\boldsymbol{\sigma}^*)]^{-1}\lambda\mathbf{R}^*.$$

Substituting into this equation conditions (6.51) we arrive at

$$\mathbf{r}(\lambda) = \mathbf{X}[\mathbf{I} - \lambda\boldsymbol{\Lambda}^{-1}]^{-1}\mathbf{X}^T\lambda\mathbf{R}^* \tag{6.52}$$

or

$$\mathbf{r}(\lambda) = \mathbf{X}[\mathbf{I}\lambda^{-1} - \boldsymbol{\Lambda}^{-1}]^{-1}\mathbf{X}^T\mathbf{R}^*.$$

Let us have a closer look at the last relationship. The matrices $\boldsymbol{\Lambda}$ and $\mathbf{X}$ may be formed only after eigenproblem (6.50) has been solved. To solve this problem we need in turn the stress distribution $\boldsymbol{\sigma}^*$ which appears as the argument in the matrix function $\mathbf{K}^{(\sigma)}$. In accordance with the simplifications introduced at the beginning of this section the stresses $\boldsymbol{\sigma}^*$ may be calculated from the solution $\mathbf{r}^*$ to the linear problem

$$\mathbf{K}^{(el)}\mathbf{r}^* = \mathbf{R}^*, \tag{6.53}$$

by using equations of linear theory of elasticity and the appropriate shape functions employed in the finite element formulation. We conclude that in order to form the matrix coefficients in eq. (6.52) we have to consecutively solve:

— linear problem of statics (6.53),

— linear eigenproblem (6.48).

Thus, having solved these two linear problems we may attempt to approximately analyse the nonlinear static response as described by eq. (6.52) by considering the gradually increase in the load parameter λ. For the given matrices $\mathbf{\Lambda}$ and $\mathbf{X}$ eq. (6.52) is analytically very simple so that it can effectively be used to assess the nonlinear displacement state corresponding to loads characterized by $\lambda < \lambda_1$ (for $\lambda = \lambda_1$ the matrix $\mathbf{I}\lambda^{-1} - \mathbf{\Lambda}^{-1}$ becomes singular).

The computational effort needed for the eigenvalue solution with respect to all the eigenvalues and eigenvectors, and an approximate character of our starting equation (6.46) raise the natural question as to the practical significance of the above formulation. However, as it turns out, only a very small number of eigenvalues and eigenvectors is in fact needed to achieve acceptable accuracy of the computations. This fact seems to be intuitively convincing as it is quite clear that the structure tends to develop strong nonlinearities in the way which should be consistent with only first few buckling modes. Thus, assuming that merely K first buckling modes contribute to the nonlinear structural behaviour, $K \ll N$, we may define the matrices

$$\begin{aligned} \mathbf{X}^{(K)}_{N\times K} &= [\mathbf{v}_1 \ \ \mathbf{v}_2 \ \ \dots \ \ \mathbf{v}_K], \\ \mathbf{\Lambda}^{(K)}_{K\times K} &= \operatorname{diag}[\lambda_1 \ \ \lambda_2 \ \ \dots \ \ \lambda_K] \end{aligned} \tag{6.54}$$

rewriting eq. (6.52) in the form

$$\mathbf{r}_{N\times 1}(\lambda) = \mathbf{X}^{(K)}_{N\times K}[\lambda^{-1}\mathbf{I}_{K\times K} - \mathbf{\Lambda}^{(K)^{-1}}_{K\times K}]^{-1}\mathbf{X}^{(K)^T}_{K\times N}\mathbf{R}^*_{N\times 1} \tag{6.55}$$

or

$$\mathbf{r}(\lambda) = \sum_{\alpha=1}^{K} \lambda\lambda_\alpha(\lambda_\alpha - \lambda)^{-1}\mathbf{v}_\alpha\mathbf{v}_\alpha^T\,\mathbf{R}^*.$$

According to the above expression, in order to find the nonlinear relation $\mathbf{r}(\lambda)$ one has to:

— find the solution $\mathbf{r}^*$ to the linear problem (6.53) for any reference value of the external load $\mathbf{R}^*$,

— using $\mathbf{r}^*$ determine the stress distribution $\boldsymbol{\sigma}^*$,
— assume a small number K (say, $K = 3$ or $K = 4$) of buckling modes to be accounted for in the analysis to follow,
— find K first eigenvalues and eigenvectors of the problem (6.48) in which the stresses $\boldsymbol{\sigma}^*$ are taken as the arguments of $\mathbf{K}^{(\sigma)}$,
— form the matrices (6.54) and substitute them into (6.55).

The two illustrative examples given below concern the frame-like configurations for which the necessary matrices have been developed in Sec. 4.5.

Example I. Consider a simply supported beam for which $l = 1.0$ m, $J = 10^{-8}$ m, $A = 10^{-4}$ m^2, $E = 10^7$ N/cm^2. The beam is divided into 8 finite elements of equal length. The nonlinear behaviour is analysed for two sets of external loadings $\{P, Q\}$ and $\{P, M\}$ shown in Figs. 104(b) and104(c),

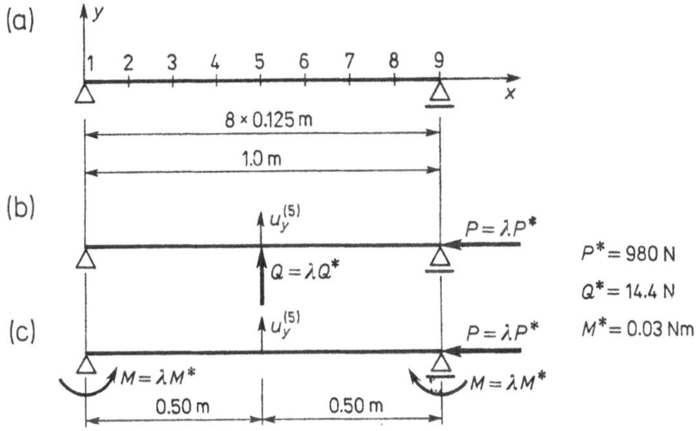

Fig. 104 Nonlinear analysis of simply supported elastic beam

respectively. The values of the vertical displacement $u_y^{(5)}$ at node 5 for the both loading cases are given in Tables 11 and 12. The displacements are compared against those obtained in [146] using the formulae

$$u_y^{(5)} = \frac{Ql^3}{48EJ}\left[\frac{3(\tan\alpha-\alpha)}{\alpha^3}\right] = \frac{\lambda Q^* l^3}{48EJ}\left[\frac{3(\tan\alpha-\alpha)}{\alpha^3}\right],$$

$$u_y^{(5)} = \frac{Ml^2}{8EJ}\left[\frac{2(1-\cos\alpha)}{\alpha^2\cos\alpha}\right] = \frac{\lambda M^* l^2}{8EJ}\left[\frac{2(1-\cos\alpha)}{\alpha^2\cos\alpha}\right], \tag{6.56}$$

$$\alpha = 0.5l\sqrt{\frac{P}{EJ}} = 0.5l\sqrt{\frac{\lambda P^*}{EJ}},$$

Table 11 Large displacement analysis of simply supported beam under the first set of the external forces

Numbers of subsequent buckling modes accounted for in eq. (6.55)		$u_y^{(5)}$ (cm), load $\{P, Q\}$, Fig. 104(b)				
		$\lambda = 2$	$\lambda = 4$	$\lambda = 6$	$\lambda = 8$	$\lambda = 10$
1	1	0.074653	0.197919	0.441439	1.15355	41.7350
2	1, 2	0.074653	0.197919	0.441439	1.15355	41.7350
3	1, 2, 3	0.074669	0.197986	0.441594	1.15383	41.7355
4	1, 2, 3, 4	0.074669	0.197986	0.441594	1.15383	41.7355
5	1, 2, 3, 4, 5	0.074670	0.197989	0.441601	1.15384	41.7355
6	1, 2, 3, 4, 5, 6	0.074670	0.197989	0.441601	1.15384	41.7355
7	Calculations based on eq. $(6.56)_1$	0.074670	0.197991	0.441615	1.15396	41.9282

Table 12 Large displacement analysis of simply supported beam under the second set of the external forces

Numbers of subsequent buckling modes accounted for in eq.(6.55)		$u_y^{(5)}$ (cm), load $\{P, M\}$, Fig. 104(c)				
		$\lambda = 2$	$\lambda = 4$	$\lambda = 6$	$\lambda = 8$	$\lambda = 10$
1	1, 2, 3, 4, 5, 6	0.941206×10^{-3}	0.251750×10^{-2}	0.566658×10^{-2}	0.149480×10^{-1}	0.546116
2	Calculations based on eq. $(6.56)_2$	0.941202×10^{-3}	0.251750×10^{-2}	0.566669×10^{-2}	0.149495×10^{-1}	0.548640

for each of the loading sets considered. Note that for $\{P, Q\}$ the displacement $u_y^{(5)}$ does not depend on antisymmetric buckling modes (i.e. modes numbered 2, 4 and 6). This is clearly due to the fact that for these modes $\mathbf{v}_j^T \mathbf{R}_1^* = 0$. The similar situation can be observed in the second case. We also note that a very high accuracy is achieved even at a very small number of the buckling modes.

Since both the loading cases induce the same axial load, the solution to eq. (6.48) is clearly identical. The smallest eigenvalue was obtained as

$\lambda_1 = 10.0713$ which results in the first critical load of $P_{cr} = \lambda_1 \cdot P^* = 9869.87$ N (compared to the Euler load of $P_{cr}^E = 9869.60$ N.

Example II. Consider a plane frame consisting of three identical beams with each beam being divided into four equal elements; the load is shown in Fig. 105. The nonlinear behaviour is analysed for two different sets of support

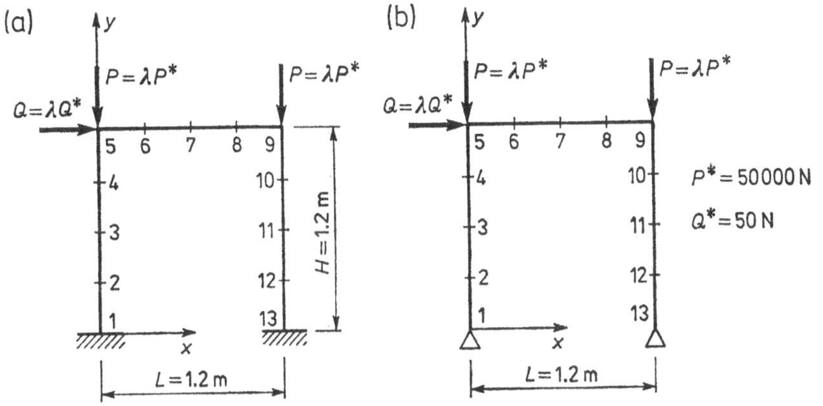

Fig. 105 Nonlinear analysis of elastic frame: two sets of boundary conditions

conditions as shown in the same figure. The horizontal displacement $u_x^{(5)}$ of the node 5 is monitored as it is supposed to best reflect the system nonlinearity. Figs. 106 and 107 illustrate the changes of $u_x^{(5)}$ with the load parameter λ for both cases considered; the broken lines are taken from [37] for compari-

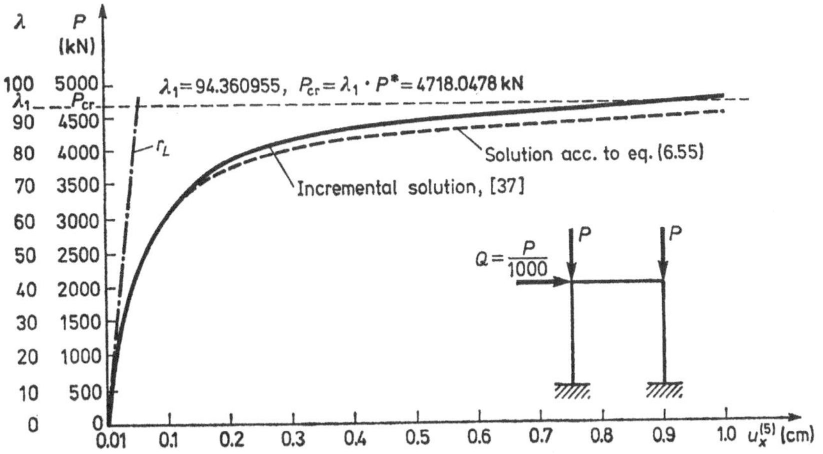

Fig. 106 Nonlinear frame analysis with hinged support

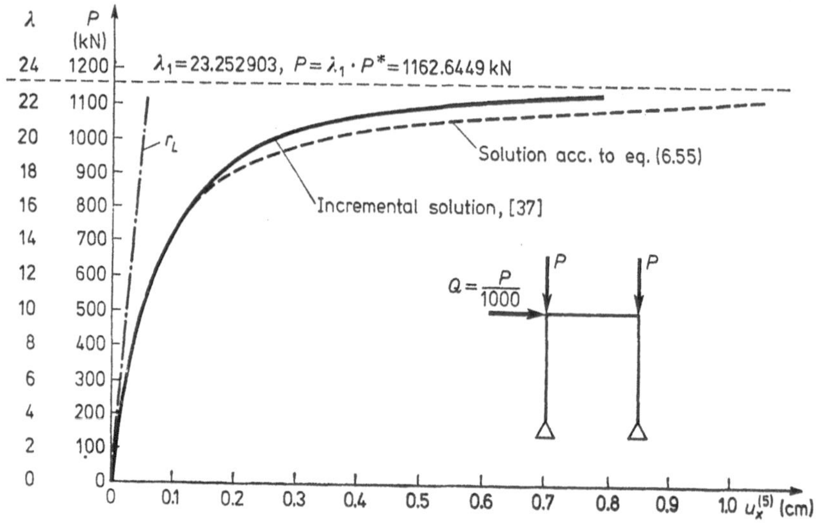

Fig. 107 Nonlinear frame analysis with clamped support

son, the linear solutions are also shown. Four buckling modes were included into the analysis but only the antisymmetric (first and third) were in fact active. A satisfactory agreement of the results up to at least 70% of the first critical load is clearly documented.

6.3 Lattice-type structures

6.3.1 Lattice shells—linear and "second-order" theories

Because of the inherent geometric complexity, the formalism used in Sec. 4.6 to derive equations of the discrete and continuous models for lattice-type structures was somewhat different from the rest of the derivations in this book. Therefore, before illustrating its use in the analysis of some nonlinear structural mechanics problems, cf. Secs. 6.3.2–6.3.5 below, in the present section we shall provide the reader with some additional insight by discussing basic aspects of the linear theory of elastic lattice-type shells showing also how a "second-order" theory can be developed.

It is assumed that for lattice shells the set of points $\boldsymbol{\chi}(\mathbf{X}, t_0)$, $\mathbf{X} \in \mathscr{X}$ (i.e. the set of lattice node centers in the undeformed configuration) can be extrapolated in a physically sound and unique manner by a certain smooth surface. Moreover, the axes of the lattice shell elements are assumed to be straight line segments which can be approximately treated as situated on the surface S_0. An example of the lattice shell is shown in Fig. 108; we have tacitly assumed here that for any time instant there exists the surface S_t which plays a similar role to that of S_0.

In order to formulate equations describing lattice shells we introduce a certain formalism based on the specification of the vector bases $\mathbf{g}_a(\mathbf{X})$,

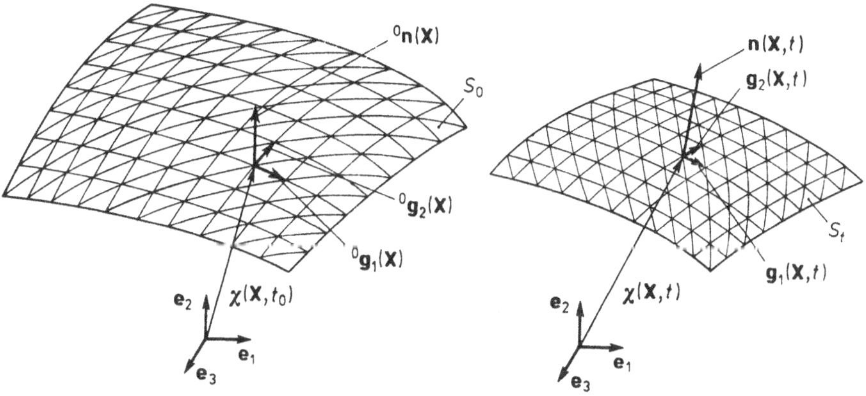

Fig. 108 Undeformed and deformed configuration of lattice shell

$\mathbf{X} \in \mathscr{X}$, $a = 1, 2, 3$, the meaning of which was explained in Sec. 4.6.1. To this aim we assume that (cf. Fig. 108)

$$\mathbf{g}_\alpha(\mathbf{X}, \tau) \equiv \delta_\alpha^A \Delta_A \boldsymbol{\chi}(\mathbf{X}, \tau); \quad \alpha = 1, 2, \quad A = \mathrm{I}, \mathrm{II}, \mathrm{III},$$

$$\mathbf{n}(\mathbf{X}, \tau) = \mathbf{g}_3(\mathbf{X}, \tau) \equiv \frac{\mathbf{g}_1(\mathbf{X}, \tau) \times \mathbf{g}_2(\mathbf{X}, \tau)}{|\mathbf{g}_1(\mathbf{X}, \tau) \times \mathbf{g}_2(\mathbf{X}, \tau)|},$$

where δ_α^A is the Kronecker symbol. Define[1]

$$a_{\beta\gamma} \equiv \mathbf{g}_\beta \cdot \mathbf{g}_\gamma, \quad a^{\beta\gamma} a_{\gamma\delta} = \delta_\delta^\beta, \quad \mathbf{g}^\alpha \equiv a^{\alpha\gamma} \mathbf{g}_\gamma,$$
$$b_{A\gamma} \equiv G_{A\gamma}^{\,3}, \quad b_A^\gamma \equiv a^{\gamma\beta} b_{A\beta}, \quad b_A \equiv G_{A3}^{\,3},$$
$$\hat{b}_{A\gamma} \equiv \hat{G}_{A\gamma}^{\,3}, \quad \hat{b}_A^\gamma \equiv a^{\gamma\beta} \hat{b}_{A\beta}, \quad \hat{b}_A \equiv \hat{G}_{A3}^{\,3},$$

where the mathematical objects $G_{Aa}^{\,b}$; $a, b = 1, 2, 3$, are those introduced in Sec. 4.6.1. Using the results there obtained we have

$$G_{A\gamma}^{\,\beta} = \mathbf{g}^\beta \cdot \Delta_A \mathbf{g}_\gamma, \quad b_{A\gamma} = \mathbf{n} \cdot \Delta_A \mathbf{g}_\gamma, \quad b_A = \mathbf{n} \cdot \Delta_A \mathbf{n},$$
$$\hat{G}_{A\gamma}^{\,\beta} = \mathbf{g}^\beta \cdot \bar{\Delta}_A \mathbf{g}_\gamma, \quad \hat{b}_{A\gamma} = \mathbf{n} \cdot \bar{\Delta}_A \mathbf{g}_\gamma, \quad \hat{b}_A = \mathbf{n} \cdot \bar{\Delta}_A \mathbf{n}.$$

[1] In all considerations concerning lattice plates and shells indices α, β, γ run over 1, 2; summation convention holds with respect to them. In formulae in which all terms depend on the same argument $\mathbf{X}$ we shall omit it. Summation convention with respect to $A = \mathrm{I}, \mathrm{II}, \mathrm{III}$ does not hold unless otherwise stated.

For an arbitrary time dependent vector field $\mathbf{w}(\mathbf{X}, \tau)$, $\mathbf{X} \in \mathscr{X}$, we may write

$$\mathbf{w} = w^\gamma \mathbf{g}_\gamma + w\mathbf{n} = w_\gamma \mathbf{g}^\gamma + w\mathbf{n},$$

and introduce the notation

$$'\delta_A w^\gamma(\mathbf{X}, \tau) \equiv \Delta_A w^\gamma(\mathbf{X}, \tau) + G^{\gamma}_{A\beta}(\mathbf{X}, \tau) w^\beta(\mathbf{X}+\Delta_A\mathbf{X}, \tau),$$
$$'\bar{\delta}_A w^\gamma(\mathbf{X}, \tau) \equiv \bar{\Delta}_A w^\gamma(\mathbf{X}, \tau) + \hat{G}^{\gamma}_{A\beta}(\mathbf{X}, \tau) w^\beta(\mathbf{X}-\Delta_A\mathbf{X}, \tau), \qquad \mathbf{X} \in \mathscr{X}^0.$$

Hence, bearing in mind the definitions of the finite differences $\delta_A w$, $\bar{\delta}_A w$ introduced in Sec. 4.6.1 and setting $'\delta_A w \equiv \Delta_A w$, $'\bar{\delta}_A w \equiv \bar{\Delta}_A w$, we obtain

$$\delta_A w^\gamma(\mathbf{X}, \tau) = {}'\delta_A w^\gamma(\mathbf{X}, \tau) + b^\gamma_A(\mathbf{X}, \tau) w(\mathbf{X}+\Delta_A\mathbf{X}, \tau),$$
$$\delta_A w(\mathbf{X}, \tau) = {}'\delta_A w(\mathbf{X}, \tau) + b_{A\beta}(\mathbf{X}, \tau) w^\beta(\mathbf{X}+\Delta_A\mathbf{X}, \tau) +$$
$$+ b_A(\mathbf{X}, \tau) w(\mathbf{X}+\Delta_A\mathbf{X}, \tau),$$
$$\bar{\delta}_A w^\gamma(\mathbf{X}, \tau) = {}'\bar{\delta}_A w^\gamma(\mathbf{X}, \tau) + \hat{b}^\gamma_A(\mathbf{X}, \tau) w(\mathbf{X}-\Delta_A\mathbf{X}, \tau),$$
$$\bar{\delta}_A w(\mathbf{X}, \tau) = {}'\bar{\delta}_A w(\mathbf{X}, \tau) + \hat{b}_{A\beta}(\mathbf{X}, \tau) w^\beta(\mathbf{X}-\Delta_A\mathbf{X}, \tau) +$$
$$+ \hat{b}_A(\mathbf{X}, \tau) w(\mathbf{X}-\Delta_A\mathbf{X}, \tau).$$

Such a finite difference formalism, [169], [174], turns out convenient in the analysis of lattice shell problems as it has certain features similar to those of the tensor calculus on smooth surfaces.

For the sake of simplicity let us adopt the following notation

$$\ddot{\boldsymbol{\chi}} = a^\beta \mathbf{g}_\beta + a\mathbf{n}, \qquad \Delta_A \boldsymbol{\chi} = l^\beta_A \mathbf{g}_\beta + l_A \mathbf{n},$$
$$\varepsilon_{\gamma\beta} = (\mathbf{g}_\gamma \times \mathbf{g}_\beta) \cdot \mathbf{n}, \qquad \varepsilon^{\gamma}_{\cdot\beta} = a^{\gamma\alpha} \varepsilon_{\alpha\beta}.$$

Assume also that for lattice shells there approximately holds the condition

$$l_A(\mathbf{X}, \tau) \simeq 0, \qquad \mathbf{X} \in \mathscr{X}^0, \qquad \tau \in [\tau_0, \tau_f].$$

Remembering the notation

$$\bar{\varphi}^A(\mathbf{X}) \equiv \varphi^A(\mathbf{X}-\Delta_A\mathbf{X}),$$

we may now represent eqs. (4.93) in the following form (summation convention holds also with respect to $A = \mathrm{I, II, III}$)

$$\begin{aligned}
&'\bar{\delta}_A P^{A\beta} + \hat{b}^\beta_A \bar{P}^A + F^\beta = m a^\beta,\\
&'\bar{\delta}_A P^A + \hat{b}_{A\beta} \bar{P}^{A\beta} + \hat{b}_A \bar{P}^A + F = m a,\\
&'\bar{\delta}_A M^{A\beta} + \hat{b}^\beta_A \bar{M}^A + \varepsilon^{\beta}_{\cdot\gamma}[\eta l^\gamma_A P^A + (1-\eta)\bar{l}^\gamma_A \bar{P}^A] + H^\beta = 0,\\
&'\bar{\delta}_A M^A + \hat{b}_{A\beta} \bar{M}^{A\beta} + \varepsilon_{\gamma\beta}[\eta l^\gamma_A P^{A\beta} + (1-\eta)\bar{l}^\gamma_A \bar{P}^{A\beta}] + H = 0.
\end{aligned} \tag{6.57}$$

It has to be emphasized that the parameter $\eta \in [0, 1]$ is assumed to be fixed a priori as the vectors $\mathbf{P}^A(\mathbf{X}, \tau)$ and $\mathbf{M}^A(\mathbf{X}, \tau)$ (cf. Fig. 17) are defined for an arbitrary but fixed value of η. Eqs. (6.57) are referred to as the equations of motion of lattice shells.

Let us observe that the decomposition

$$\mathbf{P}^A = P^{A\beta}\mathbf{g}_\beta + P^A\mathbf{n}, \quad \mathbf{M}^A = M^{A\beta}\mathbf{g}_\beta + M^A\mathbf{n}$$

determines in the case of lattice shells two kinds of the cross-sectional generalized forces: $P^{A\beta}$, M^A, which are referred to as the *membrane-type forces* and the *polar couples*, respectively, and P^A, $M^{A\beta}$, which are called the *shear-type forces* and the *bending couples*, respectively.

Structural stability problems of finding critical loads are investigated in the framework of so-called second order theories. To formulate a second order theory of lattice shells we assume the constitutive law in the linear form (4.99), we use the strain measures linearized with respect to the displacements and (small) rotations (4.98) but we retain certain nonlinear terms in the equilibrium equations of the deformed structure.

Let us first transform eqs. (6.57) to an alternative form depending explicitly on the displacements of nodes from the undeformed configuration (at the time instant $t = t_0$). To this aim for an arbitrary time dependent field $\psi(\mathbf{X}, \tau)$, $\mathbf{X} \in \mathscr{X}$, $\tau \in [\tau_0, \tau_f]$, we define its initial value setting

$${}^0\psi(\mathbf{X}) \equiv \psi(\mathbf{X}, t_0).$$

Similarly, we introduce the finite difference operators related to the initial time instant by setting

$$\begin{aligned}
{}^0\delta_A w^\gamma(\mathbf{X}) &\equiv \Delta_A w^\gamma(\mathbf{X}) + {}^0G^{\gamma}_{A\beta}(\mathbf{X})\, w^\beta(\mathbf{X}+\Delta_A\mathbf{X}),\\
{}^0\bar{\delta}_A w^\gamma(\mathbf{X}) &\equiv \Delta_A w^\gamma(\mathbf{X}) + {}^0\hat{G}^{\gamma}_{A\beta}(\mathbf{X})\, w^\beta(\mathbf{X}-\Delta_A\mathbf{X}),\\
{}^0\delta_A w(\mathbf{X}) &\equiv {}'\delta_A w(\mathbf{X}) = \Delta_A w(\mathbf{X}),\\
{}^0\bar{\delta}_A w(\mathbf{X}) &\equiv {}'\bar{\delta}_A w(\mathbf{X}) = \bar{\Delta}_A w(\mathbf{X}),
\end{aligned}$$

for an arbitrary vector field $\mathbf{w} = w^\gamma \mathbf{g}_\gamma + w\mathbf{n}$. Hence for the (time dependent) displacement field $\mathbf{u}(\mathbf{X}, \tau)$ we obtain

$$\mathbf{u}(\mathbf{X}, \tau) = \boldsymbol{\chi}(\mathbf{X}, \tau) - {}^0\boldsymbol{\chi}(\mathbf{X}) = u_\alpha(\mathbf{X}, \tau)\,{}^0\mathbf{g}^\alpha(\mathbf{X}) + \mathbf{u}(\mathbf{X}, \tau)\,{}^0\mathbf{n}(\mathbf{X}).$$

Now define the increments

$$\begin{aligned}
\Gamma^{\beta}_{A\alpha} &\equiv \hat{G}^{\beta}_{A\alpha} - {}^0\hat{G}^{\beta}_{A\alpha}, \quad & \beta_{A\beta} &\equiv \hat{b}_{A\beta} - {}^0\hat{b}_{A\beta},\\
\beta^{\beta}_{A} &\equiv \hat{b}^{\beta}_{A} - {}^0\hat{b}^{\beta}_{A}, \quad & \beta_{A} &\equiv \hat{b}_{A} - {}^0\hat{b}_{A}.
\end{aligned}$$

It can be seen that these increments depend on $\Delta_A u$, $\bar{\Delta}_B u$ and $\bar{\Delta}_A \Delta_B u$, and are equal to zero for $\Delta_B u = 0$. Since

$$
\begin{aligned}
{}'\bar{\delta}_A P^{A\beta} &= {}^0\bar{\delta}_A P^{A\beta} + \Gamma_{A\alpha}^{\ \beta} \bar{P}^{A\alpha},\\
{}'\bar{\delta}_A M^{A\beta} &= {}^0\bar{\delta}_A M^{A\beta} + \Gamma_{A\alpha}^{\ \beta} \bar{M}^{A\alpha},\\
\hat{b}_A^\beta &= {}^0\hat{b}_A^\beta + \beta_A^\beta,\\
\hat{b}_{A\beta} &= {}^0\hat{b}_{A\beta} + \beta_{A\beta},\\
\hat{b}_A &= {}^0\hat{b}_A + \beta_A,
\end{aligned}
$$

then after substituting the right-hand sides of the above formulae into eqs. (6.57) we obtain the alternative form of the equations of motion for the lattice shells as (summation convention with respect to $A = \mathrm{I, II, III}$ holds)

$$
\begin{aligned}
&{}^0\bar{\delta}_A P^{A\beta} + {}^0\hat{b}_A^\beta \bar{P}^A + \Gamma_{A\alpha}^{\ \beta} \bar{P}^{A\alpha} + \beta_A^\beta \bar{P}^A + F^\beta = m a^\beta,\\
&{}^0\bar{\delta}_A P^A + {}^0\hat{b}_{A\beta} \bar{P}^{A\beta} + {}^0\hat{b}_A \bar{P}^A + \beta_{A\beta} \bar{P}^{A\beta} + \beta_A \bar{P}^A + F = m a,\\
&{}^0\bar{\delta}_A M^{A\beta} + {}^0\hat{b}_A^\beta \bar{M}^A + \varepsilon^{\beta}_{\cdot\gamma}[\eta l_A^\gamma P^A + (1-\eta)\bar{l}_A^\gamma \bar{P}^A] + \beta_A^\beta \bar{M}^A + H^\beta = 0,\\
&{}^0\bar{\delta}_A M^A + {}^0\hat{b}_{A\beta} \bar{M}^{A\beta} + \varepsilon_{\gamma\beta}[\eta l_A^\gamma P^{A\beta} + (1-\eta)\bar{l}_A^\gamma \bar{P}^{A\beta}] + \beta_{A\beta} \bar{M}^{A\beta} + H = 0.
\end{aligned}
\tag{6.58}
$$

The linear theories of lattice shells are based on the linearized (with respect to all unknown functions) form of eqs. (6.58), i.e. on the form in which all the terms containing the increments $\Gamma_{A\alpha}^{\ \beta}$, $\beta_{A\beta}$, β_A^β, β_A are neglected and where $\varepsilon_{\gamma\beta}$, $\varepsilon^{\beta}_{\cdot\gamma}$, l_A^γ are replaced by ${}^0\varepsilon_{\gamma\beta}$, ${}^0\varepsilon^{\beta}_{\cdot\gamma}$, ${}^0l_A^\gamma$, respectively. In the second order theory we assume that the membrane-type forces $P^{A\beta}$ and the polar couples M^A can attain large values so that we retain in eqs. (6.58) the nonlinear terms $\beta_{A\beta}\bar{P}^{A\beta}$, $\beta_A^\beta \bar{M}^A$ rejecting at the same time all the other nonlinear terms. Moreover, in the calculation of $\beta_{A\beta}$ and β_A^β we neglect the differences $\mathbf{g}_\alpha - {}^0\mathbf{g}_\alpha$ and $\mathbf{n} - {}^0\mathbf{n}$ as small with respect to their finite increments $\Delta_A(\mathbf{g}_\alpha - {}^0\mathbf{g}_\alpha)$, $\Delta_A(\mathbf{n} - {}^0\mathbf{n})$; such approximation is plausible for the problems described by the second order theory.[1] Hence for the quasi-static case from eqs. (6.58) we obtain (summation convention holds also for $A = \mathrm{I, II, III}$)

$$
\begin{aligned}
&{}^0\bar{\delta}_A P^{A\beta} + {}^0\hat{b}_A^\beta \bar{P}^A + F^\beta = 0,\\
&{}^0\bar{\delta}_A P^A + {}^0\hat{b}_{A\beta} \bar{P}^{A\beta} + {}^0\hat{b}_A P^A + \beta_{A\beta} \bar{P}^{A\beta} + F = 0,\\
&{}^0\bar{\delta}_A M^{A\beta} + {}^0\hat{b}_A^\beta \bar{M}^A + {}^0\varepsilon^{\beta}_{\cdot\gamma}[\eta\, {}^0l_A^\gamma P^A + (1-\eta)\, {}^0\bar{l}_A^\gamma \bar{P}^A] + \beta_A^\beta \bar{M}^A + H^\beta = 0,\\
&{}^0\bar{\delta}_A M^A + {}^0\hat{b}_{A\beta} \bar{M}^{A\beta} + {}^0\varepsilon_{\gamma\beta}[\eta\, {}^0l_A^\gamma P^{A\beta} + (1-\eta)\, {}^0\bar{l}_A^\gamma \bar{P}^{A\beta}] + H = 0.
\end{aligned}
\tag{6.59}
$$

[1] It means that in describing deformations $S_0 \to S_t$ of the hypothetical surface S_t, cf. Fig. 108, we neglect the effects of local strains and rotations on the curvature increments of the surface.

At the same time, by the direct calculation and using the aforementioned approximation we arrive at the formulae (summation convention holds for $B = \mathrm{I}, \mathrm{II}, \mathrm{III}$)

$$\begin{aligned} \beta_{A\beta} &= \delta^{B\,0}_{\beta} n \cdot \bar{\Delta}_A \Delta_B u, \\ \beta^{\beta}_{A} &= {}^0a^{\beta\gamma} \delta^{B\,0}_{\gamma} n \cdot \bar{\Delta}_A \Delta_B u. \end{aligned} \tag{6.60}$$

Eqs. (6.59) in which $\beta_{A\beta}$, β^{β}_{A} are given by eqs. (6.60) represent the (quasi-stationary) equations of motion of the second order theory of lattice shells. Further, from eqs. (4.98) for strain measures we obtain

$$\begin{aligned} \gamma_{A\gamma} &= {}^0\delta_A u_\gamma + {}^0b_{A\gamma} u + {}^0\varepsilon_{\gamma\beta}\, {}^0l^{\beta}_{A} \{v\}_A, \\ \gamma_A &= {}^0\delta_A u + {}^0b^{\beta}_{A} u_\beta + {}^0\varepsilon_{\gamma\beta}\, {}^0l^{\gamma}_{A} \{v^{\beta}\}_A, \\ \varkappa_{A\gamma} &= {}^0\delta_A v_\gamma + {}^0b_{A\gamma} v, \\ \varkappa_A &= {}^0\delta_A v + {}^0b^{\beta}_{A} v_\beta, \end{aligned} \tag{6.61}$$

where we have used the formula $\{\varphi\}_A \equiv 0.5[\varphi(\mathbf{X}) + \varphi(\mathbf{X} + \Delta_A \mathbf{X})]$ introduced previously in Sec. 4.6.1.

The linear constitutive law (4.99) in the lattice shell setting reads

$$\begin{aligned} P^{A\gamma} &= A^{A\gamma\beta}\gamma_{A\beta} + \underline{A^{A\gamma}\gamma_A} + \underline{B^{A\gamma\beta}\varkappa_{A\beta}} + {}'B^{A\gamma}\varkappa_A, \\ P^{A} &= \underline{A^{A\beta}\gamma_{A\beta}} + A^{A}\gamma_A + {}''B^{A\beta}\varkappa_{A\beta} + \underline{B^{A}\varkappa_A}, \\ M^{A\gamma} &= \underline{B^{A\beta\gamma}\gamma_{A\beta}} + {}''B^{A\gamma}\gamma_A + C^{A\gamma\beta}\varkappa_{A\beta} + \underline{C^{A\gamma}\varkappa_A}, \\ M^{A} &= {}'B^{A\beta}\gamma_{A\beta} + \underline{B^{A}\gamma_A} + \underline{C^{A\beta}\varkappa_{A\beta}} + C^{A}\varkappa_A. \end{aligned} \tag{6.62}$$

If every cross section of the rod connecting nodes $\mathbf{X}$ and $\mathbf{X} + \Delta_A\mathbf{X}$ has an axis of symmetry which is parallel to the vector $\mathbf{n}(\mathbf{X})$ then the terms underlined in eqs. (6.62) drop out.

Eqs. (6.59) with $\beta_{A\beta}$, β^{β}_{A} given by eqs. (6.60), together with eqs. (6.61) and (6.62) constitute the governing equations of the second order theory of elastic lattice shells.

For the lattice plates we have

$$ {}^0\hat{b}^{\beta}_{A} = 0, \quad {}^0\hat{b}_{A\beta} = 0, \quad {}^0n = \text{const}. $$

Hence, eqs. (6.59) reduce to the form (summation convention with respect to $A, B = \mathrm{I}, \mathrm{II}, \mathrm{III}$ holds)

$$\begin{aligned} & {}^0\bar{\delta}_A P^{A\beta} + F^{\beta} = 0, \\ & {}^0\bar{\delta}_A P^{A} + \bar{P}^{A\beta} \delta^{B}_{\beta} \bar{\Delta}_A \Delta_B u + F = 0, \\ & {}^0\bar{\delta}_A M^{A\beta} + {}^0\varepsilon^{\beta}_{\cdot\gamma} [\eta\, {}^0l^{\gamma}_{A} P^{A} + (1-\eta)\, {}^0\bar{l}^{\gamma}_{A} \bar{P}^{A}] + M^{A\,0}a^{\beta\gamma} \delta^{B}_{\gamma} \bar{\Delta}_A \Delta_B u + H^{\beta} = 0, \\ & {}^0\bar{\delta}_A M^{A} + {}^0\varepsilon_{\gamma\beta} [\eta\, {}^0l^{\gamma}_{A} P^{A\beta} + (1-\eta)\, {}^0\bar{l}^{\gamma}_{A} \bar{P}^{A\beta}] + H = 0, \end{aligned} \tag{6.63}$$

while the strain measures (6.61) are given by

$$\begin{aligned}
\gamma_{A\gamma} &= {}^0\delta_A u_\gamma + {}^0\varepsilon_{\gamma\beta}\, {}^0l^\beta_A \{v\}_A, \\
\gamma_A &= {}^0\delta_A u + {}^0\varepsilon_{\gamma\beta}\, {}^0l^\gamma_A \{v^\beta\}_A, \\
\varkappa_{A\gamma} &= {}^0\delta_A v_\gamma, \\
\varkappa_A &= {}^0\delta_A v.
\end{aligned} \tag{6.64}$$

Eqs. (6.62), (6.63) and (6.64) represent the governing equations of the second order theory of elastic lattice plates a simple application of which will be given in the next section. For further details concerning the theories of lattice shells and plates the reader is referred to [70], [72], [73], [87]–[89], [179].

6.3.2 *Buckling of elastic grid*

To present at least one possible application of the stability equations derived in Sec. 6.3.1, cf. eqs. (6.59)–(6.61), we shall consider in this section the problem of buckling of the rectangular plane grid. The grid is shown in Fig. 109. By taking

$$\varkappa(X_1, X_2) = [l_{\rm I} X_1, l_{\rm II} X_2],$$
$$X_1 = -a, \ldots, -1, 0, 1, \ldots, a, \quad X_2 = -b, \ldots, -1, 0, 1, \ldots, b,$$

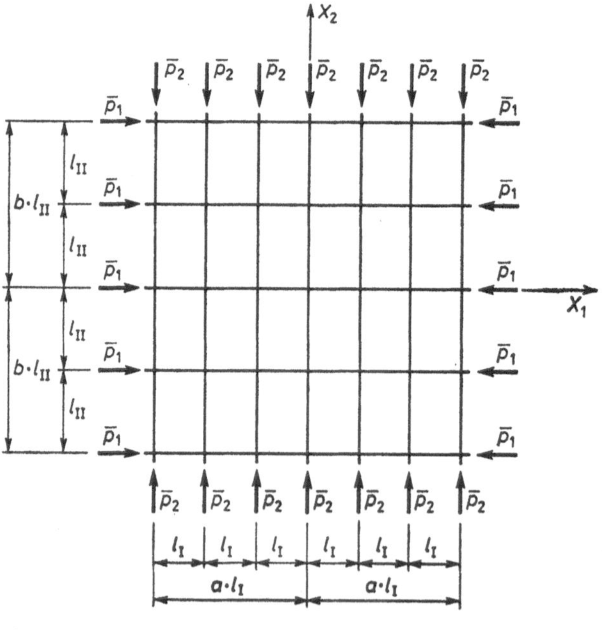

Fig. 109 Elastic grid buckling analysis

we get

$$g_1(X_1, X_2) = [l_{\text{I}}, 0], \quad g_2(X_1, X_2) = [0, l_{\text{II}}],$$
$$l^\alpha_A = \delta^\alpha_A, \quad \beta_{A\alpha} = \delta^B_\alpha \bar{\Delta}_A \Delta_B u, \quad \text{etc.}$$

and the fundamental set of equations assumes the form (cf. eqs. (6.63), (6.64) for $\eta = 1/2$, implying the generalized forces to be defined at the midsections of bars)

$$\begin{aligned}
&\bar{\Delta}_A P^{A\beta} + F^\beta = 0,\\
&\bar{\Delta}_A P^A + \delta^B_\beta \bar{\Delta}_A \Delta_B u \bar{P}^{A\beta} + F = 0,\\
&\bar{\Delta}_A M^{A\beta} + \tfrac{1}{2} \varepsilon^\beta_{.\gamma} [\delta^\gamma_A P^A + \delta^\gamma_A \bar{P}^A] + M^A \delta^{B\beta} \bar{\Delta}_A \Delta_B u + H^\beta = 0,\\
&\bar{\Delta}_A M^A + \tfrac{1}{2} \varepsilon_{\gamma\beta} [\delta^\gamma_A P^{A\beta} + \delta^\gamma_A \bar{P}^{A\beta}] + H = 0;
\end{aligned} \tag{6.65}$$

$$\begin{aligned}
&\gamma_{A\beta} = \Delta_A u_\beta + \delta^\gamma_A \varepsilon_{\beta\gamma} \{v\}_A,\\
&\gamma_A = \Delta_A u + \delta^\gamma_A \varepsilon_{\gamma\beta} \{v^\beta\}_A,\\
&\varkappa_{A\beta} = \Delta_A v_\beta,\\
&\varkappa_A = \Delta_A v;
\end{aligned} \tag{6.66}$$

$$\begin{aligned}
&P^{A\alpha} = A^{A\alpha\beta} \gamma_{A\beta}, \quad P^A = A^A \gamma_A,\\
&M^{A\alpha} = C^{A\alpha\beta} \varkappa_{A\beta}, \quad M^A = C^A \varkappa_A
\end{aligned} \tag{6.67}$$

with the elastic rigidities given by, cf. eqs. (4.101)

$$\begin{aligned}
&A^{A\alpha\beta} = \frac{E_A A_A}{l_A} d^{A\alpha}_\xi d^{A\beta}_\xi + \frac{12 E_A J_{A\eta}}{l^3_A} d^{A\alpha}_\zeta d^{A\beta}_\zeta,\\
&A^A = \frac{12 E_A J_{A\zeta}}{l^3_A} d^A_\eta d^A_\eta,\\
&C^{A\alpha\beta} = \frac{C_A}{l_A} d^{A\alpha}_\xi d^{A\beta}_\xi + \frac{E_A J_{A\zeta}}{l_A} d^{A\alpha}_\zeta d^{A\beta}_\zeta,\\
&C^A = \frac{E_A J_{A\eta}}{l_A} d^A_\eta d^A_\eta
\end{aligned} \tag{6.68}$$

$$\alpha, \beta = 1, 2, \quad A, B = \text{I}, \text{II},$$

with

$$\begin{aligned}
&d^{Ai}_\xi = d^{A\alpha}_\xi g_\alpha^{\ i} + d^A_\xi n^i,\\
&d^{Ai}_\eta = d^{A\alpha}_\eta g_\alpha^{\ i} + d^A_\eta n^i,\\
&d^{Ai}_\zeta = d^{A\alpha}_\zeta g_\alpha^{\ i} + d^A_\xi n^i, \quad i = 1, 2, 3,
\end{aligned}$$

η being normal to the grid plane.

Substituting eqs. (6.66) into eqs. (6.67) and the result into eqs. (6.65), and specifying the in-plane generalized stresses as

$$P^{I1} = -\bar{p}^1, \quad P^{II2} = -\bar{p}^2, \quad P^{I2} = P^{II1} = 0, \quad M^I = M^{II} = 0$$

we arrive at the fundamental set of equations in the form

$$\begin{aligned}
&A^I\bar{\Delta}_I[\Delta_I u+\{v^2\}_I]+A^{II}\bar{\Delta}_{II}[\Delta_{II}u-\{v^1\}_{II}]-\bar{\Delta}_I\Delta_I u\bar{p}^1-\Delta_{II}\bar{\Delta}_{II}u\bar{p}^2 = 0,\\
&C^{II1}\bar{\Delta}_I\Delta_I v^1+C^{III1}\bar{\Delta}_{II}\Delta_{II}v^1+\\
&\qquad+\tfrac{1}{2}A^{II}[\Delta_{II}u+\bar{\Delta}_{II}u-\{v^1(X^1, X^2-\Delta_{II}X^2)\}_{II}-\{v^1\}_{II}] = 0,\\
&C^{II22}\bar{\Delta}_{II}\Delta_{II}v^2+C^{I22}\bar{\Delta}_I\Delta_I v^2-\\
&\qquad-\tfrac{1}{2}A^I[\Delta_I u+\bar{\Delta}_I u+\{v^2(X^1-\Delta_I X^1, X^2)\}_I+\{v^2\}_I] = 0.
\end{aligned} \tag{6.69}$$

The solution to the so defined problem consists in finding such boundary loads $\bar{p}^1$, $\bar{p}^2$ for which the set of equations (6.69) together with the boundary conditions

$$\begin{aligned}
X^1 &= \pm a: \quad u = 0, v^1 = 0, \bar{\Delta}_I[\{v^2\}_I] = 0,\\
X^2 &= \pm b: \quad u = 0, v^2 = 0, \bar{\Delta}_{II}[\{v^1\}_{II}] = 0
\end{aligned} \tag{6.70}$$

yields solutions $u(\mathbf{X})$, $v^1(\mathbf{X})$, $v^2(\mathbf{X})$ different from zero in the region considered. (Note that only an approximate fulfillment of the static boundary conditions is implied by eqs. (6.70).)

Let the solutions to eqs. (6.69), satisfying the boundary conditions (6.70) have the form

$$\begin{aligned}
u(\mathbf{X}) &= U\sin\frac{m\pi X^1}{a}\sin\frac{n\pi X^2}{b},\\
v^1(\mathbf{X}) &= V_1\sin\frac{m\pi X^1}{a}\cos\frac{n\pi X^2}{b},\\
v^2(\mathbf{X}) &= V_2\cos\frac{m\pi X^1}{a}\sin\frac{n\pi X^2}{b},
\end{aligned} \tag{6.71}$$

where U, V_1, V_2 are constants. Substituting eqs. (6.71) into eqs. (6.69) leads to a homogeneous set of three linear algebraic equations for three constants U, V_1, V_2 in the form

$$\left[4(A^I-\bar{p}^1)\sin^2\frac{m\pi}{2a}+4(A^{II}-\bar{p}^2)\sin\frac{n\pi}{2b}\right]U-$$

$$-2A^{II}\sin\frac{n\pi}{2b}V_1+2A^I\sin\frac{m\pi}{2a}V_2 = 0, \tag{6.72}$$

$$-2A^{\mathrm{II}}\sin\frac{n\pi}{2b}U+$$
$$+\left(4C^{\mathrm{I}11}\sin^2\frac{m\pi}{2a}+4C^{\mathrm{I}11}\sin^2\frac{n\pi}{2b}+A^{\mathrm{II}}\cos^2\frac{n\pi}{2b}\right)V_1=0,$$
$$2A^{\mathrm{I}}\sin\frac{m\pi}{2a}U+ \tag{6.72}$$
$$+\left(4C^{\mathrm{I}22}\sin^2\frac{m\pi}{2a}+4C^{\mathrm{II}22}\sin^2\frac{n\pi}{2b}+A^{\mathrm{I}}\cos^2\frac{m\pi}{2a}\right)V_2=0.$$

Existence of non-zero solutions to the set (6.72) requires that its determinant vanishes. Assuming the proportional increase of the external load and setting $\bar{p}^2=\nu\bar{p}^1$ with the parameter ν being given, the zero determinant condition yields

$$\bar{p}^1=\frac{A^{\mathrm{I}}s_m^2\dfrac{C_{mn}^2}{C_{mn}^2+A^{\mathrm{I}}}+A^{\mathrm{II}}s_n^2\dfrac{C_{mn}^1}{C_{mn}^1+A^{\mathrm{II}}}}{s_m^2+\nu s_n^2} \tag{6.73}$$

where the following notation is used:

$$s_m^2=\sin^2\frac{m\pi}{2a},\qquad s_n^2=\sin^2\frac{n\pi}{2b},$$
$$C_{mn}^1=4C^{\mathrm{I}11}s_m^2+4C^{\mathrm{I}22}s_n^2, \tag{6.74}$$
$$C_{mn}^2=4C^{\mathrm{II}11}s_m^2+4C^{\mathrm{II}22}s_n^2.$$

Eq. (6.73) represents the value of the critical load, which is dependent on the buckling mode and lattice parameters.

It may be shown, cf. [73], that a reasonable approximation may be obtained by considering the components of A^A as significantly bigger than those of $C^{A\alpha\beta}$; the formula (6.73) simplifies then to the form

$$\bar{p}^1=\frac{C_{mn}^2s_m^2+C_{mn}^1s_n^2}{s_m^2+\nu s_n^2}. \tag{6.75}$$

For the square grid under one-directional compression ($\nu=0$) we get the minimum value of the load $\bar{p}^1$ for $m=n=1$. The value of the critical load reads

$$\bar{p}^1=4(C^{\mathrm{I}11}+C^{\mathrm{I}22}+C^{\mathrm{II}11}+C^{\mathrm{II}22})\sin^2\frac{\pi}{2a} \tag{6.76}$$

which, adopting additionally the approximation $\sin\frac{\pi}{2a} \approx \frac{\pi}{2a}$ valid for large a, becomes

$$\bar{p}^1 = \frac{\pi^2(C^{\mathrm{I}11}+C^{\mathrm{I}22}+C^{\mathrm{II}11}+C^{\mathrm{II}22})}{a^2}. \tag{6.77}$$

It can be easily shown that the last results is also the critical load that may be obtained by employing the continuous model, [167]. The comparison with the buckling solution for classical continuous simply supported plate is also straightforward, [153].

6.3.3 *Perfectly-plastic lattice-type plates — discrete model*

The way to effectively deal with elastic-plastic frames has been indicated in Secs. 6.2.1–6.2.3. It should be clear that a completely analogous approach can be employed to analyse inelastic plane grids. In particular, for rather dense grids (or, equivalently, for lattice-type plates) it seems plausible to combine the concept of the generalized plastic hinges with the finite difference description employed in Sec. 4.6. We shall illustrate such an approach below by discussing consecutively two its effective applications. First, we shall develop a method for solving some complex limit analysis problems by means of a purely analytical technique. Then we shall continue to describe an incremental, numerical technique effective in the step-by-step analysis up to collapse of any elastic-perfectly plastic grid.

The class of structures that we are going to deal with below is defined as follows, cf. Sec. 4.6:

— the structure is composed of slender beams,
— the axes of the beams constitute a plane mesh—for simplicity such beams layouts will be considered only for which it is possible to distinguish certain "families of beams",
— the members between any two nodes are prismatic, homogeneous, isotropic and perfectly plastic,
— the assumptions of the Euler beam theory hold,
— external load is applied to the nodes in the form of concentrated forces (with vectors normal to the grid plane) and moments (with vectors lying in the grid plane).

The equilibrium equations are obtained as, cf. eqs. $(6.57)_{2,3}$ for $\eta = 1$ and $a = 0$

$$\begin{aligned} &{}'\bar{\delta}_A P^A + F = 0,\\ &{}'\bar{\delta}_A M^{A\beta} + \varepsilon^{\beta}_{\cdot\gamma} l^{\gamma}_A P^A + H^{\beta} = 0,\\ &A = \mathrm{I}, \mathrm{II}, \mathrm{III};\quad \beta, \gamma = 1, 2. \end{aligned} \tag{6.78}$$

Summation convention holds over all the indices placed on different levels. The shear forces P^A and the bending moments $M^{A\beta}$ in eqs. (6.78) are associated with the nodal cross-sections of the beams. Since in the static approach to be employed for the assessment of ultimate load only the equilibrium equations have to be satisfied (together with the limit yield condition reached in some parts of the structure), neither the geometric nor the constitutive relationships are here recalled.

A field of generalized stresses $P^A(\mathbf{X})$, $M^{A\beta}(\mathbf{X})$ is said to be *statically admissible* if:

— for each node $\mathbf{X} \in \mathscr{X}^0$ and at the boundary the values of P^A and $M^{A\beta}$ satisfy the equilibrium equations and the static boundary conditions, respectively,

— limit state condition $\bar{F}(\mathbf{X}, P^A, M^{A\beta}) \leqslant 0$ is nowhere violated.

The limit state function is a known function of the bending and twisting moments $M^{A\beta}$, and possibly shear forces P^A at all the nodal cross-sections of the beams.

Let us now consider the grid consisting of two mutually perpendicular families of bars rigidly connected at nodes and simply supported at the outer edges, Fig. 110. The grid beams are prismatic and capable of supporting

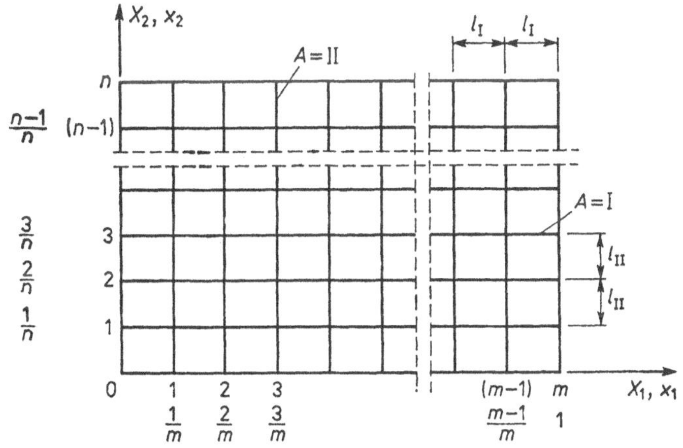

Fig. 110 Elastic-plastic rectangular grid

torque. Coordinate axes are parallel to the families of bars each of which consists of an arbitrary number of beams. The grid is subjected to the transverse uniformly distributed load of constant intensity applied at the nodes as the concentrated forces. According to Fig. 110 the following notation has been employed:

l_{I}, l_{II} — lengths of beam segments between the nodes of the first and the second family, respectively,

$N_{\rm I}, N_{\rm II}$ — number of beams in the first and second family, respectively,
X_1, X_2 — consecutive node numbers in both families,
$M_0^{\rm I}, M_0^{\rm II}$ — ultimate "beamlike" moments under pure bending in both families,
$\tilde{M}_0^{\rm I}, \tilde{M}_0^{\rm II}$ — ultimate "beamlike" moments under pure twisting in both families,
$x_1 = \dfrac{X_1}{N_{\rm I}}, x_2 = \dfrac{X_2}{N_{\rm II}}$ — discrete dimensionless coordinates.

To find the lower bound to the collapse load q we look for a statically admissible field of bending and twisting moments. In selecting this field for both the beam families we make use of "beam analogy", hence $M^{\rm I2}$ and $M^{\rm II1}$ assume parabolic distributions

$$\begin{aligned} M^{\rm I2} &= a_1 x_1^2 + b_1 x_1 + c_1, \\ M^{\rm II1} &= a_2 x_2^2 + b_2 x_2 + c_2 \end{aligned} \tag{6.79}$$

which additionally have to satisfy the following boundary conditions:

for family I —

$$M^{\rm I2}(0) = 0, \quad M^{\rm I2}(1) = 0, \quad M^{\rm I2}\left(\frac{N_{\rm I}-2+\alpha_{\rm I}}{2N_{\rm I}}\right) = M_0^{\rm I}, \tag{6.80}$$

where

$$\alpha_{\rm I} = \frac{[(-1)^{N_{\rm I}}-1]}{2},$$

for family II —

$$M^{\rm II1}(0) = 0, \quad M^{\rm II1}(1) = 0, \quad M^{\rm II1}\left(\frac{N_{\rm II}-2+\alpha_{\rm II}}{2N_{\rm II}}\right) = -M_0^{\rm II} \tag{6.81}$$

where

$$\alpha_{\rm II} = \frac{[(-1)^{N_{\rm II}}-1]}{2}.$$

Eqs. (6.79)–(6.81) imply that

$$\begin{aligned} M^{\rm I2}(x_1) &= 4M_0^{\rm I}\gamma_{\rm I}\left[\left(x_1+\frac{1}{N_{\rm I}}\right)-\left(x_1+\frac{1}{N_{\rm I}}\right)^2\right], \\ &\qquad -1/N_{\rm I} \leqslant x_1 \leqslant (N_{\rm I}-1)/N_{\rm I}, \\ M^{\rm II1}(x_2) &= -4M_0^{\rm II}\gamma_{\rm II}\left[\left(x_2+\frac{1}{N_{\rm II}}\right)-\left(x_2+\frac{1}{N_{\rm II}}\right)^2\right], \\ &\qquad -1/N_{\rm II} \leqslant x_2 \leqslant (N_{\rm II}-1)/N_{\rm II}, \end{aligned} \tag{6.82}$$

where

$$\gamma_{\mathrm{I}} = \frac{N_{\mathrm{I}}^2}{N_{\mathrm{I}}^2-\alpha_{\mathrm{I}}^2}, \qquad \gamma_{\mathrm{II}} = \frac{N_{\mathrm{II}}^2}{N_{\mathrm{II}}^2-\alpha_{\mathrm{II}}^2}.$$

Twisting moment field is postulated using the "plate analogy" so that the maximum values of moments due to the torque are assumed to occur at the corner nodes of the grid. Thus, the twisting moments assume the form

$$M^{\mathrm{I}1}(x_1, x_2) = d_1 x_1 x_2 + e_1 x_1 + f_1 x_2 + g_1,$$

$$M^{\mathrm{II}2}(x_1, x_2) = d_2 x_1 x_2 + e_2 x_1 + f_2 x_2 + g_2.$$

The coefficients $d_1, d_2, e_1, e_2, f_1, f_2, g_1, g_2$ are computed from the following boundary conditions:

for family I:

$$M^{\mathrm{I}1}\left(-\frac{1}{N_{\mathrm{I}}}, 0\right) = \tilde{M}_0^{\mathrm{I}}, \qquad M^{\mathrm{I}1}\left(\frac{N_{\mathrm{I}}-2+\alpha_{\mathrm{I}}}{2N_{\mathrm{I}}}, 0\right) = 0,$$

$$M^{\mathrm{I}1}\left(0, \frac{N_{\mathrm{II}}+\alpha_{\mathrm{II}}}{2N_{\mathrm{II}}}\right) = 0, \qquad M^{\mathrm{I}1}\left(\frac{N_{\mathrm{I}}-2+\alpha_{\mathrm{I}}}{2N_{\mathrm{I}}}, \frac{N_{\mathrm{II}}+\alpha_{\mathrm{II}}}{2N_{\mathrm{II}}}\right) = 0;$$

for family II:

$$M^{\mathrm{II}2}\left(0, -\frac{1}{N_{\mathrm{II}}}\right) = -\tilde{M}_0^{\mathrm{II}}, \qquad M^{\mathrm{II}2}\left(0, \frac{N_{\mathrm{II}}-2+\alpha_{\mathrm{II}}}{2N_{\mathrm{II}}}\right) = 0,$$

$$M^{\mathrm{II}2}\left(\frac{N_{\mathrm{I}}+\alpha_{\mathrm{I}}}{2N_{\mathrm{I}}}, 0\right) = 0, \qquad M^{\mathrm{II}2}\left(\frac{N_{\mathrm{I}}+\alpha_{\mathrm{I}}}{2N_{\mathrm{I}}}, \frac{N_{\mathrm{II}}-2+\alpha_{\mathrm{II}}}{2N_{\mathrm{II}}}\right) = 0.$$

Since the description of the twisting moment distribution by one equation in the whole region is rather akward we shall consider the first quadrant only, i.e. the region

$$x_1 \in \left[0, \frac{N_{\mathrm{I}}+\alpha_{\mathrm{I}}}{2N_{\mathrm{I}}}\right], \qquad x_2 \in \left[0, \frac{N_{\mathrm{II}}+\alpha_{\mathrm{II}}}{2N_{\mathrm{II}}}\right].$$

After some algebra the expressions for $M^{\mathrm{I}1}$ and $M^{\mathrm{II}2}$ are found to be

$$M^{\mathrm{I}1}(x_1, x_2) = \frac{4N_{\mathrm{I}}N_{\mathrm{II}}\tilde{M}_0^{\mathrm{I}}}{(N_{\mathrm{I}}+\alpha_{\mathrm{I}})(N_{\mathrm{II}}+\alpha_{\mathrm{II}})}\left[\left(x_1+\frac{1}{N_{\mathrm{I}}}\right)x_2 - \right.$$

$$\left. -\frac{N_{\mathrm{II}}+\alpha_{\mathrm{II}}}{2N_{\mathrm{II}}}\left(x_1+\frac{1}{N_{\mathrm{I}}}\right) - \frac{N_{\mathrm{I}}+\alpha_{\mathrm{I}}}{2N_{\mathrm{I}}}x_2\right] + \tilde{M}_0^{\mathrm{I}},$$

$$-\frac{1}{N_{\mathrm{I}}} \leqslant x_1 \leqslant \frac{N_{\mathrm{I}}-2+\alpha_{\mathrm{I}}}{2N_{\mathrm{I}}}, \qquad 0 \leqslant x_2 \leqslant \frac{N_{\mathrm{II}}+\alpha_{\mathrm{II}}}{2N_{\mathrm{II}}}, \tag{6.83}$$

$$M^{II2}(x_1, x_2) = -\frac{4N_I N_{II}\tilde{M}_0^{II}}{(N_I+\alpha_I)(N_{II}+\alpha_{II})}\left[x_1\left(x_2+\frac{1}{N_{II}}\right)-\right.$$

$$\left.-\frac{N_{II}+\alpha_{II}}{2N_{II}}x_1-\frac{N_I+\alpha_I}{2N_I}\left(x_2+\frac{1}{N_{II}}\right)\right]-\tilde{M}_0^{II},$$

$$0 \leqslant x_1 \leqslant \frac{N_I+\alpha_I}{2N_I}, \qquad -\frac{1}{N_{II}} \leqslant x_2 \leqslant \frac{N_{II}-2+\alpha_{II}}{2N_{II}}. \tag{6.83}$$

It should be recalled that substituting coordinates (x_1, x_2) we obtain the values of moments at the following cross-sections, cf. Fig. 110:

for A = I: on the left-hand side of the point $\left(x_1+\frac{1}{N_I}, x_2\right)$,

for A = II: below the point $\left(x_1, x_2+\frac{1}{N_{II}}\right)$.

Introducing new notation

$$x_1' = x_1+\frac{1}{N_I}, \qquad x_2' = x_2+\frac{1}{N_{II}}$$

the expressions (6.82) and (6.83) become more compact

$$M^{I2} = 4M_0^I\gamma_I(x_1'-x_1'^2), \qquad 0 \leqslant x_1' \leqslant 1,$$

$$M^{II1} = -4M_0^{II}\gamma_{II}(x_2'-x_2'^2), \qquad 0 \leqslant x_2' \leqslant 1,$$

$$M^{I1} = \frac{4N_I N_{II}\tilde{M}_0^I}{(N_I+\alpha_I)(N_{II}+\alpha_{II})}\left[x_1'x_2-\frac{N_{II}+\alpha_{II}}{2N_{II}}x_1'-\frac{N_I+\alpha_I}{2N_I}x_2\right]+\tilde{M}_0^I, \tag{6.84}$$

$$M^{II2} = -\frac{4N_I N_{II}\tilde{M}_0^{II}}{(N_I+\alpha_I)(N_{II}+\alpha_{II})}\left[x_1x_2'-\frac{N_{II}+\alpha_{II}}{2N_{II}}x_1-\frac{N_I+\alpha_I}{2N_I}x_2'\right]-\tilde{M}_0^{II}.$$

Similar twisting moment fields can be constructed for the remaining quadrants of the grid by means of suitable changes of the coordinate system and allowing for antisymmetry of twisting moments with respect to the symmetry axes of the structure. It can be readily shown that the expressions for twisting moments derived for each particular quadrant preserve the continuity while passing from a given quadrant to the adjacent one.

Now, to make use of equilibrium equations we shall reformulate the set (6.78) to obtain one equation involving moments only. After simple rearrangements we get

$$\varepsilon_{\alpha\cdot}^{\cdot A}\left(\frac{1}{l_I}\,\delta_1^A+\frac{1}{l_{II}}\,\delta_2^A\right)\bar{\Delta}_A\bar{\Delta}_B M^{B\alpha}+F = 0 \tag{6.85}$$

where

$$\delta_{\alpha}^{A} = \begin{cases} 1 & \text{if } A = \text{I and } \alpha = 1 \text{ or } A = \text{II and } \alpha = 2, \\ 0 & \text{in other cases,} \end{cases}$$

$$\varepsilon_{\alpha\cdot}^{A} = \begin{cases} 1 & \text{if } A = \text{II and } \alpha = 1, \\ -1 & \text{if } A = \text{I and } \alpha = 2, \\ 0 & \text{if } A = \text{I and } \alpha = 1 \text{ or } A = \text{II and } \alpha = 2. \end{cases}$$

Eq. (6.85) reads explicitly ($F = ql_{\text{I}}l_{\text{II}}$)

$$\bar{\Delta}_{\text{II}}^2 \frac{M^{\text{II}1}}{l_{\text{II}}} + \bar{\Delta}_{\text{I}}\bar{\Delta}_{\text{II}}\left(\frac{M^{\text{I}1}}{l_{\text{II}}} - \frac{M^{\text{II}2}}{l_{\text{I}}}\right) - \bar{\Delta}_{\text{I}}^2 \frac{M^{\text{I}2}}{l_{\text{I}}} = ql_{\text{I}}l_{\text{II}}. \tag{6.86}$$

Having calculated the second differences for the fields (6.84), the equilibrium equation (6.86) supplies the intensity of the limit load in the form

$$q = \frac{4}{l_{\text{I}}l_{\text{II}}}\left[\frac{2\gamma_{\text{I}}l_{\text{II}}M_0^{\text{I}}}{N_{\text{I}}^2} + \frac{2\gamma_{\text{II}}l_{\text{I}}M_0^{\text{II}}}{N_{\text{II}}^2} + \frac{\tilde{M}_0^{\text{I}}l_{\text{I}} + \tilde{M}_0^{\text{II}}l_{\text{II}}}{(N_{\text{I}}+\alpha_{\text{I}})(N_{\text{II}}+\alpha_{\text{II}})}\right]. \tag{6.87}$$

In the particular case when beams do not support torque ($\tilde{M}_0^{\text{I}} = \tilde{M}_0^{\text{II}} = 0$) this formula is identical with the expression given in [53]. The expression (6.87) simplifies in the case of even number of beams in both families when $\alpha_{\text{I}} = \alpha_{\text{II}} = 0$, $\gamma_{\text{I}} = \gamma_{\text{II}} = 1$ thus yielding

$$q = \frac{4}{l_{\text{I}}l_{\text{II}}}\left[\frac{2l_{\text{II}}M_0^{\text{I}}}{N_{\text{I}}^2} + \frac{2l_{\text{I}}M_0^{\text{II}}}{N_{\text{II}}^2} + \frac{\tilde{M}_0^{\text{I}}l_{\text{I}} + \tilde{M}_0^{\text{II}}l_{\text{II}}}{N_{\text{I}}N_{\text{II}}}\right]. \tag{6.88}$$

To ensure that eq. (6.87) is really a lower bound to the collapse load we must prove that the postulated moment field is in fact statically admissible. For this purpose we should find out whether the limit state condition is nowhere violated within the grid. There exist certain regions in which checking of this condition does not require any additional calculations. These are the corners where bending moments vanish and twisting ones reach their ultimate values, and the central lines of the structure where twisting moments vanish whereas bending ones achieve the ultimate values. In the remaining regions of the grid undergoing bending and torsion at the same time the check on the combined bending-twisting limit state condition must be done. For the rectangular cross-section of the beam the condition may be approximated by a circular curve, cf. Sec. 6.2.2, of the form

$$\bar{F}^A = \left(\frac{M^A}{M_0^A}\right)^2 + \left(\frac{\tilde{M}^A}{\tilde{M}_0^A}\right)^2 - 1 \leqslant 0 \tag{6.89}$$

where M^A, $\tilde{M}^A$ are the actual bending and twisting moments acting simultaneously in the family A. To check whether the condition (6.89) is satisfied at all cross-sections we combine the moment field (6.84) and eq. (6.89). Confining ourselves to the family $A = \mathrm{I}$ we get

$$[4\gamma_{\mathrm{I}}(x_1' - x_1'^2)]^2 + \left[\frac{4N_{\mathrm{I}}N_{\mathrm{II}}}{(N_{\mathrm{I}}+\alpha_{\mathrm{I}})(N_{\mathrm{II}}+\alpha_{\mathrm{II}})}\left(x_1'x_2 - \frac{N_{\mathrm{II}}+\alpha_{\mathrm{II}}}{2N_{\mathrm{II}}}x_1' - \frac{N_{\mathrm{I}}+\alpha_{\mathrm{I}}}{2N_{\mathrm{I}}}x_2\right)+1\right]^2 - 1 \leqslant 0. \quad (6.90)$$

From the character of the moment distribution it is seen that the beam with $x_2 = 0$ is critical, which results in eq. (6.90) being the function of one variable only, $\bar{F}^{\mathrm{I}}(x_1) \leqslant 0$. Further analysis does not involve any difficulties and it is straightforward to conclude that eq. (6.90) is not violated at any nodal cross-section of the first family. Since the external load is applied to the nodes only, the moments at midspan cross-sections will be at most equal to the adjacent nodal ones so there is no need to further check eq. (6.90). Moreover, the check at the nodes themselves can be done as well. Assuming that the beams of both families are monolithically connected, the nodes can be treated as plate elements, cf. Fig. 111 with the depth equal to $\max(H_{\mathrm{I}}, H_{\mathrm{II}})$ and with

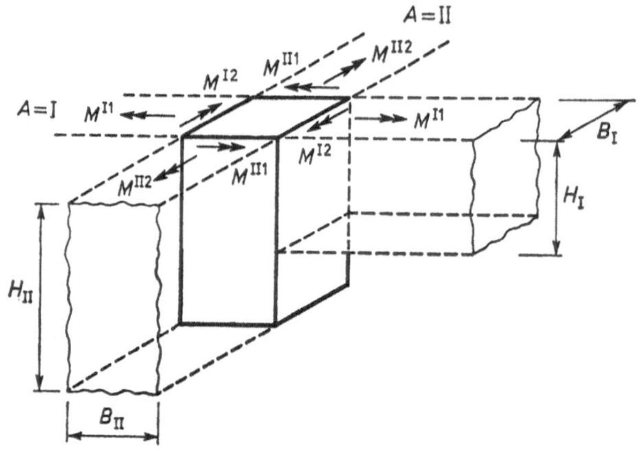

Fig. 111 Check of static admissibility of the moment fields at the grid node

the plan dimensions $B_{\mathrm{I}} \times B_{\mathrm{II}}$. Using the limit state condition for orthotropic plates it has been shown in [25] that static admissibility of the moment field is not violated at any node in the original structure.

Repeating analogous considerations we may obtain the expression for the limit load of the grid without outer (belt) beams as shown in Fig. 112. After suitable changes in the boundary conditions the moment field (6.84)

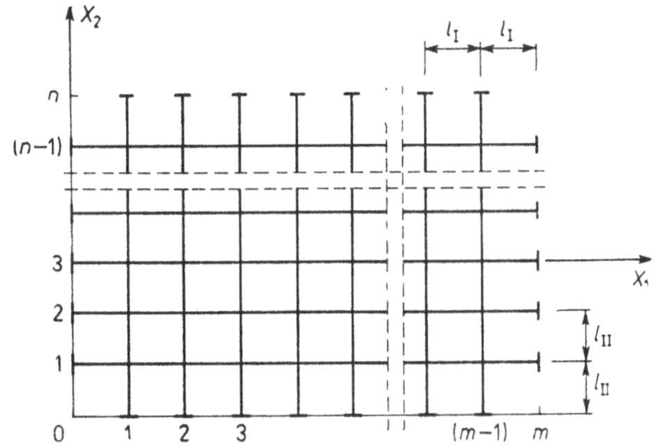

Fig. 112 Elastic-plastic rectangular grid without outer beams

is replaced by the relationships

$$M^{I2} = 4M_0^I\gamma_I(x_1 - x_1'), \quad 0 \leqslant x_1' \leqslant 1,$$

$$M^{II1} = -4M_0^{II}\gamma_{II}(x_2 - x_2'^2), \quad 0 \leqslant x_2' \leqslant 1,$$

$$M^{I1} = \frac{4N_I N_{II}\tilde{M}_0^I}{N_{II}+\alpha_{II}-2}\left[\frac{x_1' x_2}{N_I+\alpha_I} - \frac{1}{2N_{II}}\frac{N_{II}+\alpha_{II}}{N_I+\alpha_I}x_1' - \right.$$

$$\left. - \frac{1}{2N_I}x_2 + \frac{N_{II}+\alpha_{II}}{4N_I N_{II}}\right],$$

$$0 \leqslant x_1' \leqslant \frac{N_I+\alpha_I}{2N_I}, \quad \frac{1}{N_{II}} \leqslant x_2 \leqslant \frac{N_{II}+\alpha_{II}}{2N_{II}}, \tag{6.91}$$

$$M^{II2} = -\frac{4N_I N_{II}\tilde{M}_0^{II}}{N_I+\alpha_I-2}\left[\frac{x_1 x_2'}{N_{II}+\alpha_{II}} - \frac{1}{2N_{II}}x_1 - \right.$$

$$\left. - \frac{1}{2N_I}\frac{N_I+\alpha_I}{N_{II}+\alpha_{II}}x_2' + \frac{N_I+\alpha_I}{4N_I N_{II}}\right],$$

$$\frac{1}{N_I} \leqslant x_1 \leqslant \frac{N_I+\alpha_I}{2N_I}, \quad 0 \leqslant x_2' \leqslant \frac{N_{II}+\alpha_{II}}{2N_{II}}.$$

The intensity of the limit load results as

$$q = \frac{4}{l_I l_{II}}\left[\frac{2\gamma_I l_{II} M_0^I}{N_I^2} + \frac{2\gamma_{II} l_I M_0^{II}}{N_{II}^2} + \frac{\tilde{M}_0^I l_I}{(N_{II}+\alpha_{II}-2)(N_I+\alpha_I)} + \right.$$

$$\left. + \frac{\tilde{M}_0^{II} l_{II}}{(N_I+\alpha_I-2)(N_{II}+\alpha_{II})}\right]. \tag{6.92}$$

With regard to the known property of inelastic structures that the surplus of the material cannot cause the decrease in collapse load, the comparison of the limit load values for the both grids shows that the moment field for the second grid was assumed more consistently resulting in a better assessment of the actual collapse load.

Now we shall again consider the beam layout as in Fig. 112 assuming however that the beams are clamped at the boundaries. Similarly as before, in the static approach we want to obtain the best assessment of the limit load by postulating such a field of statically admissible moments that the limit state condition is reached in the largest possible number of nodal cross-sections. The "plate analogy" is used again. The bending moments are assumed to be a sum of three moment fields of the form (A = I)

$$M^{\mathrm{I}2}(x_1', x_2) = f_1(x_1')+f_2(x_1')+f_3(x_1', x_2) \tag{6.93}$$

where

$$f_1(x_1') = 4M_0^{\mathrm{I}}\gamma_{\mathrm{I}}(x_1'-x_1'^2), \quad 0 \leqslant x_1' \leqslant 1,$$

is the field taken for the simply supported grid,

$$f_2(x_1') = 4\overline{M}_0^{\mathrm{I}}\gamma_{\mathrm{I}}(x_1'-x_1'^2)-\overline{M}_0^{\mathrm{I}}, \quad 0 \leqslant x_1' \leqslant 1,$$

is a parabolic moment field satisfying the following boundary conditions

$$f_2(0) = -\overline{M}_0^{\mathrm{I}}, \quad f_2(1) = -\overline{M}_0^{\mathrm{I}}, \quad f_2\left(\frac{N_{\mathrm{I}}+\alpha_{\mathrm{I}}}{2N_{\mathrm{I}}}\right) = 0$$

and

$$f_3(x_1', x_2)$$
$$= \frac{4N_{\mathrm{I}}N_{\mathrm{II}}\overline{M}_0^{\mathrm{I}}}{N_{\mathrm{II}}+\alpha_{\mathrm{II}}-2}\left[\frac{x_1'x_2}{N_{\mathrm{I}}+\alpha_{\mathrm{I}}}-\frac{1}{2N_{\mathrm{II}}}\frac{N_{\mathrm{II}}+\alpha_{\mathrm{II}}}{N_{\mathrm{I}}+\alpha_{\mathrm{I}}}x_1'-\frac{1}{2N_{\mathrm{I}}}x_2+\frac{N_{\mathrm{II}}+\alpha_{\mathrm{II}}}{4N_{\mathrm{I}}N_{\mathrm{II}}}\right]$$

is a moment field similar to that of twisting, where $\overline{M}_0^{\mathrm{I}}$ represents negative ultimate moment in pure bending. Bending moments for family A = II are postulated similarly. Hence

$$M^{12} = 4M_0^{\mathrm{I}}\left\{\gamma_{\mathrm{I}}(x_1'-x_1'^2)+\nu_{\mathrm{I}}\left[\gamma_{\mathrm{I}}(x_1'-x_1'^2)-\frac{1}{4}+\frac{N_{\mathrm{I}}N_{\mathrm{II}}}{N_{\mathrm{II}}+\alpha_{\mathrm{II}}-2}\left(\frac{x_1'x_2}{N_{\mathrm{I}}+\alpha_{\mathrm{I}}}-\right.\right.\right.$$
$$\left.\left.\left.-\frac{1}{2N_{\mathrm{II}}}\frac{N_{\mathrm{II}}+\alpha_{\mathrm{II}}}{N_{\mathrm{I}}+\alpha_{\mathrm{I}}}x_1'-\frac{1}{2N_{\mathrm{I}}}x_2+\frac{N_{\mathrm{II}}+\alpha_{\mathrm{II}}}{4N_{\mathrm{I}}N_{\mathrm{II}}}\right)\right]\right\}\lambda_{\mathrm{I}},$$
$$0 \leqslant x_1' \leqslant \frac{N_{\mathrm{I}}+\alpha_{\mathrm{I}}}{2N_{\mathrm{I}}}, \quad \frac{1}{N_{\mathrm{II}}} \leqslant x_2 \leqslant \frac{N_{\mathrm{II}}+\alpha_{\mathrm{II}}}{2N_{\mathrm{II}}}, \tag{6.94}$$

$$M^{\text{II1}} = -4M_0^{\text{II}}\Big\{\gamma_{\text{II}}(x_2'-x_2'^2)+\nu_{\text{II}}\Big[\gamma_{\text{II}}(x_2'-x_2'^2)-\frac{1}{4}+\frac{N_{\text{I}}N_{\text{II}}}{N_{\text{I}}+\alpha_{\text{I}}-2}\Big(\frac{x_1x_2'}{N_{\text{II}}+\alpha_{\text{II}}}-$$

$$-\frac{1}{2N_{\text{II}}}x_1-\frac{1}{2N_{\text{I}}}\frac{N_{\text{I}}+\alpha_{\text{I}}}{N_{\text{II}}+\alpha_{\text{II}}}x_2'+\frac{N_{\text{I}}+\alpha_{\text{I}}}{4N_{\text{I}}N_{\text{II}}}\Big)\Big]\Big\}\lambda_{\text{II}},$$

$$\frac{1}{N_{\text{I}}}\leqslant x_1\leqslant\frac{N_{\text{I}}+\alpha_{\text{I}}}{2N_{\text{I}}},\qquad 0\leqslant x_2'\leqslant\frac{N_{\text{II}}+\alpha_{\text{II}}}{2N_{\text{II}}} \tag{6.94}$$

where

$$\nu_A = \frac{\overline{M}_0^A}{M_0^A},\qquad A = \text{I},\text{II}$$

and λ_{I}, λ_{II} are correction coefficients referring to the bending moments introduced to make the postulated moment distribution statically admissible. Twisting moments are assumed in the same way as before with the only difference being the reduction multipliers $\tilde{\lambda}_{\text{I}}$, $\tilde{\lambda}_{\text{II}}$, i.e.

$$M^{\text{I1}} = \frac{4N_{\text{I}}N_{\text{II}}\tilde{M}_0^{\text{I}}}{N_{\text{II}}+\alpha_{\text{II}}-2}\Big[\frac{x_1'x_2}{N_{\text{I}}+\alpha_{\text{I}}}-\frac{1}{2N_{\text{II}}}\frac{N_{\text{II}}+\alpha_{\text{II}}}{N_{\text{I}}+\alpha_{\text{I}}}x_1'-$$

$$-\frac{1}{2N_{\text{I}}}x_2+\frac{N_{\text{II}}+\alpha_{\text{II}}}{4N_{\text{I}}N_{\text{II}}}\Big]\tilde{\lambda}_{\text{I}},$$

$$0\leqslant x_1'\leqslant\frac{N_{\text{I}}+\alpha_{\text{I}}}{2N_{\text{I}}},\qquad \frac{1}{N_{\text{II}}}\leqslant x_2\leqslant\frac{N_{\text{II}}+\alpha_{\text{II}}}{2N_{\text{II}}}, \tag{6.95}$$

$$M^{\text{II2}} = -\frac{4N_{\text{I}}N_{\text{II}}\tilde{M}_0^{\text{II}}}{N_{\text{I}}+\alpha_{\text{I}}-2}\Big[\frac{x_1x_2'}{N_{\text{II}}+\alpha_{\text{II}}}-\frac{1}{2N_{\text{II}}}x_1-\frac{1}{2N_{\text{I}}}\frac{N_{\text{I}}+\alpha_{\text{I}}}{N_{\text{II}}+\alpha_{\text{II}}}x_2'+$$

$$+\frac{N_{\text{I}}+\alpha_{\text{I}}}{4N_{\text{I}}N_{\text{II}}}\Big]\tilde{\lambda}_{\text{II}},$$

$$\frac{1}{N_{\text{I}}}\leqslant x_1\leqslant\frac{N_{\text{I}}+\alpha_{\text{I}}}{2N_{\text{I}}},\qquad 0\leqslant x_2'\leqslant\frac{N_{\text{II}}+\alpha_{\text{II}}}{2N_{\text{II}}}.$$

All the coefficients λ_A, $\tilde{\lambda}_A$, $A = \text{I},\text{II}$ are positive and do not exceed unity, which follows directly from satisfying the limit state condition $\overline{F}^A = 0$ in the regions where only one type of moments exists. The values of reduction coefficients is selected in such a way as to maximize the limit load without violating the limit state condition at any cross-section in the grid. Having found second differences and employing the equilibrium equation we get

$$q = \frac{4}{N_{\text{I}}N_{\text{II}}}\Big[\frac{2N_{\text{I}}\gamma_{\text{II}}\lambda_{\text{II}}M_0^{\text{II}}}{N_{\text{II}}^2}(1+\nu_{\text{II}})+\frac{2N_{\text{II}}\gamma_{\text{I}}\lambda_{\text{I}}M_0^{\text{I}}}{N_{\text{I}}^2}(1+\nu_{\text{I}})+$$

$$+\frac{N_{\text{I}}\tilde{\lambda}_{\text{I}}\tilde{M}_0^{\text{I}}}{(N_{\text{II}}+\alpha_{\text{II}}-2)(N_{\text{I}}+\alpha_{\text{I}})}+\frac{N_{\text{II}}\tilde{\lambda}_{\text{II}}\tilde{M}_0^{\text{II}}}{(N_{\text{I}}+\alpha_{\text{I}}-2)(N_{\text{II}}+\alpha_{\text{II}})}\Big]. \tag{6.96}$$

Since the coefficients λ_A and $\tilde{\lambda}_A$ are functions of $\nu_A, \frac{N_I}{N_{II}}$ and $\frac{M_0^A}{\tilde{M}_0^A}$ they should be found for each specific case separately. The detailed analysis has been given in [25] and is not here repeated in full. To give an exemplary result we just cite that for $\nu_A \leqslant \frac{3}{10}$ we may set $\lambda_A = \tilde{\lambda}_A = 1.0$, for $\nu_A = \frac{1}{3} - \tilde{\lambda}_A = 0.97$, $\tilde{\lambda}_A = 1.0$, for $\nu_A = \frac{1}{2} - \lambda_A = 0.906$, $\tilde{\lambda}_A = 1.0$, for $\nu_A = \frac{7}{10} - \lambda_A = 0.905$, $\tilde{\lambda}_A = 1.0$ and for $\nu_A = 1.0 - \lambda_A = 0.86$, $\tilde{\lambda}_A = 1.0$, all for the case of $N_I = N_{II}$ and $M_0^A/\tilde{M}_0^A = 4$ (which corresponds to the rectangular cross-section with the depth twice the width, for instance).

From eqs. (6.92) and (6.96) we may try to assess the contribution of the bending as well as the twisting moments to the limit load the grid can carry. In particular, this should enable us to evaluate the common assumption that the influence of the twisting moments on the limit load is negligible. To avoid rather cumbersome calculations we restrict ourselves to the grid with the same parameters and an even number of beams in both families, $M_0^I = M_0^{II}$, $\tilde{M}_0^I = \tilde{M}_0^{II}$, $\alpha_I = \alpha_{II} = 0$, $\gamma_I = \gamma_{II} = 1$. Let us also assume that $N_I/N_{II} = 1$, $M_0^A/\tilde{M}_0^A = 4$. The percentage results for simply supported and clamped grids with square and rectangular layouts are shown in Tables 13 and 14. It follows

Table 13 **Contributions of bending and twisting moments to the collapse load for simply supported grid**

Layout	Contribution of bending moments %			Contribution of twisting moments %		
	$A = I$	$A = II$	both	$A = I$	$A = II$	both
Square $m = n$	44.5	44.5	89.0	5.5	5.5	11.0
Rectangle $2m = n$	47.0	40.0	87.0	3.0	10.0	13.0

that the contribution of twisting moments can be substantial and its neglection leads to a serious underestimation of the load carrying capacity of the grid.

To supply one more example of the application of the procedure employed in this section let us consider a polar grid with an arbitrary number of radial beams, clamped at the outer ends as shown in Fig. 113. Polar coordinates X_1, X_2 are used with the variables X_1 and X_2 taking at nodes the values: $X_1 = a_1 - 1, a_1, \ldots, a$; $X_2 = 0, 1, 2, \ldots, n$ where a_1, a denote arbitrary integers, $a_1 < a$, and $n = 2\pi/\alpha$ is the number of radial beams. The first step

Table 14 Contributions of bending and twisting moments to the collapse load for clamped grid

Layout	$\nu_{I} = \nu_{II} = \nu$	Contribution of bending moments %	Contribution of twisting moments %
Square $m = n$	0.3	91.2	8.8
	0.5	91.5	8.5
	1.0	93.3	6.7
Rectangle $m = 2n$	0.3	89.7	10.3
	0.5	90.0	10.0
	1.0	91.9	8.1

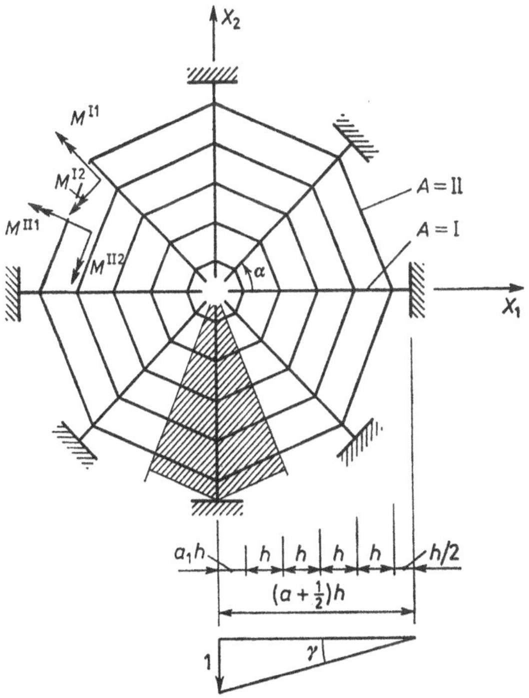

Fig. 113 Elastic-plastic polar grid

to obtain the lower bound estimate to the limit load is the transformation of the equilibrium equation (6.78) to the polar system. The derivation below is believed to be self-explanatory:

$$\varkappa(X_1, X_2) = X_1 h[\cos\alpha X_2, \sin\alpha X_2],$$

$$\Delta_{\mathrm{I}}\varkappa = h[\cos\alpha X_2, \sin\alpha X_2],$$

$$\Delta_{\mathrm{II}}\varkappa = 2X_1 h \sin\frac{\alpha}{2}\,[-\sin\alpha(X_2+\tfrac{1}{2}), \cos\alpha(X_2+\tfrac{1}{2})],$$

$$g_1 = g^1 = [\cos\alpha X_2, \sin\alpha X_2], \quad g_2 = g^2 = [-\sin\alpha X_2, \cos\alpha X_2],$$

$$\Delta_{\mathrm{II}}g_1 = 2\sin\frac{\alpha}{2}\,[-\sin\alpha(X_2+\tfrac{1}{2}), \cos\alpha(X_2+\tfrac{1}{2})],$$

$$\Delta_{\mathrm{II}}g_2 = -2\sin\frac{\alpha}{2}\,[\cos\alpha(X_2+\tfrac{1}{2}), \sin\alpha(X_2+\tfrac{1}{2})],$$

$$\Delta_{\mathrm{I}}g_1 = 0, \quad \Delta_{\mathrm{I}}g_2 = 0,$$
$$\bar{\Delta}_{\mathrm{I}}g_1 = 0, \quad \bar{\Delta}_{\mathrm{I}}g_2 = 0,$$

$$\bar{\Delta}_{\mathrm{II}}g_1 = 2\sin\frac{\alpha}{2}\,[-\sin\alpha(X_2-\tfrac{1}{2}), \cos\alpha(X_2-\tfrac{1}{2})],$$

$$\bar{\Delta}_{\mathrm{II}}g_2 = -2\sin\frac{\alpha}{2}\,[\cos\alpha(X_2-\tfrac{1}{2}), \sin\alpha(X_2-\tfrac{1}{2})],$$

$$l_{\mathrm{I}}^1 = h, \quad l_{\mathrm{I}}^2 = 0,$$

$$l_{\mathrm{II}}^1 = -2X_1 h \sin^2\frac{\alpha}{2}, \quad l_{\mathrm{II}}^2 = X_1 h \sin\alpha,$$

$$G_{\mathrm{II}1}^1 = -2\sin^2\frac{\alpha}{2}, \quad G_{\mathrm{II}2}^1 = -\sin\alpha,$$

$$G_{\mathrm{II}1}^2 = \sin\alpha, \quad G_{\mathrm{II}2}^2 = -2\sin^2\frac{\alpha}{2},$$

$$G_{1\mathrm{II}}^2 = 2\sin^2\frac{\alpha}{2}, \quad G_{2\mathrm{II}}^1 = -\sin\alpha,$$

$$G_{1\mathrm{II}}^2 = \sin\alpha, \quad G_{2\mathrm{II}}^2 = 2\sin^2\frac{\alpha}{2},$$

$$P^{\mathrm{II}} = 0,$$

$$M^{\mathrm{I}1} = 0, \quad M^{\mathrm{II}2} = M^{\mathrm{II}1}\tan\frac{\alpha}{2}$$

which finally yields

$$\bar{\Delta}_{\mathrm{I}} P^{\mathrm{I}} + F = 0,$$
$$\bar{\Delta}_{\mathrm{I}} M^{\mathrm{I}2} + \sin\alpha\left(1+\tan^2\frac{\alpha}{2}\right) M^{\mathrm{II}1} - \tfrac{1}{2} h(2P^{\mathrm{I}}+F) = 0 \tag{6.97}$$

where

$M^{\mathrm{I}2}$, $M^{\mathrm{II}1}$ are bending moments in the beam middle cross-sections ($\eta = 1/2$),

F is the load applied to the node located at (X_1, X_2),

$F(X_1) = qh^2 X_1 \sin\alpha,$

h is the constant length of radial beam segments.

By calculating the load spread on the n-th part of the grid as

$$\sum_{X_1=a_1}^{a} F(X_1) = \frac{qh^2}{2}\sin\alpha(a^2 - a_1^2 + a + a_1)$$

we have by eq. $(6.97)_1$

$$P^{\mathrm{I}}(X_1) = \frac{qh^2}{2}\sin\alpha(a_1^2 - a_1 - X_1^2 - X_1).$$

Substituting P^{I} into eq. $(6.97)_2$ we arrive at

$$\bar{\Delta}_{\mathrm{I}} M^{\mathrm{I}2} + M^{\mathrm{II}1}\left(1+\tan^2\frac{\alpha}{2}\right)\sin\alpha + \frac{qh^3}{2}\sin\alpha(X_1 - a_1^2 + a_1) = 0 \tag{6.98}$$

with the following boundary conditions

$$M^{\mathrm{I}2}(a_1 - 1) = 0, \quad M^{\mathrm{I}2}(a) = -\mu \bar{M}_0^{\mathrm{I}}$$

where

$\mu = 0$ corresponds to the simply supported grid,

$\mu = 1$ corresponds to the clamped grid.

Assuming that at collapse plastic hinges are formed in the middle cross-sections of the circumferential beams we take $M^{\mathrm{II}1} = M_0^{\mathrm{II}}\cos(\alpha/2)$ and eq. (6.98) assumes the form

$$\bar{\Delta}_{\mathrm{I}} M^{\mathrm{I}2} + 2M_0^{\mathrm{II}}\sin\frac{\alpha}{2} + \frac{qh^3}{2}\sin\alpha(X_1^2 - a_1^2 + a_1) = 0. \tag{6.99}$$

The only unknown $M^{\mathrm{I}2}$ is found to be

$$M^{\mathrm{I}2} = D - \frac{qh^3}{12}\sin\alpha[2X_1^3 + 3X_1^2 - X_1(6a_1^2 - 6a_1 - 1)] + $$
$$+ 2M_0^{\mathrm{II}} X_1 \sin\frac{\alpha}{2}. \tag{6.100}$$

On account of the boundary conditions, from eq. (6.100) first the constant D and finally the limit load q is obtained as

$$q = \frac{12\left[\mu \overline{M}_0^{\mathrm{I}} + 2M_0^{\mathrm{II}} \sin\frac{\alpha}{2}(a - a_1 + 1)\right]}{h^3 \sin\alpha [a(a+1)(2a+1) + a_1(a_1 - 1)(4a_1 - 6a - 5)]} . \tag{6.101}$$

This expression gives the lower bound to the collapse load provided the fields $M^{\mathrm{I}2}$ and $M^{\mathrm{II}1}$ are statically admissible. Without checking this we pass on to analyse the upper bound estimate. Assuming a kinematically admissible collapse mechanism as shown in Fig. 113 we write the virtual work equation for the n-th part of the grid rotating as the rigid body. The work of external load is expressed as

$$L_{\mathrm{ext}} = \sum_{X_1 = a_1}^{a} qh^2 X_1 \sin\alpha \frac{a + \frac{1}{2} - X_1}{a + \frac{1}{2}}$$

$$= qh^2 \sin\alpha \left[\sum_{X_1 = a_1}^{a} X_1 - \frac{1}{a + \frac{1}{2}}\left(\sum_{X_1 = 1}^{a} X_1^2 - \sum_{X_1 = 1}^{a_1 - 1} X_1^2\right)\right]$$

$$= qh^2 \sin\alpha \Big\{3a^2 - 3a_1^2 + 3a_1 + 3a -$$

$$- \frac{2}{2a - 1}[a(a+1)(2a+1) - a_1(a-1)(2a_1 - 1)]\Big\}$$

whereas the work dissipated in the same part of the grid can be found to be

$$L_{\mathrm{int}} = 2M_0^{\mathrm{II}} \sin\frac{\alpha}{2} \dot{\gamma}(a - a_1 - 1) + \mu \overline{M}_0^{\mathrm{I}} \dot{\gamma}$$

$$= \frac{1}{(a + \frac{1}{2})h} [2M_0^{\mathrm{II}} \sin\alpha (a - a_1 + 1) + \mu \overline{M}_0^{\mathrm{I}}]$$

where γ is shown in Fig. 113. Setting $L_{\mathrm{ext}} = L_{\mathrm{int}}$ leads to the same expression as that given by eq. (6.101) which clearly means that the exact solution to the limit load problem has been obtained.

Specifying in eq. (6.101) the values μ and a_1 we may get the limit loads for simply supported and clamped grids without and with the central hole.

We conclude this section by saying that the analytical tools developed in Sec. 4.6 for the analysis of elastic lattice-type structures within the framework of discrete modelling appear to be effective also in problems of limit analysis. Applications of the continuous model to the analysis of problems of this type will be given in Sec. 6.3.4.

So far in this section we have been dealing with plane grids modelled as discrete lattices made of rigid-perfectly plastic material. The approach will now be generalized to include the elastic phase of the material behaviour. Since such a formulation becomes analytically much more complex, only the numerically-oriented approach will be discussed.

The elastic-plastic analysis of lattice-type plates follows closely that presented earlier in this book when discussing the incremental analysis of any inelastic structural mechanics problems, cf. Secs. 4.10, 5.3, 6.2.1, for instance. In particular, the analysis of frames presented in Sec. 6.2.3 has been based on exactly the same solution algorithm. This means that we are now to employ the step-by-step analysis also for the inelastic grids assuming accordingly that plastic zones may form at one or two beam ends, cf. Fig. 76. Assuming that now the limit state conditions are functions of the bending and torsion moments, and possibly the shear forces, the sequence of transformations that leads to the derivation of the tangent stiffness matrix for the beam element

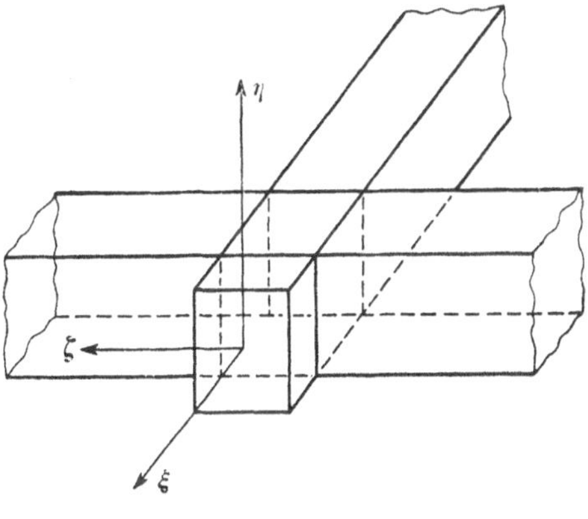

Fig. 114 Generalized displacements for grid beam

$$\mathbf{q}_{6\times 1} = \{\mathbf{q}^{(1)}_{3\times 1} \;\; \mathbf{q}^{(2)}_{3\times 1}\}, \qquad \mathbf{q}^{(\alpha)}_{3\times 1} = \{u^{(\alpha)}_{\eta} \;\; \varphi^{(\alpha)}_{\xi} \;\; \varphi^{(\alpha)}_{\zeta}\}, \qquad \alpha = 1, 2$$

with the plastic zone formed at the "left" end ($\alpha = 1$) may briefly be recalled as follows, cf. eqs. (6.16)–(6.33) and Fig. 114:

$$\Delta \mathbf{q} = \Delta \mathbf{q}^{(el)} + \Delta \mathbf{q}^{(pl)},$$

$$\Delta \mathbf{Q} = \begin{bmatrix} \Delta \mathbf{Q}^{(1)} \\ \Delta \mathbf{Q}^{(2)} \end{bmatrix} = \begin{bmatrix} \mathbf{k}^{(el)}_{11} & \mathbf{k}^{(el)}_{12} \\ \mathbf{k}^{(el)}_{12} & \mathbf{k}^{(el)}_{22} \end{bmatrix} \begin{bmatrix} \Delta \mathbf{q}^{(1)(el)} \\ \Delta \mathbf{q}^{(2)(el)} \end{bmatrix},$$

$$\Delta \mathbf{q}^{(1)(pl)} = \Delta \lambda_1 \frac{\partial f^{(1)}}{\partial \mathbf{Q}^{(1)T}} = \Delta \lambda_1 \mathbf{V}^{(1)},$$

$$\Delta \mathbf{q}^{(2)(pl)} = \mathbf{0},$$

$$\begin{bmatrix}\Delta\mathbf{Q}^{(1)}\\ \Delta\mathbf{Q}^{(2)}\end{bmatrix} = \begin{bmatrix}\mathbf{k}_{11}^{(el)} & \mathbf{k}_{12}^{(el)}\\ \mathbf{k}_{12}^{(el)} & \mathbf{k}_{22}^{(el)}\end{bmatrix}\begin{bmatrix}\Delta\mathbf{q}^{(1)}-\Delta\lambda_1\mathbf{V}^{(1)}\\ \Delta\mathbf{q}^{(2)}\end{bmatrix},$$

$$\Delta\lambda_1 = \frac{\mathbf{V}^{(1)T}\left[\mathbf{k}_{11}^{(el)}\ \ \mathbf{k}_{12}^{(el)}\right]}{\mathbf{V}^{(1)T}\mathbf{k}_{11}^{(el)}\mathbf{V}^{(1)}}\begin{bmatrix}\Delta\mathbf{q}_1\\ \Delta\mathbf{q}_2\end{bmatrix},$$

$$\Delta\mathbf{Q} = \begin{bmatrix}\mathbf{k}_{11}^{(el)} - \dfrac{\mathbf{k}_{11}^{(el)}\mathbf{V}^{(1)}\mathbf{V}^{(1)T}\mathbf{k}_{11}^{(el)}}{c_1} & \mathbf{k}_{12}^{(el)} - \dfrac{\mathbf{k}_{11}^{(el)}\mathbf{V}^{(1)}\mathbf{V}^{(1)T}\mathbf{k}_{12}^{(el)}}{c_1}\\[2ex] \mathbf{k}_{12}^{(el)} - \dfrac{\mathbf{k}_{12}^{(el)}\mathbf{V}^{(1)}\mathbf{V}^{(1)T}\mathbf{k}_{11}^{(el)}}{c_1} & \mathbf{k}_{22}^{(el)} - \dfrac{\mathbf{k}_{12}^{(el)}\mathbf{V}^{(1)}\mathbf{V}^{(1)T}\mathbf{k}_{12}^{(el)}}{c_1}\end{bmatrix}\Delta\mathbf{q},$$

$$c_1 = \mathbf{V}^{(1)T}\mathbf{k}_{11}^{(el)}\mathbf{V}^{(1)},$$

$$\Delta\mathbf{Q} = {}^1\mathbf{k}^{(e-p)}\Delta\mathbf{q}.$$

The way the incremental analysis based on the concept of updating the tangent stiffness matrix is performed was described earlier in this book. Thus, we shall confine ourselves now to just giving a few numerical illustrations, all of them based on the limit state condition in the form of a circle in the space of the bending and twisting moments, cf. eq. (6.89).

Example 1. Orthogonal grid composed of 17 single elements, simply supported at the outer edges and uniformly loaded at the nodes, Fig. 115(a). The development of successive plastic zones and the mode of failure are shown in Figs. 115(b)–115(f).

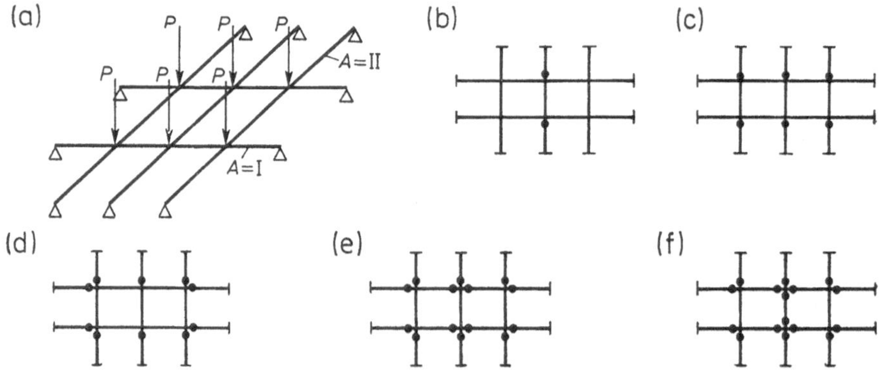

Fig. 115 Successive active zones and mode of failure for the orthogonal grid composed of 17 elements; $E_{\mathrm{I}} = 17658$ MPa, $E_{\mathrm{II}} = 17658$ MPa; $A_{\mathrm{I}} = 450\ \mathrm{cm}^2$, $A_{\mathrm{II}} = 800\ \mathrm{cm}^2$; $J_{\mathrm{I}} = 45000\ \mathrm{cm}^4$, $J_{\mathrm{II}} = 106667\ \mathrm{cm}^4$; $J_{\mathrm{I0}} = 47040\ \mathrm{cm}^4$, $J_{\mathrm{II0}} = 73280\ \mathrm{cm}^4$; ${}^+\tilde{M}_0^{\mathrm{I}} = 75.537 \times 10^5$ Ncm, ${}^+\tilde{M}_0^{\mathrm{II}} = 122.625 \times 10^5$ Ncm; ${}^-\tilde{M}_0^{\mathrm{I}} = 39.827 \times 10^5$ Ncm, ${}^-\tilde{M}_0^{\mathrm{II}} = 76.910 \times 10^5$ Ncm; $M_0^{\mathrm{I}} = 46.107 \times 10^5$ Ncm, $M_0^{\mathrm{II}} = 65.727 \times 10^5$ Ncm. (b) $P = 89.398$ kN, (c) $P = 103.525$ kN, (d) $P = 120.624$ kN, (e) $P = 120.712$ kN, (f) $P_{\mathrm{ult}} = 120.810$ kN

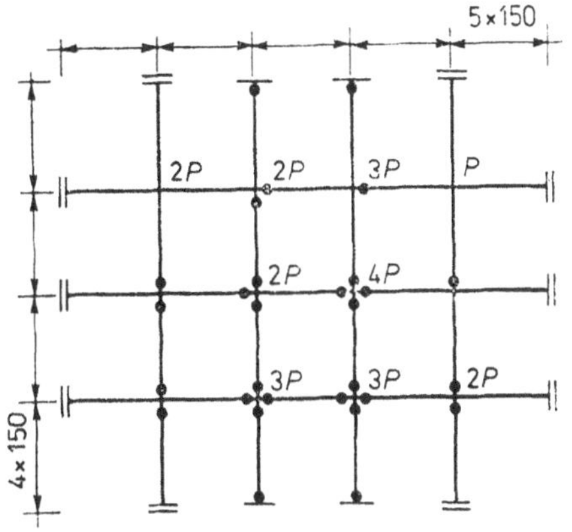

Fig. 116 Mode of failure for the unsymmetrically loaded orthogonal grid, $P_{ult.} = 0.52632 \dfrac{M_0}{P_L}$, $\dfrac{M_0}{\tilde{M}_0} = 4$

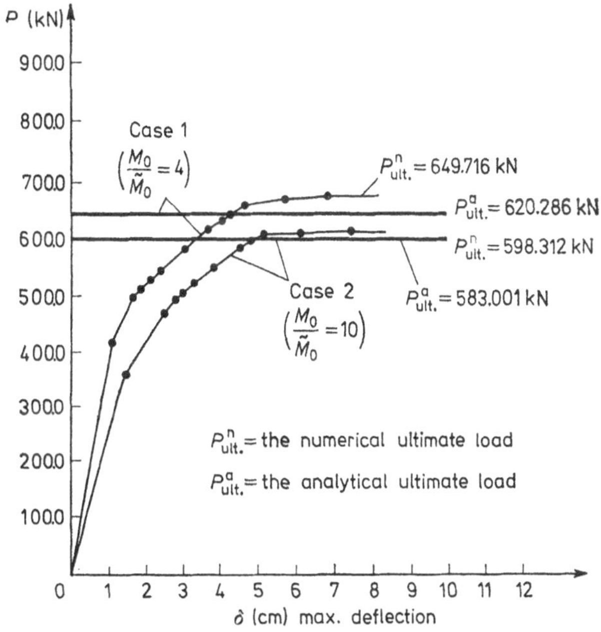

Fig. 117 Comparison of analytical and numerical results for the orthogonal grid system

Example 2. Orthogonal grid composed of 31 single elements with various boundary conditions and an unsymmetric loading. All the geometrical data and the Young modulus are the same as those given in Fig. 115 for A = II; ${}^{+}\tilde{M}_0^{\mathrm{I}} = {}^{-}\tilde{M}_0^{\mathrm{I}} = {}^{+}\tilde{M}_0^{\mathrm{II}} = {}^{-}\tilde{M}_0^{\mathrm{II}} = \tilde{M}_{\mathrm{I}0} = \tilde{M}_{\mathrm{II}0} = 98.1 \times 10^5$ Ncm. The load and the mode of failure are shown in Fig. 116.

Example 3. Orthogonal grid with 4 beams in the X^1-direction and 3 beams in the X^2-direction, simply supported and uniformly loaded. The grid was solved numerically to compare the results with those obtained analytically earlier in this section. The comparisons are shown in Fig. 117 for the case $M_0^A/\tilde{M}_0^A = 4$ and $M_0^A/\tilde{M}_0^A = 10$, $A = 1, 2$.

6.3.4 Limit load of polar grids — continuous model

In this section the limit load analysis will be presented for some dense grids subjected to bending, using the continuous model of lattice-type structures developed in Sec. 4.6. Special attention will be focused on a particular class of grids, viz. annular and circular ones with polar nets of two families of beams with constant cross-sections subjected to a transverse, uniformly distributed load of intensity q. Following the notation employed in Sec. 4.6.2 (eqs. (4.109)), cf. Problems 4.15–4.17, the load pattern is thus given by $f = q = \text{const}$, $h^\alpha = 0$.

Let us parametrize the central plane of the grid by means of a polar set of coordinates the lines of which coincide with the axes of actual ribs, $X_1 = \varrho$, $X_2 = \varphi$, Fig. 118. Due to the rotational symmetry we have $v_\varrho = 0$, $v_\varphi \neq 0$

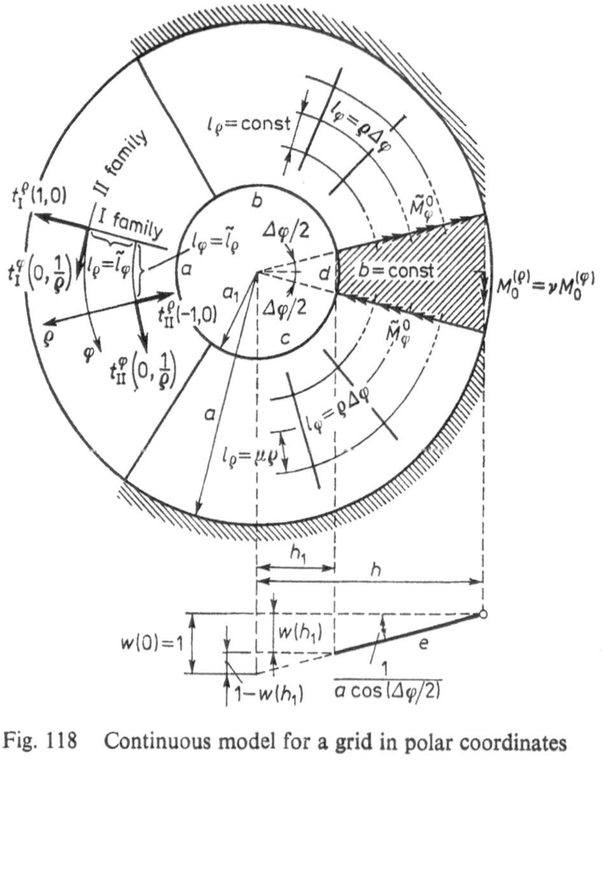

Fig. 118 Continuous model for a grid in polar coordinates

and $\frac{\partial(\cdot)}{\partial\varphi} = 0$, where v's denote the nodal rotations. It is also very easy to show that the only generalized strains that survive in the situation are $\varkappa_{\varrho\varphi}$, $\varkappa_{\varphi\varrho}$, γ_ϱ while the only non-zero components of the generalized stress state are $M^{\varrho\varphi}$, $M^{\varphi\varrho}$, P^ϱ.

The equilibrium equations of the continuous model of the plate on hand thus reduce to (cf. eqs. (4.109), Problem 4.17 and [167])

$$\begin{aligned} &T^\varrho{}_{,\varrho} + \overset{-1}{\varrho}\, T^\varrho + q = 0, \\ &M^{\varrho\varphi}{}_{,\varrho} + 2\overset{-1}{\varrho}\, M^{\varrho\varphi} + \overset{-1}{\varrho}\, M^{\varphi\varrho} - \overset{-1}{\varrho}\, P^\varrho = 0 \end{aligned} \tag{6.102}$$

where comma implies ordinary differentiation with respect to ϱ.

The bending moments and the shear forces as defined at the central cross-section of particular bars are, cf. Problem 4.17

$$\begin{aligned} M_{(\varrho)} &= \frac{\tilde{d}_{(\varphi)\alpha}\tilde{d}_{(\varrho)\beta}}{\tilde{d}_{(\varphi)\gamma}d^\gamma_{(\varrho)}}\, \tilde{l}_\varphi M^{\alpha\beta}, \\ M_{(\varphi)} &= \frac{\tilde{d}_{(\varrho)\alpha}\tilde{d}_{(\varphi)\beta}}{\tilde{d}_{(\varrho)\gamma}d^\gamma_{(\varphi)}}\, l_\varrho M^{\alpha\beta}, \\ \check{P}_{(\varrho)} &= \frac{\tilde{d}_{(\varphi)\alpha}}{\tilde{d}_{(\varphi)\beta}d^\beta_{(\varrho)}}\, \tilde{l}_\varrho P^\alpha \end{aligned} \tag{6.103}$$

the summation convention being implied over $\alpha, \beta, \gamma = \varrho, \varphi$ only, ϱ and φ denoting the families of bars. The vector bases together with their components in the polar coordinate system as well the spacings l, $\tilde{l}$ of bars are shown in Fig. 118, sector a.

Two geometrical configurations of the grid will be dealt with:

1. grid with evenly spaced circumferential ribs

$$l_\varrho = \tilde{l}_\varphi = l = \text{const},$$

$$l_\varphi = \tilde{l}_\varrho = \varrho\Delta\varphi, \quad \Delta\varphi = \text{const},$$

2. grid with geometrically similar segments

$$l_\varrho = \tilde{l}_\varphi = \mu\varrho, \quad \mu = \text{const},$$

$$l_\varphi = \tilde{l}_\varrho = \varrho\Delta\varphi, \quad \Delta\varphi = \text{const},$$

in which $l_\varrho/l_\varphi = \mu/\Delta\varphi$ remains constant throughout the grid.

In the first case we have, cf. eqs. (6.103)

$$M_{(\varrho)} = \varrho^2 \Delta\varphi M^{\varrho\varphi},$$
$$M_{(\varphi)} = l\varrho M^{\varphi\varrho}, \tag{6.104}$$
$$\check{P}_{(\varrho)} = \varrho\Delta\varphi P^{\varrho}$$

whereas in the second one we obtain

$$M_{(\varrho)} = \varrho^2 \Delta\varphi M^{\varrho\varphi},$$
$$M_{(\varphi)} = \varrho^2 \mu M^{\varphi\varrho}, \tag{6.105}$$
$$\check{P}_{(\varrho)} = \varrho\Delta\varphi P^{\varrho}.$$

For convenience the indices I, II in tensorial quantities have been replaced by ϱ, φ.

It is seen that, as far as the midsections of bars are concerned, no torque is present. Nor there are shear forces in the central cross-sections of circumferential bars.

The problem of finding a lower bound to the collapse load of the grid considered consists in solving two equilibrium equations (6.102) together with an appropriate yield condition $\bar{F}(M_{(\varrho)}, M_{(\varphi)}) = 0$, the influence of the shear force being neglected. The system of equations should be integrated under prescribed statical boundary conditions the unknown being: the distribution of the bending moments $M_{(\varrho)}$, $M_{(\varphi)}$, of the shear force $\check{P}_{(\varrho)}$ and the intensity of the limit load q.

Using the continuous plate analogy, the yield condition $\bar{F} = 0$ reduces, in the rotationally symmetric and orthogonal situation considered, to the requirement

$$|M_{(\varphi)}| = M_0^{(\varphi)} \tag{6.106}$$

to be satisfied in all the midsections of circumferential ribs, and to the requirement

$$|M_{(\varrho)}(\varrho^*)| = M_0^{(\varrho)} \tag{6.107}$$

to hold in the midsections of circumferential ribs along certain circles $\varrho = \varrho^*$, ϱ^* depending upon the geometry of the grid and the support conditions. $M_0^{(\varrho)}$ and $M_0^{(\varphi)}$ denote the yield moments.

Moreover, a check is necessary whether in any cross-section of circumferential ribs between their midspans and nodes the yield condition in simultaneous bending and twisting action is not violated,

$$\bar{F}[M_{(\varphi)}, \tilde{M}_{(\varphi)}] \leqslant 0 \tag{6.108}$$

where $M_{(\varphi)}$ denotes the bending moment and $\tilde{M}_{(\varphi)}$ stands for the twisting moment.

Let us first consider the grid with evenly spaced circumferential ribs, Fig. 118, sector b. The equilibrium equation $(6.102)_1$ under the condition

$$P^{\varrho}(a) = \frac{(a^2-a_1^2)q}{2a}$$

gives

$$P^{\varrho}_{\cdot}(\varrho) = \frac{q}{2a}(\varrho^2-a_1^2). \tag{6.109}$$

Eq. $(6.102)_2$ on substituting eq. (6.109) and denoting, from eq. (6.104), that

$$M^{\varrho\varphi} = \frac{M_{(\varrho)}}{\varrho^2\Delta\varphi} \equiv M, \quad M^{\varphi\varrho} = \frac{M_{(\varphi)}}{l\varrho}$$

takes the form

$$\frac{\mathrm{d}M}{\mathrm{d}\varrho} + \frac{2M}{\varrho} = \frac{q}{2\varrho^2}(\varrho^2-a_1^2) - \frac{M_{(\varphi)}}{l\varrho^2}. \tag{6.110}$$

The yield condition (6.106) is, under suitable sign convention,

$$M_{(\varphi)} = M_0^{(\varphi)} = \text{const} \tag{6.111}$$

which, put into eq. (6.110), yields after integrating the expression for bending moments in the radial ribs

$$M_{(\varrho)} = \Delta\varphi\left[C-\varrho\left(\frac{M_0^{(\varphi)}}{l}+\frac{qa_1^2}{2}\right)+\frac{q\varrho^3}{6}\right]. \tag{6.112}$$

The boundary conditions are

$$M_{(\varrho)}(a_1) = 0$$

and, according to eq. (6.107)

$$M_{(\varrho)}(a) = M_0^{(\varrho)} \equiv \nu M_0^{(\varphi)} \tag{6.113}$$

where ν denotes the ratio of the radial yield moment to the circumferential yield moment. Employing them in eq. (6.112), we finally get the intensity of the limit load

$$q = 6M_0^{(\varphi)}\cdot A\left(\frac{\nu}{\Delta\varphi}+\frac{a-a_1}{l}\right) \tag{6.114}$$

where

$$A = (a^3+2a_1^3-3a_1^2a)^{-1}$$

and the function for the bending moment in radial ribs

$$M_{(\varrho)} = \Delta\varphi M_0^{(\varphi)}\left[\frac{a_1-\varrho}{l} + \left(\frac{\nu}{\Delta\varphi} + \frac{a-a_1}{l}\right)R(\varrho)\right] \tag{6.115}$$

where

$$R(\varrho) = \frac{\varrho^3+2a_1^3-3a_1^2\varrho}{a^3+2a_1^3-3a_1^2 a}.$$

Eq. (6.115) together with eq. (6.111) describes the statically admissible distribution of bending moments at the limit equilibrium of the grid.

Now, the extremum of eq. (6.115) can be examined in order to find a critical value of $\Delta\varphi$ at which the yield condition (6.107) is satisfied along a certain circle with the radius ϱ^*, $a_1 < \varrho^* < a$. From $\frac{\mathrm{d}M_{(\varrho)}}{\mathrm{d}\varrho} = 0$ we get

$$\varrho^* = \left[\frac{a^3+\left(\frac{3\nu l}{\Delta\varphi}-a_1\right)a_1^2}{3\left(\frac{\nu l}{\Delta\varphi}+a-a_1\right)}\right]^{1/2}$$

and the condition (6.107) assumes the form

$$M_{(\varrho)}(\varrho^*) = \Delta\varphi M_0^{(\varphi)}\left[\frac{a_1-\varrho^*}{l} + \left(\frac{\nu}{\Delta\varphi} + \frac{a-a_1}{l}\right)R(\varrho^*)\right] \geqslant -M_0^{(\varrho)} = -\nu M_0^{(\varphi)}. \tag{6.116}$$

Denoting

$$t = \frac{\Delta\varphi}{\nu l} = \frac{2\pi s}{\nu(a-a_1)n}$$

where $n = 2\pi/\Delta\varphi$ is the total number of radial bars and $s = (a-a_1)/l$ is the total number of circumferential bar spacings, and introducing a new unknown

$$\beta = \frac{1}{ta}+1-\frac{a_1}{a}$$

the condition (6.116) leads to the third-degree equation in β:

$$3Aa^3(4Aa_1^3-1)\beta^3+6Aa^2(2Aa_1^4-2Aaa_1^3+a)\beta^2 - Aa(3a^2-4a_1^2)\beta+4/9 = 0. \tag{6.117}$$

Having found a suitable β we eventually get $t = [(\beta-1)a+a_1]^{-1}$ and

$$n_c = \frac{2\pi s}{\nu(a-a_1)t}.$$

This is the required number of radial bars as depending upon the number s of circumferential bar spacings. If $n < n_c$ the intensity of the limit load decreases and formula (6.114) no longer holds good. If $n > n_c$, this means a waste of material in radial ribs.

From the above solution some other cases can be specified:

1. Replacing eq. (6.113) by $M_{(\varrho)}(a) = 0$ we get the intensity of the limit load for a simply supported ring grid

$$q = \frac{6M_0^{(\varphi)}}{l}\bar{A} \tag{6.118}$$

where

$$\bar{A} = A(a-a_1) = (a^2+aa_1-2a_1^2)^{-1}.$$

The distribution of bending moments at limit equilibrium is expressed by eq. (6.111) and

$$M_{(\varrho)} = \frac{\Delta\varphi}{l}M_0^{(\varphi)}[a_1(1+2\bar{A}a_1^2)-\varrho(1+3\bar{A}a_1^2)+\bar{A}\varrho^3]. \tag{6.119}$$

Specification of the algebraic eq. (6.117) gives a closed form solution for the required number of radial bars

$$n_c = \frac{2\pi s}{\nu}\,\frac{(2/3\sqrt{3})(\gamma^2+\gamma+1)^{3/2}-\gamma(\gamma+1)}{(\gamma-1)(\gamma^2+\gamma-2)} \tag{6.120}$$

where $\gamma = a/a_1$. For example, assuming $\gamma = 10$, $\nu = 1$, $s = 9$, we obtain $n_c \cong 18$.

2. Assuming $a_1 \to 0$ we obtain the intensity of the limit load for a clamped, circular grid with discontinuous radial ribs at the pole

$$q = \frac{6M_0^{(\varphi)}}{a^3}\left(\frac{\nu}{\Delta\varphi}+\frac{a}{l}\right) \tag{6.121}$$

together with the distributions of bending moments

$$M_{(\varphi)} = M_0^{(\varphi)},$$

$$M_{(\varrho)} = \Delta\varphi M_0^{(\varphi)}\left[-\frac{\varrho}{l}+\left(\frac{\nu}{\Delta\varphi}+\frac{a}{l}\right)\frac{\varrho^3}{a^3}\right] \tag{6.122}$$

and the formula

$$n_c = \frac{2\pi s}{3\nu}$$

where $s = a/l$.

3. Assuming $M_{(\varrho)}(a) = 0$ and $a_1 = 0$ the results are obtained for a simply supported, circular grid with discontinuous radial ribs at the pole. These are:

$$q = \frac{6M_0^{(\varphi)}}{a^2 l},$$

$$M_{(\varphi)} = M_0^{(\varphi)}, \quad M_{(\varrho)} = \frac{\Delta\varphi}{l}M_0^{(\varphi)}\left(\frac{\varrho^3}{a^2}-\varrho\right), \tag{6.123}$$

$$n_c = \frac{4\pi s}{3\sqrt{3}\,\nu}.$$

In order to find the solution for a circular grid with radial ribs continuous over the pole, eq. (6.112) with $a_1 = 0$ must be solved under the conditions $M_{(\varrho)}(0) = -M_0^{(\varrho)}$, $M_{(\varrho)}(a) = (e-1)M_0^{(\varrho)}$ where $e = 2$ applies to the clamped edge and $e = 1$ to the simply supported edge. The solution supplies:

$$q = \frac{6M_0^{(\varphi)}}{a^3}\left(\frac{e\nu}{\Delta\varphi} + \frac{a}{l}\right),$$

$$M_{(\varphi)} = M_0^{(\varphi)}, \tag{6.124}$$

$$M_{(\varrho)} = \Delta\varphi M_0^{(\varphi)}\left[-\frac{\nu}{\Delta\varphi} - \frac{\varrho}{l} + \left(\frac{e\nu}{\Delta\varphi} + \frac{a}{l}\right)\frac{\varrho^3}{a^3}\right].$$

Since $M_{(\varrho)}$ has no extremum in the interval $0 \leqslant \varrho \leqslant a$, there is no critical number of radial ribs required for the formula (6.124) to hold true. The limit load intensity directly depends on $\Delta\varphi$ with no associated condition for n_c.

Let us consider next the grid with geometrically similar segments, Fig. 118, sector c.

Eq. $(6.102)_1$, remembering eq. (6.109) and denoting from eq. (6.105) that

$$M^{\varrho\varphi} = \frac{M_{(\varrho)}}{\varrho^2\Delta\varphi} \equiv M, \quad M^{\varphi\varrho} = \frac{M_{(\varphi)}}{\varrho^2\mu}$$

assumes the form

$$\frac{dM}{d\varrho} + \frac{2M}{\varrho} = \frac{q}{2\varrho^2}(\varrho^2 - a_1^2) - \frac{M_{(\varphi)}}{\varrho^3\mu}. \tag{6.125}$$

Using eq. (6.111), we obtain the distribution of bending moments in radial ribs

$$M_{(\varrho)} = \Delta\varphi\left(C - \frac{M_0^{(\varphi)}}{\mu}\ln\frac{\varrho}{a_1} - \frac{qa_1^2\varrho}{2} + \frac{q\varrho^3}{6}\right). \tag{6.126}$$

The boundary conditions (6.113) lead to the intensity of the limit load

$$q = 6M_0^{(\varphi)}A\left(\frac{\nu}{\Delta\varphi} + \frac{1}{\mu}\ln\frac{a}{a_1}\right) \tag{6.127}$$

while

$$M_{(\varrho)} = \Delta\varphi M_0^{(\varphi)}\left[\frac{R\nu}{\Delta\varphi} + \frac{1}{\mu}\ln\frac{a_1}{\varrho}\left(\frac{a}{a_1}\right)^R\right], \tag{6.128}$$

A and $R(\varrho)$ having been given before.

Examination of $\dfrac{dM_{(\varrho)}}{d\varrho} = 0$ results in a third-degree algebraic equation with one real root ϱ^*. Now the yield condition (6.107) is required to be satisfied in the form

$$M_{(\varrho)}(\varrho^*) = -\nu M_0^{(\varphi)} \tag{6.129}$$

which leads to the determination of the least number of radial ribs necessary for the formula (6.126) to hold good. No explicit solution to the equation (6.129) can be arrived at. However, considering a simply supported grid (which means putting $\nu = 0$ into eqs. (6.127) and (6.128)) we obtain a closed form result

$$n_c = -\frac{2\pi}{\mu\nu} \ln \frac{a_1}{\varrho^*} \gamma^{R^*}$$

where

$$\gamma = a/a_1, \qquad R^* = R(\varrho^*),$$

$$\varrho^* = a_1(\varkappa_1 + \varkappa_2),$$

$$\varkappa_1, \varkappa_2 = \left\{ \frac{\gamma^3 - 3\gamma + 2}{6 \ln \gamma} \pm \frac{1}{3} \left[\left(\frac{\gamma^3 - 3\gamma + 2}{2 \ln \gamma} \right)^2 - \frac{1}{3} \right]^{1/2} \right\}^{1/3}.$$

For example, assuming $\gamma = 2$, $\nu = 1$, $\mu = 0.05$ we obtain $\varrho^* = 1.52a_1$, $R^* = 0.235$, $n_c \cong 32$.

In order to ascertain whether or not the results of statical approach just presented are in fact lower bounds to the collapse of the grids considered, a check is necessary of the bending-twisting yield condition (6.108) being not violated along the circumferential ribs between their midsections and nodal sections. To this end, a free body diagram shown in Fig. 119 must be examined to find the bending and twisting moments $M_{(\varphi)}^{(n)}$, $\tilde{M}_{(\varphi)}^{(n)}$ acting simultaneously at the nodal cross-section of circumferential beam. Assuming the most drastic way of transmitting the loading $q = \text{const}$ to the node (i.e. that the load is acting on each node through the circumferential bar only) we obtain at the limit equilibrium

$$\begin{aligned} M_{(\varphi)}^{(n)} &= M_0^{(\varphi)} \cos \frac{\Delta\varphi}{2} - \frac{q\varrho^2 l}{8} \Delta\varphi^2 \cos \frac{\Delta\varphi}{8}, \\ \tilde{M}_{(\varphi)}^{(n)} &= \tilde{M}_0^{(\varphi)} \sin \frac{\Delta\varphi}{2} - \frac{q\varrho^2 l}{8} \Delta\varphi^2 \sin \frac{\Delta\varphi}{8}. \end{aligned} \tag{6.130}$$

Let us confine ourselves to the simply supported annular grid. For evenly spaced circumferential bars the limit load intensity is given by eq. (6.118). For geometrically similar segments the limit load amounts to the value (6.127) with $\nu = 0$ put in. Substituting those intensities of the limit load into eq. (6.130)

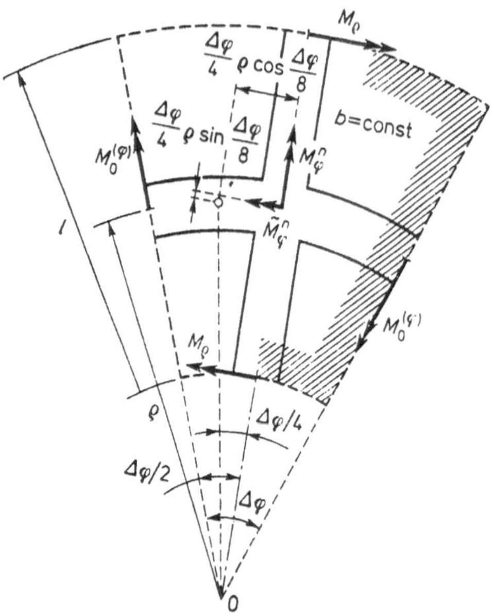

Fig. 119 Free body diagram for the polar grid

and in the latter case taking $l = \mu\varrho$, we get the relationships in terms of dimensionless moments

$$m_{(\varphi)}^{(n)} = \frac{M_{(\varphi)}^{(n)}}{M_0^{(\varphi)}} = \cos\frac{\Delta\varphi}{2} - D\Delta\varphi^2\cos\frac{\Delta\varphi}{8},$$

$$\tilde{m}_{(\varphi)}^{(n)} = \frac{\tilde{M}_{(\varphi)}^{(n)}}{\tilde{M}_0^{(\varphi)}} = \varepsilon\left(\sin\frac{\Delta\varphi}{2} - D\Delta\varphi^2\sin\frac{\Delta\varphi}{8}\right) \qquad (6.131)$$

where

$$D = D(\varrho, \gamma) = \frac{3(\varrho/a_1)^2}{4(\gamma^2+\gamma-2)}$$

applies to the evenly spaced loop bars and

$$D = D(\varrho, \gamma) = \frac{3(\varrho/a_1)^3\ln\gamma}{4(\gamma^3-3\gamma+2)}$$

corresponds to the geometrically similar segments. The coefficient $\varepsilon = \frac{M_0^{(\varphi)}}{\tilde{M}_0^{(\varphi)}}$ stands for the ratio of the yield moment in pure bending to the yield moment in pure torsion at the nodal cross-section of a circumferential rib and depends upon the cross-sectional geometry and the yield condition which is obeyed by the material.

As no confusion can arise, in the rest of this consideration the superscript n denoting nodal moments will be suppressed.

Eqs. (6.131) constitute a parametric (with respect to $\Delta\varphi$) description of a dimensionless moment profile at a generic node of the grid. The resulting relationship $f(m_{(\varphi)}, \tilde{m}_{(\varphi)}) = 0$ should now be compared with the plastic interaction curve (limit state condition) $\bar{F}(M_{(\varphi)}, \tilde{M}_{(\varphi)}) = 0$ to find out whether, in the $m_{(\varphi)}$, $\tilde{m}_{(\varphi)}$-plane, the former lies inside the latter for a sufficiently wide interval $m^c_{(\varphi)} < m_{(\varphi)} < 1$, or in terms of the geometry of the grid, $0 < \Delta\varphi < \Delta\varphi^c$. As eqs. (6.131) are the function of $D(\varrho, \gamma)$, an analysis must be made to answer at which point ϱ of the grid and for what value of $\gamma = a/a_1$ the situation is most critical. Examining the sign of slope of the function $g(D) - \tilde{m}_{(\varphi)}/(1-m_{(\varphi)})$, cf. Fig. 120, as a measure of the shortest distance between

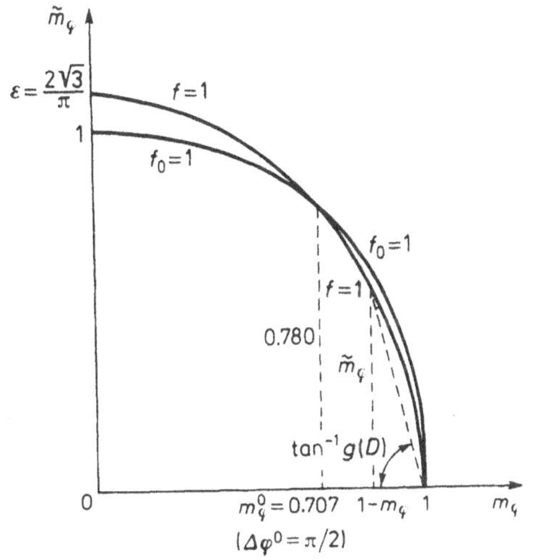

Fig. 120 Check of the bending/twisting limit state condition

the curves $f = 0$ and $\bar{F} = 0$, and finding that $\frac{\mathrm{d}}{\mathrm{d}D} g(D) < 0$ it is concluded that, for an annular grid (with finite γ) D is to be as small as possible, $\varrho = a_1$. The most drastic case is clearly a circular grid, $\gamma \to \infty$, $D \to 0$ and eqs. (6.131), on eliminating $\Delta\varphi$, yield the elliptical moment profile

$$f = m^2_{(\varphi)} + \frac{\tilde{m}^2_{(\varphi)}}{\varepsilon} - 1 = 0 \tag{6.132}$$

to be compared with the plastic interaction curve $\bar{F} = 0$.

The problem of finding $\bar{F} = 0$ in combined bending and torsion of an arbitrary cross-section is outside the scope of this book, not to mention the difficulties to be encountered, cf. Sec. 6.2.2.

Let us be specific and assume a uniform circular cross-section of loop ribs with the radius r and the material to obey the Huber–Mises yield condition. Then we have $M_0^{(\varphi)} = \frac{4}{3} r^3 \sigma_y$, $\tilde{M}_0^{(\varphi)} = \frac{2}{3} \pi r^3 \tau_y$, $\tau_y = \sigma_y/\sqrt{3}$ which combined gives $\varepsilon = 2\sqrt{3}/\pi \cong 1.105$. The limit state condition for a circular bar is found to have the form, [187]:

$$\bar{F} = 0.8845 \tilde{m}_{(\varphi)}^2 + 0.9042 m_{(\varphi)}^2 + 0.1155 \tilde{m}_{(\varphi)}^4 + 0.0958 m_{(\varphi)}^4 - 1 = 0. \tag{6.133}$$

The comparison of eqs. (6.132) and (6.133) presented diagramatically in Fig. 120 shows, the detailed calculations being omitted, that the curve $f = 0$ lies inside, however near, the curve $\bar{F} = 0$ for $0.707 < m_{(\varphi)} < 1$. (Note that since the distance between $f = 0$ and $\bar{F} = 0$ is very slight indeed the juxtaposition of the curves in Fig. 120 is an exaggeration). From eq. $(6.131)_1$ under $D = 0$ it follows that $m_{(\varphi)} = \cos\dfrac{\Delta\varphi}{2}$ hence $0 < \Delta\varphi < \varphi/2$, i.e. $4 < n < \infty$ secures the required solution. Thus, there is no danger of violating the yield condition (6.108) and the intensities of the limit load found previously are associated with statically admissible moment distributions.

Fig. 118, sector d, shows a basic, rigid element of the grid with the belonging edge yield moments (upper edge corresponds to the evenly spaced loop ribs, lower edge is loaded with the yield moments existing in the similar segment grid). Fig. 118, e, presents a conical velocity profile compatible with the moment profile $M_{(\varphi)} = M_0^{(\varphi)}$, $-M_0^{(\varrho)} \leqslant M_{(\varrho)} \leqslant M_0^{(\varrho)}$. Omitting all the necessary algebra involved in the virtual work equation, the upper bound limit load for the grid with evenly spaced loop bars is found to be

$$q = \frac{6M_0^{(\varphi)}\left(\dfrac{\nu}{2\sin(\Delta\varphi/2)} + \dfrac{a-a_1}{l}\right)}{\cos^2(\Delta\varphi/2)(a^3+2a_1^3-3a_1^2 a)}. \tag{6.134}$$

For the grid with geometrically similar segments the resultant yield bending moment acting along the radial edge of a rigid element is

$$\int_{a_1}^{a} \frac{M_0^{(\varphi)}}{\mu\varrho}\, d\varrho = \frac{M_0^{(\varphi)}}{\mu} \ln\frac{a}{a_1}$$

instead of $\dfrac{M_0^{(\varphi)}}{l}(a-a_1)$ as it was the case with the grid with $l = \text{const}$. Replacement of those terms in eq. (6.134) leads to the upper bound limit load for the grid with geometrically similar segments

$$q = \frac{6M_0^{(\varphi)}\left(\dfrac{\nu}{2\sin(\Delta\varphi/2)} + \dfrac{1}{\mu}\ln\dfrac{a}{a_1}\right)}{\cos^2(\Delta\varphi/2)(a^3+2a_1^3-3a_1^2 a)}. \tag{6.135}$$

For sufficiently dense grids we have

$$\Delta\varphi \ll 1, \quad \sin\frac{\Delta\varphi}{2} \to \frac{\Delta\varphi}{2}, \quad \cos\frac{\Delta\varphi}{2} \to 1.$$

It is readily seen that eq. (6.134) is now identical with eq. (6.114) and eq. (6.135) is the same as eq. (6.127). Thus the solutions (6.134) and (6.128) to the limit load of polar grids appear to be exact ones in the sense of limit analysis. The differences between the lower and upper assessments of collapse load (6.114) and (6.134) for a number of finite values of $\Delta\varphi$ are shown in Fig. 121 assuming $a = 10$, $a_1 = 1$, $\nu = 1$. To make the limit loads comparable

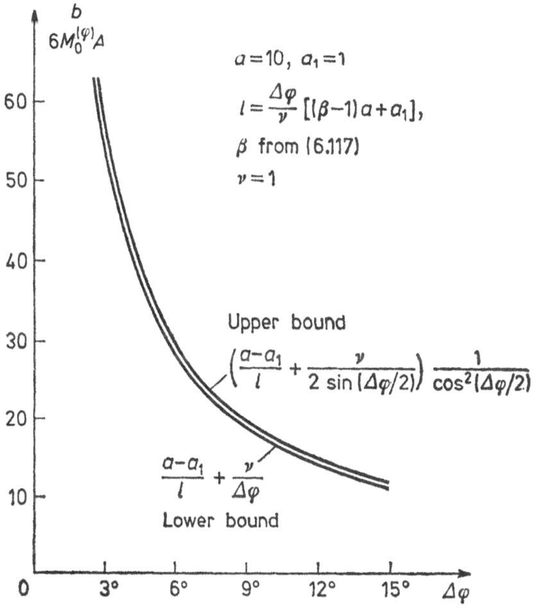

Fig. 121 Upper vs lower bounds for the polar grid

each value of $\Delta\varphi$ was first associated with suitable value of l by means of eqs. (6.117) and (6.116). The differences between the two bounds do not exceed 1.9 per cent for $\Delta\varphi \leqslant 15°$. For a simply supported grid $\dfrac{q_{\text{kin}}}{q_{\text{stat}}} = \cos^2(\Delta\varphi/2)$ not depending upon the spacing l at all.

6.3.5 *Minimum weight design of ideally plastic rectangular dense grid*

A hinge supported, rectangular, dense grid will now be studied composed of two rectilinear families of perfectly plastic bars and subjected to a transverse statical load. The material of the bars is assumed to obey either the Huber–

Mises or the Tresca yield condition. The grid bars are capable of supporting bending as well torque.

In the previous section we have applied the continuous model of dense lattice type plates to study the limit load problem of perfectly plastic grids with no necessity of introducing a large number of discrete unknowns as it is usually the case with bar structures. Following this approach, lower bound to the intensity of the collapse load is studied now by assuming reasonable statically admissible fields of bending and twisting moments. For the system of two rectilinear families of bars the magnitude of the collapse load is sought as dependent upon their inclinations to the edges of the grid, the distance between nodes being kept constant. Optimum configuration of bars will be found based on the requirement of a prescribed collapse load associated with minimum weight of the structure.

Two limiting cases of the rectangular grid will be examined: a square grid (in which the network of bars becomes orthogonal) and a strip (which appears to consist of an infinite number of parallel, vanishingly narrow ribs the ultimate bending moment per unit length of the grid remaining, however, finite and coinciding with the classical result for a solid strip under bending).

Both the continuous description of a discrete body and the assumed fields of moments in the grid are clearly approximations to the real situation. However, the finer the grid network the more exact the obtained results appear to become.

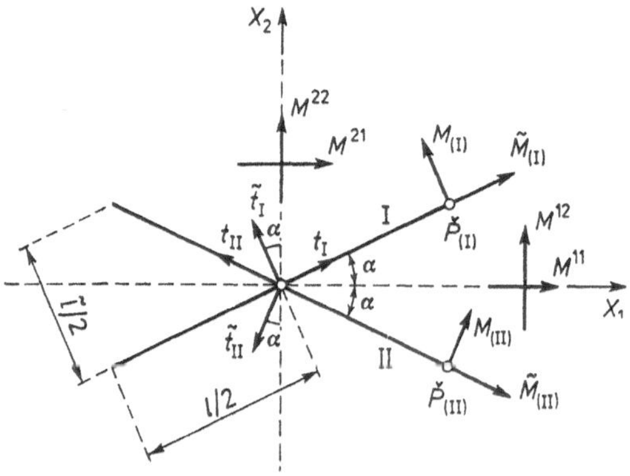

Fig. 122 Typical segment of a rectangular grid

Consider a grid whose typical segment is shown in Fig. 122. The central, principal axes of inertia of the cross-sections of all the bars are assumed to be tangent (normal) to the plane generated by the longitudinal axes of both families referred to as the central plane of the grid. The grid is subjected to the transverse, uniformly distributed load of intensity $q = \text{const}$ per unit area of the central plane.

Let us parametrize the central plane of the grid by a set of Cartesian coordinates X_1, X_2. The local vector bases associated with either family of ribs are shown in Fig. 122 and have the following components as dependent on the angle of inclination α:

$$\mathbf{d}_{(I)} = (\cos\alpha, \sin\alpha), \qquad \tilde{\mathbf{d}}_{(I)} = (-\sin\alpha, \cos\alpha),$$
$$\mathbf{d}_{(II)} = (-\cos\alpha, \sin\alpha), \quad \tilde{\mathbf{d}}_{(II)} = (-\sin\alpha, \cos\alpha)$$

(the angle α should not be confused with the index α running over 1, 2).

The relationships between the densities of generalized internal forces $M^{\alpha\beta}$, P^{α} $(\alpha, \beta = 1, 2)$ the internal beam moments and forces $M_{(A)}$, $\tilde{M}_{(A)}$, $\check{P}_{(A)}$ $(A = \mathrm{I}, \mathrm{II})$ at the midsections of particular bars, the vector bases $\mathbf{d}_{(A)}$, $\tilde{\mathbf{d}}_{(A)}$ and the spacing of the bars $\tilde{l} = l\sin 2\alpha$ have the form:

for the midspan bending moments

$$\begin{aligned} M_{(I)} &= \frac{l\sin 2\alpha}{2}\left(M^{12} - M^{21} - M^{11}\tan\alpha + M^{22}\frac{1}{\tan\alpha}\right), \\ M_{(II)} &= \frac{l\sin 2\alpha}{2}\left(M^{12} - M^{21} + M^{11}\tan\alpha - M^{22}\frac{1}{\tan\alpha}\right), \end{aligned} \tag{6.136}$$

for the midspan twisting moments

$$\begin{aligned} \tilde{M}_{(I)} &= \frac{l\sin 2\alpha}{2}\left(M^{12}\tan\alpha + M^{21}\frac{1}{\tan\alpha} + M^{11} + M^{22}\right), \\ \tilde{M}_{(II)} &= \frac{l\sin 2\alpha}{2}\left(-M^{12}\tan\alpha - M^{21}\frac{1}{\tan\alpha} + M^{11} + M^{22}\right), \end{aligned} \tag{6.137}$$

for the midspan shear forces

$$\begin{aligned} \check{P}_{(I)} &= l(P^1\sin\alpha + P^2\cos\alpha), \\ \check{P}_{(II)} &= l(-P^1\sin\alpha + P^2\cos\alpha). \end{aligned} \tag{6.138}$$

The inverse formulae read:

for the bending moment densities

$$\begin{aligned} M^{12} &= \frac{1}{2l}\left[(M_{(I)} + M_{(II)})\frac{1}{\tan\alpha} + \tilde{M}_{(I)} - \tilde{M}_{(II)}\right], \\ M^{21} &= \frac{1}{2l}\left[-(M_{(I)} + M_{(II)})\tan\alpha + \tilde{M}_{(I)} - \tilde{M}_{(II)}\right], \end{aligned} \tag{6.139}$$

for the twisting moment densities

$$M^{11} = \frac{1}{2l}\left[-M_{(I)}+M_{(II)}+(\tilde{M}_{(I)}+\tilde{M}_{(II)})\frac{1}{\tan\alpha}\right],$$
$$M^{22} = \frac{1}{2l}[M_{(I)}-M_{(II)}+(\tilde{M}_{(I)}+\tilde{M}_{(II)})\tan\alpha]. \qquad (6.140)$$

Let us assign the index zero to the considered magnitudes in the limit state of a segment of the plastic grid and assume equal plastic moduli of both families of bars $M_0^{(I)} = M_0^{(II)} = M_0$, $\tilde{M}_0^{(I)} = \tilde{M}_0^{(II)} = \tilde{M}_0$. Thus, from eqs. (6.139) and (6.140) we get the ultimate densities of bending and twisting moments as dependent on the ultimate, "beamlike" moments M_0 and $\tilde{M}_0$

$$M_0^{12} = \frac{1}{l\tan\alpha}M_0, \quad M_0^{21} = -\frac{\tan\alpha}{l}M_0,$$
$$M_0^{11} = \frac{1}{l\tan\alpha}\tilde{M}_0, \quad M_0^{22} = \frac{\tan\alpha}{l}\tilde{M}_0. \qquad (6.141)$$

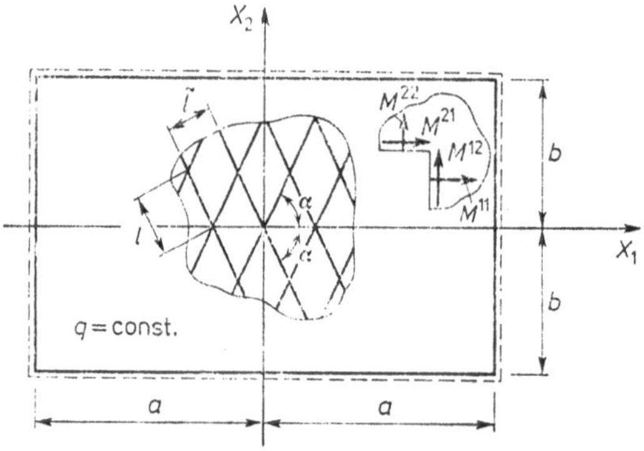

Fig. 123 Elastic-plastic rectangular grid — continuous model

Let us start with the analysis of the lower bound to the collapse load of the rectangular grid. The grid is shown in Fig. 123 in which the approximating moment densities are also shown. The equilibrium equations assume here the form

$$P^{\alpha}{}_{,\alpha}+F = 0, \quad M^{\alpha\beta}{}_{,\alpha}+\varepsilon^{\beta}_{\cdot\alpha}P^{\alpha} = 0.$$

Combined, these can be written explicitly as ($F = q = \text{const}$)

$$M^{12}{}_{,11}+(M^{22}-M^{11})_{,12}-M^{21}{}_{,22} = -q. \qquad (6.142)$$

Using again, cf. Sec. 6.3.4, the plate analogy, we assume the following statically admissible field of moment densities to apply in the considered case of regular, sufficiently dense, rectangular grid:

$$M^{12}(X_1, X_2) = A - BX_1^2, \quad M^{21}(X_1, X_2) = C - DX_2^2,$$
$$M^{11}(X_1, X_2) = EX_1X_2, \quad M^{22}(X_1, X_2) = FX_1X_2. \tag{6.143}$$

To find the six scalar coefficients, the boundary and the yield conditions must be employed. The former clearly are

$$M^{12}(\pm a, X_2) = 0, \quad M^{21}(X_1, \pm b) = 0.$$

The yield conditions, leading to a reasonably high lower bound to the limit load, are required to be satisfied in the centre of the grid

$$M^{12}(0,0) = M_0^{12}, \quad M^{21}(0,0) = -M_0^{21},$$

and at its corners, e.g.

$$M^{11}(a, b) = M_0^{11}, \quad M^{22}(a, b) = -M_0^{22}.$$

The statically admissible field of moment densities takes eventually the form

$$M^{12} = M_0^{12}\left[1 - \left(\frac{X_1}{a}\right)^2\right], \quad M^{21} = -M_0^{21}\left[1 - \left(\frac{X_2}{b}\right)^2\right],$$
$$M^{11} = M_0^{11}\frac{X_1X_2}{ab}, \quad M^{22} = -M_0^{22}\frac{X_1X_2}{ab}. \tag{6.144}$$

On substituting eqs. (6.144) into the equilibrium equation (6.142) we get the intensity of collapse load in terms of ultimate moment densities

$$q = \frac{2}{a^2}M_0^{12} + \frac{1}{ab}(M_0^{11} + M_0^{22}) + \frac{2}{b^2}M_0^{21} \tag{6.145}$$

or, on account of eqs. (6.141), expressed in terms of the two moduli of plastic strength of a constituent bar M_0, $\tilde{M}_0$

$$q = \frac{2M_0}{l}\left(\frac{1}{a^2\tan\alpha} + \frac{\tan\alpha}{b^2}\right) + \frac{\tilde{M}_0}{abl}\left(\frac{1}{\tan\alpha} + \tan\alpha\right). \tag{6.146}$$

Let the cross-sections of both families of bars be specified as narrow rectangles with constant depth $2h$ = const and variable width $2\gamma_{(\alpha)}h$ where $\gamma_{(\alpha)}$ is much smaller than unity. The ultimate moment under pure bending attains the value

$$M_0 = 2\sigma_y\gamma_{(\alpha)}h^3 = \frac{\sigma_y h}{2}A_{(\alpha)} \tag{6.147}$$

where σ_y denotes the yield point in tension or compression, while $A_{(\alpha)}$ is the cross-sectional area. The ultimate moment under pure torsion amounts to

$$\tilde{M}_0 = \frac{4}{\delta}\sigma_y\gamma_{(\alpha)}^2h^3 = \frac{\sigma_y\gamma_{(\alpha)}h}{\delta}A_{(\alpha)}. \tag{6.148}$$

The approximate formula (6.148) is arrived at by means of the so-called sand-hill analogy, the inclined planes of the "hill" near the (much) shorter sides of the rectangle being neglected.

The factor δ reflects the type of yield criterion obeyed by the material. Specifically, $\delta = \sqrt{3}$ applies to the Huber–Mises material, whereas $\delta = 2$ corresponds to the Tresca material. Let us conveniently introduce, for further reference, the ratio of ultimate moments

$$\varepsilon = \frac{M_0}{\tilde{M}_0} = \frac{\delta}{2\gamma_{(\alpha)}}. \tag{6.149}$$

We shall now pass to considering the minimum weight design problem for the grid at hand. Exact formula for the volume of the material per unit area of the central plane of the grid assumes, after some detailed calculations, the form

$$V_1 = \frac{2A_{(\alpha)}(l-\gamma_{(\alpha)}h/\sin 2\alpha)}{l^2 \sin 2\alpha}. \tag{6.150}$$

Introducing, for convenience, a reference orthogonal grid, $\alpha = \pi/4$, the cross-sectional area of a single bar is assumed to change, together with changing α, according to the rule

$$A_\alpha = A_{\pi/4} \sin 2\alpha. \tag{6.151}$$

As a consequence of the depth $2h$ being constant we have

$$\gamma_\alpha = \gamma_{\pi/4} \sin 2\alpha$$

where $\gamma_{\pi/4}$ is again the width coefficient for a bar of the reference grid. Now the volume of the material is given by the formula

$$V_1 = \frac{2A_{\pi/4}}{l^2}(l-\gamma_{\pi/4}h). \tag{6.152}$$

It can be readily seen that for narrow ribs we have $\gamma_{\pi/4}h \ll l$ and therefore the approximate formula $V_1 = 2A_{\pi/4}/l$ can be used in this case. The approximation consists in that the part of volume at a node—enclosed by the two intersecting bars—is being counted twice. On the other hand, for solid plate the relation $2\gamma_\alpha h = l \sin 2\alpha$ holds good and the volume clearly becomes $V_1 = 2h$.

The optimization problem can now be formulated as follows: Find the angle α corresponding to the prescribed collapse load associated with minimum weight, the distances between nodes being kept constant. Thus we have the condition

$$l = 2A_{\pi/4}/V_1 = \text{const} \tag{6.153}$$

applicable for the grid with narrow ribs. On combining eqs. (6.146) and (6.153) we obtain

$$V_1 = 2A_{\pi/4}q\left[2M_0\left(\frac{1}{a^2\tan\alpha}+\frac{\tan\alpha}{b^2}\right)+\frac{\tilde{M}_0}{ab}\left(\frac{1}{\tan\alpha}+\tan\alpha\right)\right]^{-1}.$$

Remembering eqs. (6.147), (6.148), (6.151), (6.152) and introducing the ratio $\beta = b/a$ of sides of the rectangular grid, we finally get the objective function in the dimensionless form

$$Z = \frac{V_1\sigma_y h}{2qb^2} = \left[\left(\beta\frac{1}{\tan\alpha}+\tan\alpha+\frac{2\beta\gamma_{\pi/4}}{\delta}\right)\sin 2\alpha\right]^{-1} \tag{6.154}$$

subject to the minimization with respect to the angle α. From $\frac{dZ}{d\alpha} = 0$ we get

$$\tan\alpha = \frac{1}{2\beta\gamma_{\pi/4}}\left[\delta(1-\beta^2)+\sqrt{\delta^2(1-\beta^2)^2+4\beta^2\gamma_{\pi/4}^2}\,\right] \tag{6.155}$$

where $\gamma_{\pi/4}$ is the reference width coefficient. The actual width coefficient can be calculated for the obtained angle α from the relation $\gamma_\alpha = \gamma_{\pi/4}\sin 2\alpha$.

The function Z is proportional to the (minimum) weight of the grid supporting a prescribed ultimate load q, whereas the function $1/Z$ is proportional to the (maximum) ultimate load associated with given volume V_1.

Let two apparent limiting cases be considered. For a square grid, $\beta = 1$, we get $\alpha = \pi/4$, and $1/Z = 2(1+\gamma_{\pi/4}/\delta)$. For an infinitely long strip with the width $2b$, $\beta = 0$, we obtain $\alpha = \pi/2$ and $1/Z = 2$.

Remembering eqs. (6.147), (6.148), (6.151), (6.152), the relationships (6.141) supply the following field of ultimate moment densities in the strip

$$M_0^{12} = 0, \quad M_0^{21} = \sigma_y h^2, \quad M_0^{11} = 0, \quad M_0^{22} = 0.$$

Moreover, it can be verified that $V_1 = 2h$. Thus we have obtained the same result as for a solid strip under cylindrical bending. The grid strip can be understood as consisting of an infinite number of parallel, vanishingly narrow ribs, touching each other.

As an example consider a rectangular grid whose perfectly plastic material obeys the Tresca yield condition, $\delta = 2$. Choose the reference width coefficient as $\gamma_{\pi/4} = 1/4$. The relevant formulae become

$$\tan\alpha = \frac{2}{\beta}\left[2(1-\beta^2)+\sqrt{4(1-\beta^2)^2+\beta^2/4}\,\right],$$

$$\frac{1}{Z} = 2\left(\beta^2\cos^2\alpha+\sin^2\alpha+\frac{\beta}{8}\sin 2\alpha\right),$$

$$\gamma_\alpha = \tfrac{1}{4}\sin 2\alpha.$$

The numerical results for $\beta = 0$; 0.5; 0.6; 0.7; 0.8; 0.9; 1.0 are shown in Table 15.

Table 15 **The inclinations of bars, width coefficients of bars and dimensionless intensities of collapse load for the Tresca rectangular grid with given weight per unit area of its central plane**

β	0	0.5	0.6	0.7	0.8	0.9	1.0
α	90°	85°15′	83°25′	80°30′	75°30′	65°	45°
γ_α	0	0.0402	0.057	0.082	0.121	0.191	0.25
$1/Z$	2	2.012	2.019	2.028	2.050	2.104	2.250

The statically admissible field of moment densities used in this analysis has the property that the yield condition is reached either in bending (the centre of the grid where the twisting moment densities are absent) or in torsion (the corners of the grid where, in turn, the bending moments densities vanish, according to the boundary conditions). Thus the combined bending-twisting limit state condition was not needed to construct the field (6.144). However, to prove that eq. (6.145) is really a lower bound to the collapse load a check is necessary on the combined condition being not violated in any cross-section of the grid undergoing bending and torsion at the same time.

The bending-twisting limit state condition for a rectangular cross-section is assumed to be reasonably well approximated by a circular interaction curve having, in the dimensionless coordinates, the equation

$$\bar{F} = \left(\frac{M}{M_0}\right)^2 + \left(\frac{\tilde{M}}{\tilde{M}_0}\right)^2 - 1 = 0 \tag{6.156}$$

where M, $\tilde{M}$ are, respectively, the current bending and twisting moments acting simultaneously on an arbitrary cross-section of the bar and M_0, $\tilde{M}_0$ are the corresponding ultimate moments under pure bending and torsion, cf. eqs. (6.147), (6.148). Consider first the midspan cross-section of a bar belonging, for instance, to the family I. Combining eqs. (6.136), (6.137), (6.144), (6.141), remembering eq. (6.149) and employing eq. (6.156) we obtain the equation for the moment profile

$$\bar{F} = \left\{\left[1-\left(\frac{X_1}{a}\right)^2\right]\cos^2\alpha + \left[1-\left(\frac{X_2}{b}\right)^2\right]\sin^2\alpha\right\}^2 + \\ + \left\{-\frac{\varepsilon}{2}\left[\left(\frac{X_1}{a}\right)^2 - \left(\frac{X_2}{b}\right)^2\right]\sin 2\alpha + \frac{X_1 X_2}{ab}\right\}^2 - 1 = 0. \tag{6.157}$$

The magnitude of the left-hand side of eq. (6.157) must not exceed zero in any midspan cross-section of the bars of the grid. This requirement is clearly satisfied in the centre and in the corners of the grid. To check eq. (6.157) at other points requires the most "unsafe" point to be found from $\frac{\partial \bar{F}}{\partial X_1} = 0$, $\frac{\partial \bar{F}}{\partial X_2} = 0$ and then eq. (6.155) to be used. This appears to be very troublesome and a numerical procedure would be necessary. The calculations made for the square grid $a = b$, $\alpha = \pi/4$ show that eq. (6.157) is nowhere violated. Moreover, in the limiting case of rectangular solid plate ($l = 2\gamma_\alpha h/\sin 2\alpha$) the yield condition has been proved not to be exceeded at any point, i.e. the field (6.144) to be safe.

Consider now the nodal cross-section of a bar. The uniformly distributed load q is here assumed to be transmitted by a floor directly to the joints of the grid, the nodal applied forces being $Q = ql^2 \sin 2\alpha$. Therefore, in the nodal section, only the bending moment generated by the midspan shear force has to be added to the internal forces $M^{(n)} = M + \check{P}\frac{l}{2}$, $\tilde{M}^{(n)} = \tilde{M}$. Having eq. (6.138) and the equilibrium equations $P^1 = M^{12}{}_{,1} + M^{22}{}_{,2}$, $P^2 = -M^{11}{}_{,1} - M^{21}{}_{,2}$, the equation for the nodal moment profile is found to be

$$\bar{F}^{(n)} = \left\{\left[1 - \left(\frac{X_1}{a}\right)^2\right]\cos^2\alpha + \left[1 - \left(\frac{X_2}{b}\right)^2\right]\sin^2\alpha - \right.$$
$$- \frac{l}{2}\left[2\left(\frac{X_1}{a}\cos\alpha + \frac{X_2}{b}\sin\alpha\right) + \right.$$
$$\left.\left. + \frac{1}{\varepsilon ab}\left(X_1\frac{\sin^2\alpha}{\cos\alpha} + X_2\frac{\cos^2\alpha}{\sin\alpha}\right)\right]\right\}^2 +$$
$$+ \left\{-\frac{\varepsilon}{l}\left[\left(\frac{X_1}{a}\right)^2 - \left(\frac{X_2}{b}\right)^2\right]\sin 2\alpha + \frac{X_1 X_2}{ab}\right\}^2 - 1 = 0. \qquad (6.158)$$

Again, the magnitude of the left-hand side of eq. (6.158) must not be positive in any nodal cross-section of the bar. This requirement, together with eqs. (6.155), (6.149) is expected to supply a condition of the type $\varphi \equiv l/a \leqslant \varphi_c(\beta, \delta, \gamma)$, i.e. the minimum "density" of the grid will be determined ensuring that the limit state condition in combined bending and torsion is nowhere violated. The critical magnitude φ_c depends on the ratio of sides, the yield condition applied, and the "narrowness" coefficient of the rectangular cross-sections of the bars. However, due to computational complexity, even in the case of a square grid the condition in the form $\varphi \leqslant \varphi_c(\delta, \gamma)$ is rather hard to obtain in a closed form.

Lastly, the question arises whether, for the nodal regions in which the bars penetrate through each other, an additional check is necessary on the plastic behaviour of the material. The answer is negative since in the situation considered a two-dimensional stress state is generated the signs of both normal stresses being the same at any point. Both the Mises and the Tresca materials are known to exhibit no "weakening" in such a state of stress. Thus, no decrease in the collapse load is to be expected due to the complex stress state in the nodal regions of the simply supported rectangular grid subjected to a distributed load acting downwards (or upwards) only.

Summary

In order to illustrate the potential of the approach advocated in the book, in this chapter we have described a number of analytical and numerical solutions to problems of nonlinear behaviour of trusses, frames and lattice-type shells. We have repeatedly emphasized that the ingenious use of available theory and numerical techniques for solving nonlinear boundary-value problems of structural mechanics makes it possible to effectively solve even very complex problems.

Apart from giving some results obtained by means of well-established algorithms, such as those based on generalized plastic hinge concept or referred to the linearized static frame buckling, we have described some novel techniques. These have been exemplified by solving some inelastic problems with non-proportionally varying loads and by thoroughly discussing solution algorithms for dealing with shell-like lattice structures.

Problems

6.1 Using the elastic and initial stress stiffness matrices find the load P which causes the truss in Fig. 124 to become unstable in its own plane. Compare the result with the exact solution contained in [136].

6.2 Perform the analysis of the influence of the axial load upon the frequency of transverse vibrations in a beam column with both ends pinned. Use one and two elements.

6.3 Using the method of Sec. 6.2.2 derive the limit state condition (as a function of the bending moment and the axial force) for the cross-section shown in Fig. 125. Compare the result against the similar condition derived for the bi-symmetric I-cross-section.

6.4 Using the approximate approach of Sec. 6.2.2 develop the limit state condition for the I-cross-section with unequal flanges.

6.5 Specify the elastic-plastic stiffness matrices for the frame element to the case of plane deformations assuming the bending moment to be the only generalized force that enters the limit state condition.

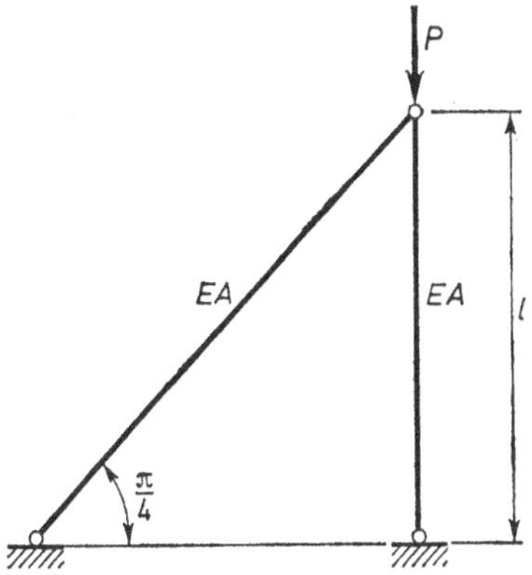

Fig. 124 Two-dimensional truss configuration

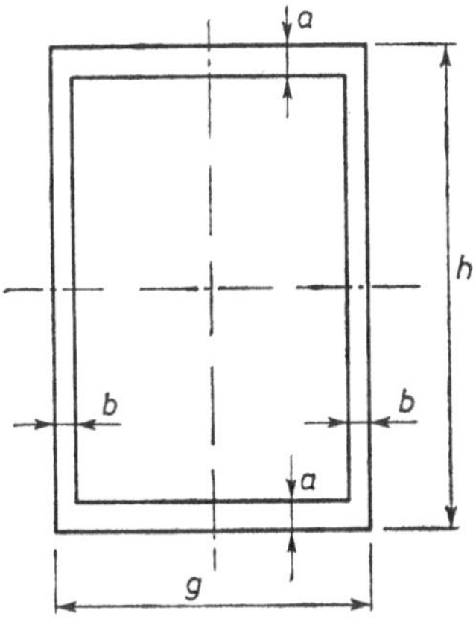

Fig. 125 Beam cross-section

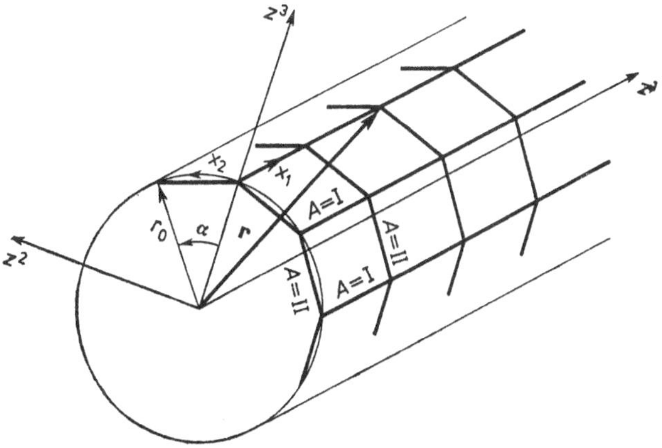

Fig. 126 Cylindrical lattice shell

6.6 Derive all the necessary geometrical relationships describing the lattice cylindrical shell, Fig. 126 within the discrete model of Sec. 6.3.1.

6.7 Derive the set of equations formulated in Problem 4.15 for the case of the lattice cylindrical shell of Fig. 126 within the framework of the continuous model.

6.8 Derive specific forms of the linear lattice shell governing equations for a shell formed on a hemisphere and consisting of two given families of meridional and parallel rigidly connected bars. Use both the discrete and continuous models.

6.9 Using plate analogy with respect to the distribution of statically admissible moment fields find a bound to the collapse load for a parallelogram grid made of two orthogonal families of rigid-perfectly plastic beams.

6.10 Using the analytical method of grid buckling analysis presented in Sec. 6.3.2 find the buckling load for the problem analyzed numerically in Sec. 6.2.6.

6.11 Find the buckling load for the grid in Fig. 109 using the similar algorithm in the framework of the grid continuous model. Compare the result with the corresponding problem of classical thin plate buckling.

6.12 Using the approach presented in Sec. 6.3.2 find the buckling load for an elastic circular grid compressed radially. The grid is composed of two families of rods as shown in Fig. 113.

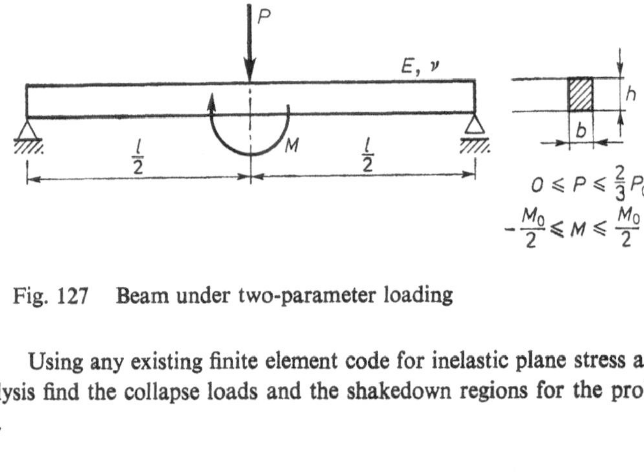

Fig. 127 Beam under two-parameter loading

6.13 Using any existing finite element code for inelastic plane stress and/or beam bending analysis find the collapse loads and the shakedown regions for the problem defined in Fig. 127.

7

Thin plates loaded in-plane

Purpose of the chapter

The main objective in this chapter is to illustrate the capabilities of the discretized approach as applied to plane stress problems. We discuss in turn small elastic-plastic deformation associated with bending of a cantilever without and with a weakening hole, the limit load analysis of a perforated plate, the behaviour of an elastic-plastic cantilever under non-proportionally varying loads and large extensional deformations in a rectangular plate with a central hole.

7.1 Elastic-plastic bending of a cantilever beam

Let us consider a cantilever beam shown in Fig. 128 and assume it is made of elastic-ideally plastic material with $E = 207$ GPa, $\nu = 0.3$, $\sigma_y = 1962$ MPa and with the Huber–Mises yield condition. The plane stress numerical analysis has been performed assuming two rather simple discretization meshes obtained by means of the triangular finite elements with quadratic displacement interpolation as presented in Sec. 4.7. The first coarse mesh employed 64 elements, cf. Fig. 128, whereas the finer one consisted of 256 elements. The calculations performed incrementally up to the limit load are shown in Fig. 128 in which the nondimensionalized load $P/P^{(el)}$ is plotted vs the nondimensionalized end displacement at the point B ($P^{(el)}$ is the maximum "elastic" load and $u_y^{(B)}$ its corresponding displacement, both obtained numerically and thus slightly different from the analytical solutions). The limit load solution obtained on the basis of the classical beam bending theory is shown as the point M; the comparison against the analytical results for upper and lower bounds to the limit load for this problem as derived in [67] are also given in the figure. A typical normal stress distribution as well as the corresponding vertical displacement distribution across the beam thickness are shown for the load level $P/P^{(el)} = 1.5$. The linear character of the displacement distribution at even highly advanced stages of the deformation process clearly supports the fundamental Euler beam assumption concerning the plane deformation of the cross-section for slender beams.

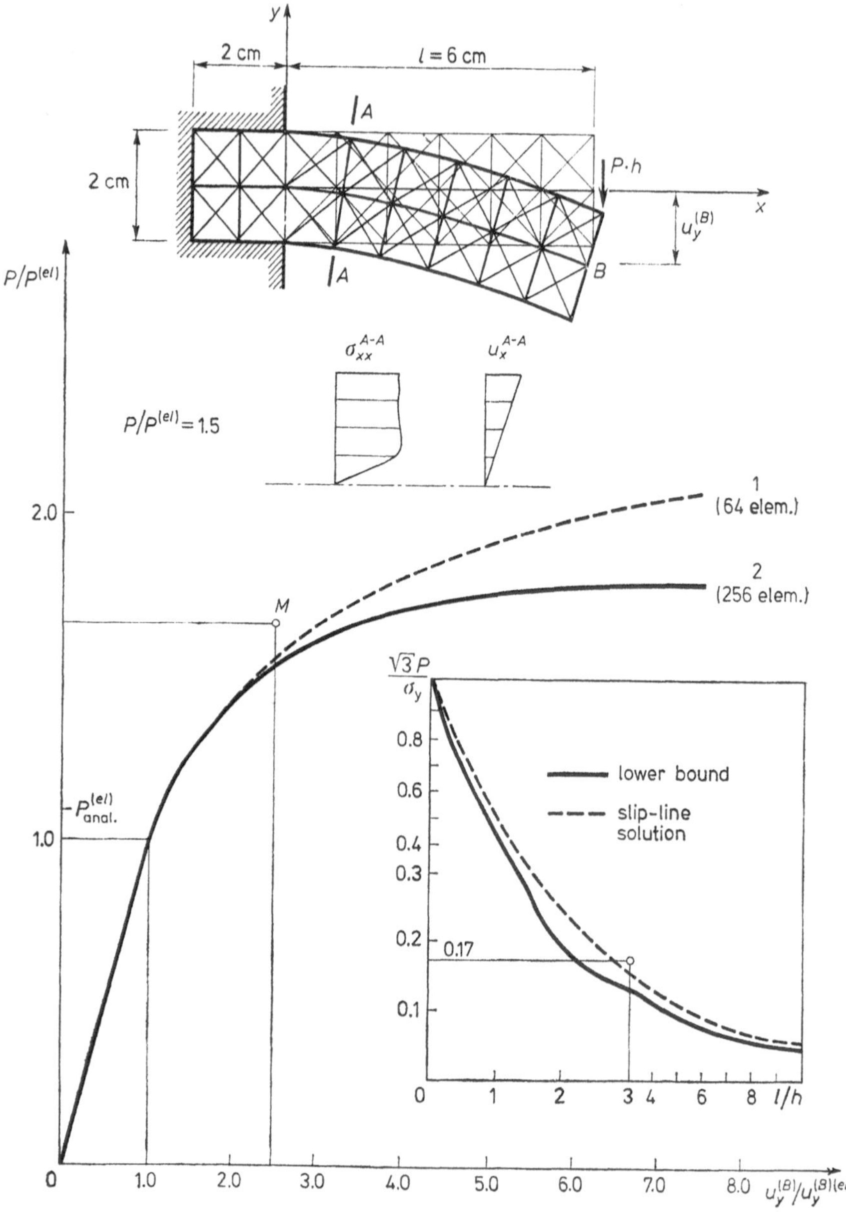

Fig. 128 Bending of a cantilever

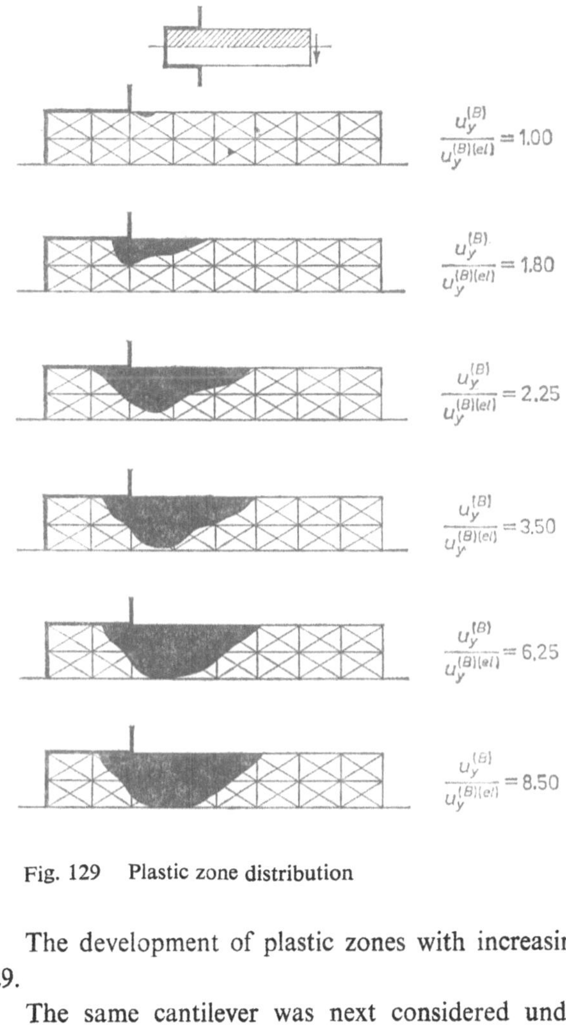

Fig. 129 Plastic zone distribution

The development of plastic zones with increasing load is shown in Fig. 129.

The same cantilever was next considered under the combined action of shear and tensile forces P and R, Fig. 130. The force R was assumed to be applied first up to the value $0.4\sqrt{3}\,\sigma_y$ and then kept constant while incrementing the force P. The results are shown in Figs. 130, 131.

As the next example we have considered bending of the same cantilever except for the square hole located as shown in Fig. 132. The results in the form of the load vs displacement curve, discretization mesh and plastic zone development are given in Figs. 132, 133, respectively. The comparison of the limit load obtained in the course of the numerical computations against the analytical results of [68] for three different hole locations are plotted in Fig. 132 as well.

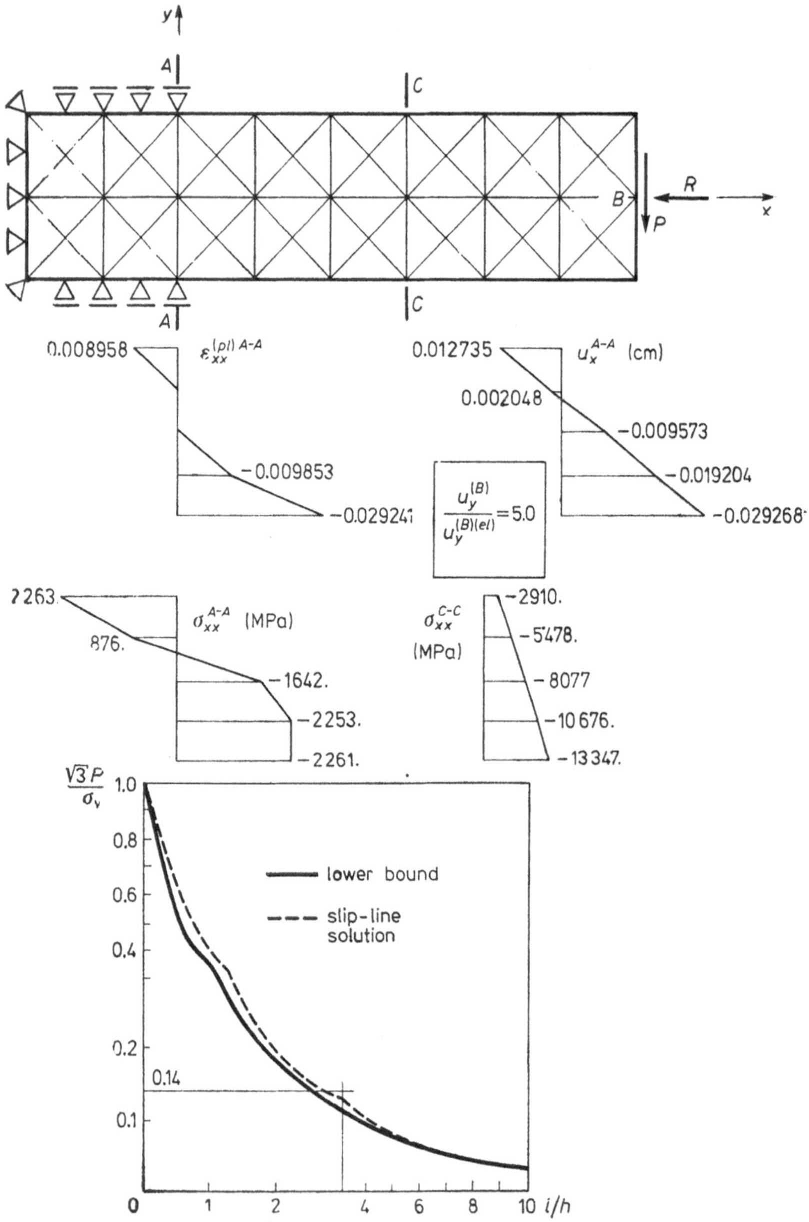

Fig. 130 Cantilever under combined bending/compression

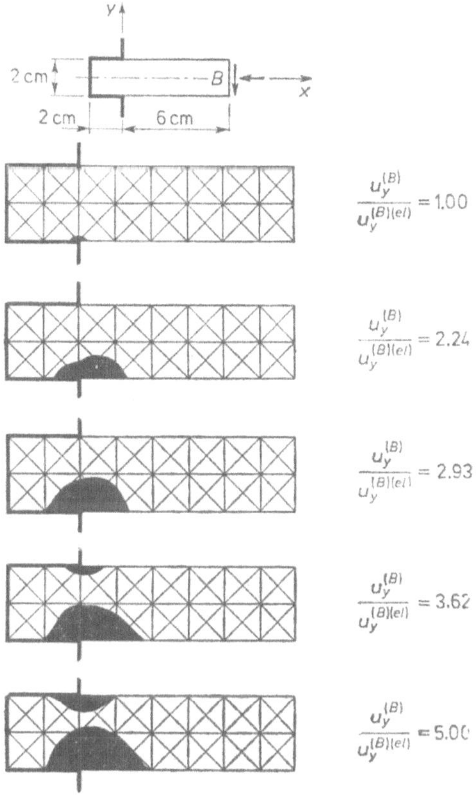

Fig. 131 Plastic zone distribution

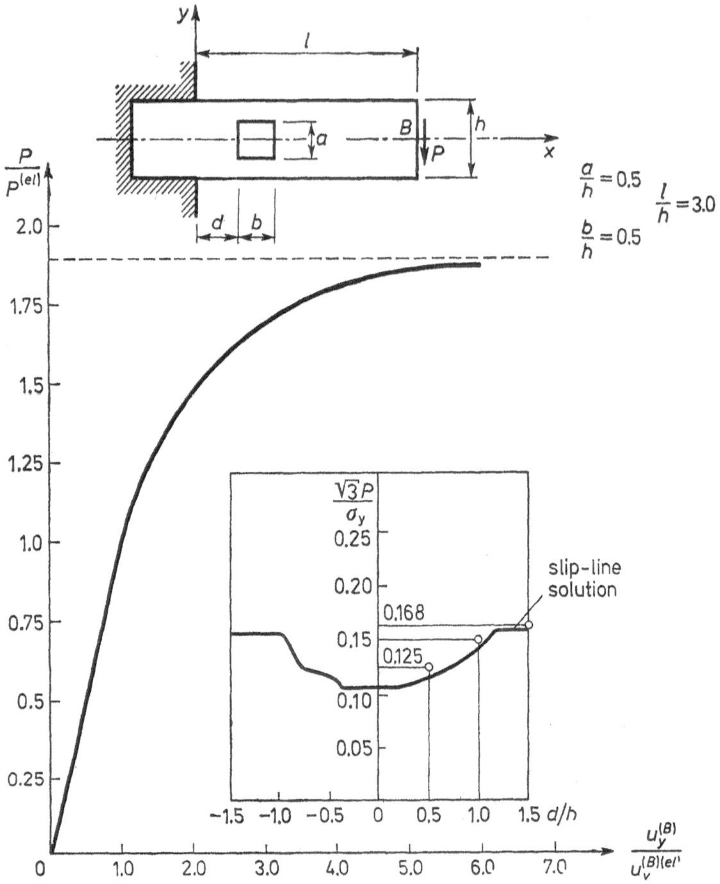

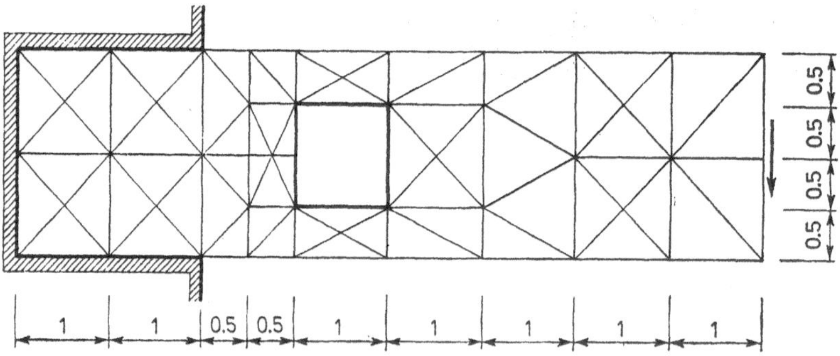

Fig. 132 Bending of a cantilever with a hole

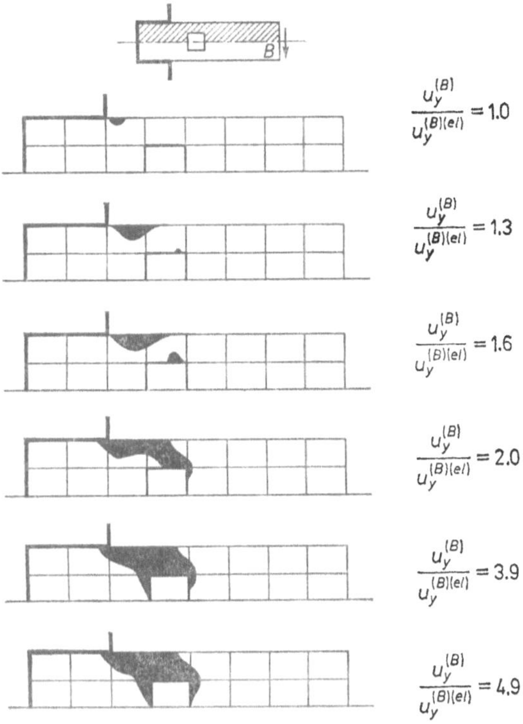

Fig. 133 Plastic zone distribution

7.2 Limit analysis of perforated plates

The knowledge of the limit state behaviour of perforated sheets subjected to in-plane biaxial load is important in many branches of mechanical and structural engineering, especially in the pressure vessels technology. The existing solutions have been as a rule obtained by using statically admissible discontinuous stress fields in order to assess the lower bound limit load for the perforated sheet. The yield curves have been constructed for the equivalent solid material under the assumptions that the distribution of normal stresses across the weakest section is uniform and that the transverse normal stresses are equal to zero. Such an uniaxial state of stress appears to be applicable merely for very low cutout coefficients $\beta = b/a$, for symbols cf. Fig. 134. Thus, if a more accurate solution to the limit analysis problem of perforated plane body is of interest the step-by-step elastic-plastic algorithm should be employed, in which clearly no necessity arises to replace the multi-connected

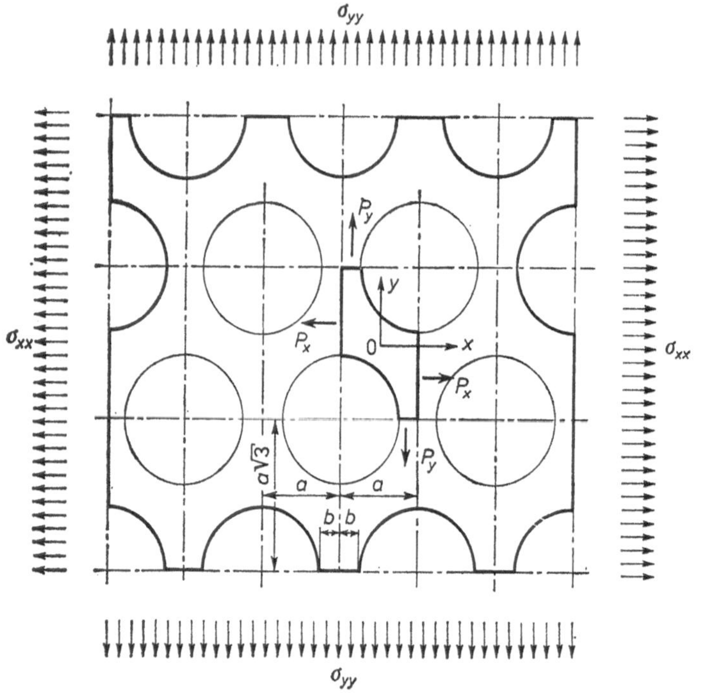

Fig. 134 Biaxial loading of perforated sheet: P_x, P_y—resultant edge loads per unit cell, σ_{xx}, σ_{yy}—uniform stresses at infinity

region by an equivalent continuous material, i.e. a solid sheet, neither does global stress uniformity need be assumed.

Using the procedure used in the previous section we shall describe now the details of the numerical analysis of a perforated sheet problem. Any other problem of this type can be solved by employing the similar algorithm.

An infinite thin sheet is considered perforated in an equilateral pattern of circular openings and subjected to biaxial two-parametric in-plane loading, Fig. 134. Perforation pitch is denoted by $2a$, width of the minimum section by $2b$. The cutout coefficient is designated by β. The material is assumed to be elastic-plastic and to obey the Mises yield condition together with the normality law. Due to its multi-symmetry the structure is split up into repeatable unit cells (subregions) as shown by heavy lines in Fig. 134. This element is subdivided into 222 triangular elements described in Sec. 4.7, Fig. 135. Owing to the regularity of the penetration pattern all points lying on horizontal and vertical edges of the unit cell undergo equal displacements. Such being the situation, the biaxial two-parameter loading of the sheet is simulated kinematically by considering incremental displacements of both pairs of edges.

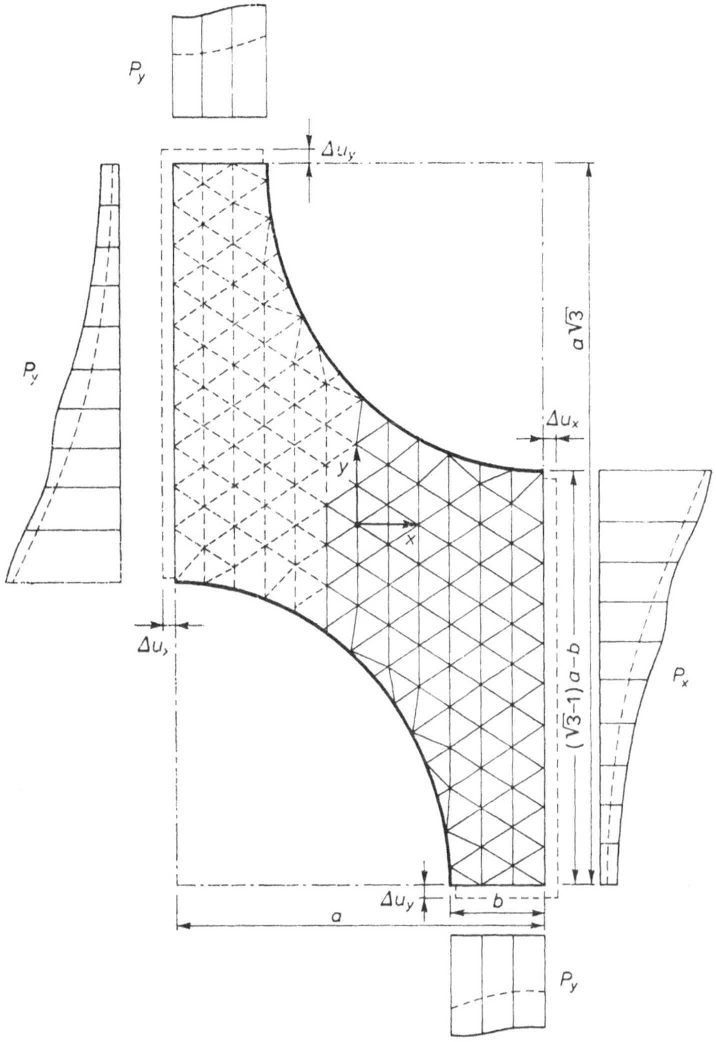

Fig. 135 Subdivision of a unit cell with the properties of its point-symmetry shown; edge displacement increments Δu_x, Δu_y and edge stresses p_x, p_y (for $\mu = \frac{\Delta u_y}{\Delta u_x} = -1$);dashed curves — elastic distribution, solid curves — plastic distribution

Corresponding forces at edge nodes are computed automatically in the program and resultant loads P_x, P_y associated with a single unit cell of Fig. 134 are calculated by a suitable summation procedure. The ratio $\mu = \frac{\Delta u_y}{\Delta u_x}$ of prescribed edge displacements is kept constant—which can be called proportional geometrical loading—and the resulting edge loads follow a certain generally curvilinear load path starting from the origin of the P_x, P_y-plane and terminating at an instant at which the limit load state is reached, for instance point B, Fig. 138. Thus the limit interaction curve (i.e. the relationship P_x vs P_y at collapse) for a unit cell can be constructed numerically by eliminating prescribed displacements from the parametric equations $P_x = P_x$ (prescribed displacements), $P_y = P_y$ (prescribed displacements).

Another procedure is also possible: starting from, for instance, point A, Fig. 138, the ratio $\mu = \dfrac{\Delta u_y}{\Delta u_x}$ can be modified at each incremental step in such a way that the $\dfrac{P_x}{P_y}$-ratio be kept constant thus resulting in the continuing proportional statical loading pattern as shown by straight line AC in this figure. However, the first procedure is numerically simpler and also in fact more useful since not only a single point C on the limit interaction curve may be found, as it is the case with proportional static loading, but a certain segment BD of this curve is, as numerical experiments show, accurately traced once the limit state is reached.

Each computational step is associated with the growing (unless local unloading occurs) elastic-plastic interface and thus the development of the plastic zone is observed until the limit state is reached and the collapse mechanism is about to form allowing for the unlimited growth of plastic strains.

It has to be noted that by employing the prescribed displacement approach the deformation process is followed (starting from the virgin state and ending at a point on the limit state curve) which clearly does not correspond to a straight path in the load plane. However, since the shape of the interaction curve does not depend on the loading path (the fact which has been repeatedly confirmed numerically in this study), the final curve obtained may be treated as the true limit state condition in the load plane.

The following data are assumed: perforation pitch $2a = 24$ cm, minimum section width $2b = 4.8$ cm, cutout coefficient $\beta = 0.2$, thickness of the sheet $t = 1$ cm, Young modulus $E = 207$ GPa, Poisson's ratio $\nu = 0.3$, yield point stress $\sigma_y = 245.25$ MPa.

Based on the numerical results, the unit cell load-displacement curves were constructed first for particular displacement paths with prescribed $\mu = \dfrac{\Delta u_y}{\Delta u_x}$. For example, in Fig. 136 the $P_x - u_x$ and $P_y - u_y$ curves are shown for the path No. 2 with $\mu = -10$. Curves corresponding to the load path No. 7 with $\mu = 2$ are shown in Fig. 137.

Elimination of displacements enables the limit interaction curve to be constructed as shown in Fig. 138. Nine displacement paths were studied to

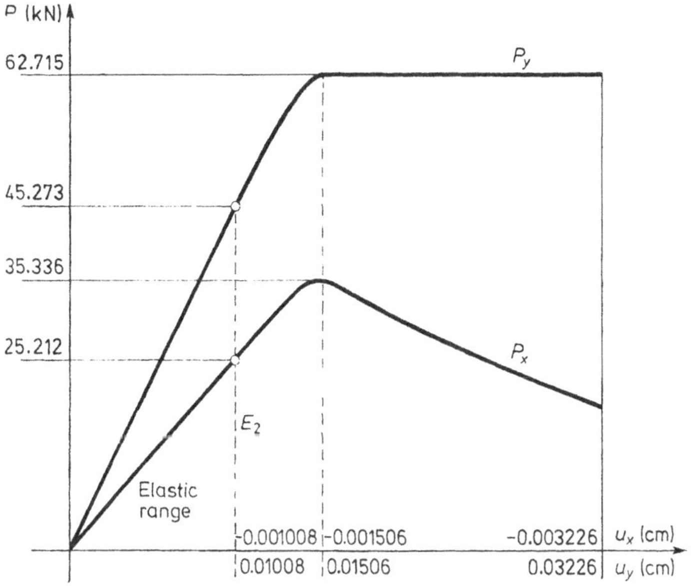

Fig. 136 Load-displacement curves for load path No. 2, $\mu = -10$

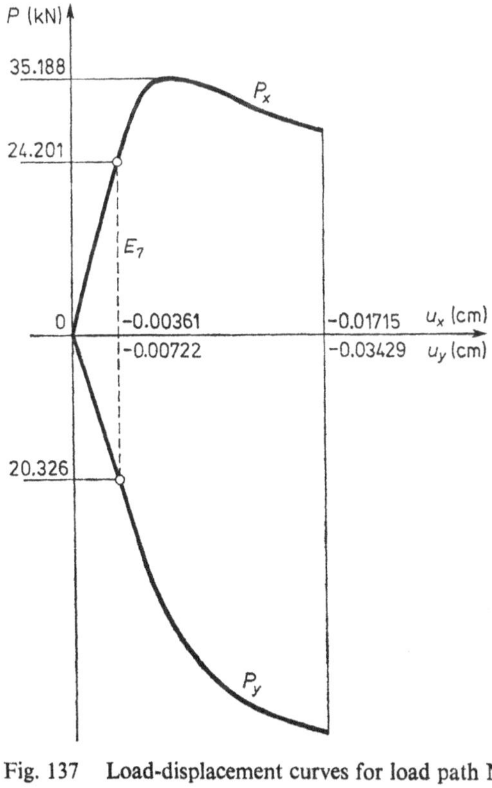

Fig. 137 Load-displacement curves for load path No. 7, $\mu = 2$

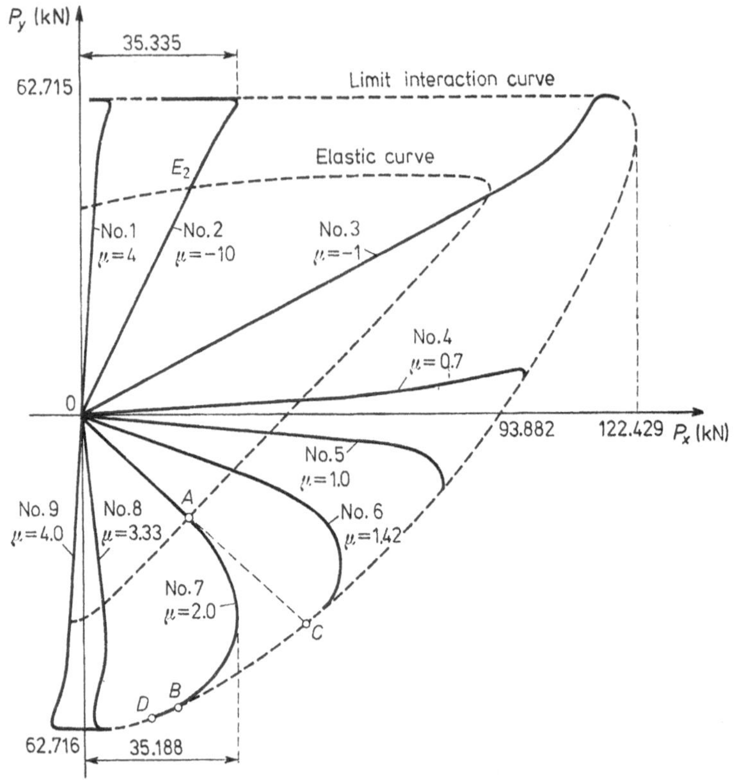

Fig. 138 Limit interaction curve for a unit cell with particular load paths shown

plot this curve in the first and second quadrant of the load plane. The elastic curve is also shown in Fig. 138 within which the load paths for μ = const form straight lines (proportional loading). Point E_2 corresponds to the ordinate shown in Fig. 136, point A to the ordinate E_7 shown in Fig. 137.

The development of plastic zones is illustrated in Fig. 139 by means of eight selected incremental steps along the path No. 3 with $\mu = -1$. The first local unloading process takes place after the 16th increment. At the 20th step both plastic regions meet and a plastic bridge begins to grow up to the 28th increment which visualizes the limit state of the weakest section and of the whole sheet as well.

To display the results in the stress (rather than load)-wise manner the actual edge stresses p_x, p_y (shown in Fig. 135 for the load path No. 3, $\mu = -1$) producing the resultant edge loads P_x, P_y, Fig. 134, can be averaged according to the formulas:

$$p_x^{av} = \frac{P_x}{(\sqrt{3}-1)a+b}, \quad p_y^{av} = \frac{P_y}{b}.$$

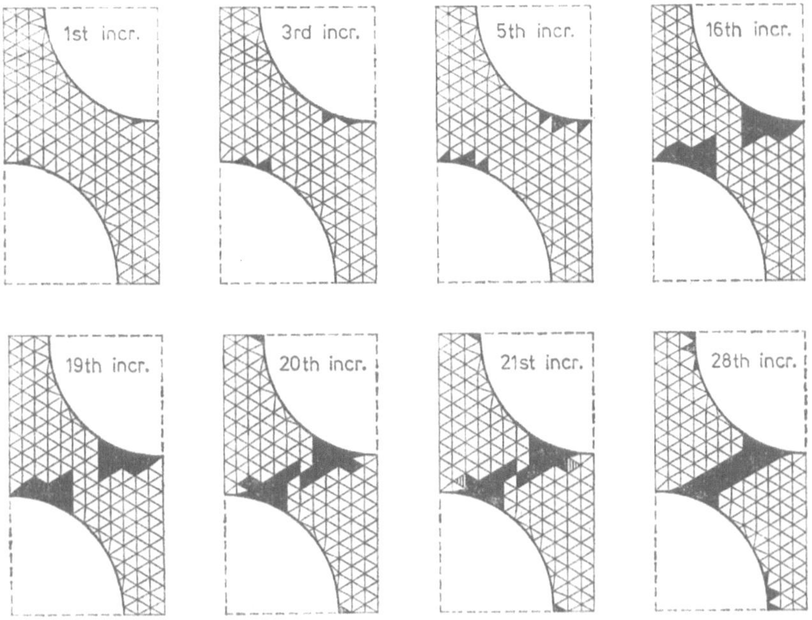

Fig. 139 Development of plastic zones (black) at particular displacement increments, path No. 3, $\mu = -1$; shaded elements have undergone unloading

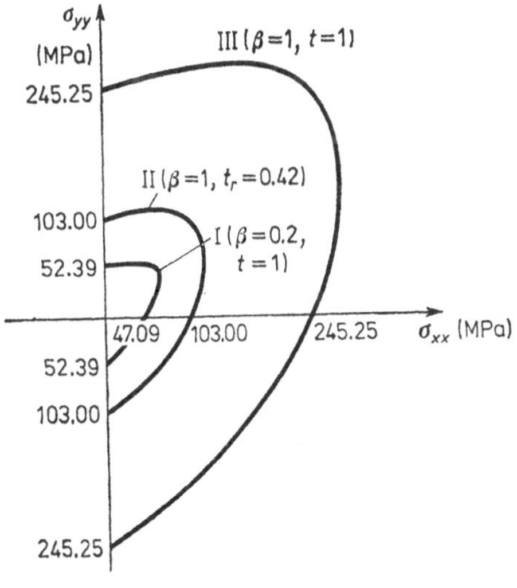

Fig. 140 Limit interaction stress curve (I) for perforated sheet with the cutout coefficient $\beta = 0.2$ and thickness $t = 1$ cm enclosed by the yield ellipse (II) for solid sheet of reduced thickness $t_r = 0.42$ cm and by the yield ellipse (III) for solid sheet with thickness $t = 1$ cm

Also the uniform stresses at infinity, constituting the biaxial load pattern of the sheet, Fig. 134, can be clearly calculated as

$$\sigma_{xx} = \frac{P_x}{\sqrt{3}a}, \qquad \sigma_{yy} = \frac{P_y}{a}.$$

Thus the limit interaction curve of Fig. 138 can conveniently be scaled down and shown in the $(\sigma_{xx}, \sigma_{yy})$-plane against the Mises yield curve for a solid, unperforated sheet, Fig. 140. Curve I refers to the sheet considered, curve II is the yield ellipse for a solid sheet of reduced thickness $t_r = t\left[1 - \frac{\pi}{2\sqrt{3}} \times (1-\beta)^2\right] = 0.42$ cm that results in the same volume of material per unit area, curve III represents the standard yield ellipse for a solid sheet of the same thickness as the perforated one.

7.3 Elastic-plastic analysis of a cantilever under non-proportionally varying loads

In Sec. 6.2.5 we have reported on the analysis of a cantilever subjected to independently varying loads in the form of a bending moment M and a shear force T, cf. Fig. 96. The cantilever was modelled in there by means of 6 beam elements. The same loading bounds as before are considered now, cf. Table 8, but the cantilever is modelled with 120 triangular elements as the plane stress problem, Fig. 141. A generic element has again 6 nodes, each

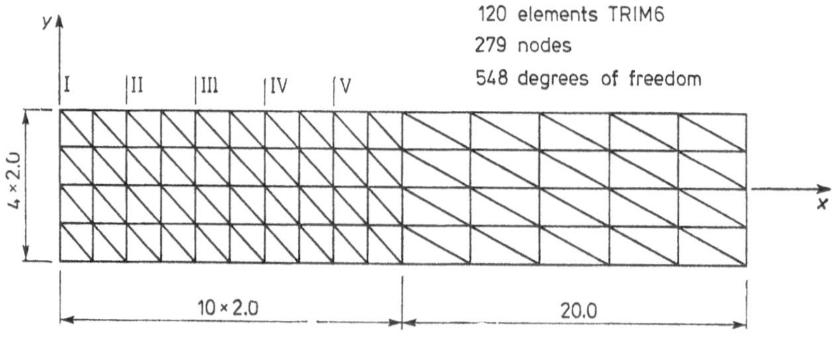

Fig. 141 Discrete model of the cantilever

of them having 2 translational degrees of freedom. The mesh shown in Fig. 141 results in 548 degrees of freedom. The shear force and the bending moment at the free end of the beam are represented by the parabolic distribution of the

Table 16 Accumulated effective plastic strains

Cycle number	Load 1	Load 2
1	0.586	0.586
2	1.172	0.586
3	1.758	0.586
4	2.344	0.586

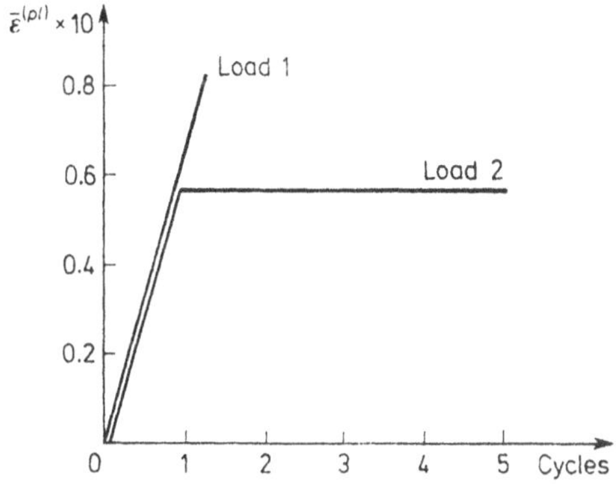

Fig. 142 Accumulation of the effective plastic strain

shear stress and linear distribution of the normal stress, respectively. The material is assumed to obey the Huber–Mises yield criterion. The results are summarized in Tables 16, 17 and Figs. 142, 143. (For the sake of comparison see the results described in Sec. 6.2.5 for the beam element model.) The adaptation to load 2 was achieved for both discrete models in the first critical cycle. The different values of the accumulated equivalent plastic strain $\bar{\varepsilon}^{(pl)}$ obtained for both models should not be interpreted as the serious discrepancy in the amount of dissipated energy. In order to find the value of $\bar{\varepsilon}^{(pl)}$ in the case of plane stress discretization an integration over element areas is clearly necessary, whereas in the computations performed only the nodal values of $\bar{\varepsilon}^{(pl)}$ were added up to simplify the coding. The value of $\bar{\varepsilon}^{(pl)}$ placed in Table 16 is thus not the real overall plastic strain making the comparison between the both models senseless. Nevertheless, within each model separately the value of $\bar{\varepsilon}^{(pl)}$ remains a rational measure of the dissipated energy.

The residual stresses σ_{xx} observed in the cross-sections I–I to V–V after the adaptation to load 2 are given in Table 17 and Fig. 143. The agreement between the beam and plane stress idealizations is very good towards the

Table 17 Residual stresses

Cross-section	I–I	II–II	III–III	IV–IV	V–V
y \ x	0.0	4.0	8.0	12.0	16.0
−4.0	905.4	812.6	718.2	629.3	540.2
−2.0	−741.0	−668.4	−591.5	−309.7	−444.3
0.0	1.560	−1.223	0.116	0.086	0.323
2.0	737.6	675.2	596.3	−630.4	448.9
4.0	−895.1	−805.8	−713.0	−624.1	−535.0

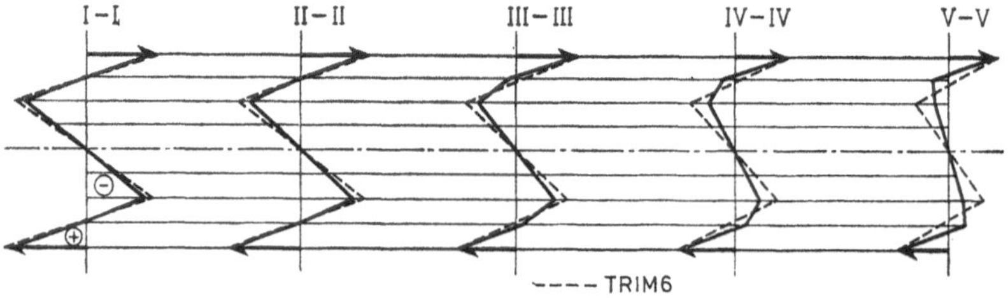

Fig. 143 Residual stresses σ_{xx} in the cross-sections I–I to V–V of the cantilever after adaptation to load 6

clamped end, where the meshes are fine. It gets worse towards the free end, where the meshes are coarse. As was to be expected for the chosen dimensions of the bar, the shear stress is of minor influence. The maximum value of the residual stress σ_{xy} found in the plane stress case was less than 3% of the maximum residual stress σ_{xx}. The residual vertical displacement of the free end of the bar was $u_y = -0.946 \times 10^{-1}$ cm which can be compared against the value $u_y = -0.636 \times 10^{-1}$ cm obtained for the beam idealization, Sec. 6.1.3.

On the basis of the examples reported here and in Sec. 6.1.3 we may generally state that the analysis of structural shakedown can be successfully accomplished by means of the incremental finite element procedures which are stable in the vicinity of the boundary of shakedown domain and exhibit sufficient accuracy for practical purposes. Also, an indication of whether shakedown occurs or not comes out clearly after a few critical cycles when the evolution of the total accumulated equivalent plastic strain is observed.

7.4 In-plane buckling of an elastic-plastic strip

Consider a thin column modelled by plane stress finite elements described in Sec. 4.9, Fig. 144. This type of modelling makes it possible to analyse only the in-plane deformations of the column. The geometry and material parameters are given in Fig. 144. The analysis aims at finding the in-plane buckling

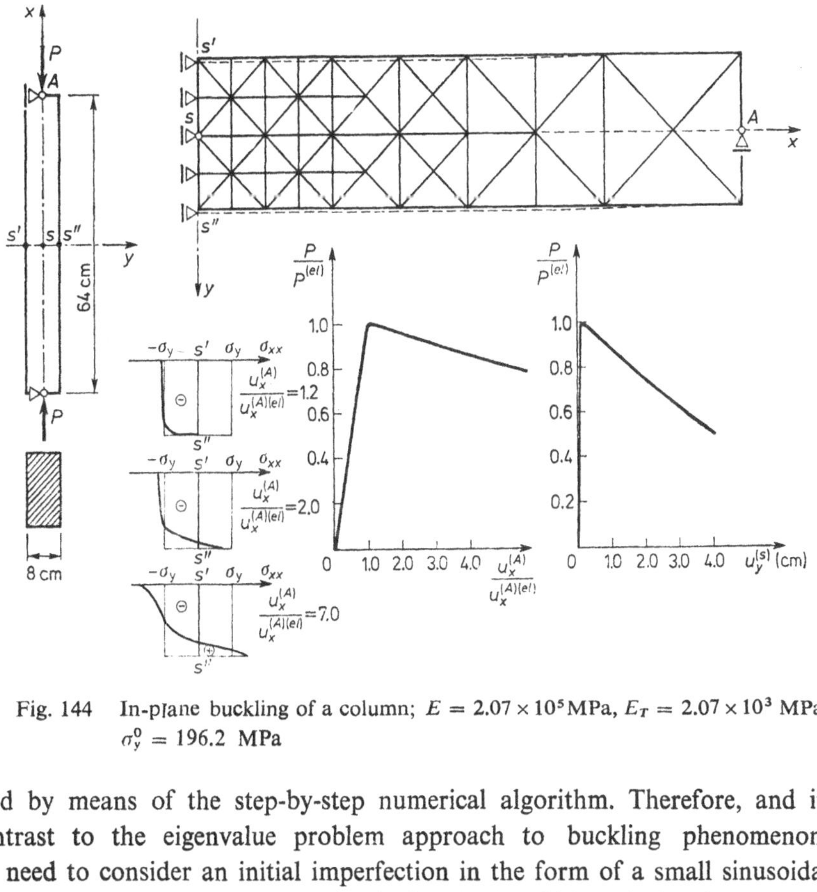

Fig. 144 In-plane buckling of a column; $E = 2.07 \times 10^5$ MPa, $E_T = 2.07 \times 10^3$ MPa, $\sigma_y^0 = 196.2$ MPa

load by means of the step-by-step numerical algorithm. Therefore, and in contrast to the eigenvalue problem approach to buckling phenomenon, we need to consider an initial imperfection in the form of a small sinusoidal eccentricity with the amplitude of 1% of the strip width and the shape corresponding to the predicted buckling mode. We note that the classical Euler beam buckling stress is given in this case by the formula

$$\sigma_{cr}^{(el)} = \frac{\pi^2 EJ}{Al^2} \cong 2660 \text{ MPa}$$

while the buckling stress for the plastic column based on the tangent modulus theory reads

$$\sigma_{cr}^{(el)} = \frac{\pi^2 E_T J}{Al^2} \cong 26.60 \text{ MPa}.$$

The average modulus theory yields the critical stress

$$\sigma_{cr}^{(el)} = \frac{4\pi^2 E E_T J}{\left(\sqrt{E}+\sqrt{E_T}\right)^2 A l^2} \cong 87.93 \text{ MPa}.$$

It may be easily observed that the above critical stresses and the value of the yield limit assumed in this problem should induce buckling right after the column mid-section goes plastic, followed by a sharp decrease of the load.

The column is compressed by imposing the series of prescribed displacements of the node A; to avoid extensive plastic straining in the vicinity of the node A artificially high yield limit is there introduced.

The results presented in Fig. 144 confirm the predictions; instability took place as soon as the midsection was plastified. The loading was continued until a typical elastic-plastic bending behaviour was observed. The distribution of normal stresses σ_{xx} at different stages of the process are also plotted in Fig. 144.

7.5 Necking of an elastic-plastic strip

In this section we shall consider a thin elastic-plastic strip subjected to tension by means of the prescribed end incremental displacements, Fig. 145. The initial length to width ratio is 4. The strip is discretized using the linear strain triangular elements of Sec. 4.9. To induce a possible necking-type bifurcation the strip is given an initial sinusoidal imperfection with the amplitude at the cross-section $x = 0$ equal to 0.5% of the strip width. The isotropic hardening of the material is given by the Ramberg–Osgood curve with $m = 8$, cf. Sec. 7.6. The analysis has been carried out by applying equal end displacement increments of $2 \cdot 10^{-4} l_0$. To obtain the final elongation of $0.2 l_0$ 1000 steps have been necessary.

The results are displayed in Fig. 145. The point of maximum load (i.e. of maximum resultant boundary reactions at nodes) is observed at the point M, Fig. 145. At the point B local unloading has occurred and the plastic zone started to localize towards the mid-section of the strip; the equilibrium curve followed the path 1. The changes in specimen geometry are shown in Fig. 145, the abrupt necking started at more or less the same time instant as the local unloading, point B. This point is thus identified as the bifurcation point which may be confirmed by additionally performing the similar calculations without any imperfections resulting in the solution curve 2, Fig. 145.

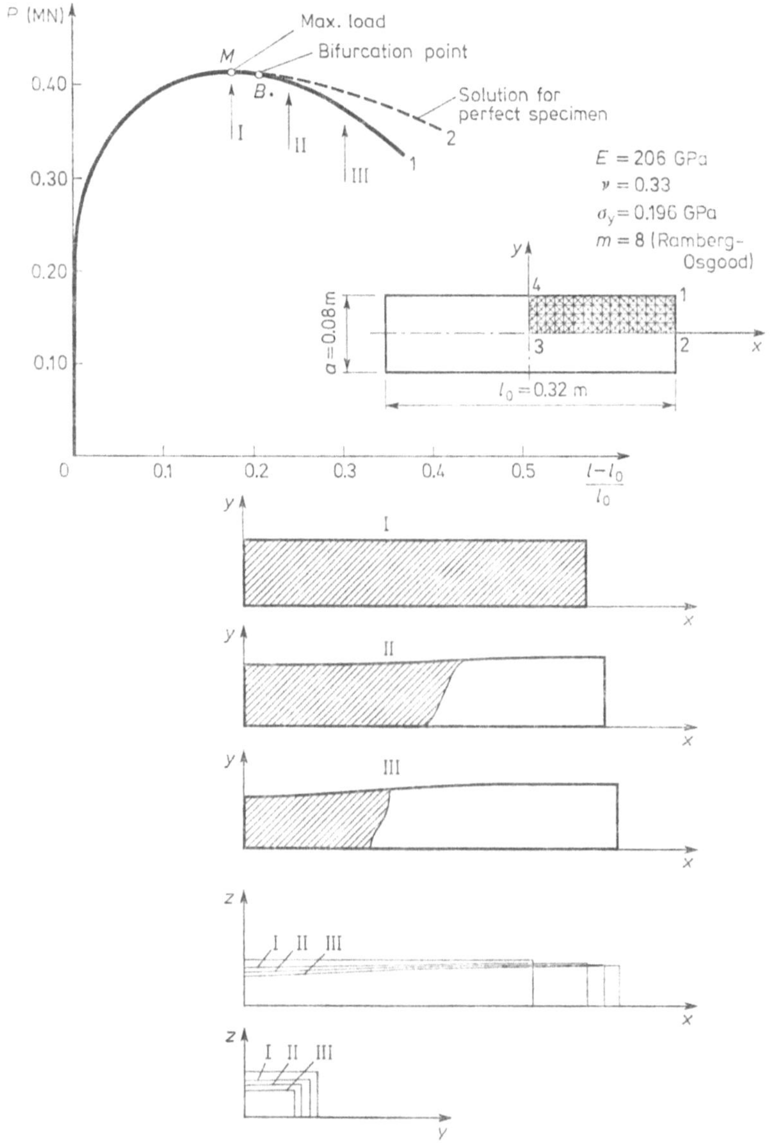

Fig. 145 Necking of an elastic-plastic strip

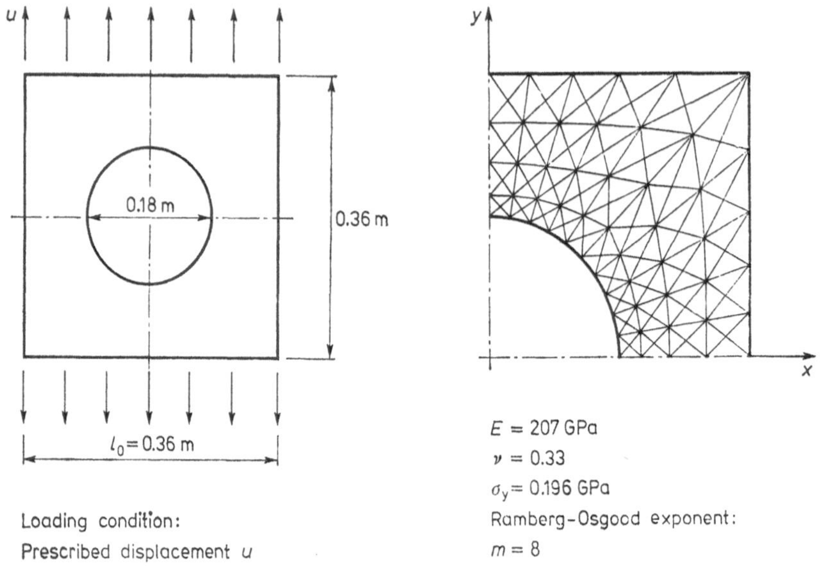

Fig. 146 Thin plate with a central hole

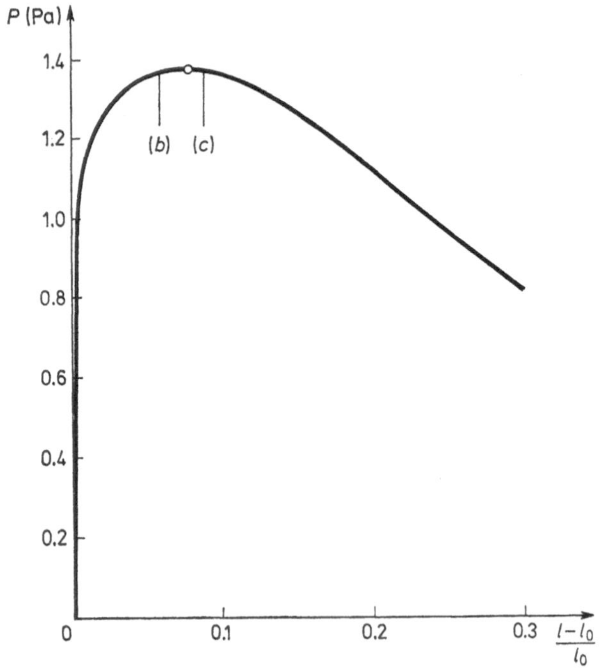

Fig. 147 Load vs extension

7.6 Extension of a thin rectangular plate with a central hole

The next plane stress problem to be reported is the one shown in Fig. 146. The loading conditions and material constants are given in the figure with the parameter $m = 8$ referring to the Ramberg–Osgood curve of the form

$$\bar{\varepsilon}^{(pl)} = \frac{1.1\sigma_y}{mE}\left[\left(\frac{\bar{\sigma}}{1.1\sigma_y^0}\right)^m - \left(\frac{1}{1.1}\right)^m\right].$$

The overall elongation of 0.30 times the initial length has been imposed in 200 equal increments of prescribed displacement. The load required for extension possesses a maximum, Fig. 147 at an overall elongation of approximately 0.08 times the initial length of the plate. After this point load instability is apparent, plastic unloading sets is in the plate, Fig. 148 and the plastic region contracts to the narrow cross-section where extensive deformations occur. The parts remaining elastic move almost rigidly while the narrow section necks up to one-half of its initial width, Fig. 149.

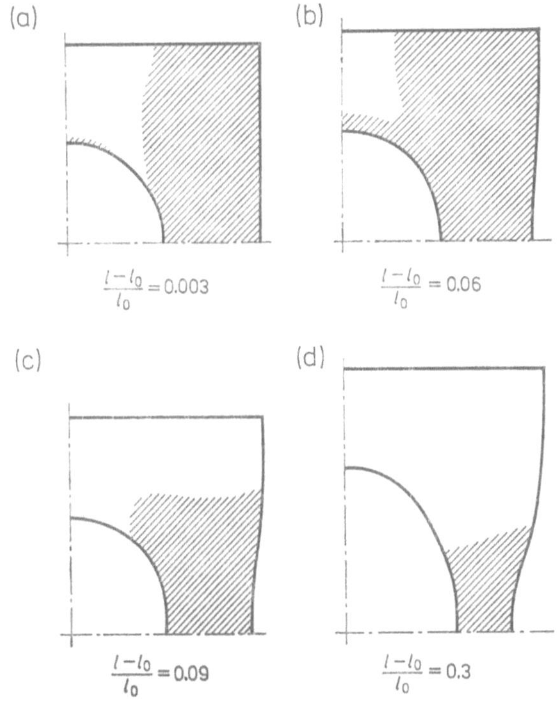

Fig. 148 Stages of deformation and plastic regions

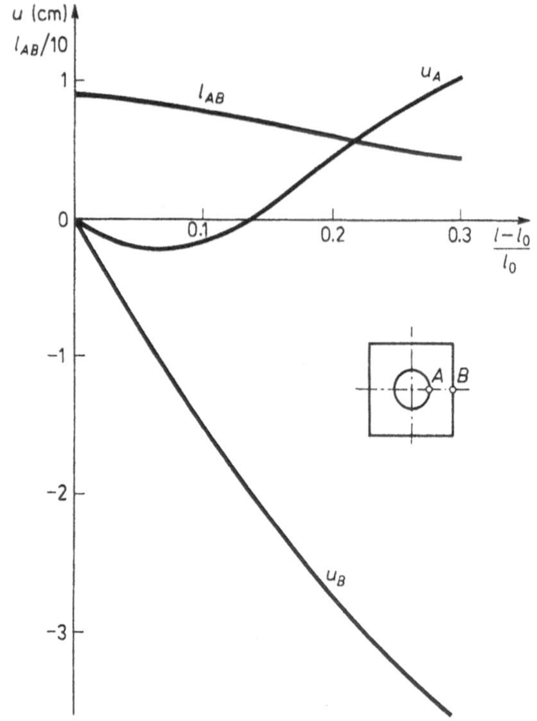

Fig. 149 Deformation of the weakened section

Summary

In this section we have exemplified the effectiveness of different formulations and algorithms by discussing the results obtained for various complex nonlinear problems under plane stress.

8

Plate and shell problems

Purpose of the chapter

We have so far discussed different theoretical aspects of the nonlinear analysis of axisymmetric thin shells, Sec. 4.7, presenting also details of the numerically oriented approach in the finite element context, Sec. 4.8. The first two sections of the present chapter are intended as the demonstration of the power of the presented approach as applied to different axisymmetric plate and shell problems. This is followed by a stability analysis of some spatial prismatic plate assemblies.

8.1 Thin elastic-plastic axisymmetric shells

Using the derivations of Sec. 4.8 we shall now present numerous illustrations of nonlinear static and dynamic analysis of thin axisymmetric shells.

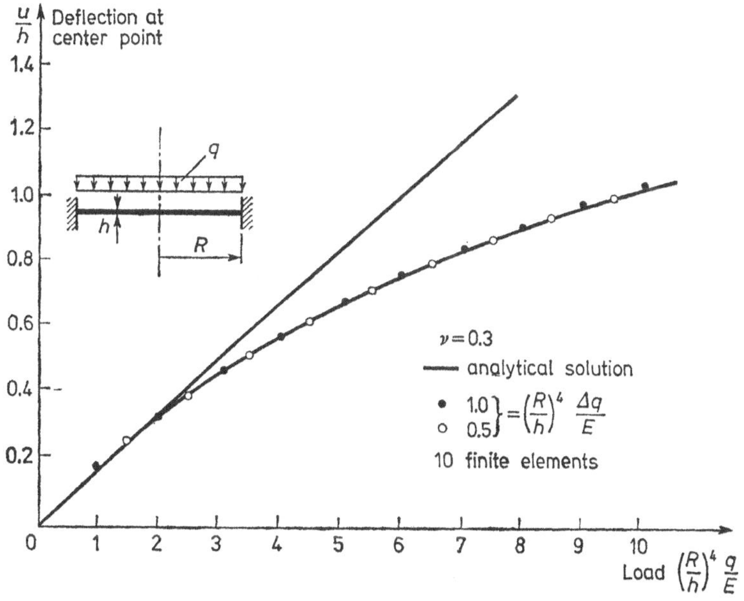

Fig. 150 Large displacements of thin elastic circular plate clamped at the boundary

Example 1. Large displacements of thin elastic circular plate clamped at the boundary.

The results of static analysis using 10 ring finite elements are given in Fig. 150. The load increment was assumed to be $\Delta q = Eh^4/R^4$; the agreement of the numerical results with those obtained in [154] by using an analytical approach is perfect.

Example 2. Limit load assessment for simply supported circular plate.

The loading, geometry and elastic-perfectly plastic material characteristics are given in Fig. 151(a). The results of a series of the step-by-step analyses

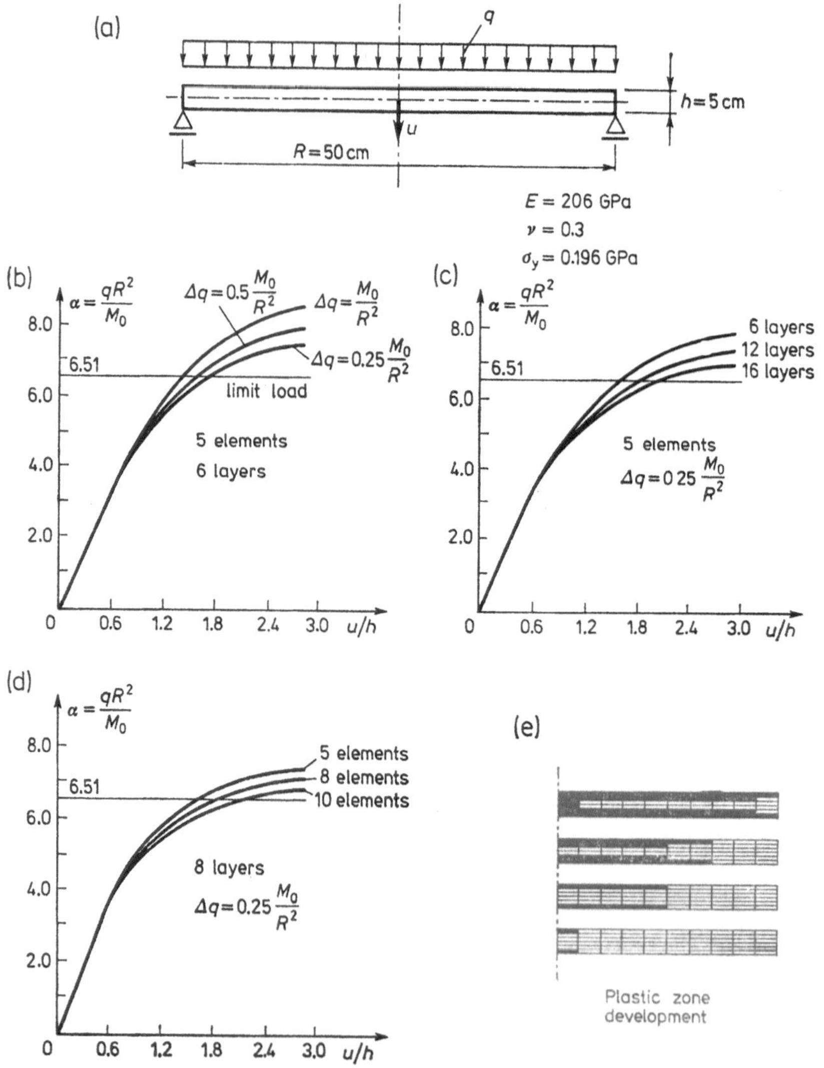

Fig. 151 Limit load assessment for simply supported circular plate

are shown in Figs. 151(b), 151(c), 151(d) and 151(e). The effects of the load step size and of the number of layers across the plate thickness are clearly visible from Figs. 151(b) and 151(c). The improvement of the solution with the increasing number of elements is illustrated in Fig. 151(d) while Fig. 151(e) displays the plastic zone development with the increasing load.

On the basis of the above experiences the plate has been analysed once more with the load step $\Delta q = 0.25 \frac{M_0}{R^2}$, 16 layers and 10 finite elements. The so obtained value of the limit load differs from the exact solution by less than 1% which proves the practical applicability of the approach. The curves of total and last (i.e. just preceding the plate collapse) incremental deflections

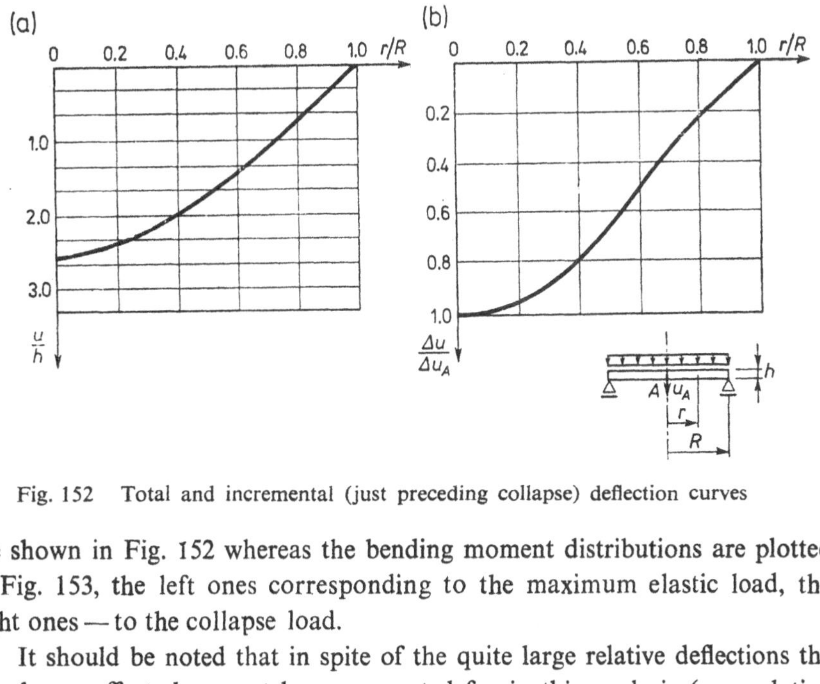

Fig. 152 Total and incremental (just preceding collapse) deflection curves

are shown in Fig. 152 whereas the bending moment distributions are plotted in Fig. 153, the left ones corresponding to the maximum elastic load, the right ones — to the collapse load.

It should be noted that in spite of the quite large relative deflections the membrane effects have not been accounted for in this analysis (no updating of the deformed plate geometry, the initial stress matrix neglected). Thus, the results may be considered as obtained in the framework of the classical limit load approach.

Example 3. Limit load assessment for clamped circular plate.

The similar analysis has been performed for the plate of the same geometry and material constants and the same type of loading but with clamped boundary conditions, Fig. 154. Because of higher gradients in the predicted functions describing plate deflections and moment distributions, 15 finite elements and 8

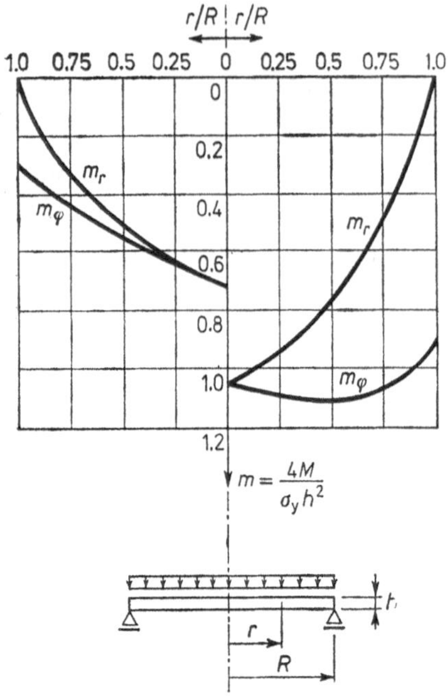

Fig. 153 Distribution of bending moments

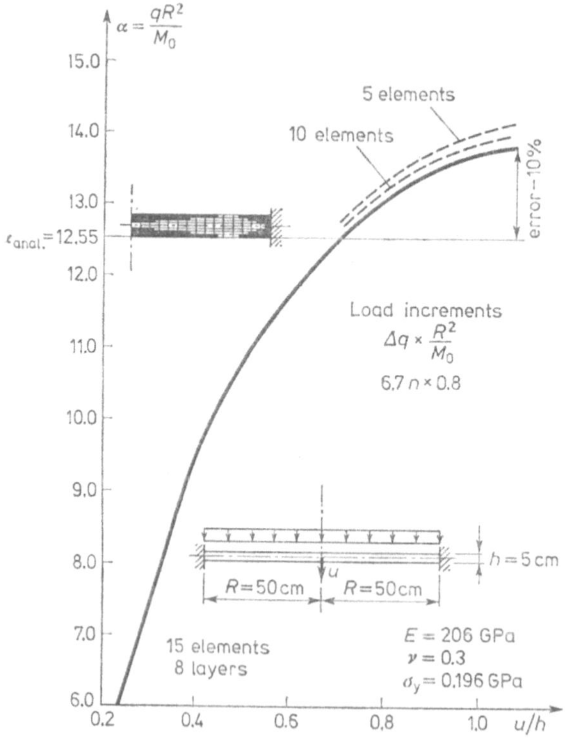

Fig. 154 Limit load assessment for clamped circular plate

layers have been considered together with the load step size given in the figure. Error of 10% in the value of the limit load has been obtained as shown in Fig. 154; the increase in number of elements to 24 and in numbers of layers to 16 has reduced the error to 2.5%. In Fig. 154 the plastic zone distribution at the load equal to the exact collapse load is also given.

Example 4. Limit load assessment for plates with varying thickness.

Simply supported plate with thickness varying according to the function given in Fig. 155 has been analysed for different value of the parameter β

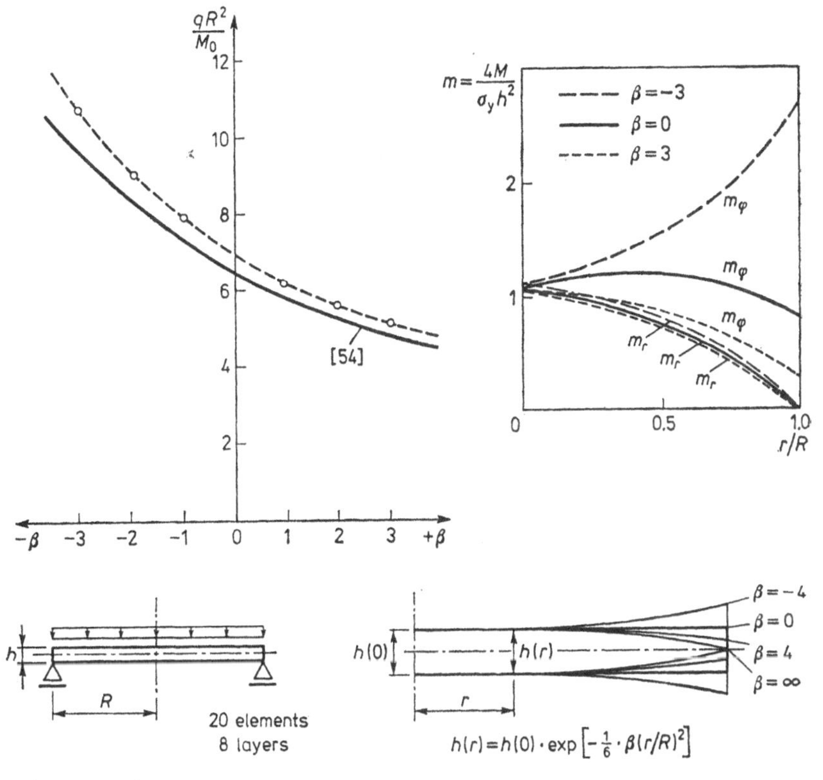

Fig. 155 Limit load assessment for plate with varying thickness

($\beta= -3, -2, -1, 1, 2, 3$). The numerically found limit loads have been compared with those obtained in [54] by using another numerical method. The radial and circumferential bending moments for different values of β are also plotted in Fig. 155.

Example 5. Circular plate of elastic-plastic material with isotropic hardening.

The piecewise linear stress-strain curve for the plate material is defined in Fig. 156(b). The plate is loaded by the load $P = 2\pi r V$ uniformly distributed

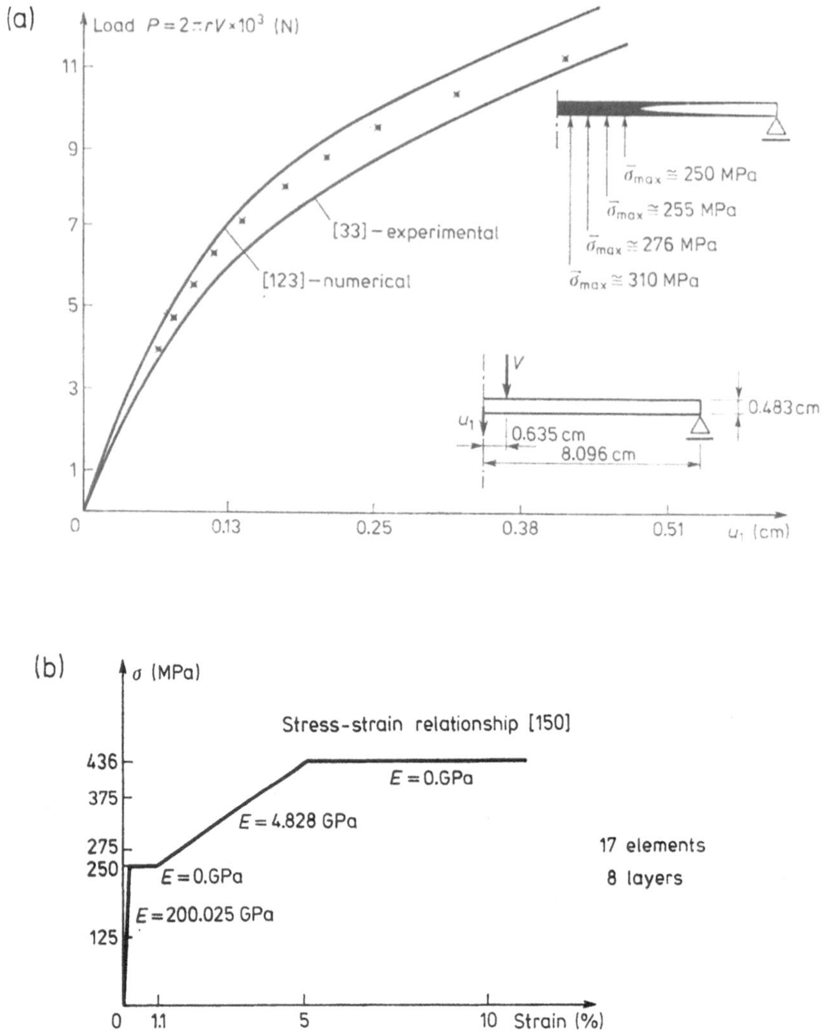

Fig. 156 Circular plate of elastic-plastic material with isotropic hardening

over the ring with the radius $r = 0.635$ cm. The results of the analysis with 17 elements and 8 layers are compared against those obtained experimentally in [33] and numerically in [123].

Example 6. Large displacement analysis of elastic-plastic circular plate.

The plate is simply supported but no radial displacements at the boundary are allowed. The geometrically nonlinear effects are fully taken into account by employing the initial stress matrix and updating the geometry at every load step, cf. Sec. 4.8. The differences in the limit load value due to the large

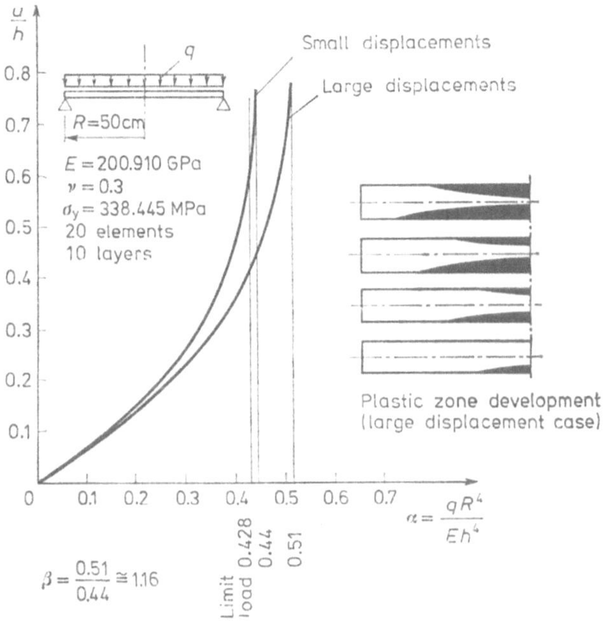

Fig. 157 Large displacement analysis of elastic-plastic circular plate

displacement effects is clearly visible from the plot in Fig. 157; accounting for geometry changes and membrane stresses made the collapse load to increase by the factor 1.16; the now asymmetric plastic zone distribution is shown in Fig. 157 as well.

Example 7. Elastic-plastic analysis of circular plate with central hole.

A plate shown in Fig. 158(a) simply supported and loaded by a uniformly distributed ring load acting at the hole boundary has been analysed in the range of large elastic-plastic displacements. The plate material has been assumed to have isotropic plastic hardening characterized by the uniaxial stress-strain diagram taken according to experimental data of [43], Fig. 158(b). In the calculation procedure 8 ring finite elements and 6 layers over the plate thickness have been assumed. The load have been applied in equal increments of 196.2 N and the hardening curve modelled by a piecewise-linear approximation consisting of 7 straight segments.

The numerical results are given in Fig. 158(c) in which the experimental curves load versus corresponding deflection of [43] are also plotted. With the greatest relative error of about 7% (for $r = 50$ mm and $P = 8829$ N) the agreement is found to be satisfactory.

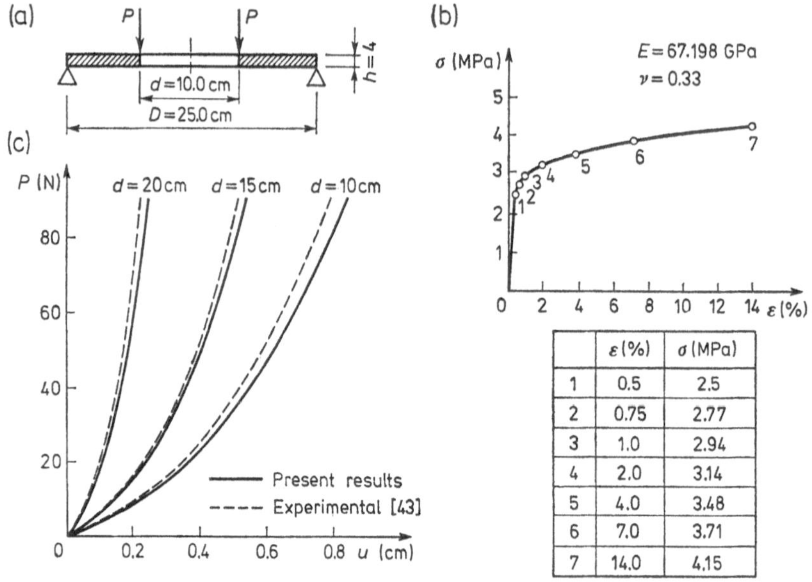

	ε (%)	σ (MPa)
1	0.5	2.5
2	0.75	2.77
3	1.0	2.94
4	2.0	3.14
5	4.0	3.48
6	7.0	3.71
7	14.0	4.15

Fig. 158 Elastic-plastic plate with a hole

Example 8. Elastic-plastic circular plate under complex load histories.

The plate of Fig. 159 has been subjected to the transverse uniformly distributed load q and to the uniform membrane tension/compression t. For different load histories shown in Fig. 159(a) the collapse loads have been determined by means of the step-by-step analysis and compared with the results of [55].

The development of plastic zones for two selected load histories is also shown.

Example 9. Elastic-plastic analysis of cylindrical shell.

The axisymmetric cylindrical shell made of elastic-perfectly plastic material and closed by two infinitely rigid diaphragms has been loaded by the tensile forces T, Fig. 160. In this figure the load/displacement curves and some characteristic plastic zone distributions are shown. Distributions of the generalized internal forces and stresses at collapse are plotted in Fig. 161.

The similar analysis has been carried out for the shell loaded additionally by the internal pressure p; the results for two selected load histories are given in Fig. 162. The findings of this analysis are summarized for seven load histories in Fig. 163 in which comparisons with upper and lower bounds to limit loads given in [59] are also plotted.

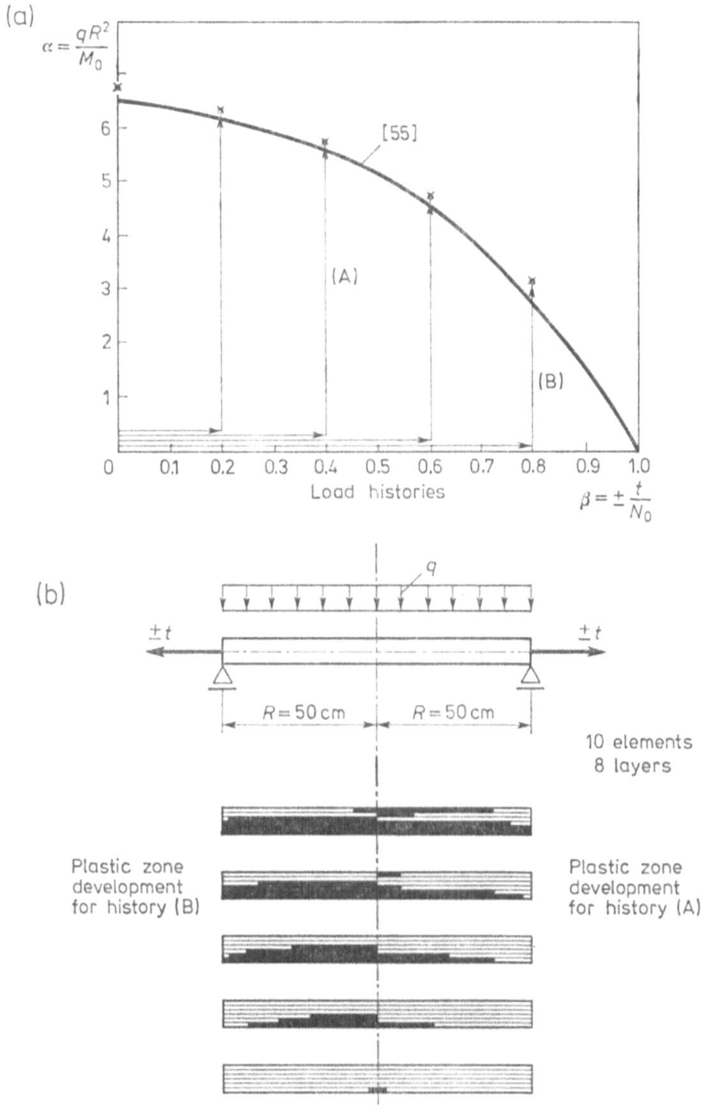

Fig. 159 Elastic-plastic plate under complex load histories

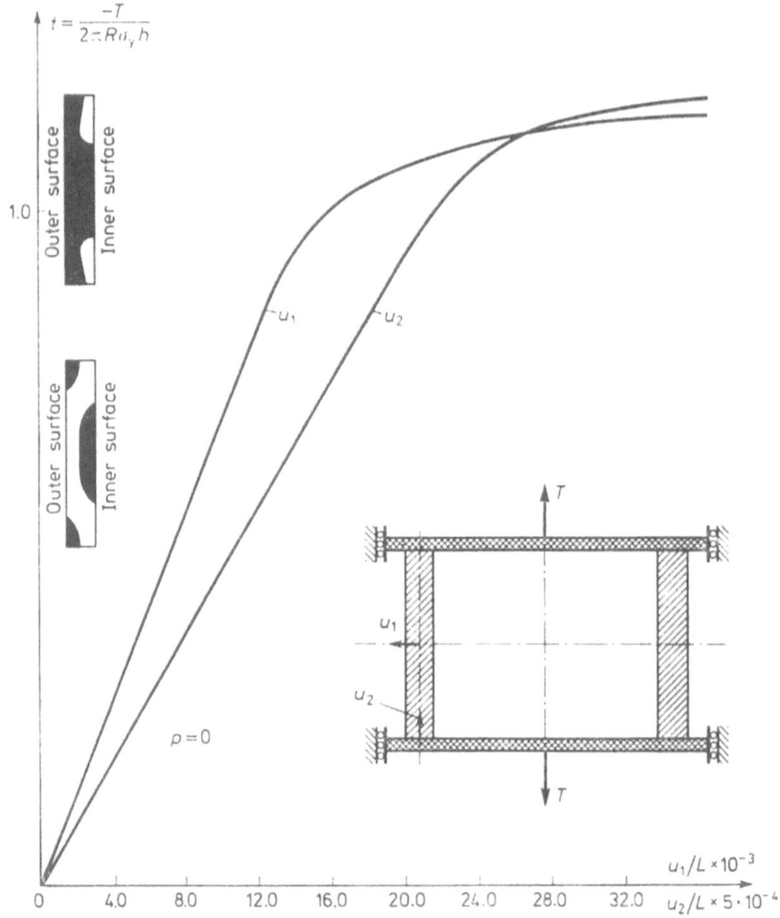

Fig. 160 Elastic-plastic analysis of a cylindrical shell

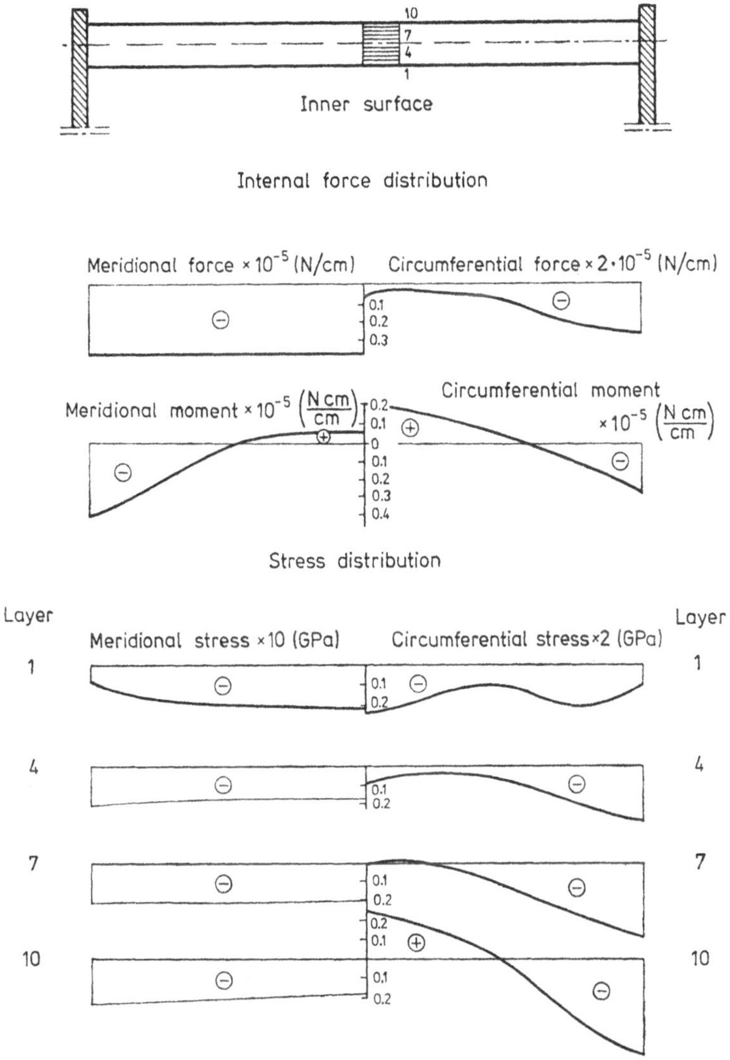

Fig. 161 Distributions of internal forces and stresses at collapse

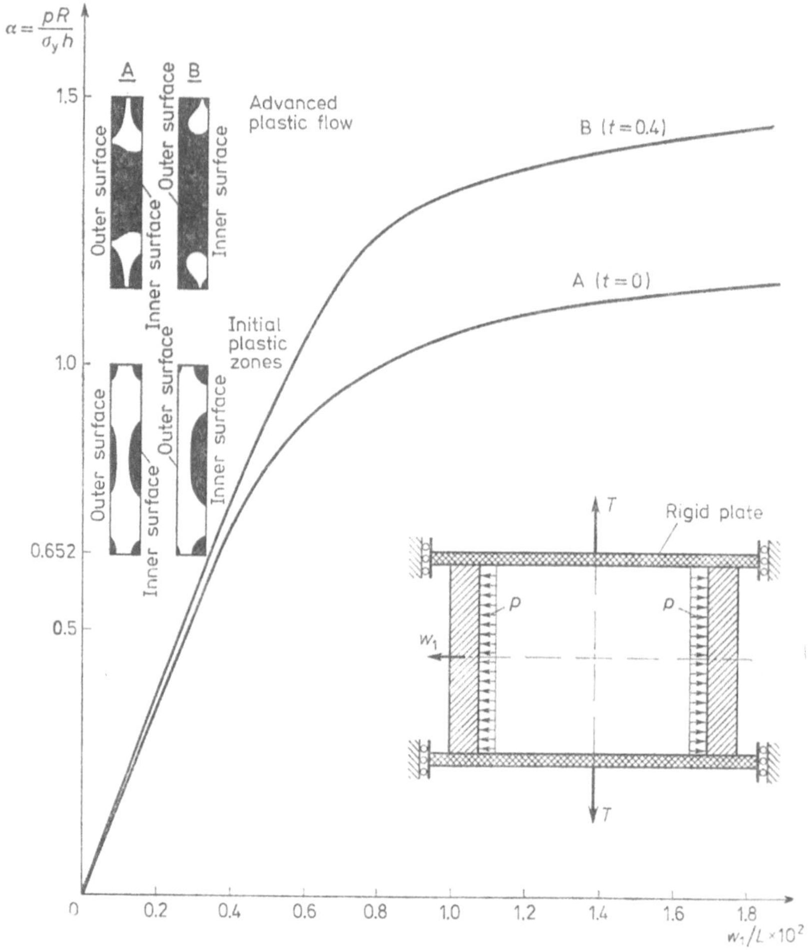

Fig. 162 Cylindrical shell under complex loading

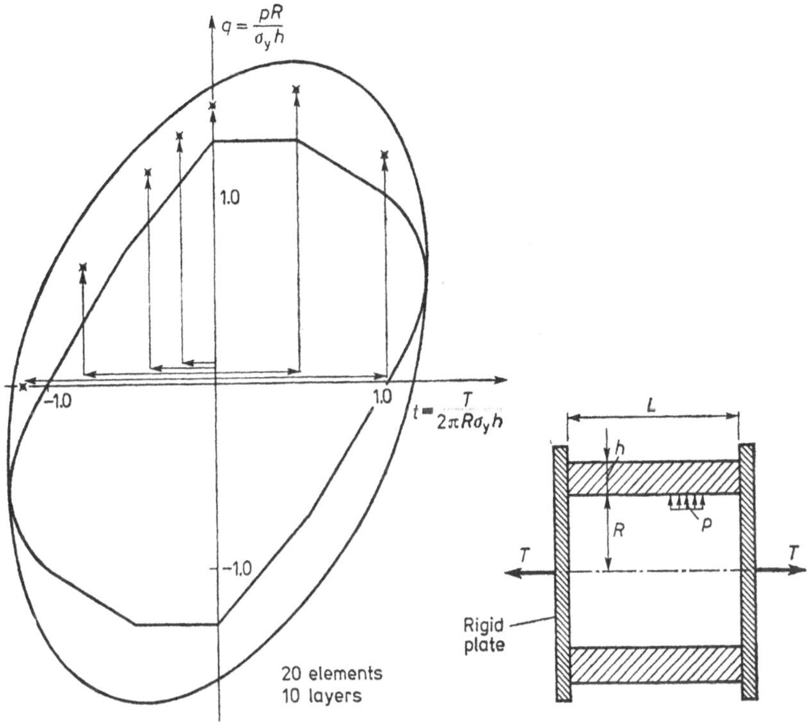

Fig. 163 Numerical results vs upper and lower bounds

Example 10. Elastic-plastic axisymmetric container under internal pressure.

The problem is defined in Fig. 164. For the pressure value of $p = 0.7863$ MPa the meridional bending moment distribution is compared with that given in [133].

Example 11. Aluminium container under external pressure.

The container geometry and its material stress-strain curve are defined in Fig. 165, the latter in accordance with the experimental results reported in [34]. The meridional strains in the vicinity of the cylindrical/conoidal surface joint for two values of the applied pressure, the magnified deformation of the container wall as well the plastic zone shape are all plotted in Fig. 166.

Example 12. Dynamic analysis of cylindrical shell.

The geometry of the shell and its support conditions are given in Fig. 167. The shell is subjected to a suddenly applied ring load acting at the middle cross-section of the shell. Only linear elastic effects are considered. Employing the discretization by means of 36 equal finite elements, the equations of motion

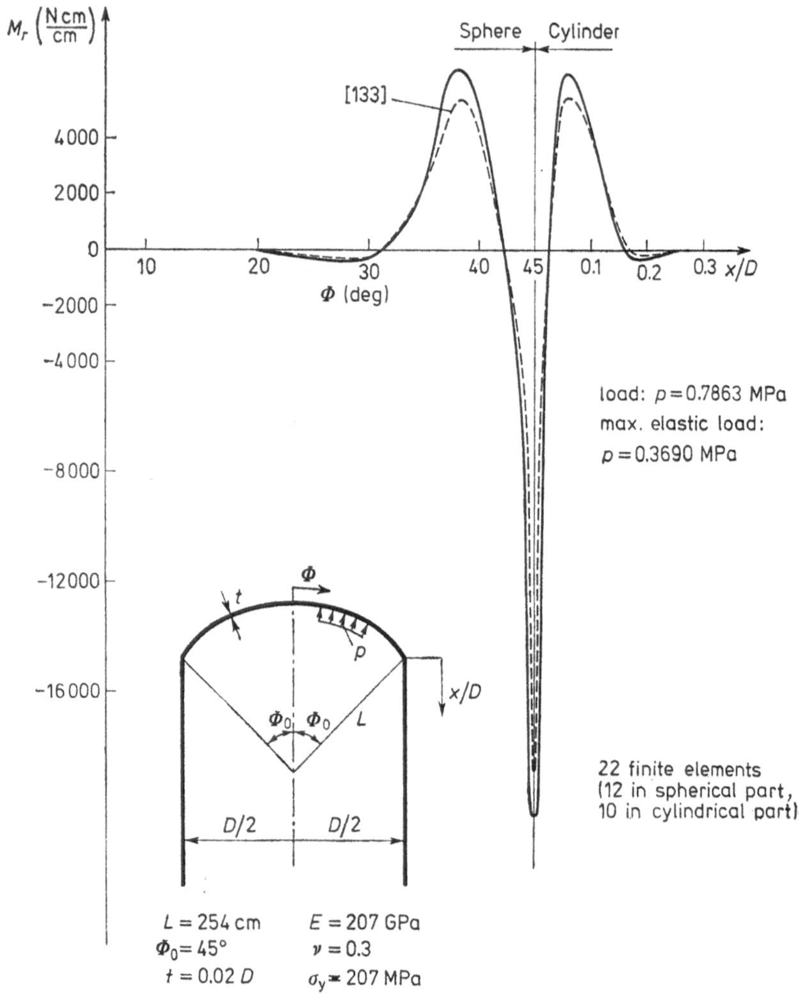

Fig. 164 Elastic-plastic container under internal pressure

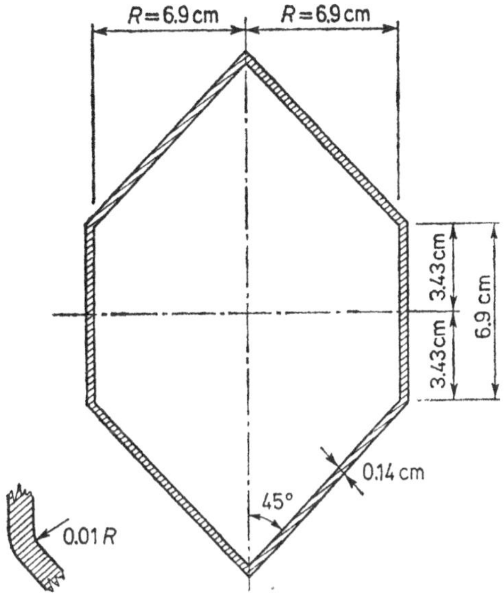

Material: Aluminium 2039	
Stress (MPa)	Strain (%)
0	0
284	0.40
316	0.45
340	0.50
354	0.55
365	0.60
379	0.70
383	0.80
386	1.00

Fig. 165 Container under external pressure

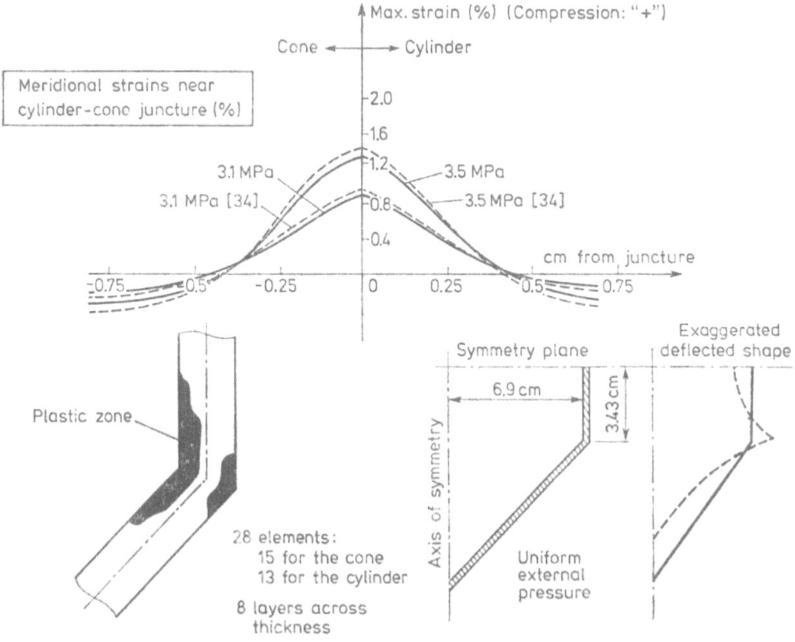

Fig. 166 Meridional strains, magnified deformations and plastic zone distribution

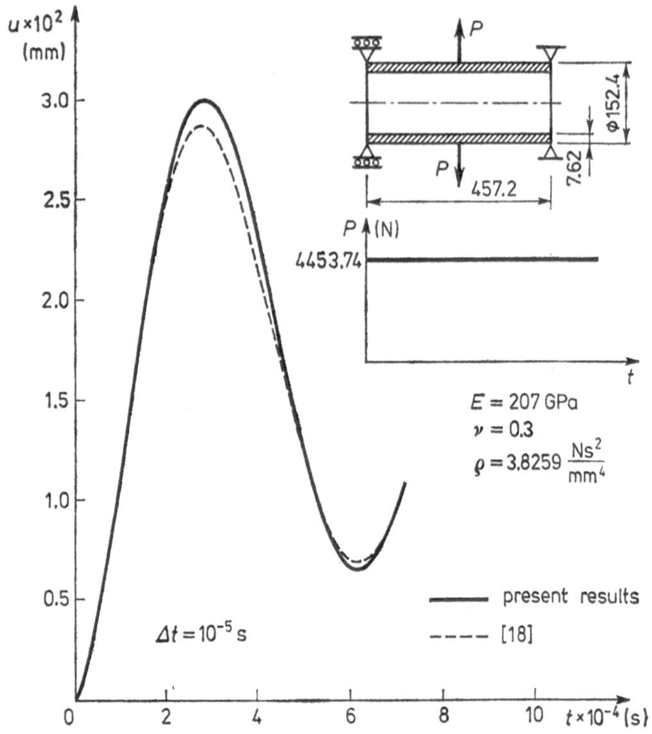

Fig. 167 Dynamic analysis of a cylindrical shell

have been integrated using the Newmark method with $\Delta t = 10^{-5}$s and $\beta = 0.25$, $\gamma = 0.5$. The central axisymmetric deflection of the shell versus time is plotted in Fig. 167. It agrees very well with the numerical results of [18].

Example 13. Nonlinear static and dynamic analysis of spherical cap under concentrated force.

The cap clamped at the entire boundary under the concentrated force P applied at the appex is considered first. The (totally inadequate) linear elastic solution and the nonlinear elastic solution obtained for two different load step sizes are all shown in Fig. 168. The successive deformed shell geometries are also plotted.

The large displacement elastic-perfectly plastic solution with the load step size of $\Delta P = 4.45$ N together with the plastic zone development are shown in Fig. 169.

The character of the solutions changes abruptly in the case of simply supported boundary conditions, Fig. 170. The typical snap-through behaviour can clearly be identified.

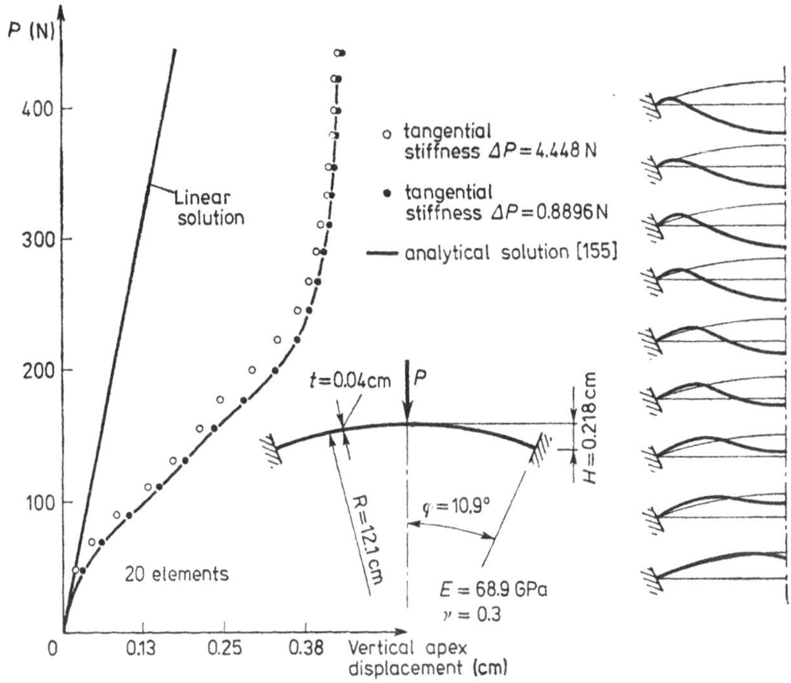

Fig. 168 Spherical cap under concentrated force

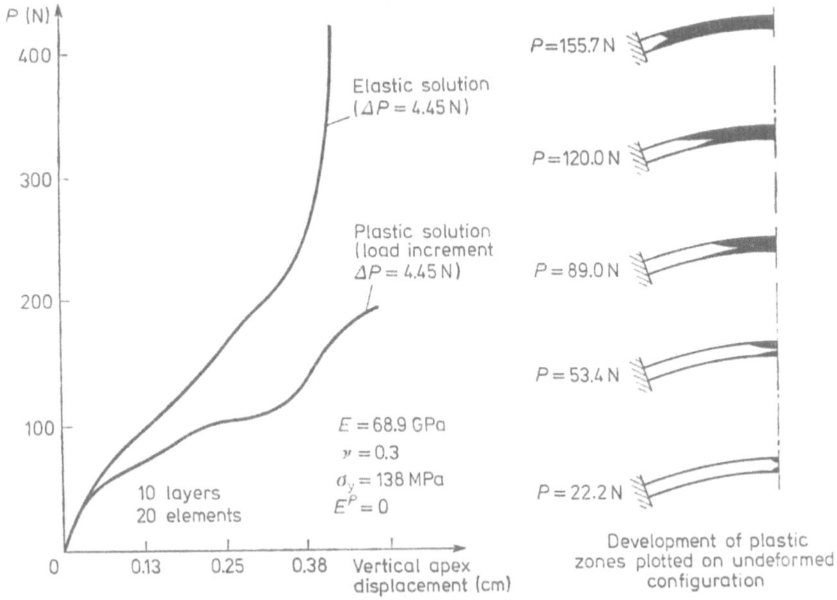

Fig. 169 Spherical cap — elastic-perfectly plastic solution and plastic zone distribution

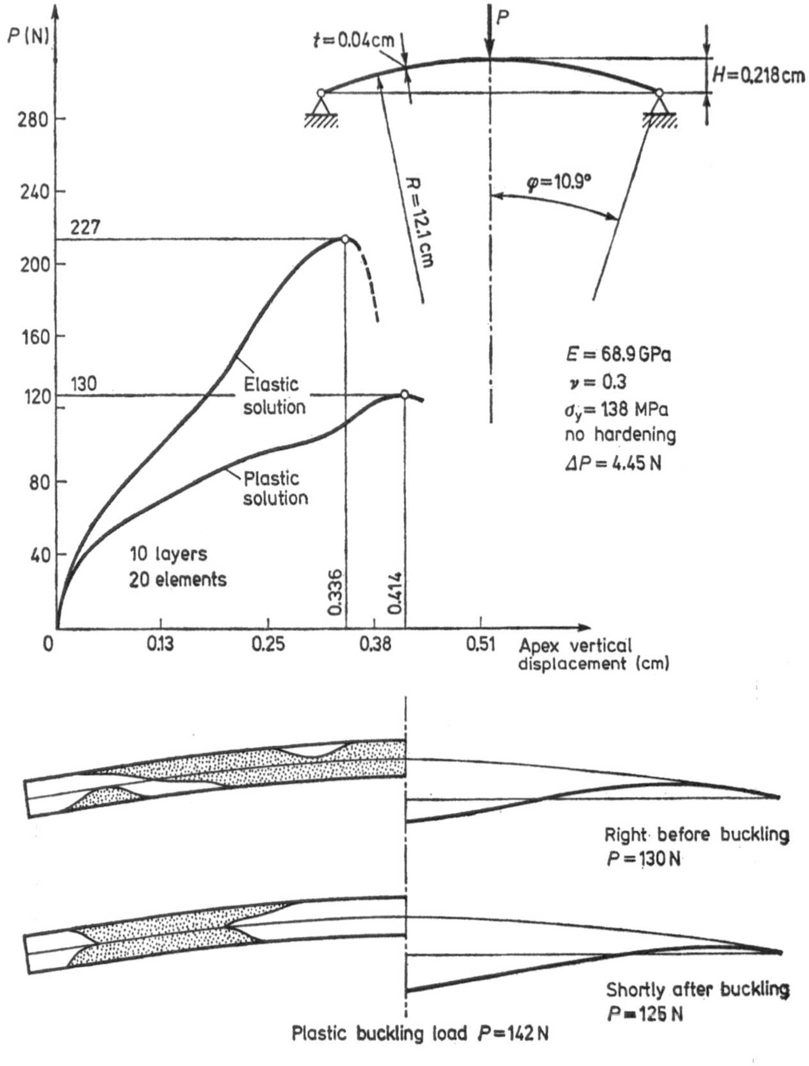

Fig. 170 Simply supported spherical cap

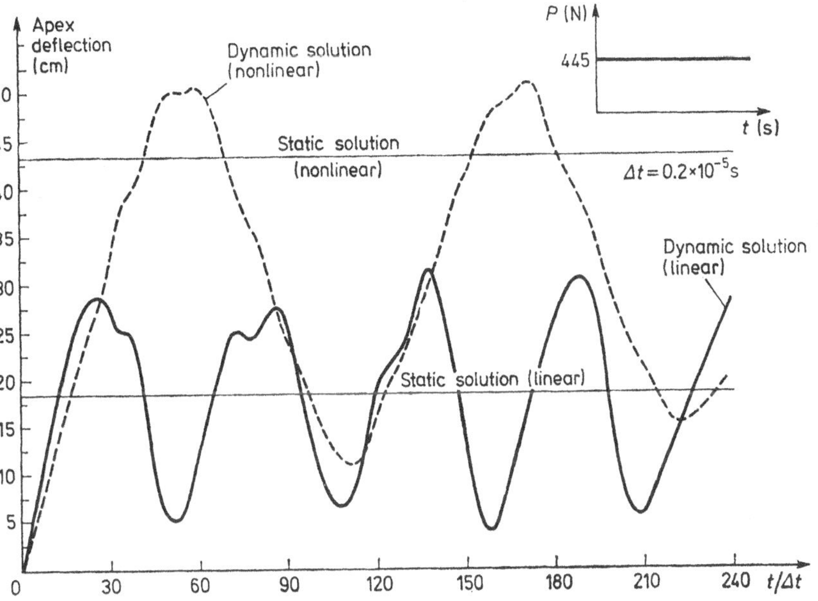

Fig. 171 Spherical cap under dynamic loading

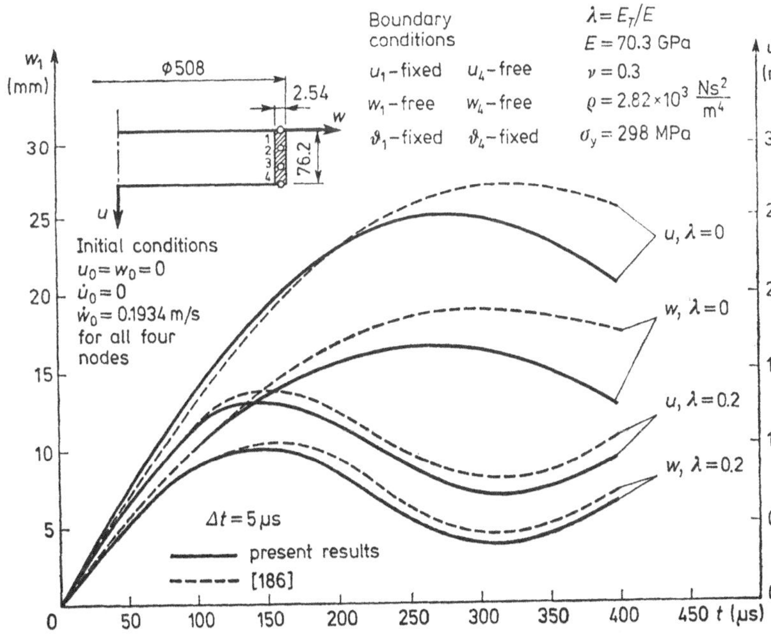

Fig. 172 Dynamic analysis of a cylindrical ring

The elastic shell was next analysed under the step-type dynamic loading, Fig. 171, using the direct integration method of Wilson, [13]. The results coincide with those reported in [16].

Example 14. Dynamic elastic-plastic analysis of cylindrical ring.

The behaviour of the cylindrical ring in Fig. 172 has been studied for two material models: elastic-perfectly plastic and elastic-plastic with linear isotropic hardening. The shell is set into motion by imposing the initial radial velocity to all its nodal points. The results are plotted in Fig. 172 and compared with those given in [186].

$E = 70$ GPa
$\nu = 0.3$
$\sigma_y = 110$ MPa
$E_T = E/200$
$\bar{\varepsilon}^{(c)} = \bar{\sigma}^n t^m$
$A = 1.88 \times 10^{-12} (\text{MPa})^{-n} \, \text{h}^{-m}$
$n = 4.43$
$m = 0.273$

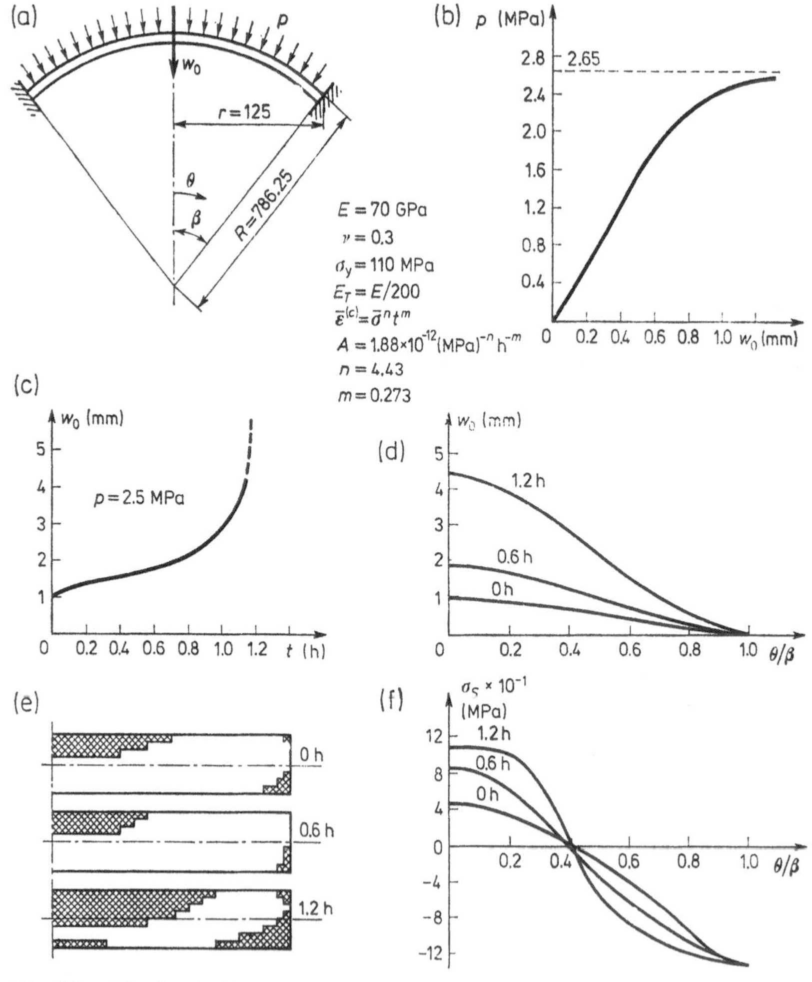

Fig. 173   Elastic-plastic-creep analysis

Example 15. Elastic-plastic-creep analysis of spherical cap.

The spherical cap clamped at the boundary is subjected to the external pressure and analyses incrementally up to the loss of stability by the snap-through. For no-creep case instability occurred at $p = 2.6$ MPa, Fig. 173. The same cap with creep effects included was simulated to creep under the pressure of $p = 2.45$ MPa which was 1.55 MPa less than the elastic-plastic critical value. Assuming the creep law in the form of eq. (4.156) with the constants A, m, n taken from [180] the numerical results are shown in Figs. 173(b)–173(f). The apex displacement versus time curve defines the creep snap-through instability taking place after 1.2 hour of the process. In Fig. 173(d) the middle surface displacements are shown at three different time instants, the last one right before the buckling. Figs. 173(e), 173(f) illustrate the inelastic zone development and the distribution of meridional stresses.

8.2 Elastic-viscoplastic analysis of axisymmetric shells

We have mentioned in Sec. 4.8 that the formulation there presented may also be employed for the analysis of time dependent problems by means of the so-called elastic-viscoplastic material model. Without going into too much theoretical details about modelling such materials we recall that the viscoplastic strain rate may be postulated in the form

$$\dot{\boldsymbol{\epsilon}}^{(vp)} = \bar{\gamma}\left\langle \Phi\left(\frac{\bar{\sigma}-\sigma_y}{\sigma_y}\right)\right\rangle \frac{\partial f}{\partial \boldsymbol{\sigma}^T} \tag{8.1}$$

where

$$\langle \Phi \rangle = \begin{cases} \Phi & \text{if} \quad \bar{\sigma}-\bar{\sigma}_y \geqslant 0, \\ 0 & \text{if} \quad \bar{\sigma}-\bar{\sigma}_y < 0, \end{cases}$$

and $\bar{\gamma}$ is referred to as the viscosity parameter. In the simplest case we may take $\bar{\sigma} = \sqrt{3J_2}$, $J_2 = \frac{1}{2}\boldsymbol{\sigma}^{D^T}\boldsymbol{\sigma}^D$ and, cf. eq. (4.157)

$$\bar{\Phi}\left(\frac{\bar{\sigma}-\bar{\sigma}_y}{\sigma_y}\right) = \frac{\bar{\sigma}-\bar{\sigma}_y}{\sigma_y} = \frac{\sqrt{3J_2}-\sigma_y}{\sigma_y}, \tag{8.2}$$

which leads to the expression for the viscoplastic strain rate in the form

$$\begin{aligned} \dot{\boldsymbol{\epsilon}}^{(vp)} &= \frac{\sqrt{3}}{2}\bar{\gamma}\frac{\sqrt{3J_2}-\bar{\sigma}}{\sqrt{J_2}\,\bar{\sigma}}\boldsymbol{\sigma}^D \quad && \text{if} \quad \sqrt{3J_2}-\sigma_y \geqslant 0, \\ \dot{\boldsymbol{\epsilon}}^{(vp)} &= 0 && \text{if} \quad \sqrt{3J_2}-\sigma_y < 0. \end{aligned} \tag{8.3}$$

Denoting

$$\gamma = \frac{\sqrt{3}}{2}\bar{\gamma}, \quad \tfrac{1}{2}\dot{\boldsymbol{\epsilon}}^{(vp)^T}\dot{\boldsymbol{\epsilon}}^{(vp)} = I_2 \tag{8.4}$$

and forming the equation

$$\tfrac{1}{2}\dot{\boldsymbol{\epsilon}}^{(vp)^T}\dot{\boldsymbol{\epsilon}}^{(vp)} = \gamma^2 \frac{\left(\sqrt{3J_2} - \sigma_y\right)^2}{\sigma_y^2 J_2} \tfrac{1}{2}\boldsymbol{\sigma}^{D^T}\boldsymbol{\sigma}^D$$

we arrive at

$$I_2 = \gamma^2 \frac{\left(\sqrt{3J_2} - \sigma_y\right)^2}{\sigma_y^2}$$

or

$$\sqrt{3J_2} - \sigma_y\left(\frac{\sqrt{I_2}}{\gamma} + 1\right) = 0. \tag{8.5}$$

The last relation may clearly be treated as the viscous counterpart to the yield condition (4.204). It may also be observed that for $\gamma \to \infty$ eq. (8.5) reduces to eq. (4.204). Also, very slow processes for which $\dot{\boldsymbol{\epsilon}}^{(vp)} \cong 0$ lead to the same result. On the other hand, for finite values of the parameter γ (and far from zero inelastic strain rates $\dot{\boldsymbol{\epsilon}}^{(vp)}$) the yield condition (8.5) cannot be considered as a real constraint imposed on stresses since only eq. (8.5), and not eq. (4.204), has to be fulfilled.

Using in eq. (8.3) the relationship (8.5) we arrive at

$$\dot{\boldsymbol{\epsilon}}^{(vp)} = \frac{\sqrt{3}}{\sigma_y} \frac{\sqrt{I_2}}{1 + \left(\sqrt{I_2}/\gamma\right)} \boldsymbol{\sigma}^D \tag{8.6}$$

which again reduces to the inviscid relation (4.205) for $\gamma \to \infty$.

The elastic-viscoplastic problems may be solved by means of elastic-plastic algorithms described in Chapter 5 by treating the viscoplastic strains as the initial strains. We also note that by means of a sequence of elastic-viscoplastic solutions we may obtain a good approximation to the corresponding elastic-plastic problem, i.e. the problem characterized by the same elastic constants, yield condition, associated flow rule and hardening law but without viscous effects. To this purpose we can either consider a limiting case of $\gamma \to \infty$ or attempt a plastic solution as a stationary value (i.e. the one obtained under load that is constant in time) of the viscoplastic problem. The first approach is possible in computing practice only through repeated application of the viscoplastic analysis with increasing values of γ so that the ultimate plastic solution may be obtained by some extrapolation technique. In the second approach the process of plastic stress redistribution is interpreted as a relaxation process during which the stress path penetrating the static yield condition

$$\left\{ A-B-B_1'-C_1-C_1'-D_1-D_1' \right\} - \text{stress history 1}$$
$$\left\{ A-B-C_2-D_2-D_2' \right\} \qquad - \text{stress history 2}$$
$$\underline{\qquad\qquad} - \text{applied stress increment}$$
$$---- - \text{viscoplastic deformation}$$
$$\qquad\qquad \text{under constant load}$$

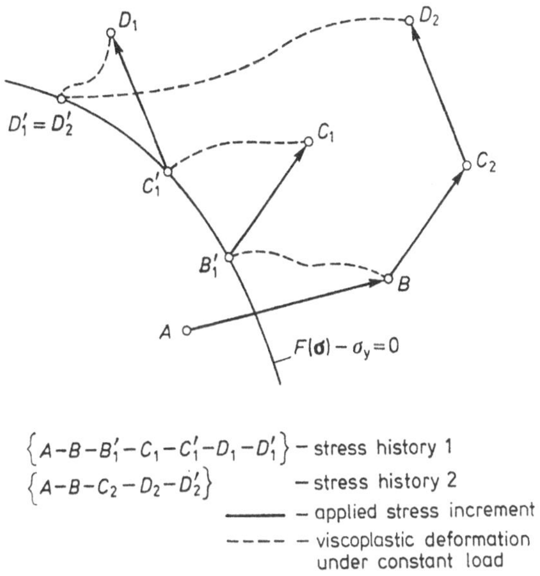

Fig. 174 Stress paths for elastic-viscoplastic behaviour

returns to the yield surface with the passage of time as illustrated for material without hardening in Fig. 174. Whichever history of viscoplastic deformations has taken place (say, $A-B-B_1'-C_1-C_1'-D_1-D_1'$ or $A-B-C_2-D_2-D_2'$ in Fig. 174), the same stationary solution $D_1' = D_2'$ should be obtained, provided the alternative external loadings differed only in their time distribution (one could be obtained from another by a mere change of time scale). However, even if this approach has been proved on numerous examples to be quite effective, [3], [111], [185], its use must be treated with a certain caution as there is no guarantee that the final stress state will be really that of the corresponding inviscid problem. In particular, this concerns statically determinate systems because of their inability to redistribute the stresses under fixed loads; in fact, stress relaxation due to viscoplastic deformation cannot occur in these structures, [4].

To illustrate these considerations we shall first consider a torispherical shell in the form of the head of a pressure vessel shown in Fig. 175. The shell is divided into 20 finite elements each of which is divided into 10 layers. An analysis of the shell is performed well into the inelastic range assuming in turn elastic-plastic and elastic-viscoplastic material properties. The step-by-step inviscid solution is given in Fig. 175(b) while those corresponding to elastic-viscoplastic material properties with different viscosity coefficients are shown in Fig. 175(c). The curves obtained for the increasing values of γ tend to the inviscid solution for which $\gamma = \infty$. In Fig. 175(d) the second approach as

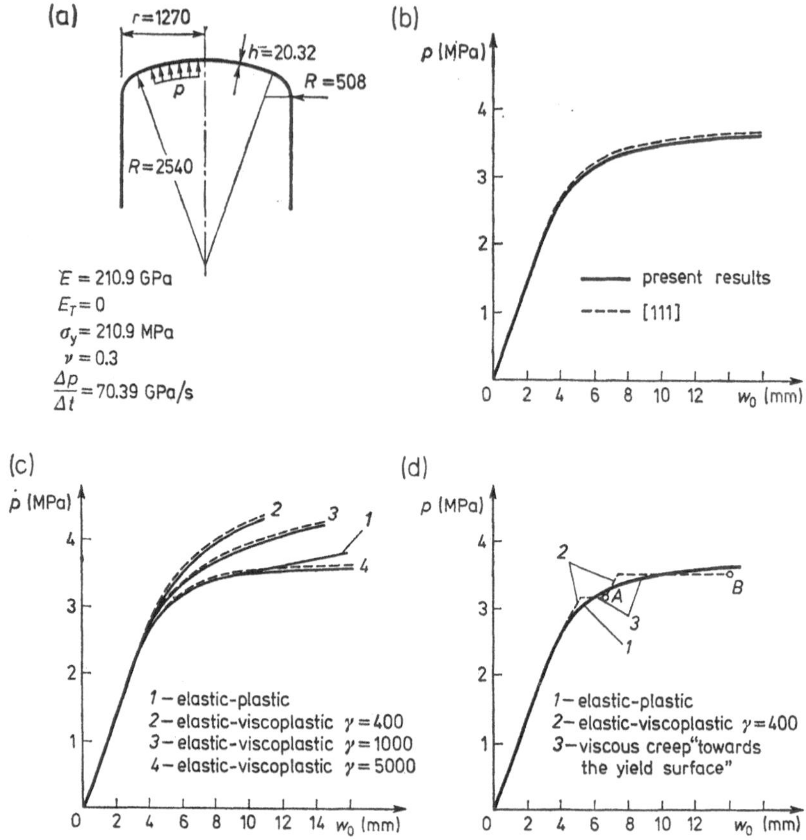

Fig. 175 Torispherical shell — elastic-viscoplastic analysis

described above is illustrated by showing that it is possible to obtain the inviscid solution by performing the viscoplastic computations and using the property of the stationarity of the viscid process. The shell for $\gamma = 400$ 1/s is loaded up to the pressure value of $p = 3.139$ MPa and in the subsequent viscoplastic flow under constant pressure a point A is obtained corresponding approximately to the elastic-plastic (i.e. inviscid) solution. For the pressure load of $p = 3.433$ MPa a point B is obtained similarly. Both elastic-plastic and elastic-viscoplastic solutions given here display a satisfactory agreement with those reported in [111].

As the second example in this section we shall consider a thick cylinder under internal pressure, Fig. 176. Since the cylinder is assumed to be infinitely long the analysis of the thin layer shown in the figure is sufficient to adequately represent the cylinder behaviour. The axisymmetric solid elements with the triangular radial cross-section in the form of the plane elements of Sec. 4.9

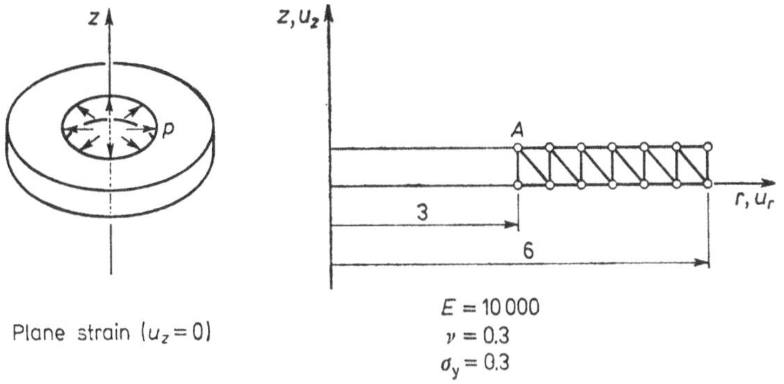

Fig. 176 Thick cylinder in plane strain

are used in the analysis—the way the axisymmetric solid elements can be constructed from the corresponding plane elements is described in every finite element textbook and is therefore omitted here.

First, the elastic-perfectly plastic material model is considered with the material constants given in Fig. 176. The well known limit load solution is given in this case by

$$p_0 = \sigma_y \frac{2}{\sqrt{3}} \ln \frac{r_2}{r_1} \cong 1594.$$

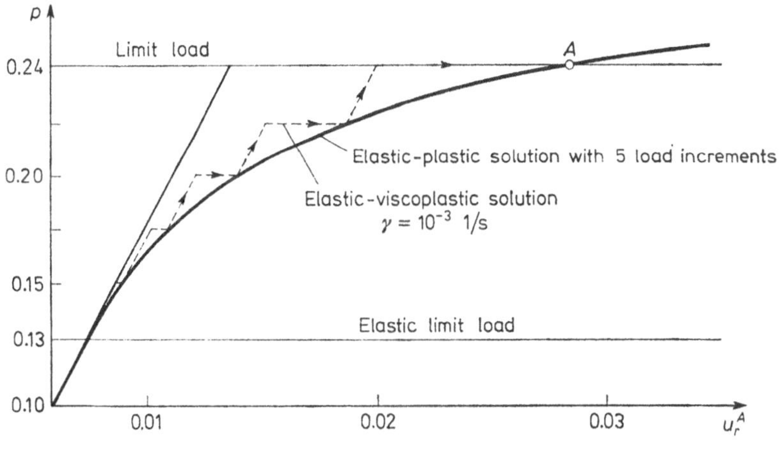

Fig. 177 Elastic-plastic and elastic-viscoplastic stationary solutions

This result is confirmed numerically by using 20 incremental steps with Newton–Raphson iterations resulting in the point A, Fig. 177.

Next, the material model is generalized to include the viscous characteristics as well. The solutions obtained for different rates of loading are shown

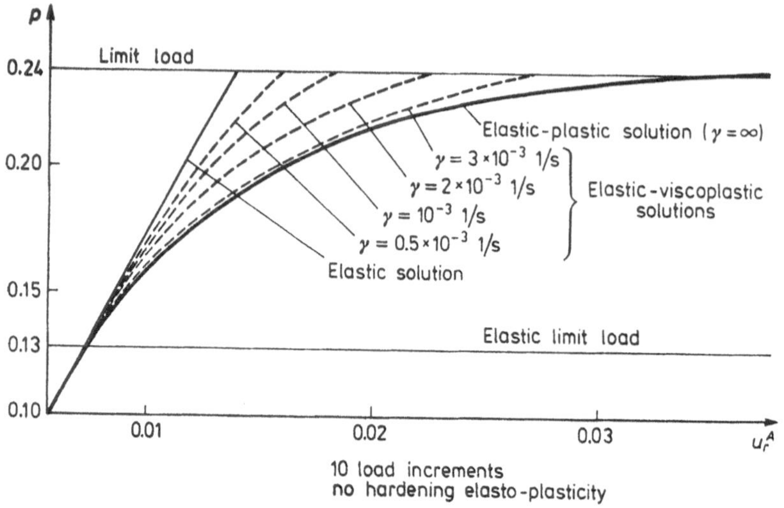

Fig. 178 Solutions for different rates of loading

in Fig. 178 whereas the variation of loading path and the stationary viscoplastic solutions are plotted in Fig. 177. It is again observed that the latter solutions coincide for each particular load level with the points on elastic-plastic curve corresponding to the same load.

8.3 Buckling of elastic-plastic spatial plate assemblies

To further illustrate the analysis of thin complex structures we shall now consider spatial structures consisting of thin, rectangular plates rigidly connected together along their longitudinal edges. A typical configuration is shown in Fig. 179. The line connections, which appear as points when a cross-section of the structure is viewed, will be referred to as nodes below.

We assume for a thin-walled cross-section that the stiffness to translation in the plane of each component flat is much greater than the stiffness to rotation about the edge lines. Consequently, no component flat is allowed to be translated in its own plane during buckling and the edge lines at the junctions between flats remain fixed in space.

Each plate may be subjected to arbitrary changing (with the axis along the transverse boundary of the structure) longitudinal stress σ_{xx} and to constant (along each element boundary) arbitrary transverse stress σ_{yy}—the load is assumed to vary proportionally to one scalar parameter.

As we know already, the initial buckling eigenvalue problem associated with various structures discretized by the finite element technique is characterized by the following matrix equation

$$[\mathbf{K}^{(con)} + \lambda \mathbf{K}^{(\sigma)}]\mathbf{v} = 0 \tag{8.7}$$

where the matrices $\mathbf{K}^{(con)}$ and $\mathbf{K}^{(\sigma)}$ of the dimension $N \times N$ are the constitutive and initial stress stiffness matrices, respectively, N is the global number of the degrees of freedom and λ is a load parameter to be determined via the eigen-

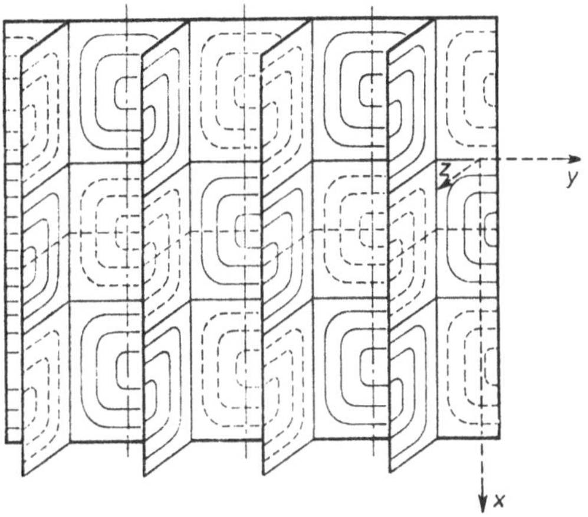

Fig. 179 Elastic-plastic spatial plate assembly

value solution. This parameter assumes values such that $\det(\mathbf{K}^{(con)} + \lambda \mathbf{K}^{(\sigma)})$ is equal to zero for the stress state used in the computation of the matrices $\mathbf{K}^{(con)}$ and $\mathbf{K}^{(\sigma)}$.

In what follows we shall consider elastic and elastic-plastic properties of the element assembly. Since the formulation (8.7) is general in nature, any suitable plasticity theory may be employed to formulate the constitutive stiffness matrix $\mathbf{K}^{(con)}$. The classical flow plasticity relations with the normality rule, smooth yield condition of the von Mises type and isotropic hardening can be assumed, for instance, in the form discussed in Sec. 4.9 with the matrix $\mathbf{K}^{(con)}$ resulting as a non-linear function of the current state of stress. However, the results reported in the existing literature prove such a formulation to be highly unsatisfactory, the structural response being much too stiff in comparison to the known experimental estimates. As a matter of fact, the problems with the application of the flow theory are in agreement with observations made by many previous researchers of the problem who stated that the generally abandoned deformation theory predicted buckling loads better than the flow theory with a smooth yield surface. Different explanations have been offered for the inaccuracy of flow theory in such problems. One of them is that the classical flow theory is not able to describe accurately enough instability phenomena because the buckling load value is quite sensitive to the direction of the normal to the loading surface and that, consequently, the shape of this surface in a local region near the stress point is very important. The smooth yield surface has also been recognized to conflict with the predictions of microstructural polycrystalline models of plasticity, which all indicate that the tangential to the yield surface total strain increment should produce inelastic strain, which is not the case for the flow theory of plasticity with a smooth yield surface. The consequence of this property is that the response for "tangential" load increments, which are typical for buckling problems, is overall much stiffer than it is for "normal" loads.

To correct this defect various forms of vertex hardening models have recently been put forward. It has also been shown that the plastic moduli which govern fully-active loading at a vertex point can be fitted to those of the deformation theory of plasticity. Put another way, for a restricted range of deformations the deformation theory of elastic-plastic material coincides with a physically acceptable incremental theory which develops a corner on its yield surface. Clearly, the necessary condition for that equivalence to hold true is that no local unloading is to occur during the whole deformation process.

This property of the deformation theory will be taken advantage of below—the so-called deformation theory will be used which accounts in a simple manner for the decreasing shear modulus with increasing plastic deformation.

In establishing relation (8.7) it was assumed that the effects of geometry up to the bifurcation load could be neglected. This assumption allows us to treat the constitutive stiffness matrix as constant in the elastic case and to neglect the initial displacement matrix. Also, in the examples to be discussed below the stress levels at bifurcation are a small fraction of the instantaneous moduli and the strains are small so that a discussion within the context of classical infinitesimal strain plasticity is justified and it is not necessary to give a precise definition to the stress measure.

The longitudinal dimension of the structural components here is many times greater than a typical cross-sectional dimension. Consequently, any standard method of matrix analysis would require a very large number of finite elements, particularly if the cross-sectional geometry is complex. To circumvent this problem the modes of buckling are assumed to vary sinusoidally in the longitudinal direction with a half-wavelength L. Thus the buckling stress obtained via (8.7) is a function of both L and the external load distributions σ_{xx}, σ_{yy}. The value of the buckling stress, minimized with respect to L, is then taken as the correct solution.

A typical configuration of the structure to be analyzed is shown in Fig.

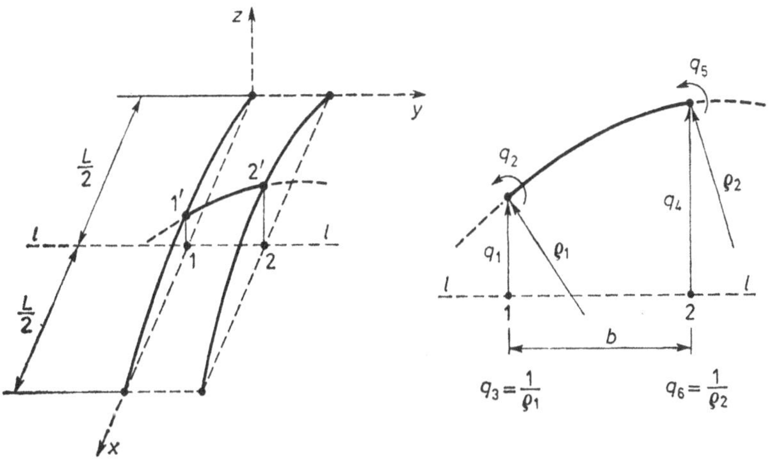

Fig. 180 One element of the plate assembly

179. One element of the assembly is shown in Fig. 180. We take the normal deflections q_1, q_4, the corresponding rotations q_2, q_5 and curvatures q_3, q_6 to represent the generalized displacement vectors at the nodes. The normal displacement distribution is taken as

$$u_z(x, y) = W(\eta)\sin\frac{\pi x}{L}, \tag{8.8}$$

where

$$\begin{aligned} &\eta = y/b, \quad W(\eta) = \mathbf{a}(\eta)^T\mathbf{q}, \\ &\mathbf{a}(\eta) = \{a_1(\eta)\ a_2(\eta)\ \dots\ a_6(\eta)\}, \quad \mathbf{q} = \{q_1\ q_2\ \dots\ q_6\}. \end{aligned} \tag{8.9}$$

As an approximation for the function $W(\eta)$ we assume a quintic polynomial whose coefficients may be determined from the compability conditions for the nodal deflections, rotations and curvatures at $\eta = 0$ and $\eta = 1$. The shape function takes then the form

$$\begin{aligned} \mathbf{a}(\eta) = \{&(1-10\eta^3+15\eta^4-6\eta^5)\ b(\eta-6\eta^3+8\eta^4-3\eta^5) \\ &b^2(0.5\eta^2-1.5\eta^3+1.5\eta^4-0.5\eta^5)\ (10\eta^3-15\eta^4+6\eta^5) \\ &b(-4\eta^3+7\eta^4-3\eta^5)\ b^2(0.5\eta^3-\eta^4+0.5\eta^5)\}. \end{aligned} \tag{8.10}$$

The components of the strain state in the element are calculated according to the standard definitions

$$\varepsilon_{xx}(x, y, z) = -z\frac{\partial^2 u_z}{\partial x^2}, \quad \varepsilon_{yy}(x, y, z) = -z\frac{\partial^2 u_z}{\partial y^2},$$

$$\varepsilon_{xy}(x, y, z) = -z\frac{\partial^2 u_z}{\partial x\,\partial y}$$

which by using eqs. (8.8) and (8.9) lead to the matrix equation of the form

$$\boldsymbol{\epsilon}(x, y, z) = \{\varepsilon_{xx} \;\; \varepsilon_{yy} \;\; \varepsilon_{xy}\} = \mathbf{B}(x, y, z)\mathbf{q} \tag{8.11}$$

where $\mathbf{B}$ is the geometric matrix, cf. eq. $(3.81)_2$, defined now as

$$\mathbf{B}_{3\times 6} = z \begin{bmatrix} \dfrac{\pi^2}{L^2}\,\mathbf{a}(\eta)^T \sin\dfrac{\pi x}{L} \\ -\dfrac{1}{b^2}\,\mathbf{a}''(\eta)^T \sin\dfrac{\pi x}{L} \\ -\dfrac{\pi}{bL}\,\mathbf{a}'(\eta)^T \cos\dfrac{\pi x}{L} \end{bmatrix} \tag{8.12}$$

and it was denoted $\mathbf{a}'(\eta) = \dfrac{\mathrm{d}\mathbf{a}(\eta)}{\mathrm{d}\eta}$.

The element stiffness matrix is formed as, cf. eq. $(3.84)_1$

$$\mathbf{k}^{(el)}_{6\times 6} = \int_0^L \int_0^b \int_{-t/2}^{t/2} \mathbf{B}^T \mathbf{C}^{(el)} \mathbf{B} \,\mathrm{d}x\,\mathrm{d}y\,\mathrm{d}z \tag{8.13}$$

where the plane stress elastic constitutive matrix is given by, cf. Sec. 2.1.5,

$$\mathbf{C}^{(el)} = \frac{E}{1-\nu^2} \begin{bmatrix} 1 & \nu & 0 \\ \nu & 1 & 0 \\ 0 & 0 & 1-\nu \end{bmatrix}, \tag{8.14}$$

E and ν denote the Young modulus and Poisson's ratio, respectively while t stands for the thickness of the element. Substituting the expressions for $\mathbf{B}$, eq. (8.12) and $\mathbf{C}^{(el)}$, eq. (8.14) into eq. (8.13) and integrating over the volume of the element we obtain the stiffness matrix in its final form as

$$\mathbf{k}^{(el)}_{6\times 6} = \mathbf{k}^{(el)}_{\mathrm{I}}\left(\frac{1}{L^3}\right) + \mathbf{k}^{(el)}_{\mathrm{II}}\left(\frac{1}{L}\right) + \mathbf{k}^{(el)}_{\mathrm{III}}(L) \tag{8.15}$$

with the matrices $\mathbf{k}^{(el)}_{\mathrm{I}}, \mathbf{k}^{(el)}_{\mathrm{II}}, \mathbf{k}^{(el)}_{\mathrm{III}}$ (which depend on the unknown half wavelength of the buckling mode in the way indicated in eq. (8.15)) given by

$$\mathbf{k}^{(el)}_{\mathrm{I}} = \frac{\pi^4 E b t^3}{24000(1-\nu^2)L^3} \times$$

$$\times \begin{bmatrix} 391.77488 & 67.31601b & 5.068542b^2 & 108.22512 & -32.68399b & 3.264791b^2 \\ & 15.00722b^2 & 1.244587b^3 & 32.68399b & -9.595952b^2 & 0.937952b^3 \\ & & 0.108225b^4 & 3.264791b^2 & -0.937952b^3 & 0.090188b^4 \\ & \text{sym.} & & 391.77488 & -67.31601b & 5.068542b^2 \\ & & & & 15.00722b^2 & -1.244587b^3 \\ & & & & & 0.108225b^4 \end{bmatrix},$$

$$\mathbf{k}_{II}^{(el)} = \frac{\pi^2 E t^3}{24000(1-\nu^2)bL} \times$$

$$\times \begin{bmatrix} 2857.1428 & (428.5715+1000\nu)b & 23.80952b^2 & -2857.1428 & 428.5715b & 23.80952b^2 \\ & 457.1428b^2 & 33.33333b^3 & -428.5715b & 28.5715b^2 & 9.523817b^3 \\ & & 3.174602b^4 & -23.80952b^2 & -9.523817b^3 & 1.587301b^4 \\ & \text{sym.} & & 2857.1428 & -(428.5715+1000\nu)b & 23.80952b^2 \\ & & & & 457.1428b^2 & -33.33333b^3 \\ & & & & & 3.174602b^4 \end{bmatrix}, \tag{8.16}$$

$$\mathbf{k}_{III}^{(el)} = \frac{ELt^3}{24(1-\nu^2)b^3} \times$$

$$\times \begin{bmatrix} 17.14286 & 8.571428b & 0.428571b^2 & -17.14286 & 8.571428b & -0.428571b^2 \\ & 5.485714b^2 & 0.314286b^2 & -8.571428b & 3.085714b^2 & -0.114286b^3 \\ & & 0.085714b^4 & -0.428571b^2 & 0.114286b^3 & 0.014286b^4 \\ & \text{sym.} & & 17.14286 & -8.571428b & 0.428571b^2 \\ & & & & 5.485714b^2 & -0.314286b^3 \\ & & & & & 0.085714b^4 \end{bmatrix}.$$

If we combine eq. (8.8) and eqs. (8.9) we shall obtain the more conventional form of the shape function approximation for the normal displacement in terms of the generalized nodal displacements as

$$u_z(x, y) = \mathbf{a}^T(\eta)\sin\left(\frac{\pi x}{L}\right)\mathbf{q} = \mathbf{\Phi q} \tag{8.17}$$

where

$$\mathbf{\Phi}(x, y) = \mathbf{a}^T(\eta)\sin\left(\frac{\pi x}{L}\right) \tag{8.18}$$

is a row vector of the shape functions.

In order to determine the initial stress stiffness matrix the same approximation for the normal displacement can be used. For constant values of the stresses σ_{xx} and σ_{yy} within the element this matrix takes the form

$$\mathbf{k}_{6\times6}^{(\sigma)} = \sigma_{xx}\int_0^L\int_0^b\int_{-t/2}^{t/2} \mathbf{\Phi}_{,x}^T\mathbf{\Phi}_{,x}\mathrm{d}x\mathrm{d}y\mathrm{d}z + \sigma_{yy}\int_0^L\int_0^b\int_{-t/2}^{t/2} \mathbf{\Phi}_{,y}^T\mathbf{\Phi}_{,y}\mathrm{d}x\mathrm{d}y\mathrm{d}z \tag{8.19}$$

where $\mathbf{\Phi}_{,x} = \dfrac{\partial\mathbf{\Phi}}{\partial x}$, $\mathbf{\Phi}_{,y} = \dfrac{\partial\mathbf{\Phi}}{\partial y}$.

Substituting the expression for $\mathbf{\Phi}$, eq. (8.18) into eq. (8.19) and integrating over the volume of the element we obtain the initial stress stiffness matrix in the form

$$\mathbf{k}_{6\times6}^{(\sigma)} = \sigma_{xx}\mathbf{k}_x^{(\sigma)}\left(\frac{1}{L}\right) + \sigma_{yy}\mathbf{k}_y^{(\sigma)}(L). \tag{8.20}$$

The matrices $\mathbf{k}_x^{(\sigma)}$ and $\mathbf{k}_y^{(\sigma)}$ depend again upon the assumed half wavelength of the buckling mode L and are given as

$$\mathbf{k}_x^{(\sigma)} = \frac{\pi^2 bt}{2000L} \times$$

$$\times \begin{bmatrix} 391.77488 & 67.31601b & 5.068542b^2 & 108.22512 & -32.68399b & 3.264791b^2 \\ & 15.00722b^2 & 1.244587b^3 & 32.68399b & -9.595952b^2 & 0.937952b^3 \\ & & 0.108225b^4 & 3.264791b^2 & -0.937952b^3 & 0.090188b^4 \\ & \text{sym.} & & 391.77488 & -67.31601b & 5.068542b^2 \\ & & & & 15.00722b^2 & -1.244587b^3 \\ & & & & & 0.108225b^4 \end{bmatrix},$$

$$\mathbf{k}_y^{(\sigma)} = \frac{Lt}{4000b} \times \tag{8.21}$$

$$\times \begin{bmatrix} 2857.1428 & 428.5715b & 23.80952b^2 & -2857.1428 & 428.5715b & -23.80952b^2 \\ & 457.1428b^2 & 33.33333b^3 & -428.5715b & -28.5716b^2 & 9.523817b^3 \\ & & 3.174602b^4 & -23.80952b^2 & -9.523817b^3 & 1.587301b^4 \\ & \text{sym.} & & 2857.1428 & -428.5715b & 23.80952b^2 \\ & & & & 457.1428b^2 & -33.33333b^3 \\ & & & & & 3.174602b^4 \end{bmatrix}.$$

The global stiffness matrices may be derived by the conventional summation of the element entries defined by eqs. (8.15) and (8.20) leading to the total stiffness of the structure which may be presented as

$$\mathbf{K}_{N\times N} = \mathbf{K}^{(el)} + \sigma_{xx}\mathbf{K}_x^{(\sigma)} + \sigma_{yy}\mathbf{K}_y^{(\sigma)} \tag{8.22}$$

where σ_{xx} and σ_{yy} now stand symbolically for the stress distributions in all elements of the assembly and care must be taken to properly account for them in building-up the overall stiffness. It is here assumed that σ_{xx} can be prescribed on the transverse boundary of the structure as an arbitrary function of the y-coordinate while σ_{yy} is constant along each element boundary in the lengthwise direction but can be different for different elements. We shall consider below two basic load cases:

(i) σ_{xx} is increasing as $\sigma_{xx} = \lambda\sigma_{xx}^*$ and $\sigma_{yy} = \sigma_{yy}^*$ is kept constant;

(ii) σ_{xx} and σ_{yy} are both proportionally increasing as $\sigma_{xx} = \lambda\sigma_{xx}^*$, $\sigma_{yy} = \lambda\sigma_{yy}^*$ so that $\sigma_{xx}/\sigma_{yy} = \text{const} = \sigma_{xx}^*/\sigma_{yy}^*$.

The starred quantities introduced above are certain prescribed stress distributions. These two cases lead to the following eigenvalue problems, respectively:

$$\begin{aligned} &[(\mathbf{K}^{(el)} + \sigma_{yy}^*\mathbf{K}_y^{(\sigma)}) + \lambda\sigma_{xx}^*\mathbf{K}_x^{(\sigma)}]\mathbf{v} = \mathbf{0}, \\ &[\mathbf{K}^{(el)} + \lambda(\sigma_{xx}^*\mathbf{K}_x^{(\sigma)} + \sigma_{yy}^*\mathbf{K}_y^{(\sigma)})]\mathbf{v} = \mathbf{0}. \end{aligned} \tag{8.23}$$

According to eqs. (8.15) and (8.20) all the stiffness matrices in eq. (8.23) are the known functions of the half wavelength L so that the standard eigenvalue problem described by eq. $(8.23)_1$ or $(8.23)_2$ contains two scalar unknowns:

L and λ (and, clearly, the eigenmode $\mathbf{v}$ corresponding to λ). To find the buckling stresses (i.e. the value of λ) we proceed as follows. For a given distribution of the reference stresses σ^*_{xx}, σ^*_{yy} the value of L is postulated as being equal to the typical width of a plate component of the cross-section, for instance. The lowest eigenvalue $\lambda = \lambda_1$ is found from eq. (8.23) and the computations are repeated for two other values of L equal to $0.90L$ and $0.95L$, for instance. Assuming next the parabolic variation of λ with L, a new value of L is computed corresponding to the lowest value of λ on this parabola. For that value of L eq. (8.23) is again solved resulting in a new value of λ. Rejecting from the set of four L's the value of L which corresponds to the highest λ a new parabola is constructed from which another value of L is found. The process is repeated until desired accuracy is obtained. We note that on the basis of an extensive numerical experimentation it is recommended to take as the first estimate of L a value too low rather than too high as it may significantly improve the convergence of the iterative procedure.

The elastic-plastic buckling analysis follows basically along the lines of the elastic analysis. We shall start with the classical J_2-flow theory for which the constitutive stiffness matrix as discussed in Secs. 2.1.4 and 4.10 reads in the case of plane stress analysis and is expressed in terms of the principal stresses as

$$\mathbf{C}^{(e\text{-}p)}_{(\text{flow})} = \frac{E}{1-\nu^2}\begin{bmatrix} 1-\dfrac{(s_x+\nu s_y)^2}{s} & \nu-\dfrac{(s_x+\nu s_y)(s_y+\nu s_x)}{s} & 0 \\ & 1-\dfrac{(s_y+\nu s_x)^2}{s} & 0 \\ \text{sym.} & & 1-\nu \end{bmatrix} \tag{8.24}$$

where

$$\begin{aligned} s_x &= \tfrac{1}{3}(2\sigma_{xx}-\sigma_{yy}), \quad s_y = \tfrac{1}{3}(2\sigma_{yy}-\sigma_{xx}), \quad s_z = -\tfrac{1}{3}(\sigma_{xx}+\sigma_{yy}), \\ s &= s_x^2+2\nu s_x s_y+s_y^2+\frac{4}{9}\,\frac{1-\nu^2}{E}\,\bar{\sigma}^2 E_T, \\ \bar{\sigma}^2 &= \tfrac{3}{2}(s_x^2+s_y^2+s_z^2) = \sigma_{xx}^2-\sigma_{xx}\sigma_{yy}+\sigma_{yy}^2. \end{aligned} \tag{8.25}$$

It will be useful to employ below the notation

$$\mathbf{C}^{(e\text{-}p)}_{(\text{flow})}(\sigma_{xx}, \sigma_{yy}, E_T) = \mathbf{C}^{(el)}-\mathbf{C}^{(pl)}(\sigma_{xx}, \sigma_{yy}, E_T). \tag{8.26}$$

The loading case (ii) is considered first, i.e. the case in which $\sigma_{xx} = \lambda\sigma^*_{xx}$, $\sigma_{yy} = \lambda\sigma^*_{yy}$ and, by eq. $(8.25)_3$, $\bar{\sigma} = \lambda\bar{\sigma}^*$ where $\bar{\sigma}^{*2} = \sigma^{*2}_{xx}-\sigma^*_{xx}\sigma^*_{yy}+\sigma^{*2}_{yy}$. We note that the following relation holds for an arbitrary α

$$\mathbf{C}^{(pl)}(\alpha\sigma_{xx}, \alpha\sigma_{yy}, E_T) = \mathbf{C}^{(pl)}(\sigma_{xx}, \sigma_{yy}, E_T) \tag{8.27}$$

which for the stress state of the type considered justifies the notation

$$\mathbf{C}^{(pl)} = \mathbf{C}^{(pl)}(\sigma_{xx}^*, \sigma_{yy}^*, E_T) = \mathbf{C}^{(pl)}(E_T). \tag{8.28}$$

The second equation of (8.23) for the elastic-plastic case takes the form

$$[\mathbf{K}^{(el)} - \mathbf{K}_*^{(pl)}(E_T) + \lambda(\sigma_{xx}^* \mathbf{K}_x^{(\sigma)} + \sigma_{yy}^* \mathbf{K}_y^{(\sigma)})]\mathbf{v} = \mathbf{0} \tag{8.29}$$

where the stiffness matrix $\mathbf{K}_*^{(pl)}$ is built up using eq. (8.13) and the element summation procedure, but replacing the matrix $\mathbf{C}^{(el)}$ in eq. (8.13) by the matrix $\mathbf{C}^{(pl)}$ of eq. (8.26) calculated at $\sigma_{xx} = \sigma_{xx}^*$, $\sigma_{yy} = \sigma_{yy}^*$.

As usual, the step-by-step approach must in general be adopted for the solution of the inelastic problem. At each loading level the value of E_T has to be determined and the stiffness matrix recomputed based on this value. The eigenvalue problem (8.29) may then be solved to determine the lowest value of λ. If such a value is smaller than the value corresponding to the step length assumed, the trial loading can be accepted as the critical one satisfying the buckling criterion. Otherwise, when λ is greater than this value, the load is incremented (i.e. a new hardening parameter value is taken) and the analysis is repeated until an adequate solution has been found.

The above algorithm can slightly be refined by taking into account the appropriate changes of the hardening parameter within each incremental interval. To this aim we note that we can write

$$E_T = E_T(\bar{\sigma})$$

and take the Taylor expansion for the matrix $\mathbf{K}^{(pl)}$ around the point $E_T^* = E_T(\bar{\sigma}^*)$ so that

$$\mathbf{K}^{(pl)}(E_T) = \mathbf{K}^{(pl)}(E_T^*) + \frac{\partial \mathbf{K}^{(pl)}(E_T)}{\partial E_T} \frac{\partial E_T}{\partial \bar{\sigma}}\bigg|_{\substack{E_T = E_T^* \\ \bar{\sigma} = \bar{\sigma}^*}} (\bar{\sigma} - \bar{\sigma}^*) + \dots \tag{8.30}$$

which can be put into a more compact form as

$$\mathbf{K}^{(pl)} = \mathbf{K}_*^{(pl)} - \bar{\sigma}^* {}'\mathbf{K}_*^{(pl)} + \lambda {}'\mathbf{K}_*^{(pl)} \bar{\sigma}^* + \dots \tag{8.31}$$

Introducing into eq. (8.29) instead of $\mathbf{K}_*^{(pl)}$ the first two terms of the expansion (8.31) we get

$$[\mathbf{K}^{(el)} - \mathbf{K}_*^{(pl)} - \bar{\sigma}^* {}'\mathbf{K}_*^{(pl)} + \lambda(\sigma_{xx}^* \mathbf{K}_x^{(\sigma)} + \sigma_{yy}^* \mathbf{K}_y^{(\sigma)} + \bar{\sigma}^* {}'\mathbf{K}_*^{(pl)})]\mathbf{v} = \mathbf{0}. \tag{8.32}$$

The procedure based upon eq. (8.32) is believed to yield slightly more accurate results than that based upon eq. (8.29). It requires some additional effort to form the gradient matrix ${}'\mathbf{K}_*^{(pl)}$, though.

There exists one situation which can easily be identified to result in much simple computations. It corresponds to the case when $E_T = E_T^0 = \text{const}$ so that the problem reduces to the solution of eq. (8.29) with the constant matrix $\mathbf{K}^{(el)} - \mathbf{K}_*^{(pl)}(E_T^0)$. Since now $\dfrac{\partial E_T}{\partial \bar{\sigma}} = 0$ which implies $'\mathbf{K}_*^{(pl)} = 0$ the standard and modified approaches given by eqs. (8.29) and (8.32), respectively, are clearly coincident. In this case we need only to check whether the structure bifurcates in the elastic range (i.e. to solve eq. (8.7)) and if not, to solve eigenvalue problem again, this time in the form (8.23).

For the loading case (i), i.e. when $\sigma_{xx} = \lambda\sigma_{xx}^*$, $\sigma_{yy} = \sigma_{yy}^* = \text{const}$, the corresponding eigenvalue problems (8.29), (8.32) take the form

$$
\begin{aligned}
&[\mathbf{K}^{(el)} - \mathbf{K}_*^{(pl)} + \sigma_{yy}^*\mathbf{K}_y^{(\sigma)} + \lambda\sigma_{xx}^*\mathbf{K}_x^{(\sigma)}]\mathbf{v} = \mathbf{0}, \\
&[\mathbf{K}^{(el)} - \mathbf{K}_*^{(pl)} + {}''\mathbf{K}_*^{(pl)}\sigma_{xx}^* + \sigma_{yy}^*\mathbf{K}_y^{(\sigma)} + \lambda\sigma_{xx}^*(\mathbf{K}_x^{(\sigma)} + {}''\mathbf{K}_x^{(pl)})]\mathbf{v} = \mathbf{0}
\end{aligned}
\tag{8.33}
$$

where

$$
{}''\mathbf{K}_*^{(pl)} = \left.\frac{\partial \mathbf{K}^{(pl)}}{\partial \sigma_{xx}}\right|_{\substack{\sigma_{xx} = \sigma_{xx}^* \\ \sigma_{yy} = \sigma_{yy}^*}}
$$

and in the calculation of the matrix $''\mathbf{K}_*^{(pl)}$ the following relations have to be consequently used

$$
E_T = E_T(\bar{\sigma}), \qquad \bar{\sigma} = \bar{\sigma}(\sigma_{xx}) = \sigma_{xx}^2 - \sigma_{xx}\sigma_{yy}^* + \sigma_{yy}^{*2}. \tag{8.34}
$$

Eq. $(8.33)_2$ is thought to improve the accuracy of the incremental solution because it approximately accounts for the influence of the stress changes upon the matrix $\mathbf{K}^{(pl)}$ within a given stress increment. The method can be considered as a special case of a general technique for nonlinear computations referred to as the perturbation approach, cf. [179].

Let us now recall that there are no physical grounds to favour the bifurcation-load predictions based upon the J_2-flow theory over those based upon a corresponding deformation theory unless a local unloading appears in a specific problem on hand. In order to exploit this observation we shall now derive equations describing the bifurcational problem for a deformation theory of plasticity known in the literature as the J_2-deformation theory, [65].

Assuming the Mises yield condition and material isotropy, the three-dimensional J_2-deformation theory of plasticity can be described by a nonlinear elasticity relation in which the total strain is expressed as a function of stress according to the relationship

$$\epsilon = \frac{1}{E} \times$$

$$
\begin{bmatrix}
1+\frac{2}{3}h_2(\bar\sigma) & -v-\frac{2}{3}h_2(\bar\sigma) & -v-\frac{1}{3}h_2(\bar\sigma) & 0 & 0 & 0 \\
 & 1+\frac{2}{3}h_2(\bar\sigma) & -v-\frac{1}{3}h_2(\bar\sigma) & 0 & 0 & 0 \\
 & & 1+\frac{2}{3}h_2(\bar\sigma) & 0 & 0 & 0 \\
 & & & 2(1+v)-h_2(\bar\sigma) & 0 & 0 \\
 & \text{sym.} & & & 2(1+v)-h_2(\bar\sigma) & 0 \\
 & & & & & 2(1+v)-h_2(\bar\sigma)
\end{bmatrix}_{6\times6} \sigma
$$

$$(8.35)$$

where

$$h_2(\bar\sigma) = \frac{3}{2}\left(\frac{E}{E_S}-1\right) \tag{8.36}$$

and $E_S = \bar\sigma/\bar\varepsilon$ is the secant modulus from the uniaxial tension/compression test. By differentiating eq. (8.35), assuming material compressibility and confining ourselves to the plane stress case the instanteneous moduli can be obtained in the form, [90]

$$
\mathbf{C}_{(\text{def})}^{(e\text{-}p)} = \frac{4}{3}E_S
\begin{bmatrix}
1-\frac{3\sigma_{xx}^2}{4\bar\sigma^2}\left(1-\frac{E_T}{E_S}\right) & \frac{1}{2}-\frac{3\sigma_{xx}\sigma_{yy}}{4\bar\sigma^2}\left(1-\frac{E_T}{E_S}\right) & 0 \\
 & 1-\frac{3\sigma_{yy}^2}{4\bar\sigma^2}\left(1-\frac{E_T}{E_S}\right) & 0 \\
 & \text{sym.} & \frac{1}{2}
\end{bmatrix}. \tag{8.37}
$$

The element stiffness matrix $\mathbf{k}_{(\text{def})}^{(e\text{-}p)}$ can be constructed by using eq. (8.13) in which $\mathbf{C}_{(\text{def})}^{(e\text{-}p)}$ replaces $\mathbf{C}_{(\text{flow})}^{(e\text{-}p)}$. In analogy to eq. (8.15) the element stiffness can be separated into three parts

$$\mathbf{k}_{(\text{def})}^{(e\text{-}p)} = {}_{\text{I}}\mathbf{k}_{(\text{def})}^{(e\text{-}p)}(1/L^3) + {}_{\text{II}}\mathbf{k}_{(\text{def})}^{(e\text{-}p)}(1/L) + {}_{\text{III}}\mathbf{k}_{(\text{def})}^{(e\text{-}p)}(L). \tag{8.38}$$

Each matrix term in eq. (8.38) can be symbolically written as

$$\mathbf{k}_{(\text{def})}^{(e\text{-}p)} = {}'\mathbf{k}_{(\text{def})}^{(e\text{-}p)}(E_S) + {}''\mathbf{k}_{(\text{def})}^{(e\text{-}p)}(E_T). \tag{8.39}$$

The element matrices defined in eq. (8.39) contribute to the global stiffness of the structure and are denoted on the assembly level by ${}'\mathbf{K}_{(\text{def})}^{(e\text{-}p)}$ and ${}''\mathbf{K}_{(\text{def})}^{(e\text{-}p)}$, respectively.

In the computer program the actual hardening curve is approximated by a piecewise linear curve with an arbitrary number of corner points. For the loading type (a), i.e. for $\sigma_{xx} = \lambda\sigma_{xx}^*$, $\sigma_{yy} = \lambda\sigma_{yy}^*$ we have the eigenvalue problem of the form

$$[{}'\mathbf{K}_{(\text{def})}^{(e\text{-}p)} + {}''\mathbf{K}_{(\text{def})}^{(e\text{-}p)} + \lambda\mathbf{K}^{(\sigma)}(\bar\sigma^*)]\mathbf{v} = 0 \tag{8.40}$$

provided the structure does not buckle elastically. Since now ${}'\mathbf{K}_{(\text{def})}^{(e\text{-}p)} \neq \text{const}$, ${}''\mathbf{K}_{(\text{def})}^{(e\text{-}p)} = \text{const}$ at the hardening line segment considered, some iterative procedure is needed to solve eq. (8.40) for a critical λ with the assumed value of the buckling half wavelength L. The algorithm requires a repetitive solution of eq. (8.40) with the updated values of E_S so that in the end the trial loading level is sufficiently close to the level assumed for the determination of E_S. The convergence can be speeded up by using stiffness expansion similar to that applied above. Thus we write out the relation

$$E_S(\lambda) = \frac{\lambda\bar{\sigma}_k(\bar{\sigma}_{k+1}-\bar{\sigma}_k)}{(\lambda-1)\bar{\sigma}_k(\bar{\varepsilon}_{k+1}-\bar{\varepsilon}_k)+\bar{\varepsilon}_k(\bar{\sigma}_{k+1}-\bar{\sigma}_k)} \tag{8.41}$$

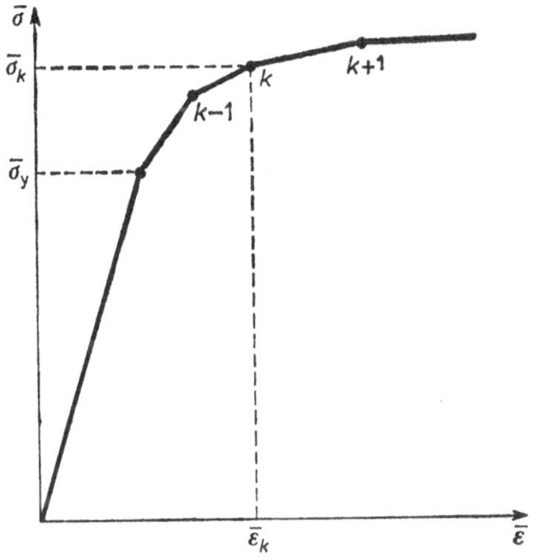

Fig. 181 Piecewise linear hardening curve

which is valid along the hardening line segment k, $k+1$, Fig. 181. It is easy to check that the secant modulus defined in eq. (8.41) takes the appropriate limiting values at the corner points of the hardening line segment as

$$E_S(\lambda)|_{\lambda=1} = \frac{\bar{\sigma}_k}{\bar{\varepsilon}_k}, \quad E_S(\lambda)|_{\lambda=\bar{\sigma}_{k+1}/\bar{\sigma}_k} = \frac{\bar{\sigma}_{k+1}}{\bar{\varepsilon}_{k+1}}.$$

We write eq. (8.41) in short as

$$E_S(\lambda) = \frac{A\lambda}{B\lambda+C} \tag{8.42}$$

and take its Taylor expansion around the point $\lambda = \lambda^*$ to get

$$\begin{aligned} E_S(\lambda) &\cong \frac{A\lambda^*}{B\lambda^*+C} + \frac{A(B\lambda^*+C)-AB\lambda^*}{(B\lambda^*+C)^2}(\lambda-\lambda^*) \\ &= \frac{A\lambda^*}{B\lambda^*+C} + \frac{AC}{(B\lambda^*+C)^2}(\lambda-\lambda^*) \\ &= \tilde{E}_S(\lambda^*)+\hat{E}_S(\lambda^*)\lambda. \end{aligned} \tag{8.43}$$

If we take into account the linearity of $'\mathbf{K}^{(e\text{-}p)}_{(\mathrm{def})}$ with respect to E_S, then eq. (8.43) implies the eigenvalue problem of the form

$$\{'\mathbf{K}^{(e\text{-}p)}_{(\mathrm{def})}[\tilde{E}_S(\lambda^*)]+{}''\mathbf{K}^{(e\text{-}p)}_{(\mathrm{def})}(E_T)+\lambda['\mathbf{K}^{(e\text{-}p)}_{(\mathrm{def})}[\hat{E}_S(\lambda^*)]+\mathbf{K}^{(\sigma)}(\sigma^*)]\}\mathbf{v}=\mathbf{0}. \quad (8.44)$$

Again, the use of eq. (8.44) proves numerically more efficient than the use of eq. (8.40).

Possible oscillations in the critical load due to the piecewise linearization of the theoretically smooth hardening curve may be handled similarly to those described above. Thus, if the critical value of λ for a subsequent linear hardening segment indicates the bifurcation to occur at the previous line segment for which no acceptable λ was found before, the relevant corner point on the piecewise linearized hardening curve is taken as the bifurcation solution.

In order to evaluate the method a number of test computations has been performed on buckling of different plates and plate assemblies with various boundary and loading conditions.

Example 1. Elastic plate of constant thickness t and width b, infinitely long in the direction of the applied uniaxial compression σ_{xx}.

The elastic buckling compressive stress can be represented as

$$\sigma_{xx} = KE(t/b)^2$$

in which the coefficient K depends upon the boundary conditions and the Poisson's ratio ν. Because of the symmetry only one half of the plate has been considered. The results of computer calculations for $\nu = 0.3$ are summarized in Table 18. The numerical results of [137] as well as the analytical results of [141] are also given. Very high accuracy is noted even for only one finite element.

Example 2. Infinitely long elastic plate of constant thickness t and width b under biaxial compression, Fig. 182.

The same plate subjected to a biaxial compressive state of stress and simply supported at the boundaries has been analysed for different ratios of σ_{xx}/σ_{yy}. The critical stresses are given in Fig. 182 in which the solid line represents the analytical results of [135]. Two finite elements were used in the analysis to represent the behaviour of half of the plate. (Symmetric modes of buckling were assumed in this and the previous example because the antisymmetric ones give, in these cases, significantly higher buckling stresses and are therefore of no practical interest). The results shown in Fig. 183 are normalized with respect to the critical stress value $\sigma^{(0)}_{xx} = 3.615E(t/b)^2$ which corresponds to the uniaxial compression with $\sigma_{yy} = 0$, cf. Table 18.

Table 18 Elastic buckling coefficients for infinitely long plate under uniaxial compression

	Simply supported plate			Clamped plate		
	Number of elements	Number of d.o.f	K	Number of elements	Number of d.o.f	K
Present solution	1	3	3.61524	1	3	6.30081
	2	6	3.61524	2	6	6.30038
	3	9	3.61525	3	9	6.30038
	.	.	.	.	.	.
[1]	4	8	3.61528	6	11	6.30075
Analytical solution			3.61524			6.29992

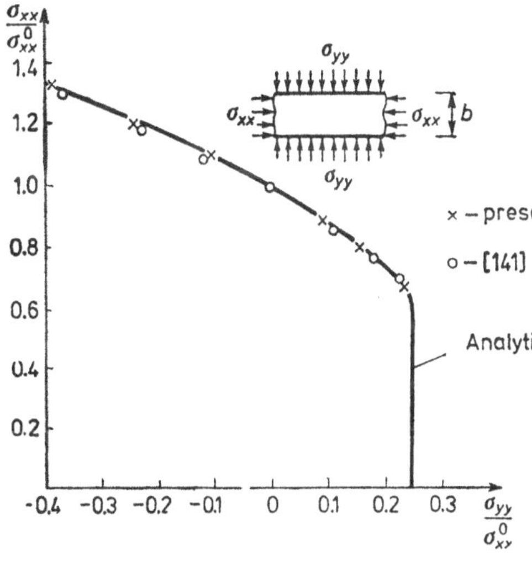

Fig. 182 Infinitely long elastic plate under biaxial compression

Example 3. Z-profile beam under uniaxial compression.

The structure analysed is defined in Fig. 183 in which the finite elements used to discretize half of the cross-section are also shown. We note that due to our assumption of fixed in space junction lines between flats the same results as for beams with the Z-cross-section would be obtained for beams with the U-cross-section, cf. Fig. 183. The values of critical stress parameter K defined by $\sigma_{xx} = KE(t_h/h)^2$ are plotted in Fig. 183 as a function of the geometri-

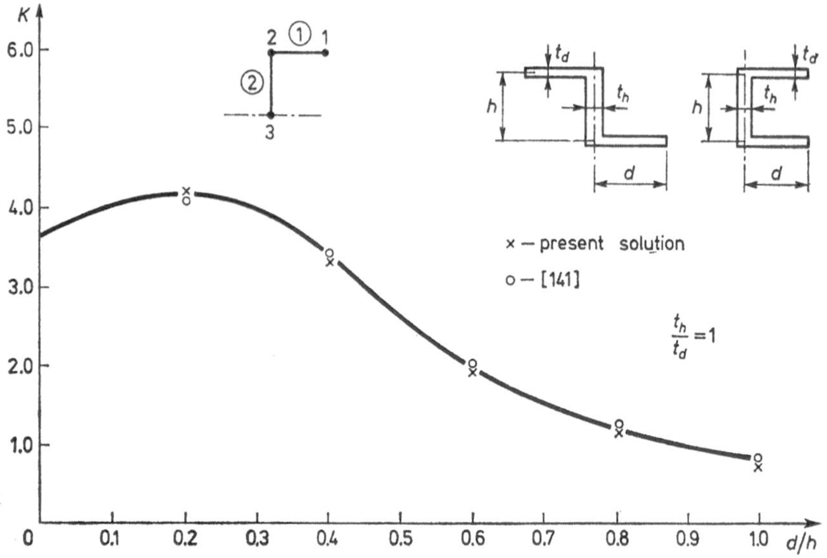

Fig. 183 *Z*-profile beam under uniaxial compression

cal parameter ratio α/h for $\nu = 0.3$. The comparison curve is taken from [141].

Example 4. I-profile beam under uniaxial compression.

The buckling coefficient K defined as in the previous example is plotted in Fig. 184 for $\nu = 0.3$. Half of the cross-section has been analysed by means of nine finite elements. The solid curve is taken from [141].

Example 5. Simply supported infinitely long plate under linearly changing uniaxial stresses, Fig. 185.

The buckling parameter K is defined as $\sigma_{xx} = KE(t/b)^2$ where σ_{xx} acts in the direction in which the plate is infinitely long while σ_{xx}^1 and σ_{xx}^2 are its limiting values. The stress σ_{xx}^1 is assumed compressive while σ_{xx}^2 is arbitrary. In Fig. 185 the curve representing the parameter K as a function of the stress parameter ratio $\sigma_{xx}^2/\sigma_{xx}^1$ is shown. The solid line is taken from [142] while the points have been obtained using the approach advocated in this section.

Example 6. Infinitely long plate with different boundary conditions under parabolically changing uniaxial stresses.

The problem considered and stress parameters are shown in Fig. 186. It is assumed that σ_{xx}^1 is compressive while σ_{xx}^2 is arbitrary. In Fig. 186 the dependence of the ratio $\sigma_{xx}^1/\sigma_{xx}^{(0)}$ upon the ratio $\sigma_{xx}^2/\sigma_{xx}^1$ is plotted where σ_{xx}^1 stands for the critical value of the stress parameter σ_{xx} while $\sigma_{xx}^{(0)}$ corresponds

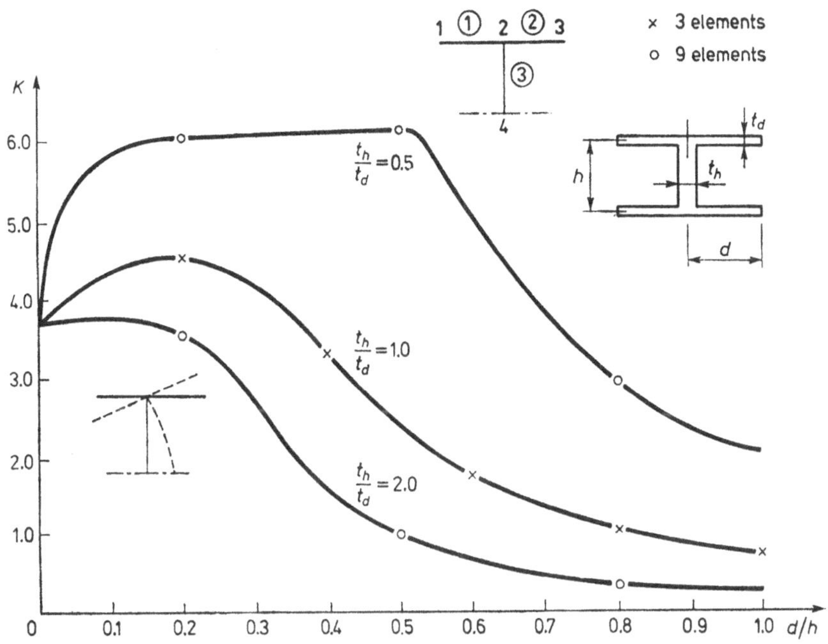

Fig. 184 I-profile beam under uniaxial compression

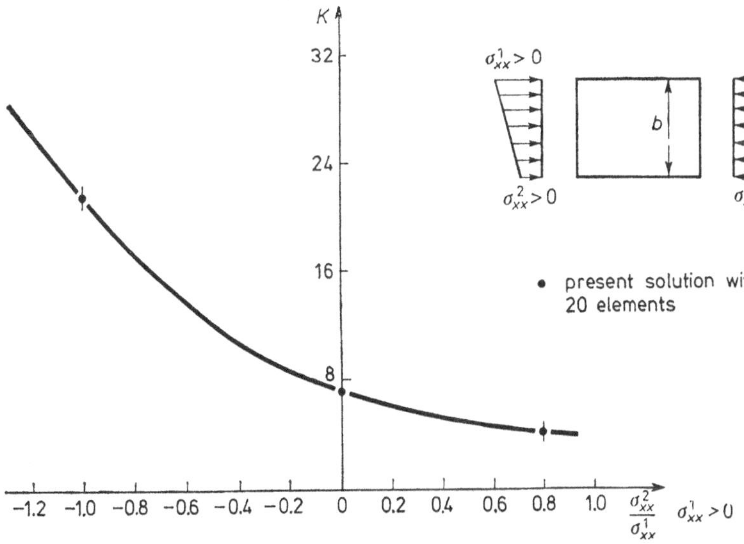

Fig. 185 Infinitely long plate under linearly changing uniaxial stresses

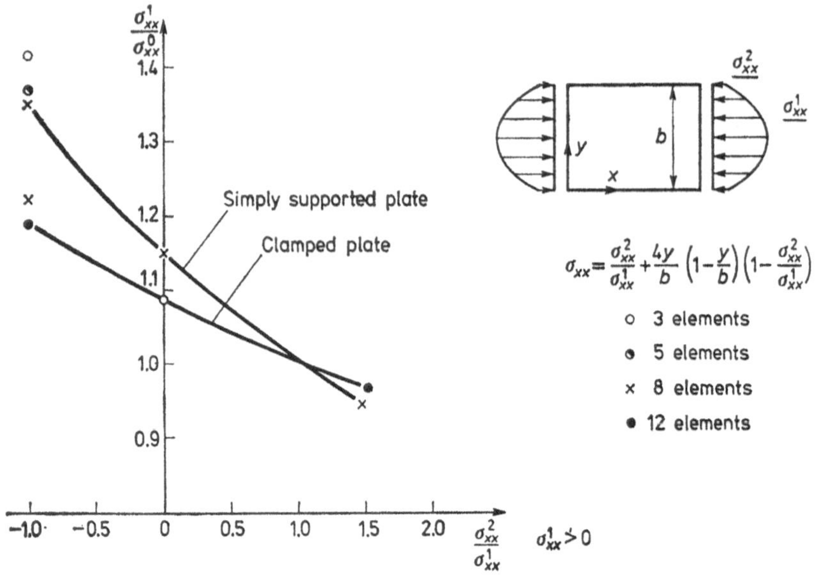

Fig. 186 Infinitely long plate under parabolically changing uniaxial stresses

to the critical uniform stress acting upon the same plate. The solid line is taken from [143] while the points are ours. The agreement was found excellent.

Example 7. Elasto-plastic buckling of simply supported rectangular plates.

After some simple modifications in the program which made it possible to analyse the buckling modes characterized by the prescribed buckling wave lengths, the computations have been carried out for some square ($a = b = 508.0$ mm) and rectangular ($a = 381.0$ mm, $b = 508.0$ mm) inelastic plates with various normal edge loadings. Assuming

$$\alpha = -\frac{\sigma_{xx}}{\sigma^*}, \quad \beta = -\frac{\sigma_{yy}}{\sigma^*}$$

where σ^* is a critical stress, three loading cases were considered: (1) $\alpha = 1$, $\beta = 0$; (2) $\alpha = 1$, $\beta = 1$; (3) $\alpha = 1$, $\beta = 0.5$. In all of the plastic buckling calculations the secant and tangent moduli were instantanously found using the Ramberg–Osgood curve of the form, [132]

$$\bar{\varepsilon} = \frac{\bar{\sigma}}{E} + \left(\frac{3\bar{\sigma}}{7E}\right)\left(\frac{\bar{\sigma}}{\sigma_{0.7}}\right)^{n-1} \tag{8.45}$$

where n is a shape parameter given by

$$n = 1 + \log\left(\frac{17}{7}\right) \Big/ \log\left(\frac{\sigma_{0.7}}{\sigma_{0.85}}\right)$$

and $\sigma_{0.7}$ and $\sigma_{0.85}$ are the stresses at which the experimental curve has secant moduli of $0.7E$ and $0.85E$, respectively. The tangent and secant moduli expressed as ratios of the elastic modulus follow as

$$\frac{E_T}{E} = \frac{1}{1+\frac{3}{7}n(\bar{\sigma}/\sigma_{0.7})^{n-1}}, \qquad \frac{E_S}{E} = \frac{1}{1+\frac{3}{7}(\bar{\sigma}/\sigma_{0.7})^{n-1}}.$$

In all the calculations the values of $E = 6.8971 \cdot 10^4$ MPa, $\nu = 0.5$, $n = 10$, $\sigma_{0.7} = 6.8971 \cdot 10^2$ MPa were used. The curve (8.45) was used in the whole range of the material behaviour so that there was no need to specify a distinct yield limit.

Table 19 Buckling stresses for elastic-plastic simply supported square plates

Thickness h (m)	σ_{xx} (MPa) exact	σ_{xx} (MPa) *FE* analysis
	$\alpha = 1,\ \beta = 0$	
0.019778	−448.33	−447.70
0.021793	−517.31	−517.09
0.024498	−586.28	−583.93
0.028453	−655.25	−653.03
0.034716	−724.23	−728.66
0.044895	−793.20	−789.68
0.060719	−862.18	−869.47
	$\alpha = 1,\ \beta = 1$	
0.028575	−448.33	−448.10
0.033015	−517.31	−514.30
0.040699	−586.28	−587.15
0.052898	−655.25	−655.68
0.070550	−724.23	−726.87
0.096157	−793.20	−788.38
0.133605	−862.18	−862.03
	$\alpha = 1,\ \beta = 0.5$	
0.024125	−448.33	−447.93
0.026387	−517.31	−515.95
0.029395	−586.28	−587.83
0.033874	−655.25	−655.05
0.040339	−724.23	−728.30
0.049202	−793.20	−791.28
0.061565	−862.18	−863.66

One sinusoidal buckling half-wave in the x-direction was assumed. Three finite elements for only half of the plate were considered. The results are given in Tables 19 and 20; the comparison data are those of [132].

Table 20 Buckling stresses for elastic-plastic simply supported rectangular plates

Thickness h (m)	σ_{xx} (MPa) exact	σ_{xx} (MPa) *FE* analysis
	$\alpha = 1,\ \beta = 0$	
0.019072	−448.33	−448.35
0.021214	−517.31	−516.13
0.024239	−586.28	−585.42
0.028628	−655.25	−654.10
0.035322	−724.23	−719.58
0.045945	−793.20	−792.73
0.062312	−862.18	−861.85
	$\alpha = 1,\ \beta = 1$	
0.024247	−448.33	−448.10
0.028014	−517.31	−514.30
0.034534	−586.28	−587.16
0.044885	−655.25	−655.67
0.059863	−724.23	−726.87
0.081591	−793.20	−788.40
0.113367	−862.18	−862.01
	$\alpha = 1,\ \beta = 0.5$	
0.021430	−448.33	−448.33
0.023510	−517.31	−518.52
0.026395	−586.28	−585.88
0.030895	−655.25	−658.25
0.037620	−724.23	−719.12
0.046921	−793.20	−791.00
0.059694	−862.18	−864.40

Example 8. Square elastic-plastic plate under triangular edge loading.

The next example is that of a simply supported square plate with a triangular distribution of load σ_{xx} on two opposite edges. The plate geometry and material properties were the same as those assumed in the previous example. The results shown in Table 21 correspond to a discretization by 10 elements and are compared with those of [132].

Table 21 Buckling stresses for square elastic-plastic plates under triangular edge loadings

Thickness (m)	Elastic critical stress (MPa)		Critical stress (MPa)	
	[132]	present solution	[132]	present solution
0.0127	−373.61	−369.31	−368.89	−369.24
0.0152	−538.00	−531.80	−528.66	−529.22
0.0178	−732.27	−723.85	−690.81	−692.06
0.0203	−956.44	−943.89	−813.36	−815.47
0.0254	−1494.44	−1485.03	−957.01	−959.13

Example 9. Infinitely long elastic-plastic plate with a beam stiffener.

To analyse the stiffened plate shown in Fig. 187 a special beam element was developed based on the same assumptions as used for the plate element.

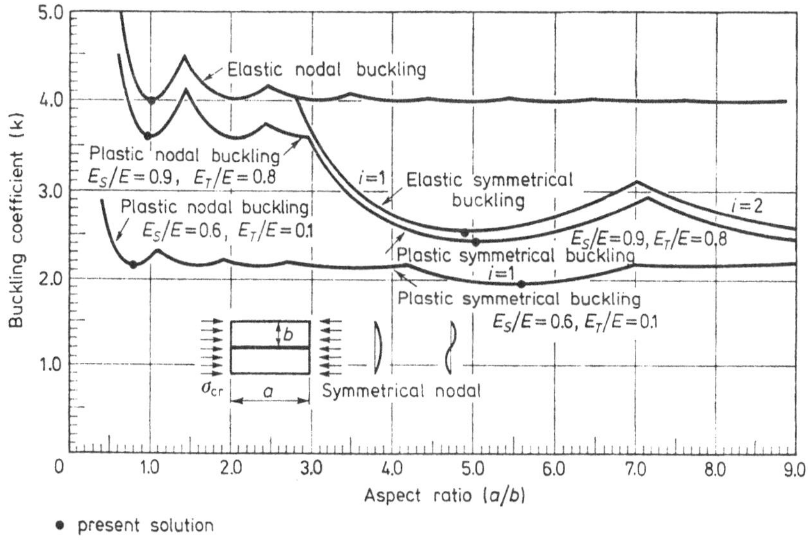

Fig. 187 Infinitely long elastic-plastic plate with a beam stiffener

The analysis of symmetric buckling modes in this example was not possible without the special stiffener element because of the assumption of the fixed edge lines between the flats. The geometry and material properties of the plate and the stiffener were taken to be

$$\gamma = \frac{E_T E_S}{bEI_p} = 40, \qquad \delta = \frac{A_s}{bt} = 0.4$$

where $I_s = \frac{1}{12} b_s h_s^3$, b_s and h_s are the width and height of the stiffener, $2b$ is the width of the plate, $I_p = \frac{1}{12} t^3$, t is the plate thickness and $A_s = b_s h_s$. The accuracy of our results as compared against those given in [151] and shown in Fig. 187 was found quite satisfactory.

Summary

In this chapter numerous examples have illustrated the effectiveness of the discretized approach in solving selected nonlinear plate and shell problems.

9

Heat conduction and thermal stress problems

Purpose of the chapter

The main goal in this chapter is to complete the presentation of various possible applications of formulations and numerical algorithms advocated in this book by giving some examples related to thermo-mechanical analysis of structures. We shall start with some illustrations of nonlinear, transient heat flow problems which will be followed by an analysis of thermal stresses in a plate deformed well into the plastic range. The chapter is completed by considering two examples of complex mechanical and thermal loads acting on a truss and on a cylindrical shell.

9.1 Heat flow in flash and friction welding

The algorithm for nonlinear heat conduction presented in Sec. 5.7 will now be used for solving two axisymmetric transient heat transfer problems. The first problem will concern the heat flow in flash welding. Flash welding of metal rods is a resistance welding process wherein coalescence is produced, simultaneously over the entire area of abutting surfaces, by heat resulting from resistance to the flow of electric current between the two surfaces and by a pressure applied to the rods after heating is essentially completed. In order to make a good flash weld it is important to be able to precisely analyse the heat treatment experienced by the metal around the weld. A common assumption in attempting to find a solution to such a problem is the postulated temperature independence of all material properties. No such simplifications have to be done in the approach reported below: the discretized approach employed in its incremental form makes it possible to account for arbitrary variations of all material characteristics during the process.

The analysis is based on the iterative Euler backward method for integrating the transient heat transfer equation (4.45), which has the form, cf. eq. (5.93)

$$\left[\frac{1}{\Delta t}\mathbf{C}_{t+\Delta t}^{(i-1)}+\mathbf{K}_{t+\Delta t}^{(i-1)}\right]\boldsymbol{\theta}_{t+\Delta t}^{(i)}=\mathbf{Q}_{t+\Delta t}^{(i-1)}+\mathbf{C}_{t+\Delta t}^{(i-1)}\frac{1}{\Delta t}\boldsymbol{\theta}_{t}. \qquad (9.1)$$

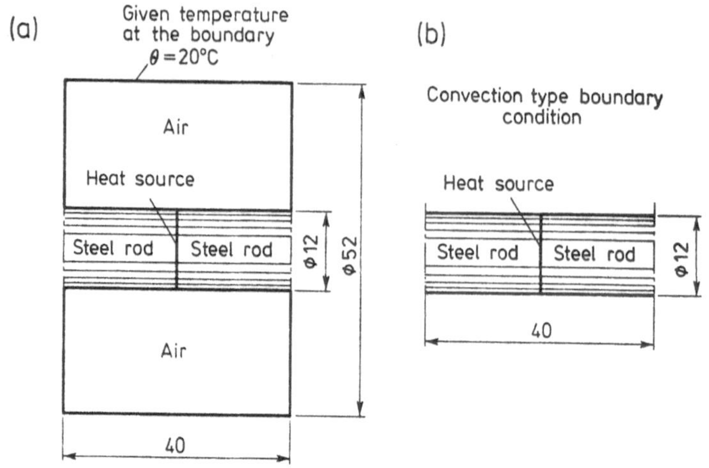

Fig. 188 Heat boundary conditions: (a) surrounding air included, (b) surrounding air excluded

Eq. (9.1) is used for the step-by-step evaluation of nodal temperatures **θ** resulting in the whole history of the process. The analysis has been based on eight node isoparametric finite elements as described in [13], for instance.

The two semi-infinite rods shown in Fig. 188 are subjected to the flash welding. At the place of abutment the heat source is given by the following equation

$$Q = J^2 \cdot R \cdot t$$

where J is the current intensity and R resistance at the place of contact. It is noted that R depends on the diameter of the rods, compressive force applied, properties of the material and time.

Steel rods of the diameter ϕ = 12 mm and density 7800 kg/m³ are considered first. The heat flux at the place of contact of the rods is assumed to be equal to $24 \cdot 10^6$ W/m² which approximately amounts to 600 cal/cm² s. The material properties are given in Table 22 for steel and Table 23 for the

Table 22

Temperature (°C)	0	500	800	900	1600
Thermal conductivity $\left(\frac{J}{m \cdot K \cdot s}\right)$	50	50	50	50	50
Specific heat $\left(\frac{J}{kg \cdot K}\right)$	510	1030	1300	680	680

Table 23

Temperature (°C)	0	1600
Thermal conductivity $\left(\frac{J}{s \cdot m \cdot K}\right)$	0.23	0.23
Specific heat $\left(\frac{J}{kg \cdot K}\right)$	1005	1100

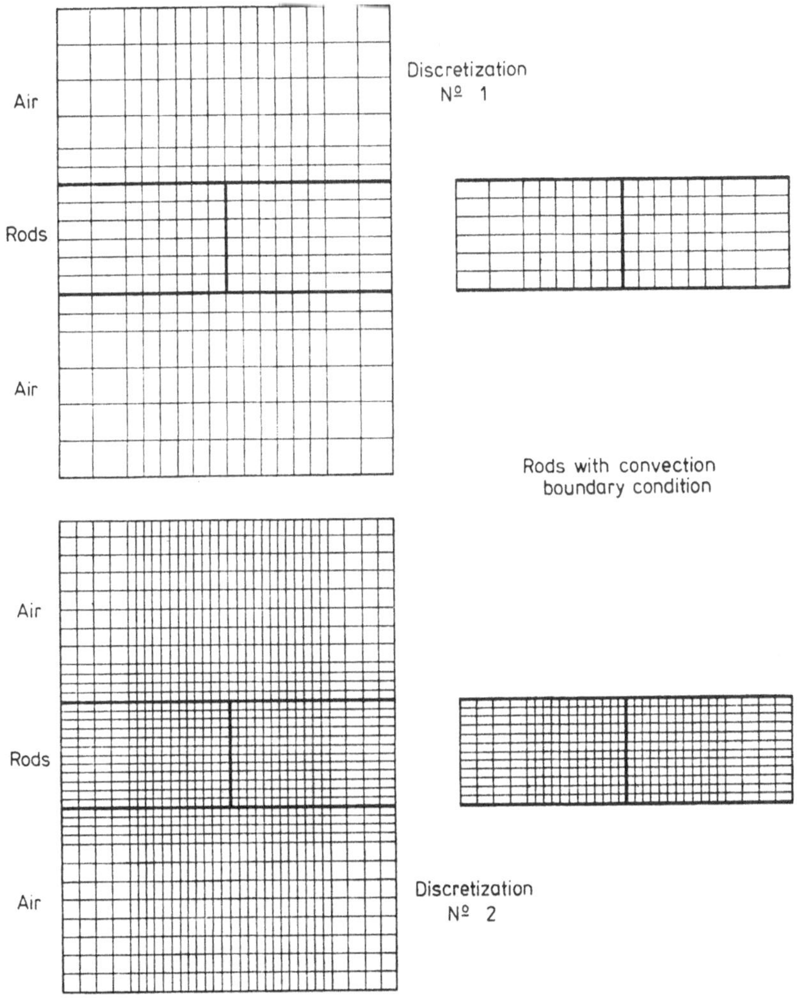

Fig. 189 Types of meshes

surrounding air (with air density taken as 1.29 kg/m³). The surface film conductance is taken as 0.25 W/m².

Two kinds of the analysis are carried out with different inclusion of the surrounding air effects, Figs. 188 and 189. The discretization meshes in the corresponding steel and air areas are shown in Fig. 189. Due to the existing symmetries, one quarter of the area is considered only for each case. The results obtained for both the coarse and fine meshes and both the boundary conditions were almost coincident and are shown in Figs. 190 and 191. The

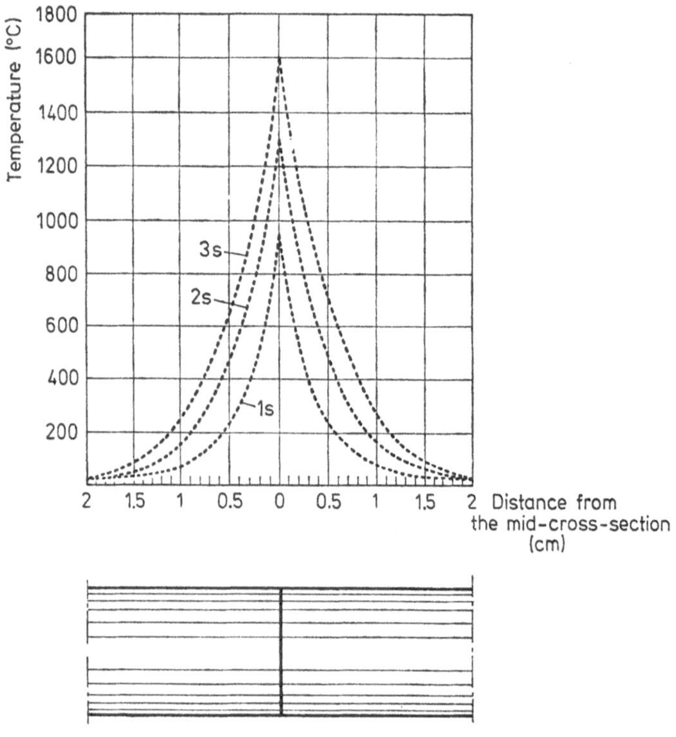

Fig. 190 Temperature distribution along the rod axis

results given in Fig. 190 refer to the time instants of $t = 1$ s, $t = 2$ s, $t = 3$ s while those shown in Fig. 191 are for $t = 3$ s.

The second example concerns the welsh welding problem of two steel pipes with the same material properties as before, Fig. 192. On the basis of the previous results one discretization mesh and the convection-type boundary conditions are assumed to yield sufficiently accurate results in the present case. The temperature distribution along the pipe axis is shown for three time instants of 1, 2 and 3 seconds.

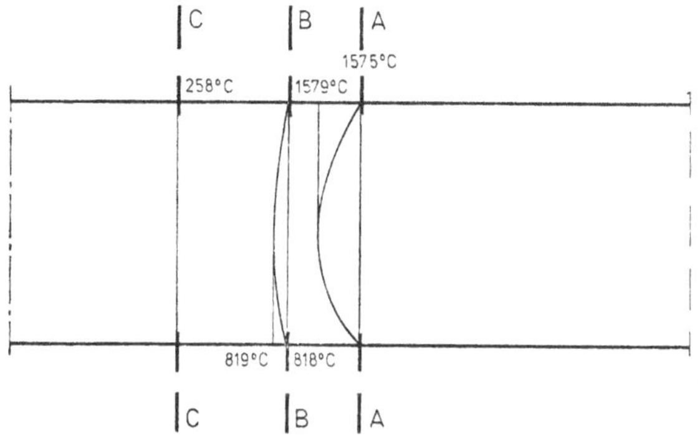

Fig. 191 Temperature distribution along the cross-section diameter (time: 3 s); A–A at the place of contact, B–B at the distance of 4 mm from A–A, C–C at the distance of 10 mm from A–A

In the next example we shall consider the problem of temperature distribution in friction welding.

Friction welding is a process in which the heat for welding is produced by direct conversion of mechanical energy to thermal energy at the interface of the workpieces without the application of electrical energy, or heat from any other sources, to the workpieces. Friction welds are made by holding a non-rotating workpiece in contact with a rotating workpiece under constant or gradually increasing pressure until the interface reaches welding temperature, and then stopping rotation to complete the weld. The frictional heat developed at the interface rapidly raises the temperature of the workpieces, over a very short axial distance, to value approaching, but below the melting range; welding occurs under the influence of a pressure that is applied while the heated zone is in the plastic temperature range. Clearly, knowing the exact temperature distributions during the whole process is of extreme significance for high quality friction weldings.

The same method as before is employed for the analysis of heat transfer in the axisymmetric specimen. Two semi-infinite rods subjected to welding are shown in Fig. 193. At the place of abutment the heat source is given by the following equation:

$$\dot{Q} = \int_A \sigma \cdot \mu \cdot \omega \cdot r \mathrm{d}A$$

where σ is a stress at the place of contact, μ is the coefficient of friction, ω is the angular speed, r is the radius, and A is the surface upon which the heat rate acts. Steel rods of the diameter $\phi = 12$ mm and density 7800 kg/m^3 are considered. The value of $\sigma \cdot \mu \cdot \omega$ is assumed to be equal to $3 \cdot 10^9$ W/m^3.

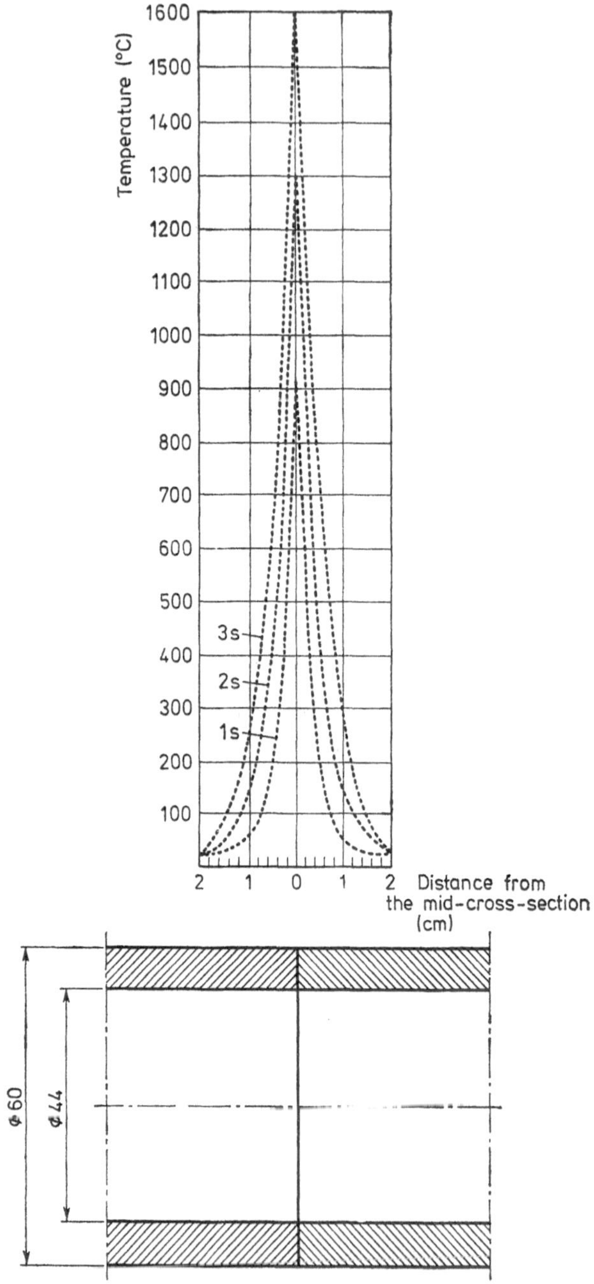

Fig. 192 Temperature distribution along the pipe axis

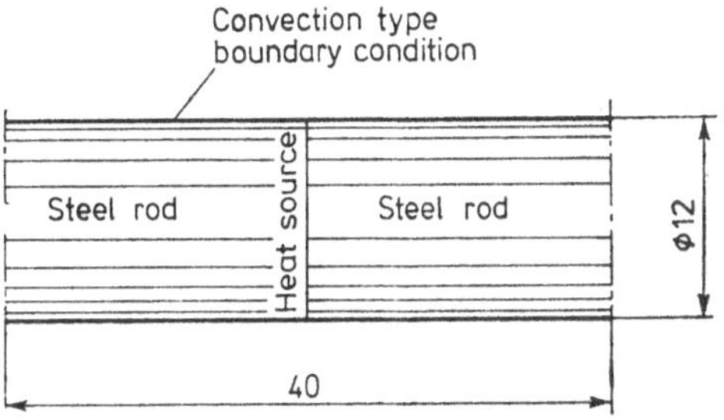

Fig. 193 Heat boundary conditions in friction welding of two rods

Table 24

Temperature (°C)	0	500	800	900	1100
Thermal conductivity $\left(\frac{J}{m \cdot K \cdot s}\right)$	50	50	50	50	50
Specific heat $\left(\frac{J}{kg \cdot K}\right)$	510	1030	1300	680	680

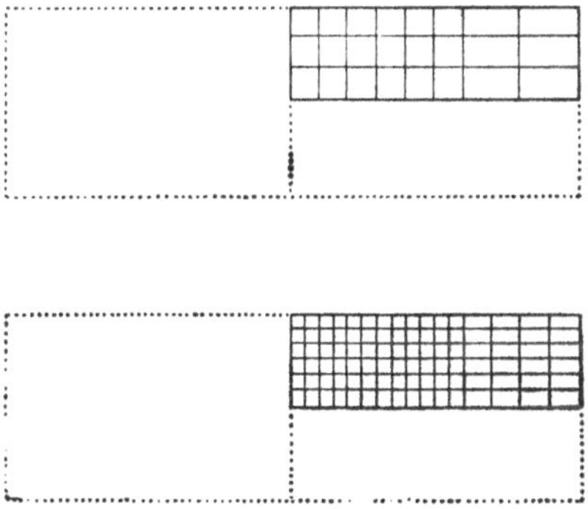

Fig. 194 Types of discretization

The material properties are given in Table 24. The surface film conductance is taken as 0.25 W/m^2. Two kinds of discretization are used for analysis, Fig. 194. The results obtained for both of them were almost coincident. The results given in Fig. 195 refer to the time instants of $t = 1, 2, 3$ s.

The next analysis concerns the friction welding problem of two different steel rods with the same material properties as before, Fig. 196. The finite element mesh used is shown in Fig. 197. The temperature distribution for the same conditions as before is given in Fig. 198.

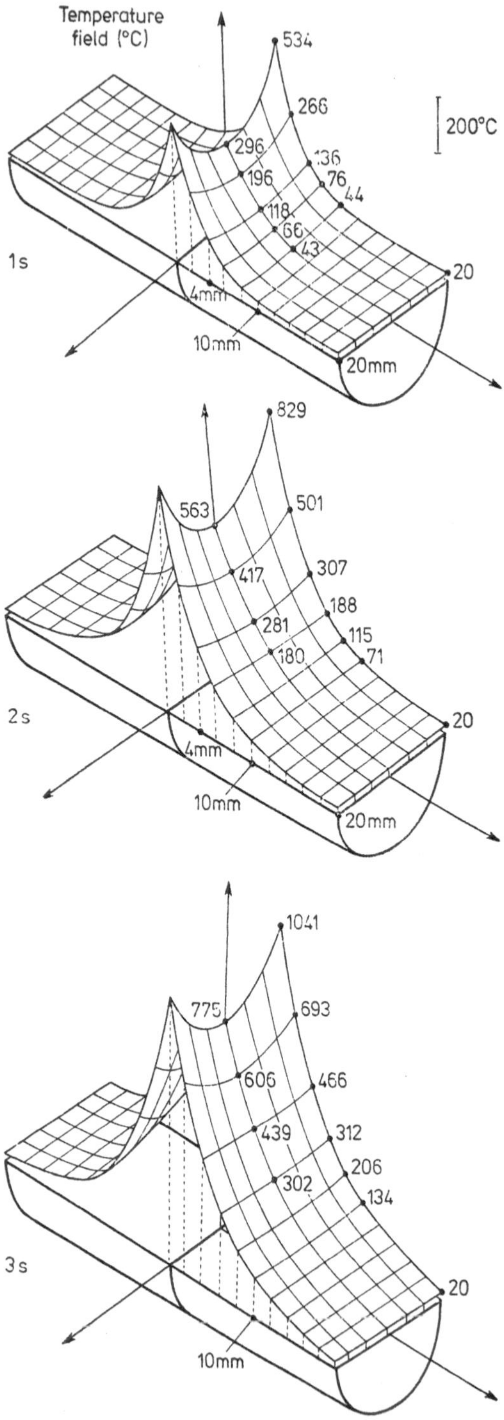

Fig. 195 Temperature distribution in friction welding

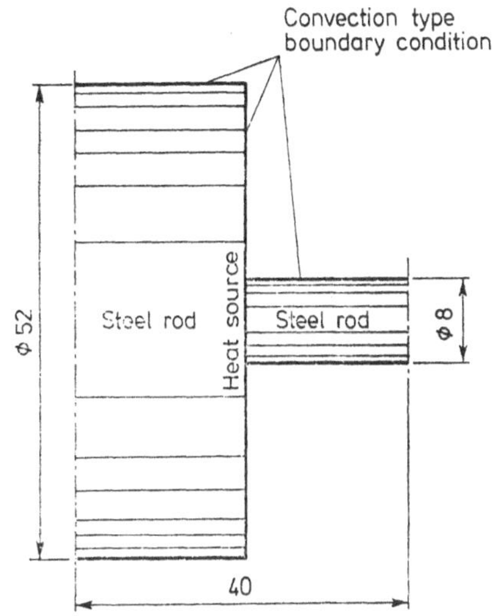

Fig. 196 Heat boundary conditions in friction welding of two rods

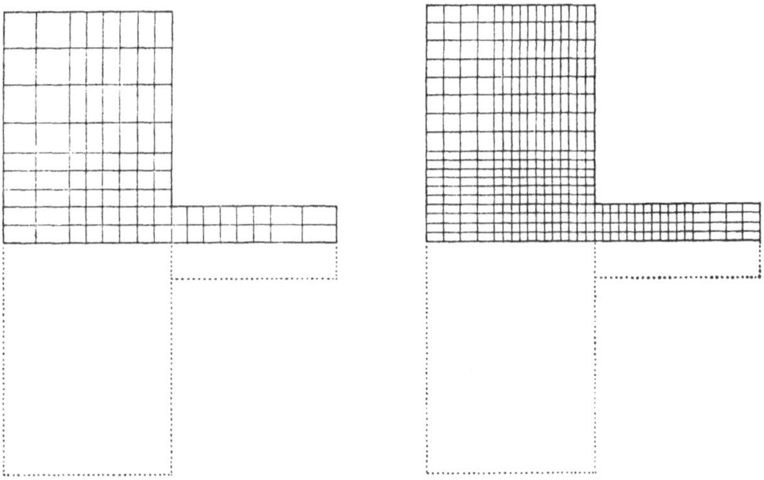

Fig. 197 Finite element meshes

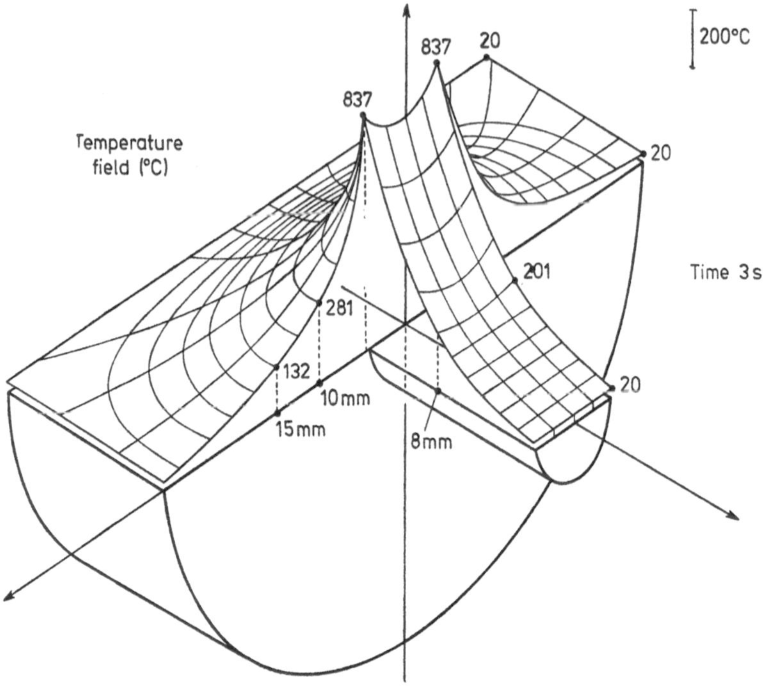

Fig. 198 Temperature field in friction welding

We may conclude that the results obtained confirm the known experimental data, [144]. As expected, the cross-section variation of temperature is significant.

9.2 Thermally induced stresses in elastic-plastic plate

We consider the plane stress problem shown in Fig. 199. The material is assumed to be elastic-ideally plastic; the material constants are: $E = 208$ GPa, $\nu = 0.3$, $\sigma_y = 235$ MPa, $E_T = 0$, thermal expansion coefficient

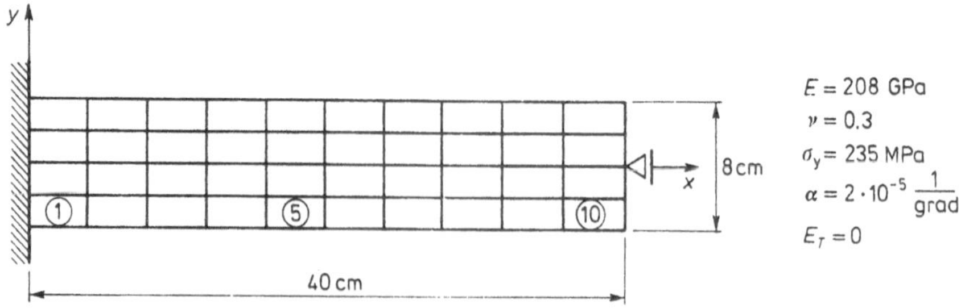

Fig. 199 Elastic-plastic plate

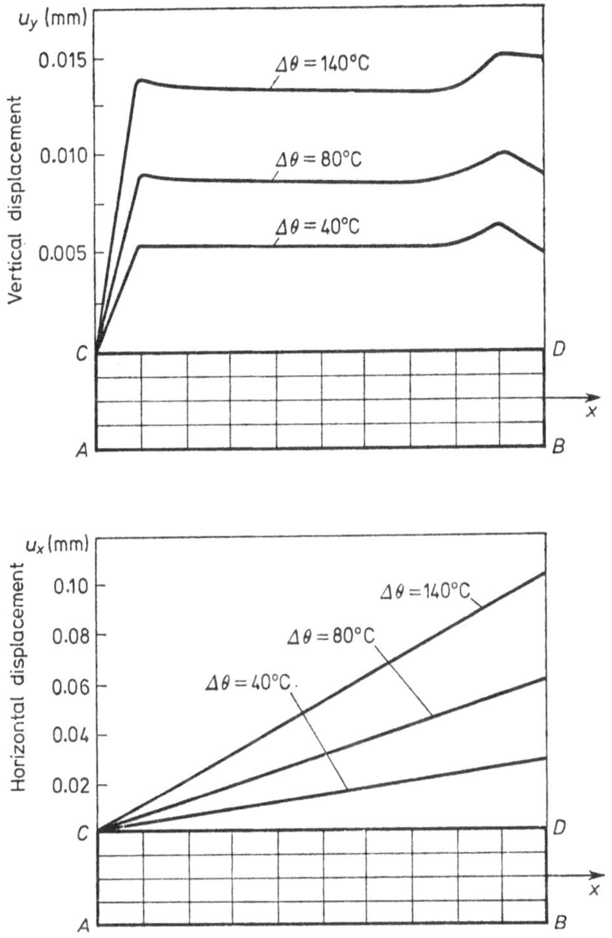

Fig. 200 Displacements of the specimen upper boundary

$\alpha = 0.00002$ grad^{-1}. The loading is assumed to consist of a uniform temperature field with the nodal temperatures increasing from 20°C to 240°C with the step of 20°C. The variation of the yield limit with temperature is assumed to be given by $\sigma_y = \sigma_y^0\left[1-\left(\frac{\Delta\theta}{800}\right)^2\right]$, $\Delta\theta = \theta - 20°$.

The x- and y-displacements of the specimen upper boundary for different temperatures are shown in Fig. 200. Fig. 201 illustrates changes in stress intensity at center points of two selected elements along the lower boundary. The development of plastic zones is shown in Fig. 202.

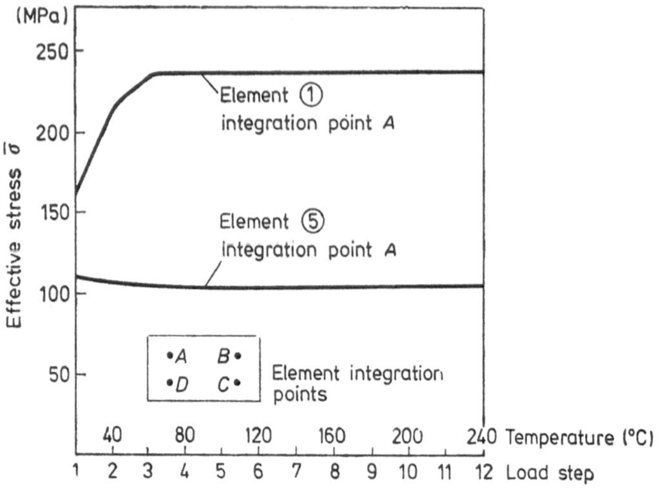

Fig. 201 Effective stresses at three selected points

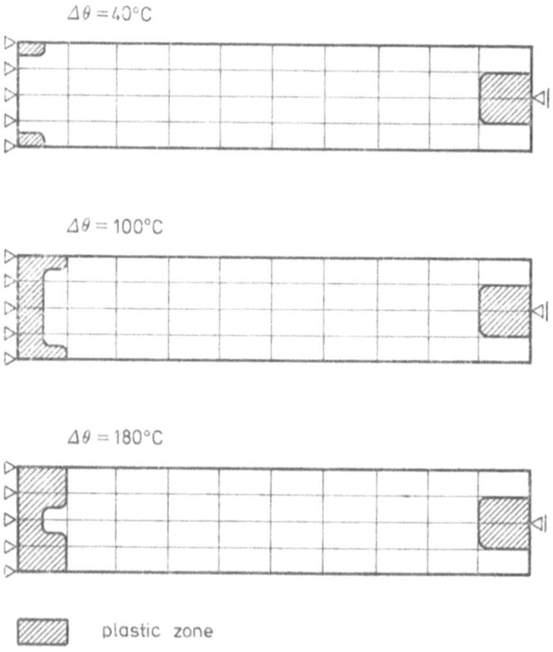

Fig. 202 Development of plastic zones

9.3 Elastic-plastic cylindrical shell under thermo-mechanical load

We consider an axisymmetric cylinder of an infinite length loaded by a uniform internal pressure and a ring of concentrated forces Q, subjected to a temperature difference $\Delta\theta$ between the inside and outside cylinder surfaces, Fig. 203. For

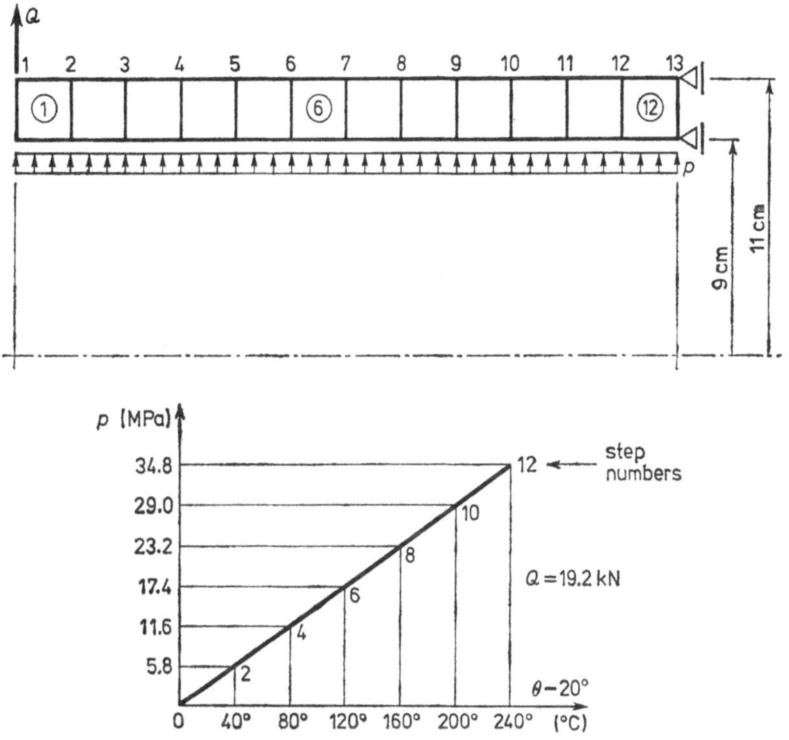

Fig. 203 Cylindrical shell under thermo-mechanical load

the sake of simplicity the linear distribution of temperature across the cylinder thickness is assumed so that no heat conduction problem needs to be solved. The material is assumed to be elastic-perfectly plastic with the material parameters as given in the previous problem. The yield limit σ_y is first assumed to be insensitive to temperature. The loading program consists of applying the forces Q in one step without exceeding the yield condition followed by a gradual proportional increase in p and ΔQ, Fig. 203. Radial displacements at selected nodes are shown in Fig. 204 for load steps 6, 8 and 12. The fast displacement increase is observed starting from the step No. 8. The step No. 13 leads to the loss of the limit load capacity of the cylinder.

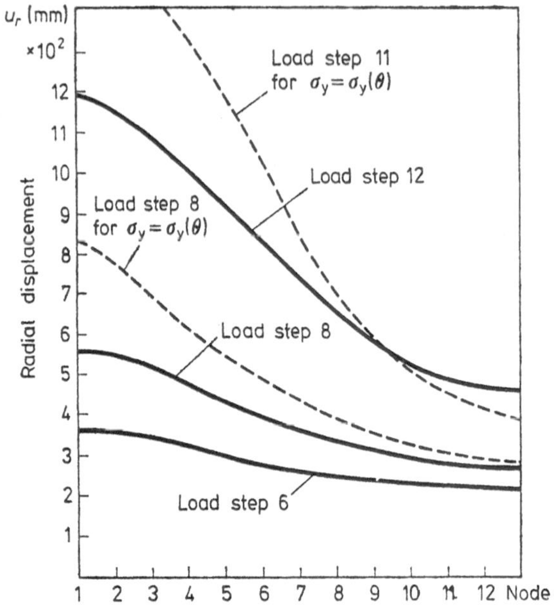

Fig. 204 Radial displacements at selected nodes

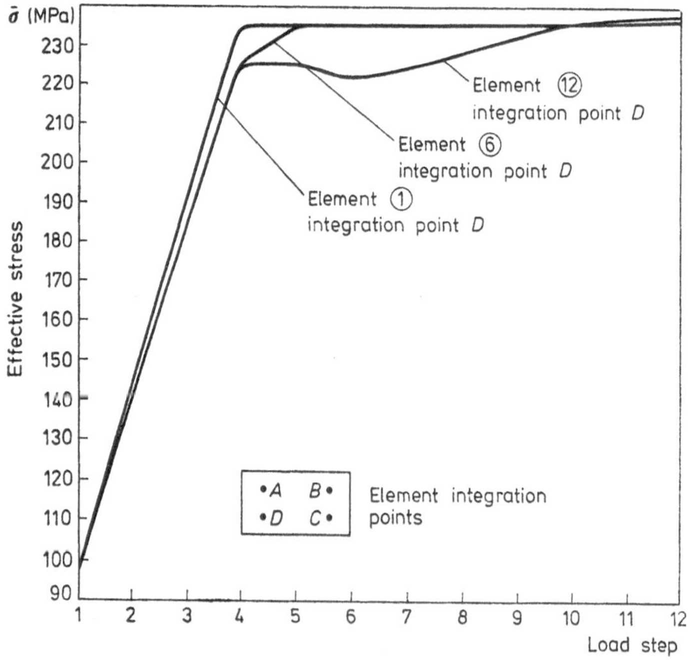

Fig. 205 Changes in effective stress during the deformation

Changes in stress intensity $\bar{\sigma}$ at points A, B and C during the process are shown in Fig. 205.

The distribution of circumferential strains (total, plastic and thermal) as well as that of equivalent plastic strain $\bar{\varepsilon}^{(pl)}$ is shown in Fig. 206 (solid lines).

Assuming the variation of the yield limit with temperature according to the formula $\sigma_y = \sigma_y^0 \left[1 - \left(\frac{\Delta\theta}{800}\right)^2\right]$ we observe the loss of limit load ca-

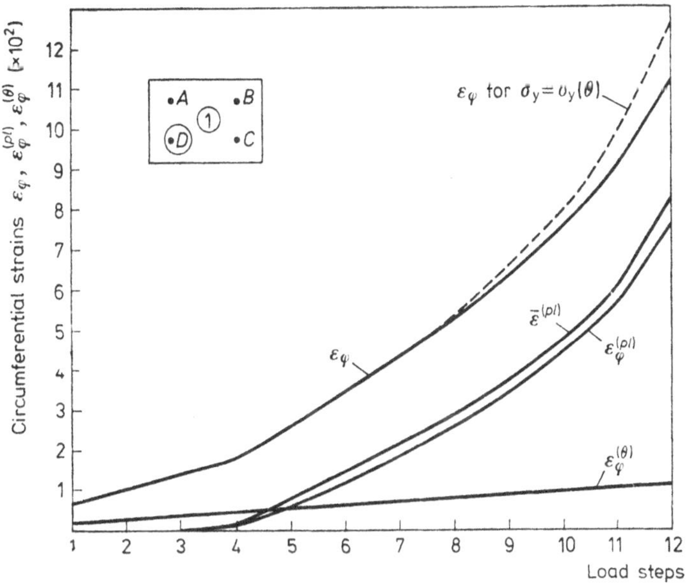

Fig. 206 Distribution of circumferential strains

pacity significantly earlier (at the step No. 12). Fig. 206 show the distribution of radial displacements at selected nodes. The broken line in Fig. 206 illustrates the distribution of total circumferential strain ε_φ in this case.

9.4 Elastic-plastic truss under nonproportionally varying temperature and mechanical load

To illustrate the theory presented in Sec. 4.11.2 we shall now again consider the steel truss shown in Fig. 70. The cross-section area of each element is $A = 10\ \text{cm}^2$, Young modulus $E = 207000$ MPa and the linear thermal expansion coefficient $\alpha = 1.1089 \times 10^{-5}\ \text{grad}^{-1}$. The temperature dependence of the

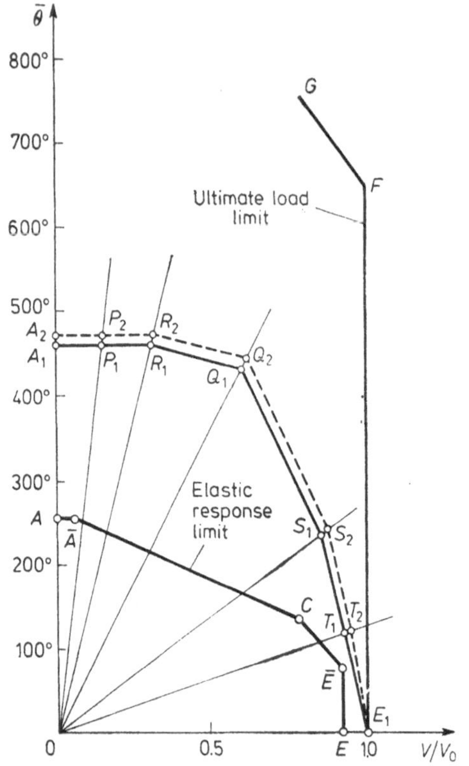

Fig. 207 Shakedown results for truss under thermo-mechanical loadings

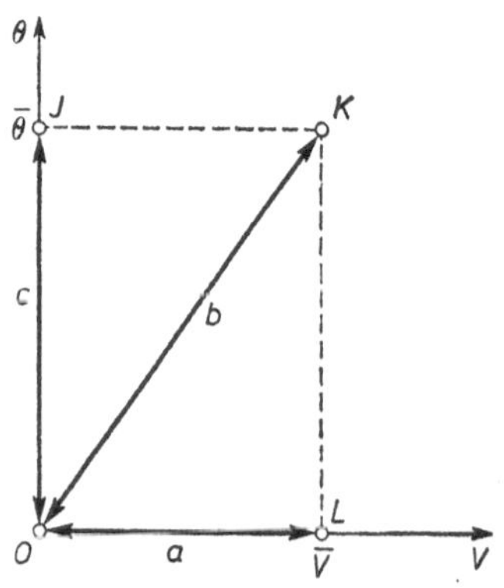

Fig. 208 Definition of cyclic paths

yield-point stress σ_y will be assumed in the following form

$$\sigma_y = \sigma_y^0 \left[1 - \left(\frac{\theta}{800}\right)^2\right] \tag{9.2}$$

where $\sigma_y^0 = 235.44$ MPa, θ being counted from 20°C. The simple relationship (9.2) describes in an acceptable manner the actual behaviour of mild steel within a fairly broad temperature range, from 0°C to about 600°C. At higher temperatures the shakedown becomes inadequate anyway, due to the influence of creep. For the sake of simplicity, buckling of compressed elements is neglected.

Loading and heating of the truss consists of the vertical force V applied at node 2, cf. Fig. 70, and of the homogeneous heating of bars 1 and 2 up to the temperature $\bar{\theta}$. These quantities may vary arbitrarily and inedependently of each other within the limits

$$0 \leqslant V \leqslant \bar{V}, \quad 0 \leqslant \theta \leqslant \bar{\theta}.$$

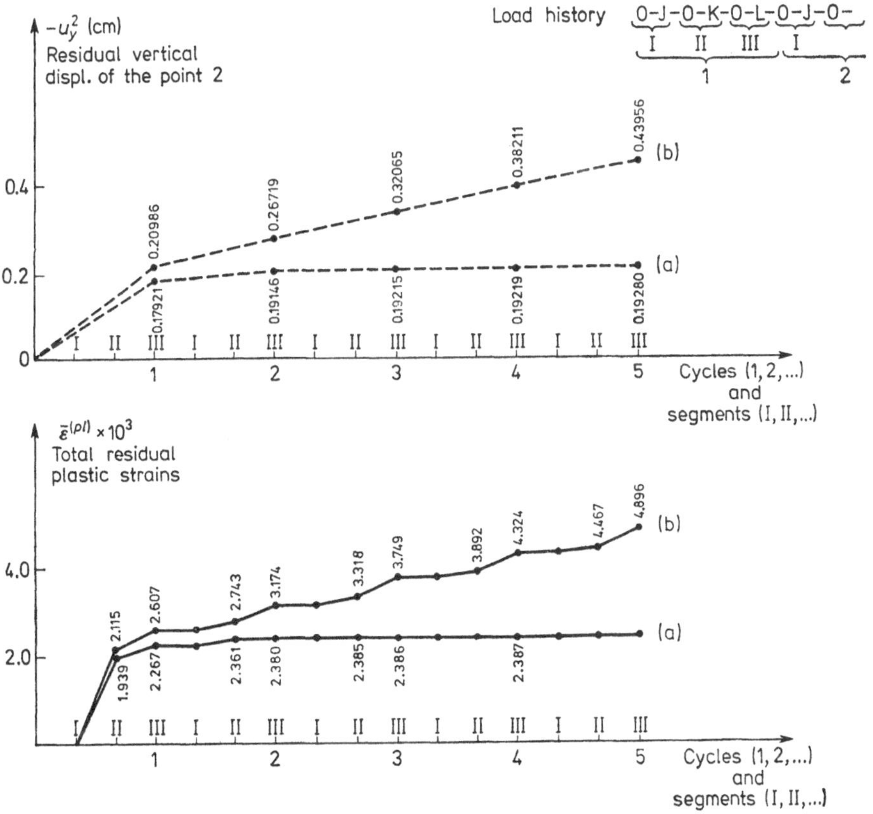

Fig. 209 Residual displacements and strains

The domains of elastic response $A\bar{A}C\bar{E}E$ of the truss, as well as the ultimate load-temperature curve GFE_1 have been determined in proportional load-temperature tests. They are shown in Fig. 207 where the results of the incremental shakedown analysis to be described below are also given. V_0 is the ultimate load for acting on its own without heating.

In the shakedown analysis the cyclic paths of Fig. 208, cf. Sec. 4.11.2, were employed. The cyclic path $0J0K0L0$ was repeated as many times as it was necessary to obtain the evidence of convergence or divergence of the truss deflections. Some results obtained are shown in Fig. 209 where the residual (permanent) part of the vertical deflection u_y^2 of the node 2 is drawn against the number of cycles. Curve (a) corresponds to the case in which the components of the point K coincide with those of point S_1 in Fig. 207. Curve (b) applies to the case of K coinciding with the point S_2. Here, the phenomenon of incremental collapse can clearly be seen.

By repeating such a cyclic procedure for various $\bar{\theta}/\bar{V}$ ratios one obtains a lower bound to the shakedown domain (line $A_1P_1R_1Q_1S_1T_1E_1$ in Fig. 207) and an upper bound (line $A_2P_2R_2Q_2S_2T_2E_1$).

The analysis documents that the incremental shakedown analysis proposed in Sec. 4.11.2 can be effectively used in practical problems of inelastic structures under cyclic thermo-mechanical loads. This may turn out very useful since, as it is clearly seen from Fig. 207 the safe shakedown loads may strongly depend of temperature changes.

Summary

The contents of this chapter, i.e. the nonlinear analysis of thermo-mechanical effects in structures, represents in a sense the most complex subject undertaken in this book. The formulations and algorithms advocated in it appear to offer effective and useful tools also for the analysis of such problems.

Bibliography

[1] Antman, S., The theory of rods, *Handbuch der Physik*, **VI a/2**, Springer 1972.

[2] Argyris, J. H., Boni, B., Hindenlang, U., Kleiber, M., Finite element analysis of 2*D* and 3*D* elasto-plastic frames—the natural approach, *Comp. Meth. Appl. Mech. Eng.* **35**, 221–248, 1982.

[3] Argyris, J. H., Doltsinis, J. St., Kleiber, M., Incremental formulation in nonlinear mechanics and large strain elasto-plasticity, *Comp. Meth. Appl. Mech. Eng.* **14**, 259–294, 1978.

[4] Argyris, J. H., Krempl, E., Willam, K. J., Constitutive models and finite element solution of inelastic behaviour, *US-German Symp. M.I.T., Boston, Aug.* 1976.

[5] Babuska, I., Rheinboldt, W. C., A posteriori error estimates for the finite element method, *Int. J. Num. Meth. Eng.* **12**, 1597–1615, 1978.

[6] Babuska, I., Rheinboldt, W. C., Error estimates for adaptive finite element computations, *SIAM J. of Num. Anal.* **15**, 1978.

[7] Babuska, I., Rheinboldt, W. C., Adaptive approaches and reliability estimations in finite element analysis, *Comp. Meth. Appl. Mech. Eng.* **17–18**, 519–540, 1979.

[8] Babuska, I., Rheinboldt, W. C., Computational error estimates and adaptive processes for some nonlinear structural problems, *Comp. Meth. Appl. Mech. Eng.* **34**, 859–937, 1982.

[9] Babuska, I., Szabo, B., On the rates of convergence of the finite element method, *Int. J. Num. Meth. Eng.* **18**, 323–341, 1982.

[10] Balmer, H. *et al.*, Elastoplastic and creep analysis with the ASKA program system, *Comp. Meth. Appl. Mech. Eng.* **3**, 87–104, 1974.

[11] Barański, W., Isotropic fibrous body as a model of shell-shaped grids, *Arch. Mech.* **18**, 553–556, 1966.

[12] Barański, W., Woźniak, Cz., The fibrous body as a simply-connected model of multihole discs, *Arch. Mech.* **18**, 273–283, 1966.

[13] Bathe, K.-J., *Finite element procedures in engineering analysis*, Prentice Hall 1981.

[14] Bathe, K.-J., Cimento, A. P., Some practical procedures for the solution of nonlinear finite element equations, *Comp. Meth. Appl. Mech. Eng.* **22**, 59–85, 1980.

[15] Bathe, K.-J., Dvorkin, E. N., On the automatic solution of nonlinear finite element equations, *Comp. Struc.* **17**, 871–879, 1983.

[16] Bathe, K.-J., Gracewski, S., On nonlinear dynamic analysis using substructuring and mode superposition, *Comp. Struc.* **13**, 699–707, 1981.

[17] Bathe, K.-J. *et al.*, Static and dynamic geometric and material nonlinear analysis, *SESM Report* 74-4, *Univ. of Calif., Berkeley.*

[18] Bathe, K.-J., Wilson, E. L., Peterson, F. E., SAP IV—A structural analysis program for static and dynamic response of linear systems, *EERC Report* 73–11, *Univ. of Calif.* 1973.

[19] BATOZ, J. L., DHATT, G., Incremental displacement algorithms for nonlinear problems, *Int. J. Num. Meth. Eng.* **12**, 1677–1696, 1978.

[20] BAUER, J., KLEIBER, M., SOSNOWSKI, W., Plane *I*-beam frames—numerical analysis accounting for inelastic material properties and local web buckling, Parts I and II, *Eng. Trans.* **33**, 353–377, 379–399, 1985.

[21] BELYTSCHKO, T., MULLEN, R., Stability of explicit-implicit mesh partitions in time integration, *Int. J. Num. Meth. Eng.* **12**, 1575–1586, 1978.

[22] BERGAN, P., Solution algorithms for nonlinear structural problems, *Comp. Struct.* **12**, 497–509, 1980.

[23] BERGAN, P. G., CLOUGH, R. W., Convergence criteria for iterative processes, *AIAA J.* **10**, 1107–1108, 1972.

[24] BOBLEWSKI, J., BOJDA, K. H., Stress equation for a pseudo-plane state of equilibrium, *Bull. Acad. Polon. Sci. Ser. Sci. Techn.* **XXIV**, 1976.

[25] BONI, B., KLEIBER, M., Limit analysis of plane bending-and- torque supporting grids *Arch. Inż. Ląd.* **XXI**, 69–87, 1975.

[26] BONI, B., KLEIBER, M., Numerical plastic collapse analysis of plane bending-and-torque supporting grids, *Comp. Meth. Appl. Mech. Eng.* **19**, 1–19, 1979.

[27] BONI, B., KLEIBER, M., SAWCZUK, A., Plastic interaction surface for non-symmetric I-cross-section, *IFTR Report No.* 29/1982 (in Polish).

[28] BORKOWSKI, A., KLEIBER, M., On a numerical approach to shakedown analysis of structures, *Comp. Meth. Appl. Mech. Eng.* **22**, 101–119, 1980.

[29] BORKOWSKI, W., KLEIBER, M., Nonlinear statics and dynamics of thin axisymmetric shells by high precision finite elements, *Arch. Mech.* **34**, 15–37, 1982.

[30] BORKOWSKI, W., KLEIBER, M., Static and dynamic analysis of spherical shell stability, *WAT Report* 1984, Warsaw (in Polish).

[31] BRENDEL, B., RAMM, E., Nicht lineare Stabilitatsuntersuchungen mit der Methode der Finiten Elemente, *Ing. Arch.* **51**, 337–362, 1982.

[32] BUDIANSKY, B., Dynamic buckling of elastic structures: criteria and estimates, in: *Dynamic Stability of Structures*, G. HERRMANN, ed., Pergamon Press 1966.

[33] BUSHNELL, D., Large deflection elastic-plastic creep analysis of axisymmetric shells, *J. Mech. Eng. ASME*, **6**, 103–138, 1973.

[34] BUSHNELL, D., Bifurcation buckling of shells of revolution including large deflections, plasticity and creep, *Int. J. Sol. Struct.* **10**, 1287–1305, 1974.

[35] CARNOY, E. G., Asymptotic study of the elastic post-buckling behaviour of structures by the finite element method, *Comp. Meth. Appl. Mech. Eng.* **29**, 147–173, 1981.

[36] CHERNYKH, K. F., *Linear theory of shells* (in Russian), Leningrad 1961.

[37] CONNOR, J. J. *et al.*, Nonlinear analysis of elastic framed structures, *J. Struc. Div.* 1526–1547, 1968.

[38] CONNOR, J. J., BREBBIA, C. A., *Finite element techniques for fluid flow*, Newnes–Butterworths 1976.

[39] COOK, R. D., *Concepts and applications of finite element analysis*, Wiley 1974.

[40] CORRADI, L., ZAVELLANI, A., A linear programming approach to shakedown analysis of structures, *Comp. Meth. Appl. Mech. Eng.* **3**, 37–53, 1974.

[41] CRISFIELD, M. A., Fast incremental/iterative solution procedure that handles "snap-through", *Comp. Struct.* **13**, 55–62, 1981.

[42] CRISFIELD, M. A., *Finite elements and solution procedures for structural analysis*, Pineridge Press 1987.

[43] DACKO, M., Elastic-plastic bending of circular-symmetric plates with large deflections (in Polish), *Biul. WAT* **9**, 1970, Warsaw.

[44] DAVIES, A. J., *The finite element method*, Clarendon 1980.

[45] DESAI, C. S., *Elementary finite element method*, Prentice-Hall 1979.

[46] DHATT, G., TOUZOT, G., *The finite element method displayed*, Wiley 1984.

[47] ERINGEN, A. C., *Mechanics of continua*, Wiley 1967.

[48] FELIPPA, C. A., *Refined finite element analysis of linear and nonlinear two-dimensional structures*, Ph.D. Thesis, Univ. of Calif. at Berkeley, 1966.

[49] GALLAGHER, R. H., *Finite element analysis: fundamentals*, Prentice-Hall 1975.

[50] GAŁKA, A., On the dynamics of elastic membranes and cords as slender bodies, *Bull. Acad. Polon. Sci. Ser. Sci. Techn.* **XXIV**, 423–427, 1976.

[51] GERARDIN, M. *et al.*, Computational strategies for the solution of large nonlinear problems via quasi-Newton methods, *Comp. Struct.* **13**, 73–81, 1983.

[52] GREEN, A. E., ADKINS, J. E., *Large elastic deformations*, Clarendon Press 1960.

[53] GRIGORIAN, M., A lower-bound solution to the collapse of uniform rectangular grids on simple supports, *Int. J. Mech. Sci.* **13**, 755–763, 1971.

[54] HAYDL, H. M., SHERBURNE, A. N., Limit loads of variable thickness circular plates in bending, *Nucl. Eng. Design*, **22**, 296–300, 1972.

[55] HAYDL, H. M., SHERBURNE, A. N., Limit loads of circular plates under combined loadings, *J. Appl. Mech.*, 799–801, 1973.

[56] HILL, R., A general theory of uniqueness and stability in elastic-plastic solids, *J. Mech. Phys. Sol.* **6**, 236–249, 1957.

[57] HODGE, P. H., *Plastic analysis of structures*, McGraw-Hill 1959.

[58] HODGE, P. H., *Limit analysis of rotationally symmetric plates and shells*, Prentice Hall 1963.

[59] HODGE, Jr. P. H., PANARELLI, J., Interaction curve for circular cylindrical shells according to the Mises or Tresca yield condition, *J. Appl. Mech.* 375–380, 1962.

[60] HÜBNER, K. H., *Finite element method for engineers*, Wiley 1975.

[61] HUGHES, T. J. R., Stability, convergence and growth and decay of energy of the average acceleration method in nonlinear structural dynamics, *Comp. Struct.* **6**, 313–324, 1976.

[62] HUGHES, T. J. R., Implicit-explicit finite element techniques for symmetric and non-symmetric systems, in: *Numerical Methods for Non-linear Problems*, C. TAYLOR *et al.*, eds., Pineridge Press 1980.

[63] HUGHES, T. J. R., Analysis of transient algorithms with particular reference to stability behaviour, in: *Comp. Methods for Transient Analysis*, T. BELYTSCHKO, T. J. R. HUGHES, eds., Elsevier 1983.

[64] HUGHES, T. J. R., LIU, W. K., Implicit-explicit finite elements in transient analysis, *J. Appl. Mech.* **45**, 371–374, 1978.

[65] HUTCHINSON, J. W., Plastic buckling, in: *Adv. in Appl. Mech.*, Vol. **14**, Acad. Press 1974.

[66] IRONS, B. M., AHMED, S., *Techniques of finite elements*, Ellis Horwood 1979.

[67] JOHNSON, W. *et al.*, Plane-stress yielding of cantilevers in bending due to combined shear and axial load, *J. Strain Anal.* **9**, 67–77, 1974.

[68] JOHNSON, W. *et al.*, Collapse loads for thin cantilevers with rectangular holes along the centre-line, *J. Strain Anal.* **11**, 84–96, 1976.

[69] KHOJASTEH-BAKHT, M., POPOV, E. P., Analysis of elastic-plastic shells of revolution, *J. Mech. Div.* 327–340, 1970.

[70] KLEIBER, M., Statics of elastic lattice-type shells, *Arch. Mech.* **25**, 179–194, 1973.

[71] KLEIBER, M., Plasticity of discretized bodies, *Bull. Acad. Polon. Sci. Ser. Sci. Techn.*, **XXI**, 369–378, 1973.

[72] KLEIBER, M., The approximate methods in the theory of elastic lattice-type shells, *Arch. Mech.* **25**, 195–211, 1973.

[73] KLEIBER, M., On stability and large deflections of shallow lattice-type shells, *Arch. Inż. Ląd.* **XIX**, 211–220, 1973 (in Polish).

[74] KLEIBER, M., SHELAX—nonlinear finite element analysis of thin axisymmetric shells, *IFTR Report No.* 49/1977 (in Polish).

[75] KLEIBER, M., PLADEP—nonlinear finite element analysis of plane stress problems, *IFTR Report No.* 48/1977 (in Polish).

[76] KLEIBER, M., An error estimation method in finite element structural analysis, *Comp. Struct.* **11**, 343–347, 1980.

[77] KLEIBER, M., Some results in the numerical analysis of structural instabilities, Part I: Statics, Part II: Dynamics, *Eng. Trans.* **30**, 327–353, 355–367, 1982.

[78] KLEIBER, M., Formulations and solution strategies for novel problems of plasticity and creep, *Nucl. Eng. Des.* **79**, 321–341, 1984.

[79] KLEIBER, M., *Incremental finite element modelling in solid mechanics*, PWN/Ellis Horwood 1989.

[80] KLEIBER, M., BREITKOPF, P., *Introduction to the finite element method with Turbo Pascal programs for the IBM PC*, PWN/Ellis Horwood, to be published.

[81] KLEIBER, M., KOTULA, W., SARAN, M., Numerical analysis of dynamic structural buckling, *Eng. Comput.* **4**, 48–52, 1987.

[82] KLEIBER, M., KONIG, J. A., SAWCZUK, A., Studies on plastic structures, *Comp. Meth. Appl. Mech. Eng.* **33**, 487–556, 1982.

[83] KLEIBER, M., SŁUŻALEC, A., Numerical analysis of heat flow in flash welding, *Arch. Mech.* **35**, 687–699, 1983.

[84] KLEIBER, M., SŁUŻALEC, A., Finite element analysis of heat flow in friction welding, *Eng. Trans.* **32**, 107–113, 1984.

[85] KLEIBER, M., SOSNOWSKI, W., Plane elastic-plastic frame analysis within the WAT-KM system, *Mech. Komp.* **6**, 83–98, 1986 (in Polish).

[86] KLEIBER, M., WIECZOREK, M., An approximate method for the nonlinear analysis of elastic frames, *Eng. Trans.* **30**, 269–281, 1982.

[87] KLEIBER, M., WOŹNIAK, CZ., On the equations of the linear theory of elastic lattice-type structures, *Bull. Acad. Polon. Sci. Ser. Sci. Techn.* **XIX**, 97–101, 1971.

[88] KLEIBER, M., WOŹNIAK, CZ., Equations of shallow lattice-type shells, *Bull. Acad. Polon. Sci. Ser. Sci. Techn.* **XIX**, 143–147, 1971.

[89] KLEIBER, M., WOŹNIAK, CZ., Edge effect in lattice-type shells, *Bull. Acad. Polon. Sci. Ser. Sci. Techn.* **XIX**, 149–152, 1971.

[90] KLEIBER, M., ZACHARSKI, A., Numerical analysis of local instabilities in elastic and elastic-plastic prismatic plate assemblies, *Comp. Meth. Appl. Mech. Eng.* **31**, 141–168, 1982.

[91] KLEMM, P., WOŹNIAK, CZ., Perforated circular plates under large deflections, *Arch. Mech.* **19**, 45–56, 1967.

[92] KLEMM, P., WOŹNIAK, CZ., Dense elastic lattices of regular structure, I and II, *Bull. Acad. Polon. Sci. Ser. Sci. Techn.* **XVIII**, 435–444, 445–453, 1970.

[93] KLEMM, P., KONIECZNY, S., WOŹNIAK, CZ., Dense elastic lattices of regular structure, III, *Bull. Acad. Polon. Sci. Ser. Sci. Techn.* **XVIII**, 497–501, 1970.

[94] KOITER, W. T., General theorems for elastic-plastic solids, in: *Progress in Solid Mechanics* **1**, North-Holland 1960.

[95] KOITER, W. T., A systematic simplification of the general equations in the linear theory of thin shells, *Proc. Neder. Acad. Wetesch.*, **B64**, 612–619, 1961.

[96] KÖNIG, J. A., *Shakedown of elastic-plastic structures*, PWN/Elsevier 1987.

[97] KÖNIG, J. A., KLEIBER, M., On a new method of shakedown analysis, *Bull. Acad. Polon. Sci. Ser. Sci. Techn.* **XXVI**, 165–171, 1978.

[98] KÖNIG, M., NAGY, D. A., STREINER, P., Buckling analysis with the ASKA program system, *Comp. Meth. Appl. Mech. Eng.* **16**, 185–212, 1978.

[99] KOTUŁA, W., KLEIBER, M., Stability and imperfection sensitivity of some 3D truss, *IFTR Report No.* 39/1983 (in Polish).

[100] KUFEL, W., On the optimal control of the discretization problems for elastic bodies, *Arch. Mech.* **28**, 3–12, 1976.

[101] KWIECIŃSKI, M., KLEIBER, M., Limit load of polar grids treated as plane fibrous bodies, *Arch. Inż. Ląd.* **XVII**, 224–237, 1971.

[102] KWIECIŃSKI, M., KLEIBER, M., Minimum weight design of ideally plastic rectangular dense grid, *Arch. Inż. Ląd.* **XVIII**, 175–183, 1972.

[103] LEE, L. H. N., On dynamic stability and quasi-bifurcation, *Int. J. Non-lin. Mech.* **16**, 79–87, 1981.

[104] LEE, L. H. N., Dynamic buckling of an inelastic column, *Int. J. Sol. Struct.* **17**, 271–279, 1981.

[105] LEVY, A., PIFKO, A. B., On computational strategies for problems involving plasticity and creep, *Int. J. Num. Meth. Eng.* **17**, 747–771, 1982.

[106] LEVY, L., LESSMAN, F., *Finite difference equations, I*, Pitman 1964.

[107] MASSONNET, H., SAVE, M. A., *Plastic analysis and design*, Blaisdell 1965.

[108] MATTHIES, H., STRANG, G., The solution of nonlinear finite element equations, *Int. J. Num. Meth. Eng.* **14**, 1613–1623, 1979.

[109] MAZUR-ŚNIADY, K., Some problems of torsion of prismatic rods as bodies with internal constraints, *Bull. Acad. Polon. Sci. Ser. Sci. Techn.* **XXII**, 389–397, 1974.

[110] MORRIS, N. F., The use of modal superposition in nonlinear dynamics, *Comp. Struct.* **7**, 65–72, 1977.

[111] NAGARAYAN, S., POPOV, E. P., Plastic and viscoplastic analysis of axisymmetric shells, *Int. J. Sol. Struct.* **11**, 1–19, 1975.

[112] NAGARAYAN, S., POPOV, E. P., Nonlinear dynamic analysis of axisymmetric shells, *Int. J. Num. Meth. Eng.* **9**, 535–550, 1977.

[113] NAGHDI, P. M., The theory of plates and shells, *Handbuch der Physik*, **VI a/2**, Springer 1972.

[114] NAGY, D. A., Modal representation of geometrically nonlinear behaviour, *Comp. Struct.* **10**, 683–688, 1979.

[115] NAGY, D. A., KÖNIG, M., Geometrically nonlinear finite element behaviour using buckling mode superposition, *Comp. Meth. Appl. Mech. Eng.* **19**, 447–484, 1979.

[116] NANIEWICZ, Z., Some examples of the functional formation of equations of constrained material continua, *Bull. Acad. Polon. Sci. Ser. Sci. Techn.* **XXVI**, 347–352, 1978.

[117] NICKELL, E., Nonlinear dynamics by mode superposition, *Comp. Meth. Appl. Mech. Eng.* **7**, 107–129, 1976.

[118] NOOR, A. K., Recent advances in reduction methods for nonlinear problems, *Comp. Struct.* **13**, 31–44, 1981.

[119] NOOR, A. K., PETERS, J. K., Recent advances in reduction methods for instability analysis of structures, *Comp. Struct.* **16**, 67–80, 1983.

[120] NORRIE, D. H., DE VRIES, G., *Finite element method: fundamentals and applications*, Academic Press 1978.

[121] ODEN, J. T., *Finite elements of nonlinear continua*, McGraw-Hill 1972.

[122] ODEN, J. T., REDDY, J. N., *An introduction to the mathematical theory of finite elements*, Wiley 1978.

[123] ONAT, E. T., HAYTHORNTHWAITE, R. M., Load carrying capacity of circular plates at large deflections, *J. Appl. Mech.* **23**, 49–55, 1956.

[124] ORTIZ, M. *et al.*, Unconditionally stable element-by-element algorithms for dynamic problems, *Comp. Meth. Appl. Mech. Eng.* **36**, 223–239, 1983.

[125] OWEN, D. R. J., HINTON, E., *Finite element in plasticity: theory and practice*, Pineridge Press 1980.

[126] PADOVAN, J., TOVICHAKCHAIKUL, S., Self-adaptive predictor-corrector algorithms for static nonlinear structural analysis, *Comp. Struct.* **15**, 365–378, 1982.

[127] PARK, K. C., An improved stiffly stable method for direct integration of nonlinear structural dynamics equations, *J. Appl. Mech.* **97**, 464–470, 1975.

[128] PEANO, A., Hierarchies of conforming finite elements for plane elasticity and plane bending, *Comp. Math. with Appls.* **2**, 211–224, 1976.

[129] PEANO, A. *et al.*, Self-adaptive finite elements in fracture mechanics, *Comp. Meth. Appl. Mech. Eng.* **16**, 69–80, 1978.

[130] PERZYNA, P., The constitutive equations for rate sensitive plastic materials, *Q. Appl. Math.* **20**, 321–332, 1962.

[131] PIETRASZKIEWICZ, W., Multivaluedness of solutions of shallow shells, *Bull. Acad. Polon. Sci. Ser. Sci. Techn.* V, 15–23, 1967.

[132] PIFKO, A., ISAKSON, G. A., A finite element method for the plastic buckling analysis of plates, *AIAAJ* **7**, 1950–1957, 1969.

[133] POPOV, E. P., SHARIFI, P., Refined finite element analysis of elastic-plastic shells of revolution, *Report to the Army Research Office-Durham, DAH CO4 69C 0037, Dec.* 1969.

[134] PRENTER, P. M., *Splines and variational methods*, Wiley 1975.

[135] PRZEMIENIECKI, J. S., Buckling of rectangular plates under bi-axial compression, *J. Aeronaut. Soc.* **59**, 566–568, 1955.

[136] PRZEMIENIECKI, J., *Theory of matrix structural analysis*, McGraw-Hill 1968.

[137] PRZEMIENIECKI, J. S., Matrix analysis of local instability in plates, stiffened panels and columns, *Int. J. Num. Meths. Eng.* **5**, 209–216, 1972.

[138] RAMM, E., Strategies for tracing nonlinear response near limit points, in: *Europe–US Workshop on Nonlinear Finite Element Analysis in Structural Mechanics*, Bochum 1980.

[139] RIDHA, R. A. *et al.*, Inelastic finite deformation of planar frames, *J. Eng. Mech. Div.* **97**, 773–789, 1971.

[140] RIKS, E., An incremental approach to the solution of snapping and buckling problems, *Int. J. Sol. Struct.* **15**, 529–551, 1979.

[141] Royal Aeronautical Society, *Structure Data Sheets*, Vol. 1.

[142] Royal Aeronautical Society, *Structure Data Sheets*, Vol. 2.

[143] Royal Aeronautical Society, *Engineering Sciences Data, Item No.* 69003, *The buckling of flat plates under non-uniform compression*, 1969.

[144] RYKALIN, N. N., *Calculation of heat processes in welding*, Lect. presented before the Amer. Welding Soc., April 1961.

[145] SARAN, M., KLEIBER, M., Experiences with nonlinear analyses of trusses under static and dynamic loadings, *Space Structures*, **2**, 204–214, 1986/87.

[146] SAWCZUK, A., JAEGER, T., *Grenztragfahigkeits—Theorie der Platten*, Springer 1963.

[147] SIMO, J. C. *et al.*, Finite deformation post-buckling analysis involving inelasticity and contact constraints, *Int. J. Num. Meth. Eng.* **23**, 779–800, 1986.

[148] SNYDER, M. D., BATHE, K.-J., A solution procedure for thermo-elastic-plastic and creep problems, *Nucl. Eng. Design*, **64**, 49–80, 1981.

[149] STRANG, G., FIX, G. J., *An analysis of the finite element method*, Prentice-Hall 1973.

[150] STRICKLIN, J. A. *et al.*, Formulation, computation and solution procedures for material and/or geometric nonlinear structural analysis, *SC-CR*-72 3102, *July* 1972, *Sandia Lab, New Mexico.*

[151] SUJATA, H. L., Plastic buckling of longitudinally stiffened plates, *J. Aeronaut. Sci.* **28**, 864–871, 1961.

[152] SZILARD, R., Critical load and post-buckling analysis by fem using energy balancing technique, *Comp. Struct.* **20**, 277–286, 1985.

[153] TIMOSHENKO, S. P., GERE, J. M., *Theory of elastic stability*, McGraw-Hill 1961.

[154] TIMOSHENKO, S. P., WOINOWSKI-KRIEGER, S., *Theory of plates and shells*, McGraw-Hill 1959.

[155] TILLERSON, J. R. *et al.*, Numerical methods for the solution of nonlinear problems in structural analysis, 1973 *Winter Annual Meeting of the ASME, Nov.* 1973, *New Mexico.*

[156] TOMITA, Y. *et al.*, Bifurcation and post-bifurcation behaviour of internally pressurized elastic-plastic circular tubes under plane strain conditions, *Int. J. Mech. Sci.* **12**, 723–732, 1981.

[157] TONG, P., ROSSETTOS, J. N., *Finite element method: basic technique and implementation*, MIT Press 1977.

[158] TRUESDELL, C., NOLL, W., The nonlinear field theories of mechanics, *Handbuch der Physik*, **III/3**, Springer 1965.

[159] UTKIN, J., On the technical theory of prismatic elastic rods as bodies with internal constraints, *Bull. Acad. Polon. Sci. Ser. Sci. Techn.* **XXIII**, 1–8, 1975.

[160] VOLTERRA, Equations of motion for curved and twisted elastic bars by the use of the "Method of constraints", *Ing. Arch.* **23**, 402–409, 1955.

[161] WASZCZYSZYN, Z., Numerical problems of nonlinear stability analysis of elastic structures, *Comp. Struct.* **17**, 13–24, 1983.

[162] WIECZOREK, M., Analysis of buckling-prone structures, *WAT Report* 1985, Warsaw (in Polish).

[163] WINNICKI, L., KWIECIŃSKI, M., KLEIBER, M., Numerical limit analysis of perforated plates, *Int. J. Num. Meth. Eng.* **11**, 553–561, 1977.

[164] WOŹNIAK, Cz., Load carrying structures of dense lattice type. The plane problem, *Arch. Mech.* **18**, 581–597, 1966.

[165] WOŹNIAK, Cz., Bending and stability problems of plates with lattice structure, *Arch. Mech.* **18**, 781–796, 1966.

[166] WOŹNIAK, Cz., On the equations of the theory of lattice structures, *Arch. Mech.* **21**, 539–555, 1969.

[167] WOŹNIAK, Cz., *Lattice-type structures*, PWN 1970 (in Polish).

[168] WOŹNIAK, Cz., Discrete elasticity, *Arch. Mech.* **23**, 801–816, 1971.

[169] WOŹNIAK, Cz., Introduction to the difference geometry, *Bull. Acad. Polon. Sci. Ser. Sci. Techn.* **XIX**, 393-398, 1971.

[170] WOŹNIAK, Cz., Theory of variated states in the discrete elasticity, *Bull. Acad. Polon. Sci. Ser. Sci. Techn.* **XX**, 47–54, 1972.

[171] WOŹNIAK, Cz., Oriented and multipolar discrete elastic media, *Bull. Acad. Polon. Sci. Ser. Sci. Techn.* **XX**, 55–61, 1972.

[172] WOŹNIAK, Cz., Equations of motion and laws of conservation in the discrete elasticity, *Arch. Mech.* **25**, 155–168, 1973.

[173] WOŹNIAK, Cz., Discrete elastic Cosserat media, *Arch. Mech.* **25**, 119–136, 1973.

[174] WOŹNIAK, Cz., Basic concepts of the difference geometry, *Annal. Polon. Math.* **28**, 24–37, 1973.

[175] WOŹNIAK, Cz., Elastic bodies with constraints imposed on deformations, stresses and moments, *Bull. Acad. Polon. Sci. Ser. Sci. Techn.* **XXII**, 407–413, 1974.

[176] WOŹNIAK, Cz., Materials with generalized constraints, *Arch. Mech.* **36**, 539–551, 1984.

[177] WOŹNIAK, Cz., Constraints in constitutive relations of mechanics, *Mech. Teor. i Stos.* **37**, 323–341, 1985.

[178] WOŹNIAK, Cz., On the modelling of materials and interactions with thermoelectro-mechanical constraints, *Bull. Acad. Polon. Sci. Ser. Sci. Techn.* **XXXIII**, 249–254, 1985.

[179] WOŹNIAK, Cz., KLEIBER, M., *Nonlinear mechanics of structures*, PWN 1982 (in Polish).

[180] YAGAWA, G., MIYAZAKI, N., ANDO, Y., An analysis of elastic-plastic creep buckling of axisymmetric shells by the finite element method, *Arch. Mech.* **27**, 869–882, 1975.

[181] YAGHMAI, S., POPOV, E. P., Incremental analysis of large deflections of shells of revolution, *Int. J. Sol. Struct.* **7**, 1375–1393, 1971.

[182] YAMADA, Y., Constitutive modelling of inelastic behaviour and numerical solution of nonlinear problems by the finite element method, *Comp. Struct.* **8**, 533–543, 1978.

[183] ZIENKIEWICZ, O. C., *The finite element method*, McGraw-Hill 1977.

[184] ZIENKIEWICZ, O. C., MORGAN, K., *Finite elements and approximation*, Wiley 1983.

[185] ZIENKIEWICZ, O. C., CORMEAU, I. C., Plasticity and creep in elastic solids—A unified numerical approach, *Int. J. Num. Meth. Eng.* **8**, 821–845, 1974.

[186] ZUDANS, Z., Implicit and explicit computational schemes in dynamic plasticity, *J. Pressure Vessel Techn.* 394–403, 1977.

[187] ŻYCZKOWSKI, M., *Complex loadings in plasticity*, PWN 1981.

Index

acceleration 4
arc-length method 221
adaptive solution refinement 103
adjacent elements 96
assembly of element contributions 113
associated flow rule 31, 226
axisymmetric shell 162
axisymmetric shell element 164

balance of energy 11
balance of momentum 10
balance of moment of momentum 10
beam theories 71
beam elements 89
bending of cantilever 364
bending/twisting limit state condition 349
BFGS method 216
bifurcation point 218
bilateral constraint 44
body force 7
boundary heat source 37
boundary heat supply 9
boundary loading 37
buckling 229, 263, 305, 318, 410
buckling mode superposition 306

Cauchy stress tensor 15
Clausius–Duhem inequality 11
collapse mechanism 274, 338
conductivity matrix 118
constitutive equations 28
constitutive stiffness matrix 93, 116, 120, 126, 179, 191, 277
constraint 27, 39
constraint function 46
constraints for heat fluxes 47
constraints for stresses 47
continuous models for lattice-type structures 141
convergence criteria 217
correction of solution 101
creep 169, 405
Crisfield's procedure 223
critical load path 203

damping matrix 117
deformation 4
deformation theory of plasticity 420
deformed body 4
discrete description 80
discretization 107
displacement field 12, 144
dissipation principle 11
dissipation inequality 22
dynamic quasi-bifurcation 238
dynamic stability 237, 259

elastic frame element in space 123
elastic lattice-type structures 131
elastic pin-jointed element in space 118
elastic-plastic beam 277
elastic-plastic grid 322
elastic-plastic plate assemblies 410
elastic strain energy 140
element connectivity table 108
energy 8
energy balance 21
engineering theories 52
engineering shell theories 147
entropy 8
equation of motion 19, 113, 117, 144, 151, 154, 182, 232, 316
equilibrium path 218

Index

Euler one-step backward method 214
Euler one-step forward method 243
evaluation of solution 99
explicit integration scheme 233
external agents 37

finite difference operators 133, 315
finite elements 107
finite element program 245
first law of thermodynamics 11
flash welding 431
frame 277
free energy 9
friction welding 431
forces 6

Green–Saint Venant strain tensor 13

hardening parameter 194
heat 9
heat absorption 9
heat capacity matrix 118
heat conduction 431
heat flux 16
Huber–Mises yield condition 36, 168
hydrostatic pressure 15
hyperelastic material 50

ideal constraint 43
implicit integration scheme 233
incremental discrete formulation 86
incremental variational formulation 55
initial buckling 229
initial displacement matrix 93, 116, 129
initial stress matrix 93, 116, 120, 127, 180, 191
integration of elastic-plastic law 225
interaction between elements 96
internal energy 8
internal force vector 117
internal heat supply 38
internal state variable 28, 30
interpolation vector 184

kinematic hardening 196
Kirchhoff plate theory 5

Lame modulae 30, 32
lattice-type shell 131, 312
layered model 176

limit point 218
limit state condition 283
linear thermo-elastic material 29
linearized buckling 229
load domain 199, 273, 301
loading 27
loading/unloading conditions 196, 199, 226
Love–Kirchhoff theory 159
LST element 182

mass 5
mass matrix 117, 121, 130
material 27
material coordinates 4
material points 4
matrix notation 16, 23, 34, 60, 91
method of constraints 61, 77, 80
midsurface of the shell 70
minimum weight design of plastic grid 351
modal superposition method 236
motion 3

necking 380
Newmark method 234
Newton–Raphson method 214, 235
nodal temperature vector 117
nodes 107
nonlinear dynamics 232
nonlinear heat transfer 242
non-proportionally varied loads 197, 272, 298, 376, 445

perforated plate 369
plane problems 74, 79
plane stress 182
plastic interaction surface 287
plastic potential 30
plastic strain rate intensity 194
plate elements 87
plate theories 65
primary equilibrium path 229
principal strain 14
principal stress 15
production of entropy 11

quasi-statics 213

reactions 42, 59
reference configuration 4
residual reactions 97

rod elements 89, 135
rod theories 71, 78

secondary equilibrium path 229
second Piola–Kirchhoff stress tensor 15
shakedown 201, 273, 298
shell analysis 385
shell elements 87
shell theories 65, 78
singular point 218
software concepts 212, 244
space 3
spatial coordinates 3
specific entropy 8
strain 12
strain rate 12
stress 14
stress deviator 15
stress rate 14
surface heat supply 9
surface traction 7

temperature 8
test function 20
thermal boundary constraints 40
thermal internal constraints 40
thermo-elastic-plastic material 30, 193
thermo-visco-elastic material 28
time 3
trial rate of stress 197, 226
truss 251

updated Lagrangian description 56

variational formulation 52
velocity 4
velocity field admissible by constraints 41
virtual displacement 42
virtual temperature increment 43
viscoplasticity 170, 405

weighted residual methods 109

yield condition 30
yield stress 36